T3-AKD-000

THE PHANEROZOIC GEOLOGY OF THE WORLD II

THE MESOZOIC, A

THE PHANEROZOIC GEOLOGY OF THE WORLD

Editors:
M. Moullade and A.E.M. Nairn

Centre de Recherches micropaléontologiques "Jean Cuvillier", Université de Nice, Parc Valrose, 06034 Nice Cédex (France)
Department of Geology, University of South Carolina, Columbia, S.C. 29208 (U.S.A.)

I. The Palaeozoic
II. The Mesozoic
III. The Cenozoic

THE PHANEROZOIC GEOLOGY OF THE WORLD II

THE MESOZOIC, A

Edited by

M. Moullade

Centre de Recherches micropaléontologiques "Jean Cuvillier"
Université de Nice
06034 Nice Cédex
France

and

A.E.M. Nairn

Department of Geology
University of South Carolina
Columbia, S.C. 29208
U.S.A.

ELSEVIER SCIENTIFIC PUBLISHING COMPANY

Amsterdam – Oxford – New York 1978

ELSEVIER SCIENTIFIC PUBLISHING COMPANY
335 Jan van Galenstraat
P.O. Box 211, 1000 AE Amsterdam, The Netherlands

Distributors for the United States and Canada:

ELSEVIER NORTH-HOLLAND INC.
52 Vanderbilt Avenue
New York, N.Y. 10017, U.S.A.

Library of Congress Cataloging in Publication Data
Main entry under title:

The Mesozoic.

(The Phanerozoic geology of the world ; 2)
Includes bibliographies and indexes.
1. Geology, Stratigraphic—Mesozoic. 2. Paleontology—Mesozoic. I. Moullade, Michel. II. Nairn, A. E. M. III. Series.
QE651.P48 vol. 2 [QE675] 551.7'08s 77-18920
ISBN 0-444-51671-4 (v. A) [551.7'6]

ISBN 0-444-41671-4 (Vol. II, Part A)
ISBN 0-444-41584-X (Series)

Printed in The Netherlands

CONTRIBUTORS TO THIS VOLUME

M. AUDLEY-CHARLES
Department of Geology
Queen Mary College
Mile End Road
London E1 4NS (Great Britain)

N.V. BEZNOSOV
All-Union Scientific Research
Institute for Petroleum Geology and Prospecting
Moscow (U.S.S.R.)

I. DE KLASZ
Department of Geology
University of Ife
Ile-Ife (Nigeria)

R.V. DINGLE
Marine Geoscience
Department of Geology
University of Cape Town
Rondebosch 7700 (South Africa)

T.N. GORBATCHIK
Department of Paleontology
Faculty of Geology
Moscow State University
Moscow (U.S.S.R.)

J. LEFELD
Institute of Geological Sciences
Polish Academy of Science
Zwirki i Wigury 93
02-089 Warszaw (Poland)

N.H. LUDBROOK
c/o Department of Mines
P.O. Box 151
Eastwood, S.A. 5063 (Australia)

T. MATSUMOTO
Department of Geology
Faculty of Science
Fukuoka 812 (Japan)

I.A. MIKHAILOVA
Department of Paleontology
Faculty of Geology
Moscow State University
Moscow (U.S.S.R.)

M. MOULLADE
Centre de Recherches micropaléontologiques "Jean Cuvillier" and Laboratoire de Géologie structurale
Université de Nice
Parc Valrose
06034 Nice Cedex (France)

A.E.M. NAIRN
Department of Geology
University of South Carolina
Columbia, S.C. 29208 (U.S.A.)

M.A. PERGAMENT
Geological Institute
U.S.S.R. Academy of Science
Moscow (U.S.S.R.)

M.F. RIDD
B.P. Research
Farburn Industrial Estate
Dyce
Aberdeen AB2 OPB (Great Britain)

P. SAINT-MARC
Centre de Recherches micropaléontologiques "Jean Cuvillier" and Laboratoire de Géologie structurale
Université de Nice
Parc Valrose
06034 Nice Cedex (France)

I.G. SPEDEN
Department of Scientific and Industrial Research
New Zealand Geological Survey
Lower Hutt (New Zealand)

G.R. STEVENS
Department of Scientific and Industrial Research
New Zealand Geological Survey
Lower Hutt (New Zealand)

CONTENTS

Chapter 1

INTRODUCTION

M. MOULLADE and A.E.M. NAIRN

The history of the evolution of the earth's surface is one of the most fascinating of all geological topics, witness the interest aroused when Wegener propounded his hypothesis or in the geological revolution accompanying the general acceptance of the plate-tectonic paradigm. The latter hypothesis provides a broad outline, the provision of the necessary detail must come from stratigraphy studied on a global rather than a regional basis. It is no exaggeration to state that stratigraphy has dismally failed to ignite the same general enthusiasm accorded to the more geophysical aspects of global evolution. This is probably because, fascinating though the evolution of global geography is, the teaching of stratigraphy from which it derives soon descends into a morasse of details in which the original objective is all too often lost to view. Yet, one cannot eschew the very facts upon which the paleogeography is based. The purpose of the present volumes is to try and kindle some of that lost interest by presenting, as far as possible on a paleogeographic basis, the history of the Mesozoic, the time during which the world as we know it began to take form. It will take into account two major advances in the earth sciences, the geophysical models indicated above and advances in sedimentation studies and the concepts of environments of deposition.

Stratigraphy is poorly represented by graduate level texts, particularly in English. There are many elementary texts, some of which give a much broader coverage than others, but very few cover regional geology at a level below monographic. Even such an excellent, if now outdated, text as Gignoux's *Géologie stratigraphique* gives scant coverage to many regions. The alternatives are the *Handbuch der Regional Geologie* series which takes a country by country approach, or the *Geological Systems* which provides system by system coverage. The objective of the present text is to provide an intermediate step by treating each era on a global basis. It was soon discovered that this ambition could not be achieved in a single volume, and the choice of contents for the present volume was based upon receipt of manuscripts consistent with a broad geological geographic presentation.

To achieve the stated goal, there was very little choice but a symposium approach with its attendant difficulties. The principal difficulty is loss of coherence of style and treatment of the material. The volume would lose some of its appeal if a too rigid adherence to style were insisted upon given the varied personalities of the distinguished geologists who have contributed. The varying state of development of geology from one part of the globe to another also poses problems in uniformity of treatment, certainly Antarctica cannot be treated in the same way as Western Europe for example. A basically paleogeographic approach was proposed to contributors in which it was suggested that much of the factual data could be presented in tables or figures leaving the text to provide the synthesis thereby eliminating the need to describe all but the more important sections. The tabled data and the bibliography provide the interested reader with the means of testing each author's interpretation for himself. A deliberate attempt was made to avoid prolonged discussion of hypothesis or to present data in terms of hypothesis. However, in attempting to summarize the geology of the myriad islands of Indonesia it was necessary to use a hypothetical model to provide some semblance of coherence to the data presentation (see Audley-Charles, pp. 165–207). Where hypothesis is involved, attention is given to the various alternatives to that preferred by the author. In the following chapters, the reader will judge the extent to which these views have prevailed. The responsability for deficiencies must rest with the editors rather than the contributors.

Historically the Mesozoic Era emerged out of the developing subject of geology as that interval of time characterized by a fauna intermediate in aspect between the ancient Paleozoic and more modern Cenozoic faunas. The limits are marked by distinctive faunal changes for which Newell (1952) does not regard the term "faunal revolution" too strong. These faunal changes did not affect all marine phyla equally, the terrestrial life forms were essentially unaffected, so that it was at one time thought that the change might be an artefact induced by the reduction of a continuum into discrete units. However, investigation has served to show that radical faunal changes did indeed occur over a relatively short time span. When the best continuous sections of the Permian–Triassic transition in the Salt Range of India and the Canadian Arctic are examined, the poverty of the basal Triassic fauna is apparent [see the *Symposium on the Permo-Triassic boundary, Calgary 1971* (1973)]. The top of the Mesozoic has also been considered to coincide with another faunal crisis and faunal extinction. Specifically, in common with most micropaleontologists and many statigraphers the top of the Mesozoic is taken at the upper limit of the Maastrichtian. Thus, the Danian is assigned as the lowest stage of the Cenozoic. An entirely satisfactory explanation for the faunal crises has not yet been found, and this probably reflects the fact that while the problem is a paleontological one, the cause may lie outside the biological realm such as the destruction of habitat or ecological niches as a result of physical or tectonic causes in the broad sense.

The same boundary questions also affect the limits of the systems composing the Mesozoic, and at every inferior level. The modern paleontologist has the advantage over his predecessor in that a greater variety of forms is available to him through advances in micropaleontology. He is on the other hand even more conscious of the problems of facies, and much of his energy and ingenuity is engaged in seeking equivalents, and in sorting out the problems of rock units, the practical mapping unit, which are time transgressive. It may be hoped that certain of the physical methods may yet provide ancillary aid, the most promising being magnetostratigraphy. The position taken concerning system boundaries is to follow the recommendations of the last International Colloquium on Stratigraphy, recommendations based essentially upon faunal considerations. Thus, the boundary of the Triassic–Jurassic lies at the limit between the Rhaetian and Hettangian stages. The boundary between the Jurassic and Cretaceous is still under debate, aggravated by the uncertainties and ambiguities concerning the relations of the Portlandian, Volgian and Tithonian, and no international position has been taken (see Moullade in Volume IIB).

The correlation of continental and marine sequences is accompanied by the usual quota of problems. Traditionally correlation has relied upon the intercalation of marine and terrestrial sequences to enable some standards to be set for marine and non-marine floras and faunas. Recent progress in more physical techniques, geochronology and, to a large extent, in paleomagnetism provide some hope that absolute correlations may eventually arise.

Yet because of the difference in the probabilities of formation and preservation of terrestrial deposits and their flora and fauna, the quality of the data available compared with marine sequences is usually inferior.

It can be noted for example, that the principal floral changes antedate the changes in marine fauna. The crises in the plant world, if the term is appropriate, were in mid-Permian and during the end-Jurassic–Early Cretaceous when the angiosperms were evolving.

It is also clear from the study of the Karroo Supergroup in South Africa that the break between the Paleozoic and Mesozoic is truly artificial. However, as soon as the sequence becomes less than complete, as for example is the case of the Karroo equivalents in Zambia and southern Rhodesia, then application of terms like Permian or Triassic seems more appropriate.

The confusion and problems introduced in attempting to interpret the imperfect record preserved in nature pale before the nomenclatural problems created by geologists. Launched with the commendable aim of providing an acceptable, and unique logical system, the literature is now bedevilled with many schemes of variable merit. While a rigid scheme is obviously undesirable, there are few who have not wished for some international nomenclatural accord such as paleontologists and biologists have been able to maintain. Most elementary text books explain the confusion of lithostratigraphic and chronostratigraphic terms, yet presumably for reasons of convenience or as a type of shorthand, terms are misapplied at a level where this should not occur. Nor is there a single system of chronostratigraphic terms, although Van Eysinga (1975) made a creditable attempt to present in a single table those which have shown the greatest survival ability. There ought to be a single stage name, defined by its stratotype, the latter by historical priority (and/or international agreement). The stratotype may be supported by supplementary reference profiles (hypo- and facio-stratotypes). The reader may judge the extent to which contributors conform to this usage.

As in the 1950's and early 1960's when paleomagnetism provided growing and internally consistent evidence of relative motion which had affected continental blocks over geological time, so in the following decade the evidence of sea-floor spreading grew. The geophysical evidence from the oceanic regions was not only internally consistent and supported the early paleomagnetic work, but also provided a series of vignettes showing the growth of the sea floor and the concomitant continental positions.

Yet there have been few serious attempts to integrate the consequences of plate activity into stratigraphy. This is perhaps because most of the effort has been understanding plate interactions, here called "plate edge effects". These interactions, due to the work of Morgan (1968), Le Pichon (1968), Dewey and Bird (1970), Dewey and Horsfield (1970), Karig (1971) and many subsequent workers, provide a remarkable dynamic interpretation of orogeny. Since the earth's surface is formed of a number of discrete plates it seems reasonable to suppose, that an interaction of two such as will produce a mountain chain will not leave the other plates unaffected. With that in mind consider the question of boundaries between geological systems. Initially these were located at significant breaks in the section, and since there has been such little change in the definition of the systems, it would seem that the units have some utility. A significant break implies at least a significant environmental change, and for cause it is normal to invoke particularly tectonism, i.e. plate interactions. The implicit assumption is that there is tectonic activity whose effect may be global, though not necessarily as precise as Stille (1944) considered, and certainly occurring at more than just system boundaries.

Of equal importance in stratigraphy are "plate surface effects". It has been rather convincingly shown, that given relatively simple assumptions concerning the thermal history of newly formed oceanic plate (McKenzie, 1969; Sleep, 1975) and the rate of ridge spreading, the bathymetry of the corresponding ocean basin may be closely calculated (Sclater et al., 1971). There therefore exists a means, variation in ridge activity, of varying oceanic depth. This might appear in geology as epeirogenic changes in sea level. Under special circumstances large-scale abrupt marine transgressions and regressions might be caused (Russell, 1968; Douglas et al., 1973). It must be stressed that alghough separated for purpose of explanation, in nature edge and surface effects are not so readily separable.

Along the trailing edge of a continent a long continued pattern of sedimentation may exist without obvious sign of plate motion, although the hypothesis would predict a gradual deepening of the basin with increasing distance from the ridge. To the editors' knowledge no one has yet compared the theoretical increase in water depth due to cooling with the accumulated sediment thickness which, with an isostatic compensation factor, might be expected to show some agreement. In contrast close to diverging or converging continental margins the sedimentation pattern may provide a precise record of the principal phases of movement. Thus, while the stratigraphical record cannot prove that movement has occurred it is equally clear that any model of movement inconsistent with the stratigraphic pattern is invalid.

There are many problems concerning the behaviour of continental lithosphere, which cannot as yet be answered, but where a knowledge of the geological history of the region, which may only be derived from stratigraphy may make a considerable contribution. Thus, while in broad terms it is possible to talk of crustal fracture and separation, little is known about the various stages in the process (see however Falvey, 1974). Discussion has centred around the process of graben formation, yet how the fractures develop and penetrate the lithosphere is not known since in the extant grabens, faults do not penetrate more than a few kilometres. However, a remarkably clear picture of the sedimentary response to graben formation in the South Atlantic region has followed from coastal and offshore exploration drilling for petroleum. Although not an issue in the Mesozoic, it is worth pointing out that zones of crust may be reactivated without separation occurring. Why this should occur, or if on the contrary, fracture and separation is the unusual event, is not decided.

The most confusing problem at the present time is

provided by the question of micro-plates, small areas deemed to have rotated based upon paleomagnetic studies, but for which the marine magnetic anomaly patterns are obscure. The introduction of spreading centres for such small regions adds a complicating factor. However, the stratigraphic history of such regions, should place clear constraints upon possible interpretations.

The choice of the Mesozoic has the advantage that as a result of many years of painstaking stratigraphical and paleontological effort a considerable volume of data is available from most parts of the world, and it is no small task to attempt to synthesize and reduce it into a presentable form. It has, however, the advantage that while the flora and fauna are different in many aspects, they begin to take on a form which permits some parallels to be drawn with modern environments. Extending back in time the parallels become less pronounced, the geological record more fragmentary and the oceanic geophysical contribution vanishes.

The areas covered in the present volume are Asia (including the European part of U.S.S.R. and excluding China), Australasia and Africa (excluding northwestern Africa). Europe, the Americas, Antarctic and geology of the ocean floor will occupy the second volume with some chapters discussing some more general characters of the Mesozoic.

ACKNOWLEDGEMENTS

The editors wish to express their gratitude to Dr. H. Frank (Elsevier, Amsterdam, Geo-Sciences Section) for his help and also wish to thank Dr. R.G. Douglas (University of Southern California, Los Angeles) for a critical review of this introductory chapter.

Nice and Columbia, 1977

REFERENCES

Dewey, J.F. and Bird, J.M., 1970. Mountain belts and the new global tectonics. *J. Geophys. Res.,* 75: 2625–2647.

Dewey, J.F. and Horsfield, B., 1970. Plate tectonics, orogeny and continental growth. *Nature,* 225: 521–525.

Douglas, R.G., Moullade, M. and Nairn, A.E.M., 1973. Causes and consequences of drift in the South Atlantic. In: D.H. Tarling and S.K. Runcorn *Implications of Continental Drift to the Earth Sciences 1.* Academic Press, London and New York, pp. 517–536.

Falvey, D.A., 1974. The development of continental margins in plate tectonic theory. *Aust. Pet. Explor. Assoc. J.,* 14: 95–106.

Gignoux, M., 1950. *Géologie stratigraphique.* Masson, Paris, 4th ed., 735 pp.

Handbücher der stratigraphischen Geologie. Enke Verlag, Stuttgart, vols. 1–14.

Karig, D.E., 1971. Origin and development of marginal basins in the western Pacific. *J. Geophys. Res.,* 76: 2542–2561.

Le Pichon, X., 1968. Sea floor spreading and continental drift. *J. Geophys. Res.,* 73: 3661–3697.

McKenzie, D.P., 1969. Speculations on the consequences and causes of plate motions. *Geophys. J.,* 18: 1–32.

Morgan, W.J., 1968. Rises, trenches, great faults and crustal blocks. *J. Geophys. Res.,* 73: 1959–1982.

Newell, N.D., 1952. Periodicity in invertebrate evolution. *J. Paleontol.,* 26: 371–385.

Russell, K.L., 1968. Oceanic ridges and eustatic changes in sea-level. *Nature,* 218: 861–862.

Sclater, J.G., Anderson, R.N. and Bell, M.L., 1971. Elevation of ridges and evolution of the central Eastern Pacific. *J. Geophys. Res.,* 76: 7888–7915.

Sleep, N.H., 1975. Stress and flow beneath island arcs. *Geophys. J.,* 42: 827–857.

Stille, H., 1944. Geotektonische Gliederung der Erdgeschichte. *Abh. Preuss. Akad. Wiss., Math.-Naturwiss., Kl;* 3: 78 pp.

Van Eysinga, F.W.B., 1975. *Geological Time Table.* Elsevier, 3rd ed., Amsterdam.

Chapter 2

SOVIET UNION

N.V. BEZNOSOV, T.N. GORBATCHIK, I.A. MIKHAILOVA and M.A. PERGAMENT

INTRODUCTION

The U.S.S.R. occupies about one sixth of the land surface of the globe and is most varied geologically. Within its limits Mesozoic deposits have accumulated in basins with different tectonic regimes, in various climatic zones, paleobiogeographical provinces and areas. The description of the Mesozoic deposits is most easily done by regions defined by large geostructures, the location of which is indicated on Fig. 1.

The subdivision of Mesozoic deposits into groups (series) and stages practiced in the U.S.S.R. sometimes does not coincide with the West-European scales which are determined both by certain specific features in the faunal and floral provinces and by historical traditions. For instance, in subdividing Lower Triassic marine deposits of the U.S.S.R. and the units of the East-Alpine Triassic, the Induan and Olenekian stages are used. These correspond approximately to the Eotriassic subdivisions of L.F. Spath. The Rhaetian stage is considered to be a part of the Triassic system and, in the majority of regions in the U.S.S.R., can not be separated from the Norian. In the U.S.S.R. the middle division of the Jurassic system consists of the Aalenian, Bajocian and Bathonian stages, while the Callovian stage is included in the upper division. The latter includes in the Boreal region not only the Tithonian, but also the Volgian stage. The Tithonian–Volgian boundary is drawn along the bottom of the Gravesia chronozone.

The main sources of information about the Mesozoic deposits of the U.S.S.R. and the sources of exhaustive references are: in the series of *Stratigraphy of the U.S.S.R., Triassic System* (1972), *Jurassic System* (1972), the volume *Cretaceous System* is now in print, and *Geology of the U.S.S.R.*, and edition in many volumes. The *Atlas of lithological and Paleogeographical maps of the U.S.S.R.* (volume 3, 1968), *Geological Structure of the U.S.S.R.* (volume 1, *Stratigraphy,* 1968) can also be recommended.

The authors take pleasure in expressing their sincerest thanks to Academician V.V. Menner because of his valuable help and advise.

EAST-EUROPEAN PLATFORM

The East-European or Russian Platform occupies the plains of European Russia, north of the Black, Azov, and Caspian seas and west of the Urals mountain range. The basement for the greater part is pre-Riphean, but is Baykalian in the northeastern corner between the Timan and northern Urals. Mesozoic deposits form part of the sedimentary cover of a platform characterized by large structural forms inherited from the Late Paleozoic and Mesozoic.

Triassic

The Triassic deposits of the East-European Platform are represented mostly by variegated beds of continental origin. Owing to scarcity of fossils, the correlation and age determination of strata in different areas is subject to dispute. On the whole, the Triassic system on the East-European platform falls naturally into two groups of deposits. The first, the Lower and, possibly, the lower part of the Middle Triassic belong to the Vetluga and Baskunchak Series, the second includes the top of the Middle (?) and Upper Triassic. In the Vetluga Series there is a predominance of red and variegated alluvial and lacustrine-bog sandstones, clays and marls. In the regions adjacent to the old uplifts of Timan and the Urals rudaceous rocks play a substantial role. In the Ural

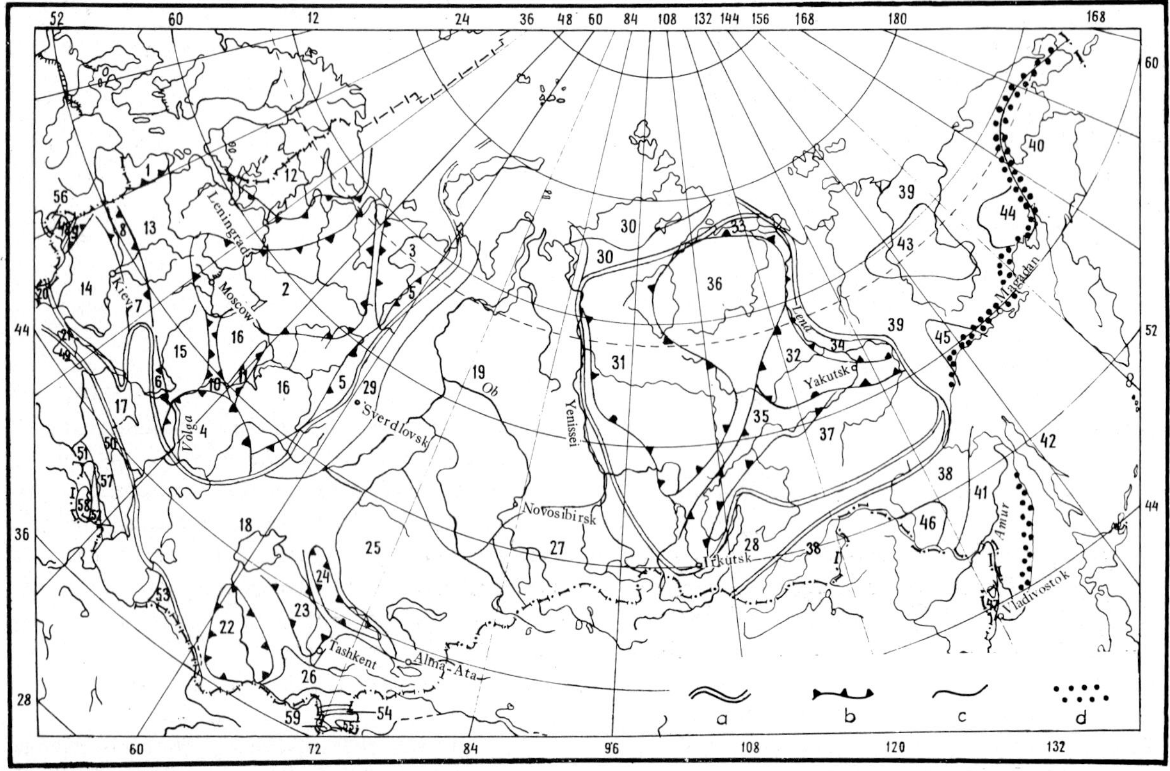

Fig. 1. Geological regionalization scheme of the territory of the U.S.S.R. Legend: *a* = boundaries of heterochronous folded areas and platforms; *b* = boundaries of syneclises, anteclises and shields; *c* = boundaries of other structures; *d* = Okhotsk volcanogene belt.

East-European old platform (*1–16*). Syneclises: *1* = Baltic; *2* = Moscow; *3* = Pechora; *4* = near-Caspian. Basins and depressions: *5* = Preuralian; *6* = Predonetz; *7* = Dnieper–Donetz (= Ukrainian syneclise); *8* = Pripyat; *9* = Lvov and Precarpathian; *10* = Pachelma; *11* = Ulyanovsk–Saratov. Shields and anteclises: *12* = Baltic; *13* = Belorussian; *14* = Ukrainian; *15* = Voronezh; *16* = Volgo-Uralian.

Central Eurasian young platform and its surrounding mountain structures (*17–30*). Plates: *17* = Scythian; *18* = Turanian; *19* = West-Siberian. Syneclises and depressions on the plates: *20* = Moldavian; *21* = north Crimean; *22* = Amu-Darya; *23* = Syr-Darya; *24* = Chu–Sarysu. Shields and active areas: *25* = Kazakh; *26* = Tien Shan–northern Pamirs; *27* = Altay-Sayansk; *28* = Selenga–Stanovoy; *29* = Uralian; *30* = Taymyr.

Siberian old platform (*31–37*). Syneclises: *31* = Tunguska; *32* = Vilyuy. Depressions: *33* = Lena–Khatanga; *34* = pre-Verkhoyansk; *35* = Angara-Vilyuy. Shields: *36* = Anabar; *37* = Aldan.

Pacific belt (*38–47*). Geosynclinal and folded areas: *38* = Mongolo-Okhotsk; *39* = Verkhoyano-Chukotka; *40* = Kamchatka–Koryak; *41* = Sikhote-Alin; *42* = Nippon. Median massifs: *43* = Kolyma; *44* = Omolon; *45* = Okhotsk; *46* = Bureya; *47* = Khankaysk.

Mediterranean fold belt (*48–59*). Geosynclinal and folded areas: *48* = East-Carpathian; *49* = mountainous Crimea; *50* = Greater Caucasus *51* = Adzharo-Trialeti; *52* = Lesser Caucasus; *53* = Kopet-Dag–Balkhan; *54* = central Pamirs; *55* = southeastern Pamirs Median massifs: *56* = Transcarpathian; *57* = Transcaucasian; *58* = Armenian; *59* = southwestern Pamirs.

foredeep the deposits of this series acquire an appearance of molasse. As a rule, the thickness of the Vetluga Series varies from a few tens of hundreds of meters, sharply increasing in the pre-Urals to 1200 m, and in the center of the near-Caspian syneclise to 1775 m. The Baskunchak Series can be clearly isolated in the near-Caspian syneclise only, for there it consists of littoral and marine, variegated and gray clays, marls, limestones, and sandstones up to 470 m thick. Over the remaining part of the East-European Platform the Baskunchak deposits are similar to Vetluga deposits and it is not always possible to separate them. In the Baltic syneclise, variegated marls (to 60 m thick) containing remains of brackish-water organisms belong to the top of the Lower and to the Middle Triassic.

In many regions the greater part of the Middle Triassic is not represented by sediments. Deposits of the Upper, and, possibly, of the Middle Triassic occur over a more restricted area. They are present in the Baltic, Pechora and the near-Caspian syneclises, in the Urals foredeep, the Dnieper–Donetz and Pripyat basins. Middle (?) and Upper Triassic deposits rest on the underlying deposits with an erosional unconformity and are represented by continental variegated and gray slightly coal-bearing sandy-clay deposits. Their thickness varies from a few tens of meters to 200–400 m, in the center of the near-Caspian syneclise to 650 m, and 700–800 m in some syneclises and grabens of the Urals foredeep (Fig. 2).

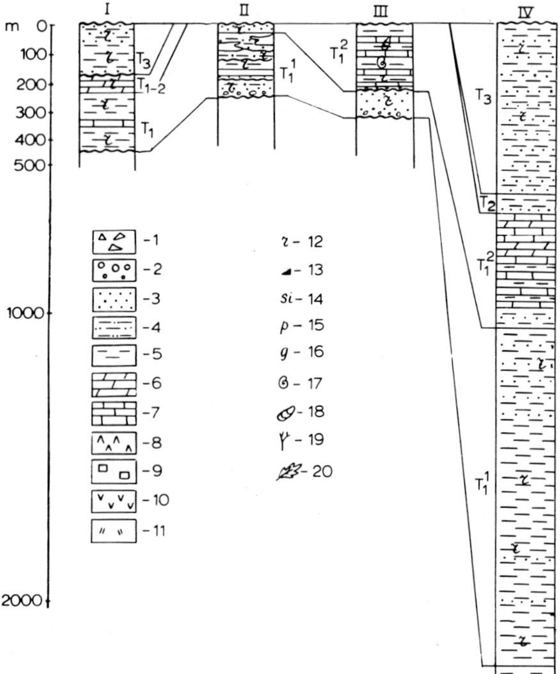

Fig. 2. Correlation of sequences in Triassic deposits on the East-European Platform. *I* = Baltic syneclise, summary sequence (after Yu.L. Kisnerius); *II* = Moscow syneclise, Vyatka River Basin (after E.M. Liutkevich); *III* = flank of the near-Caspian syneclise, Bolshoy Bogdo mountain (after V.A. Gariainov and S.P. Rykov); *IV* = central part of the near-Caspian syneclise, Aralsor borehole (after A.G. Shleifer). Legend: *1* = breccia; *2* = conglomerates, gritstones; *3* = sands, sandstones; *4* = siltstones, clay-silty rocks; *5* = clays, argillites, clay shale; *6* = marls; *7* = limestones and dolomites; *8* = gypsum, anhydrites; *9* = halite; *10* = effusive rocks; *11* = tuffogene rocks; *12* = red rocks; *13* = presence of coal; *14* = silicification; *15* = phosphorites; *16* = glauconite.
Characteristic fossil remains: *17* = ammonoidea; *18* = marine bivalves; *19* = corals; *20* = terrestrial plants.

During the Triassic the East-European Platform was poorly dissected and in the lowlands river and lacustrine-bog deposits accumulated. At the end of the Early Tirassic an arm of the sea penetrated into the northern Caspian region. It is possible that lagoons associated with the coquina sea spread over the Kaliningrad region and Lithuania. The Triassic deposits accumulated in large negative basins such as the Baltic, Moscow, Pechora, and near-Caspian syneclises, Urals and Donetz foredeeps, Dnieper–Donetz and Pripyat basins.

The Triassic fauna of the East-European Platform is mainly represented by the inhabitants of fresh-water and brackish-water bodies and terrestrial forms. Extensively known are localities with labyrinthodonts (Bentosuchidae, Capitosauridae, Mastodonsauridae, etc.) and reptiles (Procolophonidae, Pseudosuchidae, Dicynodontidae). The main role among the invertebrates belongs to the Conchostraca and Ostracoda, the range of which serves as a basis for local zonal subdivision schemes. Marine mollusks including *Ceratites, Dorikranites* and *Tirolites* are known in the near-Caspian (Bolshoy Bogdo Mountain). Early Triassic floras are poor and are represented by xerophytes, a leading role among them belonging to the lycopod *Pleuromeia.* Late Triassic floras are more abundant and varied. Oogonia of charophytes are widely used in the U.S.S.R. for the subdivision and correlation of deposits (Fig. 3).

Jurassic

The Jurassic deposits of the East-European Platform unconformably overlie Triassic and Paleozoic deposits and are represented by continental and marine, predominately gray, clay deposits. A lengthy break in sedimentation occurred at the beginning of the Jurassic period. Continental sandy-clay deposits accumulated only in the Baltic syneclise and to the east of the Dnieper–Donetz Basin. Marine Toarcian, Aalenian, Bajocian and Lower Bathonian clays and sandstones up to 200 m thick occur in the eastern part of the Dnieper–Donetz Basin. At the same time gray alluvial sandstones and conglomerates (up to 100 m thick) and overlying lacustrine-bog coal-bearing deposits (up to 250 m thick) formed in the near-Caspian syneclise. Transgressive Upper Bajocian and Bathonian marine and littoral clays and sandstones occur and are found in the center and in the west of the near-Caspian syneclise, in the Ulyanovsk–Saratov depression, on the flanks of the Voronezh and Vol-

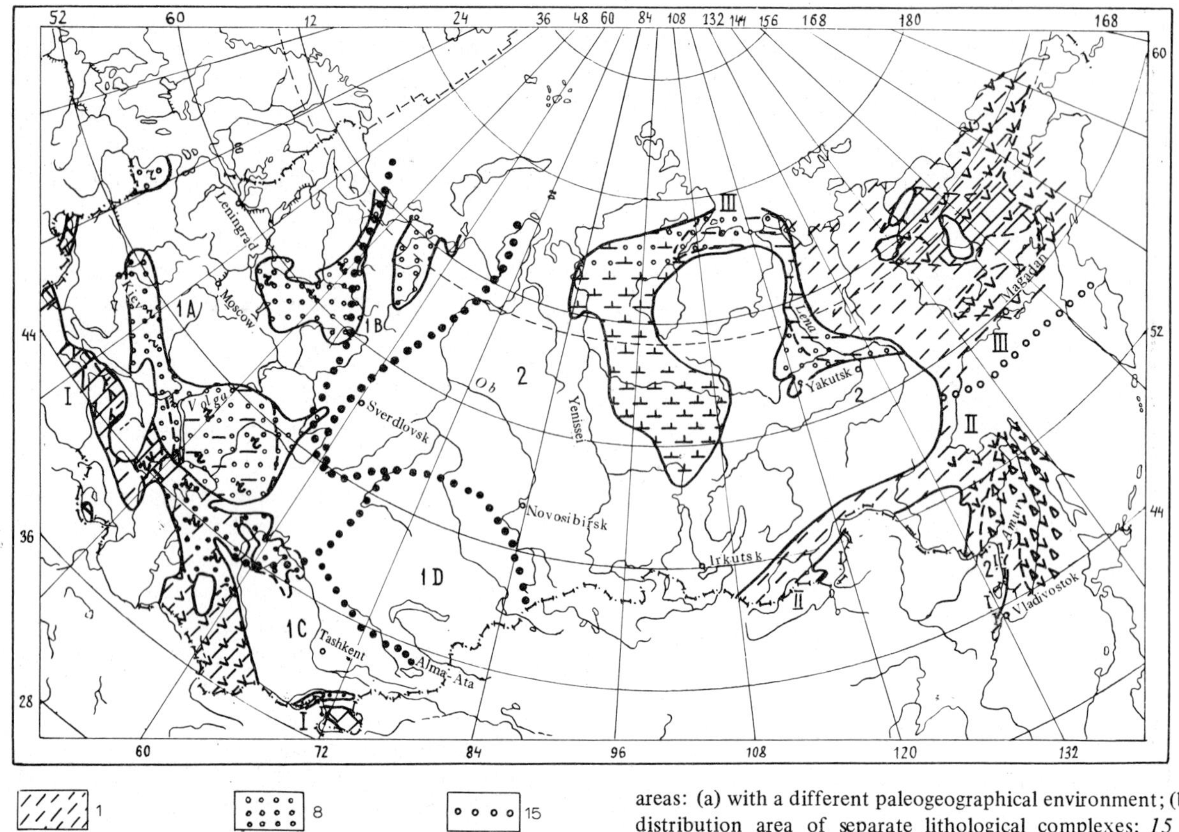

Fig. 3. Distribution scheme of the main Triassic sedimentation basins in the territory of the U.S.S.R. (compiled on the basis of the *Atlas of Lithological–Paleogeographical Maps of the U.S.S.R.*, Volume 3, 1968, simplified and changed). Legend: Sedimentation basins with a geosynclinal and orogenic type of development (*1–6*). Predominant deposits: *1* = marine terrigenous; *2* = continental terrigenous; *3* = marine carbonate; *4* = flysch; *5* = flint rocks; *6* = effusives. Sedimentation basins with a platform type of development (*7–13*). Predominant deposits: *7* = marine terrigenous; *8* = continental terrigenous; *9* = marine carbonate; *10* = salt; *11* = effusives; *12* = traps (a broken sign indicates an intermittent occurrence of the deposits, an alternation and superposition of the symbols means an alternation of deposits); *13* = red colour.
Paleobiogeographical regionalization: *14* = boundaries of areas: (a) with a different paleogeographical environment; (b) distribution area of separate lithological complexes; *15* = boundaries of paleozoogeographical areas and provinces; *16* = boundaries of paleophytogeographical areas and provinces; *17* = boundaries of distribution of Lower Jurassic deposits.

Paleozoogeographical areas and provinces

I = Mediterranean area, Alpine province. T_1: *Tompophiceras, Bernhardites, Dzulfites, Paratirolites, Flemingites, Proptychites, Owenites, Tirolites, Doricranites;* T_2: *Gymnites, Hollandites, Sturia, Laboceras, Procladiscites, Ptychites;* T_3: Arcestidae, *Nairites, Pinacoceras, Halorites, Rhabdoceras.*
II = Pacific Ocean area, Far Eastern province. T_1: *Gyronites, Euflemingites, Paranorites, Proptychites, Subcolumbites, Anasibirites, Keyserlingites, Hemiprionites;* T_2: *Acrochordiceras, Gymnotoceras, Ussurites;* T_3: *Proarcestes, Sirenites;*
III. Boreal area, Yakutsk province. T_1: *Episageceras, Otoceras, Ophiceras, Pachyproptychites, Hedenstroemia, Paranorites, Anasibirites, Sibirites, Keyserlingites, Olenekites, Hemiprionites;* T_2: *Parapopanoceras, Frechites, Nathorstites;* T_3: *Proarcestes, Sirenites, Neosirenites, Juvavites.*

Paleophytogeographical areas and provinces

1 = European–Tien Shan area: *1A* = western part of the area – T_1: Dipteridaceae, *Lepidopteris,* Cycadophyta, Ginkgoales; T_3: Dipteridaceae, *Lepidopteris,* Cycadophyta, *Yuccites,* Ginkgoales, *Podozamites, Pityophyllum. 1B* = Uralian province – T_1: *Neocalamites, Thinnfeldia, Tungussopteris, Pseudoaraucarites;* T_2: *Parasorocaulus, Danaeopsis, Aipteris, Sphenozamites, Yuccites, Tersiella;* T_3: *Aipteris,* Marratiaceae, *Lepidopteris,* Cycadophyta, *Yuccites. 1C* = Ferghana ga–Ural anteclises, in the Dnieper–Donetz and Pripyat basins. Their thickness is usually less than 100 m but reaches to 300 m in the near-Caspian. In the Pechora syneclise and in the north of the Moscow syneclise, in the basin of the Mezen River, clays, sandstones and conglomerates to 100 m thick form the top of the Middle Jurassic. They contain remains of plants and of agglutinated foraminifers. Littoral Bathonian deposits occur also in the Baltic syneclise. In the southern and central parts of the Moscow syneclise Middle Jurassic continental clays and sandstones occur intermittently and fill erosional rills in the pre-Jurassic relief. In the eastern part of the near-Caspian syneclise marine Middle Jurassic deposits become replaced by littoral and continental coal-bearing deposits (Fig. 4).

Over most of the East-European Platform various horizons of the Upper Jurassic rest transgressively on pre-Jurassic rocks. The beds consist mainly of marine clays and marls; sandstones and sand are less common. Limestones play a substantial role in southern regions in the Dnieper–Donetz and Pripyat basins, where they are associated with opokas (gaise). A characteristic feature is the occurrence of phosphorite and glauconite concretions and partings. In the Middle Callovian, oolitic rocks occur extensively. In the Kimmeridgian spongolites, in the Middle Volgian bituminous and combustible shales are common. The deposits are frequently condensed and contain numerous non-depositional horizons, so that sometimes a part of a stage or several stages are represented by a thin conglomerate parting of phosphoritic concretions. The Middle Callovian–Oxfordian deposits are the most extensive. The least are the Volgian. The thickness of Upper Jurassic deposits is usually less than 50 m. In the center of the near-Caspian syneclise it reaches 450–500 m, in the Pechora syneclise 200 m, and in the easternmost downwarped part of the Dnieper–Donetz Basin 350 m is found.

During the Early and the first half of the Middle Jurassic the greater part of the East-European platform was land. A marine gulf penetrated from the south into the eastern part of the Dnieper–Donetz Basin. In the Late Bajocian and in the Bathonian an extensive marine transgression extended from the Tethys Sea into the southern part of the platform and penetrated along the depressions up to the latitude of Kursk and Penza. A regression occurred at the end of the Middle Jurassic and the beginning of the Callovian with a corresponding break in sedimentation or accumulation of lacustrine deposits. During the Late Jurassic the sea occupied a vast part of the East-European platform, connecting the Arctic Ocean with the Tethys and, from time to time, with the Atlantic.

The Early and Middle Jurassic faunas of the East-European Platform are similar to the West-European, but are markedly impoverished, the result of a freshening of the basins. An extensive marine transgression during the Late Jurassic resulted in the abundant and varied Late Jurassic faunas described in many monographs. These faunas are characterized by the substantial role of Boreal elements (genera *Cadoceras, Cardioceras, Virgatites*, etc.).

Cretaceous

Lower Cretaceous deposits occur extensively on the East-European Platform and extend in a meridional stretch from the near-Caspian syneclise to Timan, generally transgressive over the different horizons of the Jurassic, Carboniferous, Devonian and, sometimes, even the Precambrian. All stages of the Lower Cretaceous are present, being replaced by marine and continental facies (glauconitic sands, sandstones, black clays, etc.). The general structural pattern of the East-European platform at the beginning of the Early Cretaceous period was inherited from the Late Jurassic. Throughout the Early Cretaceous the East-European Platform was subjected to numerous paleogeographical changes resulting from tectonic movements in the adjacent geosynclinal areas, the Crimean–Caucasian area in the south and the Carpathian–Balkan region to the west, as well as oscillating movements of the East-European Platform itself.

The oldest (Berriasian) deposits ("Riazan horizon") occur mainly in the near-Caspian and Moscow syneclises and are represented by a thin (0.5–12 m)

province – T_1: Dipteridaceae, Cycadophyta, *Lepidopteris*, Ginkgoales, *Albertia, Podozamites, Yuccites, Ullmannia, Pseudovoltzia;* T_3: Dipteridaceae, Matoniaceae, Marattiaceae, Cycadophyta, Ginkgoales, *Podozamites, Pityophyllum. 1D* = Kazakh province – T_1: Sphenopteridae, *Albertia, Ullmannia, Elatocladus;* T_2: *Parasorocaulus, Danaeopsis, Cladophlebis, Aipteris, Thinnfeldia, Tersiella.*
2 = Siberian area – T_1: *Pleuromeia, Neocalamites, Neokoretrophyllites, Thinnfeldia*, Ginkgoales, *Podozamites, Yuccites, Uralophyllum, Tungussopteris, Katasiopteris, Marchaella, Sphenobaiera, Elatocladus, Pseudoaraucarites;* T_2: *Thinnfeldia, Madygenia, Madygenopteris, Comsopteris;* T_3 (west): *Neocalamites, Cladophlebis*, Ginkgoales, *Yuccites, Podozamites, Pityophyllum, Uralophyllum;* T_3: (east): *Neocalamites*, Cycadophyta, *Podazamites, Pityophullum*, Dipteridaceae.

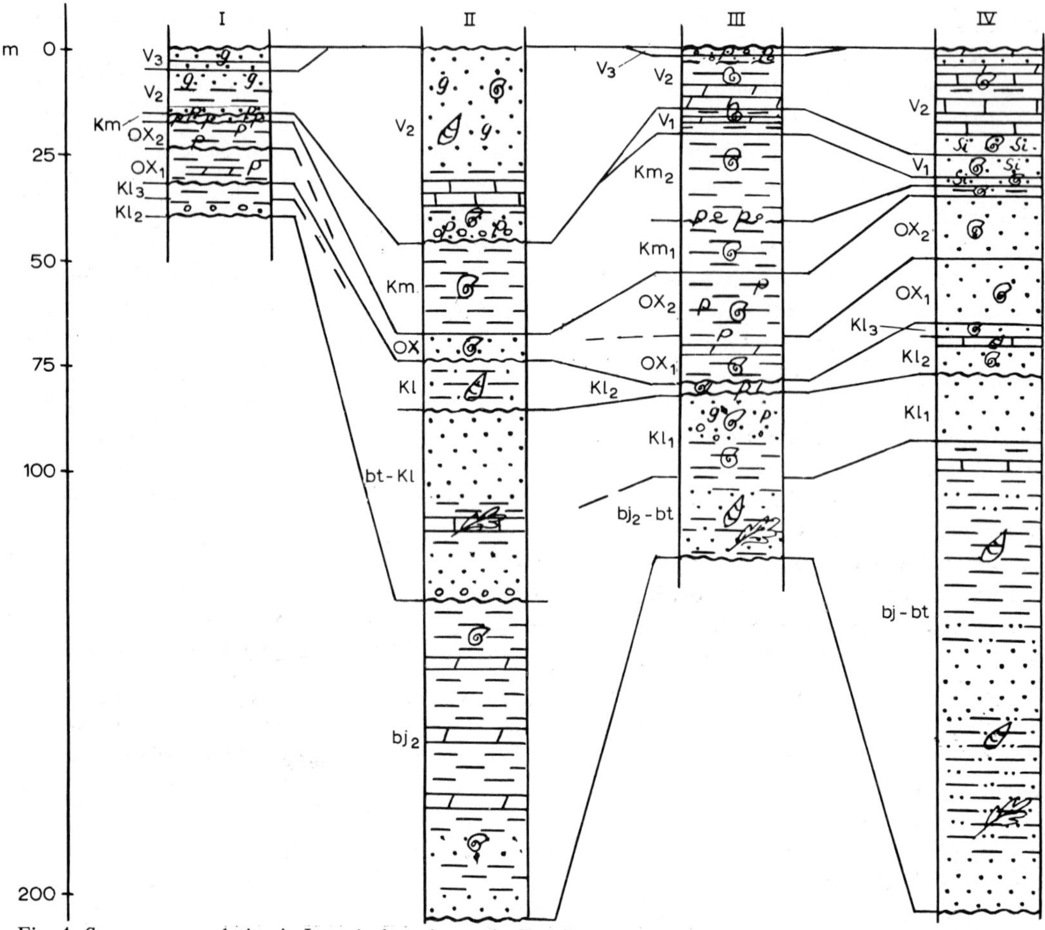

Fig. 4. Sequence correlation in Jurassic deposits on the East-European Platform. Legend on Fig. 3. *I* = Moscow syneclise, territory around Moscow (summary sequence, after P.A. Gerasimov); *II* = Voronezh anteclise, Shebeko borehole (after P.A. Gerasimov); *III* = Ulyanovsk–Saratov depression, territory around Ulyanovsk (summary sequence after Gerasimov et al., 1962); *IV* = northern flank of the near-Caspian syneclise (summary sequence, after V.P. Makridin).

packet of sandy-clay beds with *Riasanites rjasanensis.* In the same regions, as well as in the Ulyanovsk–Saratov depression, Valanginian sands and sandstones are recorded with phosphatic pebbles at the base that contain ammonites, belemnites and pelecypods. The thickness does not exceed 50 m. Hauterivian and Barremian deposits are represented by an intercalation of sands, sandstones and clays, frequently with phosphatic pebbles and containing the ammonites *Leopoldia, Simbirskites, Speetoniceras,* etc., and belemnites. The thickness ranges from a few meters to 100 m. During the Barremian continental variegated clays occur in the near-Caspian syneclise with partings of sands and sandstones containing plant remains and having a thickness up to 160 m. The fullest Aptian and Albian sequences are also recorded in the near-Caspian syneclise. These are gray clays with partings of sands, siderite and phosphatic concretions and ammonites. In the Ukrainian and Moscow syneclises in the Upper Albian, quartz-glauconitic sands and sandstones to 35 m thick predominate.

Three principal phases can be distinguished in the geological history of the platform during the Early Cretaceous. The first stage, the Berriasian to the beginning of the Early Aptian, is characterized by an alternation of transgressions from the Arctic and Mediterranean oceans and regressions, which resulted sometimes in an interruption of connections between the Boreal and the Mediterranean basins. The seas of that time were typically epicontinental, and in them accumulated shallow-water terrigenous sediments. A shallow sea (50–100 m) with numerous islands washed a very slightly elevated land area. Throughout this time various portions in the central part of the East-European Platform repeatedly became a land on which alluvial and lacustrine-bog deposits have been deposited with plant remains. The deepest part of the sea (about 200 m) was that which occupied the near-Caspian and Pechora syneclises.

The second stage, embracing from the end of the Early and the Late Aptian, is characterized by a major regression, as a result of which a substantial part of the central areas and the northwestern part of the near-Caspian syneclise became land, while the middle Russian Late Aptian sea became a gulf of the Caucasian sea. At the end of the Late Aptian the East-European Platform experienced a general uplift. The deposits of marine basins of that time are characterized by a shallow-water sandy-siltstone facies. Aptian continental deposits are represented by cross-bedded sands accumulated on a low-aggradation plain and under littoral conditions. The third stage begins with a transgression of the Albian sea from the Caucasian basin; during the Middle Albian the platform basin is connected with the Polish sea and the sea of the Carpathian foredeep, forming a vast East-European shallow, warm sea with a number of islands. By the end of the Late Albian time the sea acquired a sublatitudinal strike. The direct connection between the Mediterranean and northern basins was interrupted. The northern and, to a certain extent, the southeastern parts of the platform experienced a general uplift, whereas the southwestern part subsided and became covered by the sea.

The absence during the greater part of the Lower Cretaceous of a connection between the sea of the East-European Platform and the seas of Western Europe, as well as a close connection with the Arctic basin, resulted in the development of a specific fauna, which creates difficulties in identification here of units accepted in the West-European stratigraphic geology. Subdivision of marine deposits on the East-European Platform by ammonites and belemnites has been compiled by P.A. Gerasimov mainly on data for the Volga region (Table I).

TABLE I

Subdivision of Lower Cretaceous deposits on the East-European Platform (after Gerasimov, 1971)

Stages	Substages	Zones according to ammonites and belemnites	
ALBIAN	Upper	*Pervinquieria inflata*	
	Middle	*Hoplites dentatus*	
	Lower	*Leymeriella tardefurcata*	
		Hypacanthoplites jacobi	
APTIAN	Upper		
	Lower	*Deshayesites deshayesi*	
		Deshayesites weissi	
		Matheronites ridzewskyi	
BARREMIAN		*Oxyteuthis jasykowi*	
HAUTERIVIAN	Upper	*Simbirskites decheni*	
		Speetoniceras versicolor	
	Lower	*Dichotomites bidichotomus*	
VALANGINIAN	Upper	*Polyptychites polyptychus*	
	Middle	*Polyptychites keyserlingi*	
	Lower	*Pseudogarnieria undulato-plicatilis*	
BERRIASIAN	Upper	*Surites tzikwinianus*	Riazan horizon (Bogoslovsky, 1897)
	Lower	*Riasanites rjasanensis*	

Upper Cretaceous deposits occur extensively in the southern and central parts of the Russian Platform in a sublatitudinal stretch between the coasts of the Baltic Sea and the Urals approximately along the 40° parallel. Further north outcrops in the Pechora syneclise are associated with the ingressions from the Turonian—Senonian Boreal basin.

Among the rocks varieties of carbonate predominate (marls, chalk, limestones). In some places there are opokas, tripoli, flinty marls with radiolaria, and clays. Various sands and sandstones occur quite extensively, frequently containing glauconite and nodular phosphorites. These lithologically varied types of rocks have accumulated in a single sublatitudinal epicontinental basin closely associated with the seas of Western Europe and the Alpine geosynclinal area.

During the Late Cretaceous the general structural plan of the platform corresponded to the pattern which originated at the end of the Early Cretaceous. However, the changing intensity and differentiation of tectonic oscillations of the platform caused frequent, and locally drastic, changes in the paleogeographic environments. Stable marine conditions existed in the strongly subsiding structures in the southern part of the platform (Polish—Lithuanian, Ukrainian, near-Caspian syneclises, the near-Black Sea basin), where we find analogues of the stages belonging to the upper division of the Cretaceous system. Less complete are the sequences in the Moscow and Pechora syneclises, while on the Voronezh and Belorussian anteclises only deposits of the maximum transgressive phases are preserved. The northwestern part of the platforn remained a land area. Elsewhere the sediments were either thin or virtually completely eroded during the Cenozoic (the Ukrainian Shield, the Donetz Basin, the Volga–Ural arch, pre-Urals).

Upper Cretaceous deposits transgressively and unconformably overlie Mesozoic and Paleozoic rocks, resting on the Precambrian in the Voronezh anteclise. Only locally is a gradual transition recorded from the Upper Albian. Typical for the Cenomanian of the platform are shallow-water glauconitic sands and sandstones with phosphate partings and nodules. In the Polish–Lithuanian syneclise deposits of this age are represented by marls and limestones and in the east of the near-Caspian syneclise, where this stage is at its thickest (120 m), by clays. In the southeast and along the periphery of the Ukrainian Shield continental sands and clays containing plant remains occur.

Lower Turonian deposits are recorded locally in the southwest of the platform (chalk-like limestones and marls with *Inoceramus labiatus* Schloth.) and in other places (tripoli in the Kaluga area, sands and clays in the Uralo-Emba region). Upper Turonian deposits, on the contrary, occur extensively: these usually consist of a coarse chalk, marls, limestones (in the south with flints); in the Moscow syneclise and on adjacent structures, opokas and tripoli, in the Uralo-Emba region, clays and sands. In the majority of localities Upper Turonian deposits rest on the Lower Cenomanian and older rocks with a maximum thickness of 60 m. Lithologically monotypic deposits of the Coniacian stage (chalk and chalk-like marls 30–35 m thick in the north of the Donetz Basin) can be separated from Turonian deposits only by the fauna of sea urchins and inoceramids.

The Santonian stage in the south and southwest of the platform is represented by various marls (80–100 m thick in the Lvov trough), and more rarely by clays. In other regions it is transgressive, being represented in the Moscow syneclise and the Volga area mostly by flinty marls, clays, opokas and tripoli, and in the Uralo-Emba region by sandy marls, sands and sandstones (4–5 m thick). In the Ulyanovsk–Saratov depression a characteristic "sponge layer" (0.2–0.7 m) lies at the base of the Santonian and consists of marls with phosphorites, phosphatized mollusks and numerous sponges. The overlying marls, chalk and limestones of the Campanian stage in continuous sequence are 250–300 m thick in the southwest (Lvov trough). In the near-Caspian syneclise the corresponding deposits are represented by thick flinty and sandy-clay rocks.

Deposits of the Maastrichtian stage (consisting of layers with *Discoscaphites constrictus* Sow, and its varieties) occur irregularly and are lithologically varied. In depressions the Lower Maestrichtian is represented by marls and a white chalk replaced along the periphery of the depressions and in the Ulyanovsk–Saratov depression by sandy-clay and flinty rocks. Their average thickness is 30–40 m with a maximum of 120–150 m (Lvov trough). A smaller area is occupied by Upper Maastrichtian deposits: various marls (Polish–Lithuanian, Ukrainian syneclises, western part of the Uralo-Emba region), glauconitic and marly sands and sandstones (Volga area, east of the near-Caspian syneclise). The thickness varies from 5–10 to 50–70 m (Lvov trough).

The Danian stage (with *Hercoglossa, Echinocorys*) is recognized in the near-Caspian syneclise only. It consists of clays with partings of marls to the southwest on Obshchiy Syrt, organic-detritic limestones in the basins of the Utva River and the Shalkar Lake, and various marls (50–70 m) in the Southern Emba region. Correlated with them by microfauna are calcclayey sands (to 130 m) of the Azov–Kuban Basin, gradually replaced upwards by sands and sandstones of the Montian stage.

The distribution of fauna and different types of sediments in the Late Cretaceous basins on the platform have been controlled by climate. On this basis two paleozoogeographic provinces are distinguished: (a) a southwestern province characterized by a predominance of carbonate sedimentation, varied *Belemnitella, Gonioteuthis,* certain *Actinocamax,* sea urchins, ammonites, as well as *Hedbergella, Globotruncana,* etc.; and (b) a northeastern province with an accumulation of not only carbonate, but also sandy-clay and flinty deposits, the extensive occurrence of *Actinocamax, Belemnitella,* rare *Goniotheuthis,* ammonites and sea urchins and an occurrence of the boreal *Oxytoma tenuicostata* Roem. The subdivision of Upper Cretaceous marine deposits of the platform has been effected mainly by belemnites and inoceramids (Table II).

Several stages can be distinguished in the geological history and paleogeography of the East-European Platform during the Late Cretaceous. Downward movements at the end of the Albian continued into the Cenomanian caused a vast transgression (Fig. 5).

TABLE II

Subdivision of Upper Cretaceous deposits (Cenomanian–Campanian) of the Russian Platform (after Najdin, 1964)

Stage	Substage	Zone	Typical faunistic assemblage
Campanian	Upper	*Hoplitoplacenticeras coesfeldiense* Schlüt.	*Hoplitoplacenticeras vari* Schlüt., *Discoscaphites gibbus* Schlüt., *Belemnitella mucronata senior* Now., *Inoceramus balticus* Boehm
	Lower	*Gonioteuthis quadrata quadrata* Blv.	*Belemnitella mucronata senior* Now., *Discoscaphites binodosus* Roem., *I. balticus* Boehm, *Conulus matesovi* Moskv., *Isomicraster gibbus* Lam.
		Oxytoma tenuicostata Roem.	*Gonioteuthis quadrata quadrata* Blv., *G. granulata quadrata* Stoll., *Actinocamax laevigatus laevigatus* Arkh., *Belemnitella praecursor media* Jel.
Santonian	Upper	*Gonioteuthis granulata granulata* Blv.	*Belemnitella praecursor* Stoll. s.l., *Actinocamax verus fragilis* Arkh.
	Lower	*Inoceramus cardissoides* Goldf.	*Belemnitella propinqua* Mob., *B. praecursor* Stoll. s.l., *Actinocamax verus fragilis* Arkh., *Micraster coranguinum* Klein
Coniacian	Upper	*Inoceramus involutus* Sow.	*Inoceramus percostatus* Müll., *I. russiensis* Nik., *Goniocamax lundgreni lundgreni* Stoll., *G. lundgreni excavata* Sinz., *Actinocamax verus subfragilis* Najd., *Micraster coranguinum* Klein
	Lower	*Inoceramus wandereri* And.	*Inoceramus kleini* Müll., *I. deformis* Meek, *Micraster cortestudinarium* Goldf., *Echinocorys gravesi* Desor, *Conulus subconicus* d'Orbigny
Turonian	Upper	*Inoceramus lamarcki* Park.	*Lewesiceras peramplum* Mant., *Scaphites geinitzi* Orb., *Micraster corbovis* Forb., *M. leskei* Desm., *Conulus subconicus* Orb. below: *I. apicalis* Woods, *Conulus subrotundus* Mant. above: *Holaster planus* Mant., *Micraster cortestudinarium* Goldf.
	Lower	*Inoceramus labiatus* Schloth.	*I. hercynicus* Petr.
		Praeactinocamax plenus triangulus Najd.	
Cenomanian	Upper	*Scaphites aequalis* Sow.	*Acanthoceras rhotomagense* Defr., *Praeactinocamax plenus plenus* Blv., *Schloenbachia varians* Sow., *Chlamys aspera* Lam., *Entolium orbiculare* Sow., *Inoceramus pictus* Sow., *I. scalprum* Boehm
	Lower	*Exogyra conica* Sow.	*Entholium orbiculare* Sow., *Chlamys aspera* Lam., *Inoceramus scalprum* Boehm, *Neohibolites ultimus* Orb. below: *Parahibolites tourtiae* Weign. above: *Schloenbachia varians* Sow., *Praeactinocamax primus primus* Arkh.
Albian	Upper	*Pervinquieria* ex gr. *inflata* Sow.	

Gulfs of a latitudinal Cenomanian basin extended far towards the north, and some large islands in the basin supplied clastic material. At the end of the Cenomanian and during the first half of the Turonian the expansion of the transgression stopped. During the second half of the Turonian marine conditions expanded again in many southern and central regions. Owing to differential movements in a number of places (Donetz Basin, the north of the pre-Black Sea basin, the near-Caspian syneclise), there is a distinct interval in sedimentation between the Cenomanian and Turonian. During the Coniacian the basin shrank slightly, indicated by the sequences along the periphery of the syneclises and on the anteclises.

A new wave of subsidences began in the Santonian and the transgression reached its maximum at the beginning of the Campanian (in the southeast during the Early Maastrichtian). The sea overlapped the

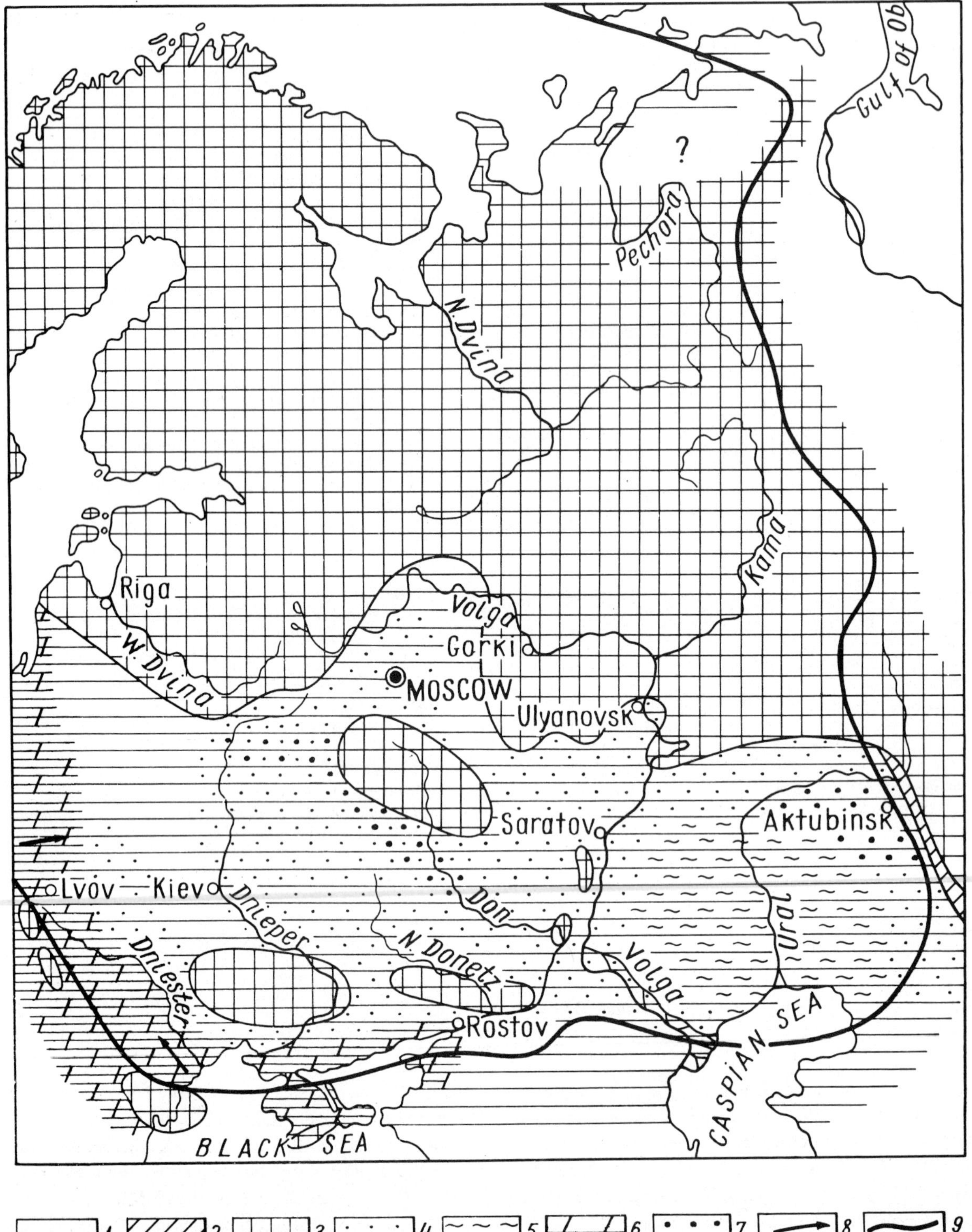

Fig. 5. Paleogeographical scheme of the East-European Platform during the Cenomanian: *1* = accumulation areas of marine sediments; *2* = accumulation areas of continental sediments; *3* = wash-out areas; *4* = sands and sandstones; *5* = clays; *6* = marls; *7* = phosphorites; *8* = directions of the transgressions; *9* = platform boundary (after D.P. Najdin).

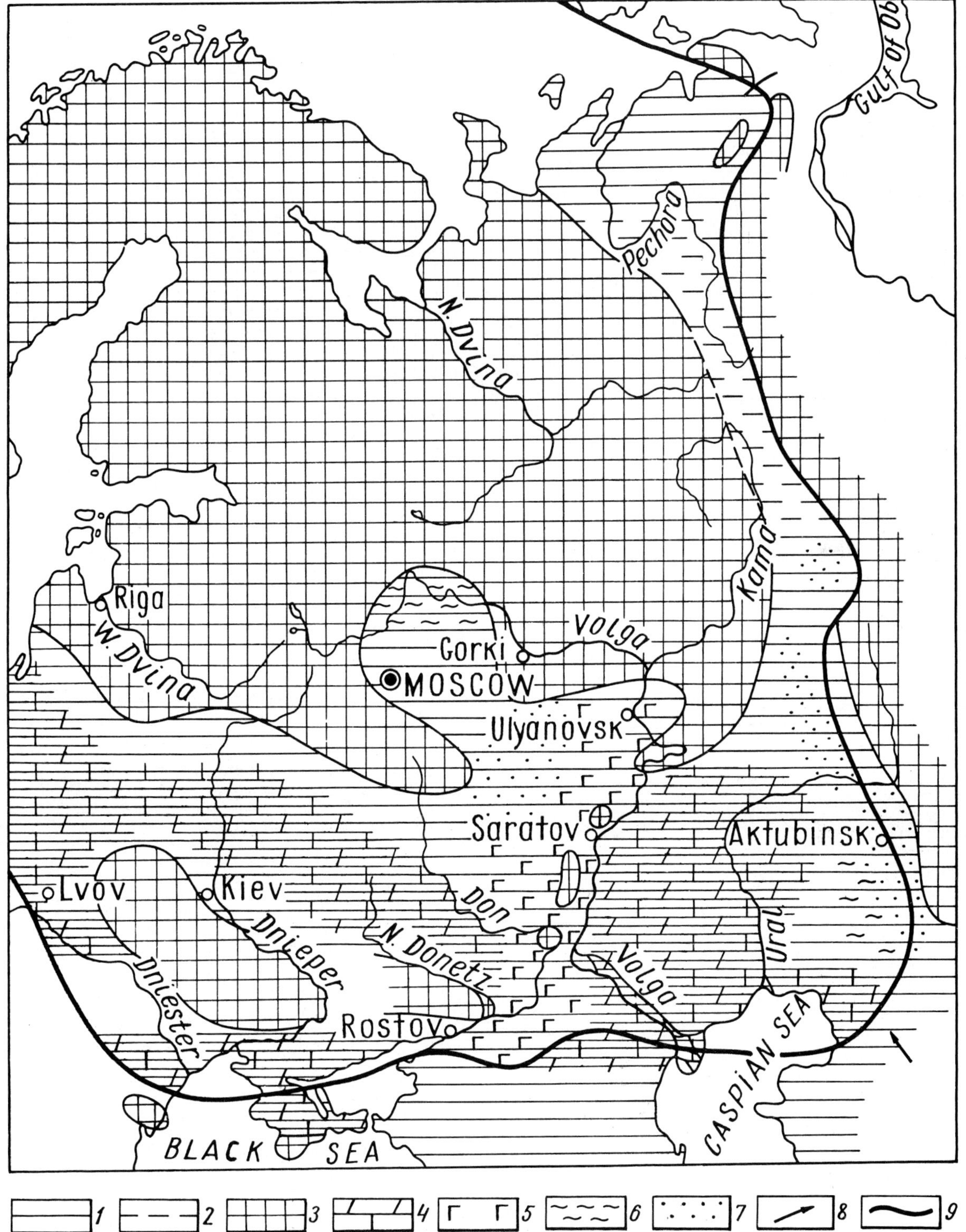

Fig. 6. Paleogeographical scheme of the East-European Platform at the end of the Santonian – beginning of the Campanian epochs [*Oxytoma (Pteria) tenuicostata* time]: *1* = accumulation areas of marine sediments; *2* = ditto, assumed; *3* = wash-out areas; *4* = marls and chalk-like marls; *5* = opokas, flint marls and clays; *6* = clays; *7* = sands; *8* = directions of the transgressions; *9* = platform boundaries (after D.P. Najdin).

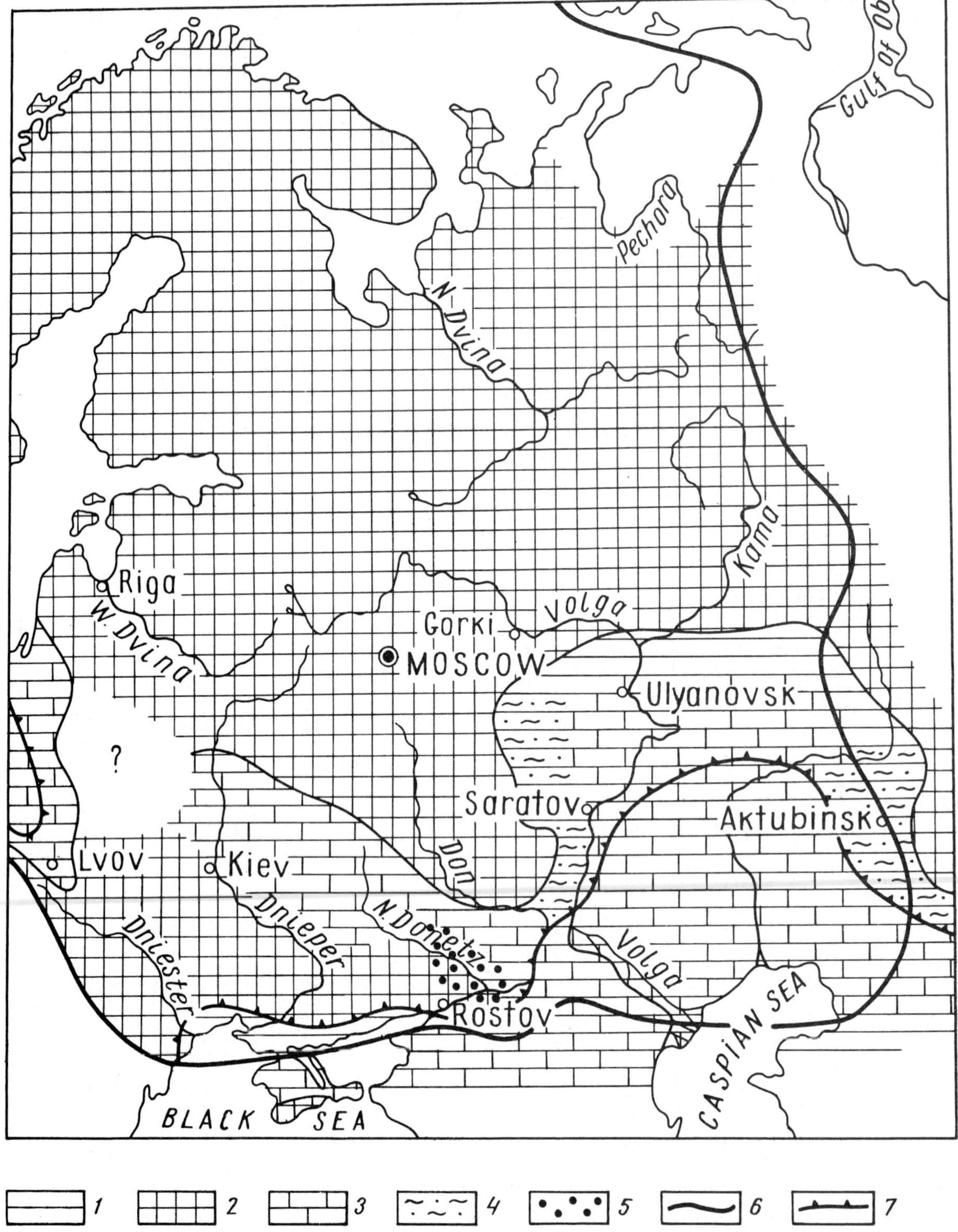

Fig. 7. Paleogeographic scheme of the East-European Platform in the Lower Maastrichtian (*Belemnella lanceolata* time): *1* = accumulation areas of marine sediments; *2* = wash-out areas; *3* = chalk and marls; *4* = sands and clays; *5* = "agglomerate"; *6* = platform boundary; *7* = boundaries of sea basins of the Danian epoch (after D.P. Najdin).

southern part of the Belorussian massif and the eastern part of the Ukrainian Shield. On the Voronezh anteclise phosphorites and brown ironstones of the "Khopersk horizon" accumulated during this time in the littoral warm-water zone. There is a definitive zonality in the distribution of sediments in the southeast of the basin, stressed by A.D. Arkhangelsky: from sands in the littoral zone to pure chalk in the open sea. In the Santonian–Campanian the eastern part of the platform experienced a more intense downwarping than the western part. The transgression developed slowly towards the Urals, so that Santonian deposits and the different horizons of the Campanian (up to the *Belemnitella langei* zone) progressively and unconformably overlie Paleozoic rocks further and further to the east (Fig. 6).

At the end of the Santonian and the beginning of the Campanian the southeastern part of the basin, in addition to the periodical connection by the Turgaysk straits with the seas of Western Siberia, could, possibly, have also had a direct connection with the sea of the Pechora syneclise, which explains the penetration far to the south of the Boreal *Oxytoma tenuicostata* Roem. and some species of *Actinocamax.*

On the shoal, the Campanian–Danian time is characterized on the platform by a gradual reduction of marine conditions. In the Moscow syneclise the basin is liquidated at the end of the Early Campanian. In the Early Maastrichtian a substantial regression is observed; sandy chalk and sands predominate in the composition of the sediments. The single platform basin shrinks and becomes differentiated, but remains connected with the seas of the Caucasus, Crimea and of the Carpathians (Fig. 7).

During the Danian the land continued to expand rapidly and in the west the coastline extended beyond the present limits of the U.S.S.R. In the southeast marine conditions contined to exist over a limited territory.

CENTRAL EURASIAN YOUNG PLATFORM AND THE SURROUNDING MOUNTAIN STRUCTURES

The Central Eurasian young platform and the epiplatform orogens include the areas of completed Paleozoic folding located between the old East-European and Siberian platforms, the Late Mesozoic and Alpine structures of the Mediterranean folded belt and the Mesozoides of the Pacific belt. The territory includes: the northern Black Sea region, the Steppe Crimea, the Precaucasian plain, part of the northern foothills of the Caucasus, the deserts of Central Asia, the Usturt Plateau, Tien Shan, northern Pamirs, Altay, Sayany, the northwestern part of the Transbaykal region, the Urals, the West-Siberian lowland and Taymyr. Common to this entire region is a development transitional from geosynclinal and orogenic to the platform type of the Mesozoic.

In a very general form the Central Eurasian Platform and their epiplatform orogens can be divided according to their geological history during the Mesozoic into the following elements (Fig.1): the Scythian, Turanian and West-Siberian plates, the Kazakh Shield, the Tien Shan–northern Pamirs, Altay–Sayansk, Selenga–Stanovoy, Urals and Taymyr active areas.

Triassic

The Triassic deposits of the Central Eurasian Platform and of the epiplatform orogens are represented by geosynclinal, orogenic and platform formations. Geosynclinal and, partly, orogenic Triassic formations of these areas are a part of the basement complex. The platform and orogenic formations in part are found in the lower structural level of the platform sedimentary cover.

On the Scythian and Turanian plates there are five areas where Triassic deposits occur. The first is the Dniester–Prut interfluve in Moldavia, where drilling has penetrated badly broken-up sandstones, argillites and limestones of Middle and Late Triassic age. The second area includes the western Precaucasus, west of the Laba River and south of Yeisk Peninsula, extending under the central part of the Sea of Azov and into the eastern Crimea. The succession consists of dislocated argillites in a flysch-like alternation with siltstones containing parts of sandstones, limestones and altered effusives. All three divisions of the Triassic are represented by an incompletely exposed thickness of 2000 m. Along the outer, northern and eastern margins of this area the flyschoid deposits become replaced by slightly disturbed terrigenous-carbonate deposits.

The third area includes the southern part of the Turanian plate. Here on the Karabilsk high, on Badghis and north of Kopet-Dag drilling located a thick mass (over 1500 m) of black dislocated argillites with lenses of effusives that contain the Carnian *Halobia.*

The fourth area of Triassic deposits is located in the eastern Precaucasus between the lower course of the Terek River and the Volga Delta. The Lower and the bottom of the Middle Triassic consist here of limestones, including reef limestone, alternating with packets of argillites towards the top. The thickness of these deposits reached 1000 m. The top of the Middle Triassic is represented by 400 m of variegated argillites and siltstones with lenses of limestone in its middle part. To the Upper Triassic belong lavas and pyroclasts, alternating with variegated conglomerates, sandstones and argillites, with a total thickness up to 1000 m.

The fifth and greatest area of Triassic deposits is in the region of the Mangyshlak, Usturt and Tuarkyr and, extending under the Sea of Aral, to the lowest courses of the Syr Darya River. In this area Triassic deposits are the thickest and most complete to west of the Mangyshlak Mountains, in the Karatau mountain range. The Lower Triassic begins with a sequence of gray and red sandstones, alternating at the top with red siltstones and argillites ranging in thickness from 500 to more than 1500m. The top of the Lower Triassic is represented by marine argillites and siltstones with partings of sandstones and limestones containing the ceratites *Tirolites, Dinarites, Nannites* and bivalves and is up to 1000 m thick. The Middle Triassic consists of variegated argillites and sandstones with partings of conglomerates with a total thickness of up to 600 m. The Upper Triassic occurs with an erosional contact and is represented by a coarsely cyclic flyschoid mass of alternating sandstones, argillites and limestones with the remains of bivalves and ichthyosauri. The thickness of the Upper Triassic comes to 2500 m.

In the southern part of the active area of the Tien Shan–northern Pamirs Triassic deposits are exposed in the southwestern Darvaza and on Peter I and Zaalaysk mountain ranges. In Darvaza the Lower Triassic comprises up to 900 m sandstones, siltstones and clays with *Meekoceras, Pseudosageceras, Dieneroceras,* etc. The Middle and Upper Triassic consists of conglomerates and sandstones with partings of tuffs and a total thickness of 600 m. In the Zaalaysk mountain range Triassic deposits [Middle and Upper (?)] are represented by 850 m of conglomerates and sandstones, alternating with tuffoconglomerates and tuffs.

On the Taymyr Peninsula Triassic deposits occur south of the Byrranga mountain range. The most complete sequence is in the east of the peninsula on Cape Tzvetkova. The lower part of the Lower Triassic and the Upper Permian consist of up to 3500 m thick basic effusives. The top part of the Lower, the Middle, and the lower part of the Upper Triassic consists of up to 900 m thick marine and terrigenous rocks with packets of coal-bearing beds of Ladinian age. The Norian and Rhaetian stages are also represented by coal-bearing deposits.

The Urals, West-Siberian plate, Kazakh Shield, Altay–Sayansk and Tien Shan–northern Pamirs (with the exception of Darvaza), active areas during the Triassic, represent a dissected land. Sedimentation occurred in narrow valleys associated with grabens, linear near-fault depressions and synclines. Despite the limited size of some of the depressions, the thickness of deposits is often quite substantial (up to 3500 m). During the Early and at the beginning of the Middle Triassic, deposition of variegated conglomerates and sandstones predominated; at the end of the Middle and during the Late Triassic, coal-bearing deposits formed. In the majority of depressions varying amounts of volcanic deposits are present, sometimes predominating over the terrigenous rocks. They are, usually, represented by basalts and tuffs, more rarely by dolerites and liparites. In the western Transbaykal region the Triassic is also represented by thick (in excess of 1000 m) effusives and pyroclastic beds of varying composition filling some of the depressions.

Thus, the greater part of the Central Eurasian Platform and its surrounding Paleozoic folded structures was a land area during the Triassic with a rather complex dissected relief. A sea extended along the margins of the area on the Scythian and Turanian plates, and in the southern part of the Tien Shan–northern Pamirs (Darvaza) and Taimyr active areas.

Jurassic

In the Jurassic deposits of the Central Eurasian Platform gray terrigenous and carbonate formations predominate with a substantial development, during the Late Jurassic, of variegated and sulphate-halite deposits. Jurassic deposits form the bottom part of the platform cover on the Scythian, Turanian and West-Siberian plates. In the active regions orogenic formations occur.

In the western and southern parts of the Scythian plate Jurassic deposits accumulated in depressions closely associated with the Mediterranean geosynclinal belt. Lower Jurassic and Aalenian marine deposits occur in the southern Precaucasus plain and in the foothills of the Caucasus. Between the Laba River and the Sea of Azov they probably form part of the flyschoid strata, the greater part of which belongs to the Triassic. East of the Laba River Lower Jurassic and Aalenian deposits rest unconformably on the Triassic and Paleozoic deposits and are represented by thick (up to 2000 m) Sinemurian–Aalenian argillites. Along the periphery of the Stavropol Massif marine deposits are partly replaced by intermittent littoral coal-bearing sandstones. Upper Bajocian and Lower Bathonian deposits occur transgressively. They consist mostly of clays and argillites, the thickness of which amounts to 1800 m in Moldavia, between the Dniester and the Prut, but only 600–800 m in the Precaucasus. In the northern Crimea and on the northeastern coast of the Sea of Azov drilling has penetrated volcanic rocks of intermediate composition under Jurassic and Cretaceous beds which may be of Middle Jurassic age. Upper Jurassic deposits are transgressive and best represented in Moldavia and in the Precaucasus. The Callovian consists of clays, sandstones and limestones and is frequently fully or partly condensed. The Oxfordian and bottom part of the Kimmeridgian is represented mostly by limestones and dolomites up to 900 m in thickness in Moldavia but not more than 300 m thick in the Precaucasus. The top part of the Kimmeridgian and the bottom part of the Tithonian consist of variegated clays, sandstones, anhydrites and salt. The latter are associated with depressions, and locally may be up to 1500 m thick. The upper part of the Tithonian occurs in the Precaucasus, where it is represented by transgressive limestones less than 100 m thick.

The northeastern part of the Scythian plate (eastern Precaucasus) and the Turanian plate from the Early–Middle Jurassic to the beginning og the Late Jurassic formed part of a sedimentary basin in the process of formation. It extended eastwards up to the Tadzhik depression. Pre-Upper Toarcian and Lower Jurassic deposits within this basin occur intermittently. They are represented by alluvial and lacustrine facies and fill the axial parts of depressions and grabens in the process of formation. Thin weathering crusts and proluvial and lacustrine deposits formed on the flanks of uplifts. Since the end of the Toarcian there has been a gradual and progressive subsidence of this entire territory (Fig. 8).

In the eastern Precaucasus there is a cyclic development of alternating packets of littoral sandstones

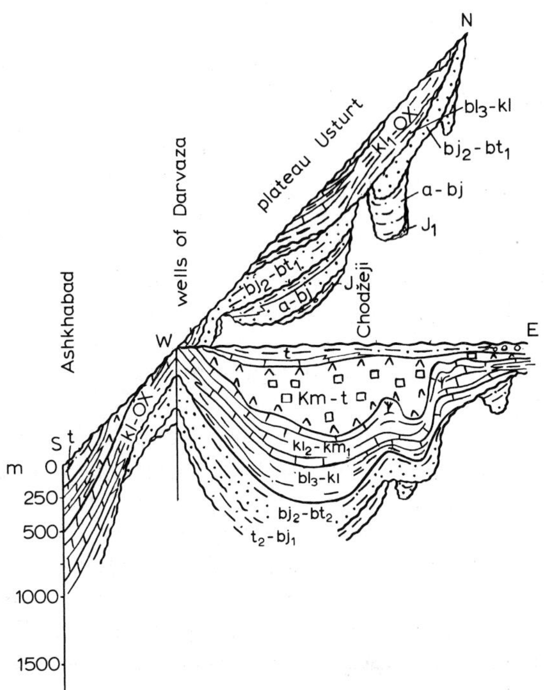

Fig. 8. Scheme of the structure of Jurassic deposits on the Turanian plate. The roof of Jurassic deposits is accepted as the zero surface. Legend on Fig. 2.

and marine clays up to 700 m thick. It covers an interval from the top of the Toarcian to the top of the Bajocian. On the Turanian plate these deposits are also cyclic. However, littoral sandstones are replaced by alluvial sediments and marine clays by littoral and lacustrine-bog coal-bearing facies. This replacement occurs gradually from the west to the east and from the south to the north. The maximum transgression is reached at the end of the Upper Bajocian and in Early Bathonian times, with the marine deposits occurring eastwards up to the foothills of the Gissar mountain range. The thickness and stratigraphic completeness of the sequence are very variable. The fullest sequences embrace an interval from the top of the Toarcian to the Middle Bathonian and are 800–1000 m thick. Over the uplifts the lower units of the sequence wedge out.

Upper Bathonian–Lower Callovian deposits in the Amu-Darya syncline and in the southwestern foothills of the Gissar mountain range are represented by a single unit of clays, marls and limestones up to 200 m thick. It is separated from the underlying deposits by a condensed horizon which changes into a washout surface. In the southern and eastern Usturt this marine rock body is replaced by variegated clays with fossil soil horizons. In the eastern Precaucasus, and the Mangyshlak, Tuarkyr and western Ustsurt Callovian deposits are represented by a sequence of transgressive sandstones, clays and limestones. Oxfordian deposits in the northern part of the sedimentary basin, in the eastern Precaucasus, south of the Kuma River and on the Turanian plate, in the southern and eastern Karakumy are represented by limestones and dolomites. In Mangyshlak and Ustsurt they are replaced by marine clays, marls and sandstones with partings of limestones, and in the east of Ustsurt by red and variegated clays and sandstones.

At the end of the Oxfordian and in the Kimmeridgian there was a regression, and the basin was divided into a number of semi-isolated basins, where salt and anhydrites accumulated. The largest of these basins is found in the Amu-Darya syneclise, where sand and anhydrite had accumulated to a thickness of 1500 m. In the Kimmeridgian the sea penetrated from the north into western Ustsurt and into the depressions of Mangyshlak. In the Tithonian the marine transgression expanded, and along the depressions the Volgian sea of the East-European Platform became connected with the Mediterranean basin through the Ustsurt, Amu-Darya syneclises and Kopet-Dag. During this transgression a packet of limestones and sandstones 50–100 m thick was deposited. At the end of the Tithonian the regression affected the entire territory reviewed and in the Amu-Darya syneclise and in adjacent regions of the southwestern foothills of the Gissar mountain range red clays and sandstones of a lagoonal and alluvial origin accumulated.

The geological history of the West-Siberian plate during the Jurassic is in many ways similar to that of the Turanian plate. During the Early and Middle Jurassic a sedimentary basin on the platform was being formed and filled by coal-bearing alluvial and lacustrine-peat deposits up to 500 m thick; thicker deposits are rare. The lowest horizons occur only in the depressions, whereas the upper beds overlie old uplifted areas. Callovian deposits are represented by littoral and marine clays and sandstones, the overlying Upper Jurassic marine clays belonging to the Oxfordian, Kimmeridgian and Volgian stages. Locally they contain packets of sandstones, spongolites and beds of organic-detritus limestones. The presence of glauconite is characteristic. The thickness of the Upper Jurassic is 300–400 m.

In the Taymyr active area Jurassic deposits occur from the lower course of the Yenisey River to the Lena Delta and extend southwards to the northern subsiding margin of the Siberian platform. The Lower–Middle Jurassic deposits are 1300 m thick and are represented by intercalated sandstones, siltstones and clays with separate partings of conglomerates and limestones. Hettangian and Sinemurian deposits are rather tentatively identified. Upper Pliensbachian, Toarcian and Aalenian deposits, along with other fossils, contain genera and species of ammonites present in Europe, such as *Amaltheus, Dactylioceras* and *Ludwigia. Normannites* has been identified in Bajocian deposits; in Bathonian deposits Oppeliidae are present along with Arctic forms of the Macrocephalitidae. Upper Jurassic deposits up to 650 m thick consist of siltstones, clays and sandstones with abundant ammonite remains which makes it possible to distinguish all the stages and provincial zones.

During the Jurassic the Ural, Tien Shan–northern Pamirs and Altay–Sayansk activate areas as well as the Kazakh Shield formed a dissected land mass. Sedimentation took place in grabens, linear depressions and isometric basins. Compared with the Triassic, the area of sedimentation was greater. In the Early and Middle Jurassic coal formation predominated frequently with rudaceous deposits, replaced during the Late Jurassic in the southern regions by an accumulation of coal-free sandstone beds and variegated deposits. Marine and littoral deposits occur at the top of the Middle Jurassic and occur in the depressions of the eastern flank of the Urals in the Upper Jurassic. Upper Jurassic salts and anhydrites associated with variegated deposits occur in Darvaza and the Zaalaysk mountain range. The role of volcanic rocks is insignificant. The latter are known in Lower–Middle Jurassic deposits of the Ferghana mountain range and the Zaalaysk range. Despite the small size of certain depressions, the thickness of deposits accumulated sometimes exceeds 2000–3000 m. In the western Transbaykal region Jurassic coal-bearing deposits fill some depressions. They differ in the completeness of the sequence and thickness of the deposits, frequently exceeding 2500 m. In addition to coal-bearing sandstones and conglomerates volcanic rocks occur extensively and may sometimes be the dominant member in a sequence.

Thus, during the Early and Middle Jurassic the greater part of the Central Eurasian Platform was a land area. The sea occupied marginal depressions of the Scythian plate and opened into the geosyncline

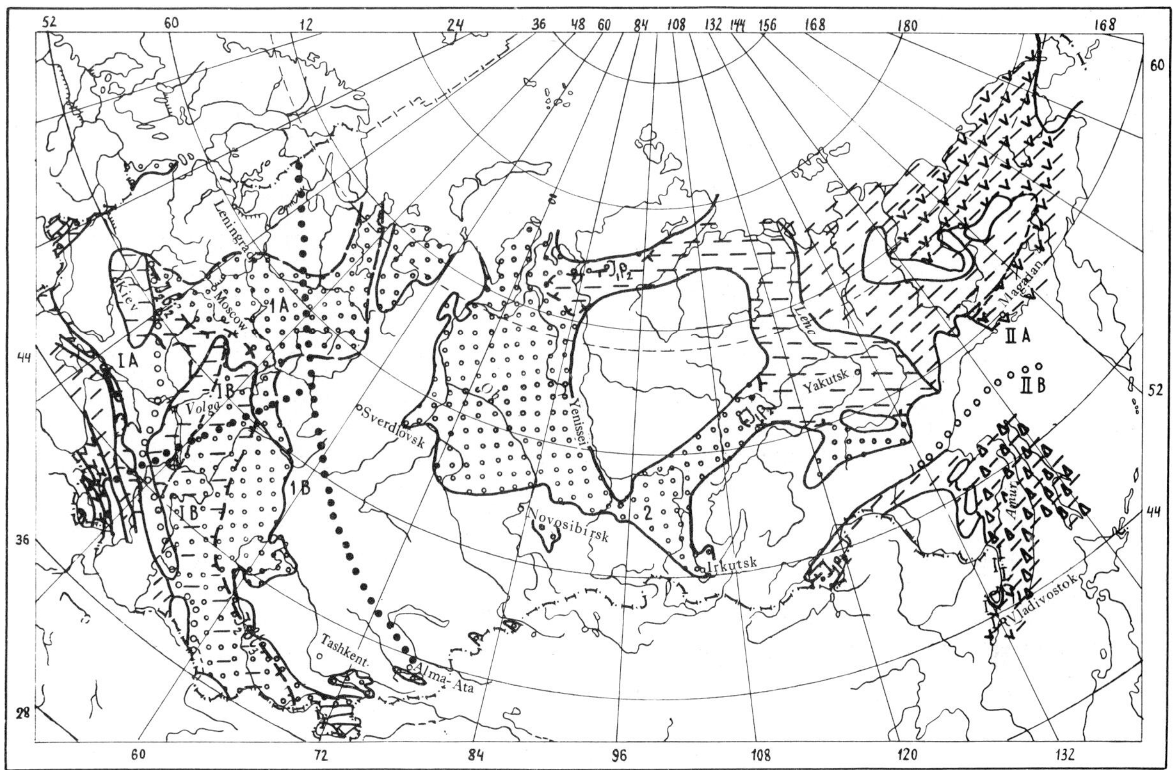

Fig. 9. Distribution scheme of the main Early and Middle Jurassic sedimentation basins in the territory of the U.S.S.R. (compiled on the basis of the *Atlas of Lithological–Paleogeographical Maps of the U.S.S.R.*, Volume 3, 1968 (simplified and changed). Legend on Fig. 3. Paleobiogeographical regionalization.

Paleozoogeographical areas and provinces

I = Mediterranean area. *IA* = Caucasian province – Hettangian–Sinemurian: *Schlotheimia, Arietites, Echioceras;* Pliensbachian: *Tragophylloceras, Uptonia, Arieticeras, Amaltheus;* Toarcian: *Grammoceras, Hildoceras, Dactylioceras, Dumortieria, Pleydellia;* Aalenian: *Leioceras, Ludwigia, Graphoceras, Hammatoceras, Tmetoceras;* Bajocian: Phylloceratida, Lytoceratida, *Hyperlioceras, Sonnina, Witchellia, Dorsetensia, Oppelia, Otoites, Stephanoceras, Leptosphinctes, Strenoceras, Parkinsonia;* Bathonian: *Oxycerites, Cadomites, Procerites, Siemiradzkia, Oraniceras. IB* = near-Caspian province (pre-Upper Bajocian deposits are represented by continental facies) – Bajocian: *Parkinsonia, Pseudocosmoceras;* Bathonian: *Procerites, Siemiradzkia, Choffatia, Tulites, Bullatimorphites, Clydoniceras, Oxycerites.*

II = Boreal area. *IIA* = East-Siberian province – Hettangian–Sinemurian: *Psiloceras, Schlotheimia, Arietites, Otapiria;* Pliensbachian: *Amaltheus, Harpax;* Toarcian: *Dactylioceras, Pseudolioceras, Harpoceras, Arctoites;* Aalenian: *Pseudolioceras, Leioceras, Ludwigia, Arctoites, Retroceramus;* Bajocian: *Arctoites, Retroceramus;* Bathonian: *Oxycerites, Arctocephalites, Cranocephalites, Retroceramus. IIB* = Far Eastern province – Hettangian–Simenurian: *Psiloceras, Schlotheimia, Arnioceras;* Pliensbachian: *Amaltheus, Harpax;* Toarcian: *Dactylioceras, Grammoceras, Phymatoceras, Pseudolioceras;* Aalenian: *Pseudolioceras, Ludwigia, Hammatoceras, Retroceramus;* Bajocian: *Stephanoceras, Retroceramus;* Bathonian: *Arctocephalites, Retroceramus.*

Paleophytogeographical areas and provinces

1 = Ind.-European area. *1A* = European province – J_1: *Marattiopsis, Dictyophyllum, Thaumatopteris, Phlebopteris, Ptilophyllum, Stachyotaxus;* J_2: *Marattiopsis, Dictyophyllum, Phlebopteris, Ptilophyllum, Otozamites, Cycadites, Brachyphyllum, Ctenozamites. 1B* = Central Asiatic province – J_1: *Marattiopsis, Dictyophyllum, Thaumatopteris, Phlebopteris, Osmundopsis, Cladophlebis, Anomozamites, Ptilophyllum, Ginkgoales, Ferganiella;* J_2: *Marattiopsis, Dictyophyllum, Phlebopteris, Ptilophyllum, Otozamites,* Ginkgoales, *Ferganiella, Brachyphyllum;*

2 = Siberian area: J_1: Ginkgoales, *Cladophlebis, Raphaelia, Ferganiella;* J_2: *Raphaelia,* Ginkgoales, *Butefia, Ferganiella.*

of Greater Caucasus, forming a strait in the southern part of the active Taymyr area and, possibly, even penetrated into the northern regions of the West-Siberian plate (Figs. 9, 10).

Cretaceous

The main features of the Early Cretaceous structure of the Central Eurasian Platform were inherited

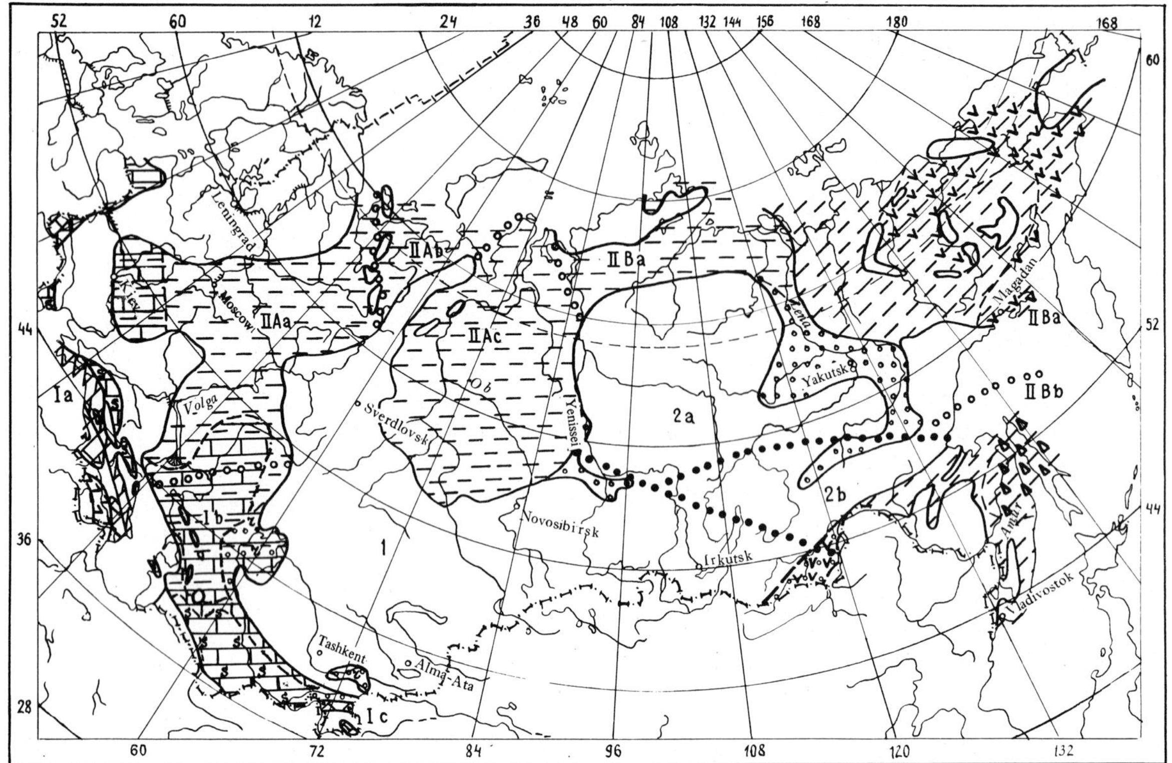

Fig. 10. Distribution scheme of the main Late Jurassic sedimentation basins in the territory of the U.S.S.R. (compiled on the basis of the *Atlas of Lithological–Paleogeographical Maps of the U.S.S.R.*, Volume 3, 1968). Legend on Fig. 3. Paleobiogeographical regionalization

Paleozoogeographical areas and provinces

I = Mediterranean area. *Ia* Caucasian province – Callovian: *Macrocephalites, Quenstedtoceras, Kepplerites, Kosmoceras, Peltoceras, Reineckeia;* Oxfordian: *Cardioceras, Perisphinctes;* Kimmeridgian: *Taramelliceras, Sutneria, Lithacoceras, Aspidoceras;* Tithonian: *Glochiceras, Neochetoceras, Lithacoceras, Virgatosphinctes,* Berriasellinae. *Ib* = near-Caspian province – Callovian: *Macrocephalites, Quenstedtoceras, Kepplerites, Kosmoceras, Peltoceras, Reineckeia;* Oxfordian: *Cardioceras, Perisphinctes, Gregoryceras, Ochetoceras;* Kimmeridgian: *Ataxioceras, Lithacoceras. Ic* = Pamirs province – Callovian: *Macrocephalites, Indocephalites, Reineckeia;* Oxfordian: *Perisphinctes.*

II = Boreal area. *IIA* = Atlantic subarea. *IIAa* = middle Russian (East-European) province – Callovian: *Cadoceras, Quenstedtoceras, Kosmoceras,* Oxfordian: *Cardioceras,* Perisphinctinae; Kimmeridgian: *Aspidoceras, Rasenia, Aulacostephanus,* Ataxioceratinae; Volgian: *Dorsoplanites, Virgatites, Craspedites. IIAb* = Pechora province – Callovian: *Arcticoceras, Cadoceras, Quenstedtoceras;* Kimmeridgian: *Aulacostephanus, Rasenia;* Volgian: *Dorsoplanites, Virgatites, Craspedites. IIAc* = West-Siberian province – Callovian: *Macrocephalites, Cadoceras, Quenstedtoceras, Kosmoceras;* Kimmeridgian: *Aulacostephanus, Rasenia;* Volgian: *Dorsoplanites, Craspedites. IIB* = Arctic subarea. *IIBa* = North-Siberian province – Callovian: *Arcticoceras, Cadoceras;* Kimmeridgian: *Euprionoceras, Haplocardioceras, Aulacostephanus, Rasenia;* Volgian: *Dorsoplanites, Chetaites, Craspedites. IIBb* = Far Eastern province – Callovian: *Cadoceras;* Tithonian (Volgian): *Virgatosphinctes, Primorites,* Berriasellinae.

Paleophytogeographical areas and provinces.

1 = Indo-European area: *Pachypteris, Otozamites, Ptilophyllum, Zamites, Brachyphyllum.*

2 = Siberian area – *2a* = Lena province: *Raphaelia,* Ginkgoales, *Heilungia,* Pinaceae. *2b* = Amur province: *Eboracia, Raphaelia,* Ginkgoales, *Heilungia, Butefia,* Pinaceae.

from the Jurassic, the principal difference being the more extensive area of subsidence and sedimentation. Marine gray detrital glauconitic and detrital-carbonate beds are predominant.

On the Scythian plate the Lower Cretaceous deposits form a virtually continuous cover. The thickest and most complete sequences occur in the marginal depressions of the plate associated with the Mediterranean geosynclinal belt, in Moldavia, in the grabens of the northern part of the Black Sea, in the southern part of the Steppe Crimea, in the western and in the eastern Precaucasus. Lower Cretaceous deposits transgressively overlie the different stages of the Jurassic, Triassic and older deposits. The transgression was intermittent and progressed from the Mediterranean area until it reached its maximum extent at the end of the Aptian and Albian times.

In the Steppe Crimea the oldest, Berriasian, deposits occur only in the zone of the present foothills, where they are represented by a flyschoid mass up to 600 m thick. Later the transgression extended northwards over the entire Steppe Crimea, and here Lower Cretaceous deposits consist of an alternation of shallow-water clays, siltstones, sandstones and limestones with partings of continental variegated rocks. Abyssal clayey sediments are characteristic of the Aptian. The maximum thickness of Lower Cretaceous deposits is up to 2000 m and this is observed in the north Crimean graben, a depression involving the Tarkhankutsk peninsula and the Dzhankoi region and separating the old and young platforms, associated with intense volcanism. The fauna consists mainly of foraminifers, rare ammonites and bivalves.

In the central and eastern Precaucasus Lower Cretaceous deposits are a part of the platform cover. They occur transgressively over different stratigraphic horizons from the Paleozoic to the Upper Jurassic. Lower Cretaceous deposits are represented by marine facies, Valanginian carbonate rocks, Hauterivian–Barremian detrital beds with rare partings of oolitic limestones and Aptian and Albian sandy-clay detrital facies. The thickness of the deposits ranges from several meters to 800–900 m.

On the Turanian plate Lower Cretaceous deposits also form a nearly continuous cover. They crop out in Gorny Mangyshlak, on Tuarkyr, in the western foothills of Tien Shan and in the Chu syneclise. Different horizons of the Lower Cretaceous transgressively overlie Jurassic deposits and, to the east of the Aral Sea, the Paleozoic rocks.

On Mangyshlak and Usturt the base of the sequence is formed by 100 m of clays or Berriasian–Lower Hauterivian sandstones and limestones replacing them. The Upper Hauterivian and Barremian are represented by variegated sandstones, clays and conglomerates up to 100 m thick. The Aptian–Albian deposits are transgressive and have a phosphoritic horizon at their base with a condensed Lower Aptian fauna. Sandstones, siltstones and clays higher in the sequence with numerous horizons of concretions contain abundant and varied complexes of ammonites, warranting a detailed subdivision of the deposits up to zones inclusively. The zonal scheme of Aptian–Albian deposits of Mangyshlak generally agress with that compiled for Kopet-Dag (Table III).

TABLE III

Unified regional stratigraphic scheme of Barremian–Albian deposits of Central Asia (1969).

Stage	Substage	Zone
ALBIAN	Upper	*Stoliczkaia dispar*
		Pervinquieria rostrata, Cantabrigites
		Pervinquieria inflata
		Hysteroceras orbignyi
		Anahoplites rossicus
	Middle	*Anahoplites daviesi*
		Anahoplites intermedius
		Hoplites dentatus
	Lower	*Douvilleiceras mammillatum*
		Leymeriella tardefurcata
APTIAN	Upper	*Hypacanthoplites jacobi*
		Acanthohoplites nolani
		Acanthohoplites prodromus
	Middle	*Parahoplites melchioris*
		Epicheloniceras subnodosocostatum
	Lower	*Dufrenoya furcata*
		Deshayesites deshayesi
		Deshayesites weissi
BARREMIAN	Upper	*Turkmeniceras turkmenicum*
		Colchidites nicortsmindensis
		Imerites giraudi
	Lower	

On Tuarkyr Lower Cretaceous deposits are associated with the limbs of anticlinal structures. In this region the Berriasian and Valanginian stages are absent. Red continental deposits, replaced in the south by organic limestones, are characteristic of the Hauterivian and Lower Barremian. All the overlying deposits are represented by sandstones, siltstones and clays. The total thickness of Lower Cretaceous deposits on Mangyshlak and Tuarkyr is 600–800 m.

To the south of the Turanian plate with the approach to Kopet-Dag, a substantial part of the Lower Cretaceous is represented by limestones which form thick packets (of several tens of meters), alternating with clays and siltstones. In the Amu-Darya syneclise Berriasian–Lower Hauterivian marine deposits are replaced by variegated red clays and sandstones, which, together with the Upper Hauterivian and Barremian, form a thick series of littoral and continental deposits. Aptian–Albian marine detrital carbonate deposits extend to the southwestern foothills of the Gissar mountain range. East of the Aral Sea, in the Kyzylkumy Desert in the Chu River Basin, the Lower Cretaceous is represented by continental variegated deposits.

In the structure of the folded area, i.e. Altay–Sayansk, Kazakhstan and Tien Shan, the role of Lower Cretaceous deposits is not important. They are represented mostly by continental deposits of a lacustrine and fluvial origin or by an alternation of marine and lagoonal forms. In the Altay–Sayansk folded area these are variegated sands and clays, in central Kazakhstan, an alternation of variegated clays, sandstones and light-colored siliceous–calcareous rocks, 60–80 m thick. The greatest thickness of Lower Cretaceous deposits (2140 m) is recorded in the Ferghana Basin, where it consists of red conglomerates, sandstones and clays with charophytes, ostracods and fresh-water fish.

On the West-Siberian plate Lower Cretaceous deposits are exposed in the southeastern part, elsewhere they are recorded in boreholes (Fig. 11). Very small outcrops are recorded on Taymyr and Novaya Zemlya. The deposits are represented by marine sandstones, siltstones and argillites, as well as by continental sandy-clay, gray and variegated rocks with bands of coal. During the entire Early Cretaceous one large basin continued to exist on the West-Siberian Platform in open communication with the Arctic Basin. The size and regime of this basin, however, were not constant, periods of normal marine conditions being replaced by conditions of a substantial freshening at the end of the Hauterivian, Barremian, Aptian and Late Albian times.

The Berriasian and Valanginian are characterized by clay–siltstone sediments with remains of the ammonites *Tollia, Polyptychites, Subcraspedites,* pelecypods, gastropods, ostracods and foraminifers which accumulated in a relatively shallow sea (less than 100 m). The regression which began in the Late Valanginian resulted in a drastic shrinking and shallowing of the sea basin during the Hauterivian. Marine Hauterivian sediments in the northern part of the West-Siberian plate are represented by argillites, siltstones and sandstones with a rich fauna of pelecypods, ammonites and belemnites. At this time lagoon-continental conditions existed in the south, with the deposition of variegated rocks with an ostracod fauna and of conglomerates. During the Barremian the region was subjected to a general uplift and the West-Siberian sea was almost completely isolated from the Arctic Basin. In the northwestern part of the plate the Barremian is represented by littoral-marine and lagoon facies, while in the eastern part it is formed of coal-bearing strata, frequently with conglomerates at the base. The thickness of the Hauterivian–Barremian deposits is 250–700 m.

The Aptian stage in the northern West-Siberian plate inherited the paleogeographical environment of the Barremian. In the south during all this time and up to the beginning of the Late Cretaceous, lagoon-continental and continental conditions with an accumulation of gray and variegated strata with the remains of a leaf flora, spores and pollen continued

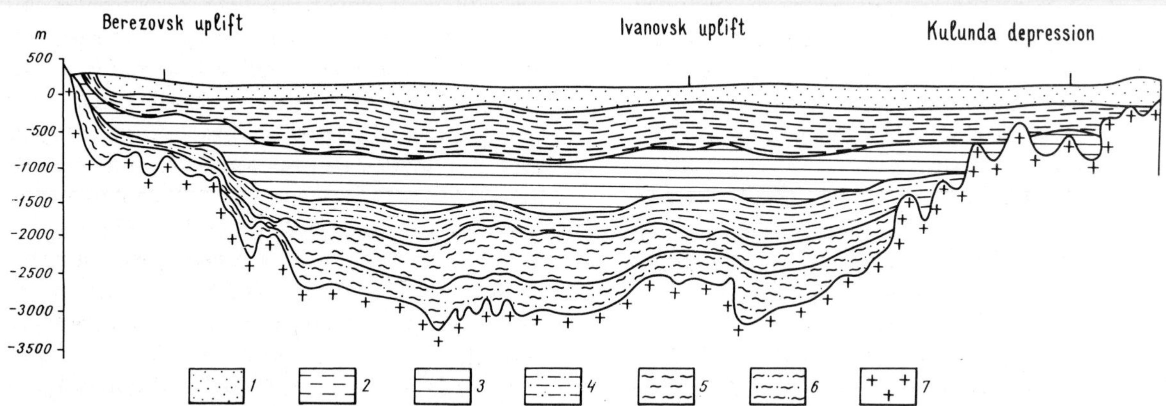

Fig. 11. Structure scheme of the West-Siberian plate (after N.N. Rostovtzev). Legend: *1* = Middle Oligocene–Anthropogen; *2* = Turonian–Eocene; *3* = Aptian–Cenomanian; *4* = Hauterivian–Barremian; *5* = Malm–Valanginian; *6* = Liassic–Dogger; *7* = basement rocks.

to exist. The thickness of the Aptian is 150–320 m. The Albian age was the time of a new transgression which reached its maximum in the Early and Middle Albian and was replaced by a regression during the Late Albian. Marine Albian deposits in the northern West-Siberian plate consist of sandy-clay strata with foraminifers, radiolaria, ostracods, rare remains of pelecypods and ammonites. However, over the greater part of the territory Aptian and Albian deposits are represented by terrigenous or continental strata with ostracods and charophytes.

Along the eastern flanks of the Urals during the Berriasian and Valanginian there was an accumulation of glauconitic sands, sandstones and argillites with *Tollia* and *Polyptychites*; in the Hauterivian and Barremian sandy-clay strata were deposited with *Speetoniceras*. The formation of continental sandy-clay and coal-bearing strata in the polar Urals was characteristic of the Aptian and was replaced during the Albian by marine siltstones with a fauna of foraminifers.

Upper Cretaceous deposits occur extensively over the Central Eurasian platform.

On the Scythian plate detrital-carbonate deposits with an abundant fauna are typical of the Upper Cretaceous in the depressions of Moldavia, western Black Sea region, the Steppe Crimea and the Precaucasus. The different pulses of the transgressions resulted in the unconformable relationship of the different horizons both between themselves and between older rocks.

In Moldavia and the western Black Sea region a mass of detrital-carbonate Cenomanian deposits up to 50–60 m thick with phosphorites at its base overlies the Silurian, Cambrian and Precambrian. The Turonian–Senonian deposits are exposed by boreholes and consist mainly of chalk-like marls. Between the Dniester and Odessa the thickness of Cenomanian–Campanian deposits does not exceed 350–500 m, while in the axial part of the northern Crimean graben-depression it reaches to 2000 m, thinning markedly towards the south.

In the Steppe Crimea Upper Cretaceous deposits are relatively uniform consisting mostly of limestones (with flints in the Turonian–Coniacian) with partings of marls, chalk-like rocks and clays. In the Maastrichtian the sandy content of deposits increases and tuff appears. Tectonic movements and the expansion of marine conditions are indicated by the unconformable relationship of Santonian deposits to the Turonian–Albian, by the transgressive relationship of the Campanian (from 50–100 m in the south, to 500 m in the Tarkhankutsk Peninsula and in the northern Crimean depression) and to the Maastrichtian (in eastern Crimea it rests on the Cenomanian and Albian). At the end of the Maastrichtian, upward movements led to an expansion of the land to the north resulting in a retreat of the sea. This explains why the carbonate Danian–Lower Paleogene strata occurring in the south are absent on the Azovo-Crimean ridge. Bluish-gray marls and limestones (340 m thick) of this age occur on the Tarkhankutsk and Kerch peninsulas.

In the Precaucasus Upper Cretaceous deposits are everywhere covered by Cenozoic deposits. In the southern regions limestones and marls with locally calcareous sandstones total from 170–620 m in the western Precaucasus, to 515–860 m in the north of the central Precaucasus (Manych depression). In the north and especially in the northwest of the Precaucasus the deposits are much more enriched in terrigenous material. In the majority of regions analogues of all stages are present, but the completeness and thickness of the sequences vary considerably. This is associated with the presence of a number of regional stratigraphic unconformities and erosion in the zones of uplifts.

In the greater part of the Precaucasus the Cenomanian stage consists of limestones, marls and siltstones with marls (Salsk region, 8–50 m thick) and clays (Beloglinsky region, 35–40 m) predominating in the north. In the monotonous overlying succession which consists mostly of silty or pure limestones with stylolites and of marls, analogues of all the remaining stages can be distinguished by their fauna. South of the Stavropol uplift Turonian–Maastrichtian carbonate deposits 340–420 m thick are, on the whole, analogues of deposits exposed on the northern flank of the Caucasus. On the Stravropol uplift their thickness drops to 10–100 m. In a number of areas here there are absolutely no Cenomanian, Campanian and Maastrichtian deposits and nearly the entire Upper Cretaceous is missing over the Armavir–Nevinnomyssky swell. Further north, in the Manych depression, the role of clastic rocks increases (clays, siltstones, sandstones).

In the Precaucasus Maastrichtian deposits occur everywhere transgressively resting on older rocks up to the Lower Cretaceous (Stavropol and Armavir regions) in age. Higher in the sequence in many places

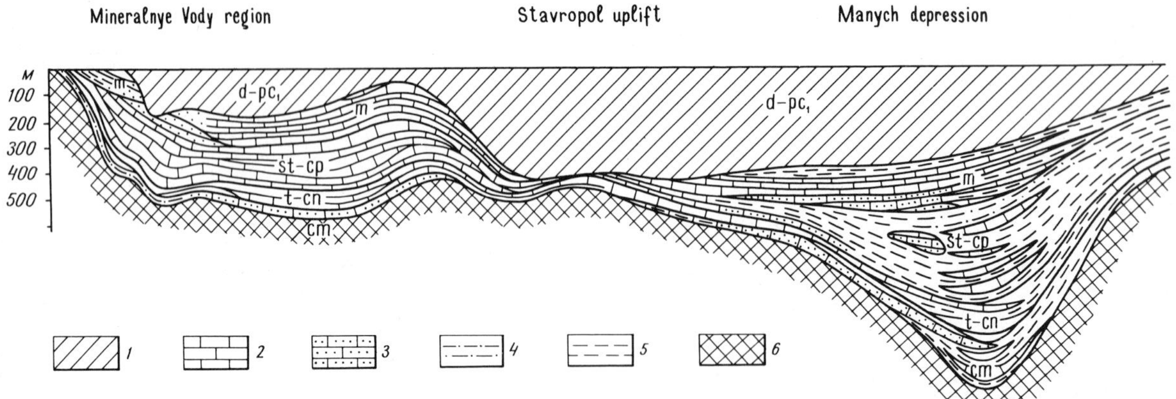

Fig. 12. Section of Upper Cretaceous deposits in the central regions of the northern Caucasus (after M.M. Moskvin). *1* = Danian and Lower Paleocene deposits of the Scythian Platform and the Rostov protrusions of the Ukrainian crystalline shield: *2* = limestones with marl partings; *3* = sandy limestones; *4* = sandy marls and clays; *5* = clay marls and clays; *6* = underlying rocks.

a sandy-clay Danian–Lower Paleogene succession (up to 400 m) overlies it conformably. Danian calcareous sandstones (25 m) in the northern central Precaucasus are conformably overlain by Lower Paleocene (Fig. 12) beds (326 m) consisting of gray sandstones. To the south of the Stavropol upland the above-mentioned sandy-clay strata are overlain by greenish-gray marls with a complex of planktonic foraminifers belonging to the Elburgansk suite of the northern Caucasus.

On the Turanian plate Upper Cretaceous deposits greatly vary lithologically and in facies. In the west of the plate (Mangyshlak, Tuarkyr, Usturt and the adjacent regions) terrigenous Cenomanian–Turonian deposits occur (290–640 m thick) with partings of phosphorites at the base. Higher up an essentially carbonate mass (including chalk) of Coniacian–Maastrichtian age overlies an erosional unconformity. The base of the Turonian is also frequently erosional, but south of Tuarkyr, beds with *Actinocamax plenus* Bl. are present in the sequence. To the east and southeast, from Mangyshlak and Tuarkyr and to the Murgab depression, carbonate deposits are replaced by detrital-carbonate, and then by exclusively sandy-clay deposits with marl partings. The thickness of Upper Cretaceous deposits decreases markedly over the central Karakumy arch (for the Coniacian, for instance, from 40–50 m to 3–4 m) and increases again in the Bayram-Ali region (to 450 m for the Santonian–Maastrichtian). A stratigraphic unconformity can be traced everywhere at the base of the Santonian.

Danian organo-clastic, melobesian and bryozoan limestones (from 150 m on the Mangyshlak to 3–4 m on the central Karakumy) transgressively overlie Maastrichtian and Upper Campanian deposits and become gradually replaced by beds of the Montian stage.

With a common fauna, the subdivisions of the Upper Cretaceous in the west of the Turanian plate corresponds to the scheme worked out for Central Asia from the Kopet-Dag sequences (Table IV).

In the east of the Turanian plate (from the Aral Sea and Amu-Darya to Ferghana) Upper Cretaceous deposits of shallow-water and littoral-marine terrigenous type predominate, but in the Kyzylkumy area the sequence (with a total thickness of 220–480 m) already includes variegated and gypsiferous rocks characteristic for the regions further east.

In a typical sequence in the western and northern Ferghana Basin gray marine beds form only the top part of the Cenomanian and the Turonian. The top of the Campanian stage is formed by the "radiolite horizon" of limestones and dolomites with rudists and ammonites. A great part of the sequence here consists of terrigenous red and variegated sediments with *Amphidonta,* etc., as in the Tashkent region, where the variegated deposits contain remains of dinosaurs, turtles, and fish. In the Maastrichtian red gypsiferous strata occur.

To the east of the Ferghana Basin, in the Chuya Basin, as well as in the depressions of the Kazakh Shield and the Altay–Sayansk area the Upper Cretaceous is mostly represented by lagoon-continental red deposits.

Over the greater part of the Tien Shan area and

TABLE IV

Subdivision of Upper Cretaceous deposits of Central Asia (Stratigraphic Conference, Samarkand, 1971)

Stages	Substages	Zones of the western areas	Zones of the eastern areas
MAASTRICHTIAN	Upper	*Inoceramus dobrovi* *Diplomoceras cylindraceum*	*Liostrea lehmani* *Biradiolites boldjuanensis*
	Lower	*Hauericeras sulcatum*	
CAMPANIAN	Upper	*Bostrychoceras polyplocum* *Hoplitoplacenticeras coesfeldiense – Stedticeras gillierone*	*Bostrychoceras polyplocum, Hoplitoplacenticeras marroti*
	Lower	*Eupachydiscus levyi – Micraster schroederi* *Offaster pomeli*	*Scaphites inflatus*
SANTONIAN	Upper	*Marsupites testudinarius*	*Asiatostantonoceras tagamense*
	Lower	*Inoceramus undulatoplicatus, Inoceramus cardissoides*	*Stantonoceras guadalupae asiaticum*
CONIACIAN	Upper	*Inoceramus involutus*	*Lewesiceras asiaticum*
	Lower	*Inoceramus wandereri*	*Horquia acrabatense*
TURONIAN	Upper	*Hyphantoceras reussianum* *Inoceramus apicalis*	*Collignoniceras intermedium*
	Lower	*Mammites nodosoides* *Inoceramus labiatus*	*Mammites nodosoides*
CENOMANIAN	Upper	*Calycoceras crassum* *Acanthoceras rhotomagense* *Euomphaloceras euomphalum*	*Placenticeras lenticulare* *Eoradiolites kugitangensis*
	Lower	*Mantelliceras mantelli* *Mantelliceras martimpreyi*	*Turkmenites gaurdakense*

northern Pamirs marine, including red, sediments have been deposited with an abundant fauna. They are closely related to the Kyzylkumy and western Ferghana deposits. In the Tadzhik depression Upper Cretaceous deposits (800–1500 m) are represented predominantly by marine and, to a lesser extent, by lagoon-continental sediments. The rocks include various clays, siltstones, sandstones, limestones, dolomites and gypsum. Gray, mainly marine strata of a similar composition form the southwestern foothills of the Gissar range.

The Late Cretaceous marine basin to the east of the Turanian plate was shallower than in the west and had specific physico-geographical conditions. A peculiar faunal assemblage developed: along with endemic forms (about 50%) the basin was inhabited by ammonites, lamellibranchs and sea urchins, which are found in the Mediterranean (North Africa), Western Europe and North America. This explains the differences in the Upper Cretaceous zones of the eastern region or the Central Asiatic (Turkestan, according to A.D. Arkhangelsky) province, from the scheme for more western regions of the plate (see Table IV).

On the West-Siberian plate Upper Cretaceous deposits crop out in Priuralie, and in the lower reaches of the Yenisey over the rest of the area they have been reached only in boreholes. They are represented by detrital marine, lagoon-lacustrine and continental rocks. During the Late Cretaceous the West-Siberian plate was covered by a shallow sea (to 200 m) penetrating deeply into the Yenisey Estuary and Khatanga basins, into the Taimyr folded area and the northerly dipping part of the Siberian Platform. In the Cenomanian, as in the Late Albian, lacustrine-alluvial sands with partings of clay, amber and wood developed along the periphery of the freshened basin that shifted westwards. Clays and siltstones up to 350 m thick are found in the central part of the basin.

At the beginning of the Turonian a boreal transgression covered the littoral plains, the basin expanding substantially across the Uralian, Taimyr and Siberian area. During the Turonian–Maastrichtian, 600–800 m of rather monotonous clay sediments with flint in the west and thickening eastwards accumulated in this basin. It contains remains of radiolaria, mollusks and foraminifers. In the southeastern and southern part of the West-Siberian plate, a total of 50–150 m of continental variegated clays accumulated. In the Chulym–Yenisey and Yenisey regions and further to the south, sands, clays, sandstones and grit with floral remains accumulated.

The maximum of the boreal transgression occurs in the Late Santonian–Campanian. In the northwest an island archipelago replaced a Uralian land mass, and the West-Siberian Basin communicated with the sea north of the Russian platform through latitudinal straits between the islands. To the south, across the Turgaisk straits, it communicated with the sea on the Turanian plate. At this time opoka-like and other flinty rocks were especially abundant on the West-Siberian plate, as well as sandy-clay sediments from 5 m (Pai-Hoi) to 100 m-150 m thick (West-Siberian lowland) and reaching 260 m (the Yenisey Estuary Basin).

At the beginning of the Maastrichtian the sea shrank and at the eastern margin of the plate and in the near-Taimyr zone a land formed at the end of the Campanian. Displaced westwards, the basin remained connected with the surrounding seas. In the southern and central regions the Maastrichtian is characterized by calcareous marly clays (from 50–100 to 200 m thick). In the west they are replaced by unstratified and opoka-like clays with partings of dolomites and brown coal, and in the east by clays with partings of sands and siderites at the top. In the Yenisey Estuary Basin shallow-water sands and sandstones contain inoceramids, belemnites and ammonites characteristic of the Maastrichtian.

On the eastern flank of the Urals Upper Cretaceous continental and lagoon-lacustrine sediments alternate with littoral-marine and marine opokas, clays, glauconitic sands and sandstones. The latter correspond to the ingressions of the West-Siberian cold-water sea (13.3° from the $^{18}O/^{16}O$ ratio in the rostra of Santonian belemnites) in the Turonian, Coniacian, Late Santonian–Early Campanian and Maastrichtian. Danian marine deposits to the east of the Urals and in Western Siberia are not known for certain. Only in the southern Transural area beds (8–10 m) are distinguished by typically Danian planktonic foraminifers and glauconitic sands with remains of nautiloids and mosasaurs.

In the Danian a regression of the West-Siberian sea resulted in the disappearance of the Turgaysk straits and a break in the connections with the Arctic Basin, which was reestablished at the beginning of the Paleogene.

OLD SIBERIAN PLATFORM

The Siberian Platform occupies Eastern Siberia between the Yenisey and Lena rivers north of Lake Baykal. Mesozoic deposits are most extensively and most completely represented in the subsiding zones along the platform margin, in the Lena–Khatanga and Verkhoyansk depressions. In addition to these depressions Triassic deposits are also well represented in the Tunguska syneclise and Jurassic deposits in the Vilyuy syneclise, the Angara–Vilyuy depression and in the junction zones of the platform with the areas of Plaeozoic folding along its southern margin.

Triassic

In the Tunguska syneclise the top part of a series of basalt sheets alternating with volcano-sedimentary rocks of a continental origin belong to the Triassic. In the trappean formation Lower and possibly Middle Triassic deposits have been recognized. They increase in thickness within the syneclise from 800 m in the south to 4000 m in the north.

In the southwestern part of the Lena–Khatanga depression, in the Kotuy and Popigay interfluve, only the Lower Triassic trappean formation occurs extending from the Tunguska syneclise. Further east, in the basin of the Anabar and Olenek rivers and on the left bank of the Lena Triassic deposits are complete. The base of the sequence consists of continental sandstone and argillites with limestone partings containing various ceratites, including *Pseudosageceras, Meekoceras, Prohungarites,* etc. The Middle Triassic (up to 400 m thick) consists of siltstones and argillites with the ceratites *Arctohungarites, Ussurites,* and *Danubites.* Upper Triassic deposits rest with an erosional contact and are represented by sandstones, conglomerates and argillites, containing Carnian bivalves in the bottom and remains of Norian–Rhaetian plants in the top. The thickness of Upper Triassic deposits is 120 m. In the Verkhoyansk depression and in the Vilyuy syneclise Triassic deposits have been reached by boreholes and are represented by argillites, siltstones and sandstones, with a total thickness of 1500 m (Fig. 13).

Jurassic

Jurassic deposits along the northern subsiding margin of the Siberian Platform, in the Lena–Khatanga

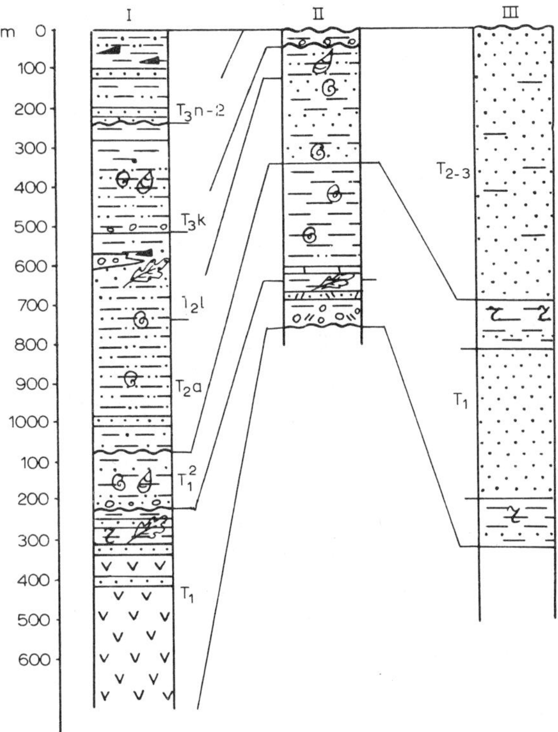

Fig. 13. Sequence correlation of Triassic deposits on Taymyr and on the northern and eastern plunges of the Siberian Platform. *I* = eastern Taymyr, Tzvetkova cape (after I.S. Gramberg); *II* = region around the mouth of the Olenek River (after Yu.N. Popov); *III* = the Vilyuy River (summary sequence, after N.S. Zabalueva). Legend on Fig. 2.

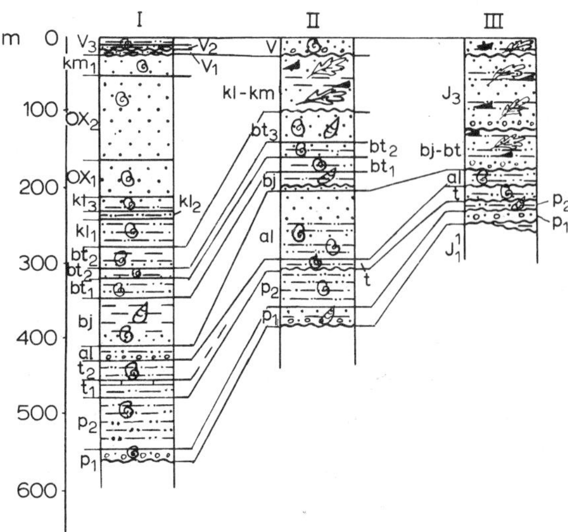

Fig. 14. Sequence correlation of Jurassic deposits on eastern Taymyr and the eastern plunge of the Siberian Platform. *I* = Khatanga depression (summary sequence, after V.A. Basov and others); *II* = northern part of the pre-Verkhoyansk depression (summary sequence, after T.I. Kirina); *III* = Vilyuy syneclise, the Markha River (after T.I. Kirina). Legend on Fig. 2.

depression, are similar in their structure to the Jurassic deposits of Taymyr discussed above. Complete and very thick Jurassic deposits occur in the pre-Verkhoyansk depression and in the Vilyuy syneclise opening into this depression (Fig. 14). Marine sandy-clay Lower–Middle Jurassic deposits to 1500 m thick predominate with coal-bearing sandy-clay Upper Jurassic deposits with a thickness of 600 m. The maximum transgression was during the Late Pliensbachian and Toarcian, the marine deposits of which penetrated into the eastern part of the Angara–Vilyuy depression. Marine deposits of the Volgian stage can be traced along the pre-Verkhoyansk depression. Callovian–Kimmeridgian marine deposits occur only in the extreme northern part of the pre-Verkhoyansk depression. In the western part of the Angara–Vilyuy depression Lower and Middle Jurassic coal-bearing continental deposits occur. In the southwestern and southeastern subsiding zones of the Siberian Platform, in certain isolated basins, coal-bearing formations have accumulated with varying stratigraphic completeness and thickness. The formations may total 2000 m.

Cretaceous

On the Siberian Platform marine Cretaceous deposits occur in the same localities as the Upper Jurassic: in the Lena–Khatanga depression and to the north of the pre-Verkhoyansk depression. Clays, siltstones and sandstones, frequently glauconitic, with numerous remains of Berriasian–Lower Valanginian ammonites (*Chetaites, Hectoroceras, Surites, Bojarkia, Neotollia, Polyptychites,* etc.) are found. A regression began at the end of the Early Valanginian. Marine deposits of the Upper Valanginian and Lower Hauterivian occur between the Yenisey and Khatanga rivers. East of Khatanga they are replaced by littoral coal-bearing strata.

In the pre-Verkhoyansk depression and in the Vilyuy syneclise Lower Cretaceous deposits are represented by continental coal-bearing deposits up to 3000 m thick. Along the flanks of the Vilyuy syneclise the accumulation of coal-bearing strata was accompanied by fissure effusions.

Upper Cretaceous deposits are known in the marginal subsiding zones to the south and east of the Siberian Platform. In the Vilyuy basin and in the pre-Verkhoyansk depression 850 to 900 m of cross-bedded, predominantly alluvial-lacustrine sands, sandstones and grits with siderites, lignites, partings of brown coal and a rich Cenomanian–Turonian flora (*Menispermites, Sassafras, Dalbergites, Cissites* and plants) and Senonian (*Trochodendroides, Viburnum, Macclintockia,* etc.) occur. On the Yenisey ridge the Senonian coaly-clay series, 85 m thick, of a lacustrine-peat origin is replaced higher in the sequence by variegated alluvial-lacustrine kaolinitic clays (up to 120 m), a product of the erosion of a lateritic weathering crust at the very end of the Late Cretaceous period.

The varied petrographic composition of the coarse clastic varieties of these fresh-water lacustrine and alluvial sediments indicates an uplift of the Verkhoyansk, Siberian and Taymyr area accompanied by a magmatic activity at the end of the Albian, the beginning of the Cenomanian, in the Coniacian–Santonian and in the Maastrichtian.

THE PACIFIC BELT

The Pacific fold belt forms the Far East and the northeastern part of the U.S.S.R. Geologically this territory is subdivided into geosynclinal areas which completed their evolution during the Mesozoic (the Verkhoyano-Chukotka and Mongolo-Okhotsk), and geosynclines which continued their development during the Cenozoic (the Sikhote-Alin, Kamchatka–Koryak and Nippon). During the Triassic and the Jurassic the Verkhoyano-Chukotka geosynclinal area included two geosynclines, the Verkhoyansk and the Chukotka separated by the Kolyma and Omolon median massifs. The Okhotsk median massif separated the eastern part of the Mongolo-Okhotsk geosyncline from the Verkhoyansk geosyncline. The Bureya median massif split the Mongolo-Okhotsk geosyncline into two branches. The western branch was separated from the southern part of the Sikhote-Alin geosyncline by the Khankaysk median massif. From the end of the Upper Jurassic the Verkhoyano-Chukotka and Mongolo-Okhotsk geosynclinal areas began their gradual transformation into an orogen.

Triassic

The greatest development of Triassic deposits in the U.S.S.R. occur within the Pacific folded belt. All three divisions are represented and frequently contain an abundant fauna. There are gradual transitions both within the system and within the overlying Jurassic deposits. The most extensive Triassic deposits in the Verkhoyano-Chukotka folded area occur where it forms the flanks of the Verkhoyansk range, the basin of the Yana River and the vast regions in the upper courses of the Indigirka and Kolyma rivers. The structural disposition of the Triassic rocks reflects in many ways the Permian pattern, and in geosynclinal depressions Triassic deposits often occur concordantly on Permian rocks. In the area discussed marine sedimentation conditions prevailed throughout the Triassic. The eroded areas, the outlines and location of which changed with time, were associated with median massifs. Terrigenous rocks, argillites, siltstones and sandstones were the predominant types of deposits. In Sikhote-Alin there are also many siliceous rocks and effusives. The latter occur also in the Okhotsk median massif and are present in the Kamchatka–Koryak geosyncline.

In the axial region of the Verkhoyansk geosyncline, which includes the eastern flank of the Verkhoyansk range and the basins of the Yana River and the upper courses of the Indigirka and Kolyma rivers to the east, the thickness of Triassic deposits reaches 4000–6500 m. They consist mainly of argillites and siltstones with concretions and contain an abundant and varied fauna of ceratites and bivalves belonging to all three divisions. In the western flank part of the geosyncline, in the Verkhoyansk range and in the Kharaulakh Mountains, the thickness drops to 1800 m. In the lower part of the Lower Triassic variegated sandstones and siltstones with *Lingula* and *Estheria* are found. In the Upper Triassic sandstones and conglomerates containing not only marine bivalves, but also terrestrial plants play an important role.

Lower Triassic and Anisian deposits thin out on the flanks of the Kolyma massif, so that over the central and northern parts Permian and older deposits are overlain by Ladinian and Upper Triassic calcareous argillites and sandstones with bands of limestones, andesites and tuffs to a total thickness of less than 500 m. On the Omolon Massif all three divisions occur and are represented by sandy-clay, and frequently calcareous rocks with a total thickness less than 500 m. In the northern part of the massif, in the basin of the Bolshoy Anyuy River, Lower Triassic and Anisian deposits thin out and the Permian rocks are transgressively overlain by Ladinian and Norian deposits.

In the Chukotka geosyncline the Lower and Middle Triassic consists of sandstones, siltstones, siliceous-chloritic schists and phyllites with bands of tuffs and sills of basic rocks. The thickness of Triassic deposits reaches 4000 m. On the Okhotsk median massif only Upper Triassic occurs as sandstones, clay shales and conglomerates with a thickness of 1000 m. They contain bands of tuffs and sheets of andesites.

In the Mongolo-Okhotsk geosyncline a Triassic marine gulf extended sublatitudinally. The Triassic deposits of this gulf are preserved in some synclines only. In the Transbaykal region, Triassic deposits are found in the Onon River Basin, where they are represented by littoral and marine sandstones, siltstones and grit with *Ophiceras, Gyronites* and *Neocalamites*. The thickness is in excess of 3000 m. In the Transbaykal region Middle Triassic deposits are absent. The Upper Triassic unconformably overlies the Paleozoic and is represented by sandy-clay rocks up to 3000 m thick. In the upper course of the Zeya River in the Tukuringra range and in the basin of the Uda River, again only Upper Triassic sandstones and argillites are found. These have a thickness of 4000 m diminishing to 1000 m at the coast of Tugursk Bay.

Triassic deposits in the southern branch of the Mongolo-Okhotsk geosyncline are exposed in the northeastern foothills of the Lesser Khingan, in the upper course of the Bureya River and in the basin of the Amgun River. Lower and Middle Triassic marine sandstones and argillites with ceratites and bivalves are found in the Lesser Khingan only, where their thickness exceeds 2000 m. Upper Triassic deposits occur in the upper course of the Bureya River and in the basin of the Amgun River. Here they are represented by conglomerates, sandstones and argillites and include a thick mass of diabasic porphyrites and siliceous schists (up to 750 m). The thickness of the Upper Triassic here totals some 3000 m. These deposits are similar to the Triassic in the Sikhote-Alin geosyncline.

Triassic deposits of the Kamchatka–Koryak geosyncline are known only in its western marginal region, where they are exposed in the basin of the Anadyr River and on the Koryak upland. Upper Triassic deposits have been recognized and consist of sandstones and siltstones over 1500 m thick as well as clay shales, siltstones and sandstones with bands and lenses of andesitic tuffs and tuffobreccia.

In the extreme southeastern part of the Sikhote-Alin geosyncline adjacent to the Khankaysk median massif (Vladivostok region and the basin of the Daubikhe River) Lower–Middle Triassic deposits unconformably overlie Paleozoic rocks. They are represented by argillites with concretions and by siltstones with bands of limestones and sandstones, and contain remains of ceratites and bivalves. The thickness of the Lower and Middle Triassic reaches 1200 m. They are unconformably overlain by Upper Triassic deposits consisting of alternations of coal-bearing and shallow-water marine sandy-clay deposits. There are localities rich in plant remains of the so-called "Mongugaisk" flora, as well as marine bivalves. Locally the thickness of the Upper Triassic exceeds 4000 m.

In the central and northern Sikhote-Alin and in the lower course of the Amur River there are extensive outcrops of volcanic beds and siliceous schist of late Middle and Late Triassic age. They contain lenses of limestones and coal-bearing shales. The thickness of these deposits is 3000–4000 m.

Jurassic

The basic Triassic structural pattern of the Pacific belt persists into the Jurassic. During the Early and Middle Jurassic the sea occupied vast expanses of the Verkhoyano-Chukotka, Kamchatka–Koryak and Sikhote-Alin geosynclinal areas and in a long gulf it advanced into the Mongolo-Okhotsk geosyncline. The median massifs and anticlinoria were represented by islands in the geosynclines, the position and size of which changed with time. The maximum transgression took place in the Late Pliensbachian and the Toarcian. During the latter half of the Middle Jurassic the size of the gulf into the Mongolo-Okhotsk geosyncline was considerably reduced and in the Late Jurassic at different times, marine sedimentation was partly replaced by continental and littoral deposition. In contrast to the Triassic period, during the Jurassic there was a much greater variety in the type of deposit and much more rapid change indicating a greater degree of differentiation of the structure. Along with marine sandy-clay deposits, volcanic siliceous and coal-bearing formations play a substantial role.

Within the Verkhoyansk geosyncline the area of maximum thickness and of the most complete sequence shifts towards the east into the regions of the upper courses of the Indigirka and Kolyma rivers and the Polousny range and between the Yana and Indigirka rivers. Lower Jurassic deposits here rest conformably on Triassic ones and are represented by clay shales and siltstones to 3000–3500 m thick. Towards the west, in the basin of the Yana River and on the flanks of the Verkhoyansk range, the role of rudaceous rocks in the composition of Lower Jurassic depostis increases. In an easterly direction, with the approach to the Kolyma Massif, tuffogene and siliceous rocks begin to appear. The deposits are comparatively rich in remains of ammonites, belemnites and bivalves of a Hettangian–Toarcian age. In the axial part of the Verkhoyansk geosyncline Middle Jurassic deposits rest conformably on Lower Jurassic deposits. In composition there is a predominance of polymict sandstones and siltstones. The thickness reaches 2500 m. Faunal remains are more rare, especially in the Bajocian and Bathonian, and are represented mostly by *Mytiloceramus* and belemnites. Upper Jurassic deposits are again conformable on Middle Jurassic beds. They consist of clayey and sandy-clay shales with beds of sandstone and grits and have a thickness of 2500 m. On the eastern flank of the Verkhoyansk geosyncline these strata are replaced by volcano-sedimentary and higher in the sequence, by coal-bearing deposits.

Over the Kolyma Massif the sequence of Jurassic deposits is thinner and less complete. The Lower Jurassic is represented by Pliensbachian and Toarcian tuffites and tuffs alternating with conglomerates totalling 300 m. The Middle Jurassic is transgressive but of a similar lithology and thickness (400 m). The Upper Jurassic on the Alazeysk Plateau, in the central part of the massif, consists of clay shales and sandstones alternating with andesites, basalts and their tuffs and has a thickness of 600 m. To the southwest of the Kolyma Massif the top part of the Upper Jurassic contains thick beds (1900 m) of clay shales and sandstones with plant remains.

On the Omolon Massif the Lower Jurassic is complete in the western part where the deposits rest conformably on Triassic and are represented by siliceous argillites and siltstones, 200 m thick. A rich fauna of Hettangian–Toarcian ammonites and bivalves is known. Towards the central part of the massif, marine argillites are replaced by littoral sandstones and wedge out. Middle Jurassic deposits rest conformably on the Lower Jurassic and are represented by volcano-mict sandstones (up to 400 m) with remains of ammonites, *Mytiloceramus* and plants. Further north, in the basin of the Bolshoy Anyuy River, Middle Jurassic deposits are transgressive, their thickness increasing to 1400 m. The Upper Jurassic is most fully represented in the basin of the Bolshoy Anyuy River. It consists of sedimentary and volcanic rocks, sandstones, argillites, basalts, andesites and their tuffs. An unconformity occurs at the base of the Oxfordian. The thickness of the Callovian reaches to 800 m; that of the Oxfordian–Volgian is slightly in excess of 2000 m. Ammonites and bivalves have been recorded, mostly *Buchia.*

In the Chukotka geosyncline, in the basin of the Maly Anyuy River, Jurassic deposits rest conformably on Triassic. At the base argillites with *Otapiria* occur and they are replaced by a thick mass of arkose sandstones (in excess of 2000 m), referred to the Lower, Middle Jurassic and to the Callovian–Oxfordian. The Volgian deposits are unconformable and are represented by 800 m of argillites and volcano-mict sandstones with lenses of intermediate and acid lavas.

In the Mongolo-Okhotsk geosyncline Jurassic deposits occur in the eastern Transbaykal region, in the middle reaches of the Zeya River, and in the basins of the Uda and Tugur rivers. In the eastern Transbaykal region marine Lower Jurassic and Aalenian deposits are found between the rivers Shilka and Argun. They accumulated in a narrow depression and are represented by gray sandstones and siltstones with subordinate conglomerates and argillites to a total thickness of 5000 m. Pliensbachian, Toarcian and Aalenian ammonites and bivalves have been recorded. On the flanks of the depression the marine deposits are replaced by rudaceous littoral and continental deposits. Bathonian and Upper Jurassic deposits which fill certain basins are of a continental sedimentary type with some volcanic rocks. Among the latter effusive series basalts, dacites, quartz porphyries are, sometimes, recorded.

In the upper Amur region Jurasssic deposits represented by all three divisions with a total thickness of 9000 m are found. Sandstones, siltstones and clay shale predominate in the sequence. The Lower Jurassic, the top of the Middle Jurassic and the bottom of the Upper Jurassic contain the remains of bivalves, belemnites and rare ammonites. The top of the Upper Jurassic is represented by coal-bearing continental deposits. In the basins of the Uda and Tugur rivers the Lower and Middle Jurassic deposits consist of sandstones and conglomerates up to 3000 m thick with the remains of marine fossils. Bathonian deposits are transgressive and unconformable and, as in the Upper Jurassic, consist mostly of sandstones with a total thickness of 4000 m and with some marine fossils. North of the Uda River the top of the sequence is replaced by coal-bearing deposits and effusives.

Jurassic deposits of the Kamchatka–Koryak geosynclinal area are exposed on the Koryak upland between the Pensha gulf and the Anadyr River. The deposits consist of tuffaceous sandstones, argillites, sandstones and clay shale with the remains of marine fossils of all three divisions.

In the Jurassic the Sikhote-Alin geosynclinal area inherited a structure which existed in the Triassic. The Lower–Middle Jurassic deposits conformably overlie Triassic rocks and are represented by sandy-shale beds, among which, in some depressions, siliceous rocks and effusives occur. The thickness of Lower Jurassic, Aalenian and Bajocian deposits varies greatly reaching a maximum of 6500 m in the lower course of the Amur River. Bathonian deposits occur unconformably and, together with the Upper Jurassic deposits, form a complex of terrigenous rocks containing lenses of siliceous and volcanic deposits. The thickness of the Bathonian–Upper Jurassic deposits reaches to 3000 m. Among the Upper Jurassic ammonites *Ataxioceras* and *Virgatosphinctes* are present. In the southeastern part of Sikhote-Alin the thickness of Jurassic deposits does not exceed 2800 m and not only marine, but also littoral and continental deposits are found.

The Jurassic deposits of the Nippon geosyncline have been recognized in Sakhalin. Here siliceous and clay slates occur with bands of diabase porphyrites to a total thickness to 2000 m. They are unconformably overlain by siliceous and tuffaceous deposits 1000 m thick. Radiolaria remains are known from the siliceous rocks and corals from the limestone partings.

Cretaceous

Cretaceous deposits occur extensively in the Pacific geosynclinal belt. In the Verkhoyano-Chukotka folded area at the end of the Jurassic and the begining of the Creteaceous, the closure of the Mesozoic geosyncline began in the northeast of the U.S.S.R. As a result the Cretaceous beds show a complex combination of marine and continental deposits both vertically and laterally. The lower part of the Cretaceous sequence consists of marine detrital deposits with an admixture of volcanic material locally unconformable on the underlying Upper Jurassic beds. A complex of *Buchia* characterizes the Berriasian–Valanginian time. The thickness reaches to 2500 m. Deposits of the Hauterivian–Aptian stages in the lower course of the Kolyma River and along the Bolshoy Anyuy River consist of sandstones and siltstones overlying, with an erosional unconformity, the subjacent beds. In the lower part *Simbirskites* are found, and in the top part various *Aucellina.* The Albian (?) deposits consist of volcanic rocks, tuffs and tuffites. To the north and west marine deposits are replaced by continental coal-bearing strata forming the Zyryansk Basin. In the Okhotsk–Chukotka volcanic belt subaerial volcanic strata are found with andesite basalts, andesites, liparites and tuffs. Their age according to the floral evidence is end of Lower to beginning of Upper Cretaceous.

In the Koryak–Kamchatka folded area Lower Cretaceous deposits are found mainly over the Koryak upland and in western Kamchatka. They are represented by marine deposits, most frequently resting conformably on volcano-siliceous or terrigenous beds of Late Jurassic age. The beds belonging to the Berriasian and Valanginian stages are the most extensive and lithologically they are similar to the Upper Jurassic deposits, differeing only in the greater number of packets of terrigenous rocks. They vary in thickness from 1500 to 2000 m. A varied complex of *Buchia* sp. including *Buchia volgensis* Lag., *B. okensis* Pavl., etc., warrants correlation of these deposits with the Riazan horizon of the East-European Platform (Fig. 15).

Deposits of the Hauterivian, Barremian and Aptian stages are not so extensive, they are known in the Mamechinsk Mountains, on the Yelistratov Peninsula, in the Pekulney range and in a number of other localities. In some places there is a basal conglomerate up to 110 m thick (Mamechinsk Mountains). At higher horizons sandstones and siltstones are recorded with partings of tuffobreccia and andesitic–basaltic tuffs (Aptian stage). The thickness reaches 2000 m. Albian deposits are often unconformable (Talovka–Maina anticlinorium), but locally there may be a gradual transition (Pensha synclinorium). They are represented by terrigenous rocks with rare partings of andesitic–basaltic tuffs and tuffobreccia. The maximum thickness is 4200 m.

The fauna of the Hauterivian stage is sparse. Barremian–Albian deposits are characterized by a com-

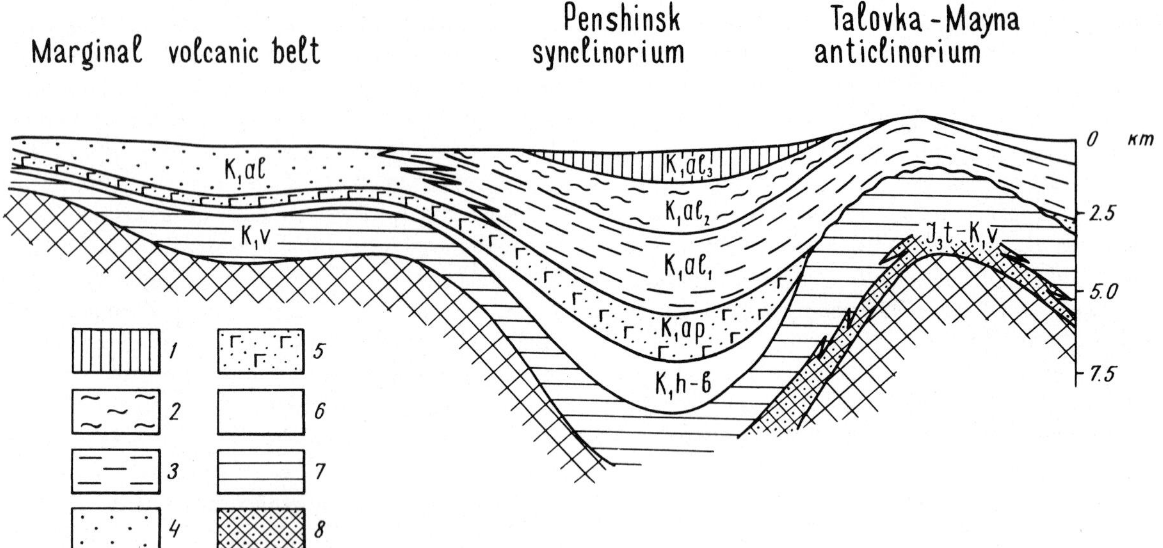

Fig. 15. Structure scheme of the southwestern part of the Koryak–Anadyr area and the marginal volcanogene belt (after G.P. Avdeiko). Legend: *1* = marine sandstones and siltstones with conglomerates at the base; *2* = marine sandstones and siltstones; *3* = marine siltstones, argillites and sandstones; *4* = continental sandstones and siltstones; *5* = marine tuffobreccia, tuffs, sandstones; *6* = marine sandstones and siltstones; *7* = marine sandstones, siltstones, partings of spilites and tuffs; *8* = spilites, tuffs, sandstones, jaspers.

plex of *Aucellina* sp. In the Albian the role of inoceramids and rare ammonites, such as *Cleoniceras, Parasilesites, Neogastroplites, Beudanticeras,* is important. Three stages are distinguished in the geological history, a transgression in the Berriasian–Valanginian, a regression in the Hauterivian–Aptian time with intense volcanism during the Aptian (Fig. 16), and a new transgression beginning at the end of the Albian.

In the Sikhote-Alin folded system vast areas are occupied by marine Berriasian–Valanginian deposits and Hauterivian and Barremian deposits are very limited, while the Aptian–Albian are represented by both marine and continental facies. The contact with underlying rocks, as a rule, is unconformable and only in the region of Komsomolsk-na-Amure and in the basin of the Tetyukhe River is there no interval or unconformity. The Berriasian–Valanginian deposits are represented by an alternation of sandstones, siltstones and argillites with partings of siliceous-clay slates and lenses of limestones with a thickness up to 4500 m. The characteristic fauna of these deposits consists of *Buchia* and rare ammonites (*Neocomites, Berriasella*) which occur only in the southern Maritime Province (Southern Primoria). Radiolaria are associated with siliceous rocks. The overlying deposits in Sikhote-Alin consist of siltstones and sandstones up to 2500 m thick. Rare ammonites (*Tetragonites, Deshayesites, Spitidiscus,* etc.) and a complex of aucellines give the approximate age of these deposits as Late Hauterivian–Early Albian. Further north packets of spilites and andesites appear among the sandstones and siltstones. Towards the south marine deposits are replaced by coal-bearing beds.

In the Mongolo-Okhotsk folded area Lower Cretaceous deposits are represented by fresh-water continental, and frequent coal-bearing deposits, the subdivision of which at the present time is possible only into local stratigraphic units. In the top part of the sequence there are dinosaur remains. The Early Cretaceous age of the deposits has been established through plant remains. Sporadic outcrops of continental Lower Cretaceous deposits are recorded in the upper Amur River region, and in the basins of the Zeya and Bureya rivers. In some regions terrigenous coal-bearing rocks (Bureya coal basin) predominate, in others volcanic rocks are more important. The age of the deposits is determined by plant remains and complexes of fresh-water bivalves, ostracods and phyllopods.

On Sakhalin Lower Cretaceous deposits are found

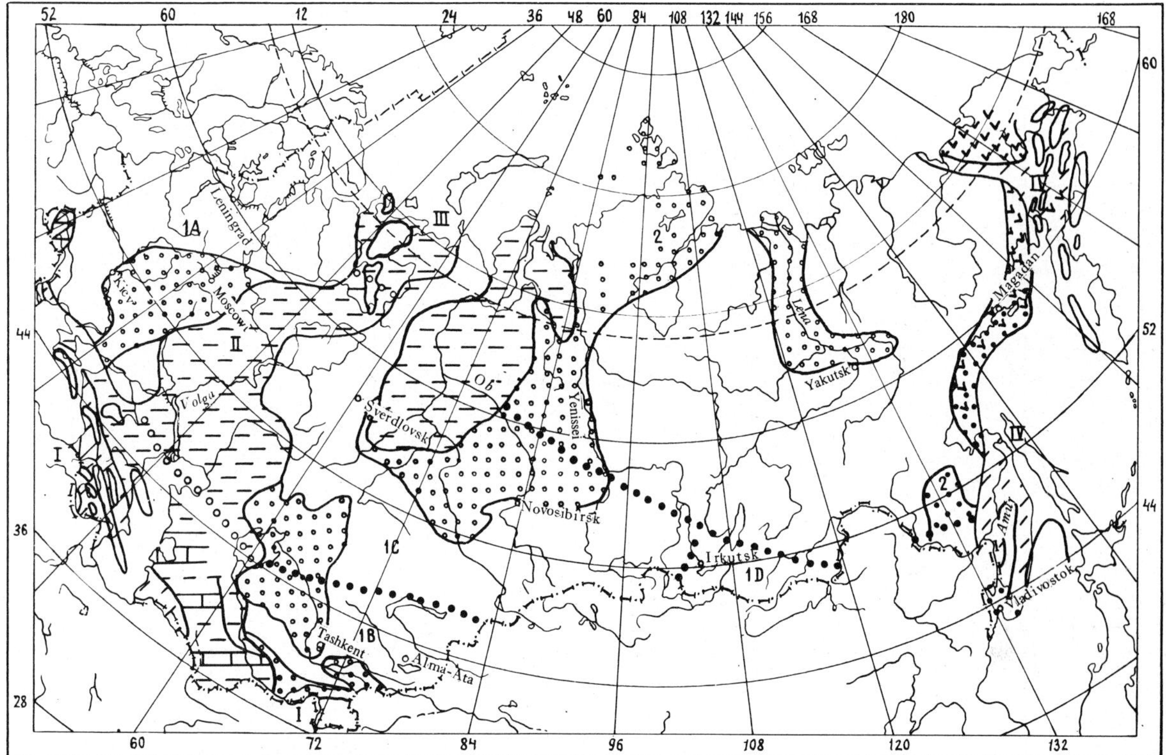

Fig. 16. Distribution scheme of the main Aptian sedimentation basins in the territory of the U.S.S.R. (after the *Atlas of Lithological–Paleogeographical Maps of the U.S.S.R.*, Volume 3, 1968). Legend on Fig. 3. Paleobiogeographical regionalization.
Paleozoogeographical areas
I = Mediterranean area: Phylloceratidae, Lytoceratidae, Deshayesitidae, Parahoplitidae, Cheloniceratidae. *II* = European area: Deshayesitidae, Parahoplitidae, Cheloniceratidae. *III* = Boreal area: Deshayesitidae, *Aucellina*. *IV* = Pacific Ocean area: Lytoceratidae, Deshayesitidae, *Aucellina*.
Paleophytogeographical areas and provinces
1 = Indo-European area – *1A* = European province: *Gleichenia, Nathorstia, Onychiopsis, Weichselia, Matonidium; 1B* = Central Asiatic province: *Gleichenia, Nathorstia, Weichselia, Brachyphyllum; 1C* = Uralo-Chulym province: *Gleichenia, Matonidium, Nathorstia, Onychiopsis, Weichselia; 1D* = East-Asiatic province: *Gleichenia, Nathorstia, Onychiopsis, Weichselia, Anomozamites, Dictyozamites, Nilssonia, Zamites, Athrotaxopsis, Phoenicopsis, Matonidium.*
2 = Siberian area: *Anozamites, Nilssonia, Sphenobaiera, Phoenicopsis, Czekanowskaia.*

extensively in the western part, where they are represented mostly by terrigenous rocks (argillites, siltstones and sandstones) conformably overlain by Cenomanian deposits and hence conditionally dated as the Albian. In the eastern part of Sakhalin volcanic and siliceous rocks are provisionally assigned to the Jurassic–Lower Cretaceous. The Lower Cretaceous deposits of the Koryak–Kamchatka folded area and of Sakhalin are a part of a geosynclinal complex of sediments.

Two fundamental lithological associations are distinguished in the Upper Cretaceous in the Cenozoic fold areas of the Pacific Ocean regions of the U.S.S.R. (from Primoria and Sakhalin to Chukotka and the Bering Sea): a detrital, predominantly marine sequence of miogeosynclinal type (the western part of Sakhalin, Kamchatka and the Koryak highland) and a flinty-volcanogenic suite of eugeosynclinal type (central and eastern parts of the same regions). The thickness of Upper Cretaceous deposits here averages from 3500 to 4000 m, although in some depressions it may reach 5000–6000 m (Fig. 17).

The detrital sequence is characterized by the frequent and rapid changes in the sequence. The sandy and clay deposits are enriched by the addition of pyroclastic material. Over the areas of uplift and in the littoral zones of islands they are replaced by rudaceous and alluvial, dominantly fresh-water continen-

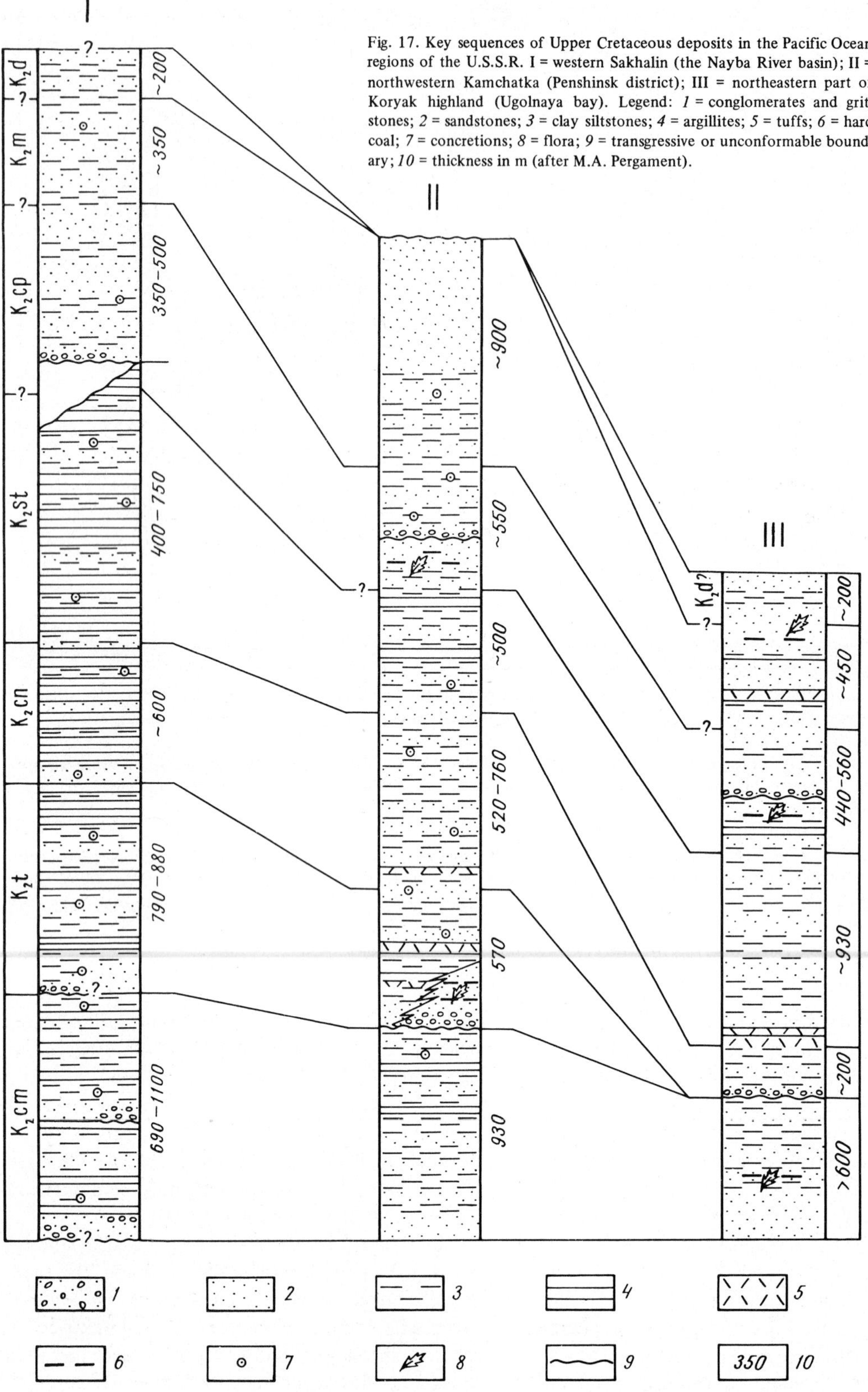

Fig. 17. Key sequences of Upper Cretaceous deposits in the Pacific Ocean regions of the U.S.S.R. I = western Sakhalin (the Nayba River basin); II = northwestern Kamchatka (Penshinsk district); III = northeastern part of Koryak highland (Ugolnaya bay). Legend: *1* = conglomerates and gritstones; *2* = sandstones; *3* = clay siltstones; *4* = argillites; *5* = tuffs; *6* = hard coal; *7* = concretions; *8* = flora; *9* = transgressive or unconformable boundary; *10* = thickness in m (after M.A. Pergament).

tal, coal-bearing deposits. The latter are found in the northern part of western Sakhalin, in the northwest of Kamchatka and in the north of the Koryak highland. The flinty-volcanogenic formation is mainly characteristic of the latter half of the Late Cretaceous. In the south of Kamchatka a thick greenstone complex of altered rocks is of Late Cretaceous age and possibly even older.

In the terrigenous deposits in the miogeosynclinal deposits all the Upper Cretaceous stages are distinguished by their fauna and flora (Fig. 17). The Cenomanian consists at the base mostly of sandstones and somewhat higher of siltstones and argillites with numerous *Desmoceras* (*Pseudouhligella*), *Anagaudryceras, Turrilites costatus* Lam., and *Inoceramus.* The thickness varies from 600 to 1100 m. In the upper reaches of the Anadyr River the Cenomanian is represented by littoral–alluvial rudaceous beds (in excess of 2000 m) with floral remains. Everywhere the Cenomanian deposits conformably and lithologically gradually replace Upper Albian sandstones and conglomerates with *Neogastroplites,* etc.

Locally Lower Turonian faunal deposits are known (sandstones with *Fagesia* in the west of Sakhalin). Usually however, the Turonian begins with "barren" sandy marine siltstones (150–200 m) or coal-bearing strata with plant remains, which transgress with a pronounced unconformity over the Cenomanian and Paleozoic and give way higher up to Upper Turonian deposits with a total thickness of 400–680 m and which contain numerous remains of inoceramids, ammonites or of thermophilous plants. In a number of regions of the Anadyr River basin and in the Ugolnaya bay Turonian deposits are entirely absent and older beds are overlain by conglomerates and basement sandstones of the Coniacian, Santonian or Campanian, or by fresh-water continental strata with a Senonian flora. In the majority of places, however, Upper Turonian–Santonian deposits are represented by a continuous sequence of clastic sediments ending with a comparatively thin (180–260 m) coal-bearing sequence with a Santonian–Campanian "Orochensk" flora of A.N. Krishtofovich. In southwestern Sakhalin (the Naiba River basin) the Turonian, Coniacian and Santonian stages are represented mostly by a clay bed. In northwestern Kamchatka the Coniacian consists of sandstones of various grain sizes and siltstones with partings of tuffs (thickness 720–760 m), while the Santonian consists mostly of tuffoargillites and siltstones (550 m). In contrast, in Ugolnaya bay consolidated clay siltstones with sandstones represent the Santonian (750 m).

The Upper Senonian is characterized by tuffaceous and coarsely clastic rocks, which transgressively overlie with a stratigraphic unconformity upon the rocks below. The Campanian in the western Sakhalin mountains, on the Koryak highland and in many other places consists of compact tuffaceous sandstones and siltstones with a total thickness from 350–520 to 540–770 m. In the lower part they include numerous remains of specific radially ribbed inoceramids (*Inoceramus schmidti* (Mich.), *I. sachalinensis* Sok., etc.); at the top *Inoceramus balticus* Boehm. and *Canadoceras* are found. The same inoceramids and a complex of radiolaria characterize the flinty-volcanogenic Senonian of the eugeosynclinal zone in the central and eastern regions of the Koryak highland and Kamchatka. Here occur tuffaceous and silicified sandstones, jaspers, jasper-quartzites, flinty-clay rocks, spilites, radiolarites, andesite-basalts, diabases and tuffs with a total thickness from 1200 to 5500 m. Locally the flinty-volcanogenic rocks are intercalated, and partly or completely replaced by terrigenous-sedimentary beds. In northwestern Kamchatka the Campanian (380–400 m) is represented by shallow-water sandstones and siltstones with oyster beds.

In the majority of regions the Maastrichtian stage consists of coarse-grained, frequently tuffaceous sandstones (350–870 m) with inoceramids and ammonites (including *Pachydiscus neubergicus, P. gollevilensis,* etc.). At the top of the stage no index fossils have been discovered nor is mention made of coal-bearing beds containing plant remains. As a rule, the eroded surface of the Maastrichtian and older rocks of the Cretaceous is overlain with a sharp unconformity by Eocene–Lower Oligocene and younger deposits. However, in a number of sequences, Maastrichtian deposits without any evident unconformity are replaced upwards by beds with spores and pollen and foraminifers of Danian age.

In the west geosynclinal depressions were bounded by the Okhotsko-Chukotka marginal volcanic belt, where a thick (up to 2000 m) complicated sequence of volcanogenic, and locally coal-bearing, rocks was formed in the Upper Cretaceous (effusives and tuffs, ash flows, tuff breccias, tuffaceous sandstones and slates with abundant plant remains). In the lower part in the southwest of the belt, paleotypical andesites, dacites and liparites are not infrequent, and in the top part kainotypic liparites, dacites and thick (to 800–

1000 m in the Arman River basin) ignimbrites, andesites and andesite-basalts occur. In the Chaunsk and east Chukotka volcanic zones Upper Cretaceous volcanogenic beds (900–1200 m) begin, usually, by tuffaceous conglomerates (to 100 m) which rest unconformably on Lower Cretaceous deposits.

The Verkhoyansk–Chukotka fold area presented in the Upper Cretaceous a mountainous area in a state of intense activity with active volcanism in the closed Mesozoic geosynclines and on the Kolyma–Omolon Massif. To the east of the Kolyma River and up to the Chukotka Peninsula Upper Cretaceous deposits are represented mainly by subaerial effusives (andesites, andesite-basalts) and pyroclastics with a thickness of 2000–3000 m. West of the Kolyma River Upper Cretaceous deposits form the top part of thick coal measures in vast intermountane basins. In the Arkagalinsk coal basin and in the Momsko-Zyryansk depression Upper Cretaceous (Cenomanian) strata (conglomerates, sandstones, siltstones, ash tuffs, coals with a total thickness from 600 to 850 m) overlie Triassic, Jurassic and Lower Cretaceous deposits with a sharp unconformity. In other regions of the Kolyma–Omolon Massif the Upper Cretaceous is represented mostly by volcanogenic deposits.

In the Mongolo-Okhotsk fold area continental and volcanogene Upper Cretaceous deposits occur mostly in the eastern regions, where they form separate basins and vast intermountane depressions. In the Amur–Zeya Basin Lower Cretaceous deposits are transgressively overlain by a thick (more than 2000 m) complex of sandstones, siltstones, argillites and clays of Cenomanian–Campanian age. Higher are sands and clays (500 m) of Maastrichtian–Danian age with dinosaur remains and the "Tsagayansk flora", and still higher the Kivdinsk Paleocene clays. Predominantly fresh-water continental deposits of Late Cretaceous age occur in the upper Priamurye and Bureya regions. On Lesser Khingan too, the Upper Cretaceous consists of volcanogenic and volcano-terrigenous deposits with a thickness of about 1000 m.

In the Sikhote-Alin area Cenomanian–Lower Senonian deposits formed during the stage of geosyncline closure and unconformably overlie Lower Cretaceous deposits. On the Sikhote-Alin and in the lower Primoria the beds are predominantly siltstones with partings of sandstones and conglomerates (with a total thickness of 2000 m), while in Primoria they consist of tuffs and tuffaceous sandstones (up to 2700 m). At the base of the Turonian–Lower Senonian (?) the deposits consist of tuffaceous sandstones (up to 2500 m) with *Actaeonella* sp. and bivalves, while at the top the beds (up to 1000 m thick) consist of porphyrites, tuffs and tuffaceous sandstones with plant remains. They are transgressed with a pronounced unconformity by a thick (up to 5000 m) volcanogenic–sedimentary post-orogenic complex of Late Senonian (and Danian) age. This is formed primarily of subaerial basic and acid effusives and tuffs, tuffites and tuffogenic-terrigenous strata accumulated in fresh-water basins.

Thus, during the Late Cretaceous the east of the U.S.S.R. was the site of a mobile geosynclinal zone, an open marine basin with an archipelago of small islands, some at least of volcanic origin. In addition to the islands, clastic material was supplied from the west from the intense erosion of a rising young Mesozoic platform fringed by a belt of intense fissure effusions and eruptive outbursts of cinder and lavas. The following stages of tectonic movements in the geosyncline have been identified: (a) at the very end of the Albian to before the beginning of the Late Cretaceous, (b) in the Early Turonian (locally occupying the entire Turonian), (c) at the beginning of the Late Senonian, and (d) at the end of the Maastrichtian into Danian times.

The marginal seas of the northwestern Pacific had

TABLE V

Subdivision of Upper Cretaceous deposits in the Pacific Ocean regions of the U.S.S.R. (after Pergament, 1973)

Stages	Zones, strata
MAASTRICHTIAN	strata with *Inoceramus* ex gr. *tegulatus – I. kusiroensis*
CAMPANIAN	*Inoceramus schmidti*
	Inoceramus patootensis – I. orientalis matsumotoi
SANTONIAN	*Inoceramus transpacificus*
	Inoceramus undulatoplicatus
CONIACIAN	*Inoceramus involutus*
	Inoceramus stantoni
TURONIAN	*Inoceramus lamarcki*
	Inoceramus labiatus
CENOMANIAN	*Inoceramus nipponicus – I. scalprum*
	Inoceramus pennatulus
	Inoceramus aff. *crippsi – Desmoceras kossmati*

extensive connections during the Late Cretaceous with the basins of the Boreal, Mediterranean and more southern areas. This is indicated by the occurrence of organic remains. Among them, along with a number of specific genera and a great number of species endemic to the Pacific Ocean, there are also forms present which occur widely including zonal species of the Upper Cretaceous in Europe, North America, Madagascar, etc. The sequence of such species and of the faunal complexes associated with them is the same everywhere, while there are many common features in the development of the phyla of mollusks, like inoceramids, for instance, in distant regions. All this warrants a subdivision of the Upper Cretaceous in the Pacific Ocean regions of the U.S.S.R. according to the West-European stage scale, though some zones are local (Table V).

MEDITERRANEAN FOLD BELT

Within the U.S.S.R. the Mediterranean fold belt includes the Alpine orogens of the eastern Carpathians, mountainous Crimea, the Caucasus, Balkhan, Kopet-Dag, central and southern Pamirs. Mesozoic deposits in this belt take part in the structure of the basement complexes, fill geosynclinal and orogenic depressions, and form the cover of median massifs. On the whole and in some of the regions the structure of the belt underwent repeated rearrangements during the Mesozoic which makes any generalization difficult.

In the Mediterranean fold belt (Fig. 1) the geosynclines of the eastern Carpathians, of the Crimea Mountains, the Greater Caucasus, and Lesser Caucasus, the Adzharo–Trialetsk, Kopet-Dag–Balkhan, of central and southern Pamirs and the median massifs of Transcaucasia, Armenia and of the southwestern Pamirs can be distinguished.

Triassic

The Triassic deposits are either geosynclinal or laid down on the median massifs. In the Crimea and in the Greater Caucasus they form part of the basement complex of younger Jurassic geosynclines, in the Pamirs they represent part of the filling of Late Paleozoic–Mesozoic geosynclinal depressions. The structure of the Mediterranean belt in the Triassic differed substantially from that in the Jurassic and Cretaceous in that it was sharply differentiated determining the variety in the type of Triassic deposits.

In the eastern Carpathians Triassic deposits form part of the structure of the basement complex of the East-Carpathian geosyncline that began to form during the Cretaceous. They are exposed in the klippen zone where they unconformably overlie Paleozoic rocks of the Marmoros and Rachov massifs. The basal part of the Triassic usually consists of conglomerates, the greater part of the sequence being limestones (marbles) and dolomites that replace them in facies. Despite a comparatively low thickness of the deposits, ordinarily not in excess of 100 m, they contain bivalves, brachiopods and foraminifers of the top of the Lower, the Middle and the Upper Triassic.

In the Crimean Mountains, Triassic deposits together with Lower Jurassic beds form the basal sequence of the geosyncline which developed in the Late Bajocian. They are represented by terrigenous flysch of the Taurian series, the apparent thickness of which (together with the Lower Jurassic) reaches 2500 m. The rocks are intensely dislocated. Middle Triassic brachiopods, *Daonella* and a varied Upper Trassic fauna is known. Fossils are mainly associated with allochthonous blocks of shallow-water limestones and sandstones.

In the Greater Caucasus Triassic deposits are a part of the basement complex of a geosyncline which began to form in the Early Jurassic. This complex is represented by deposits of three types. On the southern flank of the Caucasus, in Abkhazia, Svanetia and in northern Kakhetia in the thrust sheets clay shales, phyllites, sandstones and marmorized limestones with an apparent thickness of 600 m, and rare and uncharacteristic corals and foraminifers occur. In the northern flank of the Caucasus, between the Laba and Belaya rivers, Triassic deposits unconformably overlie the Paleozoic and are here represented by limestones, including reef limestones, marls, argillites and siltstones. The thickness of the deposits is 1500 m. An abundant and varied fauna of the Campylian, Anisian, Carnian and Norian stages has been recorded. Unconformities occur at the base of the Carnian and Norian deposits. Along the northern flank of the Caucasus, between the Kuban and Chegem rivers, variegated sandstones and conglomerates of Early Triassic age are present in some of the grabens (Fig. 18).

The Transcaucasian median massif and the Lesser Caucasus geosyncline were land in the Triassic. On

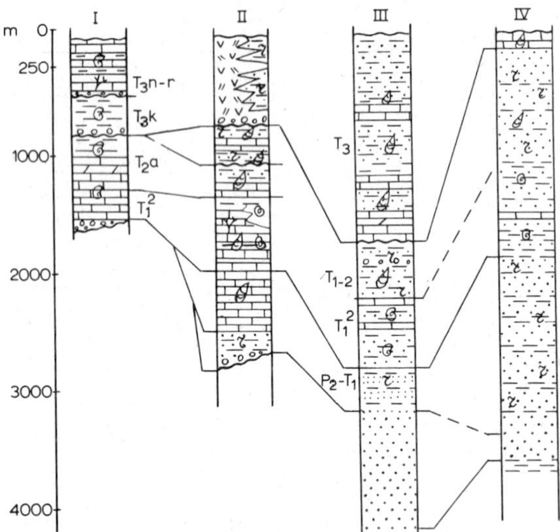

Fig. 18. Sequence correlation of Triassic deposits in the northern Caucasus and Mangyshlak. *I* = northern flank of the Caucasus, Tkhach River; *II* = eastern Pre caucasus, basin of the Kuma River (summary sequence, after Yu.N. Schwemberger); *III* = Mangyshlak, western Karatau range; *IV* = Mangyshlak, Karaschek and Karamaya Mountains (summary sequence, with the use of the data by N.V. and S.E. Petrovy). Legend on Fig. 2.

the Armenian median massif a Triassic gulf penetrated from Iran into the Dzhulfa region. Here thick (up to 1000 m) sandstones and dolomites of an Early–Middle Triassic age and, possibly earliest Late Triassic are found. In the northeastern part of this gulf the Upper Triassic deposits are represented by a paralic coal-bearing formation. In the Balkhan and on the Soviet part of the Kopet-Dag no Triassic deposits have been recorded.

In the central and southeastern Pamirs, Triassic deposits along with Upper Paleozoic fill a system of geosynclinal depressions limited in the north by an area of Paleozoic consolidation in the northern Pamirs and to the southwest by the median massif of the southwestern Pamirs. In the geosyncline of southeastern Pamirs Lower Triassic and Anisian deposits conformably overlie Permian rocks and are represented by laminated limestones and dolomites with a thickness of less than 100 m. Ladinian, Carnian and partly, Norian deposits are represented by thick (up to 1000 m) massive reef limestones and dolomites, as well as by laminated limestones, marls and siliceous rocks that are facies-equivalents, with the total thickness of the latter not exceeding 100 m. The Norian–Rhaetian deposits occur conformably and consist of an alternation of sandstones, siltstones, marls, sandstones and limestones up to 1000 m in thickness. In the marginal zone of the southeastern Pamirs geosyncline, in the Pshartsk and Rushansk ranges, and at its junction with the median massif of the southwestern Pamirs, carbonate Triassic deposits are replaced by terrigenous-carbonate and volcanic rocks, sandstones, marls, conglomerates and effusives of an intermediate and basic composition. The thickness of these deposits varies greatly but may reach 1000 m. In the geosyncline of central Pamirs, in the Muzkolsk range, Lower and Middle Triassic deposits are represented by 250 m of calcareous sandstones, marls and limestones (marbles). In the region of Rangkul Lake these strata are replaced by massive and coarsely bedded limestones which also include the Carnian stage. In the Muzkolsk range and in the western part of Rushansk range Upper Triassic deposits are represented by a thick (1500 m) mass of alternating siltstones, sandstones and conglomerates with plant remains.

Jurassic

Jurassic deposits are generally represented in the Mediterranean belt by geosynclinal and, to a lesser extent, orogenic deposits. It is in the Jurassic that the formation of the majority of geosynclinal systems began. Their development determined the fundamental features of the present Mediterranean fold belt.

In the eastern Carpathians outcrops of Jurassic deposits are limited and are, frequently, associated with klippe. Limestones and siliceous rocks in complex facies relationship predominate. The thickness does not exceed a few hundred meters. Sinemurian, Pliensbachian, Toarcian and all the stages of the Middle and Upper Jurassic are recognized. In some tectonic zones erosional intervals have been recorded at the bases of the Toarcian, Callovian and Tithonian.

The Precarpathian and Lvov depressions formed on the subsiding margin of the East-European platform, but their development is closely associated with the evolution of the East-Carpathian geosyncline. At the base of the Jurassic sequence in these depressions drilling has revealed that detrital marine rocks are replaced along the outer periphery of the depression by littoral and continental deposits. These beds, 800 m thick, belong to the Middle and Lower (?) Jurassic. The Upper Jurassic is represented by limestones replaced along the outer periphery of the depression by red beds and anhydrites. The thickness of the Upper Jurassic deposits is 500–600 m.

In the Crimean Mountains Lower Jurassic and Aalenian deposits in combination with Triassic rocks form the basement complex. During the Early Bajocian a folding took place accompanied by a marine regression. Littoral conglomerates and coal-bearing sandstones several hundred meters thick accumulated in the grabens and in some of the depressions. During the Late Bajocian a local geosynclinal depression began to form and received a fill of volcanic and sandy-clay deposits up to 1000 m thick. During the Bathonian and Early Callovian the depressions became wider and this process was accompanied by a reduction of the tectonic relief and by the accumulation of clay and flyschoid sandy-clay strata from 500 to 600 m thick. The Middle Callovian is often condensed. Upper Callovian, Oxfordian, Kimmeridgian and Tithonian deposits are represented by conglomerate, carbonate, frequently reef and flysch facies, with a distribution structurally controlled. The thickness of the Upper Jurassic deposits is highly variable coming to 3000 m and even more.

In the Greater Caucasus the onset of the geosynclinal depression began in the Early Jurassic, the structure becoming complicated as a result of Early Bajocian movements. Sinemurian and Lower Pliensbachian deposits rest unconformably on pre-Jurassic rocks. They represent deposits in shallow-water sandy-clays, with littoral coal-bearing (graphitic) deposits containing large lenses of intermediate volcanic rocks. Sometimes, as in northern Osetia for instance, volcanic rocks predominate in the sequence. The thickness of these volcanic deposits may reach 1000 m. During the Late Pliensbachian, Toarcian and Aalenian the geosynclinal depression gradually became wider and the rate of subsidence accelerated. Consequently, in the deposits shales of the "slate formation" predominate. In Domerian deposits lenses and beds of effusive rocks are regionally developed. Thick and complex Aalenian volcanic beds occur on the western slopes of the Caucasus. In Daghestan there are thick (up to 5000 m) deltaic coal-bearing sands of Late Toarcian and Aalenian age. The thickness of Upper Pliensbachian–Aalenian deposits ranges from 1500 to 3000 m in the central Caucasus, up to 10,000 m in Chechnia and Daghestan and exceeds 6000 m on the western Caucasus.

In the Early Bajocian there began the uplift of the inner part of the geosyncline of the greater Caucasus which broke up its continuity. Bajocian and Lower Bathonian deposits are mostly represented by argillites and siltstones, with a highly variable thickness which may reach 3500 m. Bajocian volcanic rocks, mostly of an intermediate composition, occur in Balkaria and on the southern flanks of the Caucasus, from Kakhetia in the east to Abkhazia in the west. At the end of the Bathonian the Caucasus was involved in folding accompanied by an uplift and regression. The Middle–Upper Bathonian littoral and shallow-water marine deposits are unconformable and occur in some depressions of Balkaria, Chechnia, Daghestan and southern Osetia (Fig. 19).

During the Late Jurassic uplift of the axial part of Greater Caucasus divided the Early–Middle Jurassic geosyncline into a system of depressions on the southern flank of the Caucasus into which flysch was poured and a system of depressions between the northern Caucasus and the Scythian platform lying to the north. In these latter depressions various limestones accumulated including the reef varieties. During the Kimmeridgian and Early Tithonian the limestones in the northern part of these depressions were

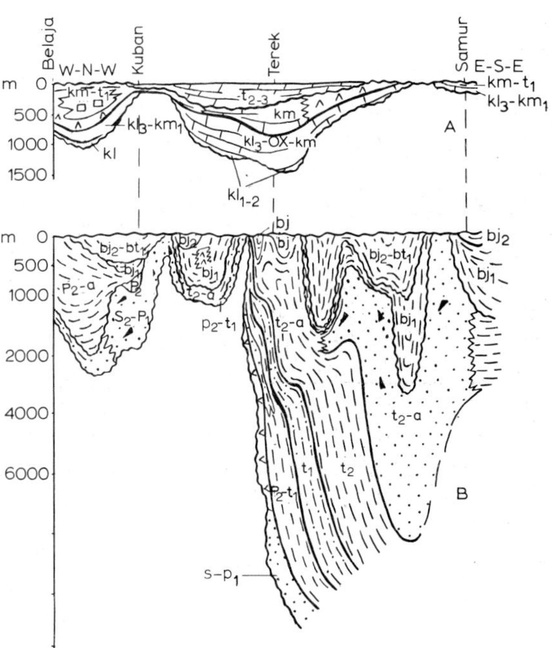

Fig. 19. Structure scheme of Jurassic deposits along the northern flank of the Caucasus. *A* = Upper Jurassic (the roof of the Upper Jurassic is accepted as the zero surface); *B* = Lower and middle Jurassic (the roof of the deposits is accepted as the zero surface). Legend on Fig. 2.

replaced by a sulphate-halite and variegated formations. Variegated terrigenous rocks of Late Jurassic age also occur in the depressions of the axial part of the eastern Caucasus. The thickness of Upper Jurassic deposits in the flysch facies reaches 3000–5000 m, in the reef facies 2000 m, and in the carbonate non-reef facies 1000 m.

Over the Transcaucasian median massif and in the geosyncline of the Lesser Caucasus Lower Jurassic and Aalenian deposits are represented by sandstones, siltstones, and limestones, not exceeding 500 m. They unconformably overlie Paleozoic and Precambrian rocks. The Bajocian is transgressive and is represented by the "porphyritic formation" up to 2000 m thick. Beginning with the Bathonian the geological history of the Lesser Caucasus geosyncline and of the Transcaucasian median massif diverges. In the geosyncline of the Lesser Caucasus thick (up to 1000 m) volcanic beds continue to form during the Bathonian. These Upper Jurassic deposits consist of porphyrites and tuffs and include several packets and lenses of limestones. Locally in the Tithonian gypsum is present with the limestone. The thickness of the Upper Jurassic deposits is 3000 m. Over the Transcaucasian median massif the sea receded during the Bathonian and littoral, terrigenous, and coal-bearing lacustrine sediments accumulated to a thickness of 500 m. Upper Jurassic deposits occur along the northern margin of the massif and in the Kolkhida (Colchis) lowland. Callovian and Oxfordian deposits are represented by marine detrital beds and limestones with reef limestones along the margin of the flysch depressions on the southern flanks of the Greater Caucasus. In the Kimmeridgian there are transgressive variegated clays and gypsum with beds of basalt. The thickness of the Upper Jurassic deposits reaches 700 m.

On the Armenian median massif Jurassic deposits accumulated in a depression of a limited size and are exposed in the Dzhulfa region. Basalts and porphyrites occur at the base with a thickness of 120 m overlain by sandstones, clays and limestones containing ammonites of Bajocian, Bathonian and Callovian age. The total thickness of Jurassic deposits does not exceed 400 m.

The oldest deposits exposed on Bolshoy Balkhan belong to the Late Bajocian and are represented by argillites with concretions, bands and packets of sandstones and allochthonous blocks of metamorphic and igneous rocks. Their thickness exceeds 1000 m. The Lower Bathonian consists of littoral sandstones and siltstones with coal lenses and includes in the middle part a mass of argillites with concretions (up to 400 m). The thickness of the Lower Bathonian comes to 1500–2000 m. Middle Bathonian–Lower Callovian deposits occur with an angular unconformity and are represented by sandstones, limestones and clays of a greatly variable thickness from a few tens of meters to 400 m. The top part of the Callovian, the Oxfordian and Kimmeridgian consist of limestones, sometimes dolomitized, up to 800 m thick. Red limestone conglomerates belong to the Tithonian.

The oldest rocks exposed and reached by drilling in the Soviet part of Kopet-Dag are Callovian limestones. Upper Jurassic deposits are represented by limestones and dolomites up to 800 m thick. In the Kimmeridgian, beds of anhydrite are present.

In the geosyncline of central Pamirs, in the Yazgulemsk and Muzkolsk mountain ranges, Jurassic deposits occur conformably on Triassic rocks. The lower part consists of clay shales and sandstones with bands of conglomerates and contains plant remains and bivalves. The thickness of these beds of the Lower and the basal Middle Jurassic is 450 m. On the southern flank of the Muzkolsk range clay shales are the predominant lithology. To the Bathonian, Callovian and Oxfordian belong thick (1000 m) limestones. These strata are conformable or rest on an erosional surface with basal coarse molasse-red conglomerates with bands of limestones of a total thickness of 1500 m. The conglomerates belong to the top of the Upper Jurassic and extend into the Lower Cretaceous.

In the geosyncline of the southeastern Pamirs Jurassic deposits are represented by limestones, including reef limestones and by marls and clay shales, which form a complex facies pattern. They sometimes conformably overlie Triassic deposits, occasionally resting on different horizons, overlapping on to the Carboniferous, Permian and Triassic deposits with an angular unconformity. Prior to the Late Oxfordian folding occurred, so that Upper Oxfordian and Kimmeridgian limestones are locally unconformable. No Tithonian deposits have yet been identified in the southeastern Pamirs.

Cretaceous

Within the Mediterranean fold belt in the U.S.S.R. the same structural elements found in the Jurassic are distinguished, namely the eastern Carpathians, the Crimean Mountains, the Caucasus, Kopet-Dag and the Pamirs. In the north these fold structures are bounded by marginal depressions. The Lower Cretaceous deposits of the region are represented by geosynclinal and subplatform marine deposits. They are best known in the Caucasus, on the Kopet-Dag and in the Crimea, where they consist mostly of detrital rocks. The presence of an abundant fauna has resulted in detailed subdivision into stages and, in some areas, into zones.

In the eastern Carpathians Lower Cretaceous deposits occur extensively, most frequently exposed in the zones of uplift. Contacts with the underlying rocks are observed only in the klippen belt (Utesovaya and Marmoros zones), where Lower Cretaceous deposits transgressively overlie Jurassic, Triassic or more rarely, Paleozoic rocks. The lithological composition is very varied, either flysch or an alternation of different, mostly detrital, beds with a total thickness of about 1000 m. In the Marmoros klippen zone organic limestones of the Urgonian facies occur extensively. The uneven faunal distribution within the sequence and its poverty means that only provisional ages assigned to isolated local units are possible. Throughout the Early Cretaceous a flysch trough existed while in the zones with a lower rate of subsidence organic rocks of the Urgonian facies were deposited. The Albian was a time of folding and the movements which followed resulted in the formation of a mountainous terrain.

The Cretaceous of the Crimean Mountains forms the second ridge extending from Feodosya to Sevastopol. Over the majority of the area, Lower Cretaceous deposits overlie transgressively and unconformably Jurassic and Triassic rocks. In the southeastern Crimea a gradual transition is observed between the Tithonian and the Berriasian. The Lower Cretaceous deposits of the Crimean Mountains are very variable both in structure and composition. The Berriasian–Valanginian deposits in some synclinoria are represented by a flysch, clay and conglomerate facies. The Hauterivian is represented by shallow-water, poorly sorted, sandy-clay rocks alternating with relatively deep-water clays and limestones. In the Barremian clays, cephalopod limestones and shallow-water arenaceous limestones are recorded. The Aptian and Lower Albian usually consists of clays with pyritized dwarf ammonites. The Upper Albian is represented by sandstones, sometimes occurring ingressively, in valleys and grabens formed during the Middle Albian, when certain anticlinoria experienced uplift (Fig. 20).

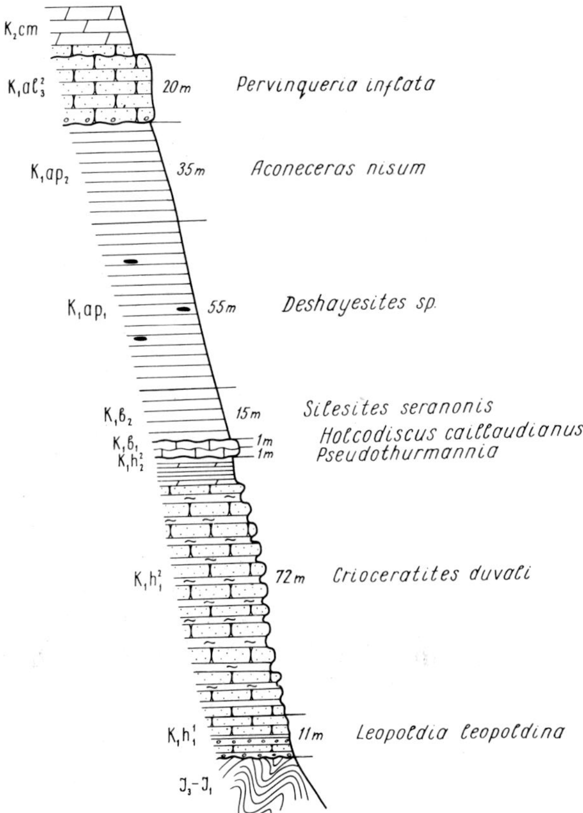

Fig. 20. Sequence of Lower Cretaceous deposits in southwestern Crimea (the Kacha River, after T.N. Gorbachik and B.T. Yanin).

In the Crimean Mountains the maximum transgression occurred in Hauterivian time, when the sea substantially expanded towards the north having flooded the plains of the Crimea. Some islands remained in the region of Belogorsk and the Old Crimea, and here conglomerates were deposited. During the Barremian the sea basin was reduced, expanding again during Aptian times. At the end of the Early Cretaceous, the Crimea was subjected to further uplift but subsidence followed in Late Albian, when the sea locally flooded the basins formed and subsequently expanded over considerable areas. Altogether Lower Cretaceous deposits are characterized by variable composition, the presence in numerous lacunae and an uneven distribution of fauna resulting from drastic changes in the sedimentation conditions from deep-sea to shallow-water.

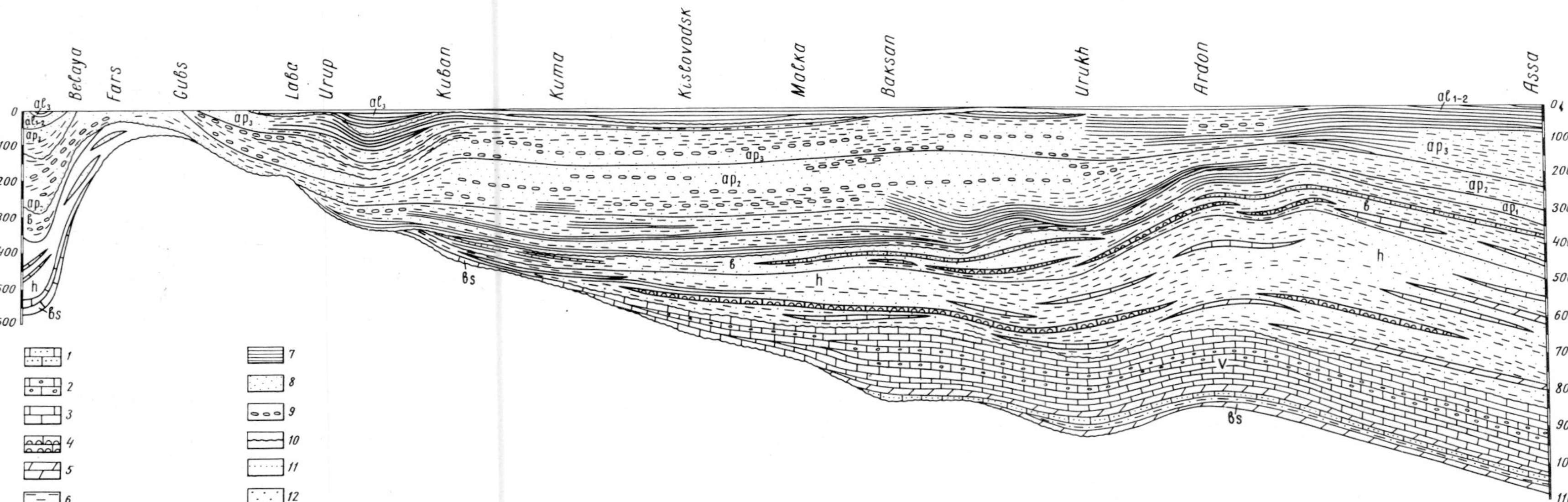

Fig. 21 Structure scheme of Lower Cretaceous deposits in the northern Caucasus (after V.V. Druschitz and I.A. Mikhailova). Legend: *1* = limestone; *2* = sandy limestone; *3* = oolitic limestone; *4* = coquina limestone; *5* = marl; *6* = siltstone; *7* = clay; *8* = sandstone; *9* = horizons of calcareous concretions; *10* = wash-out boundaries; *11* = phosphorite; *12* = conglomerate.

The Lower Cretaceous deposits of the Caucasus extend in a narrow strip along the northern flank of the Greater Caucasus and occupy extensive areas in Daghestan and the southwestern flank of the Greater Caucasus (Fig. 21). On the southern flank they are present in the Abkhazia–Racha zone, in the flysch zones and in a number of other places. Lithologically the deposits are characterized by great variety. On the northern flank carbonate rocks are observed in the lower (Berriasian–Valanginian), terrigenous-carbonate rocks in the middle (Hauterivian–Barremian) and terrigenous in the top part (Aptian–Albian). On the southern flank carbonate-flysch deposits occur (Berriasian–Valanginian) and terrigenous-flysch deposits (Hauterivian–Albian). In Abkhazia the bottom part of the sequence is calcareous, the top marly. Volcanic deposits are extensive within the Lesser Caucasus. In the majority of areas Lower Cretaceous deposits unconformably overlie Upper Jurassic rocks.

The Lower Cretaceous deposits of the northern flank of the Greater Caucasus are known in greatest stratigraphical detail. During the Early Cretaceous there was a rather shallow marine basin, in which neritic sediments were deposited and which was inhabited by a varied assemblage of cephalopods, bivalves, gastropods, brachiopods and other organisms. The lower part of the sequence (Berriasian–Valanginian) is represented by predominantly carbonate rocks containing at their base *Euthymiceras transfigurabilis* Bog., *E. euthymi* Pict., *Malbosiceras korjeli* Grig., *Riasanites rjasanensis* Nik., etc., characteristic of the Berriasian.

Deposits of the Hauterivian and Barremian stages consist of siltstones with bands of calcareous sandstones and oyster beds. There is a tendency for the amount of sandy material to increase upwards in the sequence. The Aptian stage is nearly everywhere formed by various types of unconsolidated and massive sandstones, containing numerous horizons of concretions. The concretions vary in shape and size and frequently contain shells of ammonites, bivalves and gastropods. Several phosphatic horizons of a varying extent can be traced. They contain a fauna from several stratigraphic levels, i.e., present condensed beds. The faunal distribution permits the recognition of three zones in the Lower Aptian: *Deshayesites weissi, Deshaysites dechyi* and *Dufrenoya furcata*; two zones in the Middle Aptian: *Epicheloniceras subnodosocostatum* and *Parahoplites melchioris*; two zones in the Upper Aptian: *Acanthohoplites nolani* and *Hypacanthoplites jacobi.* The Albian beds consist mainly of clays, with more rarely some siltstones. Along with the ammonites *Leymeriella tardefurcata* Leym., *Douvilleiceras mammillatum* Sow., *Hoplites dentatus* Sow., there is an important and varied assemblage of belemnites.

In the Bolshoy Balkhan, Lower Cretaceous deposits rest unconformably on Jurassic rocks and are represented by limestones in the lower part and by siltstones and sandstones in the upper part. In the Kopet-Dag, Lower Cretaceous deposits are represented by marine sediments. Lithologically they can be divided into a lower carbonate part and upper detrital beds. The boundary between them is variable, being either associated with the base of the Aptian or lying within it. The lower part of the sequence is about 1500 m thick and contains a fauna of brachiopods and bivalves with rare ammonites. *Orbitolina* limestones are very characteristic of the Lower Barremian. The Upper Barremian consists of marls containing ammonites of the *Colchidites* genus also known in synchronous deposits in Georgia and the northern Caucasus. During Aptian–Albian times there was an accumulation of clay–siltstone beds from 1200 to 2400 m thick. An abundant and varied assemblage of ammonites permits a detailed zonal subdivision of these deposits (see Table III).

In the southern Pamirs, Lower Cretaceous deposits are of limited extent. They form the top part of a coarse clastic mollase in the Muzkolsk mountain range and in the basin of the Kokubel-su River. The conglomerates contain here lenses and bands of rudistid limestones.

The Upper Cretaceous beds of the Mediterranean belt formed during a transition stage in the rearrangement of tectonic belts during the Early Mesozoic to a new Alpine plan and are extremely varied in composition.

During the Late Cretaceous in general the structural pattern of the Jurassic–Early Cretaceous time is preserved in the Greater Caucasus. Stable uplifts in the axial region of the Greater Caucasus (during the Albian–Cenomanian) separate the zone of the Upper Cretaceous volcanogenic-terrigenous and terrigenous flysch on the southern flank from the miogeosynclinal depression in the north, where only carbonate sediments formed. At the western and eastern ends of the Caucasus, flysch depressions continue their development with the deposition of a tuffaceous detrital-carbonate, locally silicified and volcanogenic flysch

(Cenomanian–lower part of the Turonian) and a younger light-colored carbonate flysch (Turonian–Danian). The thickness of these deposits in the Novorossiysk depression exceeds 4000 m. In the north these depressions are bounded by a zone of cordilleras, near which thin (to 300 m) non-flysch facies are known mainly of Upper Senonian carbonates. At the southern limits of the depressions evidence of effusive volcanism is recorded. In the Abkhazia–Racha zone and in the structures of the Transcaucasian median massif normal marine detrital (Cenomanian) deposits occur and carbonate (Turonian–Danian) deposits in which volcanogenic rocks (Turonian–Lower Senonian) play a relatively minor role.

The stratigraphy of the Upper Cretaceous is best studied in the outcrops along the northern flank of the Greater Caucasus. The most complete sequences are in Daghestan, Checheno-Ingushetia, northern Osetia and Kabardino-Balkaria. There, Upper Cretaceous deposits consist predominantly of pure limestones, the thickness of which changes sharply

TABLE VI

Subdivision of the Upper Cretaceous of the northern Caucasus (after Moskvin, 1968)

Stages	Substages	Chatacteristic fauna complexes (zonal species marked with asterisks)
DANIAN		* *Cyclaster gindrei* Seun., *Protobrissus depressus* Kongiel, *Echinocorys sulcatus* Goldf., *E. obliquus* Ravn., *E. pyrenaicus* Seun., *Galeaster carinatus* Ravn., *Coraster sphaericus* Seun., *Hercoglossa danica* Schloth.
		* *Cyclaster danicus* Schlut., *Protobrissus canaliculatus* Cotteau, *Galeaster minor* Posl., *Echinocorys edhemi* Boehm, *Homoeaster abichi* Anth., *Hercoglossa danica* Schloth.
MAASTRICHTIAN	Upper	* *Pachydiscus neubergicus* Hauer, *Pseudophyllites indra* Forb. *Discoscaphites constrictus* Sow., *Belemnella arkhangelskii* Najd., * *Inoceramus tegulatus* Hag., *Cylaster intiger* Seun.
	Lower	*Hauericeras sulcatum* Kner, *Discoscaphites constrictus* Sow., * *Belemnella lanceolata* Schloth., *Paramicraster ciplyensis* Lamb., *Stegaster chalmasi* Seun., *Echinocorys pyramidatus* Portl.
CAMPANIAN	Upper	(a) * *Micraster grimmensis* Neitsch., *Bostrychoceras schloenbachi* Favre, *Belemnitella mucronata* Now., *B. langei* Schatsk., *Ornitaster alapliensis* Lamb.; (b) * *Micraster brongniarti* Hebert, *Bostrychoceras polyplocum* Roem., *Pachydiscus koeneni* Gross., *Pseudoffaster caucasicus* Dru.
	Lower	(a) * *Micraster coravium* Posl., *Hauericeras pseudogardeni* Schlut., *Stegaster gillieroni* Lor., *Offaster pilula* Lam., *Galeola senonensis* Orb.; (b) * *Micraster schroederi* Stoll., *Eupachydiscus launayi* Gross., *Inoceramus azerbaydjanensis* Aliev, *Offaster pomeli* Mun.-Ch.
SANTONIAN	Upper	* *Marsupites testudinarius* Schloth., *Uintacrinus socialis* Grin., *Micraster rostratus* Mant., *Echinocorys turritus* Lamb., *Inoceramus haenleini* Müll.
	Lower	(a) * *Inoceramus cordiformis* Sow., *I. lesginensis* Dobr., *Micraster rostratus* Mant.; (b) * *Inoceramus undulatoplicatus* Roem., *I. cardissoides* Goldf., *Micraster coranguinum* Klein
CONIACIAN	Upper	* *Inoceramus involutus* Sow., *I. percostatus* Müll., *Micraster coranguinum* Klein
	Lower	* *Inoceramus wandereri* And., *I. deformis* Meek, *I. koeneni* Müll., *Micraster cortestudinarium* Goldf., *Echinocorys gravesi* Desor, *Conulus subconicus* Orb.
TURONIAN	Upper	* *Inoceramus lamarcki* Park., *I. seitzi* And., *Lewesiceras peramplum* Mant., *Holaster planus* Mant., *Micraster cortestudinarium* Goldf., *"M." corbovis* Forb., *"M." leskei* Desm., *Conulus subconicus* Orb.
	Lower	(a) * *Inoceramus apicalis* Woods, *Lewesiceras peramplum* Mant., *Cardiaster peroni* Lamb., *Conulus subrotundus* Mant.; (b) * *Inoceramus labiatus* Schloth., *I. hercynicus* Petr.
CENOMANIAN	Upper	*Holaster subglobosus* Leske, *Discoidea cylindrica* Lam., *Inoceramus pictus* Sow., *I. scalprum* Boehm., *Scaphites aequalis* Sow., *Turrilites costatus* Lam.
	Lower	*Neohibolites ultimus* Orb., *Mantelliceras mantelli* Sow., *Schloenbachia varians* Sow., *Inoceramus crippsi* Mant.

from the west to east from 180 to 1200–1500 m. Various fauna from all stages (Table VI) are also found in the Upper Cretaceous deposits of the Crimea.

The Cenomanian stage is separated from the Upper Albian *inflata* zone by a pronounced lithological change. The Cenomanian consists of glauconitic sandstones, sandy limestones or limestone-marls with a thickness of 1.5 m in the west to 100 m in the limestone of Daghestan. Lower Turonian deposits (15 m of white limestones with flints, marl partings) have been identified only in the inner regions of Daghestan and Checheno-Ingushetia, where they gradually replace Cenomanian deposits. In the majority of sequences the Cenomanian, and locally the Albian are overlain by the Upper Turonian (from 30 to 100 m thick) consisting of white and pink limestones with horizons of stylolites and marl partings. Lithologically similar Coniacian deposits (20–90 m), Santonian (30–150 m) and Lower Campanian deposits can be distinguished only by their fauna, but locally (the Urukh River) the Upper Santonian is separated by an unconformity. Upper Campanian white limestones and greenish marls (from 40–50 to 300–400 m thick) occur transgressively; east and south of them the amount of marls and sandy material increases. Sandy facies with phosphatic pebbles and belemnites characterize Upper Campanian deposits south of Daghestan and north of Azerbaydzhan. The Maastrichtian stage consists mostly of chalk-like limestones ranging from 45–50 m to 100–120 m thick in the Ardon River and 350–400 m in Daghestan. Conformable Danian limestones with flints (40–250 m) occur frequently and are gradually replaced by beds with *Protobrissus tercensis* Cotteau, and then by the *angulata* zone of Early Paleocene age. In the Daghestan Mountains and in other places Paleocene deposits were deposited on an eroded Danian surface.

The Upper Cretaceous in the more westerly portions of the northern flank is similar. Here, however, there are more clay-marly facies and the carbonate rocks contain locally great amounts of terrigenous material. In the central part between the Cherek and Khodz rivers the sequence is complete, a characteristic of the greater part of the Precaucasus. The thicknesses are not so variable, from 110–160 to 330–430 m. Further west, up to the Khokodz River, the Upper Cretaceous deposits are intermittent and incomplete with many gaps and the complete absence of the Lower Turonian and Santonian deposits.

In the Lesser Caucasus at the boundary between the Early and Late Cretaceous the structural plan becomes much more complicated as the result of the development of a number of new sublatitudinal depressions and uplifts. The Upper Cretaceous deposits are represented by a volcanogenic, tuffaceous, sedimentary, detrital and carbonate facies. Cenomanian, Lower and Upper Turonian, Lower and Upper Coniacian and Upper Santonian deposits are not always present; sometimes they may unconformably and transgressively overlie the basal conglomerate deposits of Early Cretaceous, Jurassic, Triassic age and crystalline rocks of the Paleozoic in the Yerevan–Ordubad and Miskhano-Zangezur zones.

In the Transcaucasian median massif, Cenomanian deposits are represented mainly by coarse tuffaceous, detrital deposits with partings of limestones and the remains of ammonites, oysters, gastropods and rudists. The maximum thicknes of the beds of this stage (400–445 m) is found in the Sevan–Karabakh and Araks zones. In the Adzharo-Trialeti geosyncline Cenomanian and Lower Turonian deposits form the top of a thick volcanogenic series. The Turonian is variable in facies, in most areas it lies at the base of a composite volcanogenic (intermediate and acid composition), volcano-sedimentary and detrital complex (to 1100 m) of Coniacian and Santonian age. On the massif and in the Yerevan–Ordubad zone the complex is replaced by detrital and carbonate rocks. At the beginning of the Late Santonian the facies became more uniform and during the Campanian and Maastrichtian limestone and marly beds accumulated over the entire Lesser Caucasus with a thickness from 200–300 to 1260 m.

A rich Upper Cretaceous fauna in the Lesser Caucasus is represented by typical Mediterranean oysters, gastropods, rudists, etc. In the northern Caucasus during the Upper Cretaceous with the expansion of the transgression, the role of the East-European Platform, Western Europe, the Transcaspian and Transcaucasian fauna increases. Typical Mediterranean forms are absent and characteristic Boreal forms appear. This may indicate a somewhat colder climate in the northern Caucasus accompanying the expansion of the transgression during the Albian and especially during Late Cretaceous time.

In the eastern Carpathians, in the central synclinal zone, a thick terrigenous flysch formed during the Albian–Danian interval. To the west of the zone variegated rocks and black argillites of Turonian–

Maastrichtian age are recorded overlain by Danian–Paleogene sandstones. In the Marmaros–Pennine zone Cenomanian sandy conglomerates followed by gray and red marls with Cenomanian–Maastrichtian foraminifers occur. In the outer part of the Carpathians (in the Klippen zone) the Upper Albian–Coniacian is represented by limestones, flinty marls and argillites, and the Santonian–Danian stage by a carbonate and sandy-clay flysch.

In the Crimea Upper Cretaceous rocks form the northern flanks of the Crimean Mountains. Cenomanian deposits (sandy and clayey marls and limestones, commonly with a basal quartz–glauconitic sandstone) transgressively overlie the Albian and Middle Jurassic

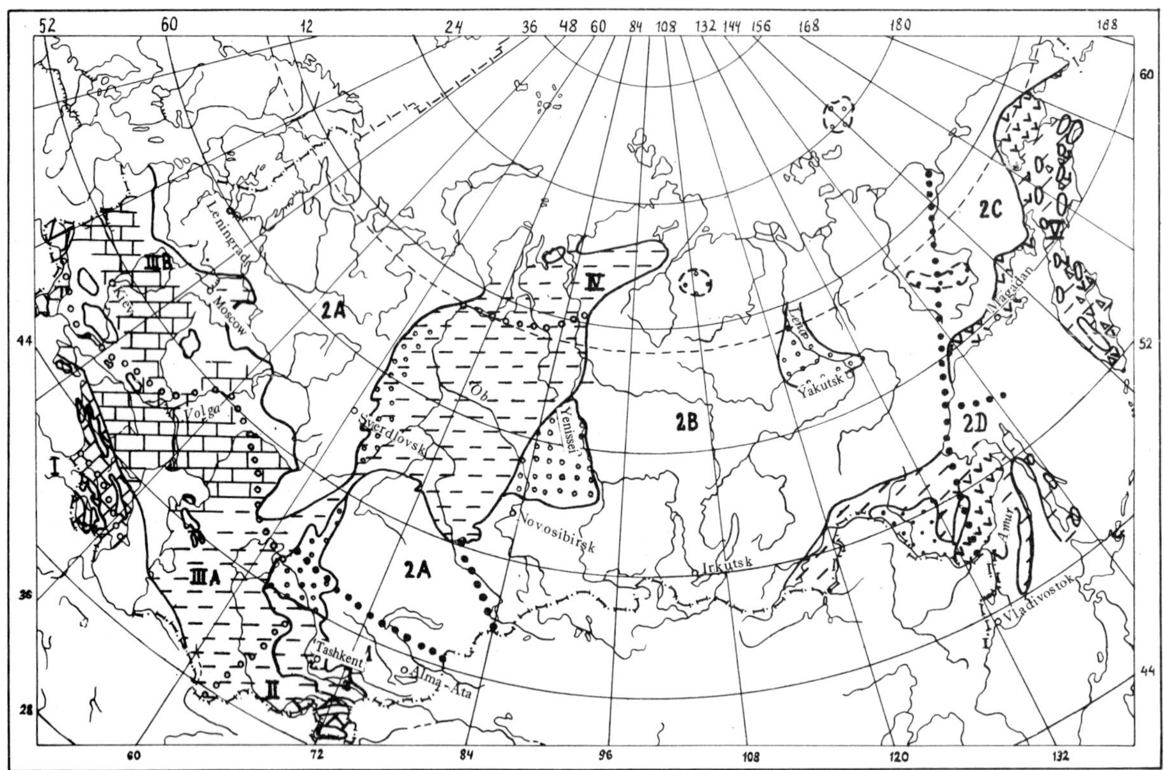

Fig. 22. Distribution scheme of the main Turonian sedimentation basins on the territory of the U.S.S.R. (compiled on the basis of the *Atlas of Lithological–Paleogeographical Maps of the U.S.S.R.*, Volume 3, 1968, simplified and changed). Legend on Fig. 3. Paleobiogeographical regionalization.

Paleozoogeographical areas and provinces

I = Mediterranean area: *Collignoniceras, Inoceramus, Exogyra, Radiolites, Vaccinites, Durania, Micraster, Infulaster, Echinocorys, Holaster, Conulus.*

II = Central Asiatic area: *Collignoniceras, Mammites, Placenticeras,* Vasoceratidae, *Inoceramus, Fatina, Exogyra, Liostrea, Korobkovitrigonia, Megatrigonia, Infulaster.*

III = European area – *IIIA* = Carpathian–Kopet-Dag province: *Arkhangelskiceras, Baculites, Beschtubeites, Borissiacoceras, Collignoniceras, Lewesiceras, Mammites, Placenticeras, Scaphites, Praeactinocamax, Inoceramus, Echinocorys, Infulaster, Holaster, Conulus, Micraster. IIIB* = Baltic–Volgian province: *Praeactinocamax, Inoceramus, Micraster.*

IV = Boreal area: *Inoceramus, Baculites.*

V = Pacific Ocean area: *Fagesia, Mammites, Inoceramus, Hemiaster.*

Paleophytogeographical areas and provinces

1 = Euro-Turkestan area, Central Asiatic province: *Brachyphyllum,* Gnetales, *Gleichenia, Normapollis.*

2 = Ural-Siberian area – *2A* = Uralian province: *Normapollis, Platanus. 2B* = East-Siberian province: *Cephalotaxopsis, Gleichenia, Aquilapollenites, Dalbergites, Sequoia, Sassafras, Protophyllum, Pseudoprotophyllum, Platanus, Trochodendroides, Quereuxia. 2C* = Kamchatka–Chukotka province: *Cephalotaxopsis,* Filicales, *Sequoia, Protophyllum, Pseudoprotophyllum, Trochodendroides, Nilssonia. 2D* = Sakhalin–Nippon province: Filicales, *Gleichenia, Cephalotaxopsis, Nilssonia, Bauhinia, Sassafras, Sequoia, Protophyllum, Pseudoprotophyllum, Trochodendroides.*

beds. Lower Turonian limestones and marls begin the Upper Cretaceous carbonate sequence. It contains a distinctive mass of white and pink limestones and marls of Late Turonian–Coniacian age. Higher, either conformably or with an erosional contact (in eastern Crimea up to the Albian), sandstones usually followed by chalk-like marls belonging to the Santonian and Campanian stages occur. Locally the Santonian–Campanian may rest on an erosional surface upon the sandstones. Extensive sandy marls and calcareous sandstones of the Maastrichtian (with numerous mollusk remains) are present. In the eastern Crimea the Maastrichtian transgressively overlies the Santonian–Albian and is conformably overlain by Danian limestones. In the central and western Crimea the Maastrichtian occurs conformably on the Campanian, but Danian sandstones and limestones containing bryozoans and crinoids rest on its eroded upper surface. The latter, in its turn, is gradually or transgressively replaced by Paleocene rocks (the Inkerman stage).

In the Kopet-Dag and Bolshoy Balkhan, Upper Cretaceous deposits are characterized by a great thickness (more than 300 m) and a complete zonal sequence (see Table IV). Minor erosional intervals are indicated by phosphatic horizons at the base of the Turonian, Coniacian, Santonian and Danian stages. Two sedimentary complexes are distinguished: the essentially detrital beds of the Cenomanian–Lower Turonian, and the carbonates of the Upper Turonian–Danian. The amount of terrigenous material increases towards the east. In the eastern Kopet-Dag terrigenous facies with oyster beds and rudists predominate from the beginning of the Late Turonian, whereas red and gypsiferous rocks occur in the Late Maastrichtian and Danian. In the carbonate complex in the western regions sandy-clay varieties are present in the Lower Coniacian, Upper Maastrichtian and make up the Danian stage of Giaursdag.

Thus, two main types of sediments accumulated in the U.S.S.R. during the Late Cretaceous: a predominantly carbonate facies in the west and south, and exclusively terrigenous beds east of the Urals and along the coast of the Pacific Ocean. Generally these were deposits laid down in vast communicating sea basins of normal salinity inhabited by varied faunas. Late Cretaceous time is characterized by pulses of an enormous transgression, which reached its maximum during the Late Senonian. At the end of the Cretaceous upward or folding movements took place in all areas. They resulted in a gradual reduction of the marine area and an expansion of the land. Sedimentation became localized and concentrated in comparatively narrow down-warping zones to the south and east. At the end of the Mesozoic the greater part of northern Eurasia acquired an outline similar to the present continent (Fig. 22).

CONCLUSIONS

The specific features of the Mesozoic geological history of the U.S.S.R. as indicated by the above data permits the recognition of three distinct tectonic areas.

(1) The Mediterranean geosynclinal fold belt.

(2) The Pacific geosynclinal fold belt.

(3) The East-European, Central Eurasian and Siberian platforms.

In the Mediterranean fold belt the Mesozoic was a time of the incipient and particularly intense development of the Alpine geosynclines. The Triassic period was characterized by a still unstable tectonic regime associated with a transition from the Variscan to the Alpine cycle. The instability of tectonic environments found its expression in numerous hiatuses and unconformities in the sequences and local phases of folding. Among the latter most extensive were the movements at the end of the Permian, at the beginning of the Triassic (eastern Carpathians, Greater Caucasus), in the Middle and at the beginning of the Late Triassic (Greater Caucasus, eastern Precaucasus, Mangyshlak).

Triassic magmatism was restricted. The geosynclinal porphyritic beds occur only in marginal zones of the geosyncline of southeastern Pamirs. The orogenic andesites occur in the eastern Precaucasus and on the Scythian plate.

Such tectonic environments, along with a geocratic nature of the Triassic period and a hot arid climate, resulted in the accumulation of a predominantly mollasoid variegated formation of a littoral origin and of marine carbonate formations.

Very extensive fold movements occurred at the end of the Triassic and the beginning of the Jurassic involving the eastern Carpathians and the Caucasus, and affecting the Scythian and Turanian plates and even Western Siberia. During the Early and Middle Jurassic the formation of the main geosynclines of the Mediterranean belt began. The climate became more humid beginning in the Late Triassic and reached its maximum during the Bajocian. The geosynclines were filled by a terrigenous (slate) and coal-bearing volcanogenic formation with spilite-keratophyric rocks dominant. The stabilization of the tectonic regime resulted in a reduction in the number of unconformities, gaps and phases of folding, but accented their regional importance. Over the greater part of the area movements began at the end of the Triassic and the beginning of the Jurassic. The next movements took place in the Early Bajocian times at which point the residual geosyncline of the Crimean Mountains and western Precaucasus ended its development. Pre-Bajocian movements are also manifest in the Caucasus, on the Scythian and Turanian plates.

At the end of the Jurassic external geosynclines of the Mediterranean folded belt, the Crimean Mountains, Greater Caucasus, Balkhan, Kopet-Dag (?), enter the late stage of their evolution, accompanied by a progressively more arid climate which began in Bathonian times. This drastically affected the nature of sediments being formed. In the Lesser Caucasus an accumulation of the spilite-keratophyric and porphyritic formations continued, to be replaced higher in the sequence by carbonate and sulphate deposits. In the Crimea and the Greater Caucasus local geosynclines were filled with detrital-carbonate and carbonate flysch replaced at higher horizons, in external depressions, by carbonates including reef deposits. The latter also extends into the depressions of the Scythian and Turanian plates and into the Donetz Basin, where they are associated with a saliferous formation. Along the periphery of the depressions variegated detrital deposits accumulated.

There were no large-scale folding movements during the Late Jurassic. The main unconformities and associated partial structural rearrangements took place only during the Kimmeridgian and Early Tithonian. A relative maximum marine regression is associated with this time interval. The fold movements in central and southeastern Pamirs that resulted in a closure of geosynclines and are onset of orogeny merits a special note.

At the end of the Late Jurassic there is a partial inversion of the Lesser Caucasus structure, ending in the early stage of geosynclinal development. Movements at the end of the Jurassic and the beginning of the Cretaceous have also been extensive on the Scythian and Turanian plates.

During the Early Cretaceous, generally speaking, the structures formed in the Late Jurassic continued their development. An intense down-warping involved the flysch-filled geosyncline of the eastern Carpathians. The gradual replacement of the arid climate by a humid one which occurred during the Early Cretaceous resulted in the replacement of carbonate formation by a detrital-carbonate and detrital sequence in the external depressions of the Mediterranean geosynclines and in the adjacent regions of the Scythian and Turanian plates. During the Early Cretaceous epoch the tectonic regime stabilized and this found its reflection in a weak magmatism and gentle folding. In addition to the late geosynclinal magmatism in the Lesser Caucasus, a limited area of intermediate lavas is known in the vicinity of Armavir on the Scythian plate. An injection of granites occurred in the orogenic area of the southeastern Pamirs.

During the Late Cretaceous the development of the Mediterranean fold belt continued, following roughly the same pattern. Carbonate rocks and flysch accumulated in the depressions of external geosynclines. Volcanic activity and the accumulation of reef and marly carbonate facies continued in the Lesser Caucasus. To the Late Cretaceous belong some granitic intrusions in the axial part of the Greater Caucasus and in the southeastern Pamirs.

The second area, the Pacific belt, consisted of Late Paleozoic-Mesozoic and Late Mesozoic-Cenozoic geosynclines and the median massifs which separated them. In the basement complex of the majority of structures in the Pacific geosynclinal fold belt, forms belonging to two cycles of tectogenesis are recorded: Sinian–Early Cambrian and Ordovician (Silurian)–Early Devonian. The precursors of the geosynclinal development during the Mesozoic developed in the Late Carboniferous and in the Permian. The closure of the geosynclines in the Soviet sector of the Pacific belt proceeded progressively from the western region adjoining the Siberian Platform, where the geosynclinal development ended in the Middle and Late Jurassic, towards the inner eastern systems, where the geosynclinal development is still in progress. Characteristic features in the structure and development of this area are the substantially larger sizes of the geosynclines, the greater duration of the intervals of continuous downwarping, the greater thickness of the deposits filling them, the much more extensive development of geosynclinal and orogenic magmatisms, the lack of synchroneity in the geocratic and thallassocratic epochs, the formation during the orogenic stage of a thick volcanogenic belt, and the extension of its links on land reaching 1500 km in the eastern Sikhote-Alin and 3000 km in the Okhotsk–Chukotka with a average width of the belt of about 100–150 km.

In the Triassic and Early Jurassic the majority of the geosynclines in the Pacific belt were still in an early stage of their development and experiencing an intense and continuous downwarping. The predominant types of deposits here were: in the Verkhoyansk geosyncline a sandy-clay (slate) formation, in the Chukotka geosyncline sandy-clay and porphyritic, in the Sikhote-Alin and Nippon geosynclines terrigenous and volcanogene-flinty. Gaps in sedimentation and unconformities are recorded only in the marginal parts of the geosynclines and over the median massifs, and are of a local importance. A special position is occupied by the Mongolo-Okhotsk geosyncline, in which distinctly pronounced folding phases occurred during the Middle Triassic and at the end of the Triassic–the beginning of the Jurassic.

During the Middle Jurassic internal uplifts occurred in the Verkhoyansk and Chukotka geosynclines, splitting them into systems of local geosynclines and associated anticlines, but downwarping continued in the geosyncline. At the end of the Bajocian – the beginning of the Bathonian the western part of the Mongolo-Okhotsk geosyncline was closed and tranformed into an orogen. In superimposed depressions in the Transbaykal region a Bathonian–Upper Jurassic andesitic formation accumulated replaced in the upper Amur region by a coal-bearing formation. The accumulation of these formations was accompanied by the injection of granitoid bodies. In the eastern part of the Mongolo-Okhotsk geosyncline and on Sikhote-Alin, Bathonian folding was very intense, but did not result in closure of the geosynclines.

During the Late Jurassic internal uplifts in the Verkhoyansk and Chukotka geosynclines developed, resulting, in a number of regions, in the replacement of geosynclinal formations by the accumulation of sandy-clay molasse associated with the porphyritic and liparite–dacitic formations. Syenitic intrusions penetrated into the uplift zone. The eastern part of the Mongolo-Okhotsk geosyncline and the Sikhote-Alin geosyncline continued their development. In the latter a thick volcanogenic flinty formation accumulated and this can also be observed in the sequences on Sakhalin.

In the Early Cretaceous the Verkhoyansk and Chukotka geosynclines and the eastern part of the Mongolo-Okhotsk geosyncline closed, accompanied by the intrusion of granitoid bodies and, to a lesser extent, of gabbros and syenites. At this time coal-bearing, sandy-clay molassic deposits accumulated in the intermontane depressions, associated with the formation of andesite. In the Okhotsk–Chukotka volcanic belt beginning at the end of the Early Cretaceous andesites, andesite-basalts and more rarely dacites were extruded. In the Kamchatka–Koryak and Sikhote-Alin geosynclines, there was especially intensive downwarping from the end of Neocomian, with the deposition of a volcanic-flinty formation, terrigenous flysch and andesites with an intrusion of granitoid bodies.

During the Late Cretaceous the development of the Okhotsk–Chukotka belt ended, the Sikhote-Alin geosyncline closed and the eastern Sikhote-Alin volcanic belt and the lower Amur depression began to be filled with acid terrestrial effusives.

In the Pacific belt, marine conditions were predominant in the Early Triassic and the Early and Middle Jurassic. Relative transgression maxima coincided with the end of the Early Triassic, the Late Triassic, the Pliensbachian, and the Late Toarcian–earliest Aalenian. Regression maxima associated with folding movements occurred during the Middle Triassic, Hettangian–Sinemurian and the Bajocian. In the Late Jurassic and Early Cretaceous the size of marine basins was gradually reduced and became limited in the Albian to the Sikhote-Alin and Kamchatka–Koryak geosynclines. A further regression occurred during the Late Cretaceous. During the Campanian the greater part of Sikhote-Alin rose above the sea level and in the Maastrichtian, the Koryak highland became land. During the Danian the present continental part of the Pacific belt became a land area.

The third and, probably, main tectonic element in the Mesozoic of the U.S.S.R. was the Central Eurasian Platform. It originated at the very end of the Paleozoic from the merging of the Siberian and East-European platforms. Owing to its extent and the heterogeneity of its basement, different parts behaved in different ways. As a relatively uplifted element, it served as a barrier separating the tropical basins of the Tethys from the Boreal-Arctic Basin, on the one hand, and the Pacific from the Atlantic West-European Basin on the other. The latter periodically flooded the peripheral, and even central, parts of the platform forming vast shallow epicontinental basins in which a distinctly climatic zonality can be observed.

During the Triassic the greater part of this vast territory remained a land area. However, the nature of continental strata in the various regions varied very much. On the East-European Platform thick variegated molasse deposits accumulated associated with the destruction of Hercynian mountain massifs of the Ural–Tien Shan fold zone and only the Caspian and Pechora syneclises were covered by shallow basins at the end of the Early, and during the Middle Triassic. In the region of Hercynian structures Triassic and Lower Jurassic deposits are represented by coal-bearing and red molasse formations of great thickness, filling narrow intermontane depressions (the Chelyabinsk and other graben-basins).

At this time the Siberian platform was a region of intense volcanic activity, where thick trapps filled the Tunguska syneclise. Triassic marine basins covered only the extreme northeastern margin and the deposits are represented exclusively by terrigenous gray sandy-clay formations. Peculiar volcanic formations similar to trapps form part of the fill of the grabens in the eastern Urals, on the Western Siberian plate and in the regions further south.

The magmatism phenomena are more varied in composition and form in the Selenga–Stanovoy area. During the Early Triassic there predominated here the accumulation of basic and intermediate effusives, and during the Middle and Late Triassic, acid and alkaline effusives played an increasingly important part. In addition to effusives Triassic granitoid bodies also occur. In the Late Triassic, against the background of a general geocratic epoch, the relative maximum transgression coincided with the end of the Early Triassic and the beginning of the Middle Triassic (Olenek and Anisian ages) and during the Carnian. The maximum regressions are associated with the beginning of the Triassic, the end of the Middle Triassic and the second half of the Late Triassic. During the Early and Middle Jurassic the geocratic epoch gradually changed into a thalassocratic one. Distinct pulses of transgressions can be identified. Trangression maxima occurred in the Late Pliensbachian, Late Toarcian, and the beginning of the Aalenian, as well as in the Late Bajocian–Early Bathonian. The depressions in the platform were filled with predominantly continental and littoral coal-bearing beds locally replaced by marine gray sandy-clay glauconitic strata.

Marine basins associated with the Alpine zone during this time covered considerable areas of the Scythian plate and Central Asia and formed a deep gulf into the regions of Mangyshlak and the Donetz Basin, where geosynclinal argillaceous beds accumulated. Vast shallow basins associated with the Verkhoyansk fold zone and the Arctic Basin at this time covered the Eastern Siberian Platform, the Khatanga depression and Western Siberia.

The climate became more arid at the beginning of the Bathonian reaching its maximum during the Kimmeridgian. However, it is clearly recognized only in the southern areas adjacent to the Mediterranean belt, on the Scythian and Turanian plates and on the East-European Platform south of the Donetz Basin, where at this time reef and nerineid limestones and red sulphate-halite strata accumulated. In the north of the platform, in contrast, a gradual marine transgression which began during the Late Bathonian flooded all the central parts of the East-European Platform including Poland and linked in the south with the Mangyshlak, Caucasian and Western Siberian basins. As a result of this transgression the climate in the northern part of the platform was temperate, and this is reflected in the dominant lithologies of the Upper Jurassic. These are in the north sandy-clay glauconitic beds, frequently with partings of nodular phosphates giving way in the marginal areas to coal measures.

Such conditions prevailed, to a great extent, during the Early Cretaceous times, with the accumulation of marine sandy-clay deposits with phosphates and glauconite continuing on the East-European Platform and on the Western Siberian plate, again being on occasion replaced along the periphery of the depressions by coal-bearing deposits. As in the Late Jurassic, the Early Cretaceous was a thalassocratic epoch. The transgression maxima occurred in the Berriasian–Valanginian, Late Hauterivian and Late Aptian–Albian, while the maxima of regressions occurred during the Early Barremian–Aptian and Lower Albian.

At the end of the Early to the beginning of the Late Cretaceous, the relief of the young Hercynian structures on the Scythian and Turanian plates is generally much reduced and this finds expression in the remarkable persistence over a wide area of Albian, Cenomanian and Turonian deposits of relative uniform thickness. From the latter part of the Late Cretaceous the isometric depressions and uplifts on the plates do not always inherit the linear form of the Jurassic–Early Cretaceous. This process of structural rearrangement is seen less distinctly on the Western Siberian plate. On the East-European Platform the area of sedimentation was reduced during the Late Cretaceous becoming limited to the southern region and the northern part of the Pechora syneclise.

Latitudinal zones can be observed in the Upper Cretaceous deposits, reflecting the climatic zonation of the Late Cretaceous epoch. In the north, on the Western Siberian plate, in the Pechora syneclise, and in the southern regions of the Moscow syneclise, there is a predominance of flinty-clay and sandy-clay deposits with glauconite. Further south, in the Ukrainian syneclise, on the flanks of the Ukrainian shield, in the near-Caspian syneclise and on the Scythian plate, chalk is present. The latter formation on the Turanian plate is gradually replaced eastwards by detrital-carbonate and terrigenous deposits, including red varieties.

At the end of the Late Cretaceous the marine basins shrank even more. In the Danian, sea covered only the near-Caspian syneclise, the Polish lowland and the very south of the Scythian and Turanian plates. The Paleocene began everywhere by a new marine transgression.

Paleobiogeographical models of the U.S.S.R. for the periods and epochs of the Mesozoic are shown in the figures (Figs. 3, 9, 10, 16, 22). Captions for these figures indicate the principal groups of fossil organisms which characterize the different paleozoogeographic areas and provinces. The paleozoogeographic models are based on the distribution of the ammonitid genera. Consideration has not been given to local peculiarities of habitat, essentially determining the appearance of the actual paleocoenoses.

On the whole the Mesozoic marine faunas of the U.S.S.R. belong to the Mediterranean, Boreal and Pacific areas. The paleophytogeographic regions of the U.S.S.R. in the Mesozoic are rather stable and are controlled by climate. Two areas existed persistently throughout the entire Mesozoic, the Indo-European (European–Tien Shan) and Siberian. It should be noted that the differences in the systematic composition of floras consecutively replacing each other during the Triassic and Jurassic are less distinct than the differences between coeval floras in different areas.

REFERENCES

Atabekian, A.A. and Likhacheva, A.A., 1961. *Upper Cretaceous Deposits of Kopet-Dag.* Gostoptekhizdat, Moscow, 201 pp. (in Russian).

Avdeiko, G.P., 1968. *Lower Cretaceous Deposits of the Circum-Pacific Belt.* Nauka, Moscow, 136 pp. (in Russian).

Baranov, A.N. (Editor), 1969. *Atlas of the U.S.S.R.* Chief Administration of Geodesy and Cartography under the Council of Ministers of the U.S.S.R., Moscow (in Russian).

Bogoslovsky, N.A., 1897. Riazan horizon. *Pap. Geol. Russ.,* 18: 157 pp. (in Russian).

Dagis, A.S. and Zokharov, V.A. (Editors), 1974. Paleobiogeography of northern Eurasia in the Mesozoic. *Tr. Inst. Geol. Geofiz., Akad. Nauk S.S.S.R., Sib. Otd.,* 80: 173 pp. (in Russian).

Drucziz, B.B. and Kudriavtzev, M.P. (Editors), 1960. *Atlas of the Lower Cretaceous Fauna of the Northern Caucasus and Crimea.* Gostoptekhizdat, Moscow, 699 pp. (in Russian).

Gerasimov, P.A., 1971. Cretaceous system. Lower section. In: *Geology of the U.S.S.R., IV.* Nedra, Moscow, pp. 416–445 (in Russian).

Gerasimov, P.A., Migacheva, E.E., Naidin, D.P. and Sterlin, B.P., 1962. Jurassic and Cretaceous deposits of the Russian platform. In: *Sketches of the Regional Geology of the U.S.S.R.,* 5. Moscow University Publishers, Moscow, 196 pp. (in Russian).

Krymholz, G.J. (Editor), 1969. Unified stratigraphic schemes of Jurassic and Cretaceous deposits of Central Asia. In: *Data for the Central Asiatic Stratigraphic Conference of 1970, Moscow,* 116 pp. (in Russian).

Milanovsky, E.E. and Khain, V.E., 1963. *Geological Structure of the Caucasus.* Moscow University Publishers, Moscow, 357 pp. (in Russian).

Moskvin, M.M. (Editors), 1959. *Atlas of the Upper Cretaceous Fauna of the Northern Caucasus and Crimea.* Gostoptekhizdat, Moscow, 501 pp. (in Russian).

Moskvin, M.M., 1968. Central and eastern regions of the monoclinal north slope (Khodz River–Tcherek River). In: *Geology of the U.S.S.R., IX.* Nedra, Moscow, pp. 288–300 (in Russian).

Najdin, D.P., 1964. *Upper Cretaceous Belemnites of the Russian Platform and Adjacent Areas.* Moscow University Publishers, 190 pp. (in Russian).

Pergament, M.A., 1973. New data on biostratigraphy and inocerami of Turonian–Coniacian deposits of northwest Kamchatka. *Izv. Akad. Nauk S.S.S.R., Ser. Geol.,* 2: 113–121 (in Russian).

Saks, V.N., Ronkina, Z.Z., Shulgina, N.I., Basov, V.A. and Bondarenko, N.M., 1963. *Stratigraphy of the Jurassic and Cretaceous Systems in the North of the U.S.S.R.* U.S.S.R. Academy of Sciences, Moscow, Leningrad, 227 pp. (in Russian).

Vassoevitch, N.B., 1961. *Unified Stratigraphic Schemes of the North-East of the U.S.S.R.* Gostoptekhizdat, Moscow, 340 pp. (in Russian).

Chapter 3

MONGOLIA

JERZY LEFELD

INTRODUCTION

The Mesozoic rocks of Mongolia are almost entirely continental – vast areas have, with few exceptions, not been covered by sea since the Early Triassic. Although Mesozoic rocks are well exposed, the Cretaceous sediments, because of their economic (oil, coal) and paleontological (unique vertebrates) interest, are much better known that the Jurassic and Triassic. Although much progress has been made in recent years in Mesozoic stratigraphy, the paleogeographical interpretation still poses many problems. Sedimentological studies are extremely rare despite the fact that most of the Mongolian Mesozoic formations lend themselves to basin analysis.

The Mesozoic sequences of Mongolia have been classified by Soviet and Mongolian geologists into "svitas"[1]. These units are stratigraphic groupings of isochronous sequences of various origins. In some cases different "svitas" may contain sequences of the same age, however, they may crop out in widely separated regions. Only a few formations have been established in the Mongolian Mesozoic so far. It should be kept in mind, however, that correlation of formations over large distances is risky until more detailed stratigraphy and basin analyses have been carried out.

Most of the topographic detail has been omitted from this review. Mongolian geography is described in greater detail by Lefeld (1971b).

GENERAL GEOLOGY

The Mongolian People's Republic occupies a central position in the Central Asiatic structural belt. It is bordered in the north by the Siberian Platform, and in the south by the North-Chinese and Tarim platforms (Fig. 1).

Mongolia forms a platform, the tectonic structure of which is mosaic, with many variously oriented structural elements limited by dislocations. The arcuate pattern of the major structural zones reflects adaptation to the shape of the southern margin of the Siberian Platform (Figs. 1 and 2). The structural units were formed either during various periods of the Paleozoic era or during Precambrian times. The Mesozoic, post-Hercynian, tectonic history of Mongolia is mainly a result of the reactivation of older, generally Paleozoic structures.

In a very general way Mongolia may be subdivided into two main tectonic areas, the northern chiefly Caledonian zone, and the southern mostly Hercynian zone.

(I) The *northern area* includes the Hövsgöl (Hubsugul) region, the western Hangay Mountains and the Great Lakes area (Fig. 3) and is subdivided into four tectonic zones (Fig. 4): (1) the North Mongolian zone; (2) the Mongol-Altay zone; (3) the Mongol-Transbaykal zone; and (4) the Central Mongolian zone. The northern area continues into Tuva and the southern Transbaykal region in the U.S.S.R. and constitutes a part of the Caledonian zone rimming the Siberian Platform to the south.

(1) The North Mongolian zone is a direct prolongation of the Early Paleozoic structures of the Altay-Sayan and Selenga-Yablon areas of southern Siberia. It comprises mainly Precambrian–Early Paleozoic formations folded during the Salair (Middle Cam-

[1] "Svita" is a local stratigraphic subdivision introduced by Russian geologists. It has no facies significance but rather groups formations of the same age.

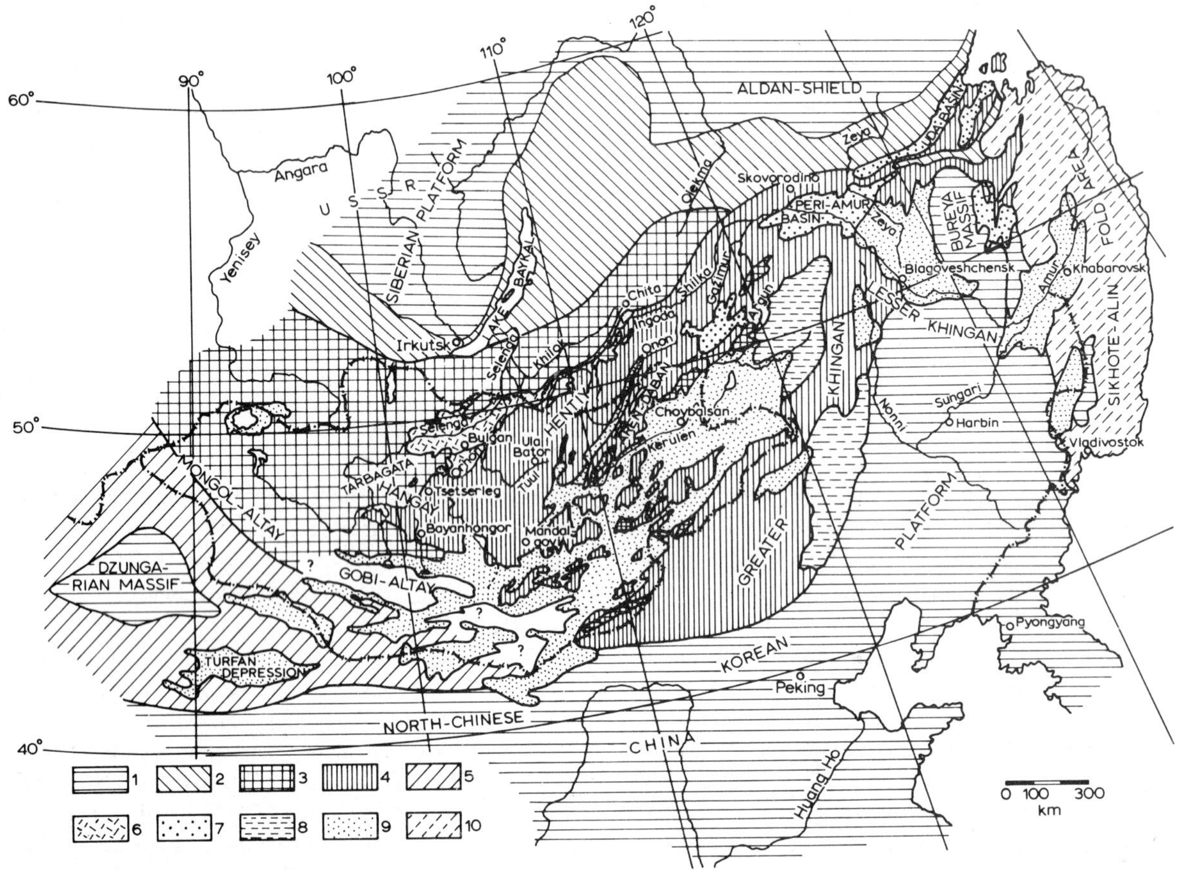

Fig. 1. Tectonic sketch map of Mongolia and surrounding areas (after Dörnfeld, 1968). Legend: *1* = Precambrian basement of the shields; *2* = Baykal fold area; *3* = Caledonian fold area; *4* = Hercynian fold area with geosynclinal, mostly terrigenous Silurian, Devonian and Lower Carboniferous formations (main orogenic phase during the Early Carboniferous); *5* = Hercynian Khazakh–Altay fold area with Ordovician and Silurian geosynclinal formations (orogenic phase during the Late Devonian); *6* = depressions with volcanic–sedimentary Upper Carboniferous, Permian and Lower Triassic formations; *7* = Mesozoic structures superposed over Hercynian substratum, containing partly continental, partly marine terrigenous formations; *8* = idem in the area of the Greater Khingan (including Upper Jurassic continental, volcanic complexes); *9* = Meso-Cenozoic continental depressions and grabens; *10* = Mesozoic (Pacific) fold area of Sikhote-Alin, geosynclinal formations of Permian, Triassic, Jurassic and Early Cretaceous age.

brian) orogenic phase. The Late Paleozoic formations form part of the cover.

(2) The Mongol-Altay zone is a southeastern prolongation of the Upper Altay and western Sayan mountain systems. It consists of thick terrigenous, highly folded formations of Middle and Late Cambrian, of Ordovician, and in places, of Silurian age. The zone divides into the Kharkhira and the Mongol-Altay branches.

(3) The Mongol-Transbaykal zone occupies the central and northeastern parts of the country. It consists of two complexes: (a) the Late Precambrian–Early Paleozoic sequences cropping out in marginal and some internal elevations of the western and eastern Hangay Mountains and in the northern and southern Hentiy Mountains, and (b) the Middle–Late Paleozoic sequences filling the synclines in the abovementioned system.

(4) The Central Mongolian zone borders the preceding zone in the south. Precambrian–Early Paleozoic sequences analogous to zone (3) are broken up into several blocks separated by narrow suture-like fractures and depressions. This zone was constantly uplifted during the middle Paleozoic.

(II) The *southern Hercynian megazone,* or megablock of Soviet geologists, includes the Mongol-Altay ranges, the Gobi-Altay ranges, the eastern part of the Hangay Mountains, the Hentiy Mountains, the

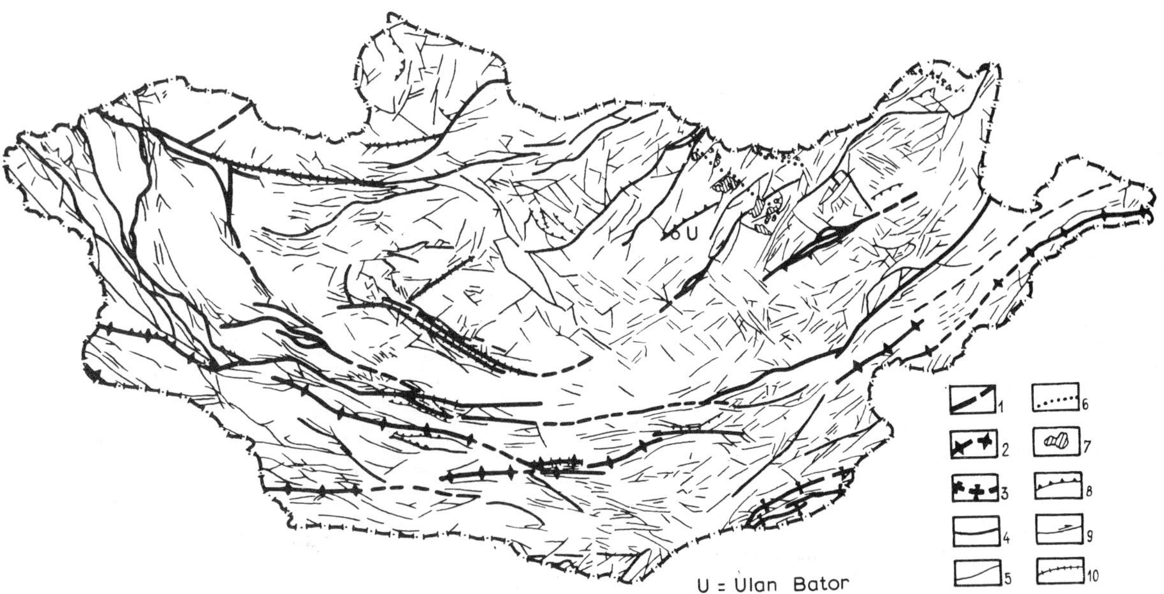

Fig. 2. Sketch map of the main fractures in Mongolia (after Zonnenshain, in Marinov et al., 1973b).
Legend: *1–3* = deep fractures controlling initial magmatism, of (*1*) Late Riphean–Cambrian age, (*2*) Middle Paleozoic age, and (*3*) Late Paleozoic age; *4* = main regional fractures; *5* = local fractures; *6* = concealed fractures traced in Mesozoic intrusions; *7* = massifs of Mesozoic granitoids; *8* = overthrusts; *9* = displacements (faults); *10* = seismically active fractures.

Gobi Tien-Shan and eastern Mongolia. It is subdivided into three tectonic zones (Fig. 4): (1) the South Mongolian zone; (2) the South Gobi zone; and (3) the Inner Mongolian zone.

(1) The South Mongolian zone forms a link between the Middle Paleozoic structures of the Greater Khingan range in the east, and the Zaysan and Dzhungaria–Balkhash (Kazakhstan) zones in the west. The Hercynian geosynclinal complex of the zone includes formations from Ordovician up to De-

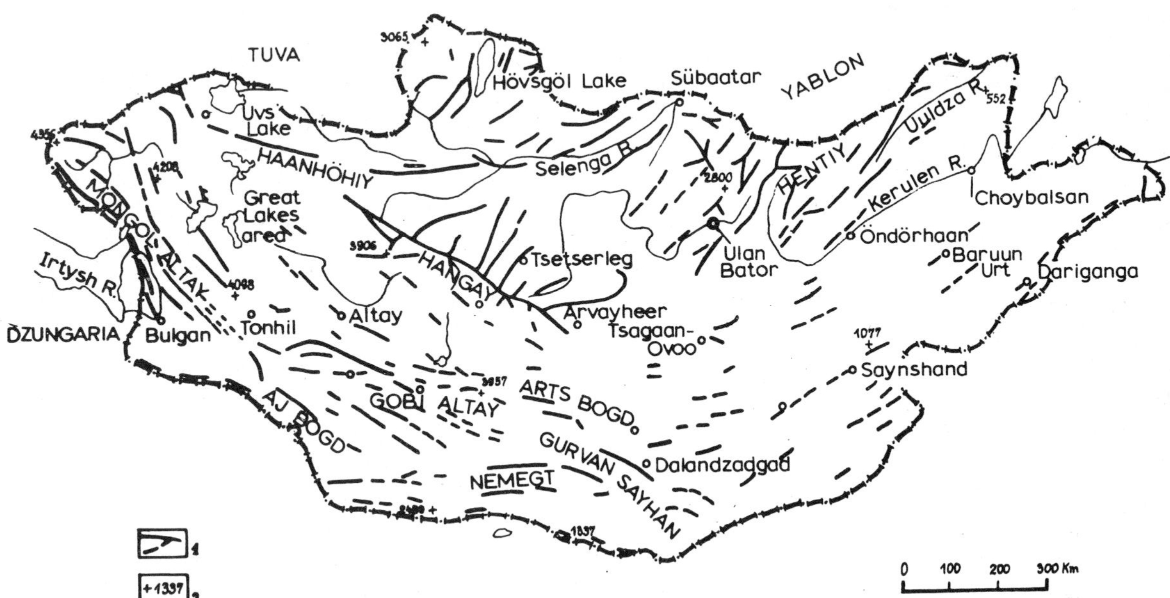

Fig. 3. Orographic scheme of Mongolia (after Marinov et al., 1973b).
Legend: *1* = mountain ranges; *2* = height in meters.

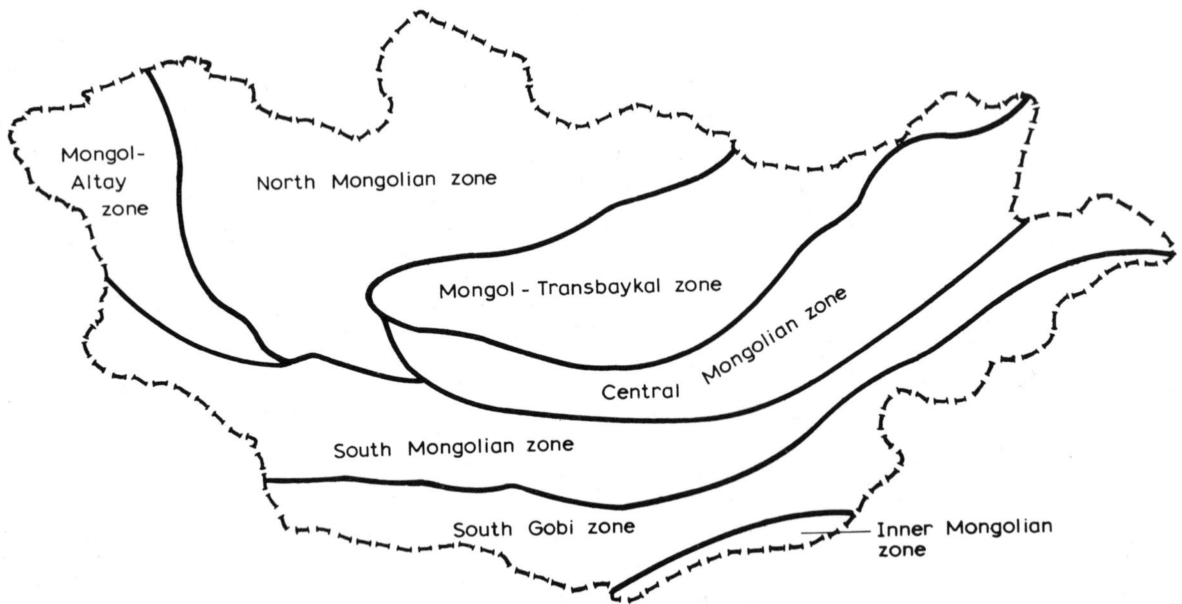

Fig. 4. Main tectonic zones of Mongolia (after Marinov et al., 1973b).

vonian and even Early Carboniferous in age.

(2) The structures found in the South Gobi zone are of Middle and Late Paleozoic age, and the result of reactivation of depressions and undoubted orogenic structures. It is divided by a system of fractures into rectangular blocks, e.g. the Gobi Tien-Shan, the Nukut-Daban, Toto-Shan, etc. The geosynclinal beds exposed are mainly Early Paleozoic (Silurian) in age.

(3) Only a small fragment of the Inner Mongolian zone lies within the Mongolian People's Republic, most of it is in China (Inner Mongolia). It consists of Paleozoic geosynclinal complexes.

In general the central and northern regions of Mongolia are characterized by the occurrence of Late Proterozoic to Early Paleozoic geosynclinal beds, uplifted during the Middle and Late Paleozoic, and block-faulted (Fig. 5). The southern regions show characteristic linear block structures, and the prevalence of Middle–Upper Paleozoic geosynclinal formations (Fig. 6). According to Amantov et al. (1970), the tectonic boundary separating the two main megablocks follows the socalled Main Mongolian Deep Fracture (Ih Bogd–Undurshilin), a feature which follows the southern boundary of the Mongol-Altay Mountains in the west, then runs south of Great Lakes area, the Hangay Mountains and the Valley of Lakes as far as the northern boundary of the East Gobi depression in the east and the Greater Khingan range in China. In both zones and especially in the southern Hercynian megazone there are many Meso-Cenozoic depressions superposed over an older, usually Paleozoic intrageosynclinal substratum where deep-water Silurian, Devonian and Lower Carboniferous facies are present (Dörnfeld, 1968). The Meso-Cenozoic complexes filling those depressions are markedly different from the Paleozoic and Precambrian rocks.

Geosynclinal conditions in Mongolia ended in Early Triassic or Late Paleozoic times. By the end of the Permian most of Mongolia was a land area with deposition restricted to a few small basins. A change in the structural plan occurred during Late Permian to Triassic times (Nagibina, 1970; Zonnenshain, 1970), the result of which being a considerable morphological diversity. One result was that the central Mongolian watershed acted as a divide between the marine basins of the Arctic and the Tethys. Many Hercynian structures were reactivated in eastern Mongolia and in the Transbaykal area as a result of Early Mesozoic tectonic movements. These affected the whole of Eastern Asia including the North- and South-Chinese platforms. The Mesozoic continental structures of Mongolia are intermediate between the platform and geosynclinal type, and are imposed upon a varied substratum. Eastern Mongolia was affected by two main tectonic phases, one during the Triassic–Jurassic and

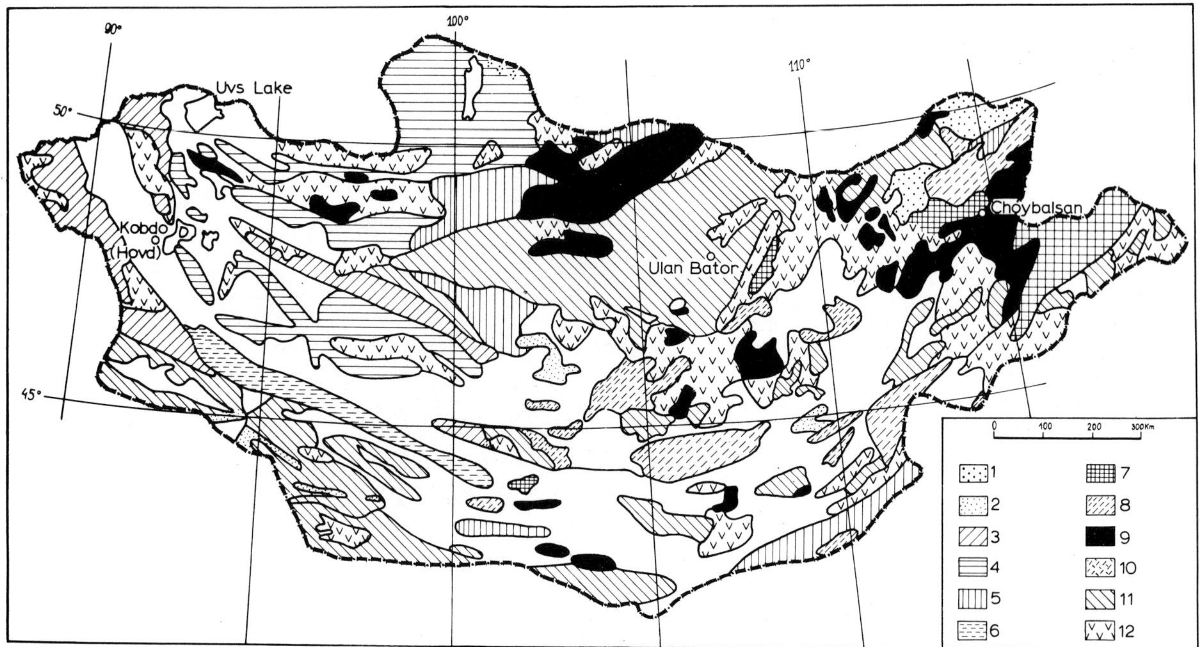

Fig. 5. Geological sketch map of Mongolia (compiled from various sources).
Legend: white = undivided Jurassic, Cretaceous, Tertiary and Quaternary and minor fields of undivided Paleozoic and volcanic rocks of various age; *1* = Permian; *2* = Carboniferous; *3* = Silurian; *4* = Cambrian; *5* = Proterozoic; *6* = Archaean; *7* = Tertiary; *8* = Cretaceous; *9* = Jurassic; *10* = Triassic; *11* = Paleozoic (undivided); *12* = volcanics.

the other during Jurassic–Cretaceous times.

As a result of block movements, the Late Jurassic–Early Cretaceous sediments are a monotonous sequence of detrital, lacustrine beds with some basic type volcanic rocks (restricted mainly to the northern regions) laid down in vast depressions. These sediments are regarded by Soviet geologists (Zonnenshain, 1975) as orogenic in type. They rest with marked discordance upon the earlier Mesozoic sequences, similar to the situation found in the eastern Transbaykal area, Peri-Amur area and in northeastern China. Lower Cretaceous tectonic movements which occurred at the Lower and the Upper Dzüün Bayan svita boundary, to some extent, controlled the Upper Dzüün Bayan and later sedimentation. The Upper Cretaceous sediments following these movements are much less disturbed than those in the Lower Cretaceous. In principle, the Upper Cretaceous sequences are of platform type.

The Mesozoic tectonic structures were controlled by movements which affected the Pacific segment of Eastern Asia. They cut obliquely across the Paleozoic structures. As a result of these movements, in vertical section several sequences are seen separated by discordance from each other, every discordance representing a phase of movement (Table I).

The block-mosaic structure of the country is the resultant of two main tectonic trends, one northwest–southeast (occasionally south), the other northeast–southwest (occasionally trending east–west). From the end of the Jurassic large parts of northern Mongolia were uplifted and eroded, while in the southern part of the country there were depositional basins. The boundary between those two regions is one of the most important lineaments in Asia and separates the Caledonian zone to the north from the Hercynian to the south. During the Mesozoic and Cenozoic the structure of the above two regions became increasingly complex due to differential block movements. Weak tectonic movements at the end of the Cenomanian (the Saynshand svita) resulted in erosion and overlying beds of the Bayan-shiré and Baruun Goyot svitas rest on the Saynshand svita with angular unconformity. Tectonic movements at the Early–Late Cretaceous boundary shown by a small unconformity between the Dzüün Bayan and Saynshand svitas resulted in the general uplift with the formation of depressions in the Gobi. In some parts of

TABLE I

Mesozoic diastrophic phases in Mongolia

		W	E
Paleocene			
Cretaceous	Late	Yen Shan orogeny	oscillatory movements in southern Mongolia
	Early	sediments partly overlie directly Paleozoic substratum	
Jurassic	Late	Great Lakes area, Mongol Altay + Gobi-Altay Mountains	N + S Gobi + Hangay Mts
	Middle		Hangay + Hentiy Mts + NE Mongolia
	Early		Onon River area
Triassic	Late		
	Middle		
	Early		

the Gobi and northern Mongolia several basins were considerably reduced, others expanded. Movements at the Cretaceous–Paleogene boundary caused further general uplift and marked the end of the lacustrine regime over most of the country.

The nature of the Mesozoic tectonic movements in Mongolia is far from being completely understood, hence it is differently interpreted (Beloussov, 1954, 1962; Sheynmann, 1955, 1959; Huan-Tsi-zin, 1961; Nagibina, 1963; Smirnov, 1963; Komarov and Khrenov, 1964).

Fractures play a very important role in the geology of Mongolia. Most of the large Mongolian fractures are rejuvenated structural lines first established during the early or middle Paleozoic or even in Precambrian times. They were active at some periods, at others inactive. In addition to the major arcuate, parallel fractures there are many smaller ones perpendicular or oblique to the main structural trends (Fig. 2).

Some deep fractures controlled the eruption of mantle ophiolites. Some of the north–south fractures originated during Mesozoic times, particularly in the Hangay and Hentiy mountain ranges where the tectonic and magmatic activities (mainly of granitic type) were most intense. In Late Jurassic times reactivation of older fractures caused morphological diversity reflected in the deposition of coarse red beds.

The role of fractures in the Mongolian tectonics is exemplified by the seismically active 1100 km long Hangay fracture. Its origin goes back to the Early Paleozoic or Precambrian. It has been reactivated several times, and is still seismically active. Several fracture zones occur in the Mongol-Altay Mountains, some of which were active in post-Tertiary times. Many of the depressions and/or grabens, which contain Mesozoic–Cenozoic beds (Fig. 5), owe their preservation to movements along these faults and major tectonic lines.

Mesozoic–Cenozoic depressions are rare in northern Mongolia and lie mainly in the northeast. They were initiated at the end of the Jurassic and Early Cretaceous times. The break-up of these depressions by wedge-like horsts begun during the Late Mesozoic and Cenozoic and was particularly intense during the Quaternary. The size of the depressions ranges from 10–15 km up to 50–70 km along their strike.

In the depressions Mesozoic beds up to 2–2.3 km thick usually overlie a Paleozoic substratum with considerable discordance. This substratum broken into

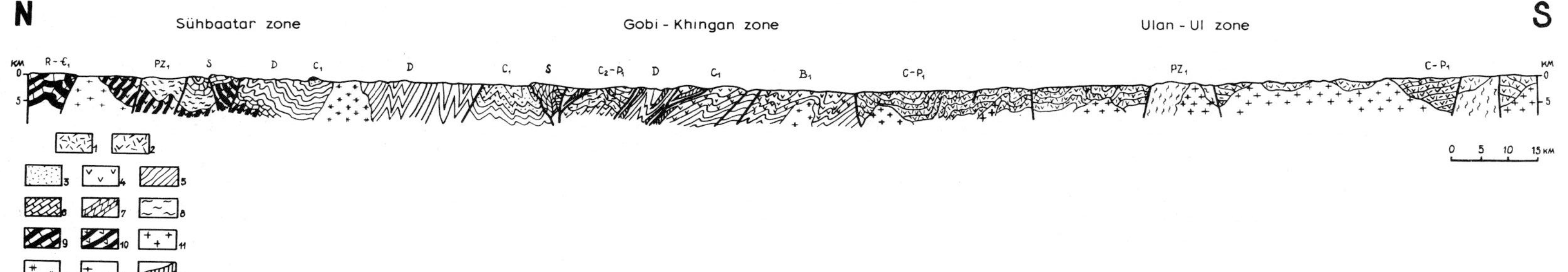

Fig. 6. Schematic cross-section through the Sühbataar–Saynshand region of southern Mongolia (after Suetenko, in Marinov et al., 1973a).
Legend: *1* = Middle Carboniferous–Lower Permian; *2* = Carboniferous–Lower Permian, Lower Carboniferous; *3* = clastic rocks; *4* = green effusive rocks predominantly of intermediate composition; *5* = Devonian, Silurian; *6* = carbonate rocks; *7* = siliceous–argillaceous rocks, effusives and limestones; *8* = Lower Paleozoic sequences: Upper Proterozoic–Lower Cambrian; *9* = carbonate–siliceous sequence; *10* = volcanic sequence; *11, 12* = Upper Paleozoic granitoids; *13* = Lower Paleozoic granitoids; *14* = ultrabasic rocks.

horst-anticlines and graben-synclines is in every case bounded by fractures. As a result of the development of horst blocks within the depressions in Middle Cretaceous times, the greatest thickness of the Cretaceous sediments is found along their flanks. These structures are the socalled "revivation" (reactivated) structures of Nagibina (1970). The original extent of the Cretaceous sediments is shown by the occurrence of remnants of the Tsagaan Tsab svita on peaks in the Arts Bogd Mountains.

THE DEEP STRUCTURE OF MONGOLIA FROM GEOPHYSICAL DATA

Geophysical data concerning Mongolia are scarce. A gravimetric survey by Stepanov and Volkhonin (1969) indicated a crustal thickness of 50–55 km in the Mongol-Altay and Hangay mountain ranges, and 45–40 km in the southern and eastern regions of the country (Fig. 7). The granitic layer is 16–20 km thick in the Mongol-Altay and Hangay mountains, 12–16 km in central and western regions, and 8–12 km in eastern Mongolia. The Mongol-Altay, Hangay and Hentiy mountains show regional gravity minima, whereas in the Selenga Basin, South Gobi and the remaining regions gravity maxima are found (Stepanov and Volkhonin, 1969). Southern Mongolia is characterized by a strong differentiation of gravitation field.

Analysis of earthquake epicenters shows that seismic areas are located within those areas the most fractured in the present day and are concentrated in the western half of the country. Earthquakes are usually shallow, less frequently from intermediate depths. A submeridional belt of earthquakes in the western Hövsgöl area is associated with the origin of the rift-type depressions of the Hövsgöl and Darhat. The pattern of seismic zones reflect a complex mosaic of relatively small angular plates, the internal parts of which show but slight seismicity, e.g. the Great Lakes area, Hangay Mountains (Zonnenshain, in Marinov et al., 1973b). It seems, however, that the Mongolian seismic zones reflect rather the repeated rejuvenation of fractures than that they mark plate boundaries, at least in western Mongolia. Recent earthquakes were recorded in Mongolia in 1905, 1957 and 1958 (Florensov and Solonenko, 1963).

The magnetic anomalies in eastern Mongolia which trend northeast–southwest coincide with geological

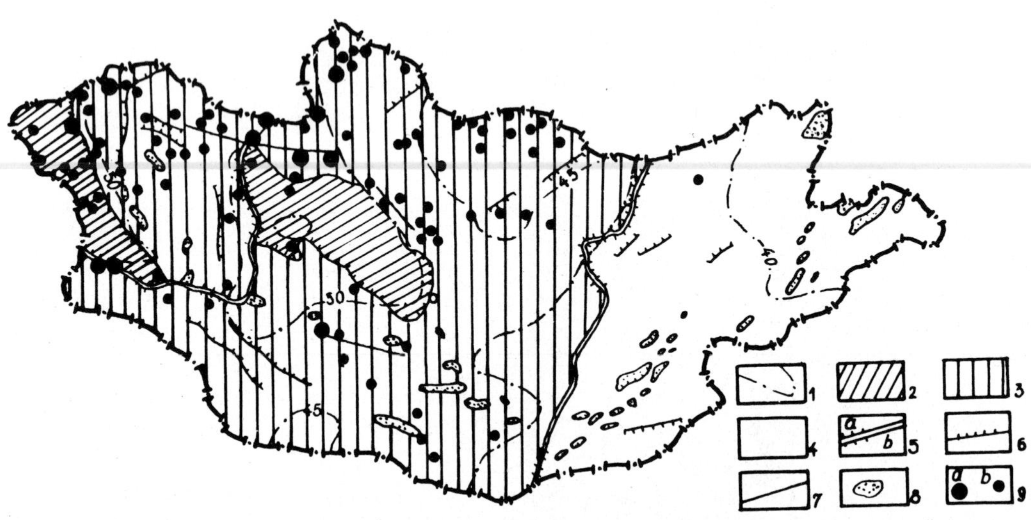

Fig. 7. Deep structure of Mongolia from geophysical data (after Stepanov and Volkhonin, 1969).
Legend: *1* = contour lines of the Moho discontinuity (in km); *2–4* = areas of granitic layers of 16–20 km thickness (*2*), of 12–16 km thickness (*3*), and of 8–12 km thickness (*4*); *5* = zones of increased gravity gradient (fractures separating main blocks), *a* = established, *b* = inferred; *6* = fractures proved by gravimetric data; *7* = recent seismically active fractures; *8* = Meso-Cenozoic depressions with sequences of over 1500 m thick; *9* = earthquake epicenters, *a* = major, *b* = minor.

structures. Other directions are submeridional and subparallel. Narrow bands of negative anomalies correspond to Meso-Cenozoic grabens. In many cases the magnetic anomalies reflect merely the rock magnetization and not the deeper structures (Blumenzvayg, in Marinov et al., 1973b). Nevertheless, many structures are easily traceable on magnetic maps. A strongly differentiated magnetic field in the Saynshand area (southeastern Mongolia) shows belts of positive (500–700 gamma) and negative (200–300 gamma) anomalies. Gradients of many tens and in places to several hundreds of gamma per kilometer have been recorded. The form of the magnetic anomalies reflects the linear style of deformations. The symmetrical pattern of negative and positive anomalies is like those of mid-ocean ridges. Stepanov and Volkhonin (1969) suggested that the thinner granitic layer and regional gravity maximum in the southeastern Gobi depressions may be explained by uprise of heavy material as the result of the spreading of the granitic layer (Zonnenshain, in Marinov et al., 1973b).

MESOZOIC MAGMATISM

Mesozoic magmatism in Mongolia was predominantly of effusive character. Granitoid intrusions of Late Triassic–Early Jurassic age (220–180 m.y.) are confined to the Hentiy Mountains (Kalenov, 1960, 1961; Hasin, 1966, 1971). In southeastern Mongolia mostly granites and granodiorites occur. A second group of intrusions, of Middle–Late Jurassic age (175–130 m.y.), consists of granitic syenites and syenites cropping out in the southernmost Gobi. Intrusions of alaskite, amazonite granites and granophyres occur in the central and eastern regions (Petrovich, 1963; Fedorova, 1968). So far no Early Mesozoic magmatic activity has been noted in western Mongolia. Since Triassic times there has been no ophiolitic magmatism. Sialic magmatism affected vast areas in the eastern and central regions. Since the end of the Jurassic only effusive rocks occur in Mongolia. Late Jurassic–Early Cretaceous basic lavas are represented mainly by basalts and trachydolerites of the Tsagaan Tsab svita. These rocks occupy vast areas in the northeastern and southeastern regions. Cenomanian basaltic rocks are assigned to the Saynshand svita.

Reactivation of tectonic movements was the main cause of Mesozoic magmatism. In northern Mongolia the Kerulen and Onon deep fractures were the main tectonic lines which controlled lava outflows (Borzakovsky, 1971). The Selenga and the eastern Mongolian volcanic belts exactly repeat the structural pattern of the Late Paleozoic. Intensification of magmatic processes occurred in the Selenga belt during early Middle Triassic, and in the eastern Mongolian belt at the end of the Permian. Proceeding from west to east the age of Late Paleozoic and Early Mesozoic volcanites becomes younger and younger in the central and eastern Mongolian belt. This is a characteristic phenomenon in the whole of Eastern Asia. Distribution of the various Early Mesozoic magmatic bodies in Mongolia shows a subparallel zonal pattern oblique to the general strike of the uplifted Pacific belt.

MESOZOIC STRATIGRAPHY AND PALEOGEOGRAPHY

Triassic

Triassic formations in Mongolia are confined to the northern and northeastern regions. Some newly discovered Triassic sediments occur near Noyon in the Trans-Altay Gobi Desert (Fig. 8). Both continental and marine facies occur but the former are more important.

The continental formations crop out in two belts more or less subparallel to the old Paleozoic fold structures. The northern belt lies in the Selenga River Basin within the socalled Selenga volcanic belt, along the southern borders of the Hentiy Mountains and in the Kerulen River Basin. In this belt the Triassic sequence is made up of a lower sequence of molasse-type sediments, and an upper sequence of volcanic rocks. The Triassic molasse is a predominantly grey, clastic sequence with coarse sandstones and grits, interbedded with siltstones, tuffites and conglomerates. Cross-bedding in sandstones and parallel fine lamination in finer fractions is typical in the molasse, pointing to fluviatile and lacustrine environments. Rocks of finer clastic material contain abundant floral remains, but no coals are found. Basal conglomerates contain mainly Permian volcanic pebbles (see type section, Table II). The thickness of the molasse varies from about 1500 to 6000 m (type section). Despite the fossil flora, the exact age of the molasse is still uncertain. Triassic floras of Mongolia show affinities to those of Kazakhstan, northern China and the south-

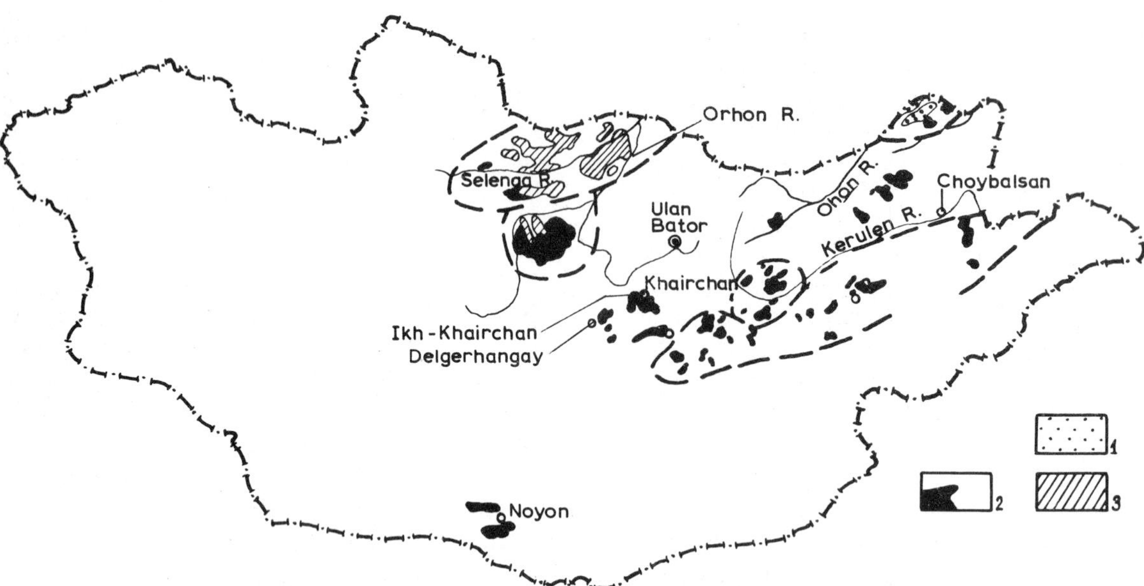

Fig. 8. Distribution of the Triassic formations in Mongolia (after Blagonravov, in Marinov et al., 1973a).
Legend: *1* = undivided Upper Permian–Lower Triassic; *2* = undivided Triassic; *3* = undivided Upper Triassic–Lower Jurassic.

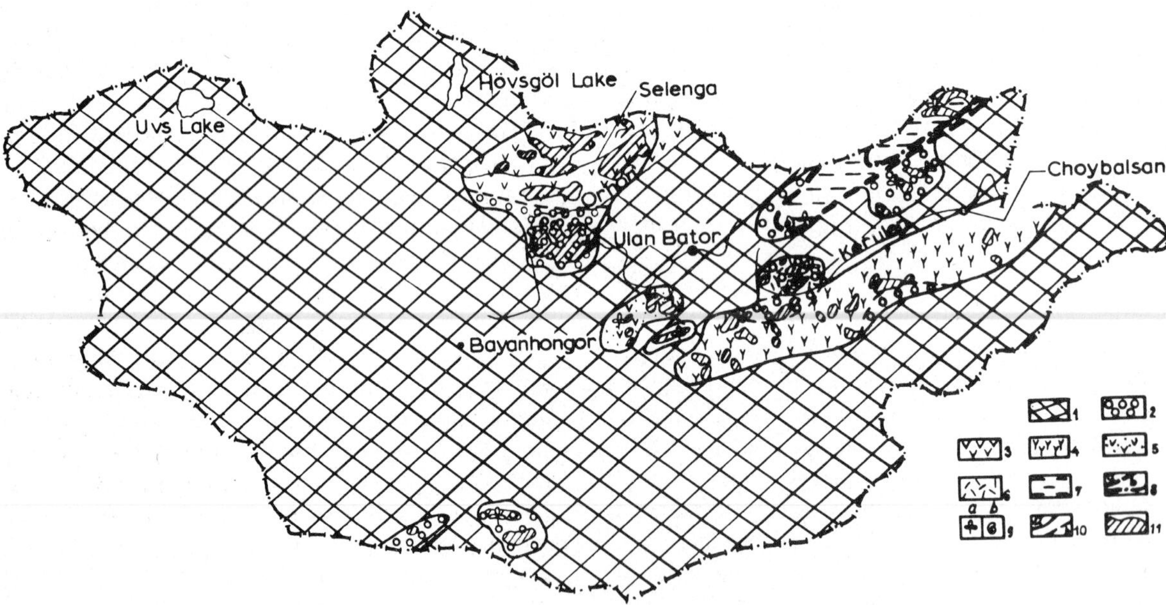

Fig. 9. Lithofacies schematic map of the Mongolian Triassic (after Blagonravov and Zonnenshain, in Marinov et al., 1973b).
Areas of continental facies: *1* = prevalence of erosion; *2* = clastic sediments; *3* = subaerial, mostly intermediate lavas and pyroclasts; *4* = same of intermediate and acid composition; *5* = subaerial intermediate lavas and pyroclasts and arenaceous and other coarse clastics; *6* = subaerial lavas of acid composition.
Areas of marine facies: *7* = clastic sediments; *8* = tentative position of ancient coast line during the Early Triassic (*a*) and during the Late Triassic (*b*); *9* = fossil localities, *a* = flora, *b* = fauna; *10,* a = boundary between erosional and depositional areas, *b* = facies boundary; *11* = actual outcrops of Triassic rocks.

ern Urals. The presence of some endemic forms, however, suggests that there may be a Mongolian Triassic subprovince. Relict *Cordaites* and other forms occurring in the lower portion of the sequence show affinities to the Late Permian flora of northern Europe. Such forms as *Cladophlebis* and *Esaylia* are common also in the Kuznetsk Basin. Other floral elements point to a Late Triassic (Keuper) or even an Early Jurassic age. The upper, volcanic sequence covers large areas in the Selenga volcanic belt, and in eastern Mongolia southwest of Choybalsan (Fig. 9). The rocks are mostly andesitic and andesite-dacitic porphyrites, orthophyres and trachyliparitic porphyries

TABLE II

Triassic type section at Dzalantu-Ula and Otson-Obo, Absog depression

No.	Thickness (m)	Description
6		effusive rocks
5	400	coarse conglomerates, variegated in color; pebble material: various volcanics, quartzites, granites, sandstones, shales
4	200	green, grey sandstone, intebedded with coaly siltstone and layers of liparite porphyries and andesite porphyrites; flora: *Neocalamites* sp., *Cladophlebis* sp., *Equisetites* sp., *Phlebopteris* sp., *Pityophyllum* ex gr. *Nordenskioldii*, etc.
3	2000	grey (sometimes red) conglomerates with interbeds of grey sandstone and gravel
2	1400	grey and green sandstones interbedded with green and dark-grey siltstones; flora: *Tersiella* sp., *Nilssonia* sp., *Cladophlebis* sp., *Neocalamites* sp., *Pseudocteni* sp.
1	1340	brown, grey conglomerates; pebble material: granites, sandstones, quartz and jaspers; sandstone intercalations
unconformity		
		Lower–Middle Carboniferous sediments

(Bobrov, 1962, 1965; Borzakovsky, 1971; Hasin, 1971). Their age is only known in terms of their position in the sequence under Jurassic and above Permian beds.

A marine Triassic sequence occurs in the northeastern part of the country, in the Dzhirgalantu River Basin (Zonnenshain et al., 1971). A type section is presented in Table III. Pelecypods and some ceratites occur in a dominantly clastic-sediments sequence 442 m thick, concordantly overlying the Permian. The distribution of the Triassic formations in northeastern Mongolia is very interesting as the marine facies which occur in the Onon River Basin link with outcrops in the eastern Transbaykal area of the U.S.S.R., and are surrounded by continental molasse sequences cropping out in the depressions. The zone of subaerial volcanism occupies still an outer position (Fig. 9).

During the Triassic Mongolia was divided into numerous elevations and intramountainous depressions. Erosion of the former produced the clastic material which filled the latter (molasse). Deposition was partly fluviatile, partly lacustrine with only one marine episode of minor importance in the north. The dominantly grey and dark colors of the rocks suggest, with a few exceptions, humid climatic conditions (Marinov et al., 1973a).

The Triassic sedimentary basins seem to have been largely mutually independent. The principal basins are the Selenga Basin, the Kerulen Basin and the area between the Kerulen and Uuldza rivers. Other basins of smaller dimensions are known. The marine ingression which continued during the marine regime of the Late Permian was connected with the Boreal zoogeographic province. During the Middle Triassic (or possibly earlier) intensive tectonic movements began in eastern Mongolia and resulted in a highly varied topography. Newly formed intramontane basins were then filled with coarse clastic sediments. The central parts of the depressions were occupied by freshwater lakes into which flowed numerous rivers from the surrounding elevations. The end of the Triassic was marked by intense volcanic activity.

Jurassic

Jurassic formations crop out over a much wider area than the Triassic. They are found in northwestern Mongolia (i.e., the Mongol-Altay Mountains, and the Great Lakes area), the Hövsgöl area, central Mongolia, eastern Gobi Desert, the Hangay and the Hentiy

TABLE III

Triassic marine type section at Sayhan-Undurin-Obo, Dzhirgalantu graben-syncline (after Zonnenshain et al., 1971)

No.	Thickness (m)	Description
7	150	pinkish-grey medium- to coarse-grained sandstones with thin interbeds of black siltstone
6	250	black shales, siltstones, dark fine-grained sandstones, and coarse-grained pinkish-grey sandstones
5	60	dark-grey sandstones with black shales and silicified dark-grey siltstones at top
4	250	black shales with interbeds of silicified siltstones and fine-grained sandstones
3	5	fossiliferous dark-grey coquina, with pelecypods: *Eumorphites* sp., *Bakewellia* (*Neobakewellia*) *reticularis, B.* (*N.*) *goldfussi, Posidonia sossunai, P.* ex. gr. *minor, Myalina* sp. indet.; and ceratites: *Euflemingites romunderi, Anasibirites* cf. *echimensis, A.* sp. indet., *Prosphingites* aff. *ovalis, Anakashmirites* sp. ind.
2	540	interbeds of dark siltstones and grey sandstones of various grain sizes, cross-bedded at the base
1	20	basal conglomerates, poorly sorted; pebble material: quartz, metamorphic rocks; Permian acid effusive rocks, and grey granites
		Permian effusive rocks and granites (eroded)
	1275 (total thickness)	

Mountains, northeastern Mongolia, southern Mongolia and the Nukut-Daban range in the extreme east-southeastern part of the country (Fig. 10). In most areas stratigraphic correlation is fairly good. In a very general way three main Jurassic type sequences may be distinguished. A continental sequence is to be found in northwestern Mongolia, in the Hövsgöl area, in central Mongolia, and in the eastern Gobi Desert. The two remaining sequences are sedimentary–volcanic and crop out in the northern and northeastern parts of the Selenga and the eastern Mongolian volcanic belts (Figs. 11, 12). In the first four regions mentioned above, Jurassic formations can be easily divided into a Lower–Middle, generally grey-colored, sequence, and a Middle–Upper Jurassic which is predominantly of red-bed type. The lower complex is coal-bearing in some areas. The upper complex unconformably overlies the lower.

Lower–Middle Jurassic (Fig. 11)

Out of many Lower–Middle Jurassic sections that in the Dzhirgalantu Mountains may be taken as an example (Table IV). The flora found in that section contains *Cladophlebis haiburnensis* L. et Heer., *Ginkgo sibirica* Heer., *G. digitata* Brogn., *Pityophyllum longifolium* Nath., *Phoenicopsis* sp., *Pterophyllum* sp., *Coniopteris burejensis* Zal. (Neiburg, 1929).

In the upper, coal-bearing part of the sequence the following mollusks were determined by Martinsson (1961): *Ferganoconcha* cf. *estherineformis* Tchern., *F. sibirica* Tchern., *Unio* sp., *Pseudocardiania* sp., etc. A similar mollusk assemblage has been found in many other Lower–Middle Jurassic sections in western Mongolia. In the Great Lakes area the Lower–Middle Jurassic beds, up to 600 m thick, crop out north of the Har Us Lake. Coal-bearing sediments of the same age in northwestern Mongolia contain plant remains similar to those of the Kuznetsk Basin, Tuva (the Erbek svita in the Ulug-khem Basin, Yenisey River) and Irkutsk Basin (Vakhrameev, 1964).

Central Mongolia and eastern Gobi desert. A type section of the Lower–Middle Jurassic sediments occurs at Bahar-Ula in the Gobi-Altay Mountains (Shuvalov, 1970). Near the Ih Bogd and Baga Bogd Mountains there are volcanic rocks in the lower part of the Bahar svita. The thickness of that svita reaches about 1000 m in places. The flora (Shuvalov, 1970; Logachev in Florensov and Solonenko, 1963) is characteristic of the Lower–Middle Jurasssic of Eurasia and Central Asia.

In the eastern Gobi Desert, the Lower–Middle Jurassic sequence is subdivided into two svitas namely the lower, Hairmot, and the upper, Hamar-Huburin. The type section occurs at Hamar-Huburin Huduk near the town of Saynshand. Abundant floras and fresh-water mollusks and ostracods are found.

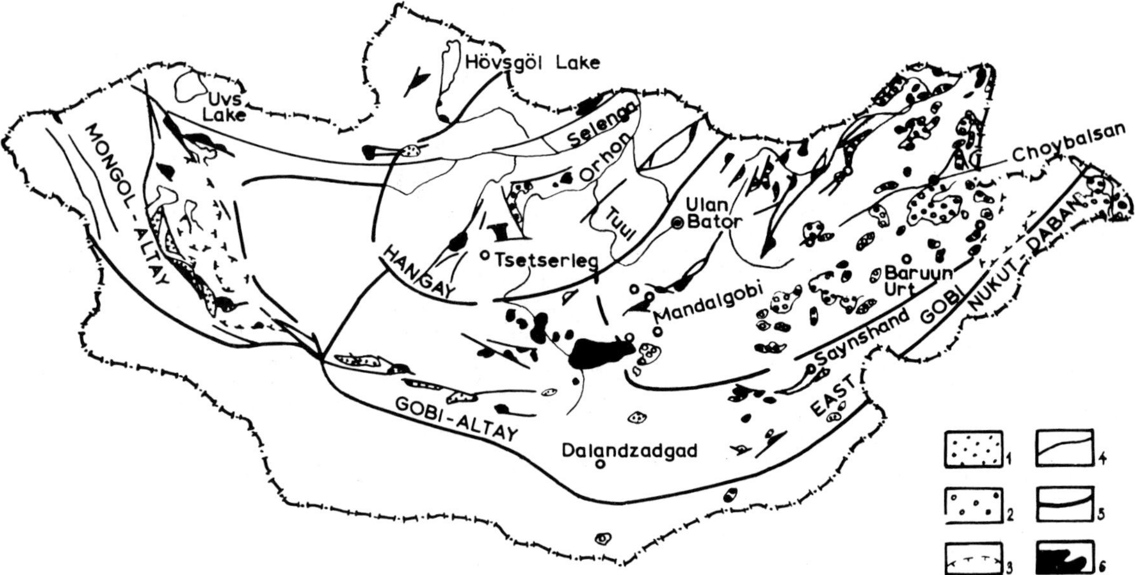

Fig. 10. Distribution of Jurassic sequences in Mongolia (after Filippova, in Marinov et al., 1973a).
Legend: *1* = Upper Jurassic sediments; *2* = Middle–Upper Jurassic sediments; *3* = inferred occurrence of Mesozoic sequences under the Cenozoic; *4* = fractures; *5* = boundaries of geological regions; *6* = Lower–Middle Jurassic sediments.

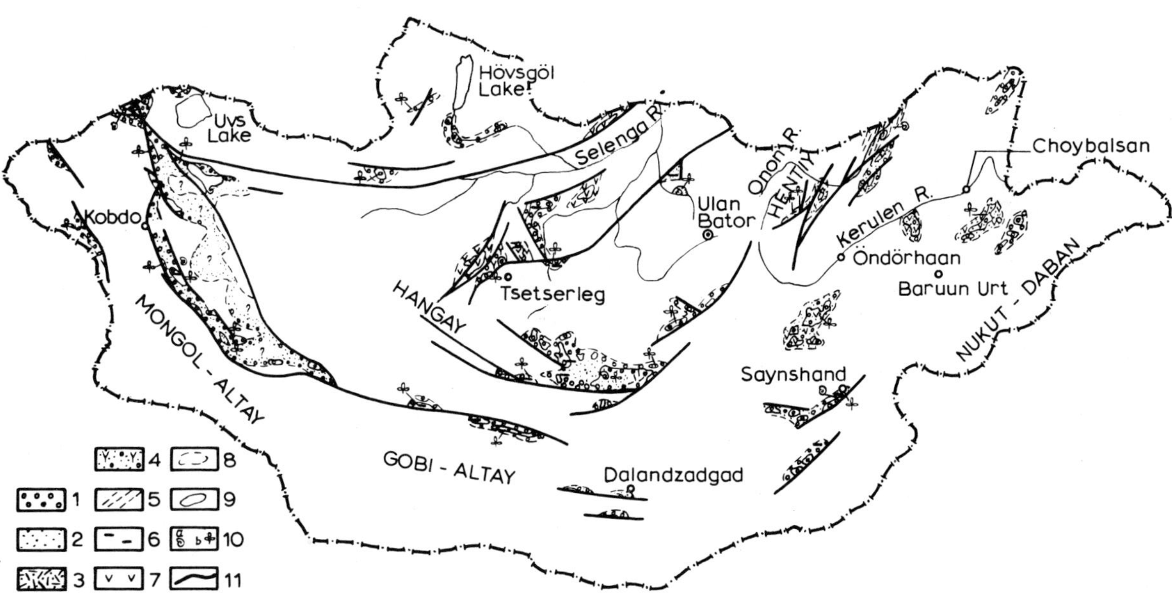

Fig. 11. Lithofacies schematic map of the Lower–Middle Jurassic in Mongolia (after Filippova et al., in Marinov et al., 1973a).
Legend: *1* = coarse clastics; *2* = fine clastics; *3* = same with subordinate lavas and pyroclasts of medium or acid composition; *4* = tuffaceous clastics with lavas of various composition; *5* = coastal lowlands covered with fine clastic sediments; *6* = coal accumulation; *7* = narrow grabens filled with lavas of trachy-andesite–basaltic composition; *8* = inferred outlines of the Lower–Middle Jurassic depressions; *9* = actual occurrence of the Lower–Middle Jurassic formations; *10* = fossil localities, *a* = fauna, *b* = flora, *11* = faults and fractures.

TABLE IV

Lower–Middle Jurassic type section in the Dzhirgalantu Mountains (after Devyatkin, in Marinov et al., 1973a)

No.	Thickness (m)	Description
3	50	coal-bearing series of grey sandstones, coaly shales with interbeds of coal (up to 3 m)
2	380	medium-grained grey to greenish-grey sandstones with fine conglomerates in lower part, and dark green siltstones and shales with coal at top
1	820	coarse conglomerates with rare interbeds of polymictic coarse-grained sandstones; granites and effusive rocks are mainly represented in pebbles
	1250 (total thickness)	

Hangay and Hentiy Mountains. In these ranges up to 3000 m of trachy-andesites, basalts and subordinate orthophyres as well as minor quantities of diabasic and andesitic porphyrites occur in the Selenga volcanic belt. A further sedimentary–volcanic sequence consisting of conglomerates, sandstones and basalts contains a flora indicating an Early–Middle Jurassic age.

Northern Mongolia. In this area the Lower–Middle Jurassic sequences are divided into a lower terrigenous series 200–250 m thick and an upper volcanic sequence which has clastic rocks at its top. The volcanics total 200–500 m, the clastic sediments about 800 m. They rest on eroded Triassic or on Precambrian limestones (Hentiy Mountains). The Early Jurassic age is based upon the flora found in both the lower and upper clastic parts of the sequence.

Marine sediments crop out in the Eren-Daban ridge along the watershed of the Onon and Uuldza rivers. There are basal black shales followed by dark-grey sandstones with shale interbeds and thin conglomerates. They contain pelecypods identical with those found in the marine Lower Jurassic in the eastern Transbaykal area.

Upper Jurassic (Fig. 12)

Upper Jurassic red beds unconformably overlie the Early–Middle Jurassic or older formations in northwestern Mongolia, the Hövsgöl area, central Mongolia and in the eastern Gobi Desert. In the

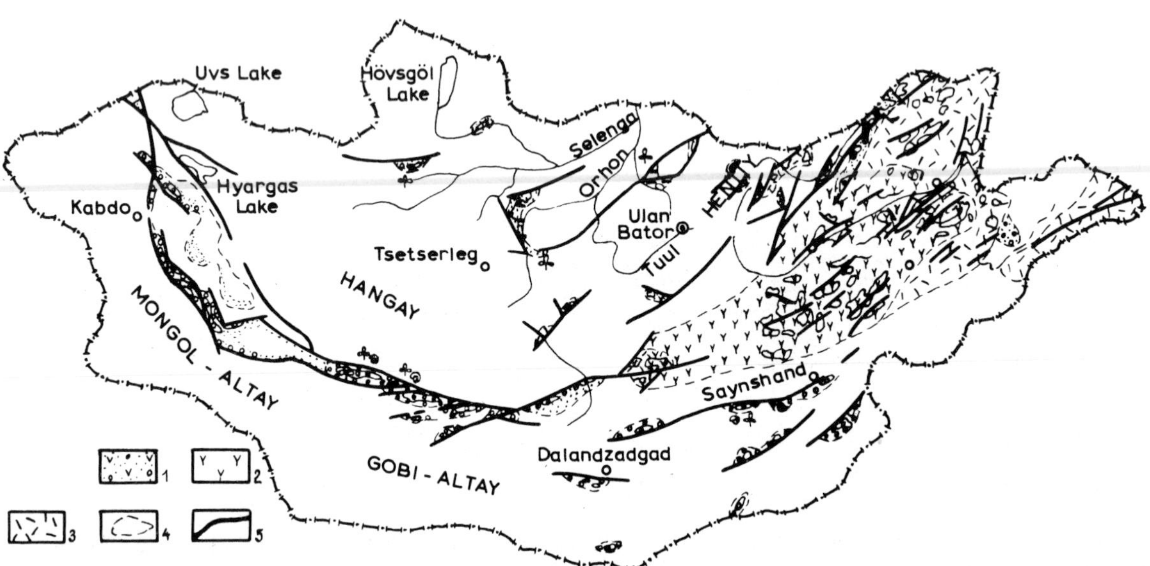

Fig. 12. Lithofacies schematic map of the Upper Jurassic in Mongolia (after Filippova et al., in Marinov et al., 1973a). Legend: *1* = tuffaceous–clastic sediments; *2* = area of volcanism of predominantly basic composition; *3* = same of mostly acid composition; *4* = inferred contours of the Upper Jurassic depressions and volcanic areas; *5* = actual occurrence of the Upper Jurassic rocks. (For other symbols see Fig. 11).

northeastern part of the country there are effusive-terrigenous rocks, and in the Hangay–Hentiy Mountains the Upper Jurassic is usually represented by grey-clastic sediments with coal.

Northwestern Mongolia. Here the Upper Jurassic red beds crop out in a broad belt in the foothills of the Mongol-Altay Mountains. They exhibit poor sorting of the clastic material which is of local derivation, and vary in thickness from 250 to 1200 m. In the Great Lakes area, sediments are finer and thinner (100–300 m) and unconformably overlie either Middle Jurassic or Paleozoic rocks. The occurrence of the pelecypods *Throacia* cf. *lata* Ag., *Pleuromya* sp. aff. *tenuistriata* Ag., *Alaria sotnikowi* Schmidt, *Tancredia subtilis* Lah., points to a Kimmeridgian age.

Central Mongolia and eastern Gobi Desert. Red beds occur in the Gobi-Altay Mountains, and in the southern part of the central Gobi Desert. They consist of conglomerates, breccias, clays and sandstones, and higher in the sequence, sandstones interbedded with conglomerates and clays of fluvial and lacustrine origin (the Tormhon svita of Shuvalov, 1969, 1970). In the Gobi-Altay Mountains a few basaltic horizons have been found in the upper part of the Tormhon svita. A type section of the Tormhon svita crops out in the upper course of the Dzhirgalantu River (Table V). A charophyte, *Aclistochara jonesi* Peck., also known from the Kimmeridgian–Tithonian of North America occurs in this section. Some of the ostracods are also found in the Purbeck beds of Europe. The thickness of the Tormhon svita varies greatly from site to site. In the eastern Gobi Desert near the Hamar-Huburin Huduk and the ruins of the Shariliin Hiid lamasery 1700 m of Upper Jurassic red-bed conglomerates contain angular or poorly rounded pebbles of porphyrite, quartzite, granite and older conglomerates in an uncemented coarse clastic matrix. They are followed by cross-bedded sandstones and arenaceous clay which complete the sequence of the Shariliin svita.

Hangay and Hentiy Mountains. In this region the Middle (in part) and Upper Jurassic sediments are exposed in rather small depressions and grabens. Here the Sayhan svita of Marinov and Petrovitch (1964) consists of 80 m of conglomerates and sandstones in the lower part, followed by finer clastic rocks (coal-bearing) 110 m thick. The flora of the coal-bearing portion contains *Raphaelia diamensis* Sew., *Cladophlebis williamsoni* Brogn., *Schizolepis* sp., *Baiera concinna* (Heer.) Kawas., *Czekanowskia setacea* Heer., *Equisetites ferganensis* Sew., *Phoenicopsis speciosa* Heer., *P. angustifolia* Heer., *Phoenicopsis* sp., *Pityophyllum nordenskioldii* (Heer.) Nath., *Sphenobaiera czekanowskiana* (Heer.) Florin, indicating the Middle–Late Jurassic age.

The coal-bearing sediments of northern Mongolia are easily correlated with the Middle–Upper Jurassic sequences of the western Transbaykal area (Kolesnikov, 1964; Nagibina, 1969).

Northeastern Mongolia. Here the Middle–Upper Jurassic sedimentary–volcanic sequence crops out. It

TABLE V

Upper Jurassic type section (Tormhon svita) in the Dzhirgalantu River Valley (after Shuvalov, in Marinov et al., 1973)

No.	Thickness (m)	Description
8	93	variegated conglomerates with interbeds of clays of the same coloration
7	131	greyish-red clay with lenses of greenish conglomerates and conglomerate-breccia (in the middle of the section)
6	20	greyish-red gravels (poorly cemented conglomerates) and clays with pebbles
5	20	greyish-red clay with white marly concretions
4	45	reddish conglomerate-breccia with interbeds of arenaceous clay
3	65	clay with pebbles, interbeds of fine conglomerates, variously grained sandstones; two clay horizons at the top
2	21	cherry-red to greenish-red conglomerate-breccia with intercalations of arenaceous greyish-red clay
1	41	greyish-red clay with intercalations of gravels and lenses of compact limestones (up to 0.5 m)
	436 (total thickness)	

broadens northeastwards continuing into the eastern Transbaykal area. The facies, both volcanic and sedimentary, are highly variable. The volcanics consist of andesites, andesite-basalts, liparite-porphyries and liparites. The lower volcanic horizons are Middle Jurassic in age according to absolute age determination (Mushnikov et al., 1966). Along the middle reaches of the Kerulen River the series is tripartite consisting of a lower sedimentary-tuffaceous sequence, a middle sequence (of basic and medium volcanics), and an upper sequence (of acid volcanics). The total thickness reaches 2000 m. To the northeast the lower member is reduced (or absent) being replaced by a thick series of predominantly acid volcanics. In general, both the composition and thickness of the different members of the series vary. Absolute age determinations of the volcanites at Tümen Tsogto on the right bank of the Kerulen River gave 154–159 m.y. Pebbles in coarse clastic sediments are mostly local igneous rocks derived from nearby elevations (Uuldza River and Eren-Daban ridge).

South Mongolia and the Nukut-Daban ridge. The volcanic rocks which unconformably overlie the Upper Paleozoic formations here are tentatively classified as Jurassic. They are light-colored liparites, tuffs and breccias 300–900 m thick. In the Toto-Shan Mountains there are andesites and basalts as well, which by analogy to similar volcanics in the adjacent Chinese territory and from absolute age determinations (150–183 m.y.) may be assigned to the Jurassic. Sedimentary rocks of Early–Middle Jurassic age are found in only a few localities.

Jurassic paleogeography

The depositional pattern of the Jurassic and the character of the volcanic activity were inherited to a considerable extent from the Triassic. Intense volcanic eruptions occurred in the northern and northeastern regions. In northeastern Mongolia, the only uplifted and denuded region during the Triassic, coarse, clastic coal-bearing and red-bed sequences were laid down during the Jurassic. The only marine ingression, one of limited extent, occurred in the extreme northeastern part of the country.

The Jurassic period in Mongolia was characterized by activation of tectonic movements. New fracture zones developed and many old ones were rejuvenated. The movements were particularly strong in the southern and western parts of the country where the Mongol- and Gobi-Altay mountain systems have been uplifted. Less pronounced uplifts were observed in the Hangay, Hentiy and the Greater Khingan belts. Many intramountainous depressions (grabens), narrow, elongate structures, were formed and were subsequently occupied by lakes. As a rule the depressions developed within reactivated zones of Paleozoic fracture zones and were filled with clastic material brought in from the surrounding massifs. Jurassic coal formed in vast, swampy lowlands (alluvial and lacustrine deposition) under warm, humid climatic conditions. Reactivation of the fracture zones was accompanied by volcanic outflows.

The red-bed character of the Upper Jurassic sediments suggests the existence of arid, warm climatic conditions. Slightly more humid conditions may have existed at the end of the Jurassic and the beginning of the Cretaceous, a suggestion based upon the increased number of floral elements in the uppermost Jurassic sediments.

Cretaceous

Cretaceous formations occupy extensive areas in Mongolia, particularly in the eastern, southern and south-central regions. There is relatively little Cretaceous in the west (Fig. 13). The Cretaceous is represented entirely by continental clastic sequences which unconformably overlie older formations, usually Early Mesozoic or Paleozoic in age. Passage beds between the Upper Jurassic and Lower Cretaceous are known in Mongolia under the name of the Tsagaan-Tsab svita (Table VI). The upper, tuffaceous, part of this svita is assigned to the Early Cretaceous (Table IX). Grey to greenish-grey clastic rocks of the Tsagaan Tsab svita are frequently intercalated with basic to intermediate volcanic rocks and tuffs. The overlying Dzüün Bayan svita (Hauterivian–Albian) consists predominantly of fine clastic sediments among which bituminous shales form a key-bed. There are also some carbonate rocks. Dark-grey colors predominate in the lower part, and greenish-grey in the upper, the latter varying in lithology and thickness. In southern Mongolia bituminous shales predominate, whereas in the eastern Gobi Desert a sandy-argillaceous series contains conglomerates at its base, with a coal-bearing sequence at a higher horizon. In relatively small-size depressions of central and eastern Mongolia the coal-bearing series show considerable extent and thickness and are of economic

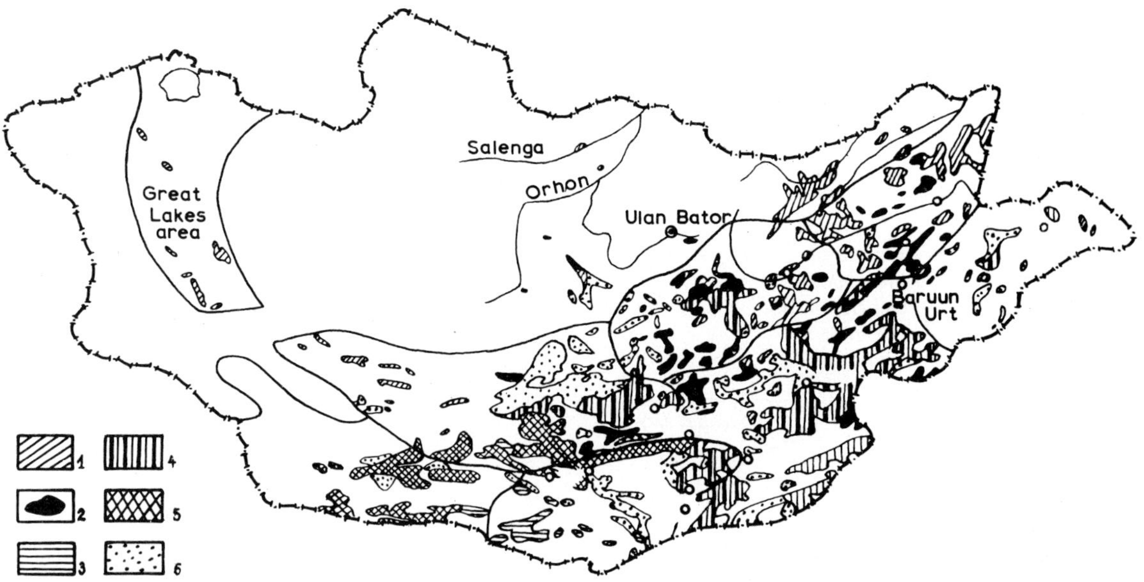

Fig. 13. Distribution of the Cretaceous formations in Mongolia (after Blagonravov, in Marinov et al., 1973a).
Legend: *1* = undivided Upper Jurassic–Lower Cretaceous (Tsagaan-Tsab svita); *2* = Dzüün Bayan svita; *3* = Lower Cretaceous undivided; *4* = Saynshand svita; *5* = Bayan-shiré svita; *6* = Baruun Goyot svita; *7* = Upper Cretaceous undivided; *8* = Upper Cretaceous–Tertiary undivided.

importance. The coal-bearing sediments of the Dzüün Bayan svita are also known to occur in the north, in the Hentiy and Hangay uplands. A brown-coal deposit in the Nalayha depression situated 35 km to the east of Ulan Bator has nine coal seams in the middle portion of the Tsagaan Tsab svita, the total thickness of which is 573 m. In other regions thickness of that svita varies from 800 to 1800 m. The thickest sequence occurs in the extreme east of the country. The uppermost horizons of the Dzüün Bayan svita contain vertebrate skeletons which include *Iguanodon orientalis* Rozdestvensky (1952). The flora contains such forms as Ginkgoales, *Podocarpus, Dacrydium, Podozamites, Abies, Picea, Cedrus, Pinus,* etc. (complete list in Marinov et al., 1973a). In the central Gobi depression Early Cretaceous sediments corresponding in age to the Dzüün Bayan svita have been described by Berkey and Morris (1927) as the Ondai Sayr Formation, a monotonous sequence of cross-bedded sandstones with intercalations of paper shales from which the reptile *Protoguanodon mongoliense* Osborn was excavated.

Undivided uppermost Jurassic and Lower Cretaceous sediments conformably overlie Jurassic beds in the Great Lakes area, and in some depressions in the foothills of the Mongol-Altay Mountains. In the central and south Gobi depressions and in the Valley of the Lakes the Lower Cretaceous clastic sediments also contain bituminous shales and basic volcanic rocks.

Cretaceous sediments younger than Albian mainly occur in the southern parts of the country. In the extreme east they disappear under Paleogene and Neogene formations. The Upper Albian–Cenomanian formations (the Saynshand svita of Soviet geologists) show a distinct red-bed character and hence differ considerably from the underlying Lower Cretaceous rocks. They are separated by erosional break from the Lower Cretaceous or pre-Cretaceous sediments. The type section of the Saynshand svita crops out at Hara-Hutul-Ula (east Gobi depression) and is presented in Table VII (Efremov, 1949; Marinov, 1957; Vasilyev et al., 1959; Barsbold, 1969). In general it consists of 50–500 m of alternating red clays, conglomerates with pebbles of various derivation, coarse sandstones, in places with blocks and slabs of crystalline rocks at the base. The Saynshand svita shows a considerable variability in facies, in some localities it is much more argillaceous than the section described above. According to Soviet geologists the deposition of the Saynshand svita took place in shallow-water lacustrine basins into which many streams brought in clastic material (poor sorting, and roundness of particles, rapid facies changes in all directions) (Marinov et al.,

TABLE VI

Type section of the Tsagaan Tsab svita (uppermost Jurassic–Lower Cretaceous) at Tsagaan Tsab Huduk (after Turistchev, in Marinov et al., 1973a)

No.	Thickness (m)	Description
7	96	grey- to greenish-grey tuffs intercalated with sandstones and clays with some interbeds of carbonates
6	89	tuffaceous light-grey sandstone interbedded with greenish-grey clays (somewhat tuffaceous)
5	50	tuffaceous sandstones and clays
4	148	clayey series with numerous horizons of white and light-grey tuffs at its lower part; fauna: *Mycetopus quadratus* Martins., *M. transbaicalensis* Martins., *M. estheriaeformis* Martins., *Estheria middendorfii* (Jones), etc.
3	59	clays, sands and sandstones, grey to dark-grey with gravels at top; fauna: *Lycoptera middendorfii* Müller, *Mycetopus estheriaeformis* Martins.
2	40	variegated clays, tuff-sandstones, interbeds of gravels and arenaceous clays
1	130	effusive-conglomeratic sequence; andesite-basalts, andesites, basalts, agglomeratic tuffs
	700 (total thickness)	

1973a). The Saynshand sediments contain characteristic ostracods and fresh-water mollusks (Galeeva, 1955; Barsbold, 1969).

Younger Cretaceous formations are assigned to the Bayan-shiré svita (uppermost Cenomanian–Santonian). They are best exposed in the east Gobi depression at Javhlant (=Uliastay), Hara-Hutul-Ula and Bayan-shiré. The type section follows above that of the Saynshand at Hara-Hutul-Ula (Table VIII) with a slight break. At the base are argillaceous–arenaceous rocks mostly of grey to light-grey color. In places the Bayan-shiré sediments rest on basalts covering the Saynshand svita. At higher horizons red-brown clays

TABLE VII

Type section of the Upper Albian–Cenomanian sediments (the Saynshand svita) at Hara-Hutul-Ula (after Barsbold, in Marinov et al., 1973a)

No.	Thickness (m)	Description
5	120	black to grey-black basalts
4	80	greenish-grey conglomerates interbedded with grey and light-brown sandstones and reddish-brown clays
3	35	greenish-grey conglomerates with interbeds of red-brown clays and variously grained sandstones
2	8	red-brown and grey sandstones and gravels with intercalations and lenses of coarse conglomerates
1	6	greenish-grey conglomerates and breccias; pebbles represented by schists, quartzites, basic effusives, and jaspers
	200–250 (total thickness)	

interbedded with sandstones occur. Among the many vertebrates an armoured dinosaur, *Talarurus plicatospinaeus* Maleev, has been excavated at Bayan-shiré (Rozdestvensky, 1971). The thickness of the Bayan-shiré svita varies between 500 and 1100–1200 m. Facial changes are small but near the margins of the depressions the sediments are coarser as a rule. The upper red argillaceous sequence is somewhat gypsiferous (as at Dadan-Hira-Obo) in the southernmost part of the east Gobi depression.

The uppermost Cretaceous sediments have been called the Baruun Goyot svita (Martinsson et al., 1969). The same name has been applied to the Campanian sediments of the Nemegt depression in the Trans-Altay Gobi Desert by Gradziński and Jerzykiewicz (1974). Their Baruun Goyot Formation, however, embrances a smaller time span than that of the svita of the same name described by Barsbold (see Table IX). According to him, the Baruun Goyot svita overlies the Bayan-shiré with slight angular unconformity. Only its lower part is exposed in the east Gobi depression where, at the base, greyish sandstones with interbeds of grits, conglomerates, and red-brownish clays, occur. In the central Gobi depression

TABLE VIII

Type section of the Turonian–Santonian sediments (the Bayan-shiré svita) at Hara-Hutul-Ula (after Barsbold, in Marinov et al., 1973a)

No.	Thickness (m)	Description
5	80	reddish-brown and greenish-grey clays alternating with grey fine-grained sands and pinkish-grey sandstones
4	200	interbeds of yellowish-grey to grey fine- to medium-grained sandstones, sands, red-brown arenaceous clays, and fine conglomerates; cross-bedding; horizons of sandy-carbonate concretions; fining upward cycles: conglomerate–sandstone–clay; mollusk fauna in conglomerates: *Plicatotrigonioides multicostatus* Barsbold
3	30	brown to grey somewhat gypsiferous clays with interbeds of grey sandstones and gravels
2	115	interbeds of grey sandstones and brownish to greenish-grey clays; inliers of fine- to medium-grained conglomerates
1	7	greenish-grey, red clays with intercalations of grey sandstones
	432 (total thickness)	

Berkey and Morris (1927) have described the Dohoin Usu Formation cropping out north of Dalandzadgad. It contains all three svitas of the Upper Cretaceous according to Soviet geologists (Efremov, 1954; Rozdestvensky, 1971; Marinov et al., 1973a). At the type locality many vertebrate remains (including *Hadrosaurus*) and fresh-water gastropods (Berkey and Morris, 1927) were found. At Olgoy-Ulan-Tsav among variegated clays, conglomerates and sandstones with carbonate admixture the largest dinosaur eggs in the world have been found by Sochava (1969). The Upper Cretaceous formations of southern Mongolia are famous because of their vertebrate remains. The Oshih locality in the northern Gobi Desert, first visited by the Expedition of the American Museum of Natural History, has yielded *Protoiguanodon* from the Lower Cretaceous and *Asiatosaurus* in sediments assigned later by Barsbold (1969) to the Bayan-shiré svita. The name "Oshih Formation", established by Berkey and Morris (1927), is obsolete as these sediments have been assigned to specific svitas by Soviet and Mongolian geologists (Sochava, 1975). The Saynshand svita at Oshih is separated from the Bayan-shiré sediments by basalts.

The Djadokhta Formation (Table X) cropping out at Bayan Dzak (= Shabarakh Usu) is famous for its dinosaurs, dinosaur eggs and mammals (Berkey and Morris, 1927; Lefeld, 1965, 1971a). Its age is probably Late? Santonian and/or Early Campanian (Gradziński et al., 1977). The sediments at Tögrög some 30 km northwest of Bayan Dzak seem to be coeval to the Djadokhta Formation although they differ both in lithology and fauna (Kielan-Jaworowska, 1974, 1975; Maryańska and Osmólska, 1975). Barsbold (1969) placed both Djadokhta and Tögrög in his Baruun Goyot svita.

In the Trans-Altay Gobi Desert the uppermost Cretaceous sediments have sedimentologically been studied at Nemegt, Altan Ula, Tsagaan-Khushu and other sites by Gradziński (1970) and Gradziński and Jerzykiewicz (1974). The lower, the Baruun Goyot Formation (Table XI) of Middle? Campanian age (Kielan-Jaworowska, 1974) is partly aeolian, partly fluviatile in origin, and the upper, the Nemegt Formation (Table XII), of Maastrichtian age is fluviatile. Vertebrates found mainly in the Nemegt Formation were excavated by the Soviet and Polish–Mongolian Paleontological expeditions (Efremov, 1954a, b; Rozdestvensky, 1971; Osmolska and Roniewicz, 1970). Both formations are red-bed clastic sequences. Sedimentological analysis has revealed that the primary basin had a tendency to aggradation and must have been much greater than the present-day Nemegt Basin (Gradziński, 1970). Thickness of both formations probably exceeds 600 m.

Cretaceous paleogeography

Late Jurassic tectonic movements considerably changed the morphology in Mongolia. Vast areas in the north of the country were uplifted. Many depressions developed in the south (e.g., Dalay Lake) in the east (Buyr Lake), in the Trans-Altay Gobi, Great Lakes area and Dzhungaria in the west. The boundary between these two major morphological zones follows the line of the Main Mongolian Deep Fracture, or north of the Valley of the Lakes, the south

TABLE IX

Correlation of the Upper Jurassic and Cretaceous formations and svitas of Mongolia (after Berkey and Morris, 1927; Martinsson, 1966; Barsbold, 1969; Marinov et al., 1973a; Gradziński and Jerzykiewicz, 1974; Kielan-Jaworowska, 1974, 1975; Gradziński et al., 1977)

Period	Stage	Svita	Formation
Cretaceous	Maastrichtian		Nemegt
	Campanian	Baruun Goyot	Baruun Goyot Djadokhta
	Santonian Coniacian Turonian	Bayan-shiré (may include some Cenomanian)	
	Cenomanian	Saynshand	
	Albian Aptian Barremian Hauterivian	Dzüün Bayan	Ondai Sayr
	Valanginian		
Upper Jurassic	Tithonian Kimmeridgian Lusitanian	Tsagaan Tsab	
	Oxfordian	Sharilin (E Gobi) / Tormhon (central Mongolia)	

Bayanhongor fracture. According to Martinsson (1955) the Lower Cretaceous lake basins were interconnected and covered vast areas. They reached from Witim in the north to China in the southeast. Some of the lakes were comparable in size to the present Caspian and Aral seas. Deposition was accompanied by oscillatory tectonic movements which tended to deepen the basins and elevate the surroundings. Diversification of morphology induced variability of deposition and thickness of sediments. The movements were accompanied by the extrusion of some lava flows.

During Valanginian times (upper part of the Tsagaan Tsab svita) the lakes are presumed to have covered most of the Gobi part of Mongolia from the Great Lakes area in the west to the rest of the territory in the east. Judging from the color of rocks the climate must have been humid and warm. The greatest subsidence was in the east Gobi depression where the thickness of the Tsagaan Tsab svita reached 1000 m. The shores of the Early Cretaceous lakes were vast, marshy lowlands covered with dense vegetation (Efremov, 1954a) in which various dinosaurs such as *Psittacosaurus* and sauropods thrived. Remains of these reptiles have been found in Mongolia, the Kuznetsk Basin and in China (Inner Mongolia). Environmental conditions were probably uniform over great areas of Asia. Among the invertebrates, fresh- and brackish-water mollusks, insects, phyllopods, and ostracods are known. The abundance of thick-shelled invertebrates suggested to some authors a possible connection of the south Mongolia lakes with the marine basins farther south (Efremov, 1954a; Martinsson, 1961). Abundant fossils are also found in the rocks of the Dzüün Bayan (Hauterivian–Albian) svita. The accumulation of vegetation and animals led to the formation of a sapropel which, in turn, formed the oil found in the east Gobi depression (Vasilyev et al., 1959). An intensification of tectonic activity at the end of the Early Cretaceous caused diversification of deposition. Some basins shrank, others changed their positions. The climate became more humid particularly in the northern and northeastern regions. Lava flows were not synchronous hence these rocks do not crop out along the same stratigraphic levels. The passage between the Dzüün-Bayan and the Saynshand svitas (Lower–Upper Cretaceous boundary according to Soviet geologists) is

TABLE X

Type section of the Djadokhta Formation (Late Santonian–Early? Campanian) at Bayan Dzak (after Lefeld, 1971a).

No.	Thickness (m)	Description
6	4.5	reddish-orange-pink sand and sandstone containing nests of dinosaur eggs and turtles
5	0.8	white marl containing little sand
4	4.0	orange-brick to reddish-orange-brown sand with calcareous nodules
3	1.0	conglomerate consisting of small, calcareous nodules with some crystalline pebbles; occasional cross-bedding
2	2.0	reddish-brown hard cavernous sandstone containing nests of dinosaur eggs
1	25.0	orange to moderate reddish-brown sand with small sandstone nodules; this horizon contains skeletons of *Protoceratops andrewsi* Granger and Gregory, *Pinacosaurus grangeri* Gregory, as well as small mammals such as *Zalambdalestes lechei* Gregory and Simpson, *Kennalestes gobiensis* Kielan-Jaworowska, *Kryptobaatar dashzevegi* Kielan-Jaworowska, *Djadochtatherium matthewi* Simpson
	37.3 (total thickness)	

TABLE XI

Type section of the Baruun Goyot Formation (Middle Campanian) (after Gradziński and Jerzykiewicz, 1974)

No.	Description
2	flat-bedded orange sandstone containing small mammals, lizards
1	massive orange sandstone forming mega-stratified units (dune sediments) intertongued with water-laid sediments; this member contains skeletons of reptiles such as: *Saychania chulsanensis* Maryanska and *Bagaceratops rozhdestvenskii* Maryańska and Osmólska, as well as small mammals: *Barunlestes butleri* Kielan-Jaworowska, *Asioryctes nemegetensis* Kielan-Jaworowska, *Chulsanbaatar vulgaris* Kielan-Jaworowska, *Djadochtatherium catopsaloides* Kielan-Jaworowska
	Total thickness: 110 m

TABLE XII

Type section of the Nemegt formation (Upper Campanian–Lower/Middle? Maastrichtian) (after Gradziński, 1970)

a repetitive sequence of fining-upward tabular and trough cross-stratified sandstone and siltstone with some levels of intraformational conglomerates; many scoured surfaces within the sequence; the formation contains reptile skeletons such as: *Tarbosaurus baatar* Maleev, *Saurolophus angustirostris* Rozhdestvenski, *Gallimimus bullatus* Osmólska and Roniewicz, etc.
Total thickness: over 450 m

marked by a slight erosional unconformity. Tectonic movements at that time caused a general uplift and formation of many mountain ridges and intradepressional uplifts in the Gobi region. In many small depressions of northern Mongolia deposition ceased or became condensed during Late Cretaceous times. In few regions, however, an expansion of the Cenomanian deposition can be observed. In general, the subsidence rate in the Late Cretaceous basins was much smaller than during earlier times. Red coloration of the Upper Cretaceous sediments and rarity of plant remains suggest hot and possibly partly arid climatic conditions. Nevertheless, the abundance of fluviatile sediments proves that the drainage systems were fairly well developed. Tectonic movements at the end of the Cenomanian (Saynshand svita) led to erosion, and the younger formations are unconformable. Lacustrine deposition of the Bayan-shiré svita is characterized by a relative abundance of carbonate sediments. Variegated gypsiferous deposits were laid down in the east Gobi depression. Considerable changes in vegetation occurred in Mongolia at the end of the Cretaceous. *Bennetites* vanished, and there were fewer Cycadaceae and Ginkgoinae. Ferns are represented only by relics. Platans, magnolias, eucalyptus, laurel and various Coniferae flourished in probably savanna-type landscape (Sinitzin, 1962). Abundance of dinosaur skeletons found in the Upper Cretaceous sediments suggests existence of extensive vegetation-covered areas of which, however, there is no record in the preserved sequences. The climatic conditions at the end of the Cretaceous must have been warm and possibly humid (at least in some areas) as indicated by some types of sediments (Gradziński, 1970). Tectonic movements at that time caused general uplift, a shrinking and even liquidation of the lacustrine and savanna (and/or semi-desert) regimes, and formation of denuded peneplains brought the life of dinosaurs to an end (Marinov et al., 1973a). Ideas of the desert conditions in Cretaceous times (Morris, 1936) must be greatly modified, although the existence of semideserts cannot be excluded. The total thickness of Cretaceous sediments in Mongolia ranges between 5000 and 6000 m. They are typical molassoid or molasse-type deposits, laid down under conditions similar to those on a young uplifted platform inheriting the structure of the Paleozoic or even Proterozoic basement.

It is hoped that future progress in Mongolian geology, in particular in sedimentological studies, will improve the paleogeographic picture here presented.

REFERENCES

Amantov, V.A., Blagonravov, Yu. A., Borsakovsky, M.V. et al., 1970. The main features of the Paleozoic stratigraphy of the Mongolian People's Republic. In: *Stratigraphy and Tectonics of the Mongolian People's Republic.* Nauka, Moscow, pp. 8–63 (in Russian).

Barsbold, R., 1969. *Stratigraphy and Fresh-Water Mollusks of the Upper Cretaceous of the Gobi Part of the Mongolian People's Republic.* Dissertation, Geological Institute, Moscow, 26 pp. (in Russian).

Beloussov, V.V., 1962. *Basic Problems in Geotectonics.* McGraw-Hill, New York, N.Y., 809 pp. (Russian edition from 1954).

Berkey, Ch. P. and Morris, F.K., 1927. *Geology of Mongolia, 2.* American Museum of Natural History, New York, N.Y.

Bobrov, V.A., 1962. Intrusive complexes of eastern Mongolia and comparison with intrusive complexes of the Transbaykal area. In: *Matieryaly po granitoidam Zabaykalya.* Gosgeoltekhizdat, Moscow, pp. 102–129 (in Russian).

Bobrov, V.A., 1965. Peculiarities of the metallogenic development of eastern Mongolia. In: *Int. Geol. Congr., 22nd Sess., Contrib. Soviet Geologists.* Nedra, Moscow, pp. 203–217 (in Russian).

Borzakovsky, Yu. A., 1971. Intrusive complexes of southeastern Mongolia. In: *Magmatism i metallogenya Mongolskoy Narodnoy Respubliki (Trudy Sovietsko-Mongolskoy ekspedicyi, 3).* Nauka, Moscow, pp. 41–59 (in Russian).

Dörnfeld, G., 1968. Zur Fortsetzung der Mongolisch-Ochotskischen Faltenzone auf das Gebiet der Mongolischen Volksrepublik. *Geologie,* 17(1): 93–98.

Efremov, I.A., 1949. Preliminary results of the works of the First Mongolian Paleontological Expedition of the U.S.S.R. Academy of Sciences in 1946. In: *Matieryaly po geomorfologii i paleontologii. Trudy Mongolskoy Kommisyi Akad. Nauk S.S.S.R.,* No. 38: 5–28 (in Russian).

Efremov, I.A., 1954a. Some remarks on the historical development of dinosaurs. *Tr. Paleontol. Inst. Akad. Nauk S.S.S.R.,* 48: 125–141 (in Russian).

Efremov, I.A., 1954b. Paleontological investigations in the Mongolian People's Republic. In: *Preliminary Results of the Expeditions of 1946, 1948, 1949. Trudy Mongolskoy Kommisyi Akad. Nauk S.S.S.R.,* No. 59: 3–32 (in Russian).

Fedorova, M.E., 1968. Upper Paleozoic and Mesozoic magmatism of the western part of the Hangay–Hentiy folded area. *Byul. Mosk. O.-Va. Ispyt. Prir., Otd. Geol.,* 43(1): 143 (in Russian).

Florensov, N.A. and Solonenko, V.P. (Editors), 1963. *The Gobi-Altay Earthquake.* Izdatelstvo Akad. Nauk S.S.S.R., Moscow, 392 pp. (in Russian).

Galeeva, L.I., 1955. *Ostracods from Cretaceous Sediments of the Mongolian People's Republic.* Gostoptekhizdat, Moscow, 98 pp. (in Russian).

Gradziński, R., 1970. Sedimentation of dinosaur-bearing Upper Cretaceous deposits of the Nemegt Basin, Gobi Desert. In: *Results of the Polish–Mongolian Palaeontological Expeditions, II. Palaeontol. Pol.,* 21: 147–229.

Gradziński, R. and Jerzykiewicz, T., 1974. Sedimentation of the Barun-Goyot Formation. In: *Results of the Polish–Mongolian Palaeontological Expeditions, V. Palaeontol. Pol.,* 30: 111–146 (plates 34–42).

Gradziński, R., Kielan-Jaworowska, Z. and Maryańska, T., 1977. Upper Cretaceous Djadokhta, Barun-Goyot and Nemegt formations of Mongolia including remarks on previous subdivisions. *Acta Geol. Pol.,* 27(3).

Hasin, R.A., 1966. Some general problems of the metallogenic position and of the Mesozoic intrusive magmatism of central and eastern Mongolia. In: *Sovremennye metody poiskov mestorozhdenii olova, volframa, molibdena. Sbornik materialov nauchno-tekhnicheskogo seminara, Moskva.* Izdatelstvo SEV, Moscow, pp. 61–69 (in Russian).

Hasin, R.A., 1971. Basic outlines of magmatic evolution in Mongolia. In: *Magmatism i metallogenya Mongolskoy Narodnoy Respubliki.* Nauka, Moscow, pp. 7–40 (in Russian).

Huan-Tsi-zin, 1961. Basin outlines of the tectonics of China. *Sov. Geol.,* No. 9: 8–56 (in Russian).

Kalenov, A.D., 1960. New data on the Mesozoic intrusions of eastern Mongolia. In: *Magmatism i svyaz z nim poleznykh iskopayemykh.* Moscow, pp. 634–636 (in Russian).

Kalenov, A.D., 1961. Mesozoic intrusions of eastern Mongolia. *Izv. Vyssh. Uchebn. Zaved., Geol. Razved.,* No. 2: 41–52 (in Russian).

Kielan-Jaworowska, Z., 1969. Preliminary data on the Upper Cretaceous eutherian mammals from Bayn Dzak, Gobi Desert. In: *Results of the Polish–Mongolian Expeditions, I. Palaeontol. Pol.,* 19: 171–197.

Kielan-Jaworowska, Z., 1974. Multituberculate succession in the Late Cretaceous of the Gobi Desert (Mongolia). In: *Results of the Polish–Mongolian Expeditions, V. Palaeontol Pol.,* 30: 23–44 (plates V–XXI).

Kielan-Jaworowska, Z., 1975. Evolution of the therian mammals in the Late Cretaceous of Asia. Part I. Deltatheridiidae. In: *Results of the Polish–Mongolian Expeditions, VI. Palaeontol. Pol.,* 33: 103–131.

Kolesnikov, E.M., 1964. Stratigraphy of the continental Mesozoic of the Transbaykal area. *Tr. Limnol. Inst., Akad. Nauk S.S.S.R., Sib. Otd.,* 4: 5–138 (in Russian).

Komarov, Yu. V. and Khrenov, P.M., 1964. On the development type of Mesozoic continental structures in Eastern Asia. In: *Skladchatye oblasti Evrazyi.* Nauka, Moscow, pp. 233–248 (in Russian).

Lefeld, J., 1965. The age of mammal-containing beds at Bayn Dzak, northern Gobi Desert. *Bull. Acad. Pol. Sci., Sér. Géol., Géogr.,* 13(1): 81–83.

Lefeld, J., 1971a. Geology of the Djadokhta Formation at Bayn Dzak (Mongolia). In: *Results of the Polish–Mongolian Palaeontological Expeditions, Part III. Palaeontol. Pol.,* 25: 101–127 (plates 19–21).

Lefeld, J., 1971b. The Gobi Desert of Mongolia. Geographic description and prospects for land use on the basis of soil, vegetation, hydrology, and climate. In: W.G. McGinnies, B.J. Goldman and P. Paylore (Editors), *Food, Fiber and the Arid Lands.* The University of Arizona Press, Tucson, Ariz., pp. 187–210.

Marinov, N.A., 1957. *Stratigraphy of the Mongolian People's Republic.* Izdatelstvo Akad. Nauk S.S.S.R., Moscow, 268 pp. (in Russian).

Marinov, N.A. and Petrovich, Yu. Ya., 1964. On the stratigraphy of the Mesozoic continental deposits of northern Mongolia. *Byul. Mosk. O.-Va. Ispyt. Prir., Otd. Geol.,* 39(6): 46–55 (in Russian).

Marinov, N.A., Zonnenshain, L.P. and Blagonravov, V.A. (Editors), 1973a. *Geology of the Mongolian People's Republic, I. Stratigraphy.* Nedra, Moscow, 583 pp. (in Russian) [1].

Marinov, N.A., Hasin, R.A., Borzakovsky, Yu. A. and Zonnenshain, L.P. (Editors), 1973b. *Geology of the Mongolian People's Republic, II. Magmatism, Metamorphism, Tectonics,* Nedra, Moscow, 751 pp. (in Russian) [1].

Martinsson, G.G., 1955. Lake basins of the geological past of Asia and their fauna. *Priroda,* No. 4: 78–82. (in Russian).

Martinsson, G.G., 1961. *Mesozoic and Cenozoic mollusks of the continental deposits of the Siberian Platform, Transbaykalia and Mongolia.* Izdatelstvo Akad. Nauk S.S.S.R., Moscow, 332 pp. (in Russian).

Martinsson, G.G., 1966. On the comparison of the Mesozoic continental deposits of Mongolia with other regions of Asia. In: *Matieryaly po geologyi Mongolskoy Narodnoy Respubliki.* Nedra, Moscow, pp. 61–64. (in Russian).

Martinsson, G.G., Sochava, A.V. and Barsbold, R., 1969. On the stratigraphic subdivision of the Upper Cretaceous deposits of Mongolia. *Dokl. Akad. Nauk S.S.S.R.,* 189(5): 1081–1085. (in Russian).

Maryańska, T., 1969. Remains of armoured dinosaurs from the uppermost Cretaceous in the Nemegt Basin, Gobi Desert. In: *Results of the Polish–Mongolian Palaeontological Expeditions, II. Palaeontol. Pol.,* 21: 23–34 (plates VI–XVII).

Maryańska, T. and Osmólska, H., 1975. Protoceratopsidae (Dinosauria) of Asia. *Palaeontol. Pol.,* No. 33: 133–182.

Morris, F.K., 1936. Central Asia in Cretaceous Time. *Geol. Soc. Am. Bull.,* 47: 1477–1533.

[1] These two volumes are the most up-to-date compilation works on the Mongolian geology in Russian. Extensive bibliography.

Mushnikov, A.F., Anashkina, K.K. and Oleynikov, B.I., 1966. Stratigraphy of the Jurassic deposits of eastern Transbaykalia. In: *Matieryaly po geologyi i poleznym iskopayemym Chitinskoy oblasti, II.* Nedra, Moscow, pp. 57–99 (in Russian).

Nagibina, M.S., 1963. *Tectonics and Magmatism of the Mongolian–Okhotsk Belt.* Izdatelstvo Akad. Nauk S.S.S.R., Moscow, 464 pp. (in Russian).

Nagibina, M.S., 1969. *Stratigraphy and Formations of the Mongolian–Okhotsk Belt.* Tr. Geol. Inst. Akad. Nauk S.S.S.R., 480 pp. (in Russian).

Nagibina, M.S., 1970. Types of Mesozoic and Cenozoic structures and principles of their development. *Geotektonika,* No. 5: 26–32 (in Russian).

Neiburg, M.F., 1929. Geological investigations in the area of the Batyr–Hairhan range in northwestern Mongolia. In: *Matieryaly Komiss, po issled. Mongolskoy i Tannu-Tuvinsk. Narodnykh Respublik i Buryat-Mongolskoy A.S.S.R., 7.* Izdatelstvo Akad. Nauk S.S.S.R., Moscow, 29 pp. (in Russian).

Osmólska, H. and Roniewicz, E., 1970. Deinocheiridae, a new family of theropod dinosaurs. In: Results of the Polish–Mongolian Palaeontological Expeditions, II. *Palaentol. Pol.,* 21: 5–19 (plates I–V).

Petrovich, Yu. Ya., 1963. Stratigraphic position of the sedimentary-effusive deposits of the Selenga depression in northern Mongolia and some features of their origin. In: *Matieryaly po geologii Mongolskoy Narodnoy Respubliki.* Gostoptekhizdat, Moscow, pp. 60–73 (in Russian).

Rozdestvensky, A.K., 1952. Discovery of an iguanodont in Mongolia. *Dokl. Akad. Nauk S.S.S.R.,* 84(6): 1243–1246 (in Russian).

Rozdestvensky, A.K., 1971. Research on dinosaurs of Mongolia and their role in the subdivision of the continental Mesozoic. In: *Mesozoic and Cenozoic Fauna of Western Mongolia.* Nauka, Moscow, pp. 21–32 (in Russian).

Sheynmann, Yu. M., 1955. Remarks on the classification of the structures of continents. *Izv. Akad. Nauk S.S.S.R., Ser. Geol.,* No. 3: 19–35 (in Russian).

Sheynmann, Yu. M., 1959. Platforms, folded belts and development of the Earth's structure. *Tr. Vses. Nauchn.-Issled. Inst. Geol.,* 49 63 pp. (in Russian).

Shuvalov, V.F., 1969. On the Upper Jurassic red-bed continental deposits of Mongolia. *Dokl. Akad. Nauk S.S.S.R.,* 189(5): 1088–1091 (in Russian).

Shuvalov, V.F., 1970. *Mesozoic Continental Stratigraphy and Mesozoic History of the Geological Development of Central Mongolia.* Dissertation, Moscow, 29 pp. (in Russian).

Sinitzin, V.M., 1962. *Paleogeography of Asia.* Izdatelstvo Akad. Nauk S.S.S.R., 267 pp. Moscow (in Russian).

Smirnov, A.M., 1963. *Joint of the Chinese Platform with the Pacific Ocean Fold Belt.* Izdatelstvo Akad. Nauk S.S.S.R., Moscow, 158 pp. (in Russian).

Sochava, A.V., 1969. Dinosaur eggs from the Upper Cretaceous in the Gobi. *Paleontol. Zh.,* No. 4: 76–88 (in Russian).

Sochava, A.V., 1975. In: *Stratigraphy of Mesozoic Deposits of Mongolia. The Joint Soviet–Mongolian Scientific Research Geological Expedition, Trans.* 13 (in Russian).

Stepanov, P.P. and Volkhonin, V.S., 1969. Present-day structure and deep structure of the Earth's crust of Mongolia according to geophysical data. *Sov. Geol.,* No. 5: 47–73. (in Russian).

Vakhrameev, V.A., 1964. *Jurassic and Early Cretaceous flora of Eurasia and Paleofloristic Provinces of Those Times.* Nauka, Moscow, 263 pp. (in Russian).

Vasilyev, B.I., Ivanov, A.H. and Marinov, N.A., 1959. *Geologic Structure of the Mongolian People's Republic.* Gostoptekhizdat, Moscow (in Russian).

Zonnenshain, L.P., 1970. *Tectonic History of the Central Asiatic Folded Belt.* Dissertation, Moscow, 44 pp. (in Russian).

Zonnenshain, L.P., 1975. Problems of global tectonics. *Bull. Am. Assoc. Pet. Geol.,* 59(1): 124–133.

Zonnenshain, L.P., Kiparisova, L.D. and Okunyeva, T.M., 1971. First find of marine Triassic deposits in Mongolia. *Dokl. Akad. Nauk S.S.S.R.,* 199(1): 164–170 (in Russian).

Chapter 4

JAPAN AND ADJOINING AREAS

TATSURO MATSUMOTO

INTRODUCTION

The area discussed in this chapter is centred on the Japanese islands but includes the Kurile Islands and Sakhalin to the north and the Ryukyu Islands and Taiwan to the south. In terms of the paleogeography of the Mesozoic this division is a little artificial, and reference will be made to geological events in mainland Asia where this is necessary to explain events within the area delimited here.

A major tectonic division of Eastern Asia is shown in Fig. 1, in which P1–P11 are ascribed to the Circum-Pacific Island Arc System with marginal seas to the west. The island arcs themselves were formed in the Cenozoic, but these areas contain Mesozoic and older rocks and the crustal movements were polycyclic. Unfortunately, there is very little information available on the Mesozoic rocks underlying the present marginal seas, the Sea of Okhotsk, the Japan Sea, the East China Sea, and the Philippine Sea. The only exception is the submarine bank in the southern Japan Sea and the northern part of the Philippine Sea.

To describe the Mesozoic geology, it is necessary to provide a tectonic framework by concisely out-

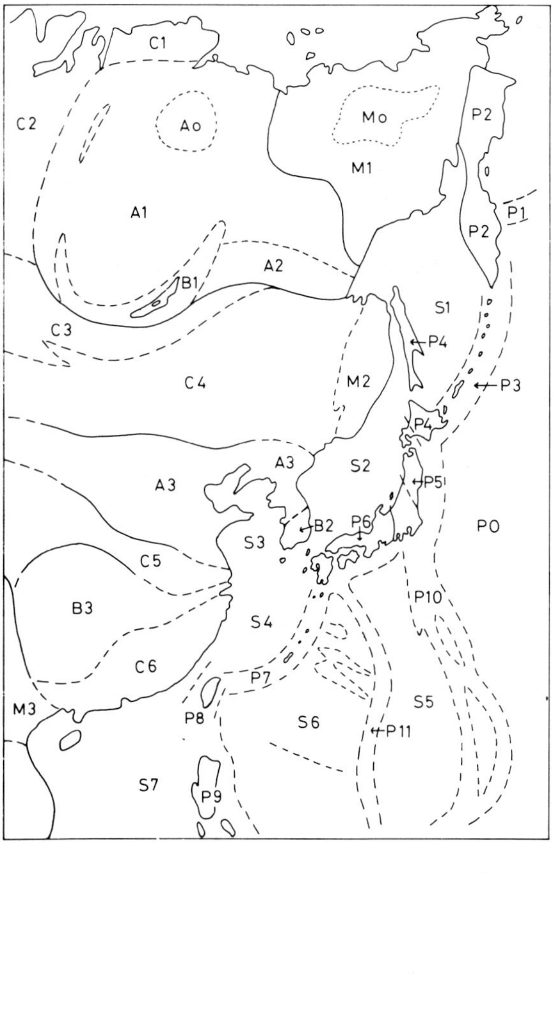

Fig. 1. Tectonic division of Eastern Asia.
Legend: *A* = shield and platform with older and Middle Precambrian basement: *A0* = Anabar Massif, *A1* = Angara or Siberian Platform, *A2* = Ardan Platform, *A3* = North China–Korean Platform; *B* = Late Precambrian orogenic belt; *B1* = Baykalian, *B2* = South Korea, *B3* = South China Platform; *C* = Paleozoic orogenic belt; *C1* = Taimyr belt, *C2* = Western Siberia, *C3* = South Siberian belt, *C4* = Central Asian or Mongolian belt, *C5* = Central China belt, *C6* = Cathaysian belt; *M* = Mesozoic orogenic belt: *M0* = Kolyma Massif, *M1* = Verkhoyansk, etc., fold systems, *M2* = Shikote-Alin fold system, *M3* = Indochina fold system; *P* = circum-Pacific mobile belt: *P1* = Aleutian arc (part), *P2* = Koryak–Kamchatka, *P3* = Kurile arc, *P4* = Hokkaido–Sakhalin, *P5* = northeast Japan, *P6* = southwest Japan, *P7* = Ryukyu arc, *P8* = Taiwan; *P9* = Philippines, *P10* = Mariana arc, *P11* = Kyushu–Palau ridge; *S* = marginal seas: *S1* = Sea of Okhotsk, *S2* = Japan Sea, *S3* = Yellow Sea, *S4* = East China Sea, *S5* = Shikoku Sea, *S6* = Philippine Sea, *S7* = South China Sea; *PO* = Pacific Ocean. (Compiled from various sources.)

TABLE I

Tectonic subdivision of Japan and adjoining areas (subdivision for the Mesozoic tectonics is primarily indicated; z = zone)

I. Kurile arc (including eastern Hokkaido)
- z1 northern zone (= Greater Kurile)
- z2 southern zone (= Lesser Kurile)

II. Yezo-Sakhalin (or Yezo) arc (= Yezo or Hidaka orogenic zone)
- z3 Tokoro–Toyokoro zone
- z4 central Hokkaido
  - a Hidaka subzone
  - b Kamuikotan subzone
  - c Ishikari subzone
  - d Kabato–Rebun subzone

III. Northeast Japan (including southwestern Hokkaido)
- z5 Kitakami zone or massif
  - a northern subzone
  - b southern subzone
- z6 Abukuma zone
- z7 Ashio–Joetsu zone (= extension of z10)
  - a Ashio massif
  - b Joetsu massif
- z8 Kwanto (Kanto) massif (= extension of z12, 13)

IV. Southwest Japan

Inner zone
- z9 Hida zone
- z10 Sangun–Yamaguchi zone
  - a northwestern province (= Chugoku subzone)
  - b Maizuru tectonic zone (narrow ENE belt)
  - c southeastern province (= Mino–Tamba subzone)
- z11 Ryoke zone

Outer zone
- z12 Sambagawa–Chichibu zone
  - a Sambagawa subzone (s.s.)
  - b northern Chichibu subzone
  - c central Chichibu subzone [(including the Yokokurayama complex (Kobayashi, 1941) or Kurosegawa tectonic zone of Ichikawa et al. (1956)]
  - d southern Chichibu subzone (= Sambosan zone)
- z13 Shimanto zone
  - a northern subzone (= Shimanto (s.s.) – Hidakagawa)
  - b southern subzone (= Nakamura – Muro – Setogawa)
- z14 Nankai submarine zone
- z15 Nankai trough

V. Ryukyu arc
- z16 Ainoshima zone
- z17 Nishisonogi zone
- z18 Motobu zone
- z19 Kunigami zone
- z20 Ishigaki zone

VI. Formosa (Taiwan) from Yen (1963), Biq (1974)
- z21 Penghu zone (Taiwan Straits shelf)
- z22 west Taiwan zone
  - a western coastal plain
  - b western foothills
- z23 central Taiwan zone
  - a western or Urai subzone
  - b eastern or Dainano subzone
- z24 Taitung zone
  - a Taitung rift valley
  - b coastal range

N.B.: Indication of better known Late Cenozoic tectonic subdivision of III and V is omitted here.

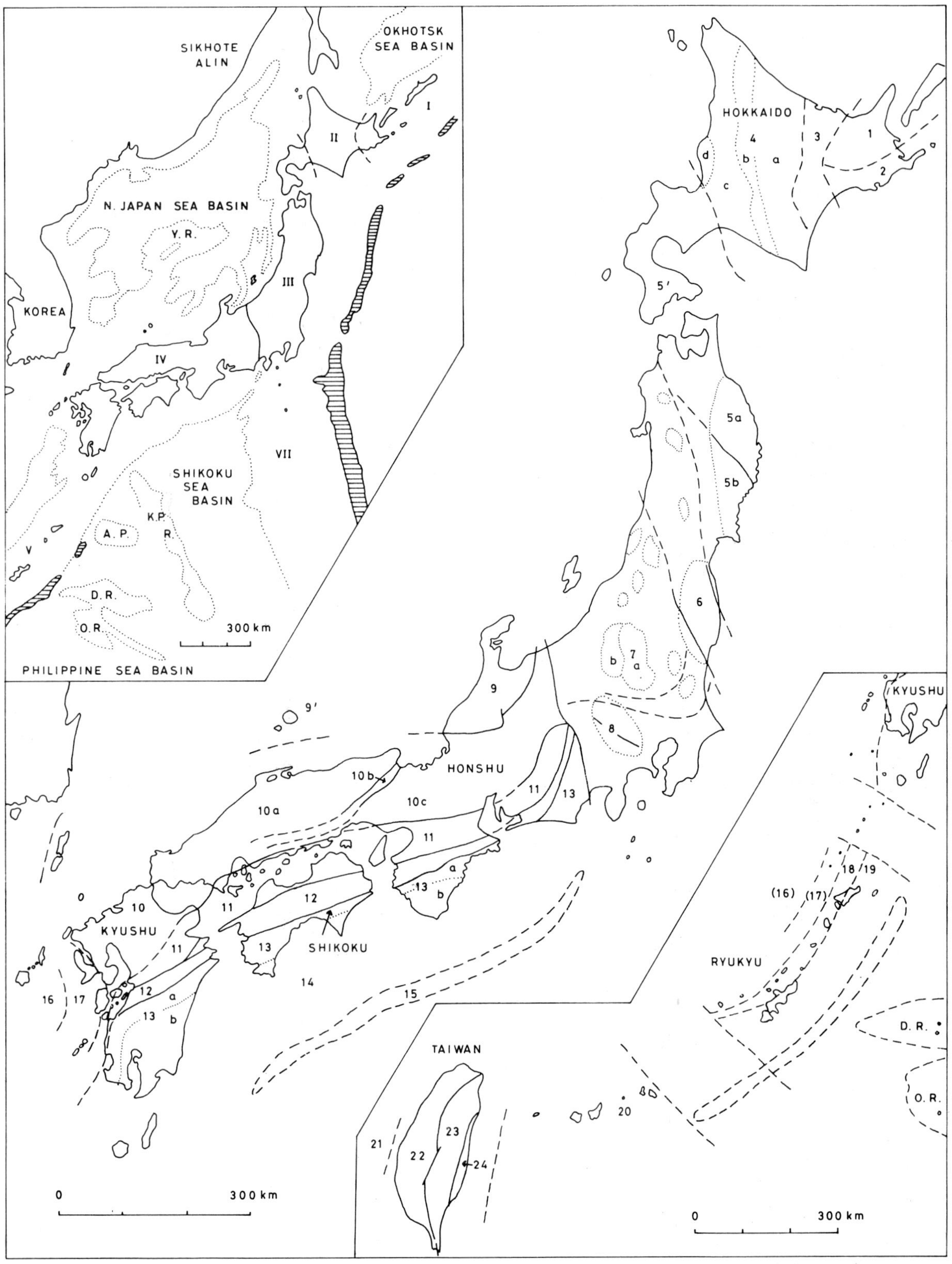

Fig. 2. Tectonic division of Japan and adjacent areas. The division applicable for the Mesozoic geology is indicated. See Table I for the explanation of zones *1–24*. In addition, letters in submarine areas from north to south mean: *Y. R.* = Yamato ridge; *K. P. R.* = Kyushu–Palau ridge; *A. P.* = Amami plateau; *D. R.* = Daito ridge; *O. R.* = Oki–Daito ridge; ruled = trench.

lining the major events which affected the area. Description will be made easier by continual reference to Table I and Fig. 2 in which the principal tectonic units are outlined and numbered. These numbers will be referred to throughout the text. The units are usually demarcated by faults and characterized by their geological constitution. In the geological development of the area four major tectonic cycles can be recognized, of which the oldest, pre-Silurian, is imperfectly known, and the youngest, the late Cenozoic Neotectonic cycle with its development of island arcs, trenches and marginal seas is still active. Of more direct concern is the Late Paleozoic–Early Mesozoic Paleotectonic cycle and the Mesozoic–Early Tertiary Mesotectonic cycle. (See also Ichikawa, 1964; Toshio Kimura, 1967, 1973; Tatsuro Matsumoto and Kimura, 1974; Toshio Kimura et al., 1975; T. Yoshida, 1975; although interpretations may differ to some extent between authors.)

During the Paleotectonic cycle the major part of Japan passed from a eugeosynclinal phase of deposition to a phase of folding, regional metamorphism and granite emplacement. The orogeny corresponds to the Akiyoshi orogeny of Kobayashi (1941) and part of the Honshu movements of Gorai (1955). The orogeny produced a mountain chain on the inner side of Japan (zones 9, part of zones 10a, b) forming the margin of the Asiatic continent, and an offshore insular belt (z12c, z8c, z5b). Parts of the geosyncline persisted as troughs in the central belt (zones 10c, 7, part of zone 10a) between the two belts and received a fill of deltaic sediments near the mountains and bedded cherts, some greenstones and turbidites in the remaining parts during the early half of the Mesozoic. They can be regarded as successor basins of the eugeosynclines, although the central belt was inadequately interpreted as a miogeosyncline by Toshio Kimura et al. (1975).

The off-shore insular belt showed characters of a narrow labile shelf during the Mesozoic and on its outer side new geosynclines developed, the Sambosan–Shimanto eugeosyncline (zones 12d, 8d, 13, 5a) and probably the Yezo (= Hidaka) geosyncline of Hokkaido and Sakhalin (zones 3, 4). The story of the Mesotectonic cycle comprises the history of this outermost belt and also various events in other belts. In Late Mesozoic times folding and thrusting occurred with the development of paired metamorphic belts (Ryoke metamorphics with granites on the inner side and Sambagawa with ultrabasics on the outer side). At the same time Late Cretaceous to Early Tertiary granite intrusions and andesitic to rhyolitic volcanism occurred within the inner zone of southwest Japan (zones 11, 10, 9, 7) and also in the northeast (zones 6, 5). A further phase of tectonic activity took place in Middle Tertiary times. Much of this overall history is simplified in Fig. 3.

The best record of the Neotectonic events is in northeastern Japan and the Kurile and Ryukyu islands and Taiwan, whereas the evidence of the Paleo- and Mesotectonic cycles is better recorded in southwestern Japan and certain restricted areas in northeastern Japan. Excluding the inevitable overlap in time and space of tectonic activity (Tatsuro Matsumoto, 1967c), there is a general outward migration of activity from the Paleotectonic to Mesotectonic cycles.

The Mesozoic history of the area essentially covers the later part of the Paleotectonic cycle and the main part of the Mesotectonic cycle. The structures of the Mesotectonic cycle are superimposed upon, but often cut obliquely across, the older trends. The Mesozoic as well as the Paleozoic rocks are sometimes concealed by Cenozoic sediments or volcanics; in other locations they have been removed by erosion. The complications introduced by structural complexities and regional metamorphism have delayed interpretation of the history. Despite these handicaps, Japan, the region studied in most detail, provides the basis for the study of the region as a whole.

TRIASSIC

General outline

The end of the Paleozoic–beginning of the Mesozoic marked a period of considerable crustal unrest in Eastern Asia. At various times during the Permian and Triassic important movements affected various regions, for example the Akiyoshi orogeny (Kobayashi, 1941) in Japan, the Indochina movements in Vietnam and southern China, evidence of movement in northern Korea and in western Sikhote-Alin. These movements profoundly affected the paleogeography (Fig. 4), with the reduction of the previously extensive seaways in Asia. Continuous marine sequences in the Triassic are restricted to three regions, the geosynclines of northeastern Siberia (Verkhoyansk, Amur and Sikhote-Alin), some parts of Japan, and

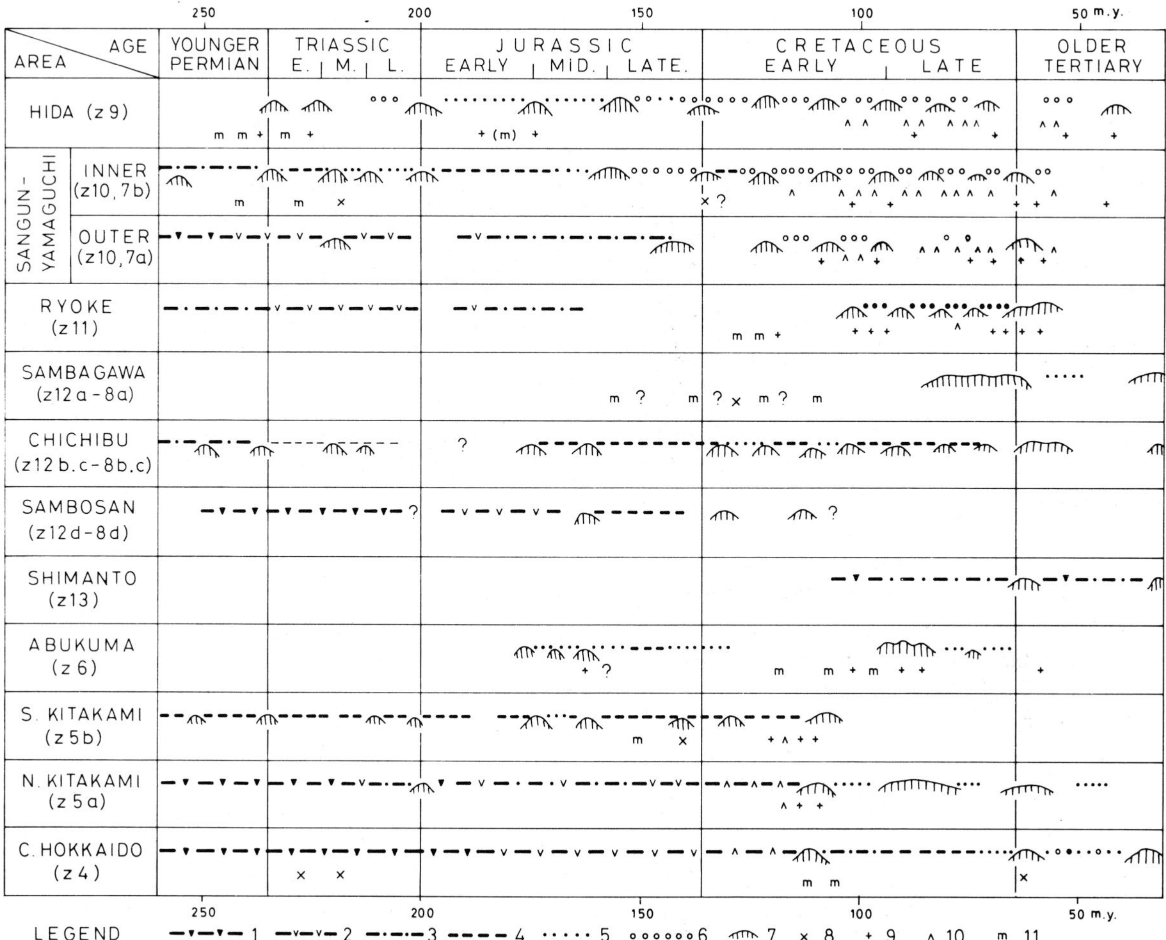

Fig. 3. Schematic diagram showing major Mesozoic events in Japan.
Legend: *1* = geosynclinal sedimentation with submarine basalt; *2* = ditto, with chert and flysch; *3* = ditto, flysch; *4* = predominantly marine muddy sedimentation of shallow to moderate depth (fossiliferous); *5* = predominantly shallow-sea sandy sediments with some conglomerate; *6* = non-marine sedimentation; *7* = tectonic phase; *8* = basic to ultrabasic intrusion; *9* = granitic intrusion; *10* = volcanism (andesitic to rhyolitic); *11* = regional metamorphism.

the eastern Tethys extending from the Himalaya through Malaya to Indonesia. The regression of the epicontinental seas over the Yangtze Basin in southern China during the Late Triassic left a wide continental area over most of China and Korea. Similarly central Siberia was also a land area, the two together forming a continent of Eastern Asia.

Japan lay at or near the eastern margin of that continental area partly covered by the marginal seas of the then Pacific Ocean, the Panthalassa of some authors. Despite the narrowness of the area a considerable variety of facies and rock types have been recognized (Fig. 5):

(1) Paralic and/or deltaic to neritic facies distributed in basins, embayments and shallow seas in front of, or between the rising mountains on the inner side of Japan (zones 10a, b).

(2) Geosynclinal chert and turbidite facies in the central belt (zone 10c, part of zones 10a, 11, 7a).

(3) Shallow, open marine facies in the outer insular belt (zones 12b, c, 8b, c, 5b).

(4) Offshore marine, eugeosynclinal facies characterized by basic submarine volcanics, chert and micritic limestone on the outermost belt (zones 12d, 18, 8d, 5a and 4).

In addition to the sedimentary facies mentioned above, some of the granitic and metamorphic rocks in zone 9 (Hida) and Sangun metamorphics and associ-

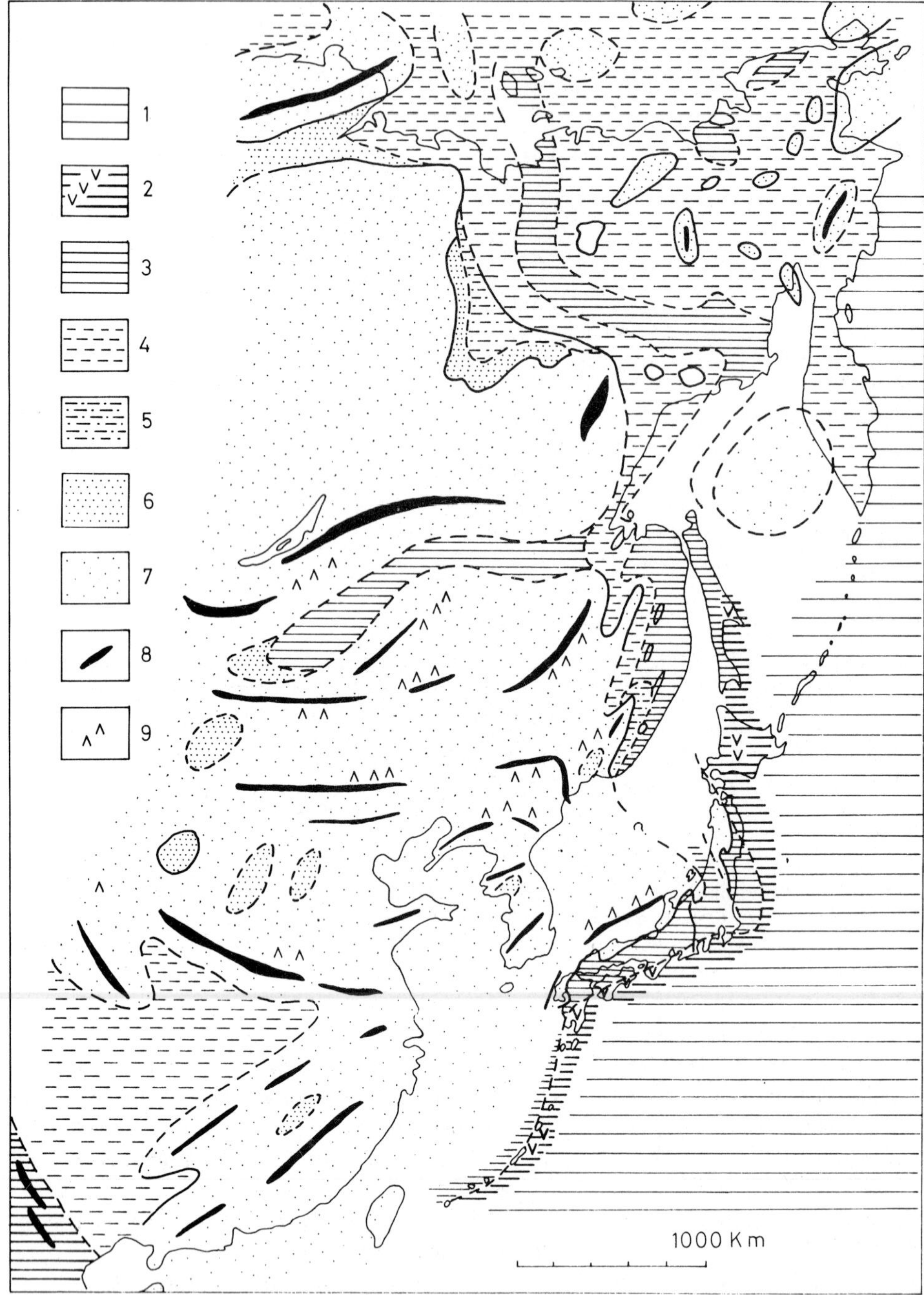

Fig. 4. Tentative paleogeographic map for the Late Triassic (Carnian) of Eastern Asia.
Legend: *1* = ocean; *2* = eugeosyncline with submarine volcanism; *3* = geosyncline with little or no volcanism; *4* = epicontinental shallow sea; *5* = paralic area; *6* = continental basin; *7* = continental area; *8* = mountain range; *9* = volcanoes on land. (Compiled from various sources.)

ated ultrabasic rocks in zone 10 have records of Permo-Triassic events.

Biostratigraphy, biogeography and paleoclimate

Biogeographically Japan is fortunately placed, for the four recognizable biofacies – non-marine, near-shore shallow marine, off-shore pelagic and deeper marine – were in contact, while the faunae and florae show affinities with those of both the northern Pacific (or Panthalassa) and the Tethys regions. While there is no continuous marine sequence, local sequences (Inai-Rifu-Saragai in zone 5b, Yakuno-Nabae in zone 10b, Nariwa and Mine-Atsu in zone 10a, Nakijin in zone 18, Kamura-Taho-Zohoin-Kochigatani in zones 12b, c, 8b, c) can be compiled into a comprehensive succession which may be correlated with that in North America, the Alps, northeastern Siberia and other regions. Only the marine Rhaetian has not been identified. The Permo-Triassic boundary problem is discussed in detail in papers by Kanmera and Nakazawa (1973) and Nakazawa et al. (1975). Triassic fossil localities are shown in Fig. 5 and Table II and the species which characterize the successive stages or substages are listed in Table III. See also correlation chart of Fig. 6.

The most recent works on ammonoid faunas and their correlation are those of Bando (1964a, b, 1966, 1967a, b, 1970), Bando and Shimoyama (1974) and Ishibashi (1970–75). Bivalves have been studied by Ichikawa (1953–54, 1958), Kambe (1951, 1957, 1963), Kobayashi and Aoti (1943), Kobayashi and Ichikawa (1949a–c, 1950), Kobayashi and Ishibashi (1970), Kobayashi and Katayama (1938), Kobayashi and Tokuyama (1959), Nakano (1957a), Nakazawa (1952–56, 1959, 1960, 1961b, 1963, 1964a, b, 1971), Tamura (1959a, b, 1965, 1972), Tokuyama (1959a, b, 1960a, b).

The molluscan faunas contain elements in common with the Arctic province (especially eastern Siberia) and others with Alpine–Himalayan affinities. There are endemic elements found in the embayments and also cosmopolitan forms.

The paralic facies on the inner side interfinger with marine beds. The plant fossils from Nariwa and Miné-Asa monographed by Oishi (1932, 1940) were correlated with the Rhaeto-Liassic of Greenland, Bornholm and Scania, however, as Kobayashi (1938, 1942) has shown, the plant beds are overlain by Upper Norian *Monotis ochotica* beds in the Nariwa area, while in the Miné-Atsu area the coal-seam and plant-bed sequence interfingers with marine Carnian, so that the fossil plants must be regarded as Carnian and Norian. Similar, if not identical stratigraphical situations are found in Ussuriysk (Maritime Provinces of eastern Siberia) and Tongking (Vietnam). Certain aspects of the evolution of the Late Paleozoic–Early Mesozoic floras were discussed by Kon'no (1944, 1949, 1961, 1962a, b, 1968), Kon'no and Naito (1960), and Tatsuaki Kimura and Asama (1975) and Kon'no (1973) added new Lower Triassic species.

An insect fauna was described by Fujiyama (1973) from the Carnian of the Miné area (zone 10a). It shows considerable affinities with the Ipswich fauna of Australia. The Lower Carnian estherian (conchostracan) fauna of southwestern Japan (zone 10a) was assigned by Kobayashi (1951, 1952) to the Taedong (Daido = Taidong) fauna which ranges up to Early Jurassic before being replaced by the Middle to Late Jurassic Jehol fauna. In a review Kobayashi (1954b) regarded the Taedong fauna as "a synorogenic fauna in the pericontinental zone", while the Jehol fauna is "an interorogenic fauna in the intercontinental terrain", thus attempting to relate the effects of orogenic movements on terrestrial life.

A number of marine faunas and floras have not been studied in detail, for example, there are few contributions to Triassic limestone biota (see however Kanmera, 1964 and Endo and Horiguchi, 1967 on corals and algae of zone 12d), and work on the radiolaria (Toshio Kimura, 1944b, c) and conodonts is in progress. Conodonts have recently been found widely distributed in ammonitic, micritic limestones, chert and siliceous shale. Descriptions are given by Nogami (1968), Hayashi (1968), Koike et al. (1970, 1971) and Igo (1972). Already their value has been demonstrated by the discovery that certain limestones assigned to the Late Jurassic (Torinosu type limestone) and certain other cherts assigned to the Permian have turned out to be of Middle to Late Triassic age.

Kobayashi (1942) suggested that the Late Triassic flora of the inner side of southwestern Japan, characterized by abundant Dipteridaceae (*Dictyophyllum,* etc) indicated high temperature and high humidity. The study of the insect fauna led Fujiyama (1973) to the same conclusion. The presence of workable coal and the absence of evaporites seems also to point to a humid climate. Red beds are rare in the Japanese Permo-Triassic (except for minor occur-

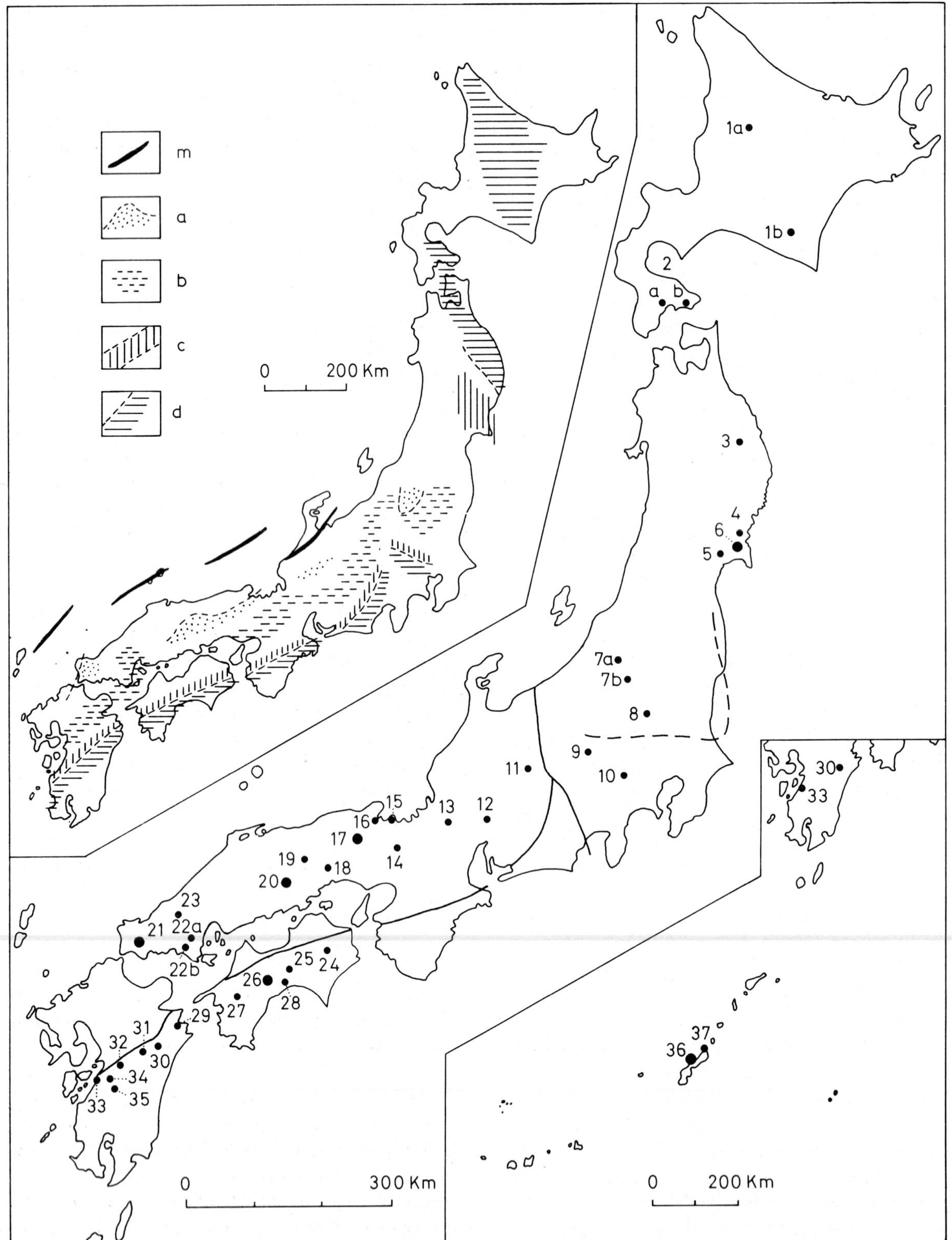

Fig. 5. Map of Japan showing the locations of Triassic fossiliferous formations. See Table II for the explanation of locality numbers. A larger solid circle indicates a location of a well-known formation where numerous localities are included. Inset at the upper left corner is an interpretation of sedimentary framework of the Triassic system in Japan.

Legend: *m* = mountain range; *a* = deltaic to shallow-bay facies; *b* = flysch facies of sandstone–shale–chert assemblage; *c* = shallow open-sea facies; *d* = eugeosynclinal facies of submarine basalt, chert and micritic limestone assemblage.

TABLE II

List of localities of Triassic fossils (see Fig. 5 for location)

The tectonic province to which the locality belongs is indicated by a numbering in square brackets (see Table I and Fig. 2). The indicated place names are mostly used as local formational names. Symbols of the fossils are: A = ammonoids; C = conodonts; M = mollusca, other than ammonoids; O = other invertebrates and marine plants; P = land plants; V = vertebrates.

0. Little Ehabi, east coast (53° 30′ N), Sakhalin) (outside Fig. 5) [z4] M
1. Hip (Hippu) near Asahikawa (*a*) and Motourakawa (*b*), central Hokkaido [z4] C, O
2. Kamiiso, southwestern Hokkaido [z5a] C
3. Kayamori, northern Kitakami [z5a] C
4. Saragai, southern Kitakami [z5c] A, M
5. Rifu, northeast of Sendai [z5c] A, M
6. Inai, including Hiraiso (*a*), Osawa (*b*), Fukoshi (*c*) and Isatomae (*d*), southern Kitakami [z5c] A, M, O, P, V
7. Kuromata-gawa (*a*) C, and Okutone (*b*) M, Joetsu [z7b]
8. Adoyama near Kuzuu, Ashio [z7a] C
9. Shionosawa, Kwanto [z8b] M
10. Iwai (*a*), Ohme (*b*), etc., Kwanto [z8c] A, M, C
11. Yabuhara and Misogawa, Kiso [z11] C
12. Inuyama (*a*), Tanikumi (*b*) Shichiso (*c*), Kami-aso (*d*) and Mino-suhara (*e*), north of Nagoya [z10c] C
13. Myogatani, Mino [z10c] M
14. Nishiyama (*a*), Shuzan (*b*) and Izuriha (*c*), west of Kyoto and in Tamba [z10c] M, C
15. Nabae (*a*) and Arakura (*b*), near Maizuru [z10b] A, M, O, P
16. Shitaka, near Maizuru [z10b] M, P
17. Yakuno, Fukumoto, etc., in the Maizuru belt [z10b] A, M, O
18. Sayo (*a*) and Nakanotani (*b*), Hyogo Prefecture [z10c] M
19. Tsuyama, north of Okayama [z10a] M
20. Nariwa (*a*) and Kyowa (*b*), northwest of Okayama [z10a] M, P
21. Mine (*a*), Asa (*b*) and Atsu (*c*), west of Yamaguchi [z10a] M, O, P
22. Mukaihata (*a*) M, and Kuga (*b*) C, east of Yamaguchi [z11]
23. Kanoashi, northwest of Yamaguchi [10a] C
24. Nagayasu, etc., eastern Shikoku [z12c] A, M
25. Kurotaki, northeast of Kochi, central Shikoku [z12b] A, M, O
26. Kochigatani (*a*) and Zohoin (*b*) near Sakawa, central Shikoku [z12b] A, M
27. Taho (*a*), Nomura (*b*) and Itadorigawa (*c*), western Shikoku [z12b–c] A, C, M, O
28. Sambosan near Kochi, Shikoku [z12d] M, O
29. Gobangadake near Tsukumi, southeastern Kyushu [z12c] M
30. Kamura near Takachiho, southeastern Kyushu [z12c] A, C, M
31. Tonegoyama near Kuraoka, Kyushu [z12c] M
32. Miyamadani, east of Yatsushiro, southwestern Kyushu [z12b] M
33. Tanoura (*a*) and Takagochi (*b*), south of Yatsushiro, Kyushu [z12c] A, M, O
34. Misaka, southeast of Yatsushiro, Kyushu [z12c] M
35. Konose, southeast of Yatsushiro, Kyushu [z12d] A, C, M, O
36. Nakijin, Motobu, Okinawa [z18] A, C, M
37. Hedo-misaki, Okinawa [z18] A

Other localities of Triassic conodonts of insufficient information are omitted here.

rence in zone 5b and that of doubtful age in zone 9). The common occurrence of conglomerates and coarse-grained sandstone suggested a heavy rainfall, active erosion and appreciable topographic relief. This is in considerable contrast to the abundant distribution of Permo-Triassic red beds and evaporites in other parts of the world (see Meyerhoff, 1970a, b; Waugh, 1973).

On the outer side of Japan, where little clastic material was supplied, two types of limestone were formed, a reef facies described by Kanmera (1964) with forms distributed in the Tethys–Indonesian region and western North America, and a more extensive micritic limestone containing planktonic microorganisms and thin-shelled bivalves. According to Kanmera (1969) the limestone is similar to the Jurassic and Cretaceous limestones of the Tethys, except in age. This is taken to suggest the occurrence of a warm current off the coast during Triassic times.

This is not in good agreement with paleomagnetic

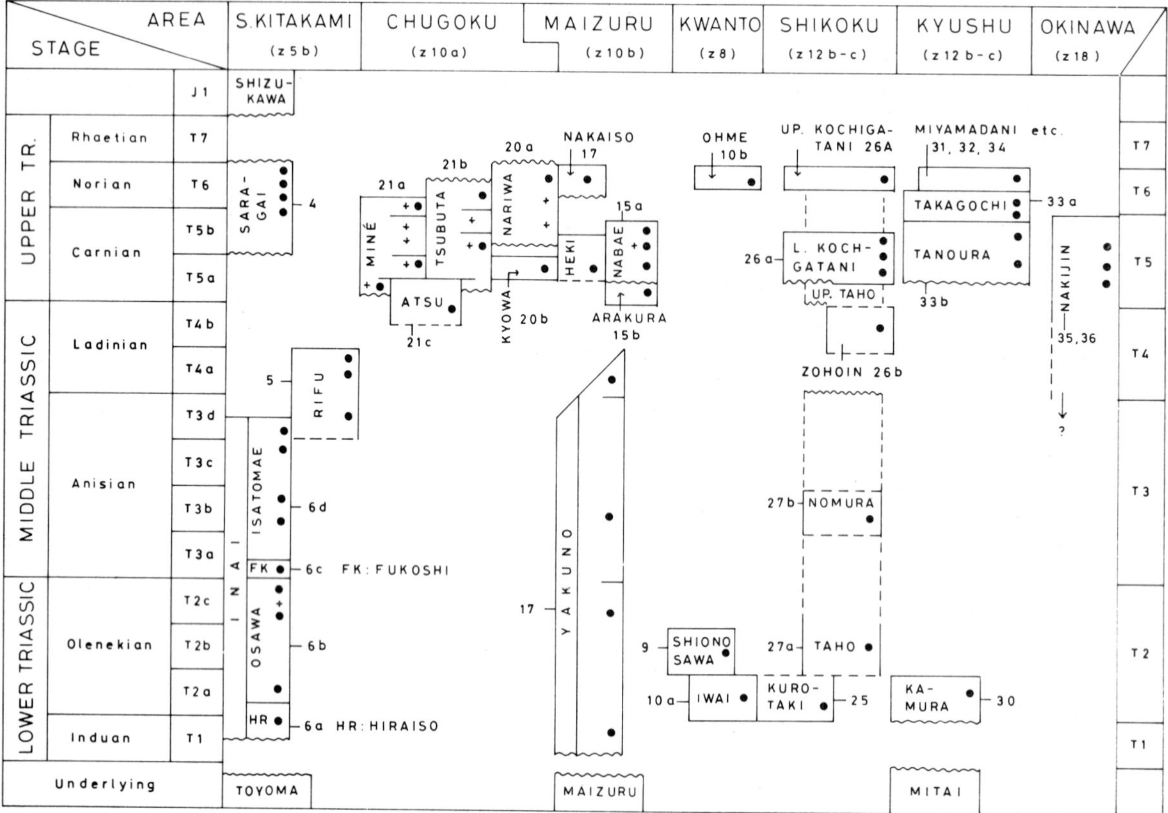

Fig. 6. Correlation of representative Triassic formations in Japan. See Table III for units T1–T7 of a biostratigraphic division, Fig. 5 and Table II for the numbered locations of indicated formations, and Fig. 2 for the tectonic units (z5b, etc.). Solid circle = stratigraphic position of index species (megafossils); cross = that of plant beds. Conodont-bearing beds are mostly excluded from this chart, except for a few (e.g. the Taho and the Nakijin).

reconstructions (Dietz and Holden, 1970; Nairn and Peterson 1973), according to which Siberia was in polar latitudes with Japan in somewhat lower latitudes. The alternative is that Arctic regions were warm during the Triassic.

Regional features

In the following section an attempt will be made to outline the general features of the rock assemblages in the various paleotectonic zones outlined above. The principal locations are listed in Table II and shown in Fig. 5 and will be referred to in the discussion. There is in each province a considerable variation in sequence, facies and rock assemblage. Local details are omitted in favour of the main features. Two kinds of correlation charts (Figs. 6 and 7) and a diagrammatic map in Fig. 5 may be helpful to understand the description.

(1) Inner belt

(a) *Possible Triassic of the Hida zone* (z9). The Hida metamorphic rocks, quartz-feldspathic gneiss, amphibolite, siliceous–calcareous gneiss and crystalline limestone, are associated with the Funatsu granitic rocks. The rocks belong to the low-pressure amphibolite facies (M. Hashimoto et al., 1970), but there is evidence of polymetamorphism. The age of the metamorphism is in some doubt but of the available radiometric ages many K–Ar determinations group around 170–190 m.y. (Kawano and Ueda, 1966). Rb–Sr whole-rock and mineral-isochron ages indicate that metamorphism and plutonic activity took place during the latest Paleozoic–Early Mesozoic. K. Shibata et al. (1970) distinguish an early event between 250 and 220 m.y. (Late Permian–Early Triassic) and a later event between 190 and 170 m.y. (Early Jurassic). The sphene and zircon ages U–Th–Pb of Ishizaka and Yamaguchi (1969) fall

TABLE III

Triassic biostratigraphic succession in Japan

Unit of division indicated with symbolized letters (with approximately correlated international stage name in parentheses): characteristic species; followed by numbers which refer to the formations or locations indicated in Table II and Fig. 5.

Lower Triassic ("Scythian" = Induan and Olenekian)	
T1	(Induan) (not yet well sorted; "Gyronitian" not yet recognized): *Glyptophiceras japonicum, G.* cf. *gracile, Claraia* aff. *decidens, Entolium discites, Eumorphotis iwanoi, Neoschizodus* aff. *ovatus, Bakevellia* cf. *exporrecta*; *6a, 17, 18b, 29*(?)
T2a	(Lower Olenekian): *Owenites shimizui, Dieneroceras iwaiense, Meekoceras* cf. *gracilitatus, Aspenites kamurensis, Clypites japonicus, Eumorphotis multiformis*; *6b, 10a, 25*(?), *30*
T2b	[Middle (?) Olenekian]: *Anasibirites multiplicatus, A. kingianus, Meekoceras japonicum, Hemiprionites katoi, H. tahoensis, Arctoprionites yeharai, Eumorphotis multiformis*; *27a, 9*(?) (T2a and T2b are represented by separate formations; unmistakable Middle Olenekian is not yet clearly recognized)
T2c	(Upper Olenekian): *Columbites parisianus, Subcolumbites perrinismithi, Arnautoceltites* sp., *Leiophyllites* cf. *pitamaha, Prenkites* cf. *timorensis*; *6b*
Middle Triassic (Anisian and Ladinian)	
T3a	(Lower Anisian): *Leiophyllites* cf. *pseudopradyumna, Gymnites watanabei*; *6c*
T3b	(Middle or main part of Anisian): *Hollandites japonicus, Balatonites gottschei, B. kitakamicus, Danubites naumanni, Cuccoceras* aff. *submarinoii, Ussurites yabei*; *6d, 17, 27b*
T3c	(Middle or Upper Anisian): *Hollandites haradai, Sturia japonica, Paraceratites* sp.; *6d*
T3d	(Upper Anisian): *Paraceratites* cf. *trinodosus, P. orientalis, Discoptychites compressus*; *5*
T4a	(Lower Ladinian): *Protrachyceras reitzei, Ptychites miyagiensis, P. japonicus, Flexoptychites matsushimaensis, Tropigastrites* aff. *halli, Epigymnites* aff. *jollyanus, Anagymnites* aff. *acutus, Daonella multistriata*; *5, 24*(?)
T4b	(Upper Ladinian): *Protrachyceras* aff. *archelaus, Daonella kotoi*; *24, 26b*
Upper Triassic (Carnian, Norian and Rhaetian)	
T5a	(Lower Carnian): *Mojvarites* (?) *arakurensis, Sirenites* cf. *nanseni, Trachyceras* (*Paratrachyceras*) sp., *Hannaoceras nusturtium, Halobia styriaca*; *15b, 36*
T5b1	(Upper Carnian): *Sandlingites* aff. *oribasus, Clydonites* aff. *daubreei, Thisbites nakijinensis, Tropiceltites* sp., *Discotropites* cf. *laurae, Paratropites* aff. *hoetzendorfi*; *36*
T5b2	(Upper Carnian): *Juvavites* cf. *kellyi, Anatomites* cf. *toulai, Hannaoceras henseli, Jovites* cf. *dacus, Discotropites quinquepunctatus, Hoplotropites* cf. *arionsis, Arietoceltites arietitoides, Tropiceltites* cf. *columbianus, Arcestes* (*Proarcestes*) aff. *carpenteri, A.* (*P.*) cf. *ausseeanus, Hypocladiscites subaratus, Placites* aff. *oldhami, Buchites kumamotoensis*; *33a, 36*
T5	(Carnian) (for near-shore shallow-sea facies): *Halobia subsedaka, H. kawadai, H. aotii, H. obsoleta, Oxytoma "zitteli", O. pulchra, Chlamys mojsisovicsi, Tosapecten suzukii, Eumorphotis* (*Asoella*) *confertoradiata, Minetrigonia hegiensis, M. katayamai, Palaeopharus maizuruensis*; *13*(?), *15a, 20b, 21, 26a, 33*
T6	(Norian): *Monotis ochotica, Monotis typica, Stenarcestes* aff. *oligosarcus, Arcestes* sp., *Placites* aff. *oxyphyllus*; *4, 7, 10b, 13, 14b, 18a, 19, 20a, 21b, 22, 26a, 27c, 31, 33b, 34, 35* (T6 is subdivisible into four subzones, *Monotis typica, M. ochotica densistriata, M. ochotica ochotica, M. zabaikalica* in ascending order; the ammonite species listed above came from the first subzone, indicating Lower Norian)
T7	(Rhaetian): index species not yet found; absent in most places

within the limits of the early event. The Funatsu granites are regarded as late-kinematic intrusions in the Hida belt and their low initial $^{87}Sr/^{86}Sr$ (0.7056) supports this (K. Shibata et al. 1970).

The red beds at one location in this area (Motodo) were regarded, without fossil evidence, as Triassic by some authors, but they could be Cretaceous.

(b) *Triassic on the inner side* (part of zones 10a, b, 7b). Comprehensive descriptions of the stratigraphy are given by Nakazawa (1958) and Tokuyama (1961a, b, 1962), the latter attempting to relate the sedimentation to the orogenic phases.

As a result of the Late Permian–Early Triassic movements part of the geosyncline was converted to a deltaic–paralic–neritic environment of deposition during the Triassic. The structural environment was that of intermontane basins or embayments surrounded by high relief. Sediments in these basins rest with marked unconformity upon folded Paleozoic rocks. There are marked unconformities under the

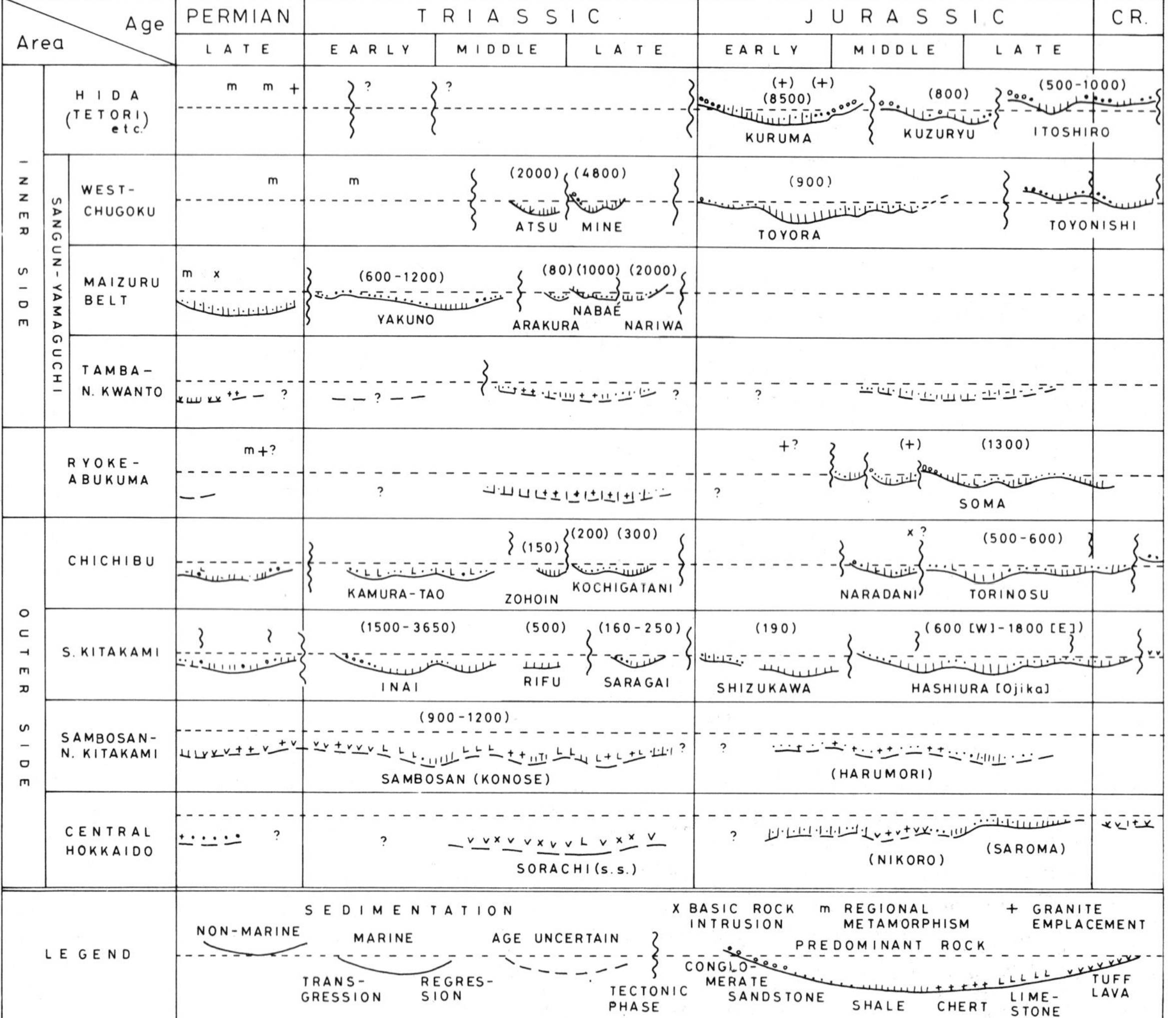

Fig. 7. Schematic diagram showing the Triassic and Jurassic geological history in Japan. The figure in parentheses means the approximate thickness (in meters) of the indicated formation or group. The diagram is not in absolute time scale.

Scythian–Anisian sequence which has a restricted occurrence in the Maizuru zone, and below the much more widely distributed Carnian and Norian deposits. The rocks of the former sequence are assigned to the Yakuno Group (loc. *17* of Fig. 5), while several local formational names, e.g. Miné and Nariwa, are assigned to the Upper Triassic sequences (loc. *15a, b, 19, 20a, b, 21a–c*).

Thick conglomeratic and coarse sandstones are common, but despite their thickness the areal distribution is, and was, probably small. In the Miné Group (loc. *21a*), 4,800 m of Carnian sediments fill a basin of about 20 X 15 km. Yet despite this restriction the facies change, both laterally and vertically, is remarkable. The fossil content changes with lithofacies change (see Fig. 8) and some charges are also seen in the petrographic character of the sandstones. Sediments range from coal measures, through deltaic to shallow-water marine.

The source material of the clastics is polygenetic though the main source in the Maizuru zone is believed to have been to the northwest. However, apart from quartz and feldspar the Miné Group received volcanic rocks from the north and chert and others from the south.

(c) *Sangun metamorphics and associated basic and ultrabasic rocks* (zone 10). These rocks are best exposed in northern Kyushu and Chugoku where

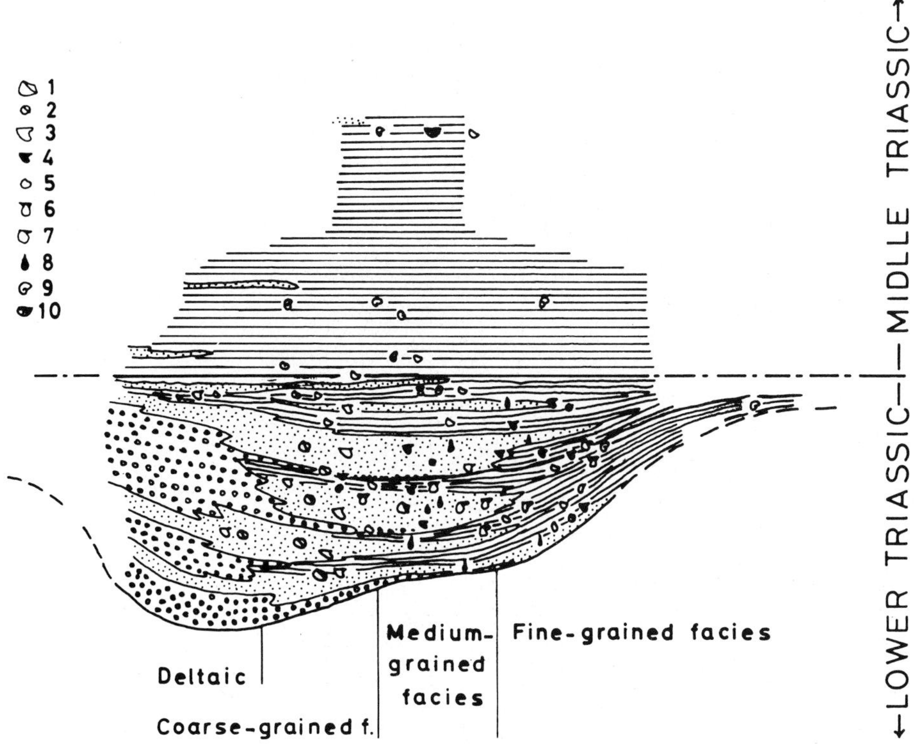

Fig. 8. Schematic profile of the Triassic of the Maizuru zone, showing the bio- and lithofacies relation.
Legend: *1* = *Neoschizodus* cf. *laevigatus* (large form); *2* = *N.* cf. *laevigatus* (small form); *3* = *Bakevellia* (*Maizuria*) *kambei*; *4* = *B.* (*M.*) *okuyamensis*; *5* = *Nuculana* and *Palaeoneilo*; *6* = *"Pecten" ussuricus*; *7* = *Claraia*; *8* = *Lingula* cf. *borealis*; *9* = *cephalopods*; *10* = *Daonella* (?) sp. (After Nakazawa, 1958.)

they are unconformably overlain by Upper Triassic beds. The rocks resembling the Sangun are found in the peri-Hida belt and parts of Zones 10c, 7b and further south in zone 20. The rocks of the typical Sangun belong to the pumpellyite–actinolite and epidote–glaucophane facies (M. Hashimoto et al., 1970), but those in the peri-Hida are of epidote amphibolite facies (Banno, 1958). The age of the metamorphism is debated. It is regarded as Permo-Triassic by Kobayashi (1941) or as older (presumably Late Carboniferous ?) but affected by later (i.e., Permo-Triassic or younger) low-grade burial metamorphism of the prehnite–pumpellyite facies by Nureki (1969) and Kuroda (1970).

The Sangun belt in a broad sense was regarded as forming a pair with the Hida metamorphic belt (Miyashiro, 1961, 1973), however it is not certain whether all the Sangun or Sangun-like metamorphics in the belt are of the same age. Some schists in zone 10a show K–Ar ages of 169 and 175 m.y. on muscovite (K. Shibata and Igi, 1969; K. Shibata et al., 1972), but other samples of metagabbro in zones 10a and 10b show ages of 248 ± 26, 297 ± 15 and 383 ± 14 m.y. on hornblende (Igi and Shibata, 1975, oral comm.). The schists at Omi in the peri-Hida belt give K–Ar ages of 309 and 324 m.y. (K. Shibata and Nozawa, 1968) and 350 ± 28 m.y. Rb–Sr age (K. Shibata et al., 1970) on micas. No radiometric age has yet been given on the Sangun-like rocks in zone 10c. Some schists in zone 20 gives a K–Ar age of 174 m.y. on muscovite.

(2) Central belt

The central belt comprising zones 10c, 11 and part of zone 7 contains a thick, intensely folded geosynclinal assemblage. Originally referred to the Upper Paleozoic it has proved to range from Permian to Upper Triassic and even in some regions to Jurassic on the basis of recently discovered fossils. Fossils however are rare, and this, together with structural and sedimentological complexities and discontinuous outcrop, means that much remains to be done.

Basic submarine effusives commonly occur in the lower part of the sequence where Permian (mostly

Lower to Middle Permian) fusulinids are found in limestones, but they occasionally occur at higher levels. The middle part of the sequence commonly consists of mudstone and sandstone with thick beds of chert, while turbidites and massive sandstones predominate in the upper part. Conodonts in the chert suggest a Middle to Late Triassic age for the series. There is little evidence for Early Triassic and at one locality of zone 7a, a distinct unconformity is seen between conodont bearing cherts (M–L. Triassic) and fusulinid (M. Permian) limestone (Yanagimoto, 1973).

Some part of the chert–turbidite series (loc. *12d*) contains cobbles and boulders of gneisses, some of which being dated as Middle Precambrian (Adachi, 1972; K. Shibata and Adachi, 1972). Olistostromes seem to be included in other parts.

Another group of intensely folded strata in z10a (loc. *23*) not far from the Miné area (loc. *21*) has also proved to range up to Triassic on the evidence of conodonts from chert. This was previously assigned to Permian and Carboniferous on some fusulinids from limestones, which may be large reworked sedimentary blocks or tectonically exotic masses (F. Toyohara, 1976, oral comm.). The relationship of these much disturbed beds with the gently folded shallow water beds of the Miné and Maizuru-Yakuno areas is still to be worked out.

(3) Outer belt

Triassic beds of shallow open-sea facies are distributed in the outer belt (comprising most of zones 12b, c, 8b, c and 5b). Although exposures are small and scattered, the succession is well known both stratigraphically and paleontologically. Lithologically the Triassic of this belt consists mainly of shale and sandy shale with subordinate sandstone and conglomerate and occasional calcareous shales and limestones. The fauna characteristically indicates shallow open marine environments. The sandstones are petrographically immature wacke ranging to mature quartzose arenite (Tokuyama, 1961b). Individual formations may be fairly thick such as the Scythian–Anisian ammonitic Inai Group, others are thinner and of shorter time-range, often with a basal unconformity, micaceous shales rich in *Daonella, Halobia* or *Monotis,* and lenticular reef-like limestones. Other limestones with ammonites and radiolaria suggest relatively off-shore facies merging to those of the outermost belt (T. Koike et al., 1974, oral comm.).

These formations probably represent deposits on a labile shelf around rising islands, i.e. in part an island shelf facies and in part miogeosynclinal facies bordering an eugeosynclinal trough on the oceanward side.

(4) Outermost belt

Triassic rocks of the off-shore geosynclinal facies

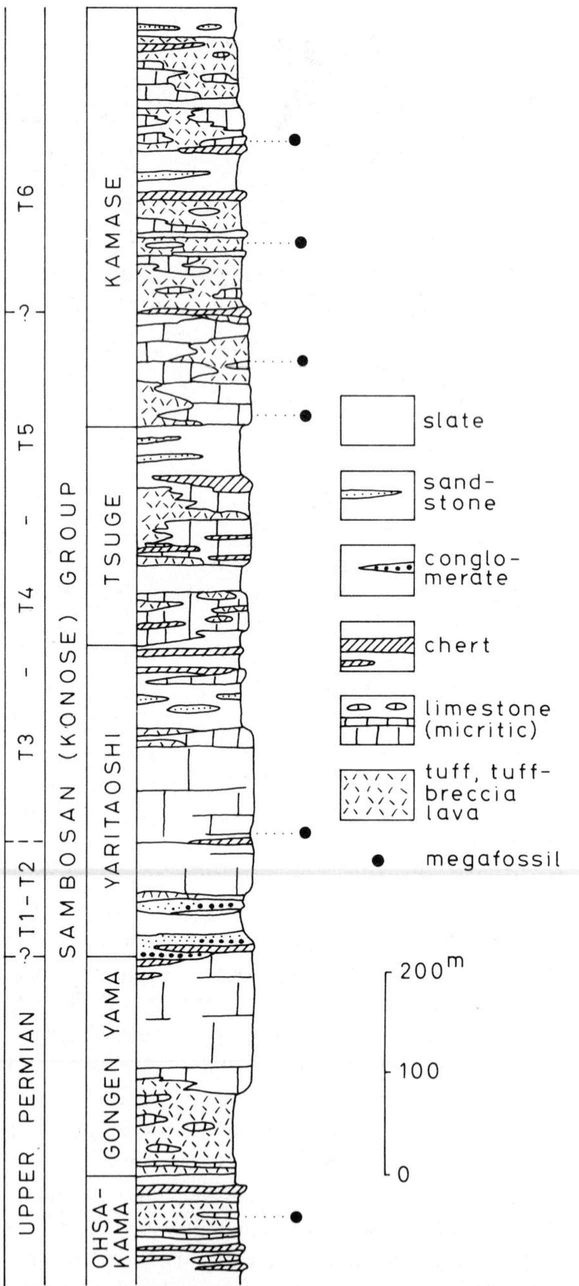

Fig. 9. Stratigraphic sequence of the Sambosan Group in the Konose area, southern Kyushu. (Adapted from Kanmera, 1969.)

are extensively exposed in the Sambosan zone (zone 12d). The characteristic Sambosan Group consists of a moderately thick conformable sequence ranging in age from late Middle Permian to Late Triassic. It is best studied in southern Kyushu (see Kanmera, 1969), where it consists of up to 1300 m of basaltic lava, tuff and tuff breccia, a number of limestones and bedded cherts with lenticular sandstones and some conglomerate (Fig. 9). The limestones are predominantly micritic with planktonic micro-organisms. In the cherts radiolaria and conodonts may be found. In some areas of the Sambosan zone cherts and sandy turbidites predominate in the upper part of the sequence. They may have been connected with those of the central belt through some submarine channels.

The Sambosan Group thus represents an off-shore eugeosynclinal facies with submarine volcanics formed in Permo-Triassic times. The rocks of this zone probably extend to zone 18 in the Ryukyu Islands where ammonites and *Halobia* are rather more common in the Carnian (Ishibashi, 1970–74).

A further possible extension into zones 8d and 5a in northeast Japan (up to southwestern Hokkaido) seems probable. Further north the Sorachi Group of central Hokkaido (zone 4) characterized by pillow basalt, serpentine, chert and subordinate limestone is, at least partly, Triassic in age and may be interpreted as a remnant of the Triassic Panthalassa ocean floor preserved in Japan.

JURASSIC

General outline

A large part of Eastern Asia, consisting of mainland China, Korea, central Siberia and a part of southeastern Asia, was a land area during the Jurassic. Marine beds are found in eastern Siberia, Japan, the Philippines, part of Vietnam and Indonesia. They represent geosynclinal deposits and sediments laid down in shallower basins around the continental margins. Apart from a Toarcian–Aalenian transgression in the Lena–Yakutsk Basin neritic epicontinental sediments are not extensive, a situation presumably related to crustal instability (Fig. 10).

Jurassic and Early Cretaceous crustal movements are known in several areas with geosynclinal folding producing the Verkhoyansk and Sikhote-Alin ranges. These are the young Kimmerian movements of some authors, but are better regarded as part of the Late Mesozoic orogeny in the Circum-Pacific belt.

Records of mobility are also found in the continental area, with the formation of a number of new basins or reactivation of Triassic basins. In the continental basins the Lower Jurassic generally follows the Upper Triassic conformably, and earth movements are reflected as unconformities or disconformities at mid-Jurassic and between the Jurassic and Cretaceous. Within the basins coal is developed, particularly in the Lower Jurassic, whereas volcanic material predominates in some Upper Jurassic basins. Late Jurassic thrusting is referred to the Yen Shan (Yenchang) movement, a term sometimes applied to all Jurassic to Cretaceous movements in China, including volcanism and plutonism. In Korea tectonic movements and granitic intrusion occurred [post-Taedong (Daedong)–pre-Kyongsong (Kyöngsang) Taebo (Daebo) disturbance]. The dominant trend of these Mesozoic structures is discordant with respect to basement trends.

Along the outer margin of the Asian continent, in Japan, there is considerable local variation due to unstable tectonic conditions. Kobayashi (1941) regarded the Jurassic of Japan as an inter-orogenic phase between the orogenies of the Permo-Triassic and Cretaceous. This interpretation may be applicable to some stratigraphic records, but if entire features are considered the Jurassic of Japan was also a period of mobility.

The Jurassic system in Japan may be treated in terms of a number of belts (Fig. 11 and Table IV):

(1) Inner belt of neritic, partly non-marine, deltaic sediments deposited in subsiding intermontane basins or bays (formations *12–16, 18–20*), sometimes called the Toyora–Tetori facies.

(2) Central belt of intensely folded beds, which may represent a continuation of the geosynclinal deposits from the Triassic.

(3) Outer belt, divided into three provinces: (a) the Chichibu zone (zone 12b–c) in which are developed sediments of shallow open-sea environments containing lenticular reef-like limestones (formations *21–40*), comprehensively called the Torinosu facies (name from loc. *30*). They are of Middle to Late Jurassic in age. (b) The eastern margin of the Abukuma massif (zone 6) where the Middle and Upper Jurassic Soma Group (loc. *10*) is developed. The beds in bio- and lithofacies closely resemble the

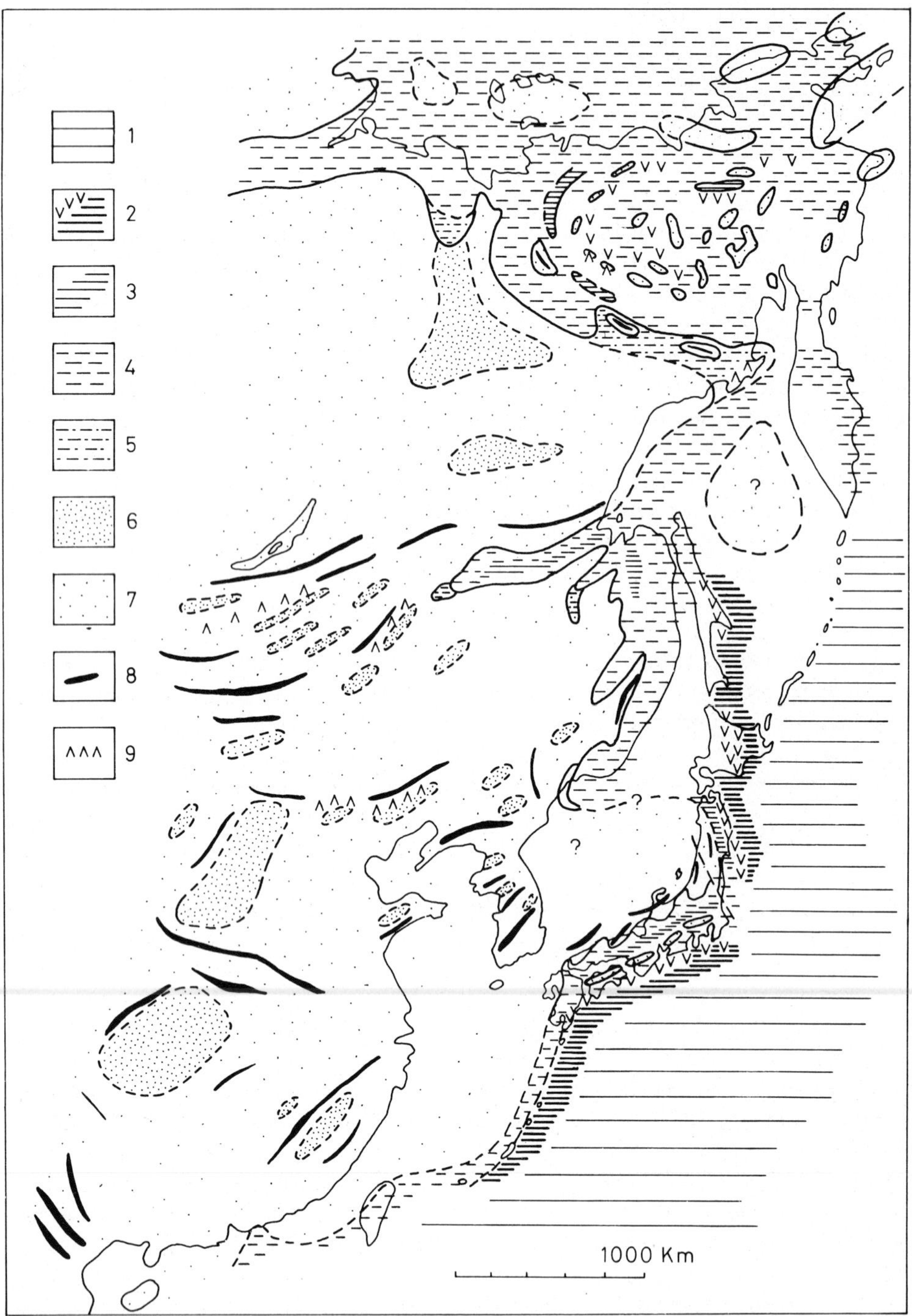

Fig. 10. Tentative paleogeographic map for the Late Jurassic (Oxfordian) of Eastern Asia.
Legend: *1* = ocean; *2* = eugeosyncline with submarine volcanism; *3* = geosyncline with little or no volcanism; *4* = shallow-sea basins, more or less labile (with some volcanism in northeastern Siberia); *5* = paralic area; *6* = continental basin; *7* = continent or island; *8* = mountain range; *9* = terrestrial volcanoes. (Compiled from various sources.)

Torinosu facies, but show some affinity with the Kitakami. (c) The Jurassic Kitakami facies of the southern Kitakami zone (5b). It consists of trigonia sandstones, ammonite shales and other sediments (formations *6–9*), showing lateral variation. (Note that this facies is distinct from that found in the northern Kitakami zone.) The sediments in the three provinces were laid down on a labile island shelf between the central trough and the outer geosyncline.

(4) Outermost belt of eugeosyncline in the Sambosan (zone 12d, 8d), northern Kitakami (zone 5a) and central Hokkaido (zones 3, 4). The Jurassic or presumed Jurassic of this belt is characterized by submarine volcanics, greywacke, chert, slate and limestone.

Biostratigraphy and biogeography

Unlike the European sequence there are many local variations and discontinuities in the Japanese succession, in common with those in other parts of the circum-Pacific belt. The fauna of the Early Jurassic contains a considerable proportion of cosmopolitan and Tethyan elements, whereas regional differentiation occurred during the Middle and Late Jurassic. In the Callovian (J8, see Table V) a few sub-Boreal elements are found in some parts in addition to the more abundant Tethyan and circum-Pacific elements.

Although neither as abundant nor as well preserved as their European counterparts, ammonites do permit a correlation with the European stages or substages. Following the classical works of Yokoyama (1902, 1904a, b), Shimizu (1930a) and Kobayashi (1935, 1947), they have been studied by Tatsuro Matsumoto (1956b), T. Sato (1954–58, 1961–62) and H. Takahashi (1969, 1973), and establishing detailed zonations has been attempted by Tatsuro Matsumoto and Ono (1947) and Hirano (1971, 1973) on the Toyora Group (*20*). Where their occurrence is sporadic, an attempt has been made to correlate with the standard zonal sequence (T. Sato, 1962a, b). Selected guide species are listed in Table V. In addition, long-ranging species of *Lytoceras, Calliphylloceras* and *Holcophylloceras* are common. Belemnoids (formations *6b, 7b, 20b, 22, 23*) and nautiloids are rare.

Bivalves are common in some facies. The most comprehensive and representative studies of them are made by Kobayashi et al. (1959), Tamura (1959a–d, 1960a–d, 1961a) and Hayami (1957a–e, 1958a–d, 1959a–d, 1960a, b, 1961a, b, 1962, 1965, 1975). Several distinctive faunal assemblages have been recognized representing biogeographic separation and ecological diversity as well as chronological difference: the Early Jurassic Toyora (*20*), Kuruma (*13, 12, 18*) and Kitakami [Shizukawa] (*6*), the Middle Jurassic Yamagami (*10*) and Kitakami [Hashiura] (*7*), and the Middle to Late Jurassic Tetori (*14*) and Soma–Torinosu (*10, 9, 8, 23–40*) assemblages. (Note that *8–9* in the eastern subbelt of southern Kitakami is better included in the Soma–Torinosu in faunal assemblage.)

In a general sense, the faunas of the inner belt (Toyora, Kuruma, Tetori and a part of Kitakami, i.e. Shizukawa and Hashiura) are biogeographically isolated from the Soma–Torinosu fauna of the outer belt. On the inner side three bivalve biofacies are recognized, an eomiodontid (*Mytilus, Bakewellia, Isognomon, Eomiodon, Crenotrapezium, Yokoyamaia, Geratrigonia*) facies in estuarine environments, the trigoniid (*Prosogyrotrigonia, Vaugania, Trigonia, Latitrigonia, Nipponitrigonia, Myophorella*) facies with pectinids and astartids in open near-shore sandstones and an inoceramid [*Inoceramus, Parainoceramus, Bositra* (="*Posidonia*")] facies with small limids and pectinids occurring in ammonitic shale of off-shore neritic environments.

The Soma–Torinosu fauna of the outer belt may be subdivided into four types in terms of bivalve assemblage and lithofacies. These are the *Grammatodon–Liostrea–Lopha–Corbula* near-shore shale facies, the *Nipponitrigonia–Haidaia* certain astartids near-shore sandstone facies, the *Neoburmesia–Pholadomya–Inoperna–Arcomytilus* off-shore shallow facies. Other forms, found in the limestone lithofacies of the outer belt, include reef corals, stromatoporoids, calcareous algae, nerineids, some brachiopods and echinoids (Yabe and Toyama, 1928; Yabe and Sugiyama, 1931, 1935; Eguchi, 1951; Tokuyama, 1957–59; W. Hashimoto, 1960a, b; Tamura, 1961a, b; Nishiyama, 1966, 1968; Shikama and Yui, 1973). Similar limestone biota are found northwards to central Hokkaido (loc. *41–43* of zone 5).

In addition to many endemic bivalves within the Lower Jurassic Toyora and Middle to Upper Jurassic Soma–Torinosu (and also part of the Kitakami) faunas there are species identical with, or closely allied to, those from Indonesia, Kutch, eastern Africa, Caucasus and Western Europe. Several species in the Upper Jurassic Tetori fauna, and a few in the Middle

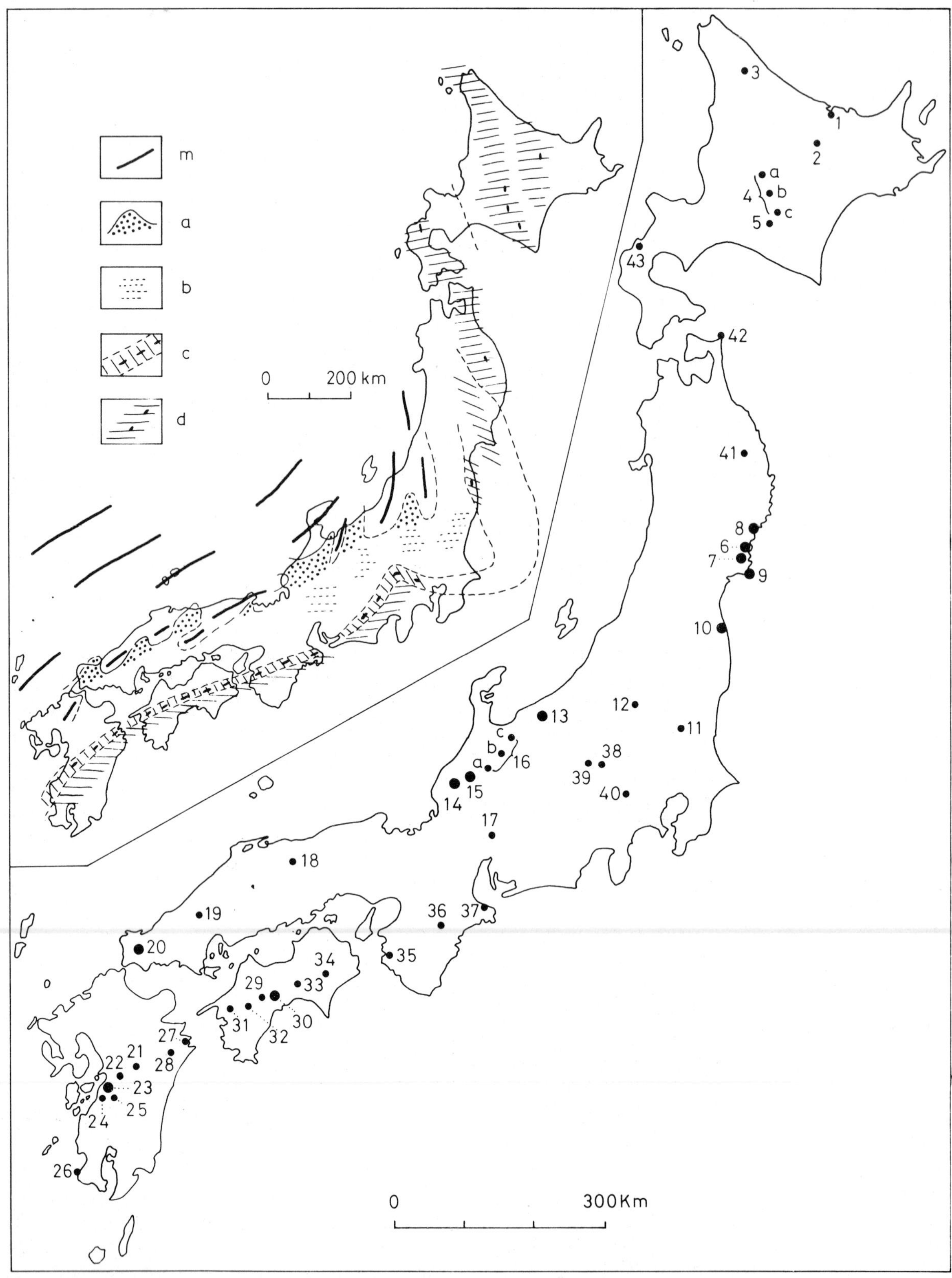

Fig. 11. Map of Japan showing the locations of Jurassic fossiliferous formations. A large solid circle means a well-known formation. See Table IV for the explanation of locality numbers. Inset at the upper left corner is an interpretation of the sedimentary framework of the Jurassic system of Japan.

Legend: *m* = mountain range; *a* = deltaic to shallow-bay facies; *b* = chert–turbidite facies; *c* = shallow open-sea facies, with reef limestone (dark spot); *d* = eugeosynclinal facies.

TABLE IV

List of localities of Jurassic fossils in Japan (see Fig. 11 for location)
The listing is similar to that given for the Triassic (Table II)

1. Saroma, Tokoro area, northeastern Hokkaido [z3] M
2. Nikoro, Tokoro area, northeastern Hokkaido [z3] O
3. Shotombetsu, upper reaches of the Tombetsu River, north-central Hokkaido [z4a] O
4. West of Yamabe (*a*), Shimekap (*b*), and east of Hidaka (*c*), central Hokkaido [z4b] O
5. Iwachishi (Nitto mine), west of Hidaka, central Hokkaido [z4b] O
6. Shizukawa, including Niranohama (*a*) and Hosoura (*b*), southern Kitakami [z5b] A, M, V
7. Hashiura, including Aratozaki (*a*), Arato (*b*) and Sodenohama (*c*), southern Kitakami [z5b] A, M
8. Karakuwa (*a*) and Shishiori (*b*), southern Kitakami [z5b] A, M, P
9. Ojika, southern Kitakami [z5b] A, M, O, P
10. Soma, Abukuma [z6] A, M, O, P
11. Toriashi, Yamizo [z7a] A, O, P
12. Iwamuro, Joetsu [z7b] M, P
13. Kuruma, east of Hida [z10c] A, M, P
14. Kuzuryu of Tetori, including Shimoyama–Tochimochiyama (*a*), Kaizara (*b*) and Yambarazaka (*c*), Hida (z9) A, M, P
15. Itoshiro of Tetori, Hida [z9] M, P, V
16. Kiritani (*a*), Furukawa (*b*) and Makito (*c*) of Tetori, Hida [z9] A, M, P
17. Inuyama, north of Nagoya [z10c] A, O
18. Yamaoka, east-central Chugoku [z10a] M, P
19. Higuchi, west-central Chugoku [z10a] A
20. Toyora [= Toyoura], including Higashinagano (*a*), Nishinakayama (*b*), and Utano (*c*), and also Kiyosuye of Toyonishi (*d*), west of Yamaguchi ("Nagato"), western Chugoku [z10a], A, M, O, P, V
21. Oishi, near Kuraoka, Kyushu [z12c] A
22. Bisho (Kawamata), east of Yatsushiro, Kyushu [z12b] A, M
23. Sakamoto, southeast of Yatsushiro, Kyushu [z12c] A, M, O
24. Tsurubami, southeast of Yatsushiro, Kyushu [z12c] M, O
25. Ebirase, southeast of Yatsushiro, Kyushu [z12c] M, O
26. Noma-ike, southwestern Kyushu [z12c] O
27. Tsui, near Saheki, southeastern Kyushu [z12d] A, M, O
28. Shinkai, southeastern Kyushu [z12c] O
29. Naradani and Nishiyama, near Sakawa, Shikoku [z12c and 12d] O, M
30. Torinosu (s.s.) near Sakawa, Shikoku [z12c] A, M, O
31. "Torinosu" of Nomura, western Shikoku [z12c] M, O
32. "Torinosu" of Go-Ochimen, Shikoku [z12c] M, O
33. "Torinosu" of Monobegawa, Shikoku [z12c] A, M, O
34. Kurisaka and Miyakodani, eastern Shikoku [z12b] A, M, O
35. Yura, Kii peninsula [z12c] M, O
36. Osako, central Kii peninsula [z12d] M, O
37. Imaura, near Toba, Shima [z12c] O
38. (?) Sanchu graben, Kwanto [z8b] O
39. Kappazaka, Minamisaku, Kwanto [z8c] A, M, O
40. Itsukaichi, west of Tokyo, Kwanto [z8c] O
41. Iwaizumi, northern Kitakami [z5a] A(?), O
42. Cape Shiriya, Shimokita peninsula [z5a] O
43. Gunai, south of Suttsu, southwestern Hokkaido [z5a] O

Locs. *4, 5, 41–43* are presumably Jurassic. Some other localities of doubtfully Jurassic fossiliferous limestone of the Torinosu aspect and those of possibly Jurassic radiolarian chert are omitted.

and Lower Jurassic Kitakami and Toyora faunas show some similarities with eastern Siberian forms.

Intercalated within the marine deposits of the inner side are plant beds. The Early Jurassic flora (found in formations *6, 12, 13, 18, 20*) is intimately related to the Late Triassic. Both are included in the *Dictyophyllum* series of Oishi (1940), who grouped the Middle–Late Jurassic with the Early Cretaceous as the *Onichiopsis* series (formations *14, 8, 9, 10, 20c*). There is a remarkable biogeographic differentiation in the latter flora (see Fig. 21) (Tatsuaki Kimura, 1973; Tatsuaki Kimura and Asama, 1975). On the

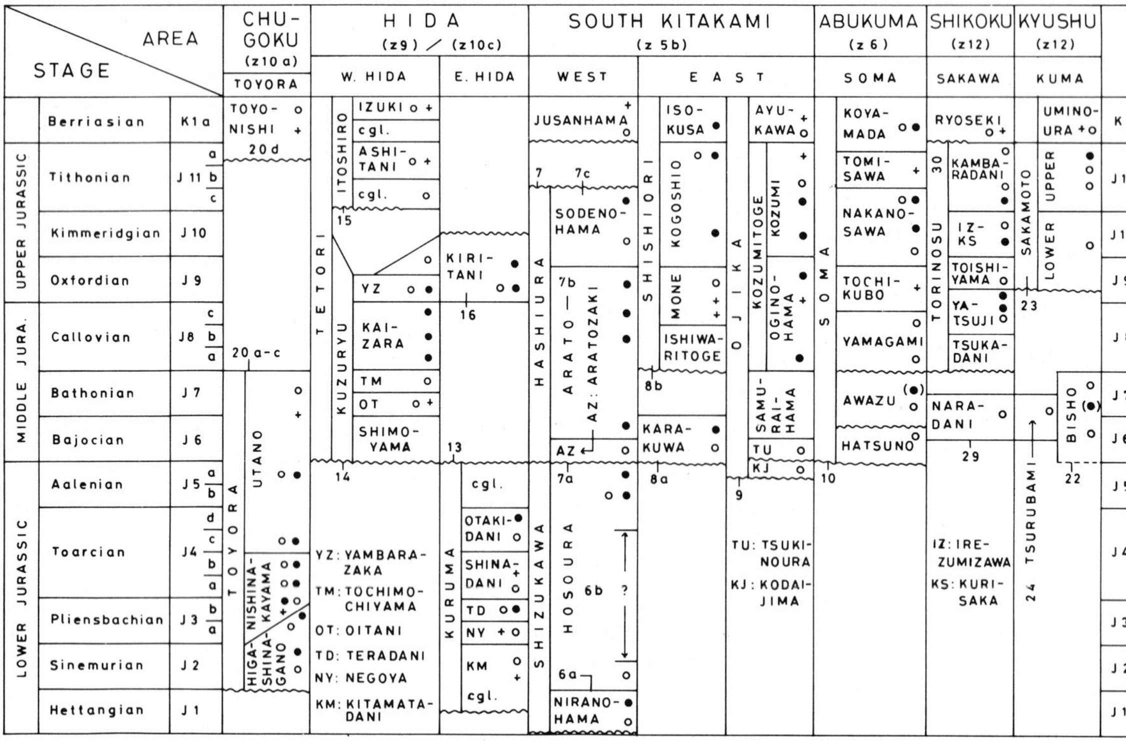

Fig. 12. Correlation of representative Jurassic formations in Japan. See Table V for units J1–J11 of a biostratigraphic division, Fig. 11 and Table IV for the numbered locations of indicated formations or groups, and Fig. 2 for the tectonic units (e.g. z10a, etc.). The stratigraphic position of guide ammonites is indicated by a solid circle (in parentheses when the identification is uncertain), other invertebrates and calcareous algae by an open circle, land plants by a cross mark, and an unconformity by a wavy line. Some formations without distinct guide species are excluded from this chart.

inner side *Coniopteris,* characteristic of the Siberian province, is found, yet in the inner province of Japan *Otozamites, Zamites* and *Dictyozamites,* not found in Siberia, are known while the characteristic Siberian forms *Heilungia, Butefia,* and *Bureja* have never been recorded. *Zamiophyllum* is characteristic of the outer province of Japan. Giant *Equisetites* from Kuruma (*13*) and *Xenoxylon* from Itoshiro (*15*) and other formations are known.

Reptilia are scarce in Japan, although a *Tedorosaurus* is known from the Tetori area (*14*) (Shikama, 1969) and an ichthyosaurian is recorded in southern Kitakami (*6b*). A new blattoid has recently been found in the Lower Jurassic of western Japan (*20b*) (Fujiyama, 1974).

Paleoclimate and paleomagnetism

There are no relatively undisturbed lavas or red beds of Jurassic age in Japan and no paleomagnetic datum is available.

The marine and terrestrial biotas indicate a generally warm climate. The boundary between the Boreal (including sub-Boreal) and Tethyan provinces must have lain at some distance north of Japan. The provincial differences in the Middle to Late Jurassic may reflect world climatic zones but may also be related to physiographic conditions.

The extensive distribution of coral reef biota on the outer side of Japan and its affinity with the Tethys suggest the existence of a warm, north-flowing current.

Conglomerate and coarse sandstones common in the Jurassic sequence on the inner side of Japan suggest strong erosion, high topographic relief and adequate rainfall consistent with the existence of a warm, humid climate flora and a blattoid record. Conditions on occasion favoured the development of coal seams.

TABLE V

Jurassic biostratigraphic division and guide ammonites in Japan (numbers in parentheses refer to formation or locality in Table IV and Fig. 11)

Lower Jurassic

J1	(Hettangian): *Alsatites onoderai* (*6a*)
J2	(Sinemurian): *Arnioceras yokoyamai* (*6b*) No ammonite of the Upper Sinemurian
J3	(Pliensbachian)
J3a	(Carixian): uncertain, except for *Deroceras* sp. (*13*)
J3b	(Domerian): *Amaltheus* sp., *Canavaria* sp. (*13* – Teradani); *Canavaria japonica, Arieticeras* cf. *pseudocanavarii, Fontanelliceras fontanellense, Paltarpites toyoranus, Fuciniceras primordium, F.* cf. *normanianum* (*20b* – *F. fontanellense* zone); *Amaltheus* cf. *stokesi* (*20a*)
J4	(Toarcian)
J4a	(Uppermost Domerian? – Lower Whitbian): *Protogrammoceras nipponicum, Lioceratoides yokoyamai, Harpoceras chrysanthemum, H. okadai, H.* aff. *exaratum, H.* (*Harpoceratoides*) *nagatoensis, Fuciniceras nakayamense, Dactylioceras helianthoides* (*20b* – *P. nipponicum* zone)
J4b	(Lower Toarcian = Witbian, main part): *Dactylioceras helianthoides, Peronoceras subfibulatum, Harpoceras chrysanthemum, H. inouyei, H. okadai, Lioceratoides yokoyamai* (*20b* – *D. helianthoides* zone)
J4c	(Upper Toarcian = Yeovilian): *Grammoceras* aff. *obesum, Phymatoceras toyoranum, Pseudolioceras* sp. (*20c* – Beds Up and Ub)
J4d	(Upper Toarcian, younger than J 4c?): *Pseudogrammoceras muelleri* (*13* – Otakidani)
J5	(Aalenian) [Some authors include this in the Middle Jurassic]
J5a	(Lower Aalenian): *Hosoureites ikianus, Tmetoceras recticostatum, Harpoceras okadai* (*6b*)
J5b	(Upper Aalenian): *Planammatoceras hosourense, P. kitakamiense, Tmetoceras recticostatum* (*6b*; *20c* – Bed Uh)

Middle Jurassic

J6	(Bajocian): *Sonninia* sp., *Stephanoceras* cf. *plicatissimus, S. hosourense, Leptosphinctes* (*Vermisphinctes*) sp., *Garantiana* sp., *Parkinsonia* aff. *parkinsoni, Cadomites bandoi* (*7b, 8a, 9*)
J7	(Bathonian): no distinctive species, except for doubtful *Bigotites* sp. (*10*); *Cadomites* sp. and *Planisphinctes* sp. (*22*) and a bivalve *Retroceramus utanoensis* (*6c*)
J8	(Callovian)
J8a	(Lower Callovian): *Neuqueniceras yokoyamai* (*14b*), *Choffatia* (?) *oginohamaensis* (9)
J8b	(Middle Callovian): *Grossouvria* cf. *subtilis, G. laeviradiata, Kepplerites* (*Seymourites*) *japonicus* (*14b*); *Kepplerites* (*Gowericeras*) *oyamai, Obtusicostites hataii* (*7b*)
J8c	(Upper Callovian): *Oppelia* aff. *subradiata, Oxycerites* sp., *Oecotraustes* sp., *Camphyllites* aff. *delmontanus, Bomburites* aff. *globuliforme* (*14b*); *Oecotraustes* cf. *bakalovi* (*7b*); *Hecticoceras japonicum* (?), *Horioceras mitodaense* (*27*?, *30*)

Upper Jurassic

J9	(Oxfordian): *Kranosphinctes matsushimai, Peltoceratoides* sp. (*14c, 7b, 9, 11*); *Perisphinctes ozikensis* (*9*); *Dichotomosphinctes kiritaniensis* (*16a*), *Perisphinctes* sp. (*17*); *Properisphinctes* aff. *bernensis* (*30*)
K10	(Kimmeridgian): *Discosphinctes* sp., *Lithacoceras onukii* (*9*), *Lithacoceras* sp. (*16a*), *Lithacoceras tarodaense* (*30, 39*) *Ataxioceras kurisakensis* (*34*); *Ataxioceras* sp., *Aspidoceras* sp., *Taramelliceras* sp. (*8b, 9, 10*); *Aulacostephanus* (?) sp. (*21*)
J11	(Tithonian):
J11a	(Lower Tithonian): *Aulacosphinctoides steigeri* (*7c, 9*?, *10, 30*); *Virgatosphinctes* aff. *communis* (*9*)
J11b	(Middle Tithonian): *Corongoceras* sp. (*34*)
J11c	(Upper Tithonian): *Substeuroceras* sp. (*8b*); *Spiticeras* [? *Kilianiceras*] cf. *eucomphalum, Aulacosphinctes* sp. (*23*)

Regional features

Following the general outline, certain of the salient features of the stratigraphy will be emphasized here. Two kinds of correlation charts (Figs. 7, 12) will clarify time relationships.

(1) Inner belt

The inner belt comprises 9, the northern part of zone 10, and part of massif 7b (Figs. 2 and 11). It was the site of the Permo-Triassic orogeny and during the Jurassic deposition was in intermontane basins with neritic to paralic sediments. The Jurassic rests directly upon metamorphosed or folded Paleozoic rocks, not directly upon Triassic. The basins were formed on or near boundary faults of tectonic units. These two facts imply that the Jurassic basins were newly formed independently of the Triassic basins as the result of tectonic activity at the end of the Triassic and continuing into the Jurassic.

The tectonic activity was not uniform and this results in variations in facies and thickness of the Jurassic formations from basin to basin. The Toyora Group (*20*) (Fig. 13), for example, ranges in age from J1 to J7 and consists of about 1000 m of clastic sediments forming a sedimentary cycle. The beds were laid down in a quiet bay with a phase of inundation (J4a–b) represented by ammonitic shales. The Kuruma Group (*13*) is made up of extremely thick clastics (nearly 10,000 m) but ranges in age from J1 to J4 (or possibly J5). The conglomerates are poorly sorted with abundant schist boulders in the lower and granite in the upper part. The gneiss and granite of zone 9 (Hida area) immediately north-west of the area of outcrop of the Kuruma Group give ages of 190–170 m.y. on micas and hornblende, suggesting that the uplift and erosion of the Hida mountains was contemporaneous with the Kuruma sedimentation.

The Tetori Group (*14–16*) (Fig. 14) is younger than J6 and ranges up into the Early Cretaceous. The beds rest with marked unconformity upon the Hida gneiss, granite, schists and unmetamorphosed Paleozoic rocks, deposited in a basin formed since mid-Jurassic times along the tectonic boundary between zones 9 and 10. It is somewhat variable in terms of thickness and facies; the Kuzuryu (*14*) and Itoshiro (*15*) subgroups amount to 1500–2000 m in the type section. Conglomerate at the base contains a variety of granite and metamorphic rocks and also orthoquartzite, and quartz-feldspathic arenites (i.e.,

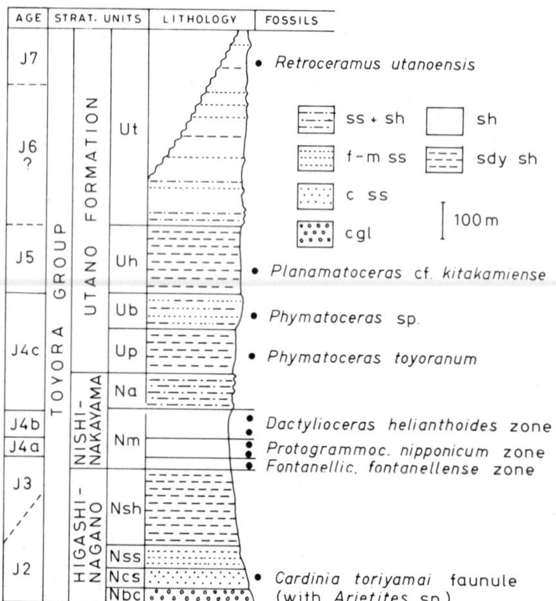

Fig. 13. Stratigraphic sequence of the Jurassic Toyora Group. Lithologic symbols: *sh* = shale; *sdy sh* = sandy shale; *ss* + *sh* = sandstone and shale in alternation; *f* – *m ss* = fine- to medium-grained sandstone; *c ss* = coarse-grained sandstone; *cgl* = conglomerate. (Compiled from the data in Tatsuro Matsumoto and Ono, 1947; Arkell, 1956; Hirano, 1971–73.)

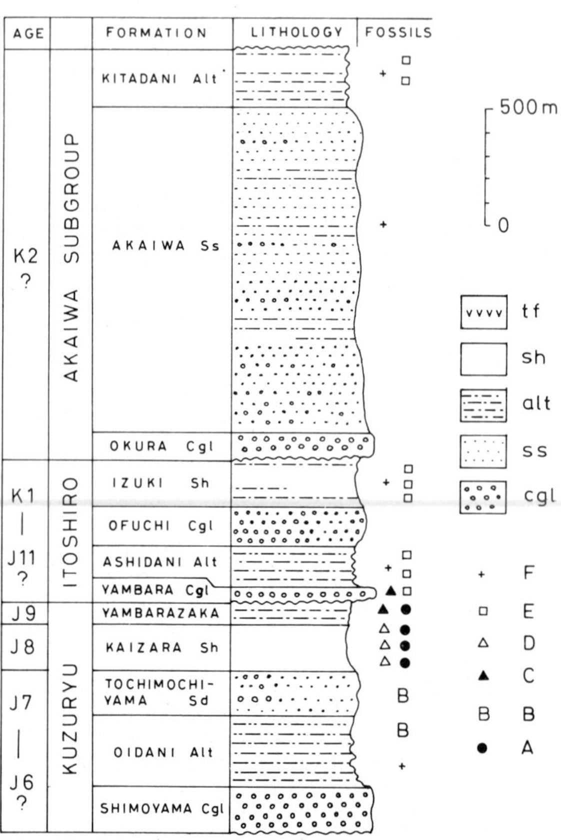

Fig. 14. Generalized stratigraphic sequence of the Tetori Group. Lithology: *tf* = tuff; *sh* = shale; *alt* = alternating sandstone and shale; *ss* = sandstone; *cgl* = conglomerate. Fossils: *A* = ammonites; *B* = belemnites; *C* = trigonians; *D* = other marine mollusca; *E* = brackish- to fresh-water mollusca; *F* = land plants. (After Maeda, 1961.)

arkoses) are common at many horizons. The source area is presumed similar to that of the Hida massif or Korea where Jurassic granite batholiths occupy a considerable area in addition to Precambrian rocks.

Within the inner belt are scattered outcrops of beds which can be correlated in part with the Toyora and Kuruma groups.

(2) Central belt

The central belt comprises the southern part of zone 10, southeastern part of zone 7, and zone 11 (Fig. 2 and 11). The major part of zones 12a and 6 could be included, but the original rocks now exposed as metamorphosed rocks may have been deep seated during the Jurassic; their supra-structure is unknown.

The record of the central belt is incomplete. Formerly it was recorded as an embryonic uplift along the axial core of Mesozoic Japan ("Eonippon Cordillera" of Kobayashi, 1941), separating the northern (inner) from the southern (outer) marine faunas. The discovery of Jurassic fossils in parts of the belt (J9 age found at localities *11* and *17*), and the thick sequence above Upper Triassic conodont beds (Triassic localities *7a, b, 8, 11, 12, 14, 22b, 23* of Fig. 5) may be presumed to range into the Jurassic. The upper part of this sequence has the characteristics of a flysch sequence and so may be the time equivalent of the coarse deltaic clastics found in the inner belt. Better age information is necessary, however, as a working hypothesis, it may be supposed that geosynclinal sedimentation continued from the Permo-Triassic into the Jurassic with an increasing clastic content. Folding and thrusting may have been contemporaneous with sedimentation, and some islands may have appeared. The hinterland lay to the north probably as part of the inner belt in the site of the present Japan Sea and Korea.

Regional metamorphism and granite emplacement began in the Jurassic in the Ryoke–Abukuma zone (zones 11 and 6), though there are few direct radiometric dates because of Cretaceous resetting. The high P/T regional metamorphism of the Sambagawa zone (12a) may likewise have begun in the Jurassic.

(3) Outer belt

The outer belt is best described in terms of three provinces, each with its own particular features, but with enough in common to separate them from the other belts. These provinces are the Chichibu zone, the eastern margin of the Abukuma massif, and the southern Kitakami zone. They are all free of volcanics, except for a few tuff layers, and have a miogeosynclinal character representing deposits laid down on a narrow insular unstable shelf facing the eugeosyncline of the outermost belt.

In the Chichibu zone (12b, c, 8b, c), the Lower Jurassic has nowhere been positively identified; the Middle and Upper Jurassic sediments are now exposed at isolated outcrops (locations *21–40*) separated by folding and thrusting and subsequent erosion, but were once part of a continuous belt. The stratigraphy is best known in the Sakawa area (loc. *29–30*) in central Shikoku and the Sakamoto area (loc. *23–24*) in western Kyushu (Toshio Kimura, 1956a; Tamura, 1961a, b). In both areas (locations *29, 24*) there are narrow outcrops of the Naradani Formation, consisting of a 150 m sequence of sandstone, sandy shale, and impure limestone, the last with corals, echinoids and brachiopods of J6–J7 ages.

The Torinosu Group rests disconformably upon the Naradani and oversteps on to the Upper Triassic and Permian. It ranges in age from J8c to J11c. It has basal conglomerate which contains chert debris and some igneous rock fragments. The sequence, about 500–600 m thick, is primarily shale laid down in shallow marine conditions but contains intercalations of sandstone and some lenticular reef-like limestones. A north to south facies change can be observed with the offshore facies to the south. At loc. *25,* southern Kyushu, a reduced thickness and a *Neoburmesia* molluscan assemblage suggests a local offshore shallowing with the development of less subsiding sub-belts existing near the boundary with the outer eugeosynclinal belt.

Along the faulted eastern margin of the Abukuma massif (zone 6) where high P/T metamorphic rocks crop out (Hashimoto et al., 1972), rocks of the Middle–Late Jurassic (J6 to J11 and K1) Soma Group (*10*) are found. These are about 1400 m of quartz-feldspathic arenite with three conglomerate horizons, some shales, calcareous sandstones and biohermal limestone. There is a disconformity between the lower two members (J6) from the rest of the Soma Group, whose age ranges up to Early Cretaceous (K1). The sediments are mainly of neritic environment, but a member in the lower part contains plant fossils and some coals. Granite pebbles within the clastics suggest either the existence of older granites in the Abukuma massif now masked by Cretaceous granites (some Jurassic radiometric ages have been recorded) or a more distant origin which may be more appropriate given the maturity of the sandstones, presence of orthoquartzite pebbles (Okami et al., 1976) and the general geological environment.

The Jurassic of the southern Kitakami zone (5b) is relatively well known (T. Sato, 1962a,b; H. Takahashi, 1969; and others), for the beds are fossiliferous and the fold structures less complex. It is found in two synclinoria, an eastern and western, some 15 km apart. The folding was completed by Aptian times and the same cleavage system is found in Permian–Neocomian shales. Folding proceeded during the Jurassic and Triassic as seen from the stratigraphy and distribution of the beds (see Fig. 15).

Two groups separated by a disconformity, the Shizukawa (J1–J5) and Hashiura (J6–J9 to J10a in places) are found in the western belt. The former averages less than 200 m thick and the latter totals about 600 m. In each a trigonian sandstone is followed by sandy shale or shale and may be capped by sandstone. In the eastern belt (loc. *8, 9*) the Lower Jurassic is absent, the beds range in age from J6 to J11 (and K1 in places), and have a much greater thickness, up to 1500 m, than in the western belt. Although there are lateral facies variations, the proportion of coarse clastics is also higher, cobbles of granite are found and the sandstones are often arkosic. The main sandstone horizons are in lower J6, J8a and J10c (?). Ammonites and open-sea molluscs are found but there are also plant horizons. The sandstone-shale alternation has often a flysch-like feature and contains *Zoophyte* trace fossils (Takizawa, 1975). The beds of the eastern belt are more strongly folded than the western.

The Jurassic formations of the outer belt are miogeosynclinal in character, without or with little volcanic debris. They suggest a narrow unstable insular shelf facing a eugeosyncline of the outermost belt.

(4) Outermost belt

The outermost belt contains the eugeosynclinal assemblage but the Jurassic there is poorly known. Eugeosynclinal sediments are exposed in three provinces which are roughly aligned along the outermost belt of Jurassic Eastern Asia.

(a) *Sambosan zone.* In this zone (12d, 8d) the Sambosan Group has been described as an Upper Permian and Triassic sequence of offshore eugeosynclinal facies which may include the Jurassic with uncertain stratigraphic relationships. The fossils which suggest the Jurassic are mostly of Torinosu type biota

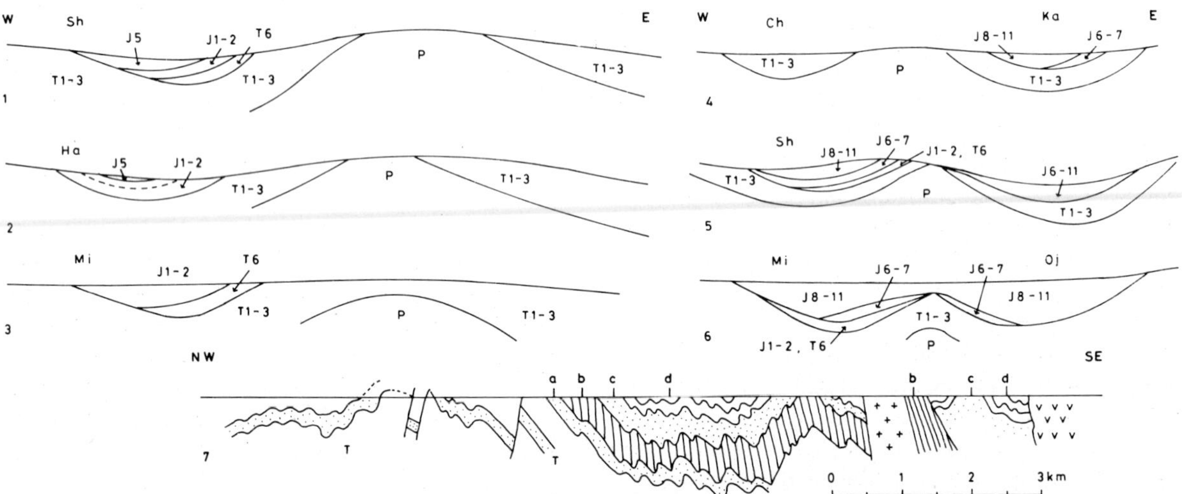

Fig. 15. Schematic profiles showing the sedimentation on Jurassic formations in south Kitakami (*1–6*) (after T. Sato, 1962a) and a geologic structure (7) (after Tokuyama, 1965). Locations of profiles: *1* = across Shizukawa (*Sh*); *2* = Hashiura (*Ha*); *3* = Mizunuma (*Mi*) (*1–3* for the Lower Jurassic from north to south); *4* = Chonomori (*Ch*) to Karakuwa (*Ka*); *5* = Shizukawa (*Sh*); *6* = Mizunuma (*Mi*) (*4–6* for the Middle to Upper Jurassic). P = Permian; T1–3 = Lower to Middle Triassic; T6 = Upper Triassic (Norian); J1–11 = subdivisions of the Jurassic defined in Table V. *7* = Structural profile of Ojika peninsula, in which the Lower to Middle Triassic (T) is in the western half, with a key bed of sandstone in the middle, and the Middle Jurassic to basal Cretaceous in the eastern half, comprising (*a*) Tsukinoura–Kodaijima, (*b*) Samuraihama, (*c*) Oginohama and (*d*) Kozumi–Ayukawa formations: quartz–diorite is shown by cross marks and porphyrite by V marks.

and partly of undescribed molluscan faunules.

(b) *Northern Kitakami zone.* The Jurassic sequence, and down to the Permian of the northern Kitakami zone (5a), is remarkably different from that of the southern Kitakami zone (5b), from which it is separated by a narrow, arcuate thrust belt, the Hayachine tectonic belt or septum. In this thrust belt basic and ultrabasic rocks are found. The northern Kitakami zone can be subdivided from west to east by major thrusts into the Kuzumaki, Iwaizumi and Taro subzones. In the central and in part of the eastern subzone intensely folded Triassic and Jurassic (and even Neocomian) are found, whereas in the western subzone only Permian slate and chert occur (see Fig. 32).

According to Sugimoto (1974), the presumed Jurassic–Neocomian sequence, the Iwaizumi Group, unconformably overlies Triassic cherts and slates which have been dated by conodonts. It consists of a thick series of andesitic lavas and pyroclastics and limestone in the lower part, siliceous shale, slate and sandstone in the upper part. Poorly preserved fossils of Torinosu aspect in the lower part suggest a Late Jurassic age.

Rocks with a similar assemblage crop out in the Shimokita Peninsula and in southwestern Hokkaido. At locations *42* and *43*, hexacorals, stromatoporoids and calcareous algae of Torinosu aspect are found (Murata, 1962; W. Hashimoto and Igo, 1962). An intraformational conglomerate near Shimokita (*42*) has, in addition to local material, occasional rare pebbles of granodiorite.

(c) *Central Hokkaido.* Over most of Hokkaido (zones 3 and 4) there are wide exposures of pre-Aptian eugeosynclinal rocks, the stratigraphy of which is incompletely known because of the intensity of folding and metamorphism, poor fossil content and discontinuity of outcrop.

The Sorachi Group of zone 4 underlies the Aptian (and younger) Yezo Group, and in the Yubari Mountains it has been subdivided by W. Hashimoto (1952) into a lower (pillow basalt) and an upper (dominantly chert, tuff and andesitic greywacke) subgroup. It was assigned a Jurassic age based upon radiolaria and the Torinosu aspect of the calcareous horizons, however, work on conodonts suggests it ranges down into the Triassic, while some of the rocks resembling the Sorachi contain Neocomian fossils (loc. *5, 6*). The location and nature of boundaries however are unsolved.

The pillow basalts of the Sorachi Group are associated with diabase and gabbro while serpentine is common in sub-belt 4b. The greenstones of the Sorachi Group geochemically belong to the high-alkali tholeiite to alkali basalts (W. Hashimoto, 1973; Sawada and Kanmera, 1973). They are weakly metamorphosed (pumpellyite–actinolite facies) and merge into the glaucophane schist facies of the Kamuikotan metamorphics in subzone 4b. There is a group of sandstones and slates to the east in subzone 4a in fault contact with the Sorachi Group. This is called the Hidaka Group (not to be confused with the Hidaka metamorphics) and it is of uncertain age.

The limestones of the Sorachi Group (and also Hidaka Group) are mostly micritic with a small amount or devoid of bioclasts, and radiolarian micrite suggest an offshore deep-sea environment (Kanmera and Obata, 1972). However, shallow-water conditions in other regions, perhaps on submarine volcanoes, is suggested by the Torinosu aspect. Slump masses of such limestone in bathyal sediments have been recorded by the same authors.

In zone 3, the Nikoro Group extending from Tokoro (near Kitami) to Toyokoro (near Obihiro) contains basalt, diabase and radiolarian chert. It resembles the Sorachi Group and also has lenticular limestones with a fauna of Torinosu aspect. It is followed by the 1300 m of the Saroma Group which has a *Buchia* faunule resembling that of the Late Jurassic in its upper part. The Nikoro is underlain by a thick clastic flysch sequence of uncertain age (Yubetsu Group). All three groups have been metamorphosed as a result of deep burial to the prehnite-pumpellyite-metagreywacke facies (see Teraoka et al., 1973).

Although zones 3 and 4 are separated by a Cenozoic fault, they must have belonged to the same eugeosynclinal belt in Triassic to Jurassic times. Both the Sorachi and Nikoro groups have an ophiolite assemblage but it is not certain whether they represent older oceanic floor. Similar rocks are found in the eastern belt of Sakhalin and seem to extend northeastwards to the Koryak belt (Avdeiko, 1970).

CRETACEOUS

General outline

During the Cretaceous the extent of the continental area in Eastern Asia was greater than that during

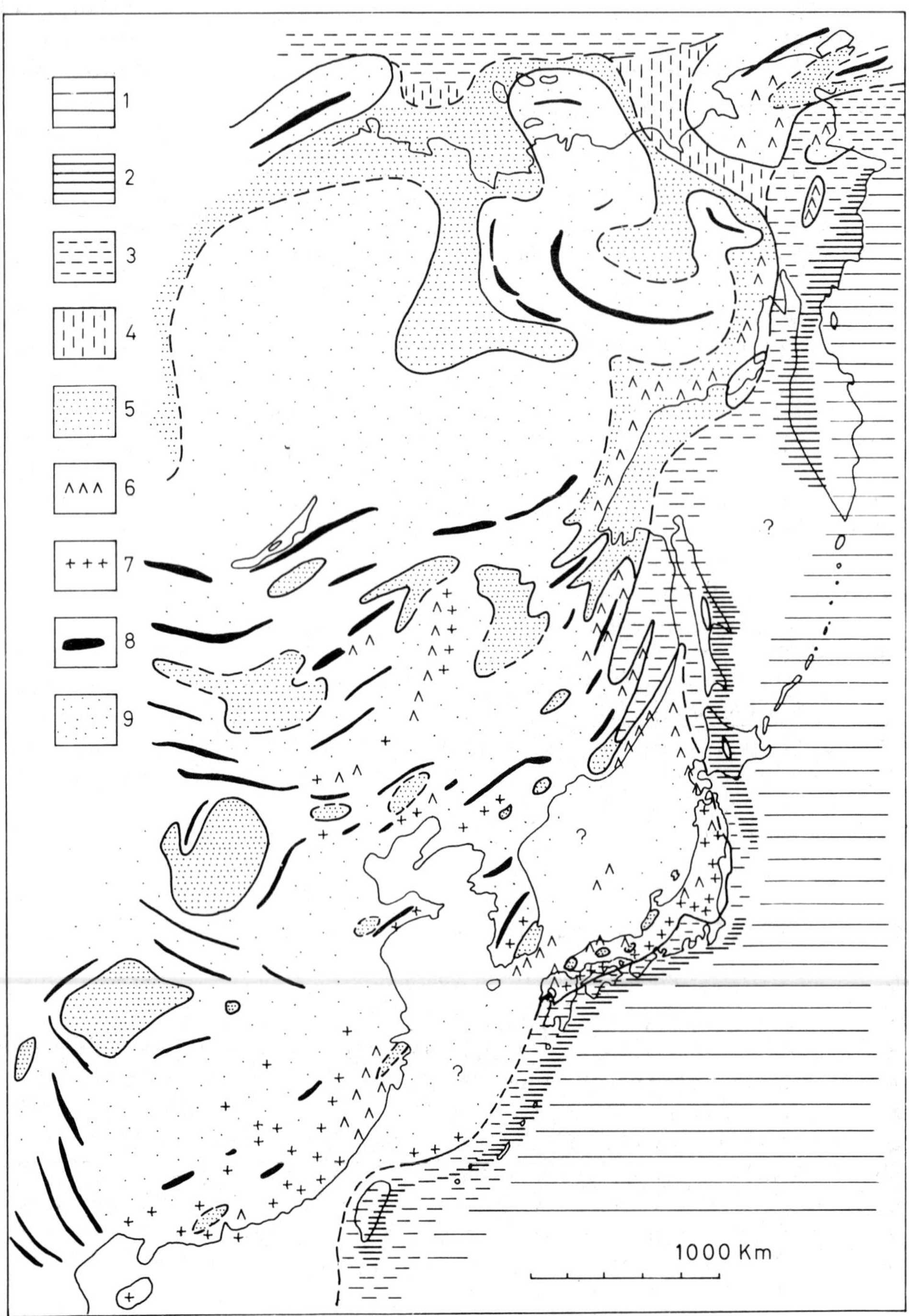

Fig. 16. Tentative paleogeographic map for the mid-Cretaceous of Eastern Asia.
Legend: *1* = ocean; *2* = geosyncline (later folded belt); *3* = shallow sea; *4* = shallow sea retreated since Albian; *5* = continental basin; *6* = volcanoes; *7* = granitic intrusion at depth; *8* = mountain range; *9* = continental area. (Compiled from various sources.)

the Jurassic. The land encompassed most of central and eastern Siberia, most of China, Korea, the inner parts of Japan, Indochina, Thailand and Malaya. Along the Pacific margin a geosynclinal belt extended through Anadyr–Koryak–Kamchatka, Sikhote-Alin, Hokkaido–Sakhalin, the Kurile arc, the outer belts of the Japanese and Ryukyu arcs, Taiwan, most of the Philippines and the outer part of Indonesia (Fig. 16). Progressive orogeny during the Cretaceous resulted in the formation of islands and mountain systems, and at the end of the Cretaceous–beginning of the Tertiary the land area, in common with many other parts of the world, was much expanded.

On the continental side of the geosynclines or along the continental margin acid magmatism with the development of extensive andesitic to rhyolitic volcanic ejecta and the emplacement of granite bodies took place. One belt may be traced from northeastern Siberia through the Dzhugdzhur Mountains to inner Sikhote-Alin (the outer volcanic belt of Russian geologists), another is the inner belt of Japan to which southern Korea may be related, the third is found in the coastal ranges of southeastern China. Paired metamorphic belts, a continental side low P/T with granitic intrusions and a high P/T with ultrabasics on the oceanic side, continued to develop and became exposed in Cretaceous times. The Ryoke–Sambagawa belts of Japan are one example, others occur in Hokkaido, Kamchatka and elsewhere, though not necessarily synchronous.

Within the continental area there were zones of instability: in the Lena depression there was heavy deposition of clastics from the rising Verkhoyansk Mountains, Early Cretaceous granites were intruded around the Kolyma massif and elsewhere and thick sediments were laid down in basins in the massif itself. A seaway, via the Anyuy synclinal zone which connected the Arctic with the Pacific, disappeared in Late Cretaceous times. Late Jurassic and Cretaceous crustal movements (i.e., Yen Shan movements) resulted in ranges and basins of various dimensions in China, Korea and Indochina. In general the basins received conglomerates and lacustrine shales and marls, often accompanied by tuff breccias, lavas and granitic emplacement. It is these mobile conditions which account for the predominance of clastic and volcanogenic rocks and intrusions during the Cretaceous which on occasion reach a great thickness. The contrast with the tectonically quiet conditions with the prominence of chalk, limestone and other fine-grained sediments in Europe and elsewhere is remarkable. In Eastern Asia petroleum and gas resources of Cretaceous age are restricted but various metallic ores were engendered.

The Cretaceous of the Japanese islands and adjacent areas can be considered in the following framework (Fig. 17): (1) inner belt (zones 9–10, 7] of non-marine deposits, volcanic series and granitic bodies; (2) central belt of regional metamorphic rocks and associated plutonics in paired zones, the Ryoke–Abukuma (zones 11 and 6) and the Sambagawa (zones 12a, 8a) where Upper Cretaceous clastic sedimentary groups are also distributed; (3) Chichibu terrain (zones 12b, c, 8b, c) of the outer belt, where Cretaceous formations of mainly shallow-sea and partly non-marine facies are distributed, without volcanic activity; (4) Shimanto zone (zones 13, 19) of the outermost belt, where geosynclinal Shimanto Group is distributed; (5) Kitakami zone (zone 5), where Cretaceous formations are similar to those in (3) but partly with volcanic rocks, and Cretaceous granites are also distributed; (6) central zone of Hokkaido (zone 4), with geosynclinal Cretaceous deposits, extending to Sakhalin; (7) eastern Hokkaido to the Lesser Kurile (zone 2), where Upper Cretaceous rocks are exposed; (8) Taiwan (zone 22, 23); (9) submarine plateau of northern Philippine Sea, where probably Cretaceous rocks crop out.

Biostratigraphy and biogeography

Fossiliferous Cretaceous beds are more widely distributed in Japan and adjacent regions than their Triassic and Jurassic counterparts. Their locations are indicated in Fig. 18 and Table VI.

A scheme of the biostratigraphic divisions and characteristic species is given in Table VII. The use of Japanese names was proposed by Yabe (1927) and modified by Tatsuro Matsumoto (1943, 1954c) until they may be replaced by international stage names. The Japanese names and approximate correlation are given below with the letter nomination proposed by Tatsuro Matsumoto (1954c, emended here [1]):

[1] Matsumoto (1954): $K3\gamma = K4a1$ (this article); $K3\gamma + K4\alpha = K4a$; $K5\gamma = K6a1$; $K5\gamma + K6\alpha = K6a$.

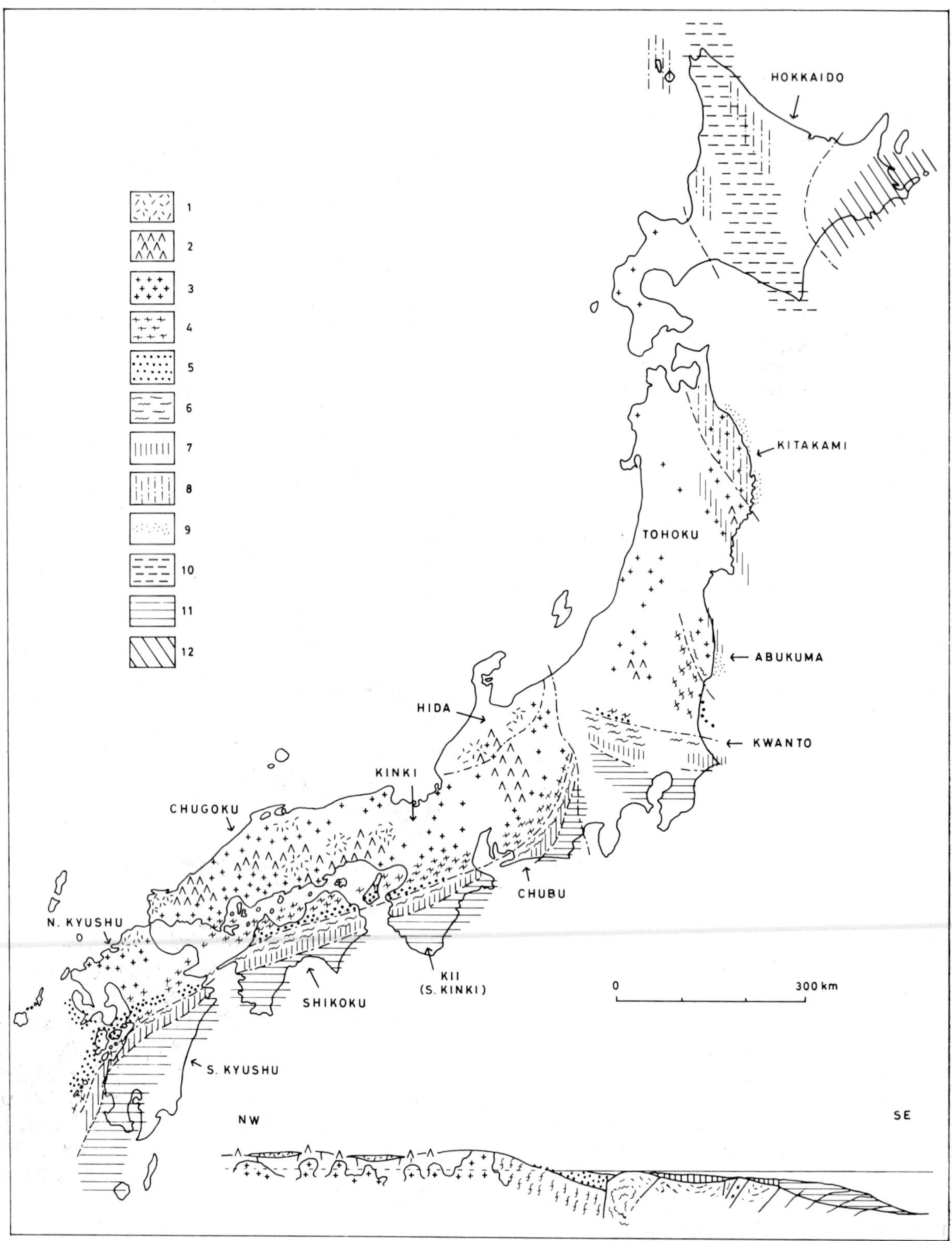

Fig. 17. Framework of the Cretaceous system in Japan. Schematic geological profile in the Late Cretaceous at the bottom. Legend: *1* = non-marine deposits; *2* = volcanoes; *3* = granites at depth and partly exposed; *4* = Ryoke–Abukuma metamorphics and granites; *5* = Upper Cretaceous clastics along the Median Tectonic Line; *6* = Sambagawa metamorphics (at depth and partly uprising); *7* = deltaic to marine deposits in the Chichibu zone and southern Kitakami (moderately folded); *8* = Neocomian geosynclinal deposits in northern Kitakami and Hokkaido; *9* = Upper Aptian–Lower Albian and Senonian littoral to neritic deposits in Kitakami and Abukuma (less deformed); *10* = Aptian–Maastrichtian deposits of the Yezo geosyncline; *11* = Cretaceous of the Shimanto geosyncline; *12* = Nemuro Group (K6 + Paleocene). (Reproduced by permission of the Nat. Res. Counc. Can.)

Kochian	K1	K1a Berriasian K1b Valanginian	Lower Neo- comian
Aridan [=Aritan]	K2	K2a Hauterivian K2b Barremian	Upper Neo- comian
Miyakoan	K3	K3a Aptian K3b Albian	
Gyliakian	K4	K4a Cenomanian K4b Turonian	
Urakawan	K5	K5a Coniacian K5b Santonian	Lower Sen- onian
Hetonaian	K6	K6a Campanian K6b Maastrichtian	Upper Sen- onian

It should be noted that Gyliakian in the above definition is not synonymous with the Gilyak Series of Sakhalin used by Russian geologists (see Nalivkin, 1973).

The literature for the lower Cretaceous ammonoids from Japan and Taiwan consists of a number of shorter papers besides Shimizu's (1931) summary of older date. They are T. Sato (1961a, b), Noda (1972), and H. Takahashi (1973) for K1; Yabe and Shimizu (1925a), Shimizu in Yabe et al. (1926), and Tatsuro Matsumoto (1947) for K2; Tatsuro Matsumoto et al. (1952, 1965), Tatsuro Matsumoto and Harada (1964), Nakai and Hada (1966), Nakai and Matsumoto (1968), Tatsuro Matsumoto and Hirata (1969), Tatsuro Matsumoto and Tashiro (1975) and Obata (1967–75), and Obata et al. (1975) for K3.

The literature for the Upper Cretaceous ammonites from Japan and Sakhalin is plentiful. Selected references among others are Schmidt (1873), Yokoyama (1890), Jimbo (1894), Yabe (1901–2, 1903–4, 1910, 1914, 1915), Yabe and Shimizu (1921, 1924, 1925b), Shimizu (1926, 1932, 1935a, b), Wright and Matsumoto (1954), Tatsuro Matsumoto (1938a, b, 1942a–d, 1953, 1954a, b, 1955a, b, 1956a, 1957a, 1959b, 1960, 1965a, b, 1967a, 1965–71, Parts I–V, 1970, 1973, 1975b), Tatsuro Matsumoto and Obata (1955, 1963a, b, 1966), Obata (1965), Tatsuro Matsumoto and Hashimoto (1953), Tatsuro Matsumoto and Saito (1954, 1956), Tatsuro Matsumoto et al. (1957, 1964, 1966, 1969, 1972a, b), Matsumoto and Ueda (in Ueda, 1962), Matsumoto and Obata (1963a, b), Matsumoto and Kanie (1967), Matsumoto and Muramoto (1967), Matsumoto and Inoma (1975), Matsumoto and Kawano (1975), Saito (1958–59, 1961–62), Druczic and Pergament (1963), Hashimoto (1973), Hirano (1975) and Tanabe (1975). These works have made clear the taxonomy and the stratigraphic succession of several important groups and the results may serve as a base for further advances (Figs. 19, 20). There are still groups awaiting monographic descriptions.

The Lower Cretaceous ammonites of Japan are mostly cosmopolitan forms and contain no Boreal elements. The Upper Cretaceous ammonites belong to the Indo-Pacific province with many species common to Japan and the Pacific coast of North America (see Tatsuro Matsumoto, 1959–60). Some species common in Japan have been recorded in southern India, Madagascar and Grahamland. There are also some cosmopolitan forms and also some endemic ones. Nautiloids occur occasionally but few have been described (Yabe and Shimizu, 1924b; Yabe and Ozaki, 1953; Tatsuro Matsumoto and Amano, 1964; and Matsumoto, 1967b). Belemnites sometimes present in the Lower Cretaceous (Hanai, 1953) are almost absent in the Upper Cretaceous where straight cephalopod shells occur (probably of a new genus allied to *Groelandibelus*).

Because of their stratigraphic utility there is a large literature devoted to the Inoceramidae (Yokoyama, 1890; Jimbo, 1894; Michael, 1899; Sokolow, 1914; Yehara, 1924; Nagao and Matsumoto, 1939–40, T. Matsumoto, 1957b; Pergament, 1965, 1966, Tatsuro Matsumoto and Noda, 1968, 1975; Tanabe, 1973; Noda, 1974, 1975; Dundo and Efremova, 1974). They are rare in the Lower Cretaceous with the exception of the Albian. They are also known from the Pacific coast of Siberia.

The Lower Cretaceous marine bivalves were monographed by Hayami (1965–66), who noted that the Berriasian fauna differed slightly from the Tithonian and that marked changes in faunal assemblage occur between K1 and K2 + K3 and again between K3 and K4. The Berriasian forms include the genera *Grammatodon, Pteroperna, Camptonectes, Ctenostreon, Vaugonia, Myophorella, Coelastarte, Eocallista, Pleuromya* and *Homomya,* while in K2 and K3 *Nanonavis, Trigonarca, Amygdalum, Pterinella, Gervillaria, Amphidonta, Rutitrigonia, Pterotrigonia, Anthonya, Pachythaerus, Ptychomya, Panopea* are common. In K1 brackish-water bivalves (called the Ryoseki fauna) occur commonly, such as *Bakewellia, Protocardia, Protocyprina, Eomiodon, Neomiodon, Isodomella* and *Tetoria.* Some of them also occur in the same facies of K3 age. Of the 180 Lower Cretaceous species 33 are closely allied to, or identical with,

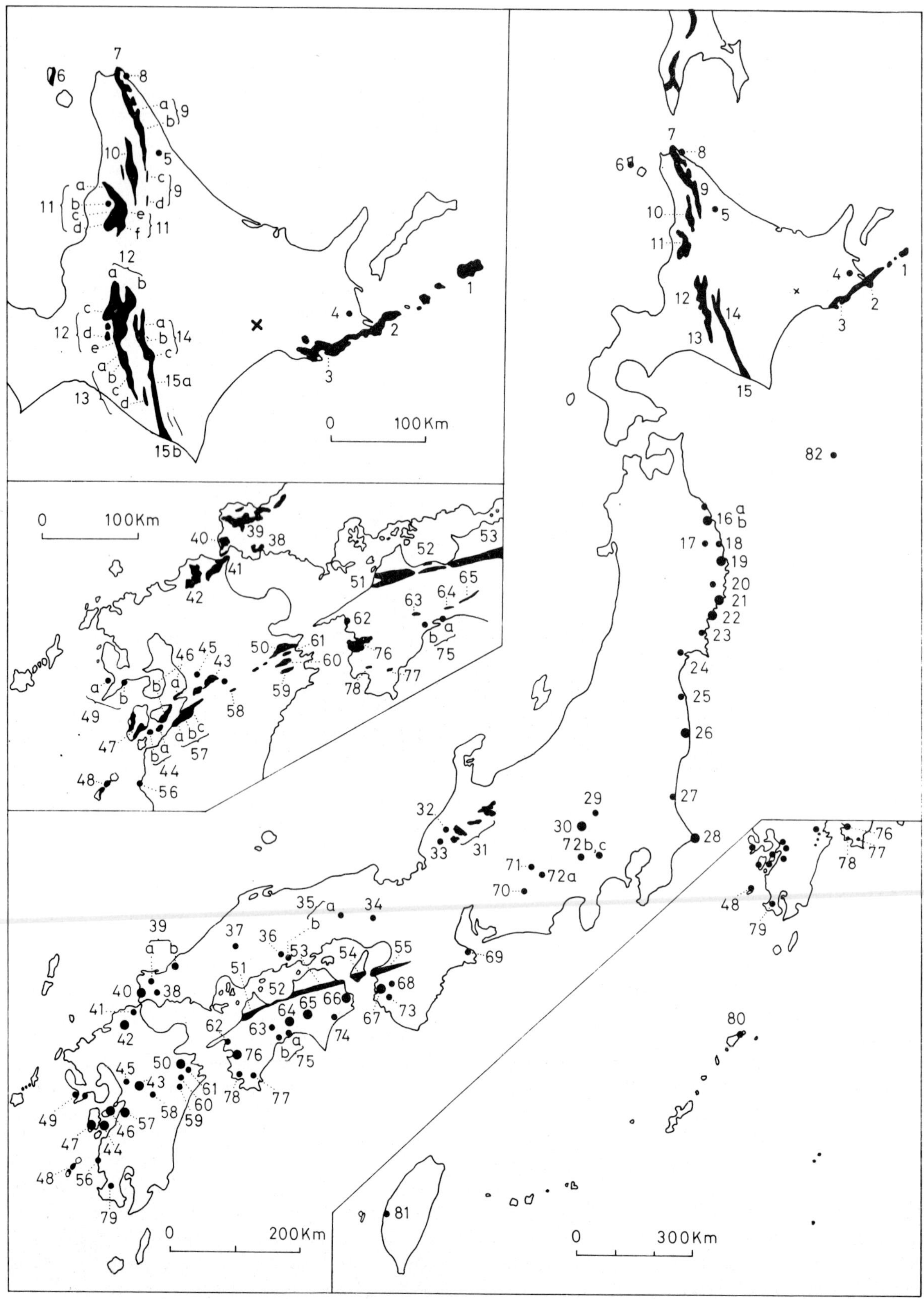

Fig. 18. Map of Japan and adjacent areas, showing the locations of Cretaceous fossiliferous formations. Black area = extensively outcropping area; large solid circles = other well-known areas; small solid circles = other localities; cross = presumably subsurface locality. See Table VI for the explanation of the numbered localities.

TABLE VI

List of localities of fossiliferous Cretaceous formations in Japan and adjacent islands (see Fig. 18 for their location)

The tectonic province to which a locality belongs is indicated with an abbreviation in square brackets (see Table I and Fig. 2 for the abbreviation). The groups of fossils hitherto reported are indicated as follows: A = ammonoids; F = foraminifera. M = mollusca, other than ammonoids; N = nannoplankton; O = other invertebrates and/or calcareous algae; P = plants, excluding S; S = spores and pollen; V = vertebrates

1. Shikotan island, southern Kurile islands [z2] M
2. Nemuro peninsula, eastern Hokkaido [z2] A, F. M, O
3. Coastal area of Akkeshi [z2] A, F, M, O
4. Nishibetsu (boring core) [z2] S
5. Horonbetsu, west of Esashi, northern Hokkaido [z4b] M
6. Rebun island, off northwest coast of Hokkaido [z4d] A, M, O
7. Soya hills and Cape Soya, northern central Hokkaido [z4] A, F, M, O, P
8. Mineoka, east of Soya [z4c] A, M, O
9. Tombetsu area, including Sarufutsu (*a*) and Nakatombetsu (*b*), and scattered southern extension at Onnenai (*c*) and Chiebun (*d*), northern central Hokkaido [z4b or c] A, F, M, O, P
10. Saku–Abeshinai area [= Teshio Nakagawa area], northern central Hokkaido [z4c] A, F, M, N, O, P. S, V
11. Rumoi district, including Chikubetsu (*a*), Haboro (*b*), Kotanbetsu (*c*), Obira [= Obirashibetsu] or Tappu [= Tap] (*d*), Shumarinai (*e*) and Soeushinai (*f*), northwestern Hokkaido [z4d] A, F, M, N, O, P, V
12. Cretaceous area adjacent to the Ishikari coal-field, including Ashibetsu (*a*), Soashibetsu (*b*), Ikushumbetsu, Pombetsu and Manji near Mikasa (*c*), Yubari (*d*), and Oyubari [= Shiyubari] (*e*), central Hokkaido [z4c] A, F, M, O, P, S, V
13. Southern extension of *12,* including Hobetsu (*a*), Tomiuchi [= Hetonai] (*b*), Furenai (*c*), and Yamamombetsu (*d*), southern central Hokkaido [z4c] A, M, O, P, V
14. Upper Sorachi valley, including Furano (*a*), Yamabe (*b*) and Kanayama (*c*), central Hokkaido [z4b] A, F, M, O, P
15. Western foot of the Hidaka range, including Hidaka (*a*) and Urakawa (*b*), southern central Hokkaido [z4b] A, M, O, P, V
16. Kuji (*a*) and Taneichi (*b*), Iwate Prefecture [z5a] A, M, O, P, S
17. Iwaizumi, Iwate Prefecture [z5a] A
18. Rikuchu, including Omoto, Iwate Prefecture [z5a] M, O, P
19. Miyako, Iwate Prefecture [z5a] A, F, M, O, P, S
20. Kamihei (near Omine mine), Iwate Prefecture [z5b] M
21. Ofunato and Ryori, Iwate Prefecture [z5b] A(?), M, O, P
22. Oshima (off Karakuwa), Miyagi Prefecture [z5b] A, M, O, P
23. Jusanhama, near Shizukawa, Miyagi Prefecture [z5b] M, P
24. Ayukawa, Ojika peninsula, Miyagi Prefecture [z5b] A, M, O, P
25. Koyamada, Soma, Fukushima Prefecture [z6] A, M
26. Futaba, Fukushima Prefecture [z6] A, F, M, O, P, S, V
27. Nakaminato, Ibaragi Prefecture [z6] A, M, O, P
28. Choshi, Chiba Prefecture [z8b?] A, F, M, O, P, S
29. Shimonita, Gunma Prefecture [z8] M
30. Sanchu graben, Gunma–Nagano Prefectures [z8b] A, M, O, P
31. Itoshiro of the Tetori Group, Hida massif [z9] M, P
32. Akaiwa of the Tetori Group, Hida massif [z9] M, P
33. Asuwa–Ohmichidani, Hida massif [z9] M, P
34. Sasayama, Hyogo Prefecture [z10] O, P
35. Ikuno, Hyogo Pref. (*a*) and Kamogata (*b*), Okayama Prefecture [z10] P
36. Inakura, north of Kasaoka, Okayama Prefecture [z10] O
37. Sakugi (Suritaki), northwest of Miyoshi, Hiroshima Prefecture [z10a] P (may be Paleocene)
38. Asa, Yamaguchi Prefecture [z10a] M
39. Tawarayama near Nishiichi, (*a*) M, and Fukuga near Abu (*b*) P, Yamaguchi Prefecture [z10a]
40. Yoshimo, Toyonishi, Yamaguchi Prefecture [z10a] M
41. Moji, Kokura and Yahata, Kitakyushu, Fukuoka Prefecture [z10] M, O, V
42. Wakino, 20 km east of Fukuoka, northern Kyushu [z10] M, O
43. Mifune, Kumamoto Prefecture, central Kyushu [z11] A, M, P
44. Goshonoura [Goshora], Kumamoto Prefecture (*a*) and Shishijima, Kagoshima Prefecture (*b*), central Kyushu [z11] A, M, O
45. Himenoura, eastern Amakusa [= Kamishima] (*a*) and Uto peninsula (*b*), Kumamoto Prefecture, central Kyushu [z11] A, M, O, P
46. Kumamoto, central Kyushu [z11] M

47. Western Amakusa [= Shimojima], Kumamoto Prefecture, central Kyushu [z17] A, M, O
48. Koshiki-jima, Kagoshima Prefecture, central Kyushu [z17] A, M, O
49. Takashima (*a*) and Mogi (*b*), Nagasaki Prefecture, central Kyushu [z17] A, M, V
50. Onogawa, Oita Prefecture, central Kyushu [z11–12] A, M, O
51. Matsuyama, Ehime Prefecture, northwestern Shikoku [z11] A, M, O
52. Hotokezaki, near Niihama, Ehime Prefecture, northern Shikoku [z11] A, M
53. Asan mountain range, Kagawa and Tokushima Prefectures, northeastern Shikoku [z11] A, M, O
54. Island of Awaji, Hyogo Prefecture [z11] A, M, O, P
55. Izumi mountain range, Osaka and Wakayama Prefectures [z11] A, M, O, P
56. Gumizaki, Kagoshima Prefecture, southern Kyushu [z12] M
57. Hinagu (*a*), southeast of Yatsushiro (*b*) and Tomochi (*c*), Kumamoto Prefecture, southern Kyushu [z12] A, M, O, P
58. Shibanomoto, near Kuraoka, Miyazaki Prefecture, southern Kyushu [z12] M, P
59. Yamabu, Oita Prefecture, southern Kyushu [z12] A, M, P
60. Haidateyama, Oita Prefecture, southern Kyushu [z12] A, M, O, P
61. Tano, Oita Prefecture, southern Kyushu [z12] A, M, O
62. Nigyu, near Mikame and Kikunotani, Ehime Prefecture, western Shikoku [z12] M
63. Ochi near Sakawa, Kochi Prefecture, central Shikoku [z12] A, M, O, P
64. Kochi and Ryoseki, Kochi Prefecture, central Shikoku [z12] A, M, O, P
65. Monobegawa, Kochi Prefecture, central Shikoku [z12] A, F, M, O, P
66. Nakagawa and Katsuuragawa, Tokushima Prefecture, eastern Shikoku [z12] A, F, M, O, P
67. Yuasa, Wakayama Prefecture [z12] A, M, O, P
68. Toyajo hills and Aridagawa [= Aritagawa] valley, Wakayama prefecture [z12] A, M, O, P
69. Shima peninsula, Miye Prefecture [z12] M, O, P
70. Misakubo, Shizuoka Prefecture, Akaishi Mountains [z12] M, O, P
71. Todai, Nagano Prefecture, Akaishi Mountains [z12] A, M, O, P
72. Shirane, Akaishi Mountains (*a*) [z13] M, Yamanashi (*b*) O, Kobotoke (*c*) [z8] M
73. Terasoma, Wakayama Prefecture [z13] M
74. Mugi, Tokushima Prefecture [z13] A
75. Kitayama, Azai and Nii, southwest of Kochi (*a*) and Doganaro, Hayama and Susaki, south of Sakawa (*b*), Kochi Prefecture, southern Shikoku [z13] A, M, O
76. Uwajima, Ehime Prefecture, southwestern Shikoku [z13] A, F, M, O, P
77. Nakamura–Arioka, Kochi Prefecture, southwestern Shikoku [z13] A, M, O
78. Johen–Ipponmatsu, Ehime Prefecture, southwestern Shikoku [z13] A, M, O
79. Kawanabe, Kagoshima Prefecture, southern Kyushu [z13] A, M
80. Ohgachi [Tatsugo], Amami Oshima, Kagoshima Prefecture, Nansei islands [z19] A
81. Peikang, western Taiwan [z22a] A, F, M, P
82. Erimo (Sysoev) seamount, off Cape Erimo, southern Hokkaido, M, O
X. Shiranuka hills, east of Honbetsu, eastern Hokkaido [1b]: inferred subsurface formation

Remarks: Localities in Sakhalin are excluded from this list.

species known in Europe and extra-Japanese provinces (Hayami, 1966).

Although studied by many paleontologists (e.g., Yabe and Nagao, 1928; Nagao, 1930, 1932, 1938; Nagao and Otatume, 1938; Tatsuro Matsumoto, 1938a; Okada, 1958; Nakano, 1957b, 1958a, b, 1960, 1961a–c; Ichikawa and Maeda, 1958, 1963; Maeda and Kitamura, 1964; Maeda and Kawabe, 1967; Tamura, 1961c, 1973a, 1975; Tamura and Tashiro, 1967; Tamura et al., 1968; Tamura and Matsumura, 1974; Tashiro, 1971, 1972, 1976; Hayami, 1975) more work is needed for Upper Cretaceous bivalves. In general species of the northern Pacific province seem to become more distinct in the Upper Cretaceous. Faunal change appeared near the K3–K4 boundary, with some change between K4 and K5, K5 and K6 and K6a and K6b, which seems to be related to the history of marine transgression and regression.

Gastropod faunas have received relatively little attention although Nagao (1932, 1934, 1939) described some species and Kanie (1975) has recently shown an evolutional series of *Anisomyon.* The non-marine molluscan faunas are noteworthy for their endemic forms (Kobayashi and Suzuki, 1936, 1937, 1939, 1941; Hoffet, 1937; Tatsuro Matsumoto, 1938a; Yabe and Hayashi, 1938; K. Suzuki, 1940, 1943, 1949; Kobayashi, 1956b, 1963, 1968; Hase, 1960; Ota, 1959a–c, 1960a, b, 1964, 1965, 1973, 1974, 1975; Maeda, 1962b, c, 1963b, Martinson,

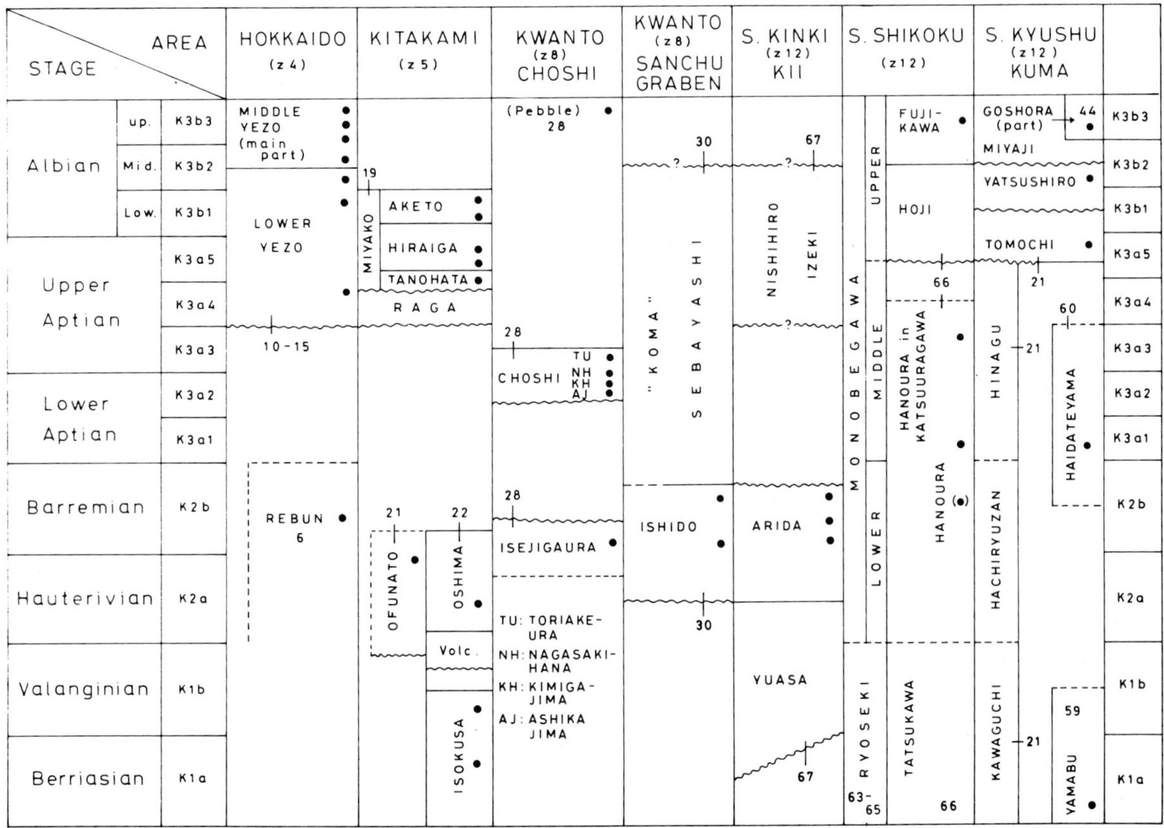

Fig. 19. Correlation of representative Lower Cretaceous formations in Japan. The stratigraphic position of guide ammonite species is indicated by a solid circle. Numbers refer to locations in Fig. 18 and Table VI. See Table VII for the biostratigraphic subdivisions (K1–K3). (Prepared by Obata and Matsumoto.)

1965; Hayami and Ichikawa, 1965; Hayami and Nakai, 1965; Tamura, 1970; Yang, 1974, 1975). The most diagnostic genera are *Nagdongia, Plicatounio, Nippononaia, Wakinoa,* and *Trigonioides* (including subgenera *Kumamotoa* and *Hoffetrigoinioides*). Correlation between non-marine and marine formations is fairly difficult, but at several places, as in central Kyushu for example, they interdigitate with one another.

Estherians and ostracods occur at some horizons in non-marine beds (Kobayashi and Huzita, 1942; Kobayashi and Kido, 1947a, b; Hanai, 1951; Kobayashi and Kusumi, 1953a, b; Kusumi, 1960, 1961). Vertebrate remains are rare in continental deposits of Japan, although some reptilian, bird and fish remains have been found in marine deposits.

Micro- and nannofossils are sometimes abundant if poorly preserved. Orbitolinids have been studied by Yabe and Hanzawa (1926), and smaller foraminifera by Asano (1950a, b), Takayanagi (1960a, b), Takayanagi and Iwamoto (1961), Yoshida (1958, 1961, 1963, 1969) and oil company geologists. No zonal planktonic succession has been firmly established, but the work is now in progress. Certain planktonic species have been used in discussions of the Cretaceous–Tertiary boundary (Yoshida, 1961; Asano, 1962) in eastern Hokkaido (loc. *2, 3*). Long-ranging species seem to give some paleoecological information. Nannoplankton from the Senonian of northeastern Japan (loc. *26*) (Takayama and Obata, 1968) and those from the Upper Cretaceous of Hokkaido are being studied by Okamura (1975, oral comm.) to attempt to correlate the nannofossil scale with the ammonite–inoceramid zonal scheme.

Tatsuaki Kimura (1973) distinguished two major floral provinces in the Lower Cretaceous K1 of Japan (Fig. 21): an outer Ryoseki flora with matoniaceous, gleicheniaceous and schizaeaceous ferns and cycade-

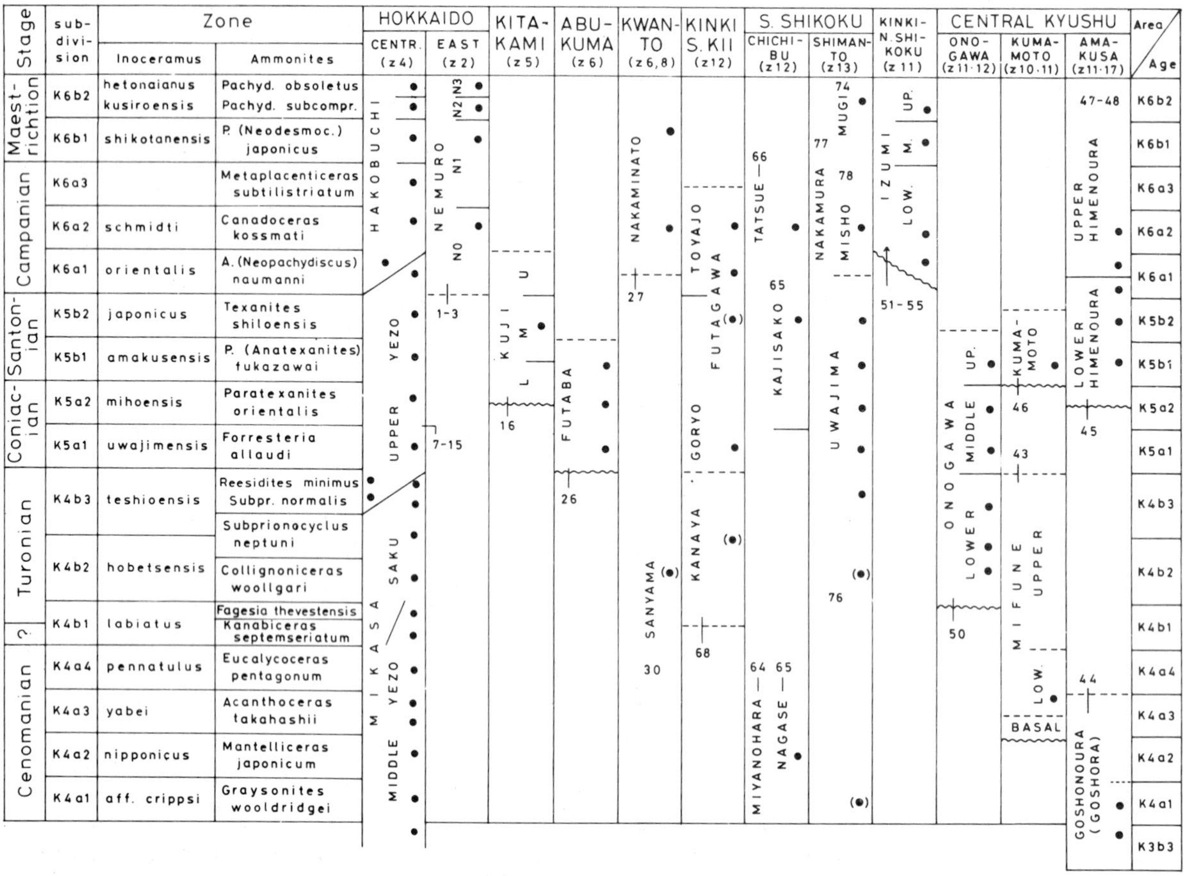
Stage
sub-division
Zone
Inoceramus
Ammonites
HOKKAIDO
CENTR. (z4)
EAST (z2)
KITA-KAMI (z5)
ABU-KUMA (z6)
KWAN-TO (z6,8)
KINKI S.KII (z12)
S. SHIKOKU
CHICHI-BU (z12)
SHIMAN-TO (z13)
KINKI-N.SHI-KOKU (z11)
CENTRAL KYUSHU
ONO-GAWA (z11·12)
KUMA-MOTO (z10·11)
AMA-KUSA (z11·17)
Area
Age
Maest-richtion
Campanian
Santon-ian
Coniac-ian
Turonian
Cenomanian
K6b2 hetonaianus kusiroensis Pachyd. obsoletus Pachyd. subcompr.
K6b1 shikotanensis P. (Neodesmoc.) japonicus
K6a3 Metaplacenticeras subtilistriatum
K6a2 schmidti Canadoceras kossmati
K6a1 orientalis A. (Neopachydiscus) naumanni
K5b2 japonicus Texanites shiloensis
K5b1 amakusensis P. (Anatexanites) fukazawai
K5a2 mihoensis Paratexanites orientalis
K5a1 uwajimensis Forresteria allaudi
K4b3 teshioensis Reesidites minimus Subpr. normalis
Subprionocyclus neptuni
K4b2 hobetsensis Collignoniceras woollgari
K4b1 labiatus Fagesia thevestensis Kanabiceras septemseriatum
K4a4 pennatulus Eucalycoceras pentagonum
K4a3 yabei Acanthoceras takahashii
K4a2 nipponicus Mantelliceras japonicum
K4a1 aff. crippsi Graysonites wooldridgei
K3b3
HAKOBUCHI
UPPER YEZO
MIKASA SAKU
MIDDLE YEZO
NEMURO
NO
N1
N2 N3
1-3
7-15
KUJI
U
M
L
16
FUTABA
26
NAKAMINATO
SANYAMA
27
30
TOYAJO
FUTAGAWA
GORYO
KANAYA
68
TATSUE
KAJISAKO
MIYANOHARA
NAGASE
66
65
64 65
NAKAMURA
MISHO
MUGI
UWAJIMA
74
77
78
76
IZUMI
UP.
M.
LOW.
51-55
ONOGAWA
UP.
MIDDLE
LOWER
50
KUMA-MOTO
MIFUNE
UPPER
LOW.
BASAL
46
43
UPPER HIMENOURA
LOWER HIMENOURA
GOSHONOURA (GOSHORA)
47-48
45
44

Fig. 20. Correlation of representative Upper Cretaceous formations in Japan. The stratigraphic position of guide ammonite and/or inoceramid species is indicated by a solid circle. Numbers refer to locations in Fig. 18 and Table VI. For the biostratigraphic subdivisions see Table VII.

oids such as *Ptilophyllum* and *Zamiophyllum* allied to the Wealden flora of Europe, India and southern Primoria, and an inner province with its Oguchi (Itoshiro) flora of the Tetori (Hida) area.

The latter flora is characterized by Osmundaceae and Dicksoniaceae forms, *Dictyozamites, Ctenis, Nilssonia, Nilssoniocladus, Podozamites* and various ginkgophytes. This flora has some affinity with the contemporary Siberian flora and may imply an environmental difference between the two Japanese floras: a warm dry climate for the Ryoseki flora and a warm moderately humid climate for the Tetori flora, although apparently contradictory with the general paleogeography (compare Fig. 16 with Fig. 21). The Kiyosuye (Toyonishi) flora of western Japan and the Naktong (= Nagdong) flora of southern Korea seem to form an intermediate province but have been inadequately studied.

Although there are local differences in the K2 and K3 floras from Akaiwa, Tamodani (loc. 32), Choshi (loc. 28) and Yatsushiro (loc. 57b) and some differences between them and the K1 flora, the general distinction of two paleofloristic provinces is maintained (Tatsuaki Kimura, 1961, 1973, 1975b; Tatsuaki Kimura and Sekido, 1965, 1966, 1967, 1972, 1974, 1975; Tatsuaki Kimura and Asama, 1975; Nishida, 1962, 1965-1967, 1973).

Upper Cretaceous plants have been reported in intermontane lake deposits [e.g., Asuwa and Ohmichidani formations (loc. *33*), and correlatives (locs. *35, 37*)] and in near-shore deposits [e.g., Hakobuchi Group (loc. *12e*), Kuji Group (loc. *16*), Mifune Group (loc. *43*), Mitsuse Formation (loc. *49*), Izumi Group (loc. *55*)] (Stopes and Fujii, 1910; Endo, 1925; Koriba and Miki, 1958; Oyama, 1960–61; Oyama and Matsuo, 1964; Matsuo, 1954, 1960, 1962, 1964,

TABLE VII

Biostratigraphic division and selected index species of the Cretaceous system in Japan and adjacent islands (localities shown by numbers referring to Table VI and Fig. 18)

Lower Cretaceous

K1a (Berriasian): *Protacanthodiscus akiyama* (*22*), *Berriasella* aff. *patula* (*59*)
K1b (Valanginian): *Thurmanniceras isokusense* (*22*), *Sarasinella* aff. *hyatti* (*25*)
K2a (Hauterivian): *Crioceratites ishiwarai* (*22*), *Holcodiscus* sp. (*21*), *Pseudothurmannia hanourensis* (*67*)
K2b (Barremian): *Pulchellia ishidoensis* (*30, 66*), *P.* (*Caicedea*) sp. (*6*), *Emericiceras emerici* (*28*), *Shasticrioceras nipponicum* (*67*), *Barremites* cf. *strettostoma, B.* sp. (*28, 30, 67*), *Silesites* aff. *seranonis* (*67*)
K3a (Aptian)
K3a1 (lower Lower Aptian): not yet well identified. *Ancyloceras* sp. from loc. *60* may be K2b or Upper Barremian
K3a2 (upper Lower Aptian): *Australiceras* cf. *gigas, Tropaeum* aff. *bowerbanki, Cheloniceras* (*Cheloniceras*) *myendorfi* (*28*), *Ch.* (*Ch.*) *shimizui* (*66, 76*), *Dufrenoyia* cf. *discoidalis* (*81*)
K3a3 (lower Upper Aptian): *Cheloniceras* (*Epicheloniceras*) cf. *martinoides* (*66*), *Ch.* (*E.*) sp. (*28, 81*)
K3a4 (middle Upper Aptian): *Hypacanthoplites subcornuerianus, Nolaniceras yaegashii, Miyakoceras tanohataense, Valdedorsella akuschaensis* (*19*), *Parahoplites* aff. *maximus* (*10*)
K3a5 (upper Upper Aptian): *Diadochoceras nodosocostatiforme, Eodouvilleiceras matsumotoi, Valdedorsella akuschaensis, V. getutina, Melchiorites yabei, Uhligella matsushimaensis* (*19, 57c*)
K3b (Albian)
K3b1a (lower Lower Albian): *Pseudoleymeriella hataii, Valdedorsella* sp., *Hulenites* sp. (*19*)
K3b1b (upper Lower Albian): *Douvilleiceras mammillatum* (19, 12?) *Valdedorsella* sp., *Desmoceras* sp. (*19, 12*)
K3b2 (Middle Albian): *Oxytropidoceras* sp. (*11*), *Adkinsites* aff. *belknapi, Dipoloceras* aff. *pseudaon* (*12*), *Dipoloceras* aff. *fredericksburgense, D.* (*Diplasioceras*) *tosaense* (*75*)
K3b3 (Upper Albian): *Mortomiceras* (*Mortoniceras*) *rostratum, M.* (*Cantabrigites*) *imaii, M.* (*Deiradoceras*) sp., *Dipoloceras* (*Diplasioceras*) cf. *tosaense, Desmoceras latidorsatum, D.* (*Pseudouhligella*) *dawsoni shikokuense, Puzosia subcorbarica, Pachydesmoceras denisoni, Hypophylloceras yeharai, Anagaudryceras sacya, Ammonoceratites ezoensis, Mariella* aff. *cantabrigiensis, Pseudohelicoceras* sp., *Inoceramus sulcatus, Inoceramus anglicus* (*11, 12, 14, 15; 44, 66*)

Upper Cretaceous

K4a1 (lower Lower Cenomanian): *Graysonites wooldridgei, G.* cf. *fountaini, G.* aff. *adkinsi, Euhystrichoceras* cf. *nicaisei, Prionocycloides* cf. *proratum, Stoliczkaia* (*Stoliczkaia*) *amonoi, Stoliczkaia* (*Shumarinaia*) *hashimotoi, Desmoceras kossmati, Marshallites* cf. *cumshwaensis, Eogunnarites unicus, Anagaudryceras sacya, Parajaubertella kawakitana, Zelandites inflatus, Z. odiensis, Anisoceras* sp., *Inoceramus* aff. *crippsi* (*8, 10, 11, 12, 13; 44*)
K4a2 (upper Lower Cenomanian): *Mantelliceras japonicum, Sharpeiceras kongo, Forbesiceras* aff. *obtectum, Desmoceras* (*Pseudouhligella*) *japonicum, D.* (*P.*) *ezoanum, Marshallites compressus, Eogunnarites unicus, Mikasaites orbicularis, Anagaudryceras sacya, Kossmatella* (*Murphyella*) *enigma, Parajaubertella imlayi, Zelandites inflatus, Hypoturrilites* cf. *gravesianus, H. tuberculatus, Ostlingoceras* aff. *bechei, Turrilites costatus, Sciponoceras baculoides, Inoceramus concentricus nipponicus* (*10, 11, 12, 13; 63, 65*)
K4a3 (Middle Cenomanian): *Acanthoceras amphibolum, A. takahashii, Euomphaloceras meridionale, Calycoceras asiaticum, C. orientale, Eucalycoceras* aff. *spathi, Desmoceras* (*Pseudouhligella*) *japonicum, D.* (*P.*) *ezoanum, Puzosia nipponica, Austiniceras* sp., *Pachydesmoceras* cf. *denisonianum, Marshallites compressus, M. olcostephanoides, Eogunnarites unicus, Anagaudryceras sacya, Turrilites costatus, Sciponoceras kossmati, Inoceramus concentricus costatus, Inoceramus yabei* (*8, 10, 11, 12, 13; 44*)
K4a4 (Upper Cenomanian): *Calycoceras* cf. *naviculare, Eucalycoceras pentagonum, Desmoceras* (*Pseudouhligella*) *japonicum, D.* (*P.*) *ezoanum, Damesites laticarinatus* (?), *Marshallites olcostephanoides, Sciponoceras kossmati, Inoceramus pennatulus* (*10, 11, 12, 15*)
K4a4 or K4b1: *Kanabiceras septemseriatum, Pseudocalycoceras* aff. *dentonense, Sciponoceras kossmati* (*12*)
K4b1 (Lower Turonian): *Mammites* sp., *Pseudaspidoceras sorachiense, Fagesia thevestensis, Fagesia* n. sp., *Yubariceras* aff. *japonicum, Y. pseudomphalum, Shuparoceras yagii, Vascoceras* aff. *durandi, Tragodesmoceroides subcostatum, Damesites* aff. *laticarinatus, Puzosia intermedia orientalis, Gaudryceras* aff. *varagurense, Allocrioceras* sp., *Sciponoceras kossmati, S. orientale, Inoceramus* (*Mytiloides*) *labiatus* (*10, 11, 12, 13, 14, 15; 73?*) *Collignoniceras* (*Selwynoceras*) sp. (*80*)
K4b2 (Middle Turonian): *Collignoniceras* (*Collignoniceras*) *woollgari, Subprionocyclus bravaisianus, Yubariceras yubarense, Y.* aff. *ornatissimum, Y. japonicum, Y. otatumei, Shuparoceras abei, Romaniceras deverioide, R. yezoense, Desmoceras* (*Pseudouhligella*) n. sp., *Tragodesmoceroides subcostatus, Mesopuzosia pacifica, M. indopacifica, M. yubarensis* (?), *Pachydesmoceras pachydiscoide, Jimboceras planulatiforme, Gaudryceras denseplicatum, G. intermedium, Anagau-*

dryceras limatum, Hypophantoceras sp., *Eubostrychoceras japonicum, Scalarites scalaris, Nipponites mirabilis, Muramotoceras yezoense, Scaphites planus, Otoscaphites puerculus, Sciponoceras orientale, Inoceramus hobetsensis, I. (Mytiloides) teraokai (10, 11, 12, 13, 14, 15; 50, 61, 76)*

K4b3 (Upper Turonian): *Subprionocyclus neptuni, S. normalis, Reesidites minimus, Prionocyclus wyomingensis, P. aberrans, Lymaniceras planulatum, Damesites ainuanus, Tragodesmoceroides subcostatus, Mesopuzosia pacifica, M. indopacifica, M. yubarensis, Pachydesmoceras pachydiscoide, Yokoyamaoceras kotoi, Gaudryceras denseplicatum, G. intermedium, Anagaudryceras limatum, Hyphantoceras oshimai, Eubostrychoceras* aff. *woodsi, Eubostrychoceras japonicum, Nipponites bacchus, Madagascarites ryu, Scalarites scalaris, S. mihoensis, S. densicostatus, Scaphites planus, Otoscaphites puerculus, Sciponoceras intermedium, Baculites undulatus, Inoceramus teshioensis, I. tenuistriatus, I. iburiensis, I. (Mytiloides) incertus, I. (M.) pedalionoides (8, 10, 11, 12, 13, 14; 50, 76)*

K5a1 (Coniacian, probably Lower and Middle Coniacian): *Pseudobarroisiceras nagaoi, Prionocycloceras sigmoidale, P. wrighti, Sornayceras wadae, Peroniceras* sp., *Barroisiceras* aff. *onilahyense, B. (Basseoceras) inornatum, Forresteria (Forresteria) allaudi, F. (Muramotoa) yezoensis, Yabeiceras orientale, Hourcquia hataii, Damesites damesi, D. semicostatus, Mesopuzosia yubarensis, Kossmaticeras theobaldianum, Yokoyamaoceras kotoi, Gaudryceras denseplicatum, G. intermedia, G. tenuiliratum, Anagaudryceras limatum, Eubostrychoceras indopacificum, E. muramotoi, Yezoceras nodosum, Scalarites mihoensis, Polyptychoceras obstrictum, Scaphites pseudoequlis, S. formosus, Otoscaphites klamathensis, O. matsumotoi, Baculites yokoyamai, Inoceramus uwajimensis, Didymotis akamatsui (10, 11, 12, 13, 14, 15; 26, 50, 67, 76)*

K5a2 (Upper Coniacian): *Paratexanites (Paratexanites) orientalis, Paratexanites (Parabevahites) serratomarginatus, Protexanites (Protexanites) planatus* (?), *Sornayceras proteus, S. omorii, Mesopuzosia yubarense, Kossmaticeras* aff. *japonicum, Yokoyamaoceras kotoi, Anagaudryceras limatum, Scalarites mihoensis, Baculites schencki, B. boulei, Inoceramus (Cremnoceramus) mihoensis, I. (Platyceramus) yubarensis, I. (Sphenoceramus) yokoyamai, I. (Cordiceramus)* sp. *(10, 11, 12, 13, 15, 26, 50, 76, 79)*

K5b (Santonian): *Protexanites (Protexanites) bontanti shimizui, P. (Anatexanites) fukazawai, Texanites (Texanites) oliveti, T. (T.)* aff. *quinquenodosus, T. (Plesiotexanites) kawasakii, T. (P) pacificus, T. (P.) stangeri, T. (P.) shiloensis, Pseudoschloenbachia* sp., *Hauericeras (Gardeniceras) angustum, Mesopuzosia densicostata, Kitchinites (Neopuzosia) japonica, K. (N.) ishikawai, Yokoyamaoceras jimboi, Anapachydiscus fascicostatus, A. sutneri, A. yezoensis, Eupachydiscus haradai, E. teshioensis, Menuites japonicus, M. pusillus, Anagaudryceras yokoyamai, Saghalinites nuperus, Hyphantoceras oshimai, H. orientale, Neocrioceras spinigerum, Eubostrychoceras densicostatum, Bostrychoceras* (?) *otsukai, Didymoceras* sp., *Baculites uedae, B. princeps, B. Baylyi, B. capensis, B. boulei, Inoceramus (Cordiceramus)* sp., *Inoceramus* (s.l.) *amakusensis, I. ezoensis, I. (Cladoceiomus?) japonicus, I. (Sphenoceramus) naumanni, I. (S.) orientalis nagaoi, I. (Cataceramus) balticus, Anisomyon cassidarius (8, 9, 10, 11, 12, 13, 15, 16, 17, 44, 45, 49, 50, 68, 72?, 73, 76)*

K6a1 (approximately Lower Campanian): *Bevahites* aff. *lapparenti (52), Submortoniceras* sp., *Desmophyllites diphylloides, Hauericeras angustum, Kitchinites (Neopuzosia) ishikawai, Anapachydiscus (Neopachydiscus) naumanni, Eupachydiscus haradai, Canadoceras yokoyamai, Menuites* aff. *sturi, Pachydiscus kobayashii* (?), *Gaudryceras striatum, Tetragonites popetense, Ainoceras kamuy, Didymoceras awajiense, Baculites chicoensis, B. tanakae, B. regina* (?), *Inoceramus (Sphenoceramus) orientalis, I. (S.) elegans, I. (Cataceramus) balticus, I. vanuxemiformis, Anisomyon transformis (7, 9, 10, 11, 13, 15, 27, 47, 48, 52, 54)*

K6a2 (approximately Middle Campanian if Campanian is tripartite): *Desmophyllites diphylloides, Natalites* sp., *Canadoceras kossmati, C. mysticum, Pachydiscus* aff. *subcompressus, Teshioites ryugasense, T. teshioensis, Urakawites rotalinoides, U.* cf. *binodatum, Gaudryceras striatum, Gaudryceras crassicostatum, Tetragonites popetense, Saghalinites nupersus, Ryugasella ryugasensis, Didymoceras awajiense, Baculites inornatus, B. occidentalis, B. subanceps, Inoceramus (Sphenoceramus) schmidti, I. (S.) sachalinensis, I. (S.) elegans, I. (Cataceramus) balticus toyajoanus, I.* (s.l.) *vanuxemiformis, I. (Cremnoceramus) yuasai, Anisomyon giganteus (1, 7, 9, 10, 12, 13, 14, 15, 27, 47, 48, 51, 53, 54, 55, 68)*

K6a3 (Upper Campanian): *Metaplacenticeras subtilistriatum, Desmophyllites diphylloides, Canadoceras multicostatum, Pachydiscus* aff. *neevesi, Pseudomenuites* sp., *Saghalinites* sp., *Tetragonites popetense, Schlueterella* (?) n. sp., *Baculites subanceps (7, 10, 15, 47?, 48?, 53?, 54, 55?)*

K6b1 (probably Lower Maastrichtian; could be uppermost Campanian): *Damesites hetonaiensis, Pachydiscus (Neodesmoceras) japonicus, Patagiosites compressus, Hauericeras* cf. *rembda, Pseudophyllites* cf. *indra, Nostoceras* sp., *Epiphylloceras subtuberculatum, Baculites rex, Inoceramus (Cataceramus) shikotanensis, Anisomyon problematicus (1, 9, 10, 13, 15, 54, 57, 58)*

K6b2 (Maastrichtian): *Pachydiscus subcompressus, P. (Neodesmoceras) obsoletum, Gaudryceras* n. sp., *Zelandites varuna japonica, Inoceramus hetonaianus, I. kusiroensis (1, 9, 13, 54)*

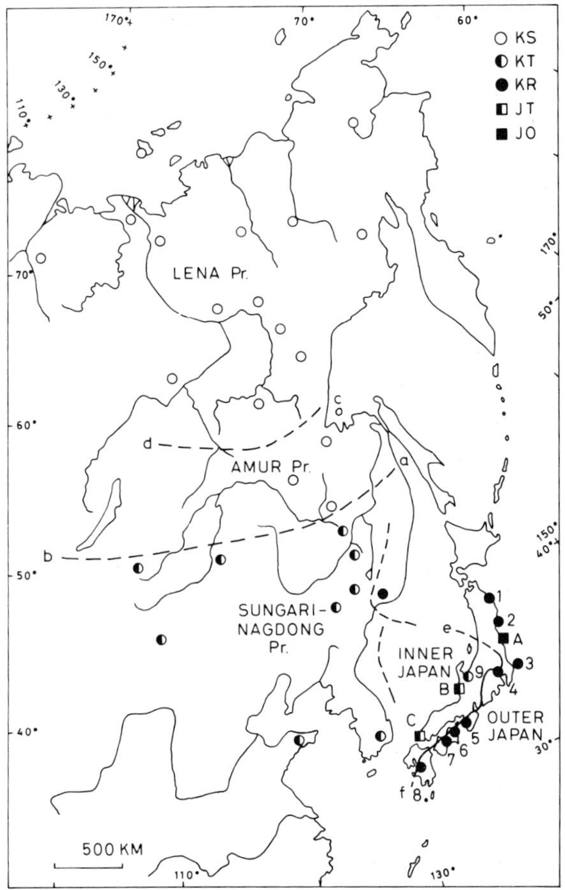

Fig. 21. Early Cretaceous floral province of Eastern Asia. Legend: *KS* = Siberian flora; *KT* = Cretaceous Tetori (Oguchi) flora; *KR* = Ryoseki flora; *JT* = Jurassic Tetori flora; *JO* = Jurassic Soma flora. Boundary of the flora provinces: *a–b* = that between the Siberian and the Indo-European provinces by Vakhrameev (1971); *c–d* = that between the Lena and the Amur provinces by Vakhrameev (1971); *e–f* = that between inner and outer Japan by Kimura. Locations: *1* = Omoto; *2* = Oshika (= "Ojika"); *3* = Choshi; *4* = Kwanto Mountains; *5* = Yuasa; *6* = Katsuuragawa; *7* = Kochi (Ryoseki); *8* = Yatsushiro; *9* = Itoshiro; *A* = Soma; *B* = Kuzuryu; *C* = Kiyosuye. (Courtesy of Kimura.)

1966, 1967, 1970, 1975). The floras suggest, according to Matsuo (1970), mixed cycad and coniferous forests with some dicotyledonous intermingling. A distinction can be made between the early (Asuwa–Mifune) flora of K4 and the late K5–K6 flora of Ohmichidani, Kuji and Hakobuchi. The former may correlate with the Gilyakian and the latter with the Orokkian [=Orochenian] floras of Sakhalin (Kryshtofovich and Baikovskaya, 1960). There are florules such as Oharai and Suritaki (Sakugi) whose age is probably Tertiary although once considered as Cretaceous.

As a result of palynological work (S. Sato, 1961; K. Takahashi, 1964, 1970, 1974; Miki, 1972a, b, 1973) various microfloras have been recognized in Hokkaido and northeast Japan. The Miyako (K3) is characterized by abundant *Classopollis*. The Saku (K4), Kuji and Futaba (K5) have been correlated with the Lower and Upper Zabitinsk floras of the Zeya–Bureya basin of eastern Siberia and the Lower Hakobuchi (K6a) and Upper Hakobuchi (K6b) with the *Aquilapollenites* province of Zaklinskaya (1962).

Paleoclimate and paleomagnetism

Some of the climatic interpretation of the paleontological data has been mentioned above. The remarkable floral change from Lower to Upper Cretaceous is paralleled by a neat contemporaneous change in the marine fauna. A paleotemperature of 18°C was recorded on a belemnite of K3 age from loc. *19* (Lowenstam and Epstein, 1954) in a fauna which includes *Orbitolina* and hippuritoids.

The predominance of quartz-feldspathic or feldspathic arenite ("arkose") and the occurrence of red beds at several horizons in the Lower Cretaceous may be consistent with the generally warm and arid climate suggested by the flora, but in detail the situation is much more complex. The presence of thick conglomerates and lithic sandstones in the Upper Cretaceous suggests stronger relief and abundant fluvial activity. Red beds still occur as high as K4 in southwest Japan, but are otherwise rare in the Upper Cretaceous.

At least three reversed epochs, Nemuro, Tenka and Akoh, have been found in the paleomagnetic investigation of Cretaceous rocks (Nagata et al., 1959; Kawai et al., 1961, 1971; Kang, 1962; Fujiwara and Nagase, 1965; Fujiwara, 1966, 1967; Kato and Muroi, 1965; Sasajima and Shimada, 1966; Sasajima et al., 1968; Sasajima, 1969; Yasukawa and Nakajima, 1972; Ito and Tokieda, 1974). Sasajima and Shimada (1966) and Sasajima et al. (1968) proposed, as a working hypothesis, that during Cretaceous time southwest Japan may have been located at 133°E 26.6°N to 125°E 29.1°N, the migration northeastwards being accompanied by a clockwise rotation of about 30° until about the end of the Miocene. The difference in the rotation angle in northeast Japan (about 50°) may have resulted from apparent bending of the Japanese arc (Kawai et al., 1961). Yasukawa and Nakajima's hypothesis (1972) of relative southerly movement of southwest Japan with respect to Korea is not necessarily inconsistent with Sasajima's.

Tsuchi and Kuroda (1973) report a warm shallow-sea fauna in the Senonian limestone on Erimo (=Sysoev) seamount (loc. *82*) which has thick-shelled gastropods allied to Mexican species but foreign to

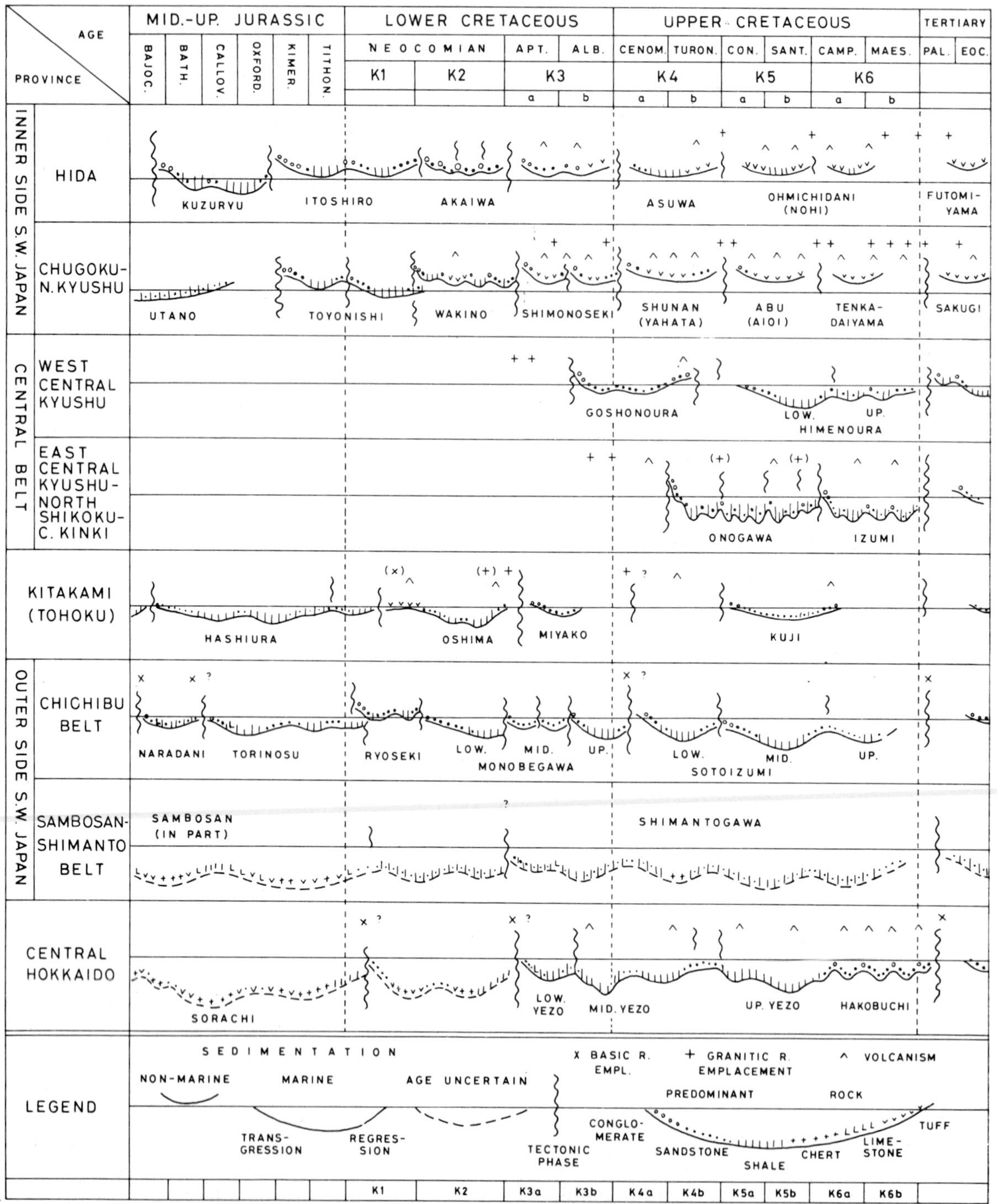

Fig. 22. Schematic diagram showing the Late Jurassic and Cretaceous geologic history in Japan. (Reproduced by permission of the Nat. Res. Counc. Can.)

Japan. The basaltic bed-rock age is 80 m.y. (Ojima et al., 1970) and it has been suggested that the seamount moved northward or northwestward with the spreading of the ocean floor. This is not inconsistent with the reconstructed history of Late Mesozoic spreading in the Pacific of Larson and Chase (1972).

Regional features

Descriptions will be given here of some of the salient features of the Cretaceous system in the various provinces or geotectonic belts briefly outlined in the preceding section. Figs. 17 and 22 may be helpful for the discussion below.

(1) Inner belt.

This is a much broader zone than in the Jurassic, comprising most of the inner zone of southwest Japan (zones 9, 10, and 11 in part) (see Fig. 23) and extends into the inner part of northeast Japan (zone 7). There is a voluminous bibliography concerning the local stratigraphy, paleontology, structures, petrography, mineralization, etc., of the area, which may be referred to in the accounts of Ichikawa et al. (1968), Nozawa (1970) and M. Murakami (1974). It was a period extending from the latest Jurassic to Paleogene, characterized by tectonic activity, block faulting and thrusting of the basement and moderate to gentle folding of the cover. Igneous activity from late Early Cretaceous was intense. The combined results of the two processes was to convert the zone to a mountainous region with non-marine deposits laid down in intermontane basins. In some basins sedimentation persisted from the Jurassic (e.g., Tetori and Toyonishi groups), but many new basins were formed. The geological structures of the area are shown in a profile (Fig. 24).

A reference sequence of non-marine deposits from northeastern Kyushu to western Chugoku leads to the recognition of five groups or series in ascending order. These are Toyonishi (probably K1), Wakino (=Lower Kwanmon) (approx. K2), Shimonoseki (=Upper Kwanmon) (? K3), Yahata–Shunan (? K4), and Abu (? K5).

The Toyonishi contains plant beds of Kiyosuye in the lower part and brackish-water shell beds in the upper. The predominant lithological types are a feldspathic arenite and conglomerate. The Wakino consists of lacustrine beds with acid tuffite. The Shimonoseki is considerably thick, about 2300 m, with

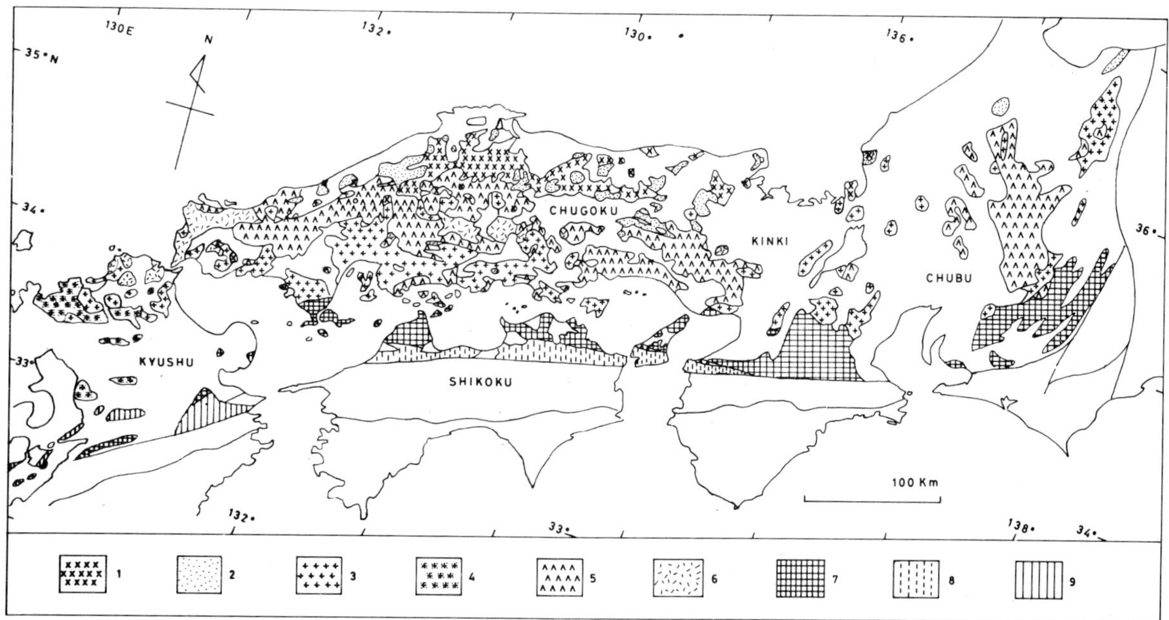

Fig. 23. Distribution of Cretaceous to Early Tertiary igneous rocks and related formations in the inner zone of southwest Japan. Legend: *1* = granite of the Sanin zone; *2* = volcanic group related to *1*; *3* = granite of the Sanyo (Hiroshima) zone; *4* = granite of deeper facies in northern Kyushu; *5* = volcanic group related to *3*; *6* = Kwanmon Group (K2 + K3); *7* = granite of the Ryoke zone; *8* = Izumi Group (K6); *9* = Upper Cretaceous (K4 + K5, partly K6) of central Kyushu. (Adapted from Murakami, 1974.)

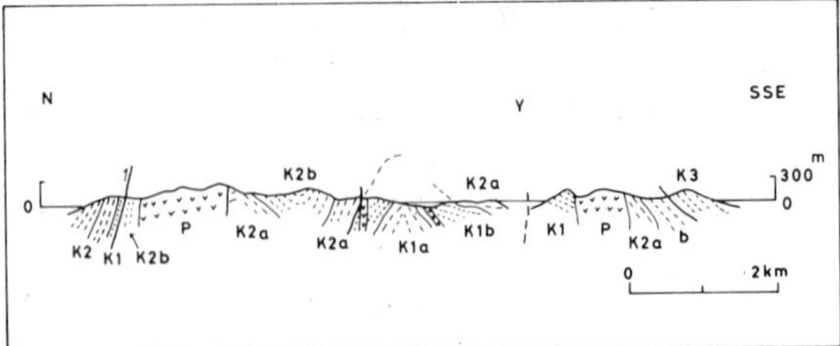

Fig. 24. Geological profile along the western coast of Yamaguchi Prefecture, near the western end of Chugoku (z10a). Legend: *K1* = Toyonishi Group; *K1a* = Kiyosuye; *K1b* = Yoshimo; *K2–K3* = Kwanmon Group; *K2* = Wakino Formation; *K3* = Shimonoseki Formation; *P* = Porphyrite; *Y* = Yoshimo. (After Tatsuro Matsumoto, 1949; Hase, 1960.)

TABLE VIII

Estimated volume of Upper Cretaceous volcanic masses (after Ichikawa et al., 1968)

Local name (from W to E)	Area (km^2)	Volume (km^3)
Abu Group	1800	4500
Takada rhyolites	6000	15000
Aioi–Arima Group	2300	5400
Ikuno Group	4000	10000
Nohi rhyolites	5200	16000
Okunikko Group	1400	>1000
Others		>1000

andesitic tuffs, tuff breccias, lavas, some rhyolite and is cut by porphyrite dykes and a small granodiorite stock. Acid volcanogenic materials make up most of the Yahata–Shunan and Abu with voluminous pyroclastic flows in the latter. Pyrophyllite deposits are associated with some of the volcanics (Kinosaki, 1965). There are considerable lateral facies and thickness changes (Hase, 1960).

Lithostratigraphic correlatives are found in central Chugoku, eastern Chugoku–western Kinki, central Kinki, Chubu (including zones 9 and 10c) and northern Kwanto (zone 7) (see Fig. 23). The most extensive acid volcanic groups comparable with the Abu have a variety of local names, e.g., the Aioi Group in western Kinki, the Nohi rhyolite in northwestern Chubu (Hida and the adjacent area), etc. The estimated volumes of volcanics are given in Table VIII, and Ichikawa et al. (1968) have compiled the chemical composition of representative rocks. A younger, presumably K6 aged group of volcanics, the Tenkadaiyama Group in western Kinki, gives Rb–Sr age of 67.5 m.y. (Seki and Hayase, 1974). Still younger rhyolitic ignimbrites, dacitic tuff and lava are found in the Hida area (northwestern Chubu). Referred to the Futomiyama Group they have a K–Ar age of 59 m.y. (N. Yamada, 1967, oral comm.), i.e., Early Paleogene.

Granitic rocks are widely distributed in the inner belt. Field relationships suggest emplacement on several occasions, possibly one associated with each of the five stages of volcanic activity indicated above, i.e. from K3 to K6 plus older Tertiary. Based upon many radiometric dates (compiled by Nozawa, 1970, 1975, and illustrated in Fig. 25), there appears to have been a general shift of predominant activity from west to east, and to some extent from south to north within the belt. A bibliography of geochronological work in Japan is compiled by Tatsuro Matsumoto (1968) and Nozawa and Shibata (1974).

Many of the granites were intruded at shallower depths as indicated by texture, minerals and contact effects, although as a result of block faulting deeper levels are sometimes seen (e.g., near Fukuoka, northern Kyushu) and here there is a tendency for the granites to become more schistose and concordant with the metamorphics (Karakida, 1969). Their chemistry has been studied by H. Shibata et al. (1960–62) (see also compilation by Ichikawa et al., 1968) and their mineralization by Ishihara (1971, 1973) (Fig. 26) and others. The origin of the granite has been discussed by Kuno (1968), Tadashi Matsumoto (1968) and Yanagi (1975). This igneous activity roughly correlates with that in southern Korea, southeastern China, inner Sikhote-Alin, Dzhugdzhur Mountains and northeasternmost Siberia. The Yamato bank in the Japan Sea appears to consist also of Cretaceous volcanic and granitic rocks.

(2) Central belt

This belt consists of paired zones of regional metamorphism, the low P/T Ryoke–Abukuma and the high P/T Sambagawa zones. They were completed, uplifted and partially eroded during the Cretaceous. Comprehensive descriptions of petrological and geological characters with essential references are to be found in Miyashiro (1961, 1967, 1972, 1973), Suwa (1961, 1973) and Hashimoto et al. (1970). It has been discussed recently that in the Ryoke–Abukuma zone an intermediate P/T type metamorphism occurred prior to the more predominant low

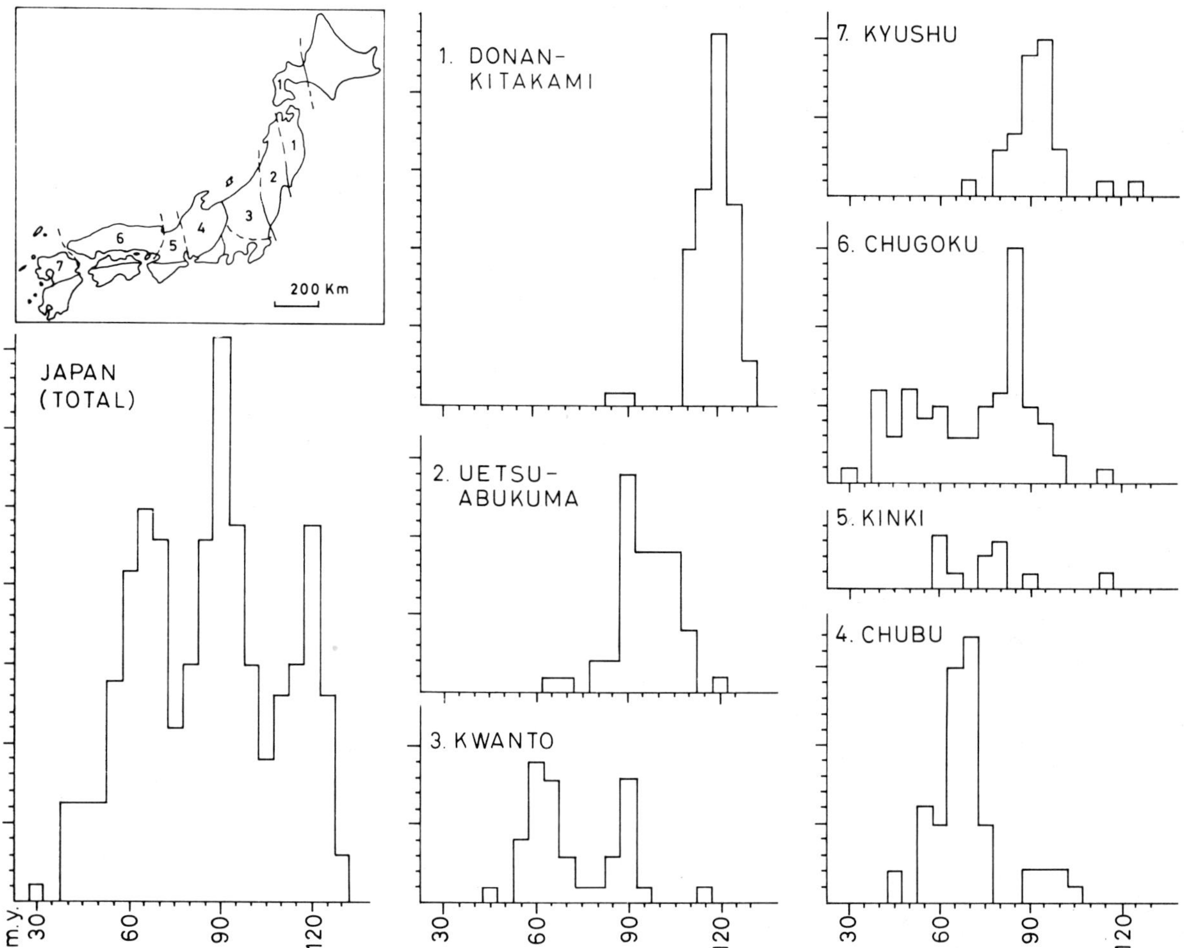

Fig. 25. Histograms illustrating the number of K–Ar age data of Cretaceous to Early Tertiary granitic rocks in Japan. (Adapted from Matsumoto, 1969 and Nozawa, 1970.) (Reproduced by permission of the Nat. Res. Counc. Can.)

P/T type and that the Median Tectonic Line cuts obliquely the thermal structure of this zone (Fig. 27). Phases of a metamorphic event have been distinguished by petrofabric analysis (Nureki, 1960; Hara, 1962). There are multiple granitic intrusions, synkinematic (related to Kashio mylonitic rocks), pre-Nohi and post-Nohi.

Sambagawa metamorphism seems to have been more complex and prolonged, and affects not only the Sambagawa belt but also the Chichibu terrain. Ophiolites of Late Carboniferous–Early Permian age are abundant in the Sambagawa belt (especially in the Mikabu subzone) and the Late Cretaceous represents the last phase of the metamorphic history, when the subduction may have ceased and the deep-seated rocks tended to rise. Isotope dating has shown an age range of 110–70 m.y. (Yamaguchi and Yanagi, 1970). It has recently been suggested that the amphibolite masses in the Sambagawa belt of central Shikoku are polymetamorphosed gabbro which had suffered granulite facies metamorphism prior to the high P/T type Sambagawa metamorphism (Banno et al., 1976).

Another characteristic feature of the central belt is the distribution of thick piles of Cretaceous clastic deposits which rest unconformably upon the Ryoke metamorphics and granites and are cut to the south by the Median Tectonic Line. The depositional basins were short-lived and shifted in time and space in central Kyushu [Goshonoura of K5b + K4a (loc. *44*), Himenoura K5 + K6 (loc. *45–49*), Mifune K4a + b (loc. *43*), Onogawa K4b + K5 (loc. *50*)] (Tatsuro

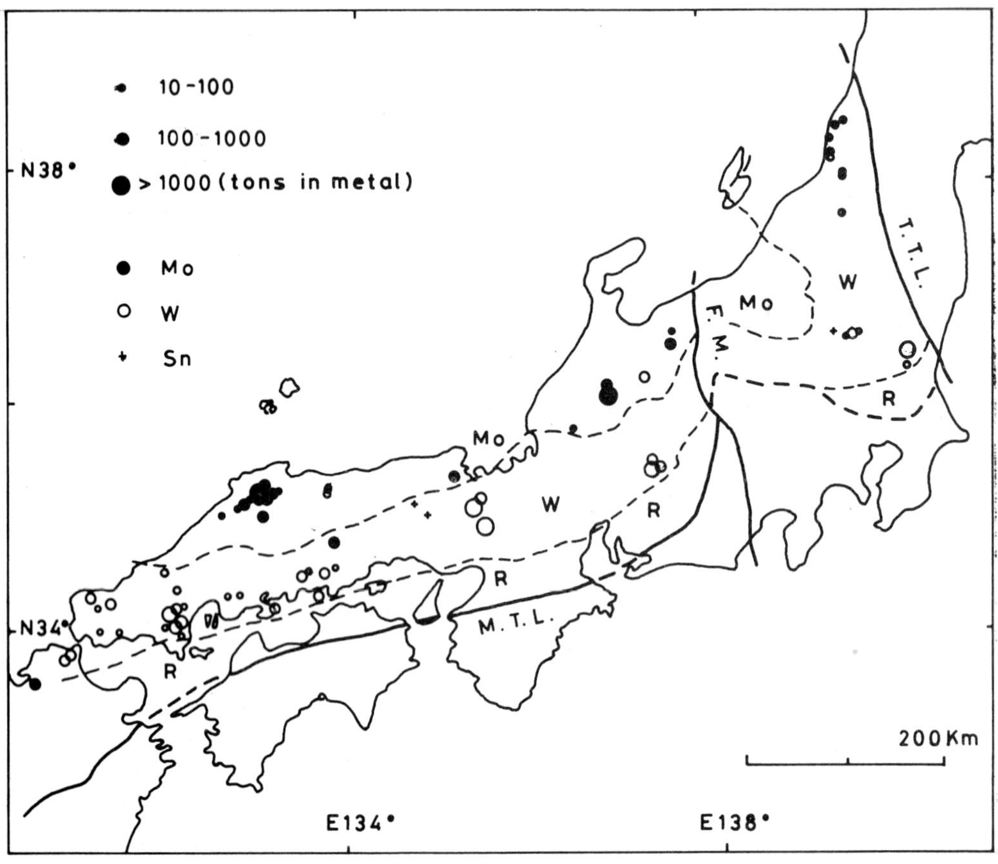

Fig. 26. Distribution of the molybdenum and tungsten deposits in western and central Japan.
Legend: *Mo* = molybdenum; *W* = tungsten; *Sn* = tin; *R* = barren Ryoke zone; *F. M.* = Fossa Magna; *M. T. L.* = Median Tectonic Line; *T. T. L.* = Tanakura Tectonic Line. (After Ishihara, 1973.)

Matsumoto, 1938a, 1954c; Okada, 1960, 1961; Yoshiro Ueda, 1962; Teraoka, 1970; Tashiro and Noda, 1973). In northern Shikoku and central Honshu (Kinki) the basin was more elongate but again of short time range [Izumi, K6 (loc. 51–55)] (Tatsuro Matsumoto, 1954c; Tanaka, 1965). It probably extended into northern Kwanto (loc. *27*) (Tanaka, 1970a). The clastics consist of conglomerate, sandstone, and shale, with some intercalated tuffite. In central Kyushu they are partly deltaic with some red beds and non-marine shell beds, otherwise the sediments were deposited in marine environments. Despite the occurrence of coarse clastics much of the Onogawa and Izumi groups have the characteristics of turbidites with current flows ENE–WSW. The clastics were derived from granitic, quartz porphyry, andesitic, rhyolitic and welded tuff source rocks. Great thickness (e.g., 7 km in K6) and the cyclic nature of the coarse clastics imply rapid sedimentation with trough subsidence keeping pace with uplift and erosion of the tectono-magmatic mountains of the inner belt. After burial they were altered at the zeolite grade and deformed into a synclinorium (Fig. 28). The Median Tectonic Line marks records of repeated movements since the Cretaceous, including a lateral displacement (Sugiyama, 1973).

(3) Chichibu zone of southwest Japan (Zones 12b, c, 8b, c)

Here shallow marine and some non-marine beds of appreciable thickness rest unconformably upon Late Jurassic or older beds. Outcrops are at present scattered occurring in a number of synclinoria within the folded Upper Paleozoic Chichibu terrain (locs. *56–71, 28, 30*), and as a result many local formational names have been applied. The main units are the Ryoseki (K1), Lower Monobegawa (K2), Upper Monobegawa (K3) (which may consist of two or

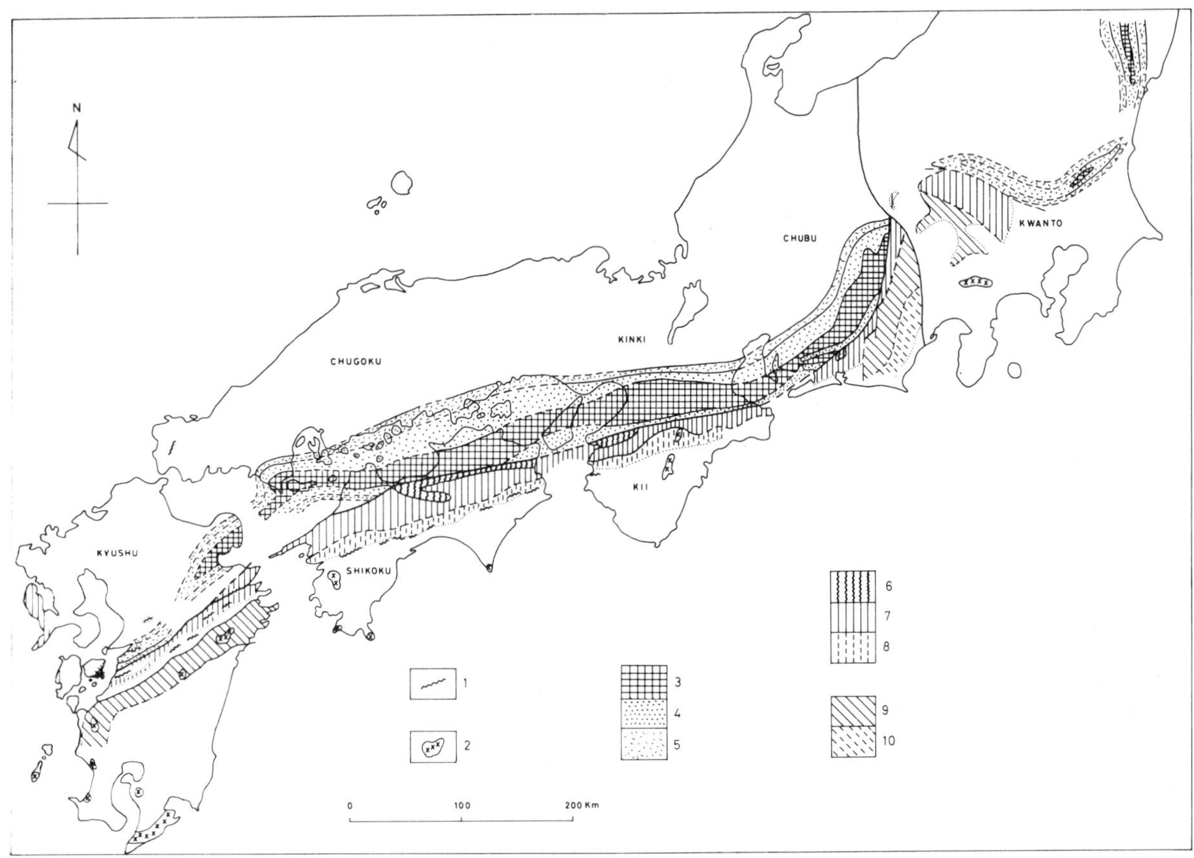

Fig. 27. Thermal structure of the Ryoke, Abukuma, Sambagawa and Shimanto metamorphic belts.
Legend: *1* = squeezed-out mass of Yokokurayama [or Kurosegawa] complex; *2* = Neogene granite; *3–5* = Ryoke–Abukuma metamorphics; low P/T facies series (partly intermediate P/T facies series), *3* = high-grade zone, *4* = intermediate zone, *5* = low-grade zone; *6–8* = Sambagawa metamorphics: high P/T facies series (classification I); *6* = high-grade zone, *7* = intermediate zone, *8* = low-grade zone; *9–10* = Shimanto metamorphics: high P/T facies series (classification II), *9* = intermediate zone, *10* = low-grade zone. The scheme of classification is principally the same as that of Zwart et al. (1967). (Adapted from Suwa, 1961, 1973; M. Hashimoto et al., 1970; and H. Yamamoto, 1975, personal communication.)

three units in some places), and the Lower, Middle and Upper Sotoizumi (K4, K5, K6) formations, each of which averages about 500 m in thickness. They indicate the expansion or shift of the sedimentary basins and the uplift and changes in the source area. Although there is little tuff, acid igneous rocks (of the foregoing two belts) constitute the main source for conglomerates and sandstones of K3–K6. Quartz-feldspathic arenite with fresh microcline is characteristic of a certain part of K1 which, together with the Ryoseki flora, may indicate a dry sub-continental climate, a time of (K1) regression.

Within the folded structures of the Chichibu terrain Cretaceous strata are usually found in synclinoria separated from each other by thrusts. Where two formations are preserved, the younger one sometimes shows a simpler structure. The general situation is illustrated in Figs. 29 and 30. Within the Chichibu belt there are narrow lenticular masses of older, presumably basement rocks squeezed out along thrusts and injected into Mesozoic strata along with serpentinite. They consist of granulites, amphibolites, biotite gneiss, granites, often with cataclastic textures, and Middle to Late Silurian fossiliferous rocks. Radiometric ages are in the 420 ± 10 m.y. range (Hayase and Ishizaka, 1967; Ishizaka, 1972).

The Cretaceous of the Chichibu zone is thus the record of tectonic movements which continued till the end of the Cretaceous with periodic phases of activity.

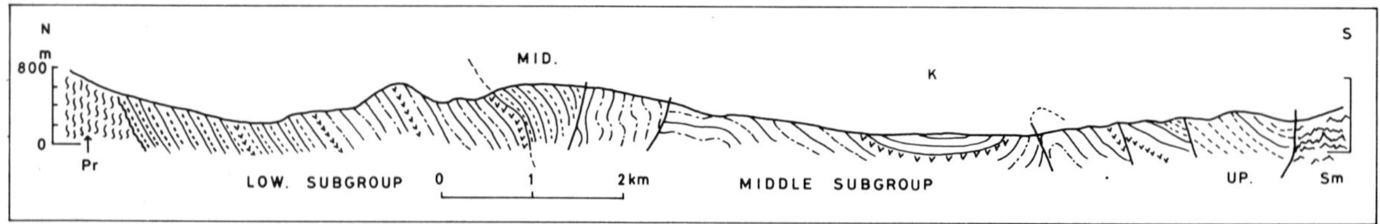

Fig. 28. Geological profile of the Izumi belt across Kawakami (*K*), 15 km east of Matsuyama, northwestern Shikoku (z11). Legend: *Pr* = pre-Izumi metamorphics of the Ryoke zone; *Sm* = Sambagawa metamorphics; *υ* = bed of predominant tuffite. (After Tatsuro Matsumoto, 1967d, by permission of the Asakura Publ. Co.)

(4) Shimanto belt

The folded beds in the Shimanto belt (zones 13, 19) are called the Shimanto or Shimantogawa Group. The group consists of a primarily Cretaceous lower division and a Lower Tertiary (here including Lower Miocene) upper division. A common generalization is that the former is exposed in the northern and the latter in the southern sub-belt consequent upon the southward migration of the sedimentary trough (Shimanto geosyncline).

The beds are much deformed and in places have suffered high P/T metamorphism (Matsuda and Kuriyakawa, 1965; Imai et al., 1971). From the imperfectly known Cretaceous there are several salient features to record. The beds are poorly fossiliferous but fossils ranging from Aptian (K3a) to Maastrichtian (K6b) have been recorded from scattered localities (locs. *72–80*). This time-range is approximately that of the Yezo Group of Hokkaido (see section 6). The predominant lithologies consist of a claystone facies in the lower unit and an alternating sandstone shale of flysch facies in the upper. Some greenstone, chert and micrite occur in the former but not so predominantly as in the Sambosan Group. Massive sandstones are usually found at the top of upwards-coarsening sequences of the latter (Fig. 31, from Okada, 1971).

The sedimentary framework of the Shimanto geosyncline may be comparable with that of the arc–trench gap (Dickinson, 1971) but there are some arguments that the environments may not have always been bathyal as generally assumed. The presence of ornate ammonites with body-chambers and/or inoceramids in shale could reflect quiet shallower-water conditions as an alternative to the floating in of the shells from shallower zones, just as the

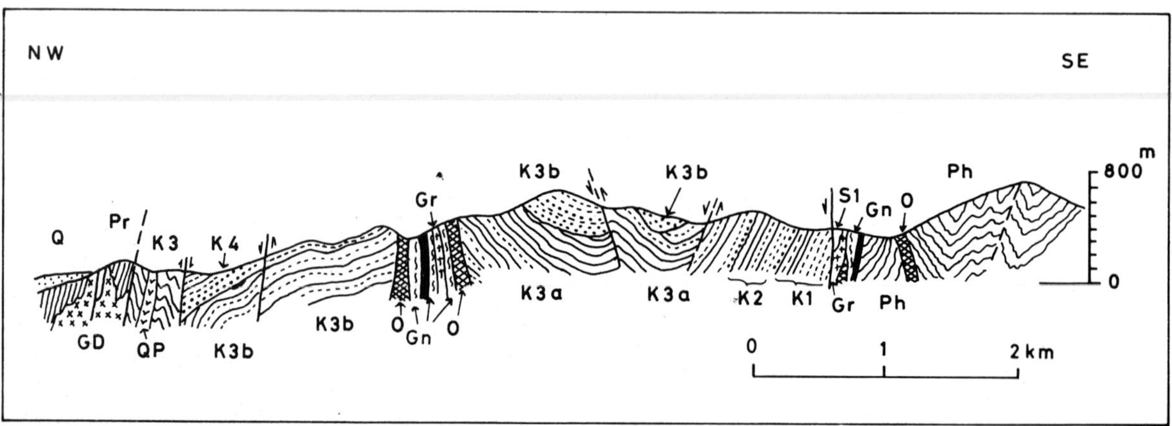

Fig. 29. Geological profile of the Cretaceous area near Yatsushiro, Kumamoto Prefecture (z12).
Legend: *K1–K4* = Lower Cretaceous, *K1* = Kawaguchi F. (= Ryoseki), *K2* = Hachiryuzan (= Lower Monobegawa), *K3a* = Hinagu F. (= Middle Monobegawa), *K3b* = Yatsushiro F. (= Upper Monobegawa), *K4* = Miyaji F. (Uppermost Monobegawa); *GD* = granodiorite of the Ryoke zone; *Gr Gn* = squeezed-out slices of older granitic and gneiss; *Sl* = Silurian; *O* = serpentinite; *Ph* = Permian; *Pr* = metamorphosed Permian (Ryuhozan Group); *QP* = Tertiary quartz–porphyry; *Q* = Quaternary. (Adapted from Matsumoto and Kanmera, 1964, in Matsumoto 1967d, by permission of the Asakura Publ. Co.)

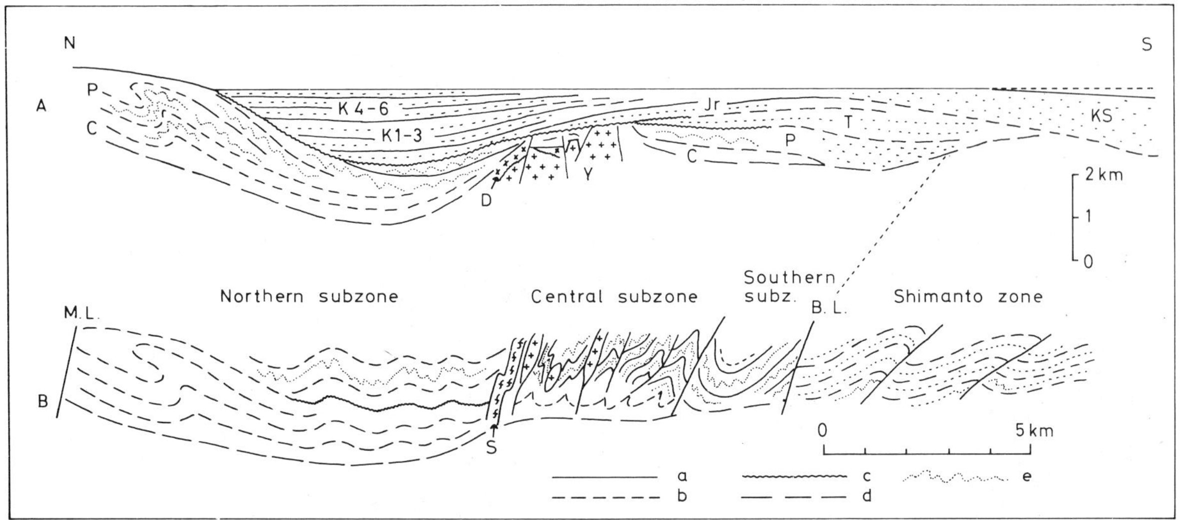

Fig. 30. Schematic geological profiles of the Chichibu terrane (z12) in eastern Shikoku.
Legend: *A* = at the time of the sedimentation of the Cretaceous; *B* = after the folding (post-Campanian). *Y* = Yokokurayama complex (granite and metamorphics); *D* = Silurian–Devonian strata; *S* = serpentinite; *C* = Carboniferous; *P* = Permian; *T* = Triassic; *Jr* = Jurassic; *K1–3* = Lower Cretaceous; *K4–6* = Upper Cretaceous; *KS* = Cretaceous of the Shimanto zone (z13); *M. L.* = Mikabu Line; *B. L.* = Butsuzo Line; *a* = boundary of major groups; *b* = ditto (presumed); *c* = unconformity; *d* = presumed boundary with the basement; *e* = minor folds (down to several meters). (Adapted from Y. Ogawa, 1974.)

presence of shallow-water bivalves including hippurids and small bodies of colonial corals and algae could avoid interpretation in terms of turbidity current transportation. Some parts of the Shimanto Group may correspond to the basin deposits inside a terrace edge (or outer high) and some other parts to the trench and slope deposits in a model of arc–trench gap (K. Kanmera, 1975, oral comm.).

In the Uwajima district (loc. *76*) of western Shikoku the Upper Cretaceous deposits are more fossiliferous and contain a larger amount of sandstone and conglomerate than found in the Shimanto Group elsewhere. These deposits, locally named the Uwajima Group, are of a facies and age (K4b–K5) similar to that of the Onogawa Group of eastern central Kyushu. Since they are not part of an exotic block, they suggest a connection of the Shimanto trough with sedimentary basins to the north. This is supported by the source material of the clastics.

(5) Tohoku region, northeast Japan

The Neocomian formations of the southern Kitakami belt (zone 5b) as represented by sequences at locs. *20–24,* like those in the Chichibu zone (see section 3), are mostly shallow-sea facies with some non-marine beds and some flysch deposits. Unlike the Chichibu, however, some volcanics occur and the folded strata are here intruded by granodiorites.

In the northern Kitakami belt (zone 5a) northeast of the Hayachine tectonic septum, a thick geosynclinal sequence seems to have continued accumulating from Jurassic into Neocomian times (see section 4b of Jurassic) with, in the Neocomian, andesitic lava, tuff and tuffaceous sandstone (Upper Rikuchu Group, loc. *18*). These beds were intensely folded, partially metamorphosed and then intruded by granodiorite (Moriya, 1972; Sugimoto, 1974). Upper Aptian–Lower Albian littoral to neritic fossiliferous sediments of the Miyako Group (loc. *19*) (Hanai et al., 1968) and Senonian beds (locs. *16, 17*) were deposited upon them with a marked unconformity.

Acid intrusives in the age range 125–110 m.y. (Kawano and Ueda, 1964–66, 1967) occur in both the northern and southern Kitakami belts (Figs. 25 and 32), and extend into southwestern Hokkaido (="Donan"). They range in composition from adamellite to gabbro with granodiorite as the commonest type. An examination of the petrographic character with respect to location by Katada (1974) shows that the K_2O content of the gabbroic rock of the same plutonic sequence increases from the outer sub-belt to the inner (Fig. 32).

The granitic rocks of the Kitakami belt have a lower SiO_2 content and a higher Fe_2O_3/FeO and

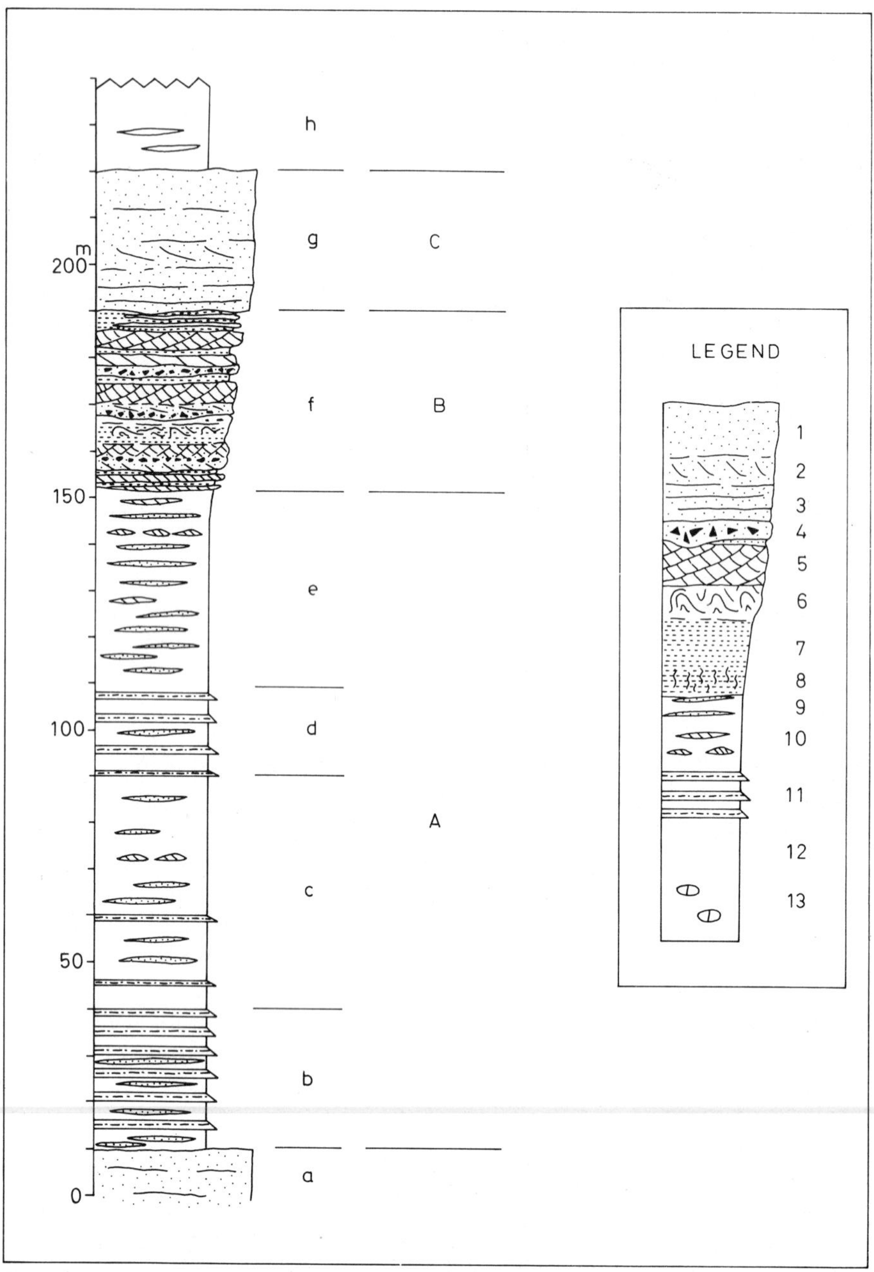

Fig. 31. Coarsening upward sedimentary succession (A–B–C) in a part of the Shimanto Group, southwestern Shikoku. Legend: *1* = thick-bedded sandstone; *2* = cross-bedded sandstone; *3* = thin and flat-bedded sandstone; *4* = shale breccia; *5* = ripple-drift cross-stratification; *6* = slump structure; *7* = siltstone; *8* = bioturbated; *9* = thin and lenticular sandstone beds; *10* = cross-laminated sandstone lenticles and small-scale ripples; *11* = graded bedding; *12* = claystone; *13* = calcareous nodules. (After Okada, 1971.)

lower Fe/Mg ratios than those of the Ryoke and adjacent belts of southwest Japan (Kanisawa, 1974a, b). The significance of these differences is not understood. Kanisawa has, however, pointed out the close relation of the extrusives to some of the intrusives constituting a volcanic–plutonic assemblage (Fig. 32).

Cretaceous granites are extensively exposed in the Abukuma massif (zone 6), where their age range is younger than that in Kitakami, 110–85 m.y. (Fig.

25). These rocks and quartz–porphyry are unconformably overlain by Senonian (K5) beds of neritic facies (loc *26*).

(6) Central zone of Hokkaido and Sakhalin (zone 4)

The mountain ranges of the main part of Hokkaido and Sakhalin form part of the Mesozoic–Cenozoic orogenic system of northeastern Asia. During the Mesozoic this orogenic zone was part of the Yezo geosyncline (Tatsuro Matsumoto, 1943), although the zone itself is called the Hidaka orogenic zone by some authors (e.g. Hunahashi, 1957).

The Sorachi ophiolitic assemblage in the lower part of the Yezo geosynclinal sequence is mainly of Triassic to Jurassic age but in some areas rocks with the same eugeosynclinal facies contain Neocomian fossils (locs. *5, 6*). Though often concealed by faulting the contact between the Yezo Group of clastic sediments and the Sorachi Group is marked by an unconformity. The Yezo Group ranges from Aptian to Maastrichtian in age and is about 5 km thick at the type locality (loc. *12*). There are four lithostratigraphic subgroups, viz., lower (K3a4–K3b2), middle (K3b3–K4b), upper (mostly K5, beginning at K4b3 in places) and the terminal Hakobuchi (K6), which follow one another essentially conformably (Fig. 33), although there are minor disconformities. The lithological sequence may be generalized with predominantly flysch in K3, muddy sediments with ammonite and *Inoceramus* nodules during K4 and K5 and coarse, partly neritic and partly deltaic sediments in K6 times. There are marked lateral facies changes in K4 and K6 times with the coarser, shallow-water facies occurring to the west. Even within essentially quiet environments of deposi-

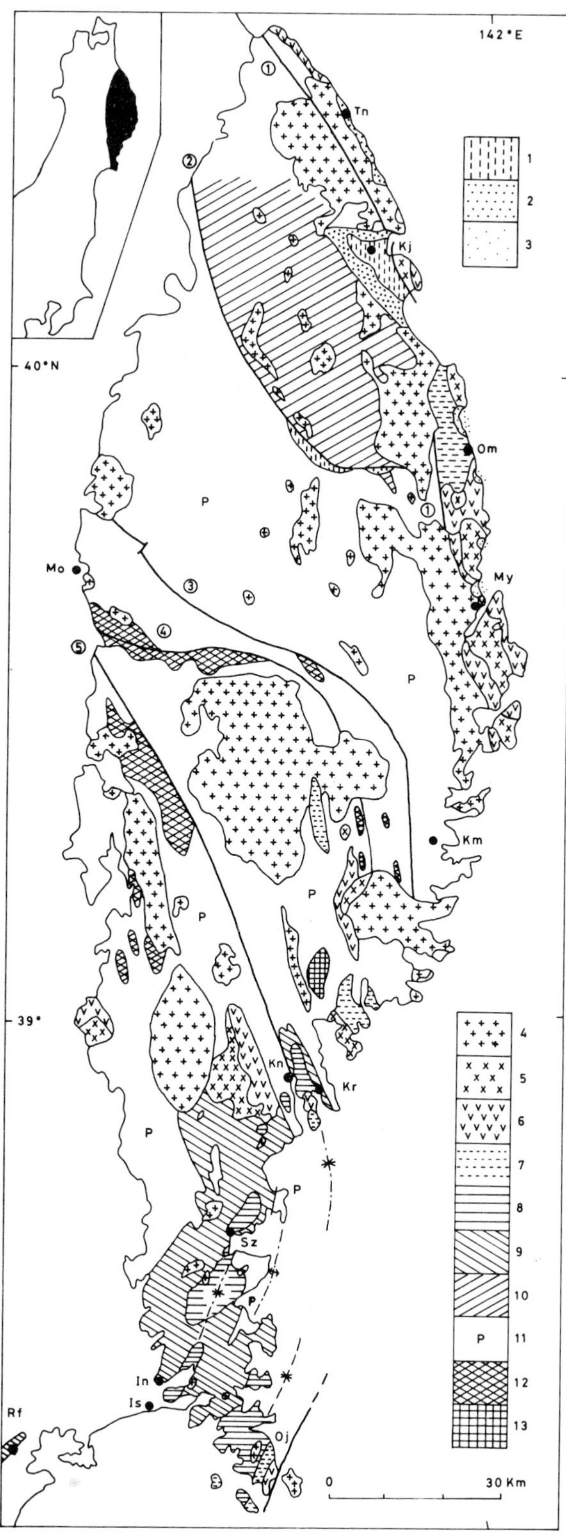

Fig. 32. Generalized geological map of the Kitakami Massif, northeast Japan.

Legend: *1* = Lower Tertiary; *2* = Upper Cretaceous (Senonian); *3* = Miyako Group (Upper Aptian–Lower Albian); *4* = granodiorite (Early Cretaceous); *5* = granite of shallower facies (Early Cretaceous); *6* = Lower Cretaceous (Neocomian) porphyrite and volcanics; *7* = Lower Cretaceous (Neocomian); *8* = Jurassic; *9* = Triassic; *10* = presumed Triassic and Jurassic of northern area; *11* = Paleozoic; *12* = serpentinite (Early Cretaceous or older); *13* = older granite of Hikamiyama. Major tectonic lines: ① Taro; ② Kuzumaki; ③ Morioka–Goyosan; ④ Hayachine; ⑤ Kesennuma. Place names (from north to south): *Tn* = Taneichi; *Kj* = Kuji; *Om* = Omoto; *Mo* = Morioka; *My* = Miyako; *Km* = Kamaishi; *Kn* = Kesennuma; *Kr* = Karakuwa; *Sz* = Shizukawa; *In* = Inai; *Is* = Ishinomaki; *Oj* = Oshika (= "Ojika"); *Rf* = Rifu. (Compiled from various sources, especially from Katada, 1974; and Kanisawa, 1974, for the igneous rocks).

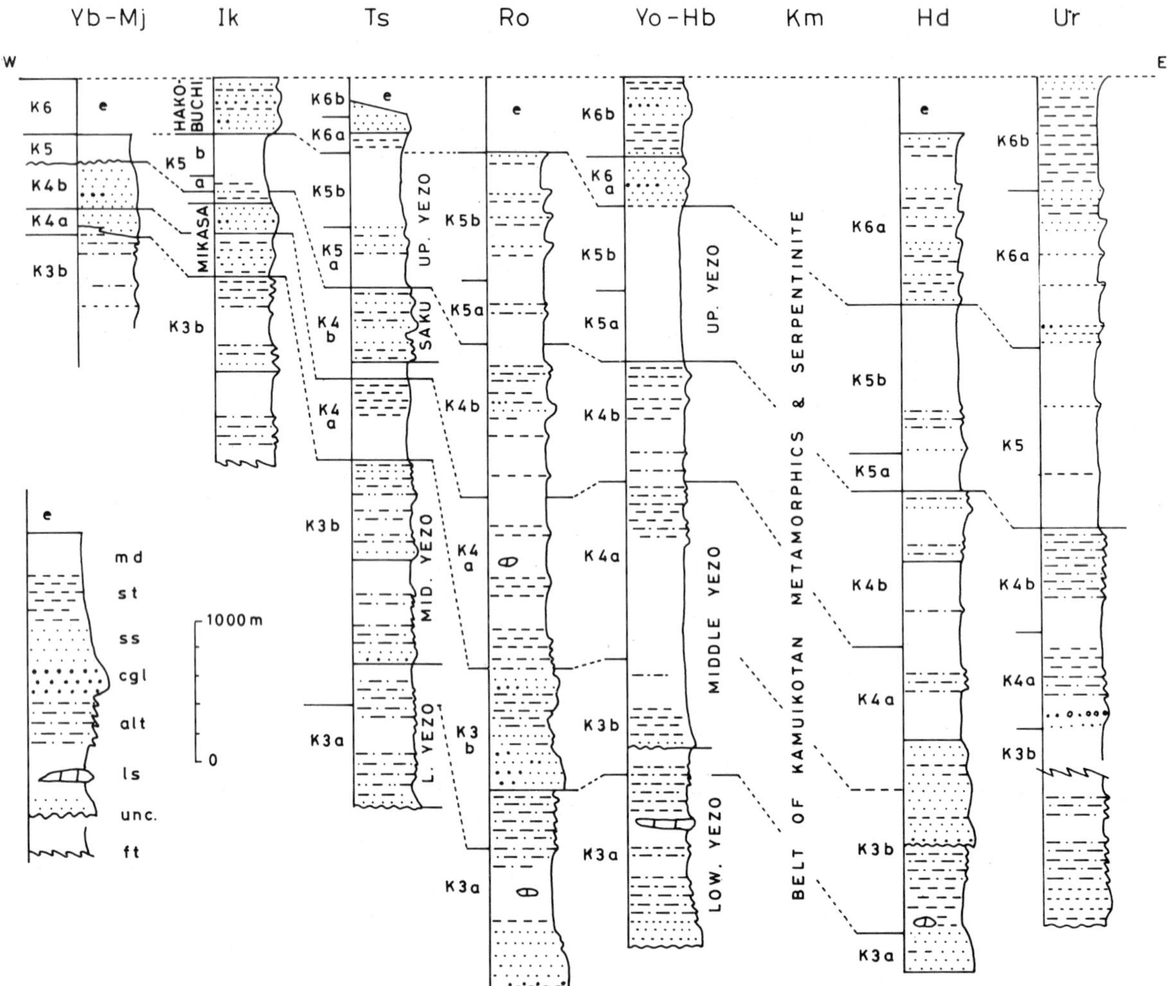

Fig. 33. Simplified stratigraphic sequences of the Cretaceous deposits in the central zone of Hokkaido, showing the correlation and the changes in facies and thickness between selected areas.
Areas: *Yb–Mj* = Yubari (Hatonosu)–Manji; *Ik* = Ikushumbetsu; *Ts* = Saku (mid-valley of the Teshio); *Ro* = Obira [= Obirashibets] valley, near Rumoi; *Yo–Hb* = Oyubari–Hobetsu; *Km* = Kamuikotan belt; *Hd* = Hidaka (Chiroro); *Ur* = Urakawa. Predominant rocks, etc.: *e* = eroded; *md* = mudstone; *st* = fine-sandy siltstone; *ss* = sandstone; *cgl* = conglomerate; *alt* = alternating sandstone and shale; *ls* = limestone; *unc* = unconformity; *ft* = faulting. [Data mostly from my own works; for Hidaka and Urakawa depending much on Obata et al. (1973), Kanie (1966), and recent pers. comm.]

tion during K4 and K5 the influx of flysch with slump structures occurred occasionally, and more frequently in K4b, when there was shallowing in western parts.

The presence of continuous exposure has led to detailed biostratigraphic zoning (Fig. 20) and intensive study of sedimentary facies and structures (see Tatsuro Matsumoto, 1942–43, 1954c, 1959b; Tanaka, 1963, 1970b, 1971; Okada, 1965, 1971; Tatsuro Matsumoto and Okada, 1971, 1973 among others). It can be shown that the Cretaceous sediments in the Yezo geosyncline were derived from an elevated zone of volcanic, granitic, hornfelsic and older sedimentary rocks including products of contemporaneous andesitic to rhyolitic volcanism. The site of this volcano-tectonic range may have lain in the region of southwestern Hokkaido and the present northern Japan Sea or in the outer volcanic belt of eastern Siberia. It is also possible that erosion of the Sorachi Group may have provided some of the sedimentary material and that the high P/T Kamuikotan metamorphic rocks may reflect the existence of a subduction zone with the tectono-volcanic arc to the west of the Yezo Group. The granites and metamorphics of the Hidaka range, once regarded as forming a pair with the Kamuikotan, are of Tertiary origin and not of concern here. The fossiliferous Cretaceous (zone 4c) on the west side of the Kamuikotan (zone 4b) may be comparable with the fore-arc basin deposits inside an outer high in a trench–slope model (Seely et al., 1974).

The Cretaceous is overlain with apparently parallel unconformity by a coal-bearing Tertiary sequence and the strata of the two systems are altogether folded and thrust. The most remarkable orogenic phase is generally assigned to the Late Tertiary.

The Cretaceous beds of Hokkaido extend northwards to Sakhalin where similar geological features are observed, but with some facies variation. In northern Sakhalin there are important continental beds with plants, intercalating with marine beds.

(7) Eastern Hokkaido and the Kurile arc (zone 2)

Although the Kurile arc–trench system is Late Cenozoic in age, the belt does contain Late Cretaceous (K6) to Paleocene marine sediments. These can be observed in the southern or Lesser Kurile belt from the island of Shikotan through the Nemuro peninsula to the coast of Akkeshi in eastern Hokkaido (locs. *1–4*). The beds, called the Nemuro Group, range in age from Campanian to Paleocene, and are overlain unconformably by a Late Eocene–Oligocene coal-bearing group.

The Nemuro Group, about 3 km thick, may be regarded as geosynclinal. It consists of a flysch facies of alternating sandstone and shale, with some basaltic to andesitic lava, tuff and tuff breccia. Slump structures found in the upper member suggest a paleoslope extending E–W (Kiminami, 1975). Near the top conglomeratic horizons are found. The observable sequence has a short time range and the beds are relatively little folded being affected mostly by block faulting.

Gabbroic masses occur and are known from borings (loc. *4*) and in some minor alkaline dolerite sills, pillow structures are found suggesting shallow intrusion in wet sediments.

(8) Taiwan (Formosa)

Although the island consists mainly of Tertiary and Quaternary rocks and Late Cenozoic movements are important, there are records of Mesozoic formations (see Biq, 1974).

Permian fusulinids have been found in rocks of the Tananao metamorphic group of the central zone (zone 22). The age of the folding and regional metamorphism may be assigned then to the Late Permian–Middle Mesozoic interval and correlated with the Indosinian or the Akiyoshi orogeny. These beds are overlain unconformably or are in fault contact with a group of phyllitic slate and sandstone ranging in age to the Early Miocene. The Cretaceous age is based on the dating of corals in lenticular limestones of the Pihou Formation, although the full sequence is imperfectly known. Within the slate are contemporaneous basaltic lavas and tuffs and some intrusive diabases. Thus, there is a similarity in lithology and age with the Shimanto Group in the outer zone of southwest Japan and the Ryukyu arc.

In the western coastal plain (zone 22a) Cretaceous has been found in the subsurface in the Peikang area (loc. *81*) where 500 m of relatively undeformed fossiliferous beds, initially non-marine passing to marine (Aptian), have been described (Tatsuro Matsumoto et al., 1965). It has been suggested that the Early Cretaceous fossils are derived fossils and that on micropaleontological grounds the formation should be younger, but judging from the condition of the megafossils this seems unlikely. An early Late Aptian marine transgression does occur.

Unconformably overlying these fossiliferous beds, and underlying Miocene, there is a formation of acid tuff and tuff breccia, which may belong to the Late Cretaceous.

(9) The submarine ridges of the northern Philippine Sea

The Philippine Basin (s.s.) is limited to the east be the Kyushu–Palau ridge and to the northwest by the Ryukyu trench. In the northern part are the Amami submarine plateau, the Daito and Oki-Daito ridges (Figs. 1, 2). The existence of Cretaceous rocks in the plateau and ridge area of the northern Philippine Basin, north of the fault along the southern margin of the Okidaito ridge was demonstrated during the GDP-11 cruise (1975) and Geol. Surv. cruise (Mizuno et al., 1975, oral comm.). They consist of green schist, serpentinite, apparently old basalt and andesite (ages 82–85 m.y. recorded), tonalite and granodiorite (age 72 m.y. recorded). They are unconformably overlain by *Nummulites boninensis* beds. Conglomerate blocks in the ridge area may be Tertiary or older.

Whether these rocks should be correlated with the Mesozoic rocks of the Philippines or not is an open question, and certainly closely related to the genesis of the Philippine Basin.

CONCLUDING REMARKS

To summarize the above descriptions the Mesozoic rocks of the Japanese islands and adjacent areas mark records of complex history of various geological events. For each system the biostratigraphy with paleobiogeography has been concisely shown, since the data are basic for constructing the various events into chronological order. Figs. 5, 6, 11, 12, 18–20, and Tables II–VII may be useful for this account. Radiometric age data on igneous and metamorphic rocks have been shown from the same viewpoint in the sections on regional features as they record thermal events.

In the descriptions of regional features the author has endeavoured to show various features, in appropriate parts of a model of geological evolution, which occurred at or near the continental margin of Mesozoic Eastern Asia facing the Pacific Ocean. Figs. 3, 4, 5 (upper left), 7, 10, 11 (upper left), 16, 17 and 22 may be useful to understand the historical development, although much remains to be worked out for further refinement.

The Cretaceous history can be interpreted in terms of plate tectonics (e.g., Uyeda and Miyashiro, 1974), but there are many questions under debate. The Triassic and Jurassic are intervening periods during which important changes occurred from the Late Paleozoic eugeosynclinal conditions to the Cretaceous continental margin. To understand the significance of these changes the nature of the socalled Late Paleozoic eugeosyncline in Japan and adjacent areas should be made clear. Recently the knowledge of the Triassic and Jurassic in Japan has been much revised and the author has attempted to outline it as a step for further improvement.

In the circum-Pacific belt orogenies took place most actively during the Mesozoic era, with some differences in details between regions and between provinces (Tatsuro Matsumoto, 1975a). This chapter has given an example from the northwestern part of the Mesozoic Pacific Ocean.

ACKNOWLEDGEMENTS

In addition to previously published works, the author owes much to Drs. Toshio Kimura, Sadao Sasajima, Kametoshi Kanmera, Keiji Nakazawa, Keisaku Tanaka, Koichiro Ichikawa, Ikuwo Obata, Toshiaki Kimura, Hakuyu Okada, Akira Tokuyama, Yujiro Ogawa, Nobuhide Murakami, Tamotsu Nozawa, Yoshimasu Kuroda, Mitsuo Hashimoto, Hirosato Yamamoto, Kanemori Suwa, Masato Katada, Ken Shibata, Shunzo Ishihara, and Satoshi Kanisawa, who have offered him up-to-date information including some results in press. Editors have helped the author in rephrasing the first draft. Miss Mutsuko Hayashida patiently assisted him in drawing illustrations and typewriting.

REFERENCES *

Adachi, M., 1972. Pelitic and quartzo-feldspathic gneisses in the Kamiaso conglomerate. *J. Geol. Soc. Jap.*, 79: 181–203.

Arkell, W.J., 1956. *Jurassic Geology of the World.* Oliver and Boyd, Edinburgh, 806 pp. (46 plates).

Asano, K., 1950a. Upper Cretaceous foraminifera from Japan. *Pac. Sci.*, 6(4): 158–163 (pl. 1).

Asano, K., 1950b. Cretaceous foraminifera from Teshio, Hokkaido. *Tohoku Univ., Inst. Geol. Paleontol., Short Pap.*, No. 2: 13–22 (pl. 3).

Asano, K., 1962. Japanese Paleogene from the view-point of foraminifera, with descriptions of several new species. *Contrib. Inst. Geol. Paleontol., Tohoku Univ.*, No. 57: 1–32 (pl. 1) [J + E].

Avdeiko, G.P., 1970. Evolution of geosynclines on Kamchatka. *Pac. Geol.*, 16(3): 1–13.

Bando, Y., 1964a. The Triassic stratigraphy and ammonite fauna of Japan. *Sci. Rep. Tohoku Univ., Sendai*, 2, 36(1): 1–137 (pls. 1–15).

Bando, Y., 1964b. On some Middle Triassic fossil cephalopods from Japan, with a note on the Middle Triassic formations in Japan. *Jap. J. Geol. Geogr.*, 35(2–4): 123–137 (pl. 5).

Bando, Y., 1966. A note on the Triassic ammonoids of Japan. *Mem. Fac. Lib. Arts Educ., Kagawa Univ.*, 2, No. 138: 1–19.

Bando, Y., 1967a. Study on the Triassic ammonoids, and stratigraphy of Japan. Part 2: Middle Triassic. *J. Geol. Soc. Jap.*, 73(3): 151–162 (pl. 1) [J + E].

Bando, Y., 1967b. A new Triassic ammonite from the Inai Group in Japan. *Mem. Fac. Lib. Arts Educ., Kagawa Univ.*, 2, No. 158: 1–7 (pls. 1–2).

* [J + E]: in Japanese with abstracts of European language. [J]: in Japanese.

Bando, Y., 1970. Lower Triassic ammonoids from the Kitakami massif. *Trans. Proc. Palaeontol. Soc. Jap., N.S.*, No. 79: 337–354 (pls. 37–38).

Bando, Y., and Shimoyama, S., 1974. Late Scythian ammonoids from the Kitakami massif. *Trans. Proc. Palaeontol. Soc. Jap., N.S.*, No. 94: 293–312 (pls. 40–42).

Banno, S., 1958. Glaucophane schists and associated rocks in the Omi district. *Jap. J. Geol. Geogr.*, 29: 29–44.

Banno, S., 1964. Petrologic studies on the Sambagawa crysstalline schists in the Bessi-Ino district, central Shikoku, Japan. *J. Fac. Sci., Univ. Tokyo*, 2, 15: 203–319.

Banno, S., Yokoyama, K., Iwata, O. and Terashima, S., 1976. Genesis of epidote amphibolite masses in the Sambagawa metamorphic belt of central Shikoku. *J. Geol. Soc. Jap.*, 82(3): 199–210 [J + E].

Biq, Ch., 1974. Taiwan. In: A.M. Spencer (Editor), *Meso-zoic–Cenozoic Orogenic Belts. Geol. Soc. Lond., Spec. Publ.*, No. 4: 501–511.

Dickinson, W.R., 1971. Clastic sedimentary sequences deposited in shelf, slope, and trough settings between magmatic arcs and associated trenches. *Pac. Geol.*, 3: 15–30.

Dietz, R.S. and Holden, J.C., 1970. Reconstruction of Pangea: Breakup and dispersion of continents, Permian to Present. *J. Geophys. Res.*, 75: 4939–4956.

Druczic, B.B. and Pergament, M.A., 1963. The genus *Nipponites* from the Upper Cretaceous of Kamchatka and Sakhalin. *Paleontol. Zh.*, 1963(2): 38–42 (in Russian).

Dundo, O.P. and Efremova, V.I., 1974. *Field Atlas of the Senonian Index Fauna in the Northeastern Part of the Koryak Highland.* Sci. Res. Inst. Arct. Geol., Leningrad 28 pp. (12 plates) (in Russian).

Eguchi, M., 1951. Mesozoic hexacorals from Japan. *Sci. Rep. Tohoku Univ.*, 2, 24: 1–96 (pls. 1–28).

Endo, R., 1961. Calcareous algae from the Jurassic Torinosu limestone of Japan. *Sci. Rep. Saitama Univ., B*, Commun. Vol. for Prof. Endo, pp. 53–75 (pls. 1–17).

Endo, R. and Horiguchi, M., 1967. Calcareous algae from the Konose Group in Kyushu, I. *Bull. Tokyo Coll. Domest. Sci.*, No. 7: 1–8 (pls. 1–3).

Endo, S., 1925. *Nilssonia*-bed of Hokkaido and its flora. *Sci. Rep. Tohoku Imp. Univ.*, 2, 7: 57–72 (pls. 11–17).

Ernst, W.G., Seki, Y., Onuki, H. and Gilbert, M.C., 1970. Comparative study of low-grade metamorphism in the California Coast Ranges and the outer metamorphic belt of Japan. *Geol. Soc. Am., Mem.*, 124: 1–276.

Fujiwara, Y., 1966. Palaeomagnetic studies on some Mesozoic rocks in Japan. Pt. 1. *J. Fac. Sci., Hokkaido Univ.*, 4, 13(3): 293–300.

Fujiwara, Y., 1967. Changing of the palaeolatitude in the Japanese islands through the Palaeozoic and Mesozoic. *J. Fac. Sci. Hokkaido Univ.*, 4, 14(2): 159–168.

Fujiwara, Y. and Nagase, M., 1965. Palaeomagnetic studies of the Cretaceous rocks in the Nemuro Peninsula, Hokkaido, Japan. *Chikyu Kagaku* [Earth Sci.], No. 79: 42–46.

Fujiyama, I., 1973. Mesozoic insect fauna of East Asia. Part I. Introduction and Upper Triassic faunas. *Bull. Natl. Sci. Mus., Tokyo*, 16 (2): 331–386 (pls. 1–5).

Fujiyama, I., 1974. A Liassic cockroach from Toyora, Japan. *Bull. Natl. Sci. Mus., Tokyo*, 17: 311–314.

Fukada, A., 1950. On the occurrence of *Perisphinctes* (s.s.) from the Ozika peninsula in the southern Kitakami mountainland. *J. Fac. Sci., Hokkaido Univ.*, 4, 7(3): 211–216 (1 pl.).

Fukada, A., 1953. A new species of *Nerinea* from central Hokkaido. *J. Fac. Sci., Hokkaido Univ.*, 4, 8(3): 211–216 (pl. 13).

Gorai, M., 1955. *Igneous Petrogenesis, 1.* Assoc. Geol. Coll., Tokyo, pp. 1–128 [J].

Gorai, M., 1965. Distinction of palaeo-, meso- and neo-tectonic provinces in Japan and the neighbouring areas and its bearing on the development of island arcs in the western Pacific. In: *Upper Mantle Symp. New Delhi 1964,* I.U.G.S., Copenhagen, pp. 60–64.

Hanai, T., 1951. Cretaceous non-marine ostracoda from the Sungari Group in Manchuria. *J. Fac. Sci., Univ. Tokyo,* 2, 7(9): 403–430 (pls. 1–2).

Hanai, T., 1953. Lower Cretaceous belemnites from Miyako district, Japan. *Jap. J. Geol. Geogr.*, 23: 63–80 (pls. 5–7).

Hanai, T., Obata, I. and Hayami, I., 1968. Notes on the Cretaceous Miyako Group. *Mem. Natl. Sci. Mus., Tokyo,* No. 1: 20–28, (pls. 1–4) [J + E].

Hara, I., 1962. Studies on the structure of the Ryoke metamorphic rocks of the Kasagi district, Southwest Japan. *J. Sci. Hiroshima Univ.*, C, 4(2): 163–224 (pls. 18–20).

Hase, A., 1960. The late Mesozoic formations and their molluscan fossils in west Chugoku and north Kyushu, Japan. *J. Sci. Hiroshima Univ.*, C, 3(2): 281–342 (pls. 31–39).

Hashimoto, M., 1973. Recent studies on Paleozoic and Mesozoic greenstones in Japan. *Bull. Natl. Sci. Mus., Tokyo,* 16(4): 739–750.

Hashimoto, M., Igi, S., Seki, Y., Banno, S. and Kojima, G., 1970. *Notes on Metamorphic Facies Map of Japan* (1 : 2,000,000). Geol. Surv. Japan, Tokyo, 18 p. (1 map).

Hashimoto, W., 1952. Jurassic stratigraphy of Hokkaido. *Rep. Geol. Surv. Jap., Spec.* No., B: 1–64 [J + E].

Hashimoto, W., 1960a. *Yezoactinia,* a new hydrozoan fossil from Shotombetsu, Nakatombetsu-machi, Esashi-gun, Hokkaido (Yezo), Japan. *Sci. Rep. Tokyo Kyoiku Daigaku,* C, 7(60): 95–97 (pls. 1–4).

Hashimoto, W., 1960b. Stromatoporoids from the Ainonai limestone, Kitami province, Hokkaido. *Sci. Rep. Tokyo Kyoiku Daigaku* 7(65): 195–203 (pls. 1–3).

Hashimoto, W., 1970. Some problems concerning the older deposits prior to the Middle Yezo Group of Hokkaido. *Sci. Rep. Tohoku Univ.*, 2, Spec. Vol., 4: 437–447 [J + E].

Hashimoto, W., 1973. *Hourcquia hataii* Hashimoto, a new species of ammonite from the Upper Cretaceous system of the Abeshinai region, Teshio province, Hokkaido, Japan. *Sci. Rep. Tohoku Univ.*, 2, Spec. Vol., 6: 315–318 (pl. 35).

Hashimoto, W. and Igo, H., 1962. Discovery of Mesozoic fossils from southwestern Hokkaido. *Proc. Jap. Acad.*, 38: 502–507.

Hayami, I., 1957a. Liassic *Bakevellia* in Japan. *Jap. J. Geol. Geogr.*, 28: 47–59 (pls. 2–3).

Hayami, I., 1957b. Liassic *Gervillia* and *Isognomon* in Japan. *Jap. J. Geol. Geogr.*, 28: 95–106 (pls. 6–7).

Hayami, I., 1957c. On the occurrence of *Cardinoides* from the Liassic Kuruma Group in central Japan. *Trans. Proc. Palaeontol. Soc. Jap., N.S.* No. 26: 69–73 (pl. 12).

Hayami, I., 1957d. *Radulonectites,* a new pectinid genus, from the Liassic Kuruma Group in central Japan. *Trans. Proc. Palaeontol. Soc. Jap., N.S.,* No. 27: 89–93 (pl. 16).

Hayami, I., 1957e. Liassic *Chlamys, "Camptonectes"* and other pectinids from the Kuruma Group in central Japan. *Trans. Proc. Palaeontol. Soc. Jap., N.S.,* No. 28: 119–127 (pl. 20).

Hayami, I., 1958a. Liassic *Volsella, Mytilus* and some other dysodont species in Japan. *Trans. Proc. Palaeontol. Soc. Jap., N.S.,* No. 29: 155–165 (pls. 23–24).

Hayami, I., 1958b. Supplementary descriptions of the Liassic pelecypods from the Kuruma and Shizukawa Groups in Japan. *Trans. Proc. Palaeontol. Soc. Jap., N.S.,* No. 30: 193–200 (pl. 28).

Hayami, I., 1958c. A review of the so-called Liassic "cyrenoids" in Japan. *Jap. J. Geol. Geogr.,* 29: 11–27 (pls. 2–3).

Hayami, I., 1958d. Taxonomic notes on *Cardinia* with description of a new species from the Lias of western Japan. *J. Fac. Sci., Univ. Tokyo,* 2, 11(2): 115–130.

Hayami, I., 1959a. Pelecypods of the Mizunuma Jurassic in Miyagi Prefecture, with some stratigraphical remarks. *Trans. Proc. Palaeontol. Soc. Jap., N.S.* No. 34: 66–78 (pl. 7).

Hayami, I., 1959b. Late Jurassic hipodont, taxodont and dysodont pelecypods from Makito, central Japan. *Jap. J. Geol. Geogr.,* 30: 135–150 (pl. 12).

Hayami, I., 1959c. Late Jurassic isodont and myacid pelecypods from Makito, central Japan. *Jap. J. Geol. Geogr.,* 30: 151–167 (pl. 13).

Hayami, I., 1959d. Lower Liassic lamellibranch fauna of the Higashinagano formation in west Japan. *J. Fac. Sci., Univ. Tokyo,* 2, 12(1): 31–84 (pls. 5–8).

Hayami, I., 1960a. Pelecypods of the Jusanhama Group (Purbeckian or Wealden) in Hashiura area, Northeast Japan. *Jap. J. Geol. Geogr.,* 31: 13–22 (pl. 3).

Hayami, I., 1960b. Jurassic inoceramids in Japan. *J. Fac. Sci., Univ. Tokyo,* 2, 12(2): 277–328 (pls. 15–18).

Hayami, I., 1961a. On the Jurassic pelecypod faunas in Japan. *J. Fac. Sci. Univ. Tokyo,* 2, 13(2): 243–343.

Hayami, I., 1961b. Jurassic pelecypods from the Awazu and Yamagami formations in Northeast Japan. *Trans. Proc. Palaeontol. Soc. Jap., N.S.* No. 43, 119–125.

Hayami, I., 1962. Jurassic pelecypod faunas in Japan, with special reference to their stratigraphical distribution and biogeographical provinces. *J. Geol. Soc. Jap.,* 68: 96–108 [J + E].

Hayami, I., 1965. Transition of Jurassic and Cretaceous marine pelecypods. *Kwaseki* [Fossils], No. 9: 13–23.

Hayami, I., 1965–66. Lower Cretaceous marine pelecypods of Japan. Pt. I. *Mem. Fac. Sci., Kyushu Univ.,* D, 15(2): 221–349 (pls. 27–52) (1965); Pt. II. *Ibid.,* 17(2): 73–150 (pls. 7–21) (1965); Pt. III. *Ibid.,* 17(3): 151–249 (pls. 22–26) (1966).

Hayami, I., 1975. A systematic survey of the Mesozoic Bivalvia from Japan. *Univ. Mus., Univ. Tokyo, Bull.,* No. 10: 240 pp. (10 pls).

Hayami, I., and Ichikawa, T., 1965. Occurrence of *Nippononaia ryosekiana* from the Sanchu area, Japan. *Trans. Proc. Palaeontol. Soc. Jap., N.S.,* No. 60: 145–155 (pl. 17).

Hayami, I., and Nakai I., 1965. On a Lower Cretaceous pelecypod, *"Cyrena" naumanni,* from Japan. *Trans, Proc. Palaeontol. Soc. Jap., N.S.,* No. 59: 114–125 (pls. 13–14).

Hayami, I., Sugita, M. and Nagumo, Y., 1960. Pelecypods of the Upper Jurassic and lowermost Cretaceous Shishiori Group in Northeast Japan. *Jap. J. Geol. Geogr.,* 31: 85–98 (pl. 8).

Hayase, I. and Ishizaka, K., 1967. Rb–Sr dating on the rocks in Japan (I). – Southwest Japan. *J. Jap. Assoc. Min. Petrol. Econ. Geol.,* 58(6): 201–212.

Hayashi, S., 1968. On conodonts from the Adoyama formation, Kuzuu-machi, Tochigi Prefecture. *Chikyu Kagaku* [Earth-Sci.], 22(2): 63–77 (pls. 1–4) [J + E].

Hirano, H., 1971. Biostratigraphic study of the Jurassic Toyora Group. Part I. *Mem. Fac. Sci., Kyushu Univ.,* D, 21(1): 93–128 (pls. 14–20).

Hirano, H., 1973. Biostratigraphic study of the Jurassic Toyora Group. Part II. *Trans. Proc. Palaeontol. Soc. Japan, N.S.,* No. 89: 1–14 (pls. 1–4); Part III. *Ibid., N.S.,* No. 90: 45–71 (pls. 9–10).

Hirano, H., 1975. Ontogenetic study of Late Cretaceous *Gaudryceras tenuiliratum. Mem Fac. Sci., Kyushu Univ.,* D, 22(2): 165–192 (pls. 24–26).

Hoffet, J.H., 1937. Les lamellibranchs saumâtres du Senonien de Muong Phalane (Bas-Laos). *Bull. Serv. Géol. Indochine,* 24(2): 4–25 (pls. 1–5).

Hunahashi, M., 1957. Alpine orogenic movement in Hokkaido. *J. Fac. Sci., Hokkaido Univ.,* 4, 9: 415–465.

Ichikawa, K., 1953–54. Late Triassic pelecypods from the Kochigatani group in the Sakuradani and Kito areas, Kokushima Prefecture, Shikoku, Part 1. *J. Inst. Polytech., Osaka City Univ.,* G, 1: 35–55 (pls. 1–2) (1953); Part 2. *Ibid.,* 2: 53–70 (pls. 3–4) (1954).

Ichikawa, K., 1958. Zur Taxonomie und Phylogenie der Triadischen "Pteriidae" (Lamellibranch.), mit besonderer berücksichtigung der Gattungen *Claraia, Eumorphotis, Oxytoma* und *Monotis. Palaeontographica,* A, 111(5–6): 131–212 (pls. 21–24).

Ichikawa, K., 1964. Tectonic status of the Honshu major belt in Southwest Japan during the Early Mesozoic. *J. Geosci., Osaka City Univ.,* 8(3): 71–107.

Ichikawa, K. and Maeda, Y., 1958. Late Cretaceous pelecypods from the Izumi Group. Part I. *J. Inst. Polytech., Osaka City Univ.,* G, 3: 61–74 (pls. 1–2); Part II. *Ibid.,* 4: 71–112 (pls. 3–7)

Ichikawa, K. and Maeda, Y., 1963. Late Cretaceous pelecypods from the Izumi Group. Part III. *J. Geosci., Osaka City Univ.,* 7(5). 113–145 (incl. pls. 8–11).

Ichikawa, K. and Maeda, Y., 1966. *Clisocolus* (Bivalvia, Late Cretaceous) from the Izumi Group of the Kinki district, Japan. *Professor Susumu Matsushita Mem. Vol., Kyoto,* pp. 233–241 (incl. pl. 7).

Ichikawa, K., Ishii, K., Nakagawa, C., Suyari, K. and Yamashita, N., 1956. Die Kurosegawa-Zone. *J. Geol. Soc. Jap.,* 62: 82–103 [J + E].

Ichikawa, K., Murakami, N., Hase, A. and Wadatsumi, K., 1968. Late Mesozoic igneous activity in the inner side of Southwest Japan. *Pac. Geol.,* 1: 97–118.

Ichikawa, K., Fujita, S. and Shimazu, M., 1970. *Tectonic Development of the Japanese Islands.* Tsukiji-shokan, Tokyo 232 pp. [J].

Igo, H., 1972. Conodonts, as a new index fossil in Japan. *J. Geogr. Soc. Tokyo,* 81(3): 142–151 [J + E].

Igo, H., and Koike, T., 1975. Geological age of the Kamiaso conglomerate and new occurrence of Triassic conodonts in the Mino mountains. *J. Geol. Soc. Jap.,* 81: 197–198 [J].

Igo, H., Koike, T., Igo, H. and Kinoshita, T., 1974. On the occurrence of Triassic conodonts from the Sorachi Group. *J. Geol. Soc. Jap.,* 80: 135–136 [J].

Imai, I., Teraoka, Y. and Okumura, K., 1971. Geologic structure and metamorphic zonation of the northeastern part of the Shimanto terrane in Kyushu. *J. Geol. Soc. Jap.,* 77: 207–220 [J + E].

Imanishi, S., 1956. On the occurrence of *Trigonia* bearing sandstone at Horonbetsu, Utanobori-mura, Eashi-gun, north Hokkaido. *Kumamoto J. Sci., B, Geol.,* 2(1): 49–53.

Ishibashi, T., 1970–75. Upper Triassic ammonites from Okinawa-jima. Part I. *Mem. Fac. Sci., Kyushu Univ.,* D, 20(2): 195–223 (pls. 26–29) (1970); Part II. *Ibid.,* 22(1): 1–12 (pls. 1–3) (1973); Part III. *Ibid.,* 22(2): 193–213 (pls. 27–28) (1975).

Ishibashi, T., 1972. Upper Triassic cephalopods from the Tanoura district, Kumamoto Prefecture, Japan. *Trans. Proc. Palaeontol. Soc. Jap.,* N.S., No. 88: 447–457 (pl. 54).

Ishihara, S., 1971. Modal and chemical compositions of the granitic rocks related to the major molybdenum and tungsten deposits in the Inner Zone of Southwest Japan. *J. Geol. Soc. Jap.,* 77: 441–452.

Ishihara, S., 1973. Molybdenum and tungsten provinces in the Japanese islands and North American Cordillera: An example of asymmetrical metal zoning in Pacific type orogeny. *Bur. Min. Res., Geol. Geophys. Bull.,* 141: 173–189.

Ishizaka, K., 1972. Rb–Sr dating on the igneous and metamorphic rocks of the Kurosegawa tectonic zone. *J. Geol. Soc. Jap.,* 78(11): 569–575 [J + E].

Ishizaka, K. and Yamaguchi, M., 1969. U–Th–Pb ages of sphene and zircon from the Hida metamorphic terrain, Japan. *Earth Plan. Sci. Lett.,* 6: 179–185.

Ito, H. and Tokieda, K., 1974. Paleomagnetism of Cretaceous granites in south Korea. *Rock Magn. Paleogeophys.,* 2: 59–61.

Jimbo, K., 1894. Beiträge zur Kenntniss der Fauna der Kreideformation von Hokkaido. *Paläontol. Abh.,* N.F., 2(6): 149–194 (pls. 17–30).

Kambe, N., 1951. On the myophorians from Kyoto Prefecture. *Trans. Proc. Palaeontol. Soc. Jap., N.S.,* No. 2: 49–56 (pl. 4).

Kambe, N., 1957. On the myophorians from the Miharaiyama Group in Hyogo Prefecture. *Rep. Geol. Surv. Jap.,* No. 173: 1–21 (pl. 1).

Kambe, N., 1963. On the boundary between the Permian and Triassic Systems in Japan. *Rep. Geol. Surv. Jap.,* No. 198: 1–66.

Kang, Y., 1962. Role of pyrrhotite in rock magnetism. *Mem. Coll. Sci., Univ. Kyoto,* B, 28: 489–526.

Kanie, Y., 1966. The Cretaceous deposits in the Urakawa District, Hokkaido. *J. Geol. Soc. Jap.,* 72: 315–328 [J + E].

Kanie, Y., 1975. Some Cretaceous patelliform gastropods from the northern Pacific region. *Sci. Rep. Yokosuka City Mus.,* No. 21: 1–44 (pls. 1–20).

Kanisawa, S., 1974a. Igneous activity and metamorphic history in Northeast Japan. *Mem. Geol. Soc. Jap.,* No. 10: 5–19 [J + E].

Kanisawa, S., 1974b. Granitic rocks closely associated with the Lower Cretaceous volcanic rocks in the Kitakami mountains, Northeast Japan. *J. Geol. Soc. Jap.,* 80: 355–367.

Kanmera, K., 1964. Triassic coral faunas from the Konose Group in Kyushu. *Mem. Fac. Sci., Kyushu Univ.,* D, 15(1): 117–147 (pls. 12–19).

Kanmera, K., 1969. Litho- and biofacies of Permo-Triassic geosynclinal limestone of the Sambosan belt in southern Kyushu. *Palaeontol. Soc. Jap., Spec. Pap.,* No. 14: 13–39 (pls. 4–8).

Kanmera, K. and Nakazawa, K., 1973. Permian–Triassic relationships and faunal changes in the eastern Tethys. In: A. Logan and L.V. Hills (Editors), *The Permian and Triassic Systems and Their Mutual Boundary. Can. Soc. Petrol. Geol., Mem.,* 2: 100–119.

Kanmera, K. and Obata, I., 1972. A preliminary study on litho- and biofacies of limestones of the Hidaka and Sorachi Groups in the Hidaka Mountains, Hokkaido. *Mem. Natl. Sci. Mus.,* No. 5: 203–212 (pl. 4) [J + E].

Kano, K., 1975. Stratigraphy of the Upper Paleozoic–Mesozoic strata in the northern Kiso–Azusagawa district, Nagano Prefecture. *J. Geol. Soc. Jap.,* 81: 285–300 [J + E].

Kanomata, N., 1961. The geology of the Yamizo, Torinoko and Toriashi mountain blocks and their geologic age. *J. Coll. Arts Sci., Chiba Univ.,* 3(3): 351–367.

Karakida, Y., 1969. On the relationship of the schistose granodiorite and massive ore in northern Kyushu. *Stud. Litt. Sci. Seinan Gakuin Univ.,* 9(2): 75–85 [J].

Katada, M. (Editor), 1974. Cretaceous granitic rocks in the Kitakami mountains. *Rep. Geol. Surv. Japan,* No. 251: 1–139 [J + E].

Kato, Y. and Muroi, I., 1965. Paleomagnetic studies of the Cretaceous granitic rocks in northeastern Japan. *Ann. Rep. Rock Magn. Res. Group Jap., Kyoto,* 1965: 179–187.

Kawai, N., Ito, H. and Kume, S., 1961. Deformation of the Japanese islands as inferred from rock magnetism. *Geophys. J. R. Astron. Soc.,* 6: 124–130.

Kawai, N., Nakajima, T. and Hirooka, K., 1971. The evolution of the island arc of Japan and formation of granites in the circum-Pacific belt. *J. Geomag. Geoelectr.,* 23: 267–293.

Kawano, Y. and Ueda, Y., 1964–66. K–Ar dating on the igneous rocks in Japan. Part I. *Sci. Rep. Tohoku Univ.,* 3, 9(1): 99–122; Part II. *Ibid.,* 9(2): 199–215 (1965); Part III. *Ibid.,* 9(3): 513–523 (1966); Part IV. *Ibid.,* 9(3): 52–539 (1966).

Kawano, Y. and Ueda, Y., 1967. Periods of the igneous activities of the granitic rocks in Japan by K–Ar dating method. *Tectonophysics*, 4(4–6): 523–530.

Kiminami, K., 1975. Sedimentology of the Nemuro Group. *J. Geol. Soc. Jap.*, 81: 215–232.

Kimura, Tatsuaki, 1958a. Mesozoic plants from the Tetori series, central Honshu, Japan (Part I). *Trans. Proc. Palaeontol. Soc. Jap., N.S.*, No. 29: 166–168 (pl. 25).

Kimura, Tatsuaki, 1958b. On the Tetori flora (Part I). Mesozoic plants from the Kuzuryu subgroup, Tetori group, Japan. *Bull. Senior High School, Tokyo Univ. Educ.*, 2(2): 1–47 (pls. 1–4).

Kimura, Tatsuaki, 1959a. Mesozoic plants from the Iwamuro Formation (Liassic), Tone-gun, Gunma Prefecture, Japan. *Bull. Senior High School, Tokyo Univ. Educ.*, 3: 1–59.

Kimura, Tatsuaki, 1959b. Mesozoic plants from the Kotaki coal-field, the Kuruma group, central Honshu, Japan. *Bull. Senior High School, Tokyo Univ. Educ.*, 3: 61–83.

Kimura, Tatsuaki, 1959c. Preliminary notes on the Liassic floras of the Japanese Islands. *Bull. Senior High School, Tokyo Univ. Educ.*, 3: 85–133.

Kimura, Tatsuaki, 1959d. On the Tetori flora (Part II). Addition to the Mesozoic plants from the Kuzuryu subgroup, Tetori Group, Japan. *Bull. Senior High School, Tokyo Univ. Educ.*, 3: 103–121 (pls. 1–2).

Kimura, Tatsuaki, 1961. Mesozoic plants from the Itoshiro subgroup, the Tetori group, central Honshu, Japan. Part II. *Trans. Proc. Palaeontol. Soc. Jap., N.S.*, No. 41: 21–32 (pls. 4–6).

Kimura, Tatauaki, 1973. Distribution of the Mesozoic plants. *Kwaseki* [Fossils], No. 25–26: 9–44 [J].

Kimura, Tatsuaki, 1975a. Middle–late Early Cretaceous plants newly found from the Upper course of the Kuzuryu River area, Fukui Prefecture, Japan. *Trans. Proc. Palaeontol. Soc. Jap., N.S.*, No., 98: 55–93 (pls. 5–8).

Kimura, Tatsuaki, 1975b. Notes on Early Cretaceous floras of Japan. *Bull. Tokyo Gakugei Univ.*, 4, 27: 218–257.

Kimura, Tatsuaki and Asama, K., 1975. On the Paleozoic and Mesozoic land floras of Japan, with special reference to the stratigraphical distribution of early Cretaceous plants in Kochi Prefecture, in the Outer Zone of Southwest Japan. *Mem. Natl. Sci. Mus.* No. 8: 91–114 [J + E].

Kimura, Tatsuaki and Hirata, M., 1975. Early Cretaceous plants from Kochi Prefecture, Southwest Japan. *Mem. Natl. Sci. Mus.* No. 8: 67–90 (pls. 10–13).

Kimura, Tatsuaki and Sekido, S., 1965. Some interesting ginkgoalean leaves from the Itoshiro subgroup, the Tetori group, central Honshu, Japan. *Mem. Mejiro Gakuen Woman's Jun. Coll.*, 2: 1–4 (pls. 1–2).

Kimura, Tatsuaki and Sekido, S., 1966. Mesozoic plants from the Itoshiro subgroup, the Tetori group, central Honshu, Japan. Part III. *Mem. Mejiro Gakuen Woman's Jun. Coll.*, 3: 1–7 (pls. 1–4).

Kimura, Tatsuaki and Sekido, S., 1967. Some Mesozoic plants from the Itoshiro subgroup, the Tetori group, central Honshu, Japan. *Professor Hidekata Shibata Mem. Vol.*, 416–419 (pls. 1–3) (Tokyo).

Kimura, Tatsuaki and Sekido, S., 1972. *Ctensis* species from the Itoshiro subgroup (Lower Cretaceous), the Tetori group, central Honshu, Japan. *Trans. Proc. Palaeontol. Soc. Jap.*, N.S., No. 86: 360–368 (pls. 44–45).

Kimura, Tatsuaki and Sekido, S., 1974. Bipinnate cycadean fronds newly found from the Lower Cretaceous Itoshiro subgroup, the Tetori group, central Honshu, Japan. In: *Symposium on morphological and Stratigraphical Paleobotany. Birbal Sahni Inst. Palaeobot., Spec. Publ.*, No. 2: 23–27 (pls. 1–3).

Kimura, Tatsuaki and Sekido, S., 1975. *Nilssoniocladus* n. gen. (Nilssoniaceae n. fam.), newly found from the early Lower Cretaceous of Japan. *Palaeontographica*, B, 153: 111–118 (pls. 1–2).

Kimura, Toshio, 1944a. The radiolarian fauna of the Naradani formation in the Sakawa basin in the Prov. of Tosa. *Jap. J. Geol. Geogr.*, 19(1–4): 273–279 (pl. 29).

Kimura, Toshio, 1944b. A study on the radiolarian chert at Hukuda on the southeastern border of the Sakawa basin in the province of Tosa. *Jap. J. Geol. Geogr.*, 19(1–4): 281–284 (pl. 30).

Kimura, Toshio, 1944c. Some radiolarians in Nippon. *Jap. J. Geol. Geogr.*, 19(1–4): 285–288 (pl. 31).

Kimura, Toshio, 1951. Some pectinids and limids from the Jurassic Torinosu Group in Japan. *J. Fac. Sci., Univ. Tokyo*, 2, 7(7): 337–350 (pl. 1).

Kimura, Toshio, 1954. The sandstone and limestone of the Nakanosawa Formation. *J. Geol. Soc. Jap.*, 60: 67–80 [J + E]

Kimura, Toshio, 1956a. The Torinosu Group and the Torinosu limestone in the Togano and Go basins, Kochi Prefecture. *J. Geol. Soc. Jap.*, 62: 515–526 [J + E]

Kimura, Toshio, 1956b. Some pelecypods from the Upper Jurassic Torinosu Group in Kochi Prefecture, Japan. *J. Earth Sci., Nagoya Univ.*, 4(2): 80–89 (pl. 1).

Kimura, Toshio, 1960. On the geologic structure of the Paleozoic group in Chugoku, West Japan. *Sci. Rep. Coll. Gen. Educ., Univ. Tokyo*, 10: 109–124.

Kimura, Toshio, 1967. Structural division of Japan and the Honshu Arc. *Jap. J. Geol. Geogr.*, 38: 117–131.

Kimura, Toshio, 1968. Some folded structures and their distribution in Japan. *Jap. J. Geol. Geogr.*, 39: 1–26.

Kimura, Toshio, 1973. The old "inner" arc and its deformation in Japan. In: P.J. Coleman (Editor), *The Western Pacific.* University of Australia Press, Canberra, A.C.T., pp. 255–273.

Kimura, Toshio, 1974a. Mesozoic folds and foldings in Japan. *J. Geogr., Tokyo*, 83: 143–156 [J + E].

Kimura, Toshio, 1974b. The ancient continental margin of Japan. In: C.A. Burk and C.L. Drake (Editors), *The Geology of Continental Margins.* Springer-Verlag, New York, N.Y., pp. 817–829.

Kimura, Toshio and Tokuyama, A., 1971. Geosynclinal prisms and tectonics in Japan. *Mem. Geol. Soc. Jap.*, No. 6: 9–20.

Kimura, Toshio, Yoshida, S. and Toyohara, E., 1975. Paleogeography and earth movements of Japan in the late Permian to early Jurassic Sambosan stage. *J. Fac. Sci., Univ. Tokyo*, 2, 19(2): 149–177.

Kinosaki, K., 1963. The pyrophyllite deposits in the Chugoku province, west Japan. *Geol. Rep. Hiroshima Univ.*, No. 12: 1–35 [J + E].

Kinosaki, K., 1965. The pyrophyllite deposits in Japan. *J. Min. Soc. Japan*, 7: 185–199 [J + E].

Kobayashi, T., 1935. Contributions to the Jurassic Torinosu Series in Japan. *Jap. J. Geol. Geogr.,* 12: 69–91 (pls. 12–13).

Kobayashi, T., 1938. On the Noric age of the Nariwa flora of the Rhaeto-Liassic aspect. *Jap. J. Geol. Geogr.,* 15: 1–12.

Kobayashi, T., 1939. The geological age of the Mesozoic land flora in Western Japan discussed from stratigraphic standpoint. *Jap. J. Geol. Geogr.,* 16: 75–103.

Kobayashi, T., 1941. The Sakawa orogenic cycle and its bearing on the origin of the Japanese Islands. *J. Fac. Sci., Univ. Tokyo,* 2, 5(7): 219–578 (pls. 1–4; 10 maps).

Kobayashi, T., 1942. On the climatic bearing of the Mesozoic floras in eastern Asia. *Jap. J. Geol. Geogr.,* 18(4): 157–196.

Kobayashi, T., 1947. On the occurrence of *Seymourites* in Nippon and its bearing on the Jurassic palaeogeography. *Jap. J. Geol. Geogr.,* 20: 19–31 (pls. 7–8).

Kobayashi, T., 1951. Older Mesozoic *Estherites* from Eastern Asia. *J. Fac. Sci., Univ. Tokyo,* 2, 7(10): 431–440 (pl. 1).

Kobayashi, T., 1952. Two new estherians from the province of Nagato in west Japan. *Trans. Proc. Palaeontol. Soc. Jap., N.S.,* No. 6: 175–178 (pl. 16).

Kobayashi, T., 1954a. Fossil estherians and allied fossils. *J. Fac. Sci., Univ. Tokyo,* 2, 9(1): 1–192.

Kobayashi, T., 1954b. Estherian evolution and orogenic cycle. In: *Congr. Géol. Int., C.R. 1ère Sess., Alger, 1952, Union Paléontol. Int.,* 19: 71–80.

Kobayashi, T., 1956a. Studies on the Jurassic trigonians in Japan. Part V. Some Jurassic trigonians from central and west Japan. *Jap. J. Geol. Geogr.,* 27(1): 1–8 (pl. 1).

Kobayashi, T., 1956b. On the dentition of *Trigonoides* and its relation to similar pelecypod genera. *Jap. J. Geol. Geogr.,* 27(1): 79–92 (pl. 5).

Kobayashi, T., 1963. On the Cretaceous Ban Na Yo fauna of East Thailand with a note on the distribution of *Nippononaia, Trigonioides* and *Plicatounio. Jap. J. Geol. Geogr.,* 35(1): 35–43 (pl. 3).

Kobayashi, T., 1968. The Cretaceous non-marine pelecypods from Nam Phung damsite in the northeastern part of the Khorat plateau, Thailand, with a note on the Trigonioididae. *Geol. Palaeontol. Southeast Asia,* 4: 109–138 (pls. 20–23).

Kobayashi, T. and Aoti, K., 1943. Halobiae in Nippon. *J. Sigen. Ken.,* 1(2): 241–255 (pls. 24–25).

Kobayashi, T. and Fukada, A., 1947a. A new species of *Ataxioceras* in Nippon. *Jap. J. Geol. Geogr.,* 20: 45–48 (pl. 11).

Kobayashi, T. and Fukada, A., 1947b. On the occurrence of *Katroliceras* in the Tetori Series. *Jap. J. Geol. Geogr.,* 20: 49–53 (pl. 12).

Kobayashi, T. and Fukada, A., 1947c. On the occurrence of *Discosphinctes* in the Kitakami Mountains in Nippon. *Jap. J. Geol. Geogr.,* 20: 55–57 (pl. 13).

Kobayashi, T. and Huzita, A., 1942. Estheriae in the Cretaceous Sungari series in Manchoukou. *J. Fac. Sci., Univ. Tokyo,* 2, 6(7): 107–128 (pls. 1–2).

Kobayashi, T. and Ichikawa, K., 1949a. *Tosapecten,* gen. nov. and other Upper Triassic Pectinidae from the Sakawa basin in Shikoku, Japan. *Jap. J. Geol. Geogr.,* 21(1–4): 163–175 (pl. 5).

Kobayashi, T. and Ichikawa, K., 1949a. *Myophoria* and other Upper Triassic pelecypods from the Sakawa basin in Shikoku, Japan. *Jap. J. Geol. Geogr.,* 21(1–4): 177–192 (pl. 6).

Kobayashi, T. and Ichikawa, K., 1949c. Late Triassic *"Pseudomonotis"* from the Sakawa basin in Shikoku, Japan. *Jap. J. Geol. Geogr.,* 21(1–4): 245–262 (pl. 9).

Kobayashi, T. and Ichikawa, K., 1950. On the Upper Triassic Kochigatani series in the Sakawa basin in Japan, and its pelecypod faunas. *J. Fac. Sci., Univ. Tokyo,* 2, 7(3–7): 179–256 (pls. 1–5).

Kobayashi, T. and Ishibashi, T., 1970. *Halobia styriaca,* Upper Triassic pelecypod, discovered in Okinawa-jima, the Ryukyu islands. *Trans. Proc. Palaeontol. Soc. Jap.,* N.S., No. 77: 243–248 (pl. 26).

Kobayashi, T. and Katayama, M., 1938. Further evidence as to the chronological determination of the so-called Rhaeto-Liassic floras with a description of *Minetrigonia,* a new subgenus of *Trigonia. Proc. Imp. Acad., Tokyo,* 14(5): 184–189.

Kobayashi, T. and Kido, Y., 1947a. Cretaceous *Estherites* from the Kyongsang group in the Tsushima basin. *Jap. J. Geol. Geogr.,* 20(2–4): 83–90 (pl. 18).

Kobayashi, T. and Kido, Y., 1947b. Cretaceous *Estherites* from the province of Chientao, Manchuria. *Jap. J. Geol. Geogr.,* 20(2–4): 91–95 (pl. 19).

Kobayashi, T. and Kusumi, H., 1953a. A study on *Estherites middendorfii. Jap. J. Geol. Geogr.,* 23: 1–24 (pls. 1–2).

Kobayashi, T. and Kusumi, H., 1953b. Younger Mesozoic estherians from Tungha region in South Manchuria. *Jap. J. Geol. Geogr.,* 23: 25–35 (pls. 1–2).

Kobayashi, T. and Nakano, M., 1957. On the Pterotrigoniinae. *Jap. J. Geol. Geogr.,* 28(4): 219–238 (pls. 16–17).

Kobayashi, T. and Suzuki, K., 1936. Non-marine shells of the Naktong-Wakino series. *Jap. J. Geol. Geogr.,* 13(3–4): 234–257 (pls. 27–29).

Kobayashi, T. and Suzuki, K., 1937. Non-marine shells of the Jurassic Tetori series in Japan. *Jap. J. Geol. Geogr.,* 14(1–2): 33–51 (pls. 4, 5).

Kobayashi, T. and Suzuki, K., 1939. The brackish Wealden fauna of the Yoshimo beds in province Nagato, Japan. *Jap. J. Geol. Geogr.,* 16(3–4): 213–224 (pls. 13, 14).

Kobayashi, T. and Suzuki, K., 1941. On the occurrence of *Trigonioides* in southeastern Manchoukuo. *Bull. Geol. Inst. Manchoukuo,* No. 101: 77–81.

Kobayashi, T. and Tokuyama, A., 1959. *Daonella* in Japan. *J. Fac. Sci., Univ. Tokyo,* 2, 12(1): 1–26 (pls. 1–4).

Kobayashi, T., Mori, K. and Tamura, M., 1959. The bearing of the trigoniids on the Jurassic stratigraphy of Japan. *Jap. J. Geol. Geogr.,* 30: 273–292.

Koike, T. and Ishibashi, T., 1974. Upper Triassic conodonts from Okinawa-jima. *Trans. Proc. Palaeontol. Soc. Jap., N.S.,* No. 96: 433–436 (pl. 57).

Koike, T., Watanabe, K. and Igo, H., 1970. New information of Triassic conodont biostratigraphy in Japan. *J. Geol. Soc. Jap.,* 76: 267–269 [J + E].

Koike, T., Igo, H., Takizawa, S. and Kinoshita, T., 1971. Contribution to the geological history of the Japanese

Islands by the conodont biostratigraphy. Part II. *J. Geol. Soc. Jap.*, 77(3): 165–168.

Konishi, K., 1963. Pre-Miocene basement complex of Okinawa and the tectonic belts of the Ryukyu islands. *Sci. Rep. Kanazawa Univ.*, 8(2): 569–602.

Kon'no, E., 1944. Contribution to our knowledge of *Swedenborgia*. *Jap. J. Geol. Geogr.*, 19(1–4): 27–66 (pls. 1–5).

Kon'no, E., 1949. *Equisetites* from Triassic and Jurassic formations in Yamaguchi prefecture. *J. Geol. Soc. Jap.*, 55: 124–125 [J].

Kon'no, E., 1961. Some *Cyadocarpidium* and *Podozamites* from the Upper Triassic formations in Yamaguchi Prefecture, Japan. *Sci. Rep. Tohoku Univ.*, 2, 32(2): 195–211 (pls. 22–24).

Kon'no, E., 1962a. Some coniferous male fructifications from the Carnic formation in Yamaguchi Prefecture, Japan. *Sci. Rep. Tohoku Univ., Spec. Vol.*, 5: 9–19 (pls. 1–8).

Kon'no, E., 1962b. Some species of *Neocalamites* and *Equisetites* in Japan and Korea. *Sci. Rep. Tohoku Univ., Spec. Vol.*, 5: 21–47 (pls. 9–18).

Kon'no, E., 1968. Some Upper Triassic species of Dipteridaceae from Japan and Borneo. *J. Linn. Soc. (Bot.)*, 61 (384): 93–105 (pls. 1, 2).

Kon'no, E., 1973. New species of *Pleuromeia* and *Neocalamites* from the Upper Scythian beds in the Kitakami massif, Japan. *J. Geol. Soc. Jap.*, 2, 32(2): 99–115 (pls. 8–12).

Kon'no, E. and Naito, G., 1960. A new *Neocalamites* from the Carnic formation in Japan with brief notes on *Lobatannularia* in Asia. *Trans. Proc. Palaeontol. Soc. Jap., N.S.*, 40: 339–351 (pls. 40, 41).

Koriba, K. and Miki, S., 1931. On "*Archaeozostera*" from the Izumi Sandstone. *Chikyu*, 15(3): 165–204 (pls. 4–5) [J].

Koriba, K. and Miki, S., 1958. *Archaeozostera*, a new genus from Upper Cretaceous in Japan. *Palaeobotanist*, 7(2): 107–110 (pls. 1–2).

Kryshtofovich, A. and Baikovskaya, T.N., 1960. *Mesozoic Floras of Sakhalin*. Izdatelstvo Akad. Nauk S.S.S.R., Moscow, 122 pp. (21 pls.) (in Russian).

Kummel, B. and Sakagami, S., 1960. Mid-Scythian ammonites from Iwai Formation, Japan. *Breviora, Mus. Comp. Zool.*, No. 126: 1–11 (pls. 1–3).

Kuno, H. (compiler), 1968. Lead and strontium isotopes in basaltic and granitic rocks of the Pacific Ocean basin. *I.U.G.S. Geol. Newslett.*, 1968(3): 67–87.

Kuroda, Y., 1970. Glaucophane–schist metamorphism in the northwestern part of the Pacific Ocean and the problem on ultrabasic intrusion. *Kozan Chishitsu* (Min. Geol.), 20: 43–52 [J + E].

Kuroda, Y. and Tazaki, K., 1969. Origin of ultrabasic rocks in the metamorphic zones in Japan. *Mem. Geol. Soc. Japan*, No. 4: 99–108 [J + E].

Kusumi, H., 1960. On the occurrence of Cretaceous estherids in north Kyushu. *J. Sci. Hiroshima Univ.*, C, 3(1): 15–24 (pl. 4).

Kusumi, H., 1961. Studies on the fossil estherids, with special reference to recent estherids. *Geol. Rep. Hiroshima Univ.*, No. 7: 1–87 (pls. 1–9) [J + E].

Larson, R.L. and Chase, C.G., 1972. Late Mesozoic evolution of the western Pacific ocean. *Geol. Soc. Am. Bull.*, 83: 2627–2644.

Lowenstam, H.A. and Epstein, S., 1954. Paleotemperature of the post-Aptian Cretaceous as determined by the oxygen isotope method. *J. Geol.*, 62: 207–248.

Maeda, S., 1961. On the geologic history of the Mesozoic Tetori Group in Japan. *J. Coll. Arts, Sci., Chiba Univ.*, 3(3): 369–426 [J + E].

Maeda, S., 1962a. Trigoniid from the Tetori Group in the Furukawa district, central Japan. *Trans. Proc. Palaeontol. Soc. Jap., N.S.*, No. 47: 273–276 (pl. 42).

Maeda, S., 1962b. Some Lower Cretaceous pelecypods from the Akiwa Subgroup, the upper division of the Tetori Group in central Japan. *Trans. Proc. Palaeontol. Soc. Jap., N.S.*, No. 48: 343–351 (pl. 53).

Maeda, S., 1962c. On some *Nipponitrigonia* in Japan. *J. Coll. Arts Sci., Chiba Univ., Nat. Sci.*, 3(4): 503–514 (pls. 1–9).

Maeda, S., 1962d. On some species of Jurassic trigoniids from the Tetori Group in central Japan. *J. Coll. Arts Sci., Chiba Univ., Nat. Sci.*, 3(4): 515–518 (pls. 1–2).

Maeda, S., 1963a. Some Jurassic trigoniids from the Tetori Group in the Kuzuryu district, central Japan. *Trans. Proc. Palaeontol. Soc. Jap., N.S.*, No. 49: 1–7 (pl. 1).

Maeda, S., 1963b. *Trigonioides* from the Late Mesozoic Tetori Group, central Japan. *Trans. Proc. Palaeontol. Soc. Jap., N.S.*, No. 51: 79–85 (pl. 12).

Maeda, S. and Kawabe, T., 1967. *Apiotrigonia* from the Futaba Group in the Joban district, north Japan. *Professor H. Shibata Mem. Vol.*, Tokyo, pp. 420–425 (pl. 1).

Maeda, S. and Kitamura, T., 1964. Lower Cretaceous trigoniids from the Todai Formation, central Japan. *J. Coll. Arts Sci., Chiba Univ., Nat. Sci.*, 4(2): 45–57 (pls. 1–4).

Martinson, W.B., 1965. Cretaceous lamellibranchs of the family Trigonioididae and their classification. *Paleontol. Zh.*, 1965(4) (in Russian).

Masatani, K. and Tamura, M., 1959. A stratigraphic study on the Jurassic Soma Group on the eastern foot of the Abukuma mountains, Northeast Japan. *Jap. J. Geol. Geogr.*, 30: 245–257.

Matsuda, T. and Kuriyakawa, S., 1965. Lower grade metamorphism in the eastern Akaishi mountains, central Japan. *Bull. Earthquake Res. Inst. Univ. Tokyo*, 43: 209–235 [J + E].

Matsumoto, Tadashi, 1968. A hypothesis on the origin of the Late Mesozoic volcano-plutonic association in East Asia. *Pac. Geol.*, 1: 77–84.

Matsumoto, Tatsuro, 1938a. Preliminary notes on some of the more important fossils among the Gosyonoura fauna. *J. Geol. Soc. Jap.*, 45: 13–26 (pls. 1, 2).

Matsumoto, Tatsuro, 1938b. *Zelandites*, a genus of Cretaceous ammonites. *Jap. J. Geol. Geogr.*, 15(3–4): 137–148 (pl. 14).

Matsumoto, Tatsuro, 1942a. A note on the Japanese Cretaceous ammonites belonging to the subfamily Desmoceratinae. *Proc. Imp. Acad., Tokyo*, 18: 24–29.

Matsumoto, Tatsuro, 1942b. A note on the Japanese ammonites belonging to the Gaudryceratidae. *Proc. Imp. Acad., Tokyo*, 18: 666–670.

Matsumoto, Tatsuro, 1942c. A note on the Japanese ammonoid species belonging to the Tetragonitidae. *Proc. Imp. Acad., Tokyo,* 18: 671–673.

Matsumoto, Tatsuro, 1942d. A short note on the Japanese Cretaceous Phylloceratidae. *Proc. Imp. Acad., Tokyo,* 18: 674–676.

Matsumoto, Tatsuro, 1942–43. Fundamentals in the Cretaceous stratigraphy of Japan. *Part I. Mem. Fac. Sci., Kyushu Imp. Univ.,* D, 1(1): 129–280 (pls. 5–20) (1942); Parts II & III. *Ibid.,* 2(1): 97–237 (1943).

Matsumoto, Tatsuro, 1947. On some interesting ammonites from the Paleocretaceous of the Yuasa district, Southwest Japan. *Sci. Rep. Dep. Geol., Kyushu Univ.,* 2(1): 13–19.

Matsumoto, Tatsuro, 1949. The late Mesozoic geological history in the Nagato province, Southwest Japan. *Jap. J. Geol. Geogr.,* 21: 235–243.

Matsumoto, Tatsuro, 1953. The ontogeny of *Metaplacenticeras subtilistriatum* (Jimbo). *Jap. J. Geol. Geogr.,* 23: 139–150 (pl. 13).

Matsumoto, Tatsuro, 1954a. Selected Cretaceous leading ammonites in Hokkaido and Saghalien. In: T. Matsumoto (Editor) *Cretaceous System in the Japanese Islands.* Jap. Soc. Promot. Sci., Tokyo, 243–313 (pls. 17–36).

Matsumoto, Tatsuro, 1954b. Family Puzosiidae from Hokkaido and Saghalien. *Mem. Fac. Sci., Kyushu Univ.,* D, 5(2): 69–118 (pls. 9–23).

Matsumoto, Tatsuro (Editor), 1954c. *The Cretaceous System in the Japanese Islands.* Jap. Soc. Promot. Sci., Tokyo, 324 pp. (26 pls.).

Matsumoto, Tatsuro, 1955a. The bituberculate pachydiscids from Hokkaido and Saghalien. *Mem. Fac. Sci., Kyushu Univ.,* D, 5(3): 153–184 (pls. 31–37).

Matsumoto, Tatsuro, 1955b. Family Kossmaticeratidae from Hokkaido and Saghalien. *Jap. J. Geol. Geogr.,* 26(1–2): 115–164 (pls. 8–10).

Matsumoto, Tatsuro, 1956a. Further notes on the kossmaticeratids from Hokkaido. *Jap. J. Geol. Geogr.,* 27(2–4): 173–187 (pls. 14–16).

Matsumoto, Tatsuro, 1956b. *Yebisites,* a new Lower Jurassic ammonite from Japan. *Trans. Proc. Palaeontol. Soc. Jap., N.S.,* No. 23: 205–212 (pl. 30).

Matsumoto, Tatsuro, 1957a. A Turonian *Damesites* from Hokkaido, Japan. *Trans. Proc. Palaeontol. Soc. Jap., N.S.,* No. 27: 86–88 (pl. 15).

Matsumoto, Tatsuro, 1957b. *Inoceramus mihoensis* n. sp. and its significance. *Mem. Fac. Sci., Kyushu Univ.,* D, 6(2): 65–68 (pl. 21).

Matsumoto, Tatsuro, 1959a. Cretaceous ammonites from the Upper Chitina valley, Alaska. *Mem. Fac. Sci., Kyushu Univ.,* D, 8(3): 49–90 (pls. 12–29).

Matsumoto, Tatsuro, 1959b. Zonation of the Upper Cretaceous in Japan. *Mem. Fac. Sci., Kyushu Univ.,* D, 9(2): 55–93 (pls. 6–11).

Matsumoto, Tatsuro, 1960. *Graysonites* (Cretaceous Ammonites) from Kyushu. *Mem. Fac. Sci., Kyushu Univ.,* D, 10(1): 41–58 (pls. 6–8).

Matsumoto, Tatsuro (Editor), 1963. *A Survey of Fossils from Japan Illustrated in Classical Monographs.* Palaeontol. Soc. Jap., Tokyo, 57 pp. (68 pls.).

Matsumoto, Tatsuro, 1965–71. A monograph of the Collignoniceratidae from Hokkaido. Part I. *Mem. Fac. Sci., Kyushu Univ.,* D, 16(1): 1–80 (pls. 1–18) (1965); Part II. *Ibid.,* 16(3), 209–243 (pls. 36–43) (1965); Part III. *Ibid.,* 19(3): 297–330 (pls. 39–45) (1969); Part IV. *Ibid.,* 20(2): 225–304 (pls. 30–47) (1970); Part V. *Ibid.,* 21(1): 129–162 (pls. 21–24) (1971).

Matsumoto, Tatsuro, 1967a. Evolution of the Nostoceratidae (Cretaceous heteromorph ammonoids). *Mem. Fac. Sci., Kyushu Univ.,* D, 18(2): 331–347 (pls. 18–19).

Matsumoto, Tatsuro, 1967b. A Cretaceous nautiloid from Urakawa, Hokkaido. *Jap. J. Geol. Geogr.,* 38: 163–169 (pl. 3).

Matsumoto, Tatsuro, 1967c. Fundamental problems in the circum-Pacific orogenesis. *Tectonophysics,* 4(4–6): 595–613.

Matsumoto, Tatsuro, 1967d. Cretaceous. In: K. Asano et al., *Historical Geology, 2.* Asakura Shoten, Tokyo, pp. 408–477 [J].

Matsumoto, Tatsuro, 1968. Bibliography of geochronological data in Japan (1). *Jap. J. Geol. Geogr.,* 39(1): 1–5.

Matsumoto, Tatsuro, 1969. Geochronology and historical geology in Japan. *Mass. Spectrosc.,* 17: 434–444.

Matsumoto, Tatsuro, 1970. Uncommon keeled ammonites from the Upper Cretaceous of Hokkaido and Saghalien. *Mem. Fac. Sci., Kyushu Univ.,* D, 20(2): 305–317 (pls. 48–49).

Matsumoto, Tatsuro, 1973. Vascoceratid ammonites from the Turonian of Hokkaido. *Trans. Proc. Palaeontol. Soc. Jap., N.S.,* No. 89: 27–41 (pl. 8).

Matsumoto, Tatsuro, 1975a. Geologic history of the circum-Pacific region – with special reference to the Mesozoic and pre-Mesozoic history of Japan. *J. Geol. Soc. Jap.,* 81(7): 461–471 [J + E].

Matsumoto, Tatsuro, 1975b. Additional acanthoceratids from Hokkaido. *Mem. Fac. Sci., Kyushu Univ.,* D, 22(2): 99–163 (pls. 11–23).

Matsumoto, Tatsuro and Amano, M., 1964. Notes on a Cretaceous nautiloid from Kyushu. *Trans. Proc. Palaeontol. Soc. Jap., N.S.,* No. 53: 173–178 (pl. 26).

Matsumoto, Tatsuro and Harada, M., 1964. Cretaceous stratigraphy of the Yubari dome, Hokkaido. *Mem. Fac. Sci., Kyushu Univ., D,* 15(1): 79–115 (pls. 9–11).

Matsumoto, Tatsuro and Hashimoto, W., 1953. A find of *Pseudaspidoceras* from Hokkaido, Japan. *Trans. Proc. Palaeontol. Soc. Jap., N.S.,* No. 12: 97–102 (pls. 10).

Matsumoto, Tatsuro and Hirata, M., 1969. A new ammonite from the Shimantogawa Group of Shikoku. *Trans. Proc. Palaeontol. Soc. Jap., N.S.,* No. 76: 177–184 (pl. 20).

Matsumoto, Tatsuro and Inoma, A., 1975. Mid-Cretaceous ammonites from the Shumarinai–Soeushinai area, Hokkaido. Part I. *Mem. Fac. Sci., Kyushu Univ., D,* 23(2): 1–16 (pls. 38–42).

Matsumoto, Tatsuro and Kanie, Y., 1967. *Ainoceras,* a new heteromorph ammonoid genus from the Upper Cretaceous of Hokkaido. *Mem. Fac. Sci., Kyushu Univ.,* D, 18(2): 349–359 (pls. 20–21).

Matsumoto, Tatsuro and Kanmera, K., 1964. *Hinagu. Expl. Text, Geol. Map. Japan, 1: 50,000.* Geol. Surv. Japan, 147 + 27 pp. (1 map) [J + E].

Matsumoto, Tatsuro and Kawano, T., 1975. A find of *Pseudocalycoceras* from Hokkaido. *Trans. Proc. Palaeontol. Soc. Jap.,* N.S., No. 97: 7–21 (pl. 1).

Matsumoto, Tatsuro and Kimura, T. 1974. Southwest Japan. In: A.M. Spencer (Editor), Mesozoic-Cenozoic Orogenic Belts. *Geol. Soc. Lond., Spec. Publ.*, No. 4: 515–541.

Matsumoto, Tatsuro and Muramoto, T., 1967. Two interesting heteromorph ammonoids from Hokkaido. *Mem. Fac. Sci., Kyushu Univ.*, D, 18(2): 361–366 (pls. 22–24).

Matsumoto, Tatsuro and Noda, M., 1968. An interesting species of *Inoceramus* from the Upper Cretaceous of Kyushu. *Trans. Proc. Palaeontol. Soc. Jap., N.S.*, No. 71: 317–325 (pl. 32).

Matsumoto, Tatsuro and Noda, M., 1975. Notes on *Inoceramus labiatus* (Cretaceous Bivalvia) from Hokkaido. *Trans. Proc. Palaeontol. Soc. Jap., N.S.*, No. 100: 188–208 (pl. 18).

Matsumoto, Tatsuro and Obata, I., 1955. Some Upper Cretaceous desmoceratids from Hokkaido and Saghalien. *Mem. Fac. Sci., Kyushu Univ.*, D, 5(3): 119–151 (pls. 24–30).

Matsumoto, Tatsuro and Obata, I., 1963a. A monograph of the Baculitidae from Japan. *Mem. Fac. Sci., Kyushu Univ.*, D, 13(1): 1–116 (pls. 1–27).

Matsumoto, Tatsuro and Obata, I., 1963b. *Bevahites* (Cretaceous ammonite) from Shikoku. *Bull. Natl. Sci. Mus., Tokyo*, 6(4): 405–410 (pl. 61).

Matsumoto, Tatsuro and Obata, I., 1966. An acanthoceratid ammonite from Sakhalin. *Bull. Natl. Sci. Mus.*, 9(1): 43–52 (pls. 1–4).

Matsumoto, Tatsuro and Okada, H., 1971. Clastic sediments of the Cretaceous Yezo geosyncline. *Mem. Geol. Soc. Jap.*, No. 6: 61–74.

Matsumoto, Tatsuro and Okada, H., 1973. Saku Formation of the Yezo geosyncline. *Sci. Rep. Dep. Geol., Kyushu Univ.*, 11(2): 275–309 [J + E].

Matsumoto, Tatsuro and Ono, A., 1947. A biostratigraphic study of the Jurassic Toyora Group, with special reference to ammonites. *Sci. Rep. Dep. Geol., Kyushu Univ.*, 2(1): 20–33 (incl. 2 pls.) [J].

Matsumoto, Tatsuro and Saito, R., 1954. A nearly smooth pachydiscid from Hokkaido, Japan. *Jap. J. Geol. Geogr.*, 24: 87–92 (pls. 9–11).

Matsumoto, Tatsuro and Saito, R., 1956. A new species of *Damesites* from the Cenomanian of Hokkaido, Japan. *Trans. Proc. Palaeontol. Soc. Jap., N.S.*, No. 22: 191–194.

Matsumoto, Tatsuro and Tashiro, M., 1975. A record of *Mortoniceras* (Cretaceous ammonite) from Goshonoura island, Kyushu. *Trans. Proc. Palaeontol. Soc. Jap., N.S.*, No. 100: 230–238, (pl. 25).

Matsumoto, Tatsuro, Kimura, T. and Katto, J., 1952. Discovery of Cretaceous ammonites from the undivided Mesozoic complex of Shikoku, Japan. *Mem. Fac. Sci., Kyushu Univ., D*, 3: 179–186 (pl. 13).

Matsumoto, Tatsuro, Saito, R. and Fukada, A., 1957. Some acanthoceratids from Hokkaido. *Mem. Fac. Sci., Kyushu Univ.*, D, 6(1): 1–45 (pls. 1–18).

Matsumoto, Tatsuro, Obata, I., Maeda, S. and Sato, T., 1964. *Yabeiceras* (Cretaceous ammonites) from Futaba, northeast Japan. *Trans. Proc. Palaeontol. Soc. Jap., N.S.*, No. 56: 322–331 (pl. 48).

Matsumoto, Tatsuro, Hayami, I. and Hashimoto, W., 1965. Some molluscan fossils from the buried Cretaceous of western Taiwan. *Pet. Geol. Taiwan*, 4: 1–23.

Matsumoto, Tatsuro, Ishikawa, H. and Yamakuchi, S., 1966. A Mesozoic ammonite from Amami-Oshima. *Trans. Proc. Palaeontol. Soc. Jap., N.S.*, No. 62: 234–241.

Matsumoto, Tatsuro, Kanmera, K. and Sakamoto, H., 1968. Notes on two Cretaceous ammonites from the Tomochi Formation of Kyushu. *Jap. J. Geol. Geogr.*, 39: 139–148 (pl. 2).

Matsumoto, Tatsuro, Muramoto, T., and Takahashi, T., 1969. Selected acanthoceratids from Hokkaido. *Mem. Fac. Sci., Kyushu Univ.*, D, 19(2): 251–296 (pls. 25–38).

Matsumoto, Tatsuro, Muramoto, T. and Inoma, A., 1972. Two small desmoceratid ammonites from Hokkaido. *Trans. Proc. Palaeontol. Soc. Jap., N.S.*, No. 87: 377–394 (pl. 47).

Matsumoto, Tatsuro, Muramoto, T. and Takahashi, T., 1972. A new gaudryceratine ammonite from the Cenomanian of Hokkaido. *Mem. Fac. Sci., Kyushu Univ.*, D, 21(2): 207–215 (pl. 33).

Matsuo, H., 1954. Discovery of *Nelumbo* from the Asuwa flora (Upper Cretaceous) in Fukui Prefecture in the inner side of central Japan. *Trans. Proc. Palaeontol. Soc. Jap., N.S.*, No. 14: 155–158 (pl. 20).

Matsuo, H., 1960. On the new Nymphaeacean plant from the Omichidani bed (Cretaceous system), Ishikawa Prefecture, central Japan. *Trans. Proc. Palaeontol. Soc. Jap., N.S.*, No. 40: 329–336 (pl. 38).

Matsuo, H., 1962. A study on the Asuwa flora (Late Cretaceous age) in the Hokuriku district, central Japan. *Sci. Rep. Kanazawa Univ.*, 8(1): 177–250 (pls. 1–24).

Matsuo, H., 1964. On the late Cretaceous flora in Japan. *Ann. Sci. Coll. Lib. Arts, Kanazawa Univ.*, 1: 39–65 [J + E].

Matsuo, H., 1966. Plant fossils of the Izumi Group (Upper Cretaceous) in the Izumi mountain range, Kinki district, Japan. *Ann. Sci. Coll. Lib. Arts, Kanazawa Univ.*, 3: 67–75.

Matsuo, H., 1967. A Cretaceous *Salvinia* from the Hashima Is. (Gunkan-Jima), outside of the Nagasaki harbour, Kyushu, Japan. *Trans. Proc. Palaeontol. Soc. Jap., N.S.*, No. 66: 49–55 (pl. 5).

Matsuo, H., 1970. On the Omichidani flora (Upper Cretaceous), inner side of central Japan. *Trans. Proc. Palaeontol. Soc. Jap., N.S.*, No. 80: 371–389 (pls. 42–43).

Matsuo, H., 1975. Some evidence on the climatic conditions of the Neophyta in the inner side of Honshu, Japan. *Ann. Sci. Kanazawa Univ.*, 12: 73–90 [J].

Meyerhoff, A.A., 1970a. Continental drift: implications of paleomagnetic studies, meteorology, physical oceanography, and climatology. *J. Geol.*, 78: 1–55.

Meyerhoff, A.A., 1970b. Continental drift, II: high latitude evaporite deposits and geologic history of Arctic and North Atlantic Oceans. *J. Geol.*, 78: 406–444.

Michael, R., 1899. Ueber Kreidefossilien von der Insel Sakhalin. *Jahrb. K. Preuss. Geol. Landesanst.*, 18: 153–164 (pls. 5, 6).

Miki, T., 1972a. Palynological study of the Kuji Group in northeastern Honshu, Japan. *J. Fac. Sci., Hokkaido Univ.*, 4, 15: 513–604 (incl. pls. 1–11).

Miki, T., 1972b. Spores and pollen flora of the Upper Cretaceous Futaba Group in northeastern Japan. *J. Geol. Soc. Jap.*, 78: 241–252 (pls. 1–2) [J + E].

Miki, T., 1973. Spores and pollen flora from the Middle Yezo Group in northern Hokkaido, Japan. *J. Geol. Soc. Jap.*, 79: 205–218 (pls. 1–2) [J + E].

Miyashiro, A., 1958. Regional metamorphism of the Gosaisyo–Takanuki district in the central Abukuma plateau. *J. Fac. Sci. Univ. Tokyo,* 2, 11: 219–272.

Miyashiro, A., 1961. Evolution of metamorphic belts. *J. Petrol.*, 2: 277–311.

Miyashiro, A., 1967. Orogeny, regional metamorphism and magmatism in the Japanese Islands. *Medd. Dan. Geol. För.*, 17: 390–446.

Miyashiro, A., 1972. Metamorphism and related magmatism in plate tectonics. *Am. J. Sci.*, 272: 629–656.

Miyashiro, A., 1973. *Metamorphism and Metamorphic Belts.* G. Allen and Unwin, London, 492 pp.

Moriya, S., 1972. Low-grade metamorphic rocks of the northern Kitakami mountainland. *Sci. Rep. Tohoku Univ.*, 3, 11: 239–282.

Morozumi, Y., 1970. Upper Cretaceous *Inoceramus* from the Shimanto belt of the Kii peninsula. *Bull. Osaka Mus. Nat. Hist.*, No. 23: 19–24 (pls. 2–4).

Murakami, A. 1976. Finding of Triassic conodonts from the Kanoashi Group. *J. Geol. Soc. Jap.*, 82: 143–144 [J].

Murakami, M., 1974. Some problems concerning late Mesozoic to early Tertiary igneous activity on the inner side of Southwest Japan. *Pac. Geol.*, 8: 139–151.

Murata, M., 1962. The Upper Jurassic of Cape Shiriya, Aomori Prefecture, Japan. *Sci. Rep. Tohoku Univ., 2, Spec. Vol.*, 5: 119–126 (pls. 30–32).

Nagao, T., 1930. On some Cretaceous fossils from the islands of Amakusa, Kyushu, Japan. *J. Fac. Sci., Hokkaido Imp. Univ.*, 4, 1: 1–25 (pls. 1–3).

Nagao, T., 1932. Some Cretaceous mollusca from Japanese Saghalien and Hokkaido (Lamellibranchiata and Gastropoda). *J. Fac. Sci., Hokkaido Imp. Univ.*, 4, 2: 23–50 (pls. 5–8).

Nagao, T., 1934. Cretaceous mollusca from the Miyako district, Honshu, Japan (Lamellibranchiata and Gastropoda). *J. Fac. Sci., Hokkaido Imp. Univ.*, 4, 2: 177–277 (pls. 23–39).

Nagao, T., 1938. Some molluscan fossils from the Cretaceous deposits of Hokkaido and Japanese Saghalien. Part 1. Lamellibranchiata and Scaphopoda. *J. Fac. Sci., Hokkaido Imp. Univ.*, 4, 4: 117–142 (pls. 14–16).

Nagao, T., 1939. Some molluscan fossils from the Cretaceous deposits of Hokkaido and Japanese Saghalien. Part 2. *J. Fac. Sci., Hokkaido Imp. Univ.*, 4, 4: 213–239 (pls. 20–22).

Nagao, T. and Matsumoto, T., 1939–40. A monograpph of the Cretaceous *Inoceramus* of Japan. Part I. *J. Fac. Sci., Hokkaido Imp. Univ.*, 4, 4(3–4): 241–299 (pls. 23–34) (1939); Part II. *Ibid.*, 6(1): 1–64 (pls. 1–22) (1949).

Nagao, T. and Otatume, K., 1938. Molluscan fossils of the Hakobuti sandstone of Hokkaido. *J. Fac. Sci., Hokkaido Imp. Univ.*, 4, 4: 31–56 (pls. 1–4).

Nagata, T., Akimoto, S., Shimizu, Y., Kobayashi, K. and Kuno, H., 1959. Palaeomagnetic studies on Tertiary and Cretaceous rocks in Japan. *Proc. Jap. Acad.*, 35: 378–383.

Nairn, A.E.M. and Peterson, D.N., 1973. A review of Permian and Triassic paleomagnetic data with respect to paleogeographic conditions at, and the location of, the Permian–Triassic boundary. In: A. Logan and L.V. Hills (Editors), *The Permian and Triassic Systems and Their Mutual Boundary. Can. Soc. Petrol. Geol., Mem.*, 2: 694–713.

Nakai, I. and Hada, S., 1966. Discovery of Aptian ammonites from the Shimanto terrain, western Shikoku. *Trans. Proc. Palaeontol. Soc. Jap., N.S.*, No. 62: 242–250 (pls. 29, 30).

Nakai, I. and Matsumoto, T., 1968. On some ammonites from the Cretaceous Fujikawa Formation of Shikoku. *J. Sci. Hiroshima Univ.*, G, 6: 1–15, (pls. 1–3).

Nakano, M., 1957a. Carnian fossils from Kyushu, Okayama Prefecture, Japan. *J. Sci. Hiroshima Univ.*, C, 2: 63–67 (pl. 9).

Nakano, M., 1957b. On the Cretaceous pennatae trigonians in Japan. *Jap. J. Geol. Geogr.*, 28(1–3): 107–120 (pls. 8–9).

Nakano, M., 1958a. On some Upper Cretaceous *Steinmanella (Yeharella)* in Japan. *J. Sci. Hiroshima Univ.*, C, 2(2): 83–88, (pls. 13–14).

Nakano, M., 1958b. Scabrotigonians in Japan. *J. Sci. Hiroshima Univ.*, C, 2(3): 227–233 (pl. 29).

Nakano, M., 1960. Stratigraphic occurrences of the Cretaceous trigoniids in the Japanese islands and their faunal significances. *J. Sci. Hiroshima Univ.*, C, 3(2): 215–280 (pl. 23–30).

Nakano, M., 1961a. Note on *Heterotrigonia subovalis* (Jimbo). *Trans. Proc. Palaeontol. Soc. Jap., N.S.*, No. 42: 55–62 (pl. 9).

Nakano, M., 1961b. On some Gyliakian Pterotrigoniae from Kyushu and Hokkaido, Japan. *Trans. Proc. Palaeontol. Soc. Jap., N.S.*, No. 43: 89–98 (pl. 13).

Nakano, M., 1961c. Note on *Steinmannella (Yeharella) ainuana* (Yabe and Nagao). *Trans. Proc. Palaeontol. Soc. Jap., N.S.*, No. 44: 139–145 (pls. 20–21).

Nakazawa, K., 1952–56. A study on the pelecypod-fauna of the Upper Triassic Nabae Group in the northern part of Kyoto Prefecture, Japan. Part 1. *Mem. Coll. Sci., Univ. Kyoto,* B, 20(2): 95–106 (pls. 7–10) (1952); Part 2 *Ibid.*, 21(2): 213–222 (pls. 3–6) (1954); Part 3. *Ibid.*, 22(2): 243–260 (pls. 13–16) (1955); Part 4. *Ibid.*, 23(3): 231–253 (pls. 1–4) (1956).

Nakazawa, K., 1953. Discovery of *Claraia* and *Eumorphotis* from the Triassic Yakuno Group, Kyoto Pref., Japan. *Mem. Coll. Sci., Univ. Kyoto*, B, 20(4): 261–269 (pl. 3).

Nakazawa, K., 1958. The Triassic System in the Maizuru zone, Southwest Japan. *Mem. Coll. Sci., Univ. Kyoto,* B, 24(4): 265–313.

Nakazawa, K., 1959. Permian and Eo-Triassic Bakevellias from the Maizuru zone, Southwest Japan. *Mem. Coll. Sci., Univ. Kyoto*, B, 26(2): 193–213 (pls. 3–4).

Nakazawa, K., 1960. Permian and Eo-Triassic Myophoriidae from the Maizuru zone, Southwest Japan. *Jap. J. Geol. Geogr.*, 31(1): 49–62 (pl. 6).

Nakazawa, K., 1961a. On the so-called Yakuno intrusive rocks in the Yakuno district, Southwest Japan. In: S. Matsushita et al. (Editors), *Professor Jiro Makiyama Mem-*

orial Volume. University of Kyoto pp. 149–161 [J + E].

Nakazawa, K., 1961b. Early and Middle Triassic pelecypod-fossils from the Maizuru zone, Southwest Japan. *Mem. Coll. Sci., Univ. Kyoto,* B, 27(3): 249–291 (incl. pls. 12–14).

Nakazawa, K., 1963. Norian pelecypod-fossils from Jito, Okayama Prefecture, west Japan. *Mem. Coll. Sci., Univ. Kyoto,* B, 30(2): 47–58 (pls. 1–2).

Nakazawa, K., 1964a. Discovery of the Anisian fauna from Shikoku, Southwest Japan and its geological meaning. *Mem. Coll. Sci., Univ.Kyoto,* B, 30(4): 7–19 (pls. 1–2).

Nakazawa, K., 1971. The Lower Triassic Kutotaki fauna in Shikoku and its allied faunas in Japan. *Mem. Fac. Sci. Kyoto Univ., Geol. Min.* 38(1): 103–133 (pls. 23–25).

Nakazawa, K. and Murata, M., 1966. On the Lower Cretaceous fossils found near the Omine mine, Iwata Prefecture, Northeast Japan. *Mem. Coll. Sci., Univ. Kyoto,* B, 32(4): 303–325 (pls. 3–6).

Nakazawa, K. and Nogami, Y., 1967. Problematic occurrence of the Upper Triassic fossils from the western hills of Kyoto. *Mem. Fac. Sci., Kyoto Univ., Geol. Min.* 34(1): 9–22 (pl. 1).

Nakazawa, K. and Shimizu, D., 1955. Discovery of *Glyptophiceras* from Hyogo Prefecture, Japan. *Trans. Proc. Palaeontol. Soc. Jap., N.S.*, No. 17: 13–18 (pl. 3).

Nakazawa, K. et al. (Working Group on the Permian-Triassic Systems), 1975. Stratigraphy near the Permian–Triassic boundary in Japan and its correlation. *J. Geol. Soc. Jap.*, 81(3): 165–184 [J + E].

Nalivkin, D.V. (translated by N. Rast), 1973. *Geology of the U.S.S.R.* Oliver and Boyd, Edinburgh, 855 pp.

Nishida, M., 1962–1973. On some petrified plants from the Cretaceous of Choshi, Chiba Prefecture. I. *Jap. J. Bot.* 18: 87–96 (8 pls.) (1962); II. *Bot. Mag. Tokyo,* 78: 138–146 (5 pls.) (1965); III. *Ibid.*, 78: 226–235 (1965); IV. *Ibid.*, 80: 383–393 (1967); V. *Ibid.*, 80: 487–497 (1967); VI. *Ibid.*, 86: 189–202 (1973).

Nishiyama, S., 1966. The echinoid fauna from Japan and adjacent regions. Part I. *Palaeontol. Soc. Jap., Spec. Pap.*, No. 11: 1–277 (pls. 1–18).

Nishiyama, S., 1968. The echinoid fauna from Japan and adjacent regions. Part II *Palaeontol. Soc. Jap., Spec. Pap.*, No. 13: 1–491 (pls. 1–30).

Noda, M., 1972. Ammonites from the Mesozoic Yamabu Formation, Kyushu. *Trans. Proc. Palaeontol. Soc. Jap., N.S.*, No. 88: 462–471 (pl. 56).

Noda, M., 1974. A new species of *Inoceramus* from the Shimantogawa Group of South Shikoku. *Trans. Proc. Palaeontol. Soc. Jap., N.S.*, No. 93: 240–248 (pl. 34).

Noda, M., 1975. Succession of *Inoceramus* in the Upper Cretaceous of Southwest Japan. *Mem. Fac. Sci., Kyushu Univ., D,* 23(2): 211–261 (pls. 32–37).

Nogami, Y., 1968. Trias-Conodonten von Timor, Malaysien und Japan. *Mem. Fac. Sci., Kyoto Univ., Geol. Min.*, 34(2): 115–136 (pls. 8–11).

Nohda, S. and Setoguchi, T., 1967. An occurrence of Jurassic conodonts from Japan. *Mem. Coll. Sci., Kyoto Univ.*, B, 33: 227–237.

Nozawa, T., 1970. Isotopic ages of Late Cretaceous acid rocks in Japanese Islands; Summary and notes in 1970. *J. Geol. Soc. Jap.*, 76: 493–518 [J + E].

Nozawa, T., 1975. Radiometric age map of Japan – Granitic rocks. *Geol. Surv. Jap., 1:2,000,000 Map Ser.*, No. 16–1.

Nozawa, T. and Shibata, K., 1974. Bibliography of geochronological data in Japan (2). *Jap. J. Geol. Geogr.*, 43: 35–40.

Nureki, T., 1960. Structural investigation of the Ryoke metamorphic rocks of the area between Iwakuni and Yunai, Southwest Japan. *J. Sci. Hiroshima Univ.*, C, 3(1): 69–141.

Nureki, T., 1969. Geological relations of the Sangun metamorphic rocks to the "non-metamorphic" Paleozoic formations in Chugoku province. *Mem. Geol. Soc. Jap.*, No. 4: 23–39 [J + E].

Obata, I., 1965. Allometry of *Reesidites minimus,* a Cretaceous ammonite species. *Trans. Proc. Palaeontol. Soc. Jap., N.S.*, No. 58: 39–63 (pls. 4, 5).

Obata, I., 1967–1975. Lower Cretaceous ammonites from the Miyako Group. Pt. 1. *Valdedorsella* from the Miyako Group. *Trans. Proc. Palaeontol. Soc. Jap., N.S.*, No. 66: 63–72 (pl. 8) (1967); Pt. 2. Some silesitids from the Miyako Group. *Ibid., N.S.*, No. 67: 129–138 (pl. 11) (1967); Pt. 3. Some douvilleiceratids from the Miyako Group. *Ibid., N.S.*, No. 76: 165–176 (pls. 18, 19) (1969); Pt. 4. *Pseudoleymeriella* from the Miyako Group. *Sci. Rep. Tohoku Univ., 2, Spec. Vol.*, 6 [Hatai Memorial Vol.]: 309–314 (pl. 34) (1973); Pt. 5. *Diadochoceras* from the Miyako Group. *Bull. Natl. Sci. Mus. Tokyo,* C, 1(1): 1–10 (pls. 1–3).

Obata, I., Maehara, T. and Tsuda, H., 1973. Cretaceous deposits in the Hidaka area, Hokkaido. *Mem. Natl. Sci. Mus.*, No. 6: 131–145.

Obata, I., Hagiwara, S. and Kamiko, S., 1975. Geological age of the Cretaceous Choshi Group. *Bull. Natl. Sci. Mus.*, C, 1(1): 1–36 (pls. 1–5) [J + E].

Ogawa, Y., 1973. Tectonic development of the Chichibu terrain in eastern Shikoku, Japan, with special reference to the deformational stages. *J. Fac. Sci., Univ. Tokyo,* 2, 18(3): 475–506 (pls. 12–14).

Ogawa, Y., 1974. Geologic structure of the Chichibu terrain in eastern Shikoku, Japan. *J. Geol. Soc. Jap.*, 80: 439–455 [J + E].

Oishi, S., 1932. The Rhaetic plants from the Nariwa district, province Bitchu (Okayama Prefecture), Japan. *J. Fac. Sci., Hokkaido Imp. Univ.*, 4, 1(3–4): 257–380 (pls. 19–53).

Oishi, S., 1940. The Mesozoic floras of Japan. *J. Fac. Sci., Hokkaido Imp. Univ.*, 5(2–4): 125–480 (pls. 1–48).

Ojima, M., Kaneoka, I. and Aramaki, S., 1970. K–Ar ages of submarine basalts dredged from seamounts in the western Pacific area and discussion of oceanic crust. *Earth Planet. Sci. Lett.*, 8: 237–249.

Okada, H., 1958. *Matsumotoa*: a new prionodont pelecypod genus from the Cretaceous Mifune Group, Kyushu, Japan. *Mem. Fac. Sci., Kyushu Univ.*, D, 8(2): 35–48 (pls. 10, 11).

Okada, H., 1960. Sandstones of the Cretaceous Mifune Group, Kyushu. *Mem. Fac. Sci., Kyushu Univ.*, D, 10(1): 1–40 (pls. 1–5).

Okada, H., 1961. Cretaceous sandstones of the Goshonoura island, Kyushu. *Mem. Fac. Sci., Kyushu Univ.*, D, 11(1): 1–48 (pls. 1–2).

Okada, H., 1965. Sedimentology of the Cretaceous Mikasa Formation. *Mem. Fac. Sci., Kyushu Univ.*, D, 16(1): 81–111 (pls. 19–24).

Okada, H., 1971. A pattern of sedimentation in clastic sediments in geosynclines. *Mem. Geol. Soc. Jap.*, (6): 75–82, [J + E].

Okami, K., Masuya, H. and Mogi, T., 1976. Exotic pebbles in the eastern terrain of the Abukuma plateau, northeast Japan (Part 1). *J. Geol. Soc. Jap.*, 82: 83–98.

Okubo, M. and Matsushima, N. 1959. On a new species of Pachyodonta from the Akaishi mountains, central Japan. *Chikyu Kagaku* (Earth Sci.), 42: 1–4 [J].

Onuki, Y., 1969. Geology of the Kitakami massif, Northeast Japan. *Contrib. Inst. Geol. Paleontol. Tohoku Univ.*, 69: 1–239 [J + E].

Ota, Y., 1959a. *Plicatounio* of the Wakino Formation. *Trans. Proc. Palaeontol. Soc. Jap., N.S.*, No. 33: 15–18 (pl. 3).

Ota, Y., 1959b. *Trigonioides* and its classification. *Trans. Proc. Palaeontol. Soc. Jap., N.S.*, No. 34: 97–104 (pl. 10).

Ota, Y., 1959c. On the "*Nippononaia*" and its classification. *Trans. Proc. Palaeontol. Soc. Jap., N.S.*, No. (34): 105–110 (pl. 11).

Ota, Y., 1960a. Gastropods from the Kwanmon Group (Inkstone Series). *J. Sci. Hiroshima Univ.*, C, 3(1): 1–3 (pls. 1–3).

Ota, Y., 1960b. The zonal distribution of the non-marine fauna in the Upper Mesozoic Wakino Subgroup. *Mem. Fac. Sci., Kyushu Univ.*, D, 9(3): 187–209.

Ota, Y., 1963. Notes on the relationship of *Trigonioides* and *Plicatounio*, non-marine Mesozoic Bivalvia from eastern Asia. *Geol. Rep. Hiroshima Univ.*, No. 12: 503–512.

Ota, Y., 1964. On some Cretaceous corbulids from Japan. *Mem. Fac. Sci., Kyushu Univ.*, D, 15(1): 149–161 (pls. 20, 21).

Ota, Y., 1965. On the Corbiculidae from the lower Neocomian of Japan. *Geol. Rep. Hiroshima Univ.*, No. 14: 165–171 (pls. 12, 13).

Ota, Y., 1973. Pelecypod family Neomiodontidae from the Lower Neocomian of Japan. *Bull. Fukuoka Univ. Educ.*, 22(3): 245–273 (incl. pls. 1–4).

Ota, Y., 1974. A new bakevellid (Bivalvia) from the Lower Cretaceous of Southwest Japan. *Bull. Fukuoka Univ. Educ.*, 23(3): 79–89 (incl. pl. 1).

Ota, Y., 1975. Two new non-marine species of Bivalvia from the Lower Cretaceous of Southwest Japan. *Trans. Proc. Palaeontol. Soc. Jap., N.S.*, No. 98: 95–104 (pl. 9).

Oyama, T., 1960–61. On the conclusion of the Oarai flora from the Oarai Formation, in Oarai, Ibaraki Prefecture, Japan, Part 1. Ibaraki Univ., Fac. Lib. Arts Sci. Bull., No. 11: 75–105; Part 2. *Ibid.*, No. 12: 61–101.

Oyama, T. and Matsuo, H., 1964. Notes on palmaean leaf from the Oarai flora, Oarai machi, Ibaraki Prefecture, Japan. *Trans. Proc. Palaeontol. Soc. Jap., N.S.*, No. 55: 241–246.

Ozaki, H. and Shikama, T., 1954. On three Skytic molluscs from Gumma Prefecture, central Japan. *Bull. Natl. Sci. Mus., N.S.*, 1(2): 42–45 (pl. 19).

Pergament, M.A., 1965. *Inoceramus* and Cretaceous stratigraphy of the Pacific region. *Tr. Geol. Inst. Akad. Nauk S.S.S.R.*, 118: 1–102 (pls. 1–12) (in Russian).

Pergament, M.A., 1966. Zonal stratigraphy and *Inoceramus* of the lowermost Upper Cretaceous of the Pacific Coast of the U.S.S.R. *Tr. Geol. Inst. Akad. Nauk S.S.S.R.*, 146: 1–83 (pls. 1–36).

Research Members of the GDP-11 Cruise, 1975. *Nummulites*, and pebbles of hornblende-tonalite and other igneous rocks, collected at the Amami plateau. *J. Geol. Soc. Jap.*, 81: 269–271 [J].

Rikitake, T. (Editor), 1973. *The Crust and Upper Mantle of the Japanese Area. Part II. Geology and Geochemistry.* Japanese National Committee for Upper Mantle Project, Geol. Surv. Japan, Tokyo, 176 pp.

Saito, T., 1958–59. Notes on some Cretaceous fossils from the Nakaminato Formation, Nakaminato City, Ibaraki Prefecture, Japan, Part I. *Bull. Fac. Lib. Arts, Ibaraki Univ., Nat. Sci.*, No. 8: 83–94 (pls. 1–5); Part II. *Ibid.*, No. 9: 79–85 (pl. 1–2).

Saito, T., 1961–62. The Upper Cretaceous system of Ibaraki and Fukushima Prefectures, Japan, Part I. *Bull. Fac. Lib. Arts, Ibaraki Univ., Nat. Sci.*, No. 12: 103–144; Part 2. *Ibid.*, No. 13: 51–88 (pls. 1–8).

Sakagami, S., 1955. Lower Triassic ammonites from Iwai, Oguno-mura, Nishitama-gun, Kwanto massif, Japan. *Sci. Rep. Tokyo Kyoiku Daigaku*, C, 4(30): 131–140 (pls. 1–2).

Sasajima, T., 1969. A consideration on paleomagnetic stratigraphy, with special reference to its development and present state of study. *J. Geol. Soc. Jap.*, 75: 13–25 [J + E].

Sasajima, T. and Shimada, M., 1966. Paleomagnetic studies of the Cretaceous volcanic rocks in Southwest Japan – An assumed drift of the Honshu island. *J. Geol. Soc. Jap.*, 72: 503–514 [J + E].

Sasajima, T., Nishida, J. and Shimada, M., 1968. Paleomagnetic evidence of a drift of the Japanese main island during the Paleogene period. *Earth Planet. Sci. Lett.*, 5: 135–141.

Sato, S., 1961. Pollen analysis of carbonaceous matter from the Hakobuchi Group in the Enbetsu district, northern Hokkaido, Japan. *J. Fac. Sci., Hokkaido Univ.*, 4, 11(1): 77–93.

Sato, T., 1954a. Découverte de *Tmetoceras* dans le plateau de Kitakami au nord du Japon. *Jap. J. Geol. Geogr.*, 24: 115–121 (pl. 13).

Sato, T., 1954b. *Hammatoceras* de Kitakami, Japon. *Jap. J. Geol. Geogr.*, 25(1–2): 81–100 (pls. 7–9).

Sato, T., 1955. Les ammonites recueillies dans le groupe de Kuruma, nord du Japon central. *Trans. Proc. Palaeontol. Soc. Jap., N.S.*, No. 20: 111–118 (pl. 18).

Sato, T., 1956. Révision chronologique de la série de Karakuwa (Jurassique moyen). *Jap. J. Geol. Geogr.*, 27(2–4): 167–191 (pl. 13).

Sato, T., 1957. Biostratigraphie de la série de Shizukawa (Jurassique inférieur) du Japon septentrional. *Jap. J. Geol. Geogr.*, 29(1–3): 153–159 (pls. 1–2).

Sato, T., 1958. Supplement à la faune de la série de Shizukawa (Jurassique inférieur) du Japon septentrional. *Jap. J. Geol. Geogr.*, 29(1–3): 153–159 (pl. 18).

Sato, T., 1959. Précision du Berriasien dans la stratigraphie du plateau de Kitakami. *Bull. Soc. Géol. Fr.*, 6, 8: 585–599 (pl. 28).

Sato, T., 1961a. La limite jurassico-crétacée dans la stratigraphie japonaise. *Jap. J. Geol. Geogr.*, 32(3–4): 533–541 (pl. 12).

Sato, T., 1961b. Faune berriasienne et tithonique supérieure nouvellement découverte du Japon. *Jap. J. Geol. Geogr.*, 32(3–4): 543–551 (pl. 13).

Sato, T., 1962a. Études biostratigraphiques des ammonites du Jurassique du Japon. *Mém. Soc. Géol. Fr., N.S.*, 41, Mém. 94: 1–122 (pls. 1–10).

Sato, T., 1962b. Le Jurassiques du Japon – Zones d'ammonites. *C. R. Mém., Coll. Jurassique, Luxembourg, 1962*, 885–896.

Sato, T., 1972. Some Bajocian ammonites from Kitakami, Northeast Japan. *Trans. Proc. Palaeontol. Soc. Jap., N.S.*, No. 85: 280–292 (pl. 34).

Sato, T., 1974. A Jurassic ammonite from near Inuyama, north of Nagoya. *Trans. Proc. Palaeontol. Soc. Jap., N.S.*, No. 96: 427–432.

Sato, T., 1975. Permian–Triassic in the Kuromata-gawa area, Niigata Prefecture. *J. Geol. Soc. Jap.*, 81: 709–711 [J].

Sawada, K. and Kanmera, K., 1973. Greenstones from the Sorachi and Hidaka Groups of the Hidaka Mountains, Hokkaido. *Mem. Natl., Sci. Mus.*, No. 6: 147–161.

Schmidt, M.F., 1873. Ueber die Petrefakten der Kreideformation von der Insel Sachalin. *Mem. Acad. Imp. Sci. St. Petersbourg*, 7, 19 (3): 1–33 (pls. 1–8).

Seely, D.R., Vail, P.R. and Walton, G.G., 1974. Trench slope model. In: C.A. Burk and C.L. Drake (Editors), *The Geology of Continental Margins.* Springer-Verlag, Berlin, pp. 249–260.

Seki, T. and Hayase, I., 1974. Rb–Sr isochron of the Cretaceous acid volcanic rocks of Himeji district, Hyogo Prefecture, Japan. *Mass Spectrosc.*, 22: 55–59 [J + E].

Shibata, H. et al., 1960–1962. Chemical composition of Japanese granitic rocks in regard to petrographic provinces. Part 7. *Sci. Rep. Tokyo Kyoiku Daigaku*, C, 7: 71–94; Part 8. *Ibid.*, 7: 217–270; Part 9. *Ibid.*, 8: 19–31; Part 10. *Ibid.*, 8: 33–47.

Shibata, K., 1958. K–Ar age determinations on granitic and metamorphic rocks in Japan. *Rep. Geol. Surv. Jap.*, No. 227: 1–73.

Shibata, K. and Adachi, M., 1972. Rb–Sr and K–Ar geochronology of metamorphic rocks in the Kamiaso conglomerate, central Japan. *J. Geol. Soc. Jap.*, 78: 265–271.

Shibata, K. and Igi, S., 1969. K–Ar ages of muscovite from the muscovite-quartz schist of the Sangun metamorphic terrain in the Tari district, Tottori Prefecture, Japan. *Bull. Geol. Surv. Jap.*, 20: 707–709.

Shibata, K. and Nozawa, T., 1968. K–Ar age of Omi schist, Hida mountains, Japan. *Bull. Geol. Surv. Jap.*, 19: 243–246.

Shibata, K., Nozawa, T. and Wanless, R.K., 1970. Rb–Sr geochronology of the Hida metamorphic belt, Japan. *Can. J. Earth Sci.*, 7: 1383–1401.

Shibata, K., Wanless, R.K., Kano, H., Yoshida, T., Nozawa, T., Igi, S. and Konishi, K., 1972. Rb–Sr ages of several so-called basement rocks in the Japanese islands. *Bull. Geol. Surv. Jap.*, 23: 505–510 [J + E].

Shikama, T., 1969. On a Jurassic reptile from Miyama-cho, Fukui Prefecture, Japan. *Sci. Rep. Yokohama Natl. Univ.*, 2, 20: 25–34 (pl. 1).

Shikama, T. and Yui, S., 1973. On some nerineid gastropoda in Japan. *Sci. Rep. Yokohama Natl. Univ.*, 2, 20: 9–57 (pls. 3–8).

Shimazaki, H., 1973. Cu, Zn and Pb in the skarn ores of Southwest Japan. *G.D.P. Lett., Tectonics*, 1: 39–42 [J].

Shimizu, S., 1926. Three interesting Cretaceous ammonites recently acquired from Hokkaido and Saghalien. *Proc. Imp. Acad., Tokyo*, 2 (10): 547–550.

Shimizu, S., 1930a. Notes on two Tithonian species of *Perisphinctes* from the Torinosu Limestone of Koike, province of Iwaki. *Jap. J. Geol. Geogr.*, 7 (2): 45–48.

Shimizu, S., 1930b. On some Anisic ammonites from the *Hollandites*-beds of the Kitakami mountainland. *Sci. Rep. Tohoku Imp. Univ.*, 2, 14 (1): 63–74 (pl. 24).

Shimizu, S., 1930c. Two new species of Ladinic ammonites from the *Daonella*-beds of Rifu, province of Rikuzen. *Sci. Rep. Tohoku Imp. Univ.*, 14 (1): 75–78 (pl. 24).

Shimizu, S., 1931. The marine Lower Cretaceous deposits of Japan, with special reference to the ammonite-bearing zones. *Sci. Rep. Tohoku Imp. Univ.*, 2, 15 (1): 1–40 (pls. 1–4).

Shimizu, S., 1932. On a new type of Senonian ammonite, *Pseudobarroisiceras nagaoi* Shimizu gen. et sp. nov. from Teshio province, Hokkaido, *Jap. J. Geol. Geogr.*, 10 (1–2): 1–4 (pl. 1).

Shimizu, S., 1935a. The Upper Cretaceous cephalopods of Japan, Part I. *J. Shanghai Sci. Inst.*, 2, 1: 159–226.

Shimizu, S., 1935b. The Upper Cretaceous ammonites so-called *Hamites* in Japan. *Proc. Imp. Acad., Tokyo*, 11 (7): 271–273.

Sokolow, D.W., 1914. Kreideinoceramen des russischen Sachalin. *Mém. Com. Géol. Leningrad, N.S.*, 83: 1–95 (pls. 1–6).

Stopes, M.C. and Fujii, K., 1910. Studies on the structure and affinities of Cretaceous plants. *Philos. Trans. R. Soc. Lond., B*, 201: 1–9 (pls. 1–9).

Sugimoto, M., 1974. Stratigraphical study in the outer belt of the Kitakami massif, northeast Japan. *Contrib. Inst. Geol. Paleontol., Tohoku Univ.*, No. 74: 1–48 (pls. 1–6, folded maps) [J + E].

Sugiyaman, R. (Editor), 1973. *Median Tectonic Line.* Tokai

Suwa, K., 1961. Petrological and geological studies on the Ryoke metamorphic belt. *J. Earth Sci., Nagoya Univ.*, 9: 224–303.

Suwa, K., 1973. Metamorphic rocks occurring along the Median Tectonic Line in the Japanese Islands: Ryoke and Sambagawa metamorphic belts. In: R. Sugiyama (Editor), *Median Tectonic Line.* Tokai Univ. Press, Tokyo, pp. 221–238 [J + E].

Univ. Press, Tokyo, 401 pp. [J].

Suzuki, J., 1952. Ultrabasic rocks and associated ore deposits of Hokkaido, Japan. *J. Fac. Sci., Hokkaido Univ.*, 4, 8:

175–210.

Suzuki, K., 1940. Non-marine molluscan faunule of the Siragi Series in south Tyosen. *Jap. J. Geol. Geogr.*, 17 (3–4): 215–231 (pls. 22–24).

Suzuki, K., 1943. Restudy on the non-marine molluscan fauna of the Rakuto series in Keisyo-do, Tyosen. *J. Shigen-Kagaku Kenkyusho*, 1 (2): 189–219 (pls. 14–19).

Suzuki, K., 1949. Development of the fossil non-marine molluscan faunas in eastern Asia. *Jap. J. Geol. Geogr.*, 21 (1–4): 91–133.

Suzuki, A. and Sato, T., 1972. Discovery of Jurassic ammonites from Toriashi Mountain. *J. Geol. Soc. Jap.*, 78: 213–215 [J + E].

Takahashi, E., 1959. Floral change since the Mesozoic age in western Honshu, Japan. *Sci. Rep. Yamaguchi Univ.*, 10: 181–237 [J].

Takahashi, H., 1969. Stratigraphy and ammonite fauna of the Jurassic system of the southern Kitakami massif, northeast Honshu, Japan. *Sci. Rep. Tohoku Univ.*, 2, 41 (1): 1–93 (pls. 1–19).

Takahashi, H., 1973. The Isokusa formation and its late Upper Jurassic and early Lower Cretaceous ammonite fauna. *Sci. Rep. Tohoku Univ.*, Spec. Vol., No. 6: 319–336 (pls. 36–37).

Takahashi, K., 1964. Sporen und Pollen der oberkretazeischen Hakobuchi Schichtengruppe, Hokkaido. *Mem. Fac. Sci., Kyushu Univ.*, D, 14 (3): 159–271 (pls. 23–44).

Takahashi, K., 1970. Some palynomorphs from the Upper Cretaceous sediments of Hokkaido. *Trans. Proc. Palaeontol. Soc. Jap., N.S.*, No. 73: 265–275 (pls. 29–30).

Takahashi, K., 1974. Palynology of the Upper Aptian Tanohata Formation of the Miyako Group, northeast Japan. *Pollen Spores*, 16: 535–564.

Takai, F., Matsumoto, T. and Toriyama, R., 1963. *Geology of Japan.* Univ. Tokyo Press, Tokyo, 279 pp.

Takayama, T. and Obata, I., 1968. Discovery of nannoplanktons from the Upper Cretaceous Futaba Group. *J. Geol. Soc. Jap.*, 74: 187–189 [J].

Takayanagi, Y., 1960a. Cretaceous foraminifera from Hokkaido, Japan. Sci. Rep. Tohoku Univ., 2, 32 (1): 1–154 (pls. 1–11).

Takayanagi, Y., 1960b. Annoted bibliography of the Cretaceous foraminifera from Japan. *Sci. Rep. Tohoku Univ.*, 2, Spec. Vol., No. 4: 309–315.

Takayanagi, Y. and Iwamoto, H., 1961. Cretaceous planktonic foraminifera from the Middle Yezo Group of the Ikushumbetsu, Miruto and Hatonosu areas, Hokkaido. *Trans. Proc. Palaeontol. Soc. Jap., N.S.*, No. 45: 183–196 (pl. 28).

Takizawa, F., 1970. Ayukawa Formation of the Ojika peninsula, Miyagi Prefecture, Northeast Japan. *Bull. Geol. Surv. Jap.*, 21: 567–578.

Takizawa, F., 1975. Lower Cretaceous sedimentation in the Oshika Peninsula, Miyagi Prefecture, Northeast Japan. *Bull. Geol. Surv. Jap.*, 26: 267–305 (pls. 10–16) [J + E].

Tamura, M., 1959a. On *Kumatrigonia*, a new subgenus of *Frenguelliella*, and a *Tosapecten* from the Carnic Tanoura formation in Kyushu, Japan. *Mem. Fac. Educ., Kumamoto Univ.*, 7: 212–218 (pl. 2).

Tamura, M., 1959b. Carnic pelecypods from Matsukuma in central Kyushu, Japan. *Mem. Fac. Educ., Kumamoto Univ.*, 7: 219–224.

Tamura, M., 1959c. Trigoniidae, Ostreidae, Bakevelliidae, Pteriidae, Cardiidae and Astartidae from the Upper Jurassic Sakamoto Formation in central Kyushu, Japan. *Trans. Proc. Palaeontol. Soc. Jap., N.S.*, No. 33: 23–32 (pl. 5).

Tamura, M., 1959d. Taxodonta and Isodonta from the Upper Jurassic Sakamoto Formation in central Kyushu, Japan. *Trans. Proc. Palaeontol. Soc. Jap., N.S.*, No. 34: 53–65 (pl. 6).

Tamura, M., 1959e. Some pelecypods from the Upper Jurassic Sakamoto Formation in central Kyushu, Japan. *Trans. Proc. Palaeontol. Soc. Jap., N.S.*, No. 35: 113–120 (pl. 12).

Tamura, M., 1959f. Taxodonta and Isodonta from the Jurassic Soma Group in North Japan. *Trans. Proc. Palaeontol. Soc. Jap., N.S.*,No. 86: 168–180 (pl. 19).

Tamura, M., 1960a. Upper Jurassic pelecypods from the Torinosu Group in Shikoku, Japan. *Mem. Fac. Educ., Kumamoto Univ.*, 8: 227–244 (pl. 2).

Tamura, M., 1960b. Upper Jurassic Pteriacea from the Soma Group, Fukushima Prefecture, Japan. *Trans. Proc. Palaeontol. Soc. Jap., N.S.*, No. 37: 223–229 (pl. 26).

Tamura, M., 1960c. A note on *Neoburmesia*, a peculiar Jurassic pelecypod, with description of mitilids and myacids from the Upper Jurassic Soma Group in Japan. *Trans. Proc. Palaeontol. Soc. Jap. N.S.*, No. 38: 275–283 (pl. 32).

Tamura, M., 1960d. Heterodont and other pelecypods from the Upper Jurassic Soma Group, Japan. *Trans. Proc. Palaeontol. Soc. Jap., N.S.*, No. 39: 285–292 (pl. 33).

Tamura, M., 1961a. The Torinosu series and fossils therein. *Jap. J. Geol. Geogr.*, 32 (2): 219–251.

Tamura, M., 1961b. The geologic history of the Torinosu epoch and the Mesozoic reef-limestones in Japan. *Jap. J. Geol. Geogr.*, 32 (2): 253–277 (pl. 8).

Tamura, M., 1961c. New subgenus of pearl oyster, *Eopinctada* from the Cretaceous Mifune Group in Kumamoto Prefecture, Japan. *Trans. Proc. Palaeontol. Soc. Jap., N.S.*, No. 44: 147–151 (pl. 22).

Tamura, M., 1965. *Monotis* (*Entomonotis*) from Kyushu, Japan. *Mem. Fac. Educ., Kumamoto Univ.*, 13 (1): 42–59.

Tamura, M., 1970. The hinge structure of *Trigonioides*, with description of *Trigonioides mifunensis* sp. nov. from Upper Cretaceous Mifune Group, Kumamoto Pref., Japan. *Mem. Fac. Educ., Kumamoto Univ.*, 18 (1): 38–52 (incl. pls. 1, 2).

Tamura, M., 1972. Myophorian fossils discovered from the Konose Group, Kumamoto Prefecture, Japan, with a note on Japanese myophoriids. *Mem. Fac. Educ., Kumamoto Univ.*, 21 (1): 66–72 (pl. 1).

Tamura, M., 1973a. *Meekia hokkaidoana*, new species, from Hokkaido, Japan. *Mem. Fac. Educ. Kumamoto Univ.*, 22 (1): 101–104 (pl. 1).

Tamura, M., 1973b. Two species of Lower Cretaceous *Parvamussium* from Kyushu, Japan, and Sarawak Borneo. *Geol. Palaeontol. Southeast Asia*, 11: 119–124 (pl. 17).

Tamura, M., 1975. On the bivalvian faunas from the Gyliakian of Japan. *Mem. Fac. Educ., Kumamoto Univ.*, 24 (1): 59–62 (pl. 3) [J + E].

Tamura, M. and Matsumura, M., 1974. On the age of the Mifune Group, central Kyushu, Japan. With a description of ammonites from the Group by T. Matsumoto. *Mem. Fac. Educ., Kumamoto Univ.*, 23 (1): 47–56 (pl. 1).

Tamura, M. and Tashiro, M., 1966. Upper Cretaceous system south of Kumamoto. *Mem. Fac. Educ., Kumamoto Univ.*, 14 (1): 29–35 (incl. pl. 1) [J + E].

Tamura, M. and Tashiro, M., 1967. Cretaceous trigoniids from the Mifune Group. *Mem. Fac. Educ., Kumamoto Univ.*, 15 (1): 13–23 (incl. pl. 1).

Tamura, M., Tashiro, M. and Motojima, T., 1968. The correlation of the Mifune Group with the upper formation of the Goshonoura Group, with description of some important pelecypods from the strata. *Mem. Fac. Educ., Kamomoto Univ.*, 16 (1): 28–43 (incl. pl. 1).

Tanabe, K., 1973. Evolution and mode of life of *Inoceramus (Sphenoceramus) naumanni* Yokoyama emend., an Upper Cretaceous bivalve. *Trans. Proc. Palaeontol. Soc. Jap., N.S.*, No. 92: 163–184 (pls. 27–28).

Tanabe, K., 1975. Functional morphology of *Otoscaphites puerculus* (Jimbo), an Upper Cretaceous ammonite. *Trans. Proc. Paleontol. Soc. Jap., N.S.*, No. 99: 109–132 (pls. 10, 11).

Tanaka, K., 1963. A study of the Cretaceous sedimentation in Hokkaido, Japan. *Rep. Geol. Surv. Jap.*, No. 197: 1–122 (pls. 1–3).

Tanaka, K., 1965. Izumi Group in the central part of the Izumi Mountain range, Southwest Japan, with special reference to the sedimentary facies and cyclic sedimentation. *Rep. Geol. Surv. Jap.*, No. 212: 1–33 (pls. 1–8) [J + E].

Tanaka, K., 1970a. Upper Cretaceous turbidite formation of the Nakaminato area, Ibaraki Prefecture, Japan. *Bull. Geol. Surv. Jap.*, 21: 579–593 [J + E].

Tanaka, K., 1970b. Sedimentation of the Cretaceous flysch sequence in the Ikushumbetsu area, Hokkaido, Japan. *Rep. Geol. Surv. Jap.*, No. 236: 1–107 (pls. 1–12).

Tanaka, K., 1971. Trace fossils from the Cretaceous flysch of the Ikushumbetsu area, Hokkaido, Japan. *Rep. Geol. Surv. Jap.*, No. 242; 1–32 (pls. 1–11).

Tashiro, M., 1971. Upper Cretaceous glycymerids in Japan. *Trans. Proc. Palaeontol. Soc. Jap., N.S.*, No. 84: 225–242 (pls. 27, 28).

Tashiro, M., 1972. On the surface ornamentation of the pennate trigoniids, and on three new species of the trigoniids from the Himenoura Group, Kyushu, Japan. *Trans. Proc. Palaeontol. Soc. Jap., N.S.*, No. 86: 325–339 (pls 40, 41).

Tashiro, M., 1976. Bivalve faunas of the Upper Cretaceous Himenoura Group in Kyushu. *Palaeontol. Soc. Jap., Spec. Pap.*, No. 19: 1–102 (pls. 1–12).

Tashiro, M. and Noda, M., 1973. The geological age of the Himenoura Group, Kyushu. *J. Geol. Soc. Jap.*, 79: 465–480 [J + E].

Tatsumi, T. (Editor), 1970. *Volcanism and Ore Genesis.* Univ. Tokyo Press, Tokyo, 448 pp.

Teraoka, Y., 1970. Cretaceous formations in the Onogawa basin and its vicinity, Kyushu, Southwest Japan. *Rep. Geol. Surv. Jap.*, No. 237: 1–87 (pls. 1–18) [J + E].

Teraoka, Y., Hashimoto, M. and Okumura, K., 1973. Geology and metamorphism of the Mesozoic formations of northeastern Hokkaido. *Bull. Geol. Surv. Jap.*, 24: 385–392 [J + E].

Tokunaga, S. and Shimizu, S., 1926. The Cretaceous formation of Futaba in Iwaki and its fossils. *J. Fac. Sci., Imp. Univ. Tokyo,* 2, 1 (6): 181–212 (pls. 21–27).

Tokuyama, A., 1957a. On some Upper Triassic spiriferinoids from the Sakawa basin in prov. Tosa, Japan. *Trans. Proc. Palaeontol. Soc. Jap., N.S.*, No. 27: 99–106 (pl. 17).

Tokuyama, A., 1957b. On the Late Triassic rhynchonellids of Japan. *Jap. J. Geol. Geogr.*, 28 (1–3): 121–127 (pl. 10–11).

Tokuyama, A., 1957c. On some Jurassic rhynchonellids from Shikoku, Japan. *Trans. Proc. Palaeontol. Soc. Jap., N.S.*, No. 28: 128–136 (pl. 21).

Tokuyama, A., 1958a. On some terebratuloids from the Middle Jurassic Naradani formation in Shikoku, Japan. *Jap, J. Geol. Geogr.*, 29 (1–3): 1–10 (pl. 1).

Tokuyama, A., 1958b. On some terebratuloids from the late Jurassic Torinosu series in Shikoku, Japan. *Jap. J. Geol. Geogr.*, 29 (1–3): 119–131 (pl. 9).

Tokuyama, A., 1959a. Bemerkungen über die Brachiopodenfazies der Oberjurassichen Torinosuserie Südwestjapans, mit Beschreibungen einiger Formen. *Jap. J. Geol. Geogr.*, 30: 183–194 (pl. 15).

Tokuyama, A., 1959b. "*Bakevellia*" and "*Edentula*" from the late Triassic Mine Series in West Japan. *Trans. Proc. Palaeontol. Soc. Jap., N.S.*, No. 35: 147–155 (pl. 16).

Tokuyama, A., 1959c. Late Triassic Pteriacea from the Atsu and Mine Series, West Japan. *Jap. J. Geol. Geogr.*, 30 (1): 1–19 (pl. 1).

Tokuyama, A., 1960a. Late Triassic pelecypod fauna of the Aso Formation in West Japan. *Jap. J. Geol. Geogr.*, 31 (1): 23–38 (pl. 4).

Tokuyama, A., 1960b. On the pelecypod fauna of the late Triassic Hirabara Formation in West Japan. *Jap. J. Geol. Geogr.*, 31 (2–4): 201–217 (pl. 13).

Tokuyama, A., 1961a. Entwicklungsgeschichte der orogenen Ablagerungen. *Jap. J. Geol. Geogr.*, 32 (1): 85–110.

Tokuyama, A., 1961b. On the late Triassic sedimentary facies of Japan. *Jap. J. Geol. Geogr.*, 32 (2): 279–292 (pl. 9).

Tokuyama, A., 1962. Triassic and some other orogenic sediments of the Akiyoshi cycle in Japan, with special reference to their evolution. *J. Fac. Sci., Univ. Tokyo,* 2, 13 (3): 379–469 (pls. 15–18).

Tokuyama, A., 1965. Faltungsstockwerke in der Ojika-Halbinsel Nordostjapans. *Sci. Pap. Coll. Gen. Educ., Univ. Tokyo,* 15 (2): 217–236.

Toyohara, F., 1976. Geologic structures from Sangun-Yamaguchi zone to "Ryoke zone" in eastern Yamaguchi Prefecture. *J. Geol. Soc. Jap.*, 82: 99–111 [J + E].

Tsuchi, R. and Kuroda, N., 1973. Erimo (Sysoev) seamount and its relation to the tectonic history of the Pacific ocean basin. In: P.J. Coleman (Editor), *The Western Pacific.* Univ. W. Australia Press, Perth, W.A., pp. 57–64.

Ueda, Yoshio and Onuki, H., 1968. K–Ar dating on the

metamorphic rocks in Japan. I. *J. Jap. Assoc. Min. Petrol. Econ. Geol.*, 60: 159–166 [J + E].

Ueda, Yoshio, Yamaoka, K., Onuki, H. and Tagiri, M., 1969. K–Ar dating on the metamorphic rocks in Japan. II. *J. Jap. Assoc. Min. Petrol. Econ. Geol.*, 61: 92–99.

Ueda, Yoshiro, 1962. The type Himenoura Group, with palaeontological notes by Matsumoto, T. and Ueda, Y. *Mem. Fac. Sci., Kyushu Univ.*, D, 12 (2): 129–178 (pls. 22–27).

Uyeda, S. and Miyashiro, A., 1974. Plate tectonics and the Japanese Islands: a synthesis. *Geol. Soc. Am. Bull.*, 85: 1159–1170.

Vakhrameev, V.A., 1971. Development of the Early Cretaceous flora in Siberia. *Geophytika*, 1 (1): 75–83.

Waugh, B., 1973. The distribution and formation of Permian–Triassic red beds. In: A. Logan and L.V. Hills (Editors), *The Permian and Triassic Systems and Their Mutual Boundary. Can. Soc. Petrol. Geol., Mem.* 2.

Wright, C.W. and Matsumoto, T., 1954. Some doubtful Cretaceous ammonite genera from Japan and Saghalien. *Mem. Fac. Sci., Kyushu Univ.*, D, 4 (2): 107–134 (pls. 7–8).

Yabe, H., 1901–2. Note on the three Upper Cretaceous ammonites from Japan, outside of Hokkaido, *J. Geol. Soc. Tokyo* (*Jap.*), 8 (1901): 1–4 (English pages); *Ibid.*, 9 (1902): 5–10 (English pages) (pl. 10).

Yabe, H., 1903–4. Cretaceous Cephalopoda from the Hokkaido. Part I. *J. Coll. Sci., Imp. Univ. Tokyo*, 18 (2): 1–55 (pls. 1–7) (1903); Part II. *Ibid.*, 20 (2): 1–45 (pls. 1–6) (1904).

Yabe, H., 1909. Zur Stratigraphie und Paläontologie der oberen Kreide von Hokkaido und Sachalin. *Z. Dtsch. Geol. Ges.*, 61: 402–444.

Yabe, H., 1910. Die Scaphiten aus der Oberkreide von Hokkaido. *Beitr. Palaeontol. Geol. Österr.-Ung. Orients.*, 23: 159–174 (pl. 15).

Yabe, H., 1914. Ein neuer Ammonitenfund aus der Trigonia-sandstein-Gruppe von Provinz Tosa. *Sci. Rep. Tohoku Imp. Univ.*, 2, 1 (5): 71–74 (pl. 12).

Yabe, H., 1915. Notes on some Cretaceous fossils from Anaga on the island of Awaji and Toyajo in the province of Kii. *Sci. Rep. Tohoku Imp. Univ.*, 4 (1): 13–24 (pls. 1–4).

Yabe, H., 1927. Cretaceous stratigraphy of the Japanese Islands. *Sci. Rep. Tohoku Imp. Univ.*, 2, 11 (1): 27–100 (pls. 3–9).

Yabe, H. and Hanzawa, S., 1926. Geological age of *Orbitolina*-bearing rocks of Japan. *Sci. Rep. Tohoku Imp. Univ.*, 2, 9 (1): 13–20 (pls. 3–6).

Yabe, H. and Hayashi, Z., 1938. A Mesozoic unionid from Manchuria. *Jap. J. Geol. Geogr.*, 15: 31–33 (pl. 4).

Yabe, H. and Nagao, T., 1925. New or little known Cretaceous fossils from North Saghalin (Lamellibranchiata and Gastropoda). *Sci. Rep. Tohoku Imp. Univ.*, 2, 7 (4): 111–124 (pls. 29, 30).

Yabe, H. and Nagao, T., 1926. *Praecaprotina* nov. gen., from the Lower Cretaceous of Japan. *Sci. Rep. Tohoku Imp. Univ.*, 9 (1): 21–24 (pl. 7).

Yabe, H. and Nagao, T., 1928. Cretaceous fossils from Hokkaido: Annelida, Gastropoda and Lamellibranchiata. *Sci. Rep. Tohoku Imp. Univ.*, 9 (3): 77–96 (pls. 16–17).

Yabe, H. and Obata, T., 1930a. Discovery of *Ptychodus rugosus* Dixon from the Upper Cretaceous of the Japanese Saghalin. *Jap. J. Geol. Geogr.*, 7 (2): 43–44.

Yabe, H. and Obata, T., 1930b. On some fossil fishes from the Cretaceous of Japan. *Jap. J. Geol. Geogr.*, 8 (1–2): 1–7 (pls. 1, 2).

Yabe, H. and Ozaki, H., 1953. A new type of Cretaceous nautiloids from Tyosi peninsula, Kwanto region. *Bull. Natl. Sci. Mus. Tokyo*, No. 32: 55–61.

Yabe, H. and Shimizu, S., 1921. Notes on some Cretaceous ammonites from Japan and California. *Sci. Rep. Tohoku Imp. Univ.*, 2, 5 (3): 51–59 (pls. 8, 9).

Yabe, H. and Shimizu, S., 1924a. A new species of *Brahmaites* from the Upper Cretaceous of South Saghalin, with a remark on the genus *Brahmaites*. *Jap. J. Geol. Geogr.*, 3 (2): 77–80 (pl. 13).

Yabe, H. and Shimizu, S., 1924b. A new species of *Nautilus*, *N.* (*Cymatoceras*) *pseudo-atlas* Yabe and Shimizu, from the Upper Cretaceous of Amakusa. *Jap. J. Geol. Geogr.*, 3 (2): 41–43 (pl. 5).

Yabe, H. and Shimizu, S., 1925a. A new Lower Cretaceous ammonite, *Crioceras ishiwarai*, from Oshima, province of Rikuzen. *Jap. J. Geol. Geogr.*, 4 (3–4): 85–87 (pl. 4).

Yabe, H. and Shimizu, S., 1925b. Japanese Cretaceous ammonites belonging to Prionotropidae, I. *Sci. Rep. Tohoku Imp. Univ.*, 2, 7 (4): 125–138 (pls. 30–33).

Yabe, H. and Shimizu, S., 1927. The Triassic fauna of Rifu, near Sendai. *Sci. Rep. Tohoku Imp. Univ.*, 11 (2): 101–136 (pls. 10–14).

Yabe, H. and Shimizu, S., 1933. Triassic deposits of Japan. *Jap. J. Geol. Geogr.*, 10 (3–4): 87–98.

Yabe, H. and Sugiyama, T., 1930. Stromatoporoids and related forms from the Jurassic of Japan. *Jap. J. Geol. Geogr.*, 8 (1–2): 23–28 (folded table).

Yabe, H. and Sugiyama, T., 1931. On some spongiomorphid corals from the Jurassic of Japan. *Sci. Rep. Tohoku Imp. Univ.*, 2, 14 (2A): 103–105 (pl. 33–35).

Yabe, H. and Sugiyama, T., 1935. Jurassic stromatoporoids from Japan. *Sci. Rep. Tohoku Imp. Univ.*, 2, 14 (2B): 135–192 (pls. 40–71).

Yabe, H. and Sugiyama, T., 1939. Discovery of a Mesozoic hexacoral in "a green schistose rock of the Kamuikotan system" of Hokkaido. *Proc. Imp. Acad., Tokyo*, 15: 86–89.

Yabe, H. and Toyama, S., 1928. On some rock-forming algae from the younger Mesozoic of Japan. *Sci. Rep. Tohoku Imp. Univ.*, 2, 12 (1): 141–152 (pls. 18–23).

Yabe, H., Nagao, T. and Shimizu, S., 1926. Cretaceous mollusca from the Sanchu-graben in the Kwanto Mountainland, Japan. *Sci. Rep. Tohoku Imp. Univ.*, 2, 9 (2): 33–76 (pls. 12–15).

Yamada, N., Katada, M. et al. (Compilers), 1974. Geological map of the Ryoke belt, central Japan, Scale 1 : 200,000. *Geol. Surv. Jap., Misc. Map. Ser.* 18.

Yamaguchi, M. and Yanagi, T., 1970. Geochronology of some metamorphic rocks in Japan. *Eclog. Geol. Helv.*, 63: 371–388.

Yamamoto, H., 1962. Plutonic and metamorphic rocks along the Usuki–Yatsushiro tectonic line in the western part of central Kyushu. *Bull. Fukuoka Gakugei Univ.*, 12: 92–172.

Yanagi, T., 1975. Rubidium–Strontium model of formation of the continental crust and the granite at the island arc. *Mem. Fac. Sci., Kyushu Univ.*, D, 22 (2): 37–98.

Yanagimoto, Y., 1973. Stratigraphy and geological structure of the Paleozoic and Mesozoic formations in the vicinity of Kuzuu, Tochigi Prefecture. *J. Geol. Soc. Jap.*, 79:441–451 [J + E].

Yang, S.-Y., 1974. Note on the genus *Trigonioides* (Bivalvia). *Trans. Proc. Palaeontol. Soc. Jap., N.S.*, No. 95: 395–408 (pls. 54–55).

Yang, S.-Y., 1975. On a new non-marine pelecypod genus from the Upper Mesozoic Gyeongsang Group of Korea. *Trans. Proc. Palaeontol. Soc. Jap., N.S.*, No. 100: 177–187 (pls. 16–17).

Yasukawa, K. and Nakajima, M., 1972. Relative southward shift of Southwest Japan. *Kagaku (Tokyo)*, 42: 163–165 [J].

Yehara, S., 1924. On the Izumi-sandstone Group in the Onogawa basin (Prov. Bungo) and the same group in Uwajima (Prov. Iyo). *Jap. J. Geol. Geogr.*, 3: 27–39 (pls. 2–4).

Yen, T.P., 1963. Major geological, geophysical and geochemical features of Taiwan. *Taiwan Mining Ind.*, 15: 1–21.

Yokoyama, M., 1890. Versteinerungen aus der japanischen Kreide. *Palaeontographica,* 36: 159–202 (pls. 18–25).

Yokoyama, M., 1902. On the ammonites of Echizen. *J. Geol. Soc. Jap.*, 9: 387–390.

Yokoyama, M., 1904a. On some Jurassic fossils from Rikuzen. *J. Coll. Sci., Imp. Univ. Tokyo,* 18 (6): 1–13 (2 pl).

Yokoyama, M., 1904b. Jurassic ammonites from Echizen and Nagato. *J. Coll. Sci., Imp. Univ. Tokyo,* 19 (20): 1–17 (pl. 4).

Yoshida, S., 1958. The foraminifera fauna of the Upper Cretaceous Hamanaka and Kritappu Formations of eastern Hokkaido, Japan. *J. Hokkaido Gakugei Univ.*, 9 (1): 250–264 (incl. 3 pls.).

Yoshida, S., 1961. The Cretaceous–Tertiary boundary in eastern Hokkaido, Japan. *J. Hokkaido Gakugei Univ.*, 2B, 12 (1): 14–38.

Yoshida, S., 1963. Upper Cretaceous foraminifera from the Nemuro Group, eastern Hokkaido, Japan. *J. Hokkaido Gakugei Univ.*, 2B, 13 (2): 211–258 (incl. 17 pls.).

Yoshida, S., 1969. Upper Cretaceous foraminifera from the Izumi Group, Awaji-shima, Hyogo Prefecture. *Bull. Yamagata Univ., (Nat. Sci.)* 7 (2): 183–189.

Yoshida, T. (Editor), 1975. *An Outline of the Geology of Japan.* Geol. Surv., Japan, Tokyo, 3rd ed., 61 pp.

Yossii, M., Toyama, S. and Sugiyama, T., 1933. Cordierite slate with *Pseudomonotis* (*Claraia*) found among the shore boulders of the lagoon, Little Ehabi, on the east coast of North Saghalin. *Jap. J. Geol. Geogr.*, 10 (3–4): 99–105 (pl. 6).

Zaklinskaya, E.D., 1962. Importance of angiosperm pollen for the stratigraphy of Upper Cretaceous and Lower Paleogene deposits and botanical–geographical provinces at the boundary between the Cretaceous and Tertiary system. *Rep. Sov. Palynol. First Int. Conf. Palynol. Acad. Sci. U.S.S.R., Inst. Geol.*, pp. 105–113 [R + E].

Zwart, H.J., Corvalan, J., James, H.L., Miyashiro, M., Saggerson, E.P., Sobolev, V.S., Subramaniam, A.P. and Vallance, T.G., 1967. A scheme of metamorphic facies for the cartographic representation of regional metamorphic belts. *Geol. Newsl., IUGS,* 1967 (2): 57–74.

Chapter 5

THAILAND

M.F. RIDD

INTRODUCTION

All three of the Mesozoic systems are represented in Thailand, cropping out extensively over the north and northeast of the country and in more discontinuous areas in the west, southeast and on the Peninsula (Fig. 1). They overlie a fairly continuous Palaeozoic succession and are only locally concealed beneath Tertiary or Quaternary beds.

Broadly, the Mesozoic history of Thailand can be viewed as an orogenic cycle. Thus in the Triassic, marine flysch sedimentation took place, derived from a source in the east and passing westwards into limestone. In the Late Triassic there was a major orogeny with acid plutonism and volcanicity. Then, starting in the Late Triassic or Early Jurassic and continuing into the Cretaceous, sedimentation was of post-orogenic molasse type, mostly of non-marine facies but periodically marine. That too was derived from an easterly source. But in the opposite direction, toward the Burmese border, marine limestone deposition continued. In the Late Cretaceous or Early Tertiary there was a second period of faulting and folding, insignificant in the east of the country but intense in the west. The uplift which resulted from those earth movements produced an elevated area which separated the ocean (in Burma or beyond) from the eastern part of Thailand. In the land-locked basin which resulted, deposition of evaporites occurred, bringing to an end the Mesozoic cycle.

The Palaeozoic and Triassic rocks of Thailand lie within what Burton (1967) called the Yunnan–Malaya Geosyncline. This was a major belt of sedimentation which has been traced from the islands south of Singapore, through West Malaysia, Thailand, eastern Burma and into Yunnan Province of southern China where it meets the eastern end of the Himalayas. As might be expected the Triassic and older facies belts of Thailand trend roughly parallel with the geosyncline, as does the regional structural strike. After the Late Triassic orogeny the facies distribution was less obviously linear in so far as north–south-trending facies boundaries cannot be mapped over such long distances.

Reports written by travellers in Thailand (then called Siam) in the first four decades of this century are now largely of historical interest although some, for example Hogbom (1914), Lee (1927), Gregory (1930) and Heim and Hirschi (1939), contain valuable details of Mesozoic rocks in the north of the country.

In 1951, Brown et al. produced a detailed compilation including the work of geologists in the Thailand Department of Mineral Resources, and they accompanied their bulletin with a geological map of the country at a scale of 1 : 2,500,000. They used the name "Khorat Series" for all the Mesozoic sedimentary rocks above the Permian Rat Buri Limestone (including those of the Khorat Plateau) and underlying the Tertiary, although they did map the Triassic–Jurassic Kamawkala limestone as a distinct unit within the Khorat Series of the northwest.

The Khorat Plateau of northeast Thailand has received a lot of attention. An account, including measured sections, was written by Ward and Bunnag (1964) and the Khorat Group, as they called it, was divided into seven formations extending, they thought, from Lower Triassic to Upper Cretaceous. Two years later a comprehensive ground-water study by Haworth et al. (1966) adopted the stratigraphy of Ward and Bunnag (1964) and included an excellent geological map of the Khorat Plateau and adjacent country at a scale of 1 : 750,000. Meanwhile interest in the petroleum potential of the Khorat Group led

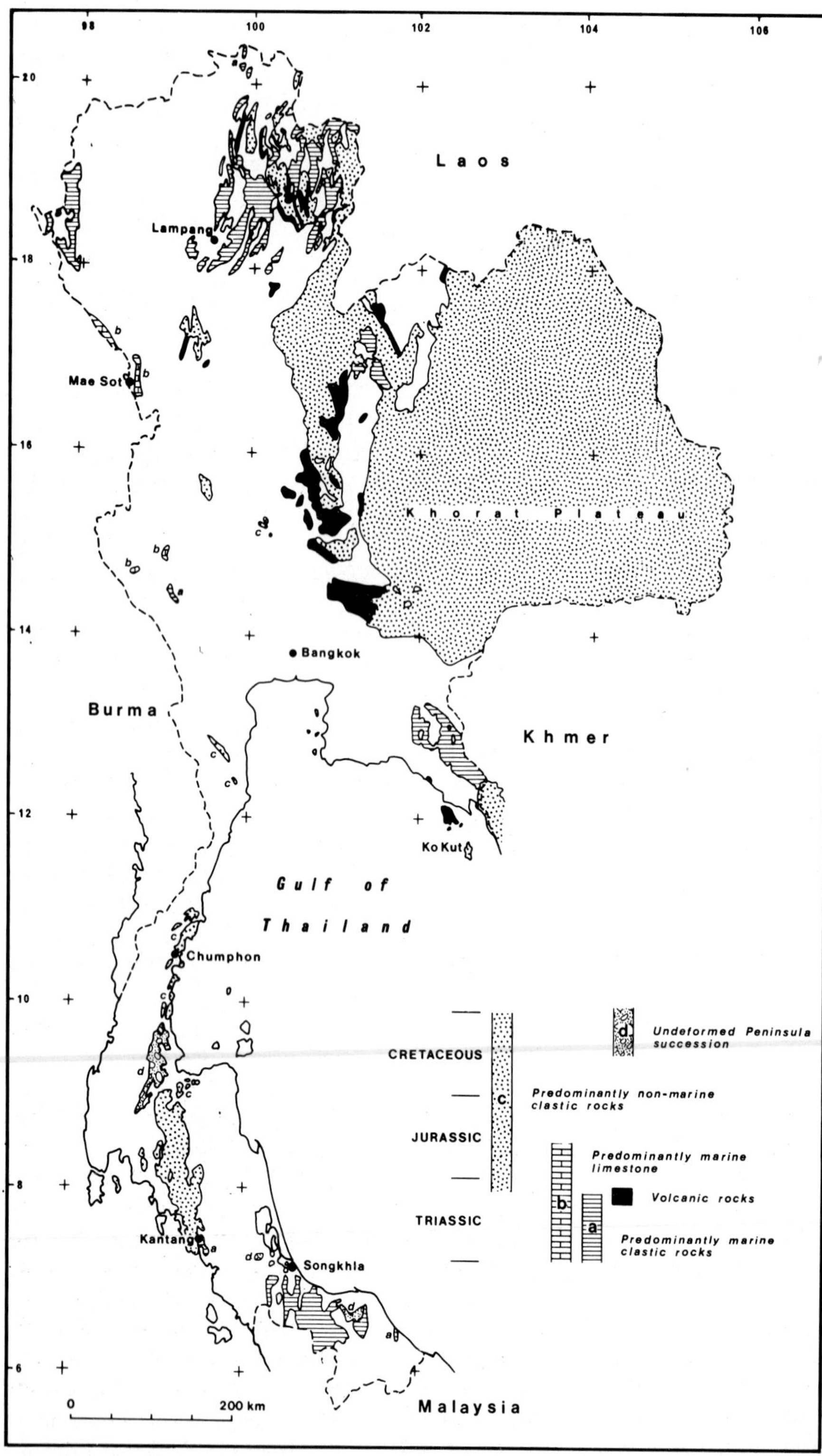

Fig. 1. Mesozoic sedimentary and volcanic rocks in Thailand. The four lithostratigraphic units *a* to *d* correspond to those shown in Fig. 2 and in the text.

to the publication by Borax and Stewart (1965) of another detailed account of the stratigraphy of the plateau.

A Japanese team under the direction of Professor Teichi Kobayashi made collecting expeditions to Thailand in the 1960's. The emphasis of their work was biostratigraphic and their results have appeared mostly in the series of volumes *Geology and Palaeontology of Southeast Asia* published by the University of Tokyo Press. Mesozoic geology was the subject of a number of their papers. Thus Triassic marine faunas were described by Kobayashi et al. (1966); the Khorat Group and various aspects of its biostratigraphy were described by Hayami (1968), Kobayashi (1968) and Iwai et al. (1966) (who established its Late Triassic to Cretaceous age); and there were several others, for example Asama (1973), Endo and Fujiyama (1966), Iwai (1968, 1972, 1973), Kobayashi and Tokuyama (1959), Kon'no and Asama (1973). Biostratigraphic reviews of the Triassic, the marine Jurassic and the non-marine Mesozoic, putting Thailand into its southeast Asian context, were written respectively by Tamura et al. (1975), Sato (1975) and Iwai et al. (1975).

Geological mapping and geochemical prospecting surveys by British teams from the Institute of Geological Sciences, working with geologists from Thailand's Department of Mineral Resources, have included some areas of Mesozoic rocks. On peninsular Thailand, Garson et al. (1975) mapped the Phuket region, Hughes and Bateson (1967) the Chanthaburi area in the southeast, and Bleakly et al. (1965) the Loei area in the north. Similar joint Thai-German teams have mapped northernmost Thailand (Baum et al., 1970) and the Sri Sawat area in the west (Koch, 1973). Mesozoic rocks were mapped in both areas but the published accounts are unfortunately brief.

A major advance has been the publication by Thailand's Department of Mineral Resources of the Lampang geological map at 1 : 250,000 scale (Piyasin, 1972); that is the type-area for the marine Triassic in the north of the country. Its accompanying text is in Thai with an English summary.

Economic deposits are associated with some Mesozoic rocks and these have stimulated certain geological investigations. Thus, the Triassic plutonic rocks are particularly well known because of their tin mineralisation (e.g., Brown et al., 1951; Garson et al., 1969, 1975; Hughes and Bateson, 1967). The water resources of the Khorat Plateau prompted the detailed work by Haworth et al. (1966), mentioned earlier, while at the top of the Khorat Group the salt deposits have been reported on by Gardner et al. (1967).

The Gulf of Thailand has been shown to be a Tertiary hydrocarbon prospect only and the drilling there has not increased our understanding of the Mesozoic. However, onshore the Mesozoic Khorat Plateau has attracted some attention and in 1972 a well was drilled to 3356 m by Union Oil Company at Kuchinarai (16°42′36″N, 104°04′07″E).

Radiometric work on the igneous rocks of Thailand has been carried out by Burton and Bignell (1969), and Braun (1969), while Haile and Tarling (1975) made the first reconnaissance palaeomagnetic measurements in the country and included samples of redbeds from the Khorat Group.

This review of the Mesozoic geology of Thailand describes the main lithostratigraphic units and discusses the evidence for their ages; this is done at some length because of the lack of any existing comprehensive accounts of the geology on which to base interpretive work. It goes on to outline the igneous rocks of Mesozoic age and concludes by interpreting the Mesozoic geological history of the country.

SEDIMENTARY ROCKS

Most geological observations in Thailand have been made on surface outcrops, subsurface data being limited to a few shallow wells on the Khorat Plateau (Fig. 1). However, natural outcrops are generally poor inland because of deep weathering and a heavy cover of vegetation; the best exposures are coastal outcrops and, inland, quarries and road-cuttings.

Macrofossil collections from the Mesozoic are not numerous and particularly few have been made in southern Thailand. Microfossil and microfloral collections are even scarcer. As a result of this poor palaeontological control, coupled with the discontinuous exposures, the stratigraphy of the Mesozoic has still not been worked out in detail. Another problem for the stratigrapher is that thicknesses of rock units can generally only be estimated and not measured. Only on the Khorat Plateau, where the structure is simple and the dry climate makes for a less dense vegetation cover, are reported thicknesses likely to be at all accurate.

The diagrammatic stratigraphic cross-sections

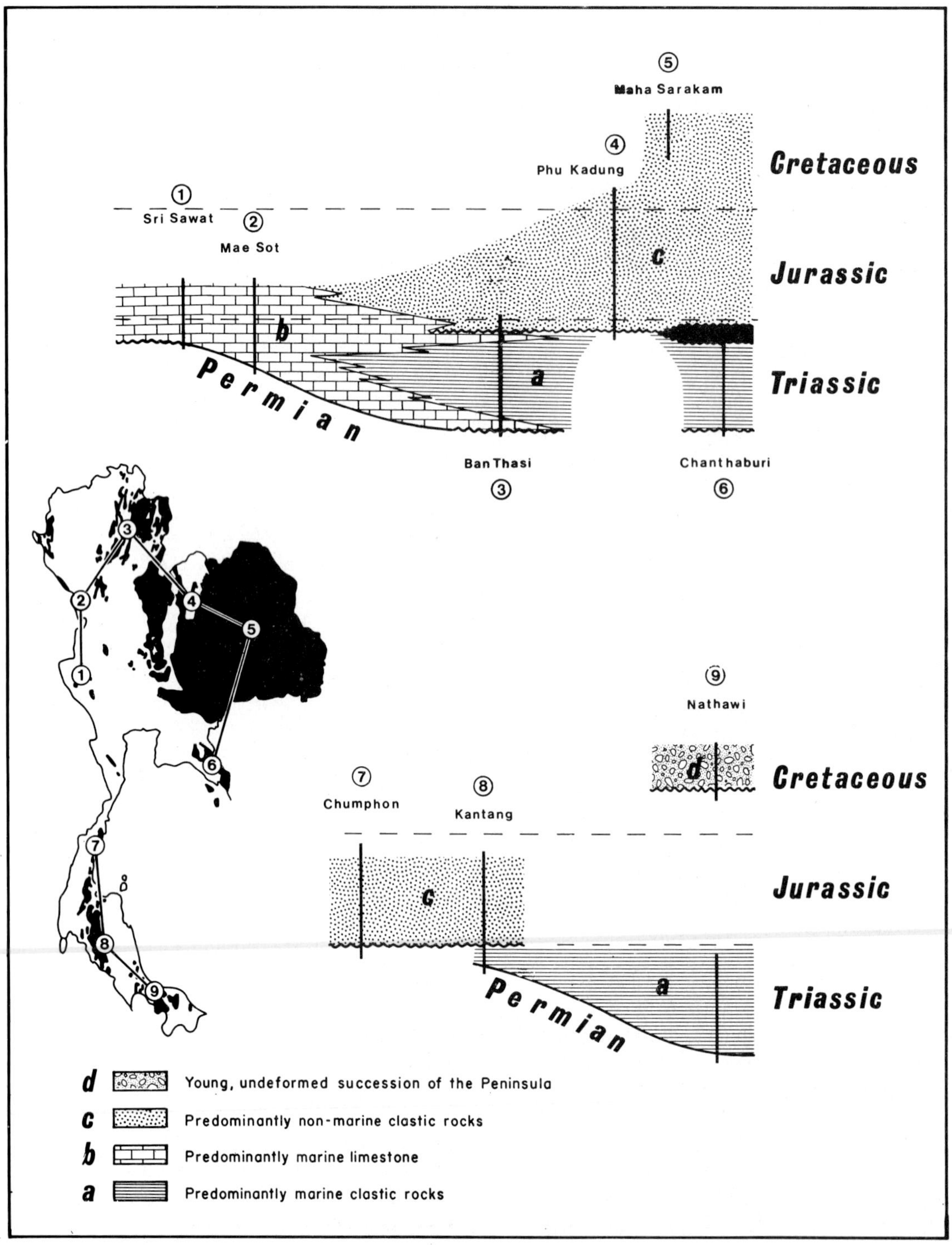

Fig. 2. Diagrammatic cross-sections showing stratigraphic relations of Mesozoic sedimentary and volcanic rocks (the latter shown black). They are not to scale. The four lithostratigraphic units *a* to *d* correspond to those shown on Fig. 1 and in the text.

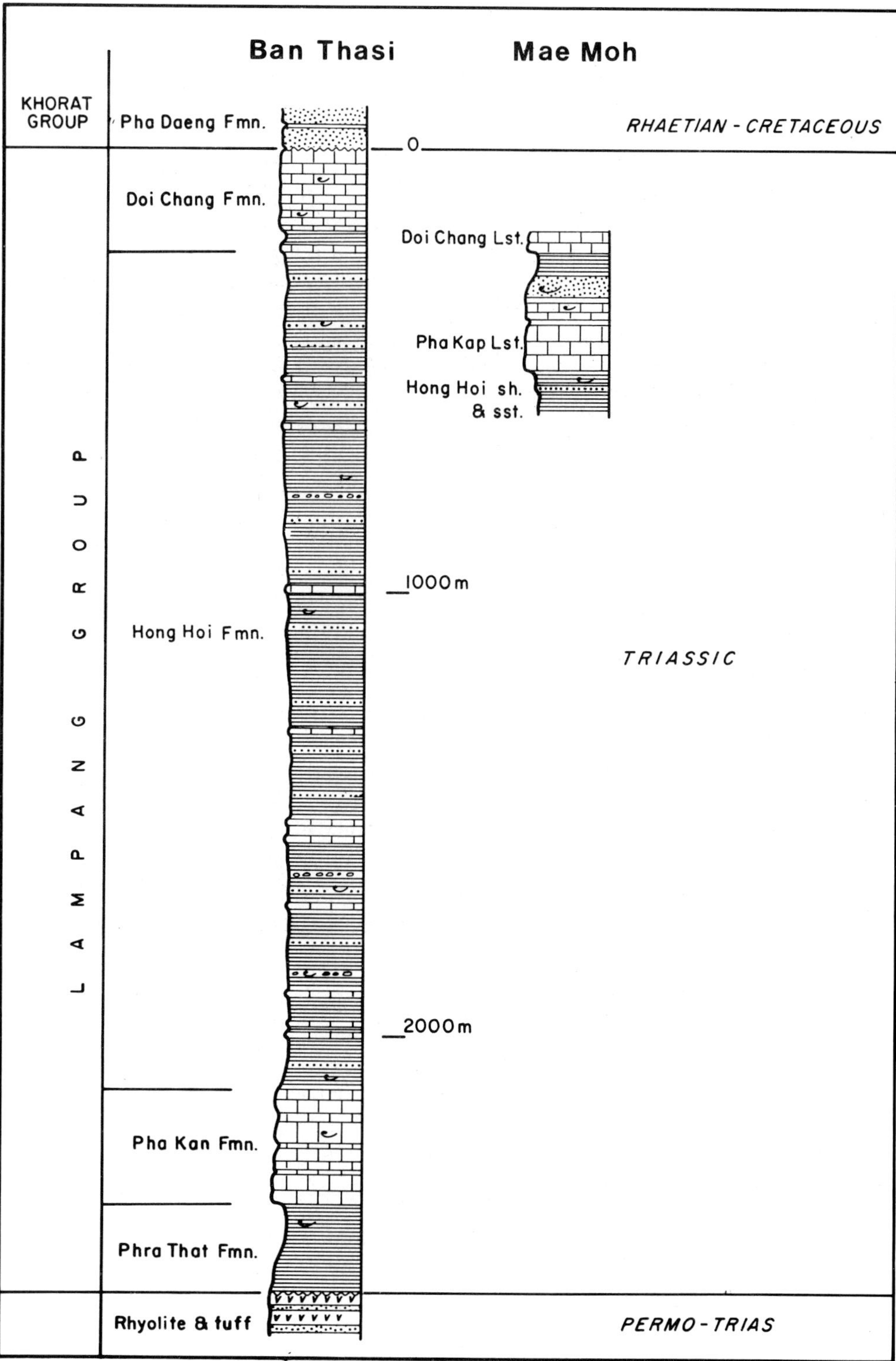

Fig. 3. Marine Triassic Lampang Group of northern Thailand. The type-section at Ban Thasi is based on the work of Chonglakmani (1972), and the Mae Moh section on Pitakpaivan (1955). For location see Fig. 2.

which form Fig. 2 show the relationships of the main lithostratigraphic units of the Mesozoic. Because of the sparse palaeontological control, and consequently the imprecise determination of chronostratigraphic boundaries, these units are described and their ages discussed one by one in approximately ascending order rather than the stratigraphy being described system by system. For ease of reference the four main units are labelled *a* to *d* on Figs. 1 and 2, and in the following text.

(a) Predominantly marine clastic rocks (Triassic)

In northern Thailand these rocks have been named the Lampang Group. Broadly similar beds occur southeast of Bangkok and on the Peninsula, close to the Malaysian border (Figs. 1, 2 and 6). The term "flysch" is used for them in the sense that they are pre- or synorogenic, marine, and consist largely of repetitious sandstone and mudstone beds.

Northern Thailand

The *Lampang Group* is a sequence of marine fossiliferous shale, sandstone, limestone and conglomerate. It attains a thickness of 2950 m and occurs in northern Thailand in the region of Lampang and the Laotian border. The type-section is at Ban Thasi, 75 km northeast of Lampang, and was first described by Piyasin (1972). However, further work there by Chonglakmani (1972) and unpublished work by Piyasin have shown that what was previously regarded as the uppermost unit, the Pha Daeng Formation (a non-marine red-brown shale and sandstone with an hiatus at the base), is more correctly included in the Khorat Group (described later) than in the Lampang Group. The four formations which comprise the Lampang Group at Ban Thasi, the type-section, are shown in Fig. 3.

At the base, the *Phra That Formation* consists of shale, sandstone, calcareous shale and conglomerate with pebbles of andesite and rhyolite. Locally there is a conformable passage from Permian sedimentary rocks beneath but commonly the base is an unconformity with a basal conglomerate resting on Permo-Triassic rhyolite and andesite or on Palaeozoic sedimentary rocks. Fossils, including ammonoids, indicate that the environment of deposition was marine. It is overlain by the *Pha Kan Formation* which is a massive, grey limestone with gastropods, brachiopods, crinoids and algae. At Ban Thasi it is 260 m thick but it thins to nothing elsewhere in the Lampang area. At Pha Kan itself, 55 km east of Lampang, Middle to Upper Triassic corals are described by Pitakpaivan et al. (1969). The thickest unit of the Lampang Group is the *Hong Hoi Formation*. It is up to 1900 m thick (Chonglakmani, 1972) and consists of flysch-type, rhythmically alternating, grey shale and graded tuffaceous sandstone or greywacke, with intervals of siltstone, mudstone, polymict conglomerate and rare argillaceous limestone. A varied marine fauna is present in the shales and mudstones at Ban Thasi and also at Mae Moh, 10 km south (Pitakpaivan, 1955; Kummel, 1960), including ammonoids, and the bivalves *Halobia* and *Daonella.* The *Doi Chang Formation* is the uppermost unit of the (revised) Lampang Group. It is a 230 m thick, massive limestone and limestone conglomerate at Ban Thasi, containing *Ostrea*-like bivalves, gastropods and brachiopods. This formation at Mae Moh contains several ammonoid genera (Pitakpaivan, 1955; Kummel, 1960) and calcareous algae (Endo, 1966).

The age of the Lampang Group at Ban Thasi ranges from Scythian (conformable on *Leptodus*-bearing Permian) through Middle Triassic to probable Norian (Piyasin, 1972; Chonglakmani, 1972). Nearby at Mae Moh, Kummel (1960) has recognised Anisian and Carnian faunas.

Rocks assignable to the Lampang Group have been mapped by Baum et al. (1970) east of Lampang nearly to the Laotian border. They describe a basal conglomerate from which a Scythian conodont fauna has been obtained, passing up into argillaceous and sandy sediments which reach Late Carnian age. In the middle of the succession are limestone and chert intercalations which increase in importance westwards to the Burmese border (Fig. 6) and have a *Daonella* and *Halobia* fauna of Carnian age. An increase in red colouration toward the east is suggested by Baum et al. (1970) to indicate continental conditions in that direction.

Koch (1973) mentions sandstone and shales of so-called normal facies, with a *Halobia* and *Daonella* fauna, in the Khwae Yai valley northwest of Bangkok. No further details are given but it would seem to correlate with the Hong Hoi Formation of the Lampang Group.

Southeast Thailand

Southeast of Bangkok, in the region of Chanthaburi (Figs. 1 and 2), a sequence of unfossiliferous rocks has been mapped by the present writer and given the informal name, Chanthaburi group. It forms a southeast-trending belt crossing the Khmer border (Gubler, 1935) and is distinctive, comprising only two rock types: graded greywacke beds and mudstone.

It is intruded by granite and is locally quartz-veined. Neither cleavage nor regional metamorphism effects were seen but the beds have been strongly folded and steep dips or overturned beds are common.

An unconformity apparently separates this group from the pre-Permian rocks as the Permian Rat Buri Limestone is locally absent. In one locality, however, on the Khmer border, a breccia of Permian fusulinid limestone is associated with extrusive igneous rock at the base of the Chanthaburi group. Although this has been interpreted as a remnant of the Permian which escaped erosion, it may be that these rocks are part of the Chanthaburi group.

Toward the southeast the group is overstepped by the little-deformed Khorat Group of latest Triassic–Cretaceous age. Areas of subhorizontal basalt of probable young Tertiary age locally overlie the group.

The poor exposure and structural complexity make it impossible to determine the thickness of the Chanthaburi group. From the width of the outcrop belt one might guess it to be at least 2000 m.

Peninsular Thailand

In the Songkhla and (50 km south) the Nathawi districts of peninsular Thailand (Figs. 1 and 2) a succession of sandstone, conglomerate, mudstone chert and siltstone rests with apparent conformity upon the Permian Rat Buri Limestone.

The structure of these beds is extremely complex and the succession has not been worked out; but it seems likely that in the Songkhla area, on the western side of what is assumed to be a synclinorium, the older beds outcrop. These are mostly sandstones. Probably somewhat younger, at the roadside to the west of Nathawi, are conglomerate beds in which the clasts are uniformly about 0.5 cm diameter, composed of grey chert, quartz, and moulds of what were probably limestone clasts. The youngest and the largest part of the succession is the thinly bedded, argillaceous sandstone and siltstone, interbedded with mudstone and thinly bedded chert. Although the rocks lack most of the typical features of turbidites they are thought, nevertheless, to be deep-water marine deposits.

The only record of fossils is that by Kobayashi and Tokuyama (1959) and Kobayashi et al. (1966) who state that *Daonella sumatrensis* has been found near Nathawi and indicates a Carnian age. Along strike to the south the same beds have been mapped in West Malaysia as the Semanggol Formation or Lipis Group (Burton, 1973). Kobayashi (1963a) has collected a marine fauna of Triassic age from those beds, comprising *Halobia, Daonella* and *Posidonia.*

Granite intrudes the succession and has resulted in locally abundant quart-veining and hornfels development. Cleavage is locally developed but no regional metamorphism was noted. The thickness, again, cannot be determined but at least 2000 m is likely.

Geographically quite separate, on the western side of the Peninsula at Kantang (Figs. 1, 2 and 4), marine Triassic rocks have been mapped by the present writer. Thin-bedded, pyritic sandstones and shales predominate but the most interesting rocks are limestones. In one quarry micritic limestone contains abundant calcisponges (?*Thalamida*) and branching stromatoporoids. In another exposure, a rhynchonellid brachiopod was found as well as a microflora containing the Triassic miospore *Ovalipollis* sp. It is likely that these limestones correlate with the Triassic Kodiang Limestone Formation 150 km southeast, in Malaysia (De Coo and Smit, 1975).

(b) Predominantly marine limestone (Triassic–Jurassic)

Near the Burmese border of northern Thailand a predominantly limestone succession occurs (Figs. 1 and 2). Cotter (1923) first described the rocks in the Mae Sot and Kamawkala area (southwest of Lampang), calling them the "Kamawkala limestone" and saying they are composed of impure limestone, red sandstone and conglomerate. Gregory (1930), Trauth (1930), and Weir (1930) described the rich coral, brachiopod, bivalve, ammonite, sponge, algal, crinoid, echinoid and gastropod fauna as Late Triassic.

The stratigraphy is confused but according to Sato (1975) the Triassic Kamawkala limestone is overlain by Jurassic beds as follows:

(3) Ban Hui Hin Fon bed: composed of dark, impure marly limestone with ammonites
(2) Ban Yang Puteh bed: composed of dark brownish grey cryptocrystalline marly limestone with ammonites
(1) Kamawkala limestone

He states that the age of the upper two "beds" is

Aalenian (topmost Early Jurassic). Since the Kamawkala limestone is apparently Triassic, an unconformity between it and the Jurassic is implied.

Sato (1975) described the Jurassic as being thin without saying how thin it is. According to Heim and Hirschi (1939), the thickness of the whole limestone (presumably Triassic and Jurassic) on the eastern side of Mae Sot Valley is more than 1000 m.

Further south, in the Sri Sawat area 185 km northwest of Bangkok, Koch (1973) has described a similar predominantly limestone succession of Triassic and probably Jurassic age (Fig. 2). An Anisian conodont fauna was found in limestone which appears to overlie conformably the Permian Rat Buri Limestone. The same author believes the Anisian limestone to be the lateral equivalent of a 200 m thick limestone conglomerate (consisting of pebbles of fossiliferous Permian limestone) which occurs also in the Sri Sawat area. The conglomerate in turn passes up into limestone, 300 m thick, from which microfossils of probable Early Jurassic age have been obtained.

The relations of the Triassic and Jurassic carbonate rocks with the terrigenous clastic rocks further east is presumably an interfingering one (Fig. 6). Cotter's description (1923) of the Kamawkala limestone includes red quartzitic sandstone and conglomerate which may be tongues of the Lampang Group while, similarly, the Triassic Lampang Group at Ban Thasi includes intervals of limestone (Fig. 3). The regional work of Baum et al. (1970) also refers to a westward increase in the proportion of limestone in the Triassic succession.

(c) Predominantly non-marine clastic rocks (?Upper Triassic–Cretaceous)

These are the most widely distributed rocks in Thailand (Figs. 1, 2 and 6). They consist predomi-

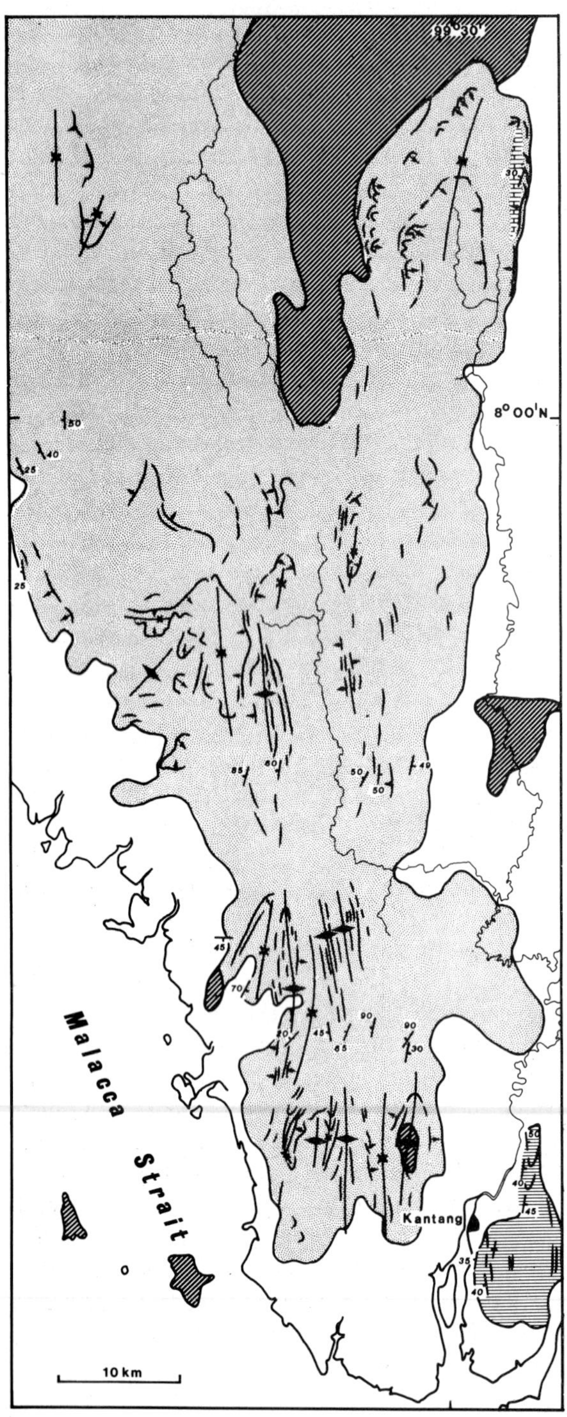

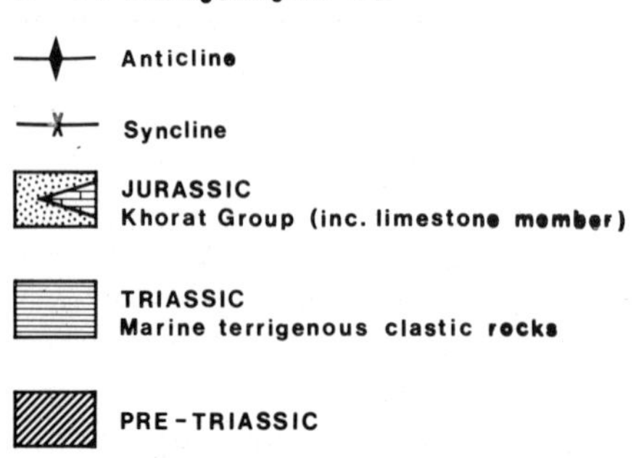

Fig. 4. Geological map of the Mesozoic and older rocks in the Kantang area of southwest peninsular Thailand. (Photogeology is based on unpublished work by W.M. Johnstone.)

nantly of sandstone, mudstone and conglomerate although minor limestone intervals are also present; red, brown and grey colours predominate. In broad terms the facies can be described as molasse or red-beds. Terrestrial plants and non-marine bivalves are the principal fossils although some marine bivalves and vertebrate fossils have also been described.

In the northeast of the country the beds are over 4500 m thick and occupy a gently downwarped structural basin which gives rise to the Khorat Plateau from which the *Khorat Group* was named. There is sufficient lithological homogeneity to allow the same name to be applied to the predominantly non-marine clastic rocks in northern, southeastern and peninsular Thailand, notwithstanding their geographical separation.

Khorat Group in the type-area

In the type-area, the Khorat Plateau, this unit has received considerable attention, partly because its simple structure has invited stratigraphical research and partly because of its economic importance as a source of subsurface water, salt and, it was hoped, of oil. The deep borehole drilled in search of oil at Kuchinarai in the northeast part of the Plateau reached 3356 m. It is shown on the map by Haworth et al. (1966) to have commenced in the Phu Kadung Formation. No details have been published but that thickness is likely to be about the thickness of the lower part of the Khorat Group there.

There has been a confusion of lithostratigraphic names for units in the group, for example those of La Moreaux et al. (1959), Ward and Bunnag (1964), Borax and Stewart (1965), Iwai et al. (1966), Haworth et al. (1966) and Gardner et al. (1967). The following succession is that of Ward and Bunnag (1964) except that in addition the salt-bearing Maha Sarakam Formation of Gardner et al. (1967) is included at the top and the Huai Hin Lat Formation of Iwai et al. (1966) at the bottom. The succession is therefore:

Maha Sarakam Formation	Cretaceous
Khok Kruat Formation	Cretaceous
Phu Phan Formation	Jurassic
Sao Khua Formation	Jurassic
Phra Vihan Formation	Jurassic
Phu Kadung Formation	Jurassic
Nam Phong Formation	Upper Triassic–Jurassic
Huai Hin Lat Formation	Upper Triassic

The basal contact of the Khorat Group in the type-area is an angular unconformity. Locally developed above this is the *Huai Hin Lat Formation*, a conglomerate up to 140 m thick. Pebbles are mostly of limestone, rhyolite, porphyry, chert and argillite. There are minor intercalations of shale, siltstone and calcareous sandstone which have yielded land-plant fossils (mostly Equisetales) indicating a Rhaetian–Early Jurassic age (Iwai et al., 1966; Kon'no and Asama, 1973).

Overlying that basal facies, and elsewhere unconformable on older rocks, is the *Nam Phong Formation* (shown on the representative stratigraphic column of part of the Khorat Group, Fig. 5). Soft, grey-red to pale red siltstone which makes up 70% of the formation is rarely exposed and only the thick resistant beds of sandstone and conglomerate crop out. In a zone 300 m thick above the middle of the formation, beds of conglomeratic sandstone and conglomerate contain pebbles of quartz, chert and red-brown siltstone. The sandstones are medium to fine grained below this conglomeratic zone and very fine grained above it. Cross-bedding is conspicuous in the conglomerates and in some of the upper sandstones. The top of the Nam Phong Formation is taken as the base of an 85 m thick interval of hard calcareous siltstone and limestone. At its type-section (Fig. 5) the thickness of the formation is 1456 m. Fossils are rare but the crustacean *Euestheria mansuyi* has been found at the base of laterally equivalent beds (Bunopas, 1971) and is believed by Kobayashi (1973) to indicate a Norian age.

The basal beds of the *Phu Kadung Formation* are pelletal, micritic, dolomitic limestones which have been described by Iwai (1972, 1973) and interpreted as of lacustrine origin. Mostly siltstone and cross-bedded sandstone occupy the remaining 900 m, or thereabouts, of the formation. Non-marine bivalves have been described by Hayami (1968) from this formation and possibly Early Jurassic marine reptilian teeth (plesiosaurs) by Kobayashi et al. (1963). The land plant *Araucarioxylon* has been reported by Kobayashi (1960) and probably was collected from this formation.

Forming a resistant and regionally extensive cap to the main Khorat Plateau escarpment is the *Phra Vihan Formation,* predominantly a cross-bedded sandstone. Some beds are conglomeratic and there are also minor siltstone and carbonaceous shale bands. Its thickness varies from 56 m to 136 m. Apart from silicified wood, the formation lacks fossils; the fossils listed by Japanese authors (e.g., Iwai et al. 1966) as

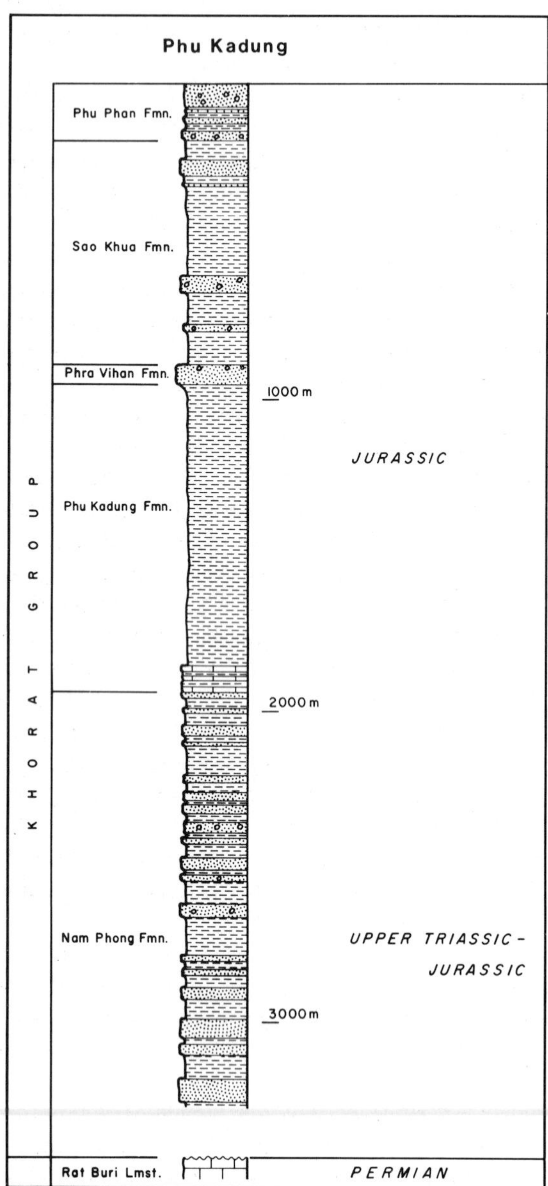

Fig. 5. Part of the Khorat Group at Phu Kadung, near the western edge of the Khorat Plateau (for location see Fig. 2). Based on Ward and Bunnag (1964).

being from this formation would be included by Ward and Bunnag (1964) in the Sao Khua Formation (see below). Northwest of the Khorat Plateau, in the Loei area, photogeological work by Bleakly et al. (1965) has brought to light a possible angular unconformity between the Phu Kadung and Phra Vihan Formations.

The *Sao Khua Formation* represents a return to a predominantly grey-red and brown siltstone succession. The thickness of the formation varies from 404 m to 702 m. Jurassic marine bivalves and ichthyosaur teeth have been described by Kobayashi et al. (1963) and non-diagnostic land plants by Kon'no and Asama (1973).

The *Phu Phan Formation* is predominantly variegated, cross-bedded, conglomeratic sandstone. Its thickness varies from 82 m to 183 m. Apart from some unidentified bone fragments in a caliche-siltstone conglomerate (Ward and Bunnag, 1964) no fossils have been found in this formation.

The *Khok Kruat Formation* comprises mostly siltstone and is poorly exposed, the best sections being in the cored boreholes described by Ward and Bunnag (1964). Thin beds of gypsum occur in the upper part and indicate the transitional boundary with the overlying gypsum- and halite-bearing Maha Sarakam Formation. The composite section compiled from boreholes by Ward and Bunnag (1964) is about 700 m thick. Non-marine bivalve faunas have been described by Kobayashi (1963b, 1968, 1973) from beds which appear to be correlatable with the Khok Kruat Formation and are considered by him to be Cretaceous. Land plants described by Endo and Fujiyama (1966) and Asama (1973) from probably correlative beds are dated by them as Late Cretaceous. In addition, indeterminate vertebrate remains are described by Ward and Bunnag (1964).

Gardner et al. (1967) introduced the name *Maha Sarakam Formation* for the uppermost unit of the Khorat Group. Its type-section is a well near Maha Sarakam at the centre of the Khorat Plateau. The well penetrated 610 m of salt-bearing beds without reaching the bottom and a maximum thickness of about 1000 m is suggested by Gardner et al. (1967). Although very poorly exposed the formation has been proved by boreholes to be very extensive over the Khorat Plateau, extending into Laos. The dominant lithology is soft, variegated, siltstone and mudstone. Halite occurs as white, pure, coarsely crystalline rock, forming beds from less than 1 cm to 250 m thick, and also disseminated through the siltstone and mudstone. Beds of gypsum up to 50 m thick occur in some wells but generally they are much thinner or present as disseminated grains and crystals or clusters of crystals. Anhydrite also occurs. The age of the

Maha Sarakam Formation is unclear since it contains only rare and indeterminate fragments of bone and wood, and some carbonaceous material. However, since it overlies and is transitional with the Cretaceous Khok Kruat Formation a Late Cretaceous age is probable, perhaps extending into the Early Tertiary.

Khorat Group elsewhere in Thailand

In the north of the country, sandstone, mudstone and conglomerate of red-bed facies overlie the Triassic marine Lampang Group (Baum et al., 1970; Piyasin, 1972). Locally the contact is disconformable above the Lampang Group (e.g., at Ban Thasi, Figs. 2 and 3) but elsewhere acid volcanic rocks intervene. No fossils have been found in these beds but their relations with dated Triassic rocks show that they are Late Triassic or younger. Thus lithostratigraphically and chronostratigraphically they correlate with the Khorat Group of the type-area. Westward tongues of Khorat Group have also been described in the marine Jurassic limestone succession (Baum et al., 1970) (Figs. 2 and 6).

On the Khmer border southeast of Bangkok, and forming the island of Ko Kut (Fig. 1), almost flat-lying Khorat Group has been mapped by the present writer. It unconformably overlies strongly folded presumed Triassic beds on the mainland but on Ko Kut its base is not seen. Sandstone is the main rock type and it occurs mostly as thick cross-bedded units, commonly with well-rounded quartz pebbles. It may also occur in thinner beds with alternations of red or grey mudstone in which case it is more grey and argillaceous and instead of large-scale cross-bedding there are ripple cross-lamination and load-cast structures. The only macrofossils (apart from worm burrows) found in the Khorat Group of southeast Thailand are vertebrate teeth and bones which occur on the east coast of Ko Kut in a half-metre thick, calcareous, carbonaceous sandstone. Dr. W.J. Clarke (pers. comm.) described the collection made by the writer as containing "stumpy peg-like teeth referable to *Lepidotes* and palatal pavement teeth of *Ptychodus* or *Coelodus* type, as well as saurian bone". A microflora was obtained from nearby grey mudstone which indicates an Early Cretaceous age (Dr. D. Batten, pers. comm.). Across the border in Cambodia, Gubler (1935) mapped the Khorat Group as Grès Supérieurs and measured 1400 m at Phnom Samkos. Its total thickness is unknown but may be of the same order as that in the Khorat Plateau.

Peninsular Thailand (Figs. 1 and 2) has a similar sandstone, mudstone and conglomerate succession which is also apparently predominantly non-marine; its basal contact is believed to be unconformable on the Permian Rat Buri Limestone (and locally on still older beds) and in places there is a coarse basal conglomerate containing cobbles of Permian limestone. However, its principal lithology is cross-bedded sandstone. An interesting pink and cream, micritic, massive limestone unit some tens of metres thick occurs in the sandstone, mudstone and conglomerate succession on the east flank of the broad syncline 90 km north of Kantang (Fig. 4). Algal or faecal pellets are present in the limestone but are not age-diagnositic. The only fossil to have been identified in the Khorat Group on the Peninsula is a bivalve from near Chumphon. Buravas (1961) called it *Astarte* and suggested a Triassic age but according to Hayami (1960) it is *Eomiodon chumphonensis* and of probable Jurassic age. On regional geological grounds and by comparison with the Jurassic Tembeling Formation of the Malay Peninsula (Burton, 1973), the latter age is preferred. Folding and faulting have affected these rocks throughout the Peninsula (Fig. 4). The thickness of the Khorat Group there is probably no more than 1000 m.

(d) Young, undeformed succession of the Peninsula (Upper Cretaceous–Tertiary)

In the low-lying country on the Gulf of Thailand side of the Peninsula is a group consisting of argillaceous sandstone, conglomerate, ripple-marked siltstone and mudstone (Figs. 1 and 2). Their colours range from white to red-brown and purple and a non-marine origin is likely. Apart from some indeterminate silicified wood no fossils have been found and an assessment of the age of the group has to be made on the basis of its lack of folding, apparent unconformity with the Khorat Group, and poor consolidation; it would thus seem to be of Cretaceous or Tertiary age. Its thickness onshore is likely to be no more than 200 or 300 m.

IGNEOUS ROCKS

The Mesozoic was a period of considerable igneous activity in Thailand, mostly of an acidic kind.

Scattered throughout the country (except in the

northeast) are stocks and elongate batholiths of granite, granodiorite, diorite, tonalite, monzonite and adamellite, generally grouped together and loosely called "granite". They trend in a generally north–south direction, are up to about 100 km long, and give rise to the highest mountains in the country including Doi Inthanon (2560 m) in the northwest. Most of the rocks contain abundant muscovite and biotite; hornblende is generally present while zircon, sphene, pyrite, apatite and tourmaline are common accessory minerals. All of the tin and tungsten deposits of Thailand are associated with these rocks and there are also deposits of gold, copper, molybdenum, iron, antimony, lead and zinc ores.

It had been thought (e.g., Brown et al., 1951; Klompé, 1962) that two different Mesozoic plutonic suites could be recognized, an older non-tin-bearing one of Triassic age and a younger tin-bearing suite of Cretaceous age. However, it now appears that with possible minor exceptions all of the batholiths and stocks are of Late Triassic age or older; moreover, apparently barren ones have since been found to have associated tin mineralization (e.g., Hughes and Bateson, 1967). Triassic sedimentary rocks have been intruded and, for example at Songkhla, have obvious metamorphic aureoles. However, nowhere can the Khorat Group be seen to have been intruded (with two possible local exceptions mapped, but not described by Baum et al., 1970); the Khorat Group of the Khorat Plateau is conspicuously lacking any granite intrusions although they intrude the pre-Khorat Group rocks around the western and southwestern margins.

Radiometric work on the granites of Thailand have produced inconclusive results. Rb/Sr (whole-rock) determinations published by Burton and Bignell (1969) gave Permian and Cretaceous ages for rocks in southeast Thailand and Jurassic–Cretaceous ages for rocks on the Peninsula. Their K/Ar (mineral) determinations on the same rocks gave conflicting ages, some older and others younger. The authors conclude that spurious results are likely to have been caused by heating or tectonism after emplacement.

Braun (1969) states that the most abundant granite in northern Thailand is Late Triassic. He cites an example which intrudes *Halobia*-bearing Middle Triassic rocks and is overlain by the Khorat Group; a K/Ar (mica) determination on it gave an age of 212 Ma, that is Late Triassic. However, a petrologically identical rock, but having less certain field-relations, gave a Rb/Sr (whole-rock) age of 266 Ma, that is, Permian. His conclusion is that those dates are probably valid and represent granite emplacement of Permian and Late Triassic age. In the same paper he describes Rb/Sr-determined ages as young as 62 Ma for some northern granites. He concludes that there must therefore also have been a phase of Early Tertiary granite intrusions, notwithstanding their petrological similarity to his "normal" Triassic granite. In the present writer's view the explanation more consistent with the regional field evidence is that the youngest granite intrusions in Thailand are Late Triassic and that any younger dates obtained are the result of later tectonism. Such re-setting of the "radiometric clock" to give spuriously young dates is widely realised to apply in the case of K/Ar determinations, while Bottino et al. (1970) have shown that, with respect to Rb/Sr also, igneous bodies do not necessarily behave as closed systems after crystallization.

Extrusive igneous rocks of Mesozoic age are principally andesite, porphyritic rhyolite and rhyolitic tuff. In northern Thailand they are of two ages: Late Permian–Early Triassic and Late Triassic.

The former locally underlies the Triassic Lampang Group (Fig. 3) and reaches a maximum thickness of about 100 m according to Piyasin (1972). He ascribes a submarine origin to the rocks because they are interbedded with marine sandstones and he states that tuffaceous debris in the Lampang Group indicates waning activity through the Triassic. Baum et al. (1970) describe acid volcanic rocks with the same field-relations elsewhere in northern Thailand. However, further south, volcanicity of that age is only known from one locality: on the Khmer border, at Khlong Pong Nam Ron, the present writer has found a red-green rock described by Dr. R. Walls (pers. comm.) as possibly a devitrified volcanic glass with variolitic texture but now extensively chloritized. It contains inclusions of fossiliferous Permian limestone and from its field-relations appears to be a Late Permian–Early Triassic submarine volcanic rock.

Andesite, porphyritic rhyolite and tuff of Late Triassic age are more widespread. In northern Thailand these rocks occur locally at the base of the Khorat Group disconformably overlying the Triassic Lampang Group (Bleakly et al., 1965; Baum et al., 1970; Piyasin, 1972). Around the western margin of the Khorat Plateau they unconformably overlie Permian and older rocks and are themselves unconformably overlain by the Khorat Group (Fig. 2). The relationships of the tuffs, agglomerates and rhyolite mapped by the writer on the coast 220 km southeast of Bangkok and forming the island of Ko Chang are less clear, however. Their contact with underlying dark grey shale and sandstone has been seen and since the sedimentary rocks are inferred as being of Triassic age, a Late Triassic age is inferred for the volcanic rocks too. The whole of Ko Chang, with hills over 700 m high, is apparently made of rhyolite and this figure can be considered a minimum thickness there.

A single isolated exposure of porphyritic rhyolite has been found by the writer in the granite mountain range on the Peninsula, 155 km northwest of Songkhla. Its age is assumed to be Late Triassic.

GEOLOGICAL HISTORY

In late Permian times shelf-carbonates were being deposited in warm seas over much of Thailand although, in the north, terrigenous clastic rocks were also laid down. That was the closing phase of a sedimentary cycle which began in the mid-Palaeozoic. The precursor of Mesozoic orogenesis is evident in the local disconformity at the base of the Triassic and the acid volcanicity in northern Thailand. Variolitic basalt on the Khmer border southeast of Bangkok is of about the same Late Permian–Early Triassic age, but in southern Thailand at that time there was no volcanic activity.

Early Triassic marine fossils (conodonts) have been found locally in the Lampang Group in the north, while in other places in the north the Lampang Group can be inferred to be of Early Triassic age because of an apparently unbroken succession up from the Permian (Fig. 6). In southeast Thailand and on the Peninsula, although no Lower Triassic fossils have been found, it is suggested that deposition of marine terrigenous clastic rocks (of Lampang Group lithofacies) had begun. Little or no detailed sedimentological work has been done but it is evident from the fauna, the apparently high rate of deposition, the repetitious graded-bedding and the pyrite which is present that deep-water marine conditions prevailed. Influxes of sand occurred at times, by turbidity currents and grain-flow. Such a suite of rocks is here described as flysch (Fig. 6).

The marine limestone of western Thailand (including the western part of the Peninsula) has not yielded any specifically Early Triassic faunas but rocks of that age cannot be ruled out (Fig. 6). The interfingering relationship of the two Triassic lithofacies, limestone and marine terrigenous clastic rocks, indicates that the source of the latter sediments must have lain in the east.

Middle Triassic fossils have been found in the Lampang Group in the north of the country, and the rocks of similar flysch facies in the southeast and at the southern end of the Peninsula are assumed to be partly of the same age. Close to the Burmese border Middle Triassic conodonts have been found in the limestone characterizing that area.

The limestone of the Kantang area of the Peninsula is Triassic but a more precise age cannot be determined; its argillaceous and pyritic content indicate a low-energy environment while the abundant calcisponge and stromatoporoid fauna suggests a shallow-water, perhaps lagoonal, facies. Its possible correlative, the Kodiang Limestone of northwest Malaysia, is believed by De Coo and Smit (1975) to be of deep-water origin and is of particular interest in that slump-bedding is evidence of a sloping sea-bed, presaging the Late Triassic orogenesis and gravity-gliding.

Late Triassic. The same bipartite facies distribution (limestone in the west and terrigenous clastic marine rocks in the east) continued into the Late Triassic (Fig. 6). Fossil collections of this age have been made from the limestones of the Burmese border, the Lampang Group in the north and similar beds at the southern end of the Peninsula.

The Late Triassic was a time of major orogenesis, perhaps the main Phanerozoic event of the Southeast Asian Peninsula, and called by writers in Laos, Khmer and Vietnam, the Indosinian Orogeny. Folding occurred along trends varying either side of north–south. It was more intense in the east where the unconformity with the overlying Khorat Group is greatest (Fig. 7). Meanwhile in the west, on the Burmese border, Late Triassic shelf-environment Kamawkala limestone was being deposited, apparently without a break. If the folding of the Triassic flysch belt had been a subaerial event accompanied by erosion the detritus produced would be expected to have extended far to the west, replacing the area of limestone deposition. The fact that the facies belts remained and were not shifted (Fig. 6) suggests that the folding was not subaerial but submarine, probably

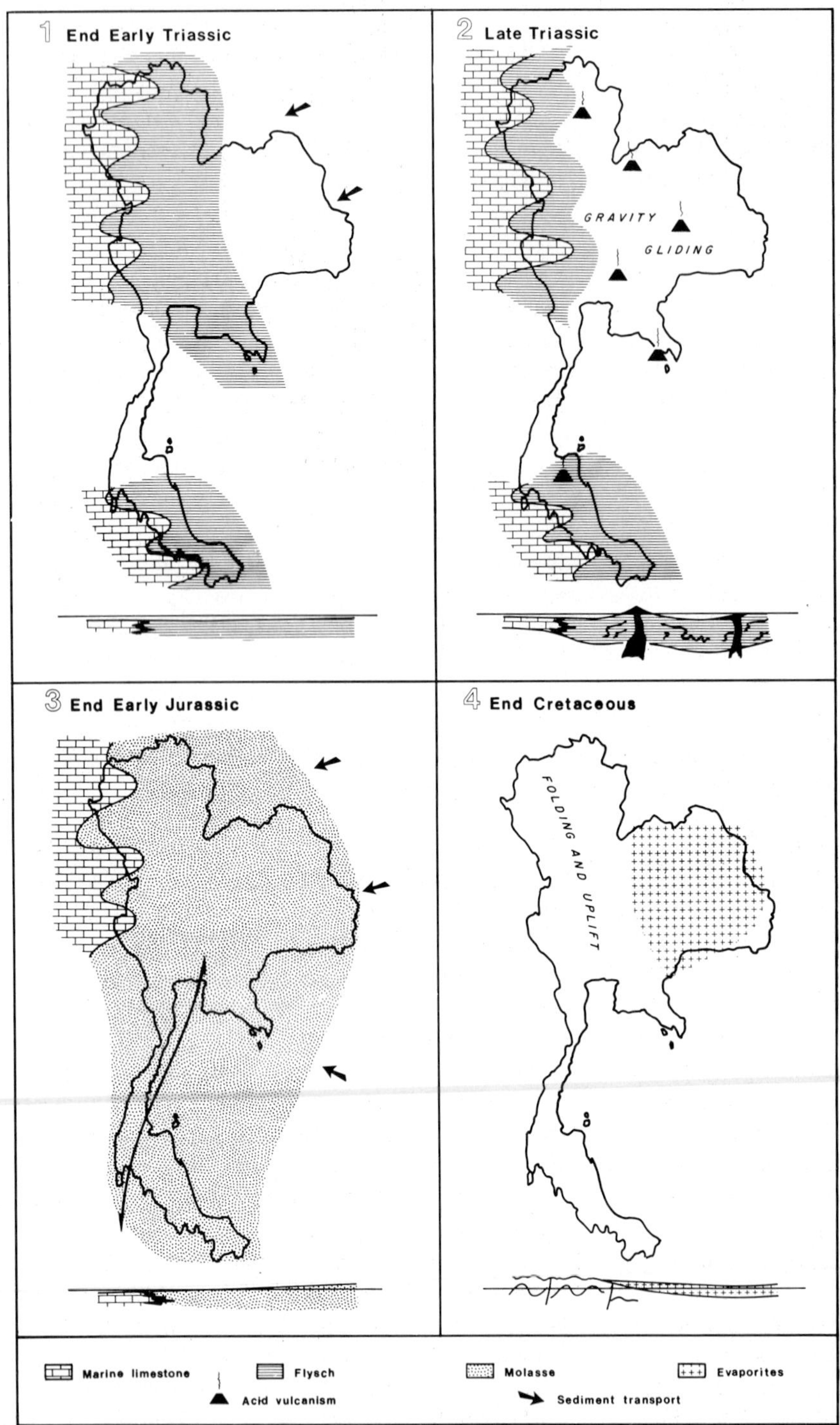

Fig. 6. Palaeogeographic reconstructions of Thailand during the Mesozoic. The section beneath each map is a schematic east–west section across northern Thailand at that time. Note that the maps are not palinspastic, i.e. movement on the wrench-faults has not been reversed.

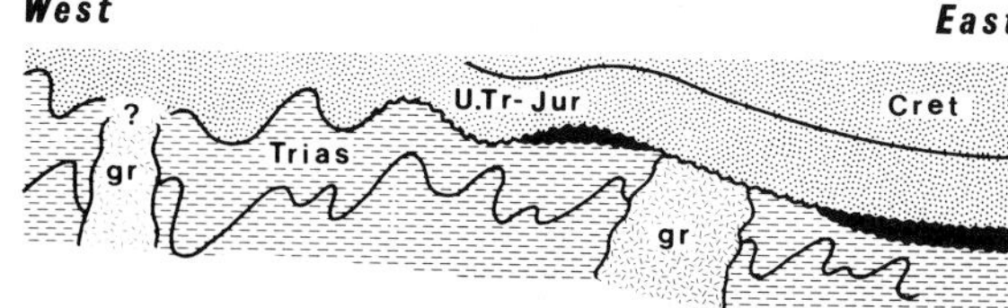

Fig. 7. Diagrammatic east–west section across northern Thailand to show the westward migration of folding. In the east there was strong folding in the Late Triassic but only slight warping thereafter. In the west there are no major unconformities in the Mesozoic succession and the folding was of Late Jurassic or younger age. (The diagram is not to scale, and the ornaments have no lithological significance.)

the result of gravity-gliding down a tilted surface. It is probably significant that in Burma gravity-gliding of about the same age has been inferred by Brunnschweiler (1970) and Garson et al. (1972). What therefore must have been essentially vertical crustal movements were accompanied by the emplacement of acid plutonic stocks and batholiths in roughly north–south aligned belts covering the length and breadth of the country. Acid lavas and tuffs were the surface manifestation of that igneous activity.

At some time after the igneous activity major wrench-faulting took place. The pronounced bend of the Thailand Peninsula is the site of a north-north-east-trending fault, named by Garson and Mitchell (1970) the Khlong Marui Fault (Fig. 8). Other similar trending faults crossing the Peninsula are the Ranong and Kapoe faults. By matching the chains of acid igneous plutons, the regional strike, and areas of probable Precambrian gneiss and schist, about 250 km of sinistral transcurrent displacement can be inferred on the fault-belt (Fig. 8).

The question of its northward continuation through Thailand is equivocal. A broad anticlinal belt of largely pre-Mesozoic outcrops along the western edge of the Khorat Plateau (Fig. 1) may be related to the presence of a fault-belt and this possibility is supported by east–west Permo-Carboniferous facies differences mapped in the Loei area by Bleakly et al. (1965). The alternative, that the Khlong Marui Fault belt has itself been offset sinistrally by the Tertiary–Quaternary northwest–southeast-trending Thai-Burma Fault Belt, was proposed by Ridd (1971).

Although an older limit on the age of the Khlong Marui belt of wrench-faulting can be established by the displacement of the Late Triassic plutonic rocks, a younger limit cannot be determined directly. The Khlong Marui, Ranong, and Kapoe Faults cannot be detected in the Oligocene–Recent rocks in the Gulf of Thailand (Woollands and Haw, 1976) which therefore narrows down the time of possible faulting. A Late Triassic age would best account for the difficulty in detecting the fault belt north of the Gulf of Thailand but the possibility remains of it being as young as Early Tertiary. On Fig. 6 it is shown as having moved in the Jurassic.

After the Late Triassic tectonic and igneous activity the Khorat Group was deposited, representing a long period of molasse sedimentation. The age of the lowest beds of the Khorat Group has been variously put at Norian to Early Jurassic. But in view of the magnitude of tectonic and igneous events which occurred between the dated Norian marine deposition and deposition of the lowest Khorat Group, an age as early as Norian for the onset of molasse deposition seems unlikely; a Rhaetian or Early Jurassic age is therefore preferred. Meanwhile in the northwest of the country limestone deposition continued, perhaps with a break at the Triassic–Jurassic boundary (Fig. 6).

Jurassic times saw predominantly non-marine sands and mudstones laid down in a lacustrine and fluviatile environment in the east of Thailand and on the Peninsula. These rocks show considerable uniformity of lithofacies, varying mainly in the proportions of sandstone, conglomerate and mudstone. Although the terrestrial plant remains and non-marine bivalves are clear indications of the predominantly non-marine environment of deposition, it is also evident from the marine bivalves, ichthyosaur and plesiosaur bones that the sea periodically transgressed the coastal plain or, at least, that marine animals entered the hinterland up the rivers.

An attempt was made by Iwai (1968) to map the cross-bedding directions in the Khorat Group of the Khorat Plateau, but no consistent pattern emerged. However, toward the Burmese border, marine limestone deposition continued until at least as late as the Aalenian. The limestone tongues in the Khorat Group of northern Thailand mapped by Baum et al. (1970) show that the boundary between the limestone facies belt in the west and the molasse belt in the east was not fixed but that it fluctuated. The palaeogeography implied by that facies distribution again indicates that the source of the great thickness of Jurassic molasse must have lain in the east (Fig. 6).

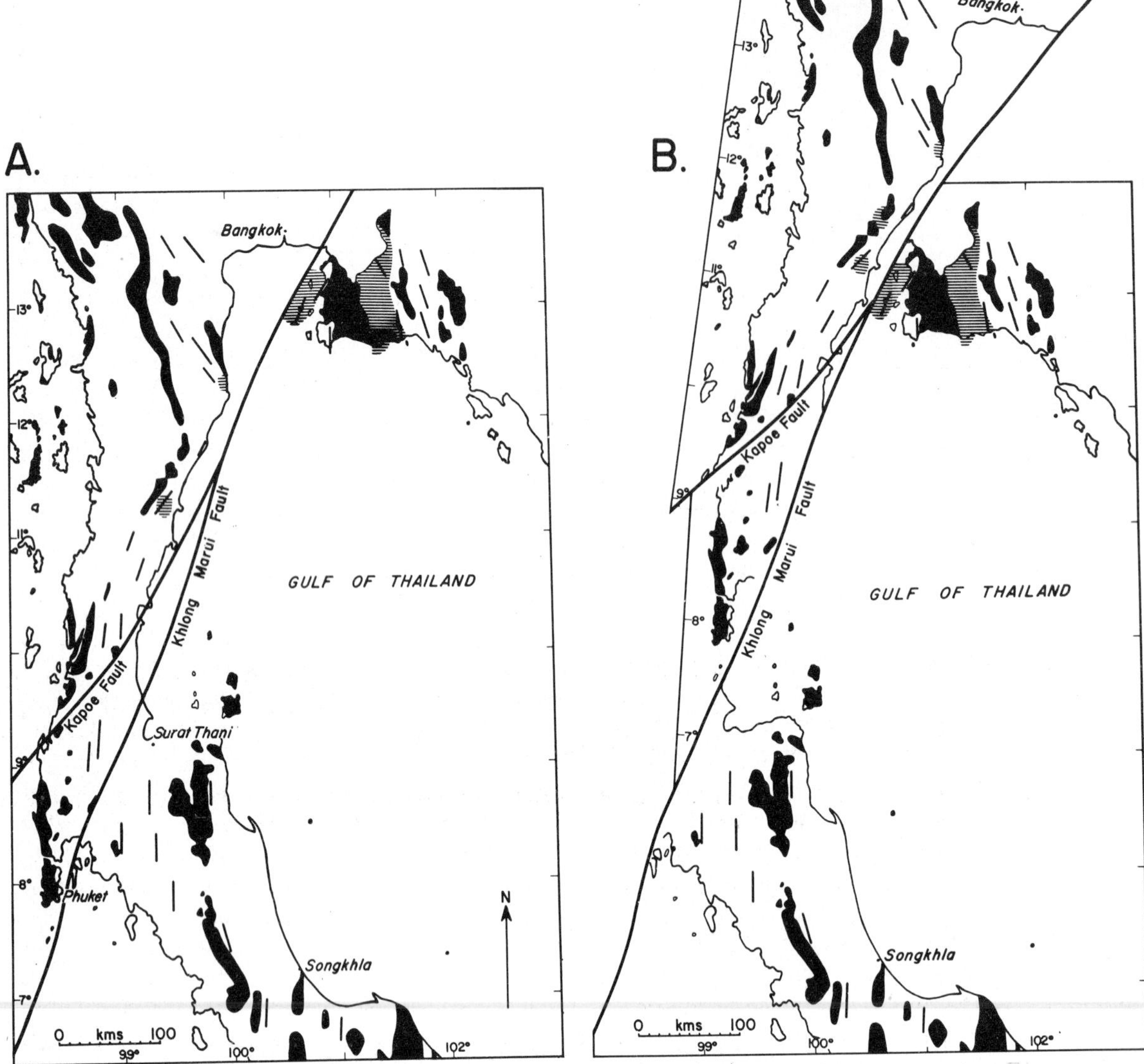

Fig. 8. A. Southern Thailand showing present-day geological relations. Acid igneous plutons are shown black. ?Precambrian regional metamorphic rocks are horizontally lined, and structural strike is shown as short lines. B. Palinspastic geological map showing relations before sinistral transcurrent movement on the Kapoe and Khlong Marui Faults, i.e. in the Late Triassic. Note the alignment of plutons and strike trends and the juxtaposition of ?Precambrian regional metamorphic rocks.

Cretaceous rocks have only been identified on the Khorat Plateau and southeast of Bangkok, near the Khmer border. In both areas molasse sedimentation (the Khorat Group) continued from the Jurassic, with the same fossil indicators of non-marine and marine conditions. On the island of Ko Kut, close to the Khmer border, the present writer noted that cross-bedding showed currents to have flowed mainly from the southeast quadrant, supporting the sediment-source deduced for the Jurassic.

It is not known with certainty when the folding took place which affected the Jurassic limestones of the northwest and the Jurassic Khorat Group of peninsular Thailand (Fig. 7). If it occurred in the Cretaceous it might account for the Cretaceous salt deposition of the Khorat Plateau by having provided an elevated barrier between the Mesozoic seas in the west and the salt basin; this interpretation is adopted in Fig. 6.

SUMMARY

To summarise the Mesozoic history of Thailand in the context of the whole of the Southeast Asian Peninsula it is useful to invoke plate-tectonic theory. The present writer believes (Ridd, 1976) that in Triassic times the Sundaland plate (i.e., essentially the Southeast Asian Peninsula with parts of Borneo, Java and Sumatra) was carried as an entity northeastwards having broken away from Gondwanaland in the mid-Palaeozoic. On its leading edge island arcs were present and provided a source for the flysch deposition over much of the plate. On the trailing edge, remote from that source, carbonate deposition prevailed. In the Late Triassic, Sundaland collided with mainland Asia along the Song Ma line which runs northwest from the Gulf of Tongking. Deformation of the plate accompanied by acid igneous activity were the result, as the plate buckled and fusion of its lower levels took place. The Jurassic and Cretaceous saw the deposition of a thick molasse wedge build out southwestwards from the collision zone. But meanwhile on the southwestern side of the plate a new compressional continent–ocean plate boundary was developing. That resulted in waves of folding entering Thailand from the west and uplifting that part of the country to produce, for the only time in the Mesozoic, a barrier between the Tethyan Ocean and the area of declining molasse deposition. In the Late Cretaceous, playa-lake conditions prevailed on what had become the land-locked Khorat Plateau basin and thick beds of halite were deposited.

ACKNOWLEDGEMENTS

Field work in southern Thailand, on which part of this account is based, was carried out on behalf of the British Petroleum Company Limited and the writer thanks the Chairman and Directors for permission to publish some of the findings. The writer was accompanied in the field by Mr. A.C.J. Wainwright and geologists of the Thai Department of Mineral Resources. Dr. G.F. Elliott of the British Museum (Natural History) and Dr. D. Batten of the University of Aberdeen kindly examined palaeontological collections.

REFERENCES

Asama, K., 1973. Some younger Mesozoic plants from the Lom Sak Formation, Thailand. *Geol. Palaeontol. Southeast Asia,* 13: 39–46.

Baum, F., Braun, E. Von, Hahn, L., Hess, A., Koch, K., Kruse, G., Quarch, H. and Siebenhuner, M., 1970. On the geology of Northern Thailand. *Beih. Geol. Jahrb.,* 102: 1–23.

Bleakly, D., Stephens, A.E., Cratchley, C.R., Workman, D.R., Newman, D., Cogger, N., Sanevong, P., Thanvarachorn, P., Intrakhao, B. and Chaungpaisal, S., 1965. *The Regional Geology of the Loei–Chiengkarn Area of Thailand and Detailed Investigations of the Phu-Khum Lead–Zinc Mineral Prospects.* U.K. Overseas Geological Surveys and Thai Royal Dep. Min. Res. Rep.

Borax, E. and Stewart, R.D., 1965. Notes on the Khorat series of Northeastern Thailand. *ECAFE Third Symp. Dev. Pet. Resources of Asia and the Far East, Tokyo, 1965.*

Bottino, M.L., Fullagar, P.D., Fairbairn, H.B., Pinson, W.H. and Hurley, P.M., 1970. The Blue Hills Complex Massachusetts: whole-rock Rb/Sr open systems. *Geol. Soc. Am. Bull.,* 81: 3739–3746.

Braun, E. Von, 1969. On the age of the granites in northern Thailand. *Proc. Second Tech. Conf. Tin, Bangkok,* pp. 151–157.

Brown, G.F., Buravas, S., Charaljavanaphet, J., Jalichandra, N., Johnston, W.D., Sresthaputra, V. and Taylor, G.C., 1951. Geologic reconnaissance of the mineral deposits of Thailand – geologic investigations in Asia. *Bull. U.S. Geol. Surv.,* 984.

Brunnschweiler, R.O., 1970. Contributions to the post-Silurian geology of Burma (Northern Shan States and Karen State). *J. Geol. Soc. Aust.,* 17 (Pt. 1): 59–79.

Bunopas, S., 1971. On the geology and stratigraphy of the Nam Phrom Dam and its vicinities. *Geol. Soc. Thailand Newsl.,* 4 (1–3).

Buravas, S., 1961. Stratigraphy of Thailand. *Proc. Ninth Pac. Sci. Congr.,* pp. 301–305.

Burton, C.K., 1967. Graptolite and tentaculite correlations and palaeogeography of the Silurian and Devonian in the Yunnan–Malaya Geosyncline. *Trans. Proc. Palaeontol. Soc. Jap., N.S.,* No. 65: 27–46.

Burton, C.K., 1973. Mesozoic. In: D.J. Gobbett and C.S. Hutchinson (Editors), *Geology of the Malay Peninsula.* Wiley-Interscience, New York, N.Y., pp. 97–141.

Burton, C.K. and Bignell, J.D., 1969. Cretaceous–Tertiary events in Southeast Asia. *Geol. Soc. Am. Bull.,* 80: 681–688.

Chonglakmani, C., 1972. Stratigraphy of the Triassic Lampang Group in Northern Thailand. *Geol. Soc. Thailand Newsl.,* 5 (5–6).

Cotter, G. de P., 1923. The oil shales of eastern Amherst,

Burma, with a sketch of the geology of the neighbourhood. *Indian Geol. Surv. Rec.*, 55 (pt. 4): 275–286.

De Coo, J.C.M. and Smit, O.E., 1975. The Triassic Kodiang Limestone Formation in Kedah, West Malaysia. *Geol. Mijnbouw*, 54: 169–176.

Endo, R., 1966. Calcareous algae from Thailand. *Geol. Palaeontol. Southeast Asia*, 2: 289–296.

Endo, S. and Fujiyama, I., 1966. Some Late Mesozoic and Late Tertiary plants and a fossil insect from Thailand. *Geol. Palaeontol. Southeast Asia*, 2: 301–304.

Gardner, L.S., Haworth, H.F. and Chiangmai, P.N., 1967. Salt resources of Thailand. *Dep. Min. Resour., Bangkok Rep. Invest.*, No. 11.

Garson, M.S. and Mitchell, A.H.G., 1970. Transform faulting in the Thai Peninsula. *Nature*, 228: 45–47.

Garson, M.S., Bradshaw, N. and Rattawong, S., 1969, Lepidolite pegmatite in the Phangnga area of Peninsular Thailand. *Proc. Sec. Tech. Tin Conf., Int. Tin Council, Bangkok*, pp. 1–14.

Garson, M.S., Mitchell, A.H.G., Amos, B.J., Hutchison, D., Soe, K., Myint, P. and Pau, N.C., 1972. Economic geology and geochemistry of the area around Neyaungga and Yengan, southern Shan States, Burma. *U.K. Inst. Geol. Sci. Overseas Div. Rep.*, No. 22.

Garson, M.S., Young, B., Mitchell, A.H.G. and Tait, B.A.R., 1975. Geology of the tin belt in Peninsular Thailand around Phuket, Phangnga, and Takua Pa. *U.K. Inst. Geol. Sci. Overseas Mem.*, No. 1.

Gregory, J.W., 1930. Upper Triassic fossils from the Burma–Siamese frontier. The Thaungyin Trias and description of corals. *Rec. Geol. Surv. India*, 63: 155–167.

Gubler, J., 1935. Etudes géologiques au Cambodge Occidental. *Bull. Serv. Geol. Indochine*, 22 (2).

Haile, N.S. and Tarling, D.H., 1975. Palaeomagnetic reconnaissance study of Mesozoic rocks from the Khorat Plateau, Thailand. *Pac. Geol.*, 10: 101–103.

Haworth, H.F., Chiangmai, P.N. and Phiancharoen, C., 1966. Ground water resources development of Northeastern Thailand. *Thailand Dep. Min. Resour., Ground Water Bull.*, 2.

Hayami, I., 1960. Two Jurassic pelecypods from West Thailand. *Trans. Proc. Palaeontol. Soc. Jap., N.S.*, No. 38: 284.

Hayami, I., 1968. Some non-marine bivalves from the Mesozoic Khorat Group of Thailand. *Geol. Palaeontol. Southeast Asia.*, 4: 100–108.

Heim, A. and Hirschi, H., 1939. A section of the mountain ranges of Northwestern Siam. *Eclog. Geol. Helv.*, 32(1): 1–16.

Hogbom, B., 1914. Contributions to the geology and morphology of Siam. *Upps. Univ., Geol. Inst. Bull.*, 12.

Hughes, I.G. and Bateson, J.H., 1967. Reconnaissance geological and mineral survey of the Chanthaburi area of Southeast Thailand. *U.K. Inst. Geol. Sci., Rep.* No. 7.

Iwai, J., 1968. The sedimentary structures observed in rocks of the Khorat Group and overlying formation. *Geol. Palaeontol. Southeast Asia*, 5: 166–172.

Iwai, J., 1972. Pelletal limestone of the Phu Kadung Formation, Mesozoic Khorat Group, Thailand. *Geol. Palaeontol. Southeast Asia*, 10: 257–263.

Iwai, J., 1973. Dolomitic limestone of the Phu Kadung Formation, Mesozoic Khorat Group, Thailand. *Geol. Palaeontol. Southeast Asia*, 12: 173–178.

Iwai, J., Asama, K., Veeraburas, M. and Hongnusonthi, A., 1966. Stratigraphy of the so-called Khorat Series and a note on the fossil plant-bearing Palaeozoic strata in Thailand. *Geol. Palaeontol. Southeast Asia*, 2: 179–196.

Iwai, J., Hongnusonthi, A., Asama, K., Kobayashi, T., Kon'no, E., Nakornsri, N., Veeraburas, M. and Yuyen, W., 1975. Non-marine Mesozoic formations and fossils in Thailand and Malaysia. *Geol. Palaeontol. Southeast Asia*, 15: 191–218.

Klompé,, H.F., 1962. Igneous and structural features of Thailand. *Geol. Mijnbouw*, 41: 290–302.

Kobayashi, T., 1960. Notes on the geologic history of Thailand and adjacent territories. *Jap. J. Geol. Geogr.*, 31: 129–148.

Kobayashi, T., 1963a. *Halobia* and some other fossils from Kedah, Northwest Malaya. *Jap. J. Geol. Geogr.*, 34: 113–128.

Kobayashi, T., 1963b. On the Cretaceous Ban Na Yo fauna of east Thailand with a note on the distribution of *Nippononaia, Trigonoides* and *Plicatounio*. *Jap. J. Geol. Geogr.*, 34: 35–43.

Kobayashi, T., 1968. The Cretaceous non-marine pelecypods from the Nam Phung Dam Site in the northeastern part of the Khorat Plateau, Thailand, with a note on the Trigonioididae. *Geol. Palaeontol. Southeast Asia*, 4: 109–138.

Kobayashi, T., 1973. A Norian conchostracan from the basal part of the Khorat Group in Central Thailand. *Proc. Jap. Acad.*, 49 (10): 825–828.

Kobayashi, T. and Tokuyama, A., 1959. The Halobidae from Thailand. *J. Fac. Sci.Univ. Tokyo*, 12: 27–30.

Kobayashi, T., Takai, F. and Hayami, I., 1963. On some Mesozoic fossils from the Khorat Series. *Jap. J. Geol. Geogr.*, 34: 187–191.

Kobayashi, T., Burton, C.K., Tokuyama, A. and Yin, E.H., 1966. The *Daonella* and *Halobia* facies of the Thai-Malay Peninsula compared with those of Japan. *Geol. Palaeontol. Southeast Asia*, 3: 98–122.

Koch, K.E., 1973. Geology of the region Sri Sawat – Thong Pha Phum – Sangkhlaburi (Kanchanaburi Province, Thailand). *Geol. Soc. Malaysia Bull.*, 6: 177–185.

Kon'no, E. and Asama, K., 1973. Mesozoic plants from Khorat, Thailand. *Geol. Palaeontol. Southeast Asia*, 13: 149–171.

Kummel, B., 1960. Triassic ammonoids from Thailand. *J. Palaeontol.*, 39: 682–694.

La Moreaux, P.E., Javanaphet, J.C., Jalichon, N., Chiengmai, P.N., Bunnag, D., Thaviori, A. and Rakprathum, C., 1959. Reconnaissance of the geology and groundwater of the Khorat Plateau, Thailand. *U.S. Geol. Surv. Water-Supply Pap.*, 1429: 407–415.

Lee, W.M., 1927. Outline of the geology of Siam with reference to petroleum. *Am. Assoc. Pet. Geol. Bull.*, 11: 144.

Pitakpaivan, K., 1955. Occurrence of Triassic Formation at Mae Moh. *Dep. Min. Resour. Bangkok, Rep. Invest.*, No. 1: 47–57.

Pitakpaivan, K., Ingavat, R. and Paritwatvorn, P., 1969. Fossils of Thailand. *Dep. Min. Resour. Bangkok, Geol. Surv. Mem.*, 3: 1–111.

Piyasin, S., 1972. Geology of Lampang Sheet NE 47-7. *Dep. Min. Resour. Bangkok, Rep. Invest.,* No. 14.

Ridd, M.F., 1971. Faults in Southeast Asia, and the Andaman Rhombochasm. *Nature,* 229 (2): 51–52.

Ridd, M.F., 1976. Southern Thailand and its place in Southeast Asian plate reconstructions. *J. Geol. Soc. Lond.,* 132: 361–362 (abstract and discussion).

Sato, T., 1975. Marine Jurassic formations and faunas in Southeast Asia and New Guinea. *Geol. Palaeontol. Southeast Asia,* 15: 151–189.

Tamura, M., Hashimoto, W., Igo, H., Ishibashi, T., Iwai, J., Kobayashi, T., Koike, T., Pitakpaivan, K., Sato, T. and Yin, E.H., 1975. The Triassic System of Malaysia Thailand and some adjacent areas. *Geol. Palaeontol. Southeast Asia,* 15: 103–150.

Trauth, F., 1930. Upper Triassic fossils from the Burma–Siamese Frontier. On some fossils from Kamawkala Limestone. *Rec. Geol. Surv. India,* 63: 174–176.

Ward, D.E. and Bunnag, D., 1964. Stratigraphy of the Mesozoic Khorat Group in Northeastern Thailand. *Dep. Min. Resour., Thailand Rep. Invest.,* No. 6.

Weir, J., 1930. Upper Triassic fossils from the Burma–Siamese frontier – brachiopoda and lamellibranchia from the Thaungyin River. *Rec. Geol. Surv. India,* 63 (pt. 1): 168–173.

Woollands, M.A. and Haw, D., 1976. Tertiary stratigraphy and sedimentation in the Gulf of Thailand. *Proc. Offshore Southeast Asia Conf., 1976, Seapex, Singapore.*

Chapter 6

THE INDONESIAN AND PHILIPPINE ARCHIPELAGOS

M.G. AUDLEY-CHARLES

INTRODUCTION

Available information

The land area of the Indonesian Archipelago is 1,950,963 km^2 and the land area of the Philippine Archipelago is 300,000 km^2. Mesozoic rocks are widely scattered among the thousands of islands that make up these archipelagos (Fig. 1) although they are not usually exposed in the very small islands where Late Cainozoic rocks predominate. Very little published information is available about the Mesozoic rocks in the offshore parts of these archipelagos. Many areas mapped as Mesozoic outcrop are known to contain other undifferentiated rocks. In most cases the age of the Mesozoic rocks is not known very

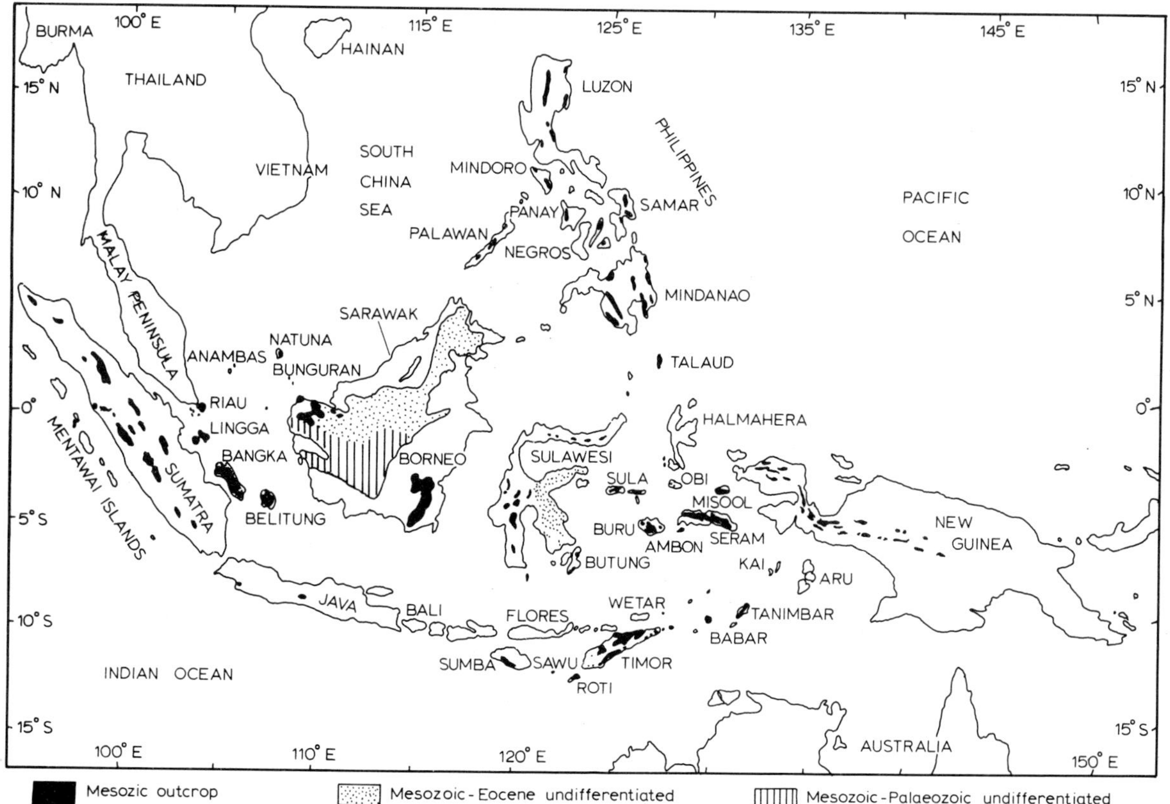

Fig. 1. Present distribution of Mesozoic outcrops in the Indonesian and Philippine archipelagos.

precisely and only rarely can the rocks be referred to a biostratigraphical stage with confidence. In some cases even the Mesozoic age of the rocks is in doubt. There is in general very little *detailed* biostratigraphical or lithostratigraphical information about the Mesozoic rocks of Indonesia as may be seen from the data provided by Van Bemmelen (1949) and Marks (1956), and the same applies to the Philippines (Teves, 1956; and Gervasio, 1973). Locally however, as for instance in Sarawak (Leichti et al., 1960) and in Misool (Van Bemmelen, 1949), the lithology and fauna have been recorded in some detail, and in Sarawak formations have been mapped over considerable distances. The reconnaissance quality of much of the information about the Mesozoic rocks of these archipelagos results partly from the scattered distribution of the numerous islands and the often poor exposures. Another factor that has led to the lack of detailed information about the Mesozoic rocks is that they have long been regarded as generally lacking economic prospects. Furthermore in Sumatra, Java and the Philippines the Mesozoic rocks have often been metamorphosed, while in the islands of the Outer Banda Arc they have been strongly folded into complex structures inhibiting the determination of their stratigraphy. As a consequence of this there are very few *detailed* stratigraphical sections on which precise correlations can be built and from which it is possible to develop palaeogeographical interpretations in which speculation can be kept within usually acceptable proportions.

Earlier palaeogeographical interpretations

Umbgrove (1938) published a map for each system of the Mesozoic on which the known facies were plotted but without attempting to distinguish autochthonous from allochthonous elements nor indicating any crustal movements. These maps were supported by a summary of the available stratigraphical information. In these maps Umbgrove pointed out the presence of what he called the Timor–East Celebes geosyncline (or Banda geosyncline), in which Mesozoic rocks very different in facies and subsequent deformation history were distinguished from Mesozoic rocks in all other parts of Indonesia.

Teichert (1939) pointed out that the facies and faunas of Umbgrove's Timor–East Celebes geosyncline are very closely related to those of the Carnarvon and Canning basins of west and northwest Australia. Teichert argued that the Westralia geosyncline and the Timor–East Celebes geosyncline formed a single long sinous belt (Fig. 2) connecting Western Australia through Timor, Misool and Seram to East Celebes (i.e., east Sulawesi). It is of particular interest to note that in a later paper Teichert (1941) remarked on the Permian faunas of Timor being greatly different from those of Western Australia. These Timor Permian faunas are very similar in some important respects to those found in Sumatra, Borneo, Malay Peninsula and the Philippines (Stauffer and Gobbett, 1972). This apparent paradox can be easily understood when it is recognised that the Timor Permian faunas, made famous by Wanner's *Paleontologie von Timor* (1914–1929), were obtained either from thrust sheets of Permian strata or from Permian olistoliths in the Bobonaro Scaly Clay olistostrome (Audley-Charles, 1968), although this was not recognised at that time. Some of the apparent difficulties in understanding the Timor Mesozoic faunas and those of other islands of the Outer Banda Arc, where faunas of the same age from very different facies have been found very close together, can now be understood in terms of the Australian faunas belonging to para-autochthonous strata and the Asian faunas belonging to the allochthonous elements forming both thrust sheets and olistoliths (Audley-Charles and Carter, 1974; Carter et al., 1976a, b; Barber et al., 1977). Runnegar (personal communication, 1976) suggested that the brachiopods from the Timor allochthon show too great an affinity with those of Western Australia for them to be regarded as belonging to the Asian province. Runnegar's argument is based on the distribution of brachiopod families and neglects the more powerful palaeographical argument based on striking features of the populations of the faunal and floral assemblages.

The interpretation of Mesozoic palaeogeography put forward in this chapter is an extension of the work of Umbgrove and Teichert, who first pointed out that the islands of the Banda Arc and Sula Spur formed a distinct palaeogeographical feature during the Mesozoic. The detailed information acquired by recent work has allowed Asian and Gondwanan elements in these islands of eastern Indonesia to be distinguished.

Most palaeogeographical interpretations of the last thirty years have tended to overlook the work of Teichert and Umbgrove. Most recent maps show all the Indonesian islands as part of Asia in a fixistic rela-

Fig. 2. Westralia–East Celebes Geosyncline as defined by Umbgrove (1938) and Teichert (1939). This Mesozoic geosyncline was identified on the basis of it containing facies and faunas uniquely different from those of western Indonesia and the Philippines. These Australian (Gondwanan) facies in eastern Indonesia form part of the Australian para-autochthonous imbricate zone in the islands of the Outer Banda Arc. Asian Mesozoic facies are also found in these Banda Arc islands but always in an allochthonous position.

tionship with each other, separated from Australia during the Mesozoic by a wide Tethys Ocean (Smith et al., 1973), probably because of the difficulty in understanding the origin of the mixture of Asian and Australian elements in the Banda Arc islands. The world maps for the Mesozoic published by Smith et al. (1973) are most useful for any study of Indonesian palaeogeography because they are based on palaeomagnetic data and so provide geophysical evidence for Australia's Mesozoic position. McElhinny et al. (1974) and Haile et al. (1977) have provided palaeomagnetic evidence for the Mesozoic position of the Malay Peninsula. By using these two sources of information the northern and southern margins of the Indonesian Archipelago can be tied to palaeomagnetic latitudes during the Mesozoic.

Ben-Avraham and Uyeda (1973), in a discussion of sea-floor magnetic anomalies in the South China Sea, indicated the extent of young Mesozoic marginal sea-floor spreading, which they interpreted as having been responsible for moving the whole island of Borneo southwards away from the Asian margin during its Late Jurassic spreading. At the southern margin of Indonesia Falvey (1972) reported magnetic anomalies on the floor of the Wharton Basin which could be interpreted as indicating a spreading system in this northeast part of the Indian Ocean. Veevers and Heirtzler (1974) discussed this Late Jurassic spreading system in terms of palaeogeographical evolution of the sea floor in the region of Sumba and Timor. Audley-Charles (1975) argued that this magnetic data implies the Late Jurassic movement of Sumba along faults postulated by Warris (1973) on the basis of Mesozoic facies distribution on the northwest Australian shelf.

Most palaeogeographical interpretations since Umbgrove (1938), for example Pupilli (1973), Ben-Avraham and Uyeda (1973) and Geiger (1963) have been concerned with local regions of the archipelago, while others have concentrated on defining palaeotectonic rather than palaeogeographical features, for example Hutchison (1973), Haile (1973), Klompé (1961), Katili (1971) and Stauffer (1974).

Present structural configuration: Its relevance to Mesozoic palaeogeography

Particular difficulties are encountered in trying to interpret the palaeogeographical evolution of the pre-Cainozoic rocks of this region because of the results of Cainozoic tectonics. These two archipelagos are among the most tectonically complicated parts of the world. In an attempt to present a coherent history of the palaeogeographical development of the region, it is proposed to consider the area in terms of two postulates, that the present configuration of the islands, seas and continental margins in this region can be attributed partly to two different processes: (1) the Late Cainozoic collision of the Australian continent with the Banda Arc in eastern Indonesia, and (2) the migration of groups of islands within the archipelago away from the continental margin of Asia during the Cainozoic (Dickinson, 1973). These two very different types and directions of movement within the archipelagos have brought about important changes in the distribution of the Mesozoic rocks, so that the interpretation of their original site of deposition and the reconstruction of Mesozoic geography can only be achieved if Cainozoic tectonics are understood. That requires not only a knowledge of Cainozoic history of the various arcs within the archipelagos, it also requires knowing something of the evolution of the marginal seas, which may have developed during the Cainozoic and possibly some seas which may have been totally or partially consumed at converging margins during the last 65 m.y. The stratigraphy and palaeogeography is then a test of the proposed model. Although this approach appears to ignore any other interpretation, with such dispersed and meagre information it does provide a coherent and simplistic picture of a very complex region.

No published palaeomagnetic data for Mesozoic rocks within the archipelagos are known and the only published magnetic data for the sea floors within this region seems to be limited to the Wharton Basin of the northeast Indian Ocean (Falvey, 1972) and the somewhat ambiguous information from the South China Sea (Ben-Avraham and Uyeda, 1973).

One of the consequences of the Late Cainozoic collision between the Australian continent and eastern Indonesia (Audley-Charles and Carter, 1974) is the presence in the islands of the Outer Banda Arc of both Asian and Gondwanan facies (Carter et al., 1976a). This is a key factor which will be discussed in detail later (see pp. 170–194) but may be summarized here by noting that in the Banda Arc islands Australian facies Mesozoic strata occupy a para-autochthonous position where they are overlain by entirely allochthonous Mesozoic strata having un-

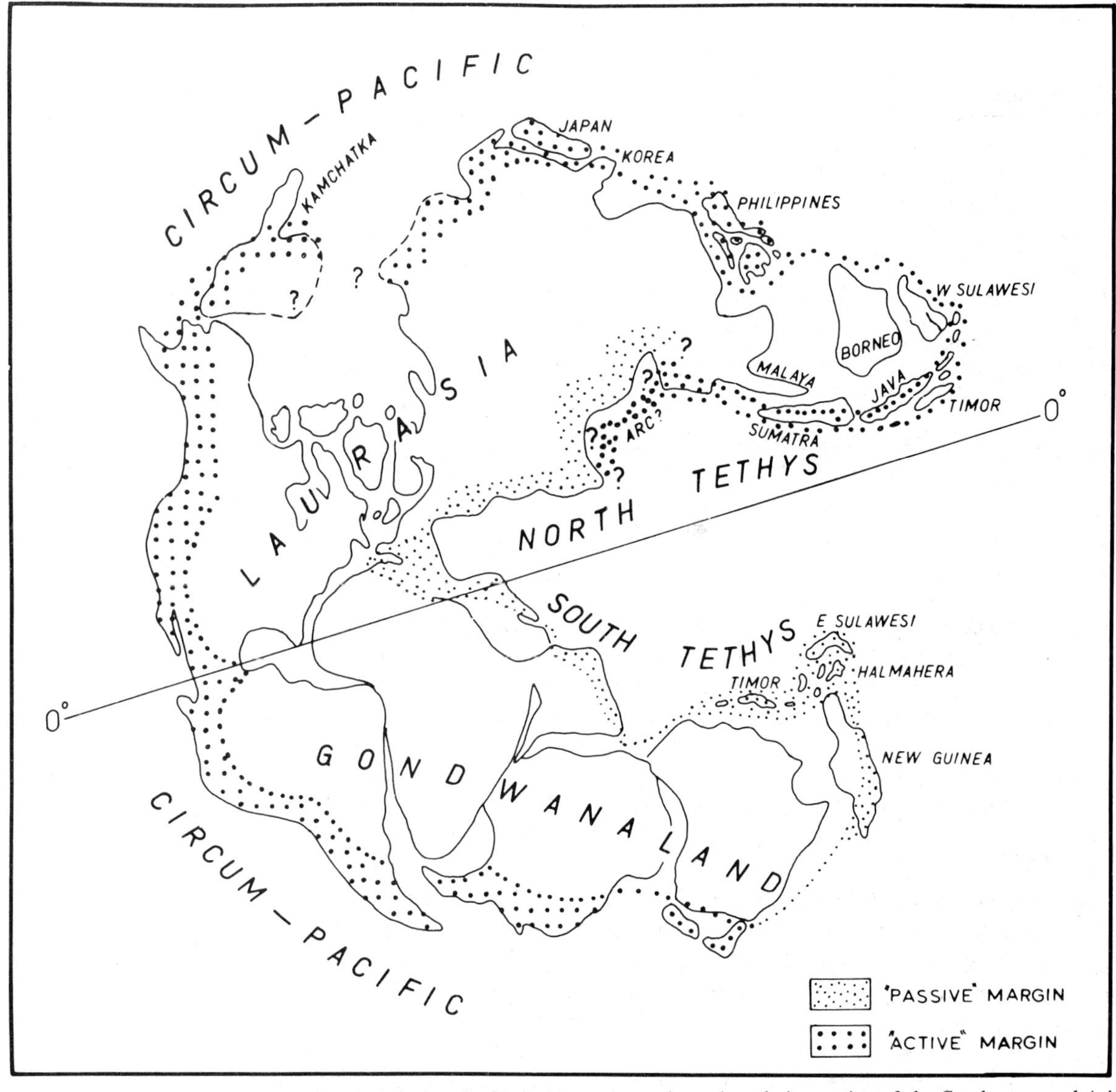

Fig. 3. Simplified reconstruction of Tethys during the Early Mesozoic to show the relative setting of the Gondwanan and Asian parts of Indonesia and the Philippines during the Mesozoic. Present-day outlines are for reference only.

doubted Asian affinities. Evidence will be set forth later to show that the Asian allochthonous facies must have been deposited in a tropical marine environment while the Australian para-autochthonous facies accumulated under much cooler conditions (Audley-Charles and Carter, 1974; Fitch and Hamilton, 1974). These observations imply that parts of Indonesia occupied very widely separated climatic zones during the Mesozoic and that the rocks representing these different zones have been brought together by converging plates during the Cainozoic. It will be argued later in this chapter that rocks in the island of Timor which are now only a few metres apart were originally separated by about 2800 km of the Tethys Ocean (Fig. 3). Unfortunately there is very little available geophysical data to constrain this interpretation, but what little there is supports what for the present is a palaeogeographical speculation built mainly on the available geological information and this forms the basis for the palaeogeographical interpretation in this chapter. This appears to be justified because any fixistic interpretation conflicts with the available faunal and lithological evidence and with the structural relations of the rocks. It is not the large-scale continental drift concept applied to this region that is in serious doubt here; it is the lack of detailed information to define more precisely the original palaeogeographical setting of the Asian rocks now found resting on the Australian strata in the Banda Arc islands.

STRATIGRAPHICAL CORRELATION OF THE MESOZOIC ROCKS OF THE INDONESIAN AND PHILIPPINE ARCHIPELAGOS

The dominating problem encountered in trying to make stratigraphical correlations of Mesozoic rocks in Indonesia and the Philippines has already been discussed in the Introduction, where it was attributed to the general lack of biostratigraphical details and lack of published details of key stratigraphical sections. This limits the preparation of Mesozoic correlation charts to a consideration of the rocks in terms only of Early, Middle and Late divisions of the three periods. In many cases even these are in doubt. Average or maximum thicknesses of the stratigraphical divisions are seldom known and many divisions do not have formal lithostratigraphical names, in some cases formal names have been applied but are considered by later workers to be ambiguous; in these instances they are not included in the charts.

In some sections of the chart the presence of particular fossils have been mentioned not for the purpose of indicating the age of the rocks but because they are a significant constituent of the facies and may indicate something of the conditions of deposition. The basis for the age of the divisions is dealt with by the list of references for each part of the chart.

The greatest source of stratigraphical information about the Mesozoic rocks of most of Indonesia remains the *Geology of Indonesia* by Van Bemmelen (1949). Some useful supplementary information has been put together by Marks (1956) in *Lexique Stratigraphique International.* In some areas great advances in our knowledge have been made, for example in Sarawak, Brunei and Sabah regions of Borneo (Leichti, 1960; and in the annual reports of the Geological Survey of Malaysia). In the Irian Jaya region of New Guinea much new information was provided by Visser and Hermes (1962). Considerable advances in the knowledge of Mesozoic stratigraphy of the Philippines has been summarised by Gervasio (1973) although it remains very much of a reconnaissance quality.

The extensive petroleum exploration in western Indonesia since about 1968 has not yet resulted in much published stratigraphical detail about the Mesozoic rocks. This is largely because the Mesozoic, often metamorphosed in this region, is widely regarded as uneconomic basement for petroleum exploration. It is unfortunate that the very large offshore area of the Sunda shelf, Java Sea and South China Sea, which has been the target for much exploration, has not provided much new detailed published information about the Mesozoic.

One very useful advance has been the radiometric dating of the Mesozoic granites and metamorphic rocks in the Sunda shelf, Java Sea and its islands (Katili, 1973a; Haile and Bignell, 1971).

One particular problem encountered in preparing these correlation charts was the very uncertain age of the basic-ultramafic igneous suites, sometimes called ophiolites. It often seems from the literature that the age attributed to these rocks is based on tectonic or petrologic concepts rather than on direct evidence. The recent summary of the ophiolites of western Indonesia (Hutchison, 1975) indicates the uncertain age of many attributed to the Mesozoic.

The preparation of the correlation charts for the islands of the Banda Arc presented special problems, resulting from the presence of allochthonous, and para-autochthonous facies of the same age; this has already been discussed briefly in the Introduction to this chapter. In the case of Timor, where some detailed studies of the Mesozoic have been carried out during recent years, the differences in the allochthonous and para-autochthonous facies are generally well known and can be attributed to the Asian or Australian provinces with some confidence (Figs. 4–6). In other islands of the Banda Arc, such as Seram, where the last published field work on these rocks was carried out in 1919, it is very difficult to be confident into which structural element or palaeogeographical province some of the rocks should be placed. In Timor five allochthonous Mesozoic elements can be clearly distinguished, some of which are thrust sheets and some olistoliths in the Bobonaro Scaly Clay olistostrome. Nevertheless, in Timor some important uncertainties remain, for instance there is insufficient evidence to be certain whether the strata immediately

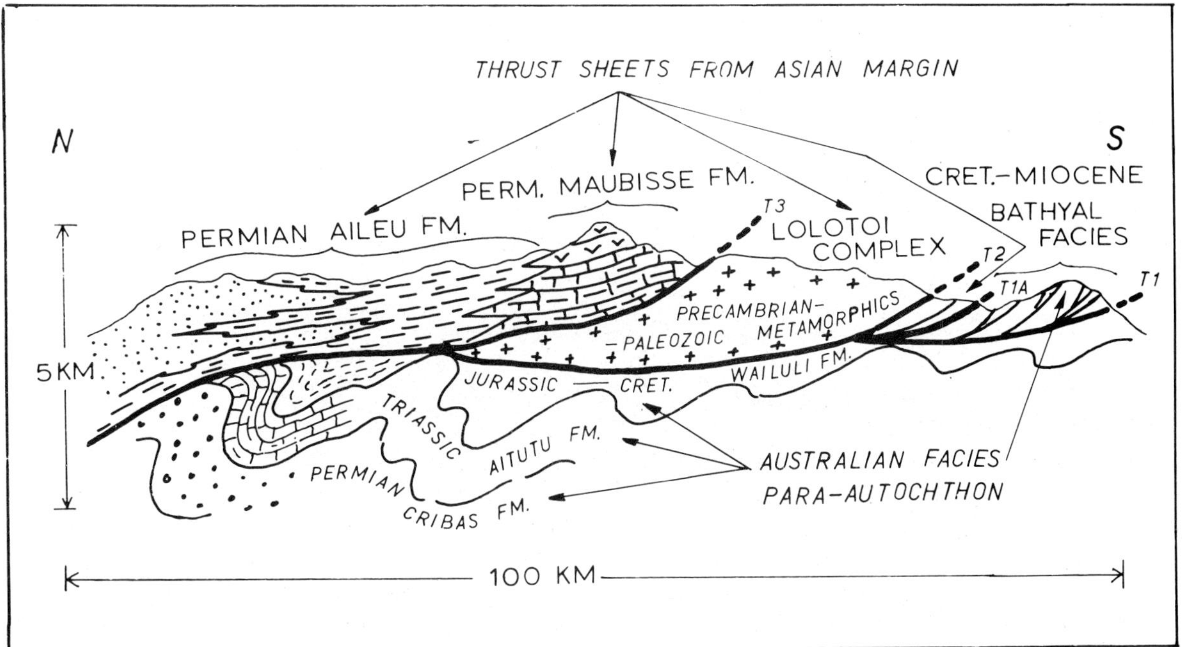

Fig. 4. Schematic section through Timor to illustrate the stratigraphy of the allochthonous Asian elements thrust over the para-autochthonous Australian facies of the Gondwana margin. (After Audley-Charles, 1976.)

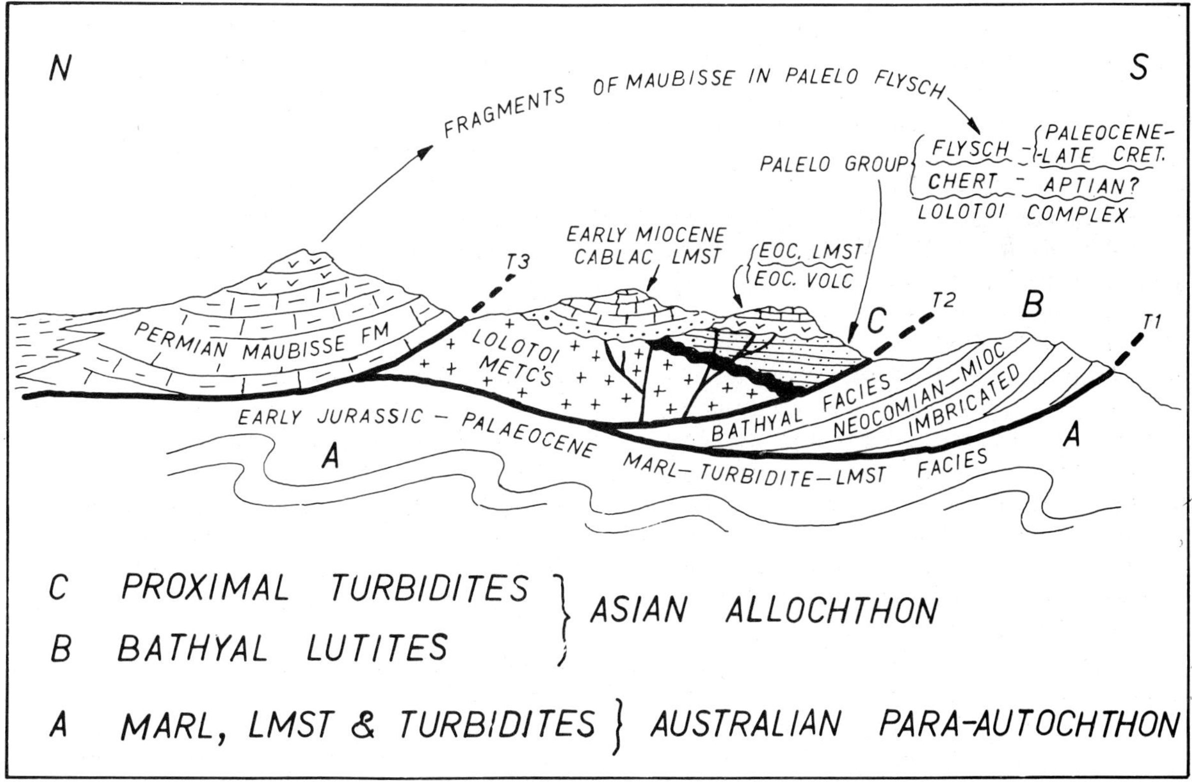

Fig. 5. Schematic section through Timor to illustrate the detailed relationships of the Asian thrust sheets. (After Audley-Charles, 1976.)

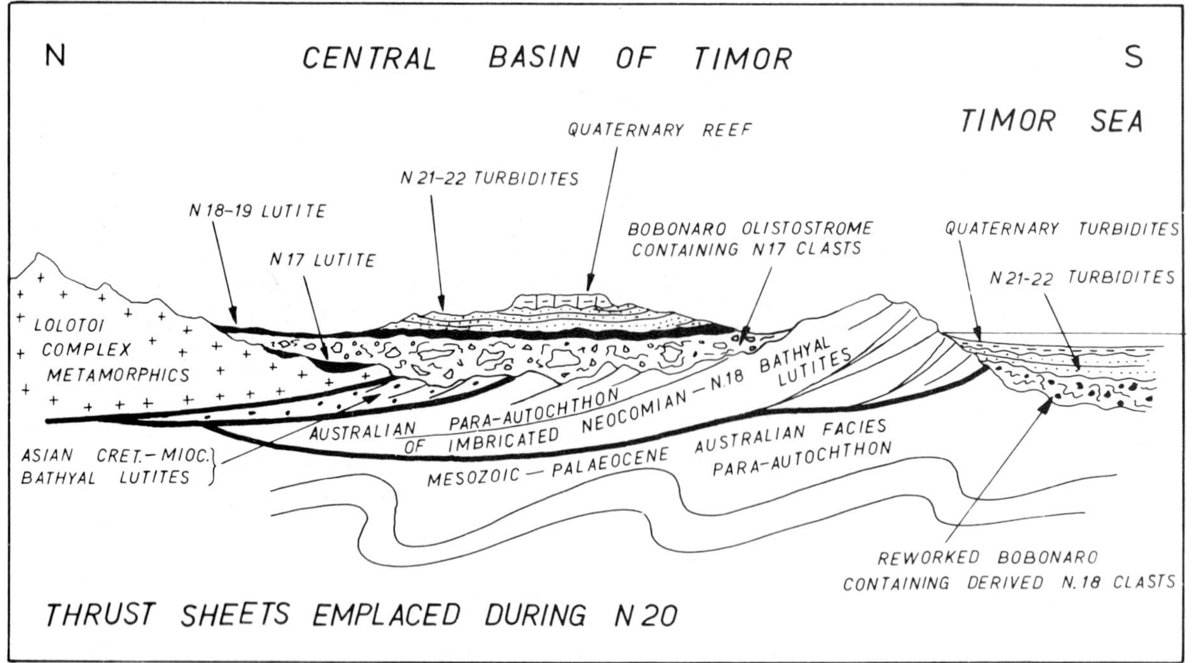

Fig. 6. Schematic section through western Timor to illustrate some details of the relationships between the Asian allochthonous elements, Australian para-autochthon and the post-thrusting autochthon. (After Audley-Charles, 1976.)

below the *Stomiosphera* and *Cadosina* bearing mudstones are allochthonous. These rocks are included in the correlation charts with the para-autochthon but they might be part of the allochthon. In Timor three distinct Mesozoic para-autochthonous elements are recognised which have been structurally condensed into a narrow geographical zone by the compressive forces of the Late Cainozoic orogenesis resulting from the collision of the northern part of the Australian margin with the Asian island arc (Carter et al., 1976a). The rocks reported from Seram can be matched with many of the different allochthonous and para-autochthonous types in Timor, but their stratigraphy and structural relations have not been worked out sufficiently to permit correlations with the detailed subdivisions established in Timor. Furthermore, the recognition of an olistostrome in Seram (Audley-Charles et al., in prep.) comparable in many respects with the Bobonaro Scaly Clay olistostrome places considerable uncertainty on any attempt to interpret the stratigraphical sequences in Seram from published accounts.

One outstanding biostratigraphical problem in Seram discussed by Van der Sluis (1950) and Van Bemmelen (1949) concerns the question whether the thick section of flysch facies containing rocks with *Lovcenipora* are Late Triassic or Late Jurassic in age; recent work suggests they are Triassic. The problem goes beyond that because what is called the Greywacke Formation and what is sometimes referred to as the flysch facies may be identical in places and very different elsewhere. The three main authors who have discussed the stratigraphy of Seram (Valk,1945; Germeraad, 1946; Van der Sluis, 1950) appear to have different views about what may be the same sequences of rocks. Recent work (Audley-Charles et al., in prep.) suggests that it is most likely as Van Bemmelen (1949) proposed that both Triassic and Jurassic flysch facies are present in Seram. Comparison with Timor suggests that some of the Greywacke Formation maybe equivalent to the Aileu Formation of Timor and be derived from the Asian continental margin of Sundaland (Barber and Audley-Charles, 1976; Carter et al., 1976a). In summary, the correlation charts for all the islands of the Outer Banda Arc must be regarded as tentative and considerable detailed stratigraphical and structural mapping will be required before the Mesozoic rocks of these islands can be precisely correlated with Timor and Misool, and there remain many unsolved problems with the Mesozoic of Timor. From the east arm of Sulawesi Kundig (1956) described Cretaceous and Eocene sections of "bathyal" sediments that may be correlated with the imbricated Kolbano sections of Timor and imbricated Nief sections of Seram. Despite the lack of detailed information of these rocks the presence of Mesozoic facies that can be correlated with very similar rocks in all or most of the islands of the Outer Banda Arc (Umbgrove, 1938; Teichert, 1939; Van Bemmelen, 1949) provides support for the basic palaeogeographic concepts presented in this chapter.

Mesozoic correlation charts (pp. 177–192)

As explained above some of the ages to which rocks have been ascribed in these charts are open to doubt. The basic data on which these charts have been constructed is to be found in the references listed below the appropriate chart.

In these charts listed rock types are not in stratigraphic sequence except where they are separated by horizontal lines which imply superposition of the rocks or groups of rocks above the horizontal line. Intrusive igneous rocks are shown in capitals, extrusive igneous rocks in lower case.

RECONSTRUCTION OF MESOZOIC PALAEOGEOGRAPHY

Palaeogeographical reconstruction of the Australian (Gondwanan) part of Mesozoic Indonesia

It was argued by Teichert (1939) that the islands of the Outer Banda Arc and the Sula Spur of eastern Indonesia contain Mesozoic rocks that were originally deposited at the margin of the Australian shelf. Subsequent work in Timor led Audley-Charles (1965) and Crostella and Powell (1975) to support Teichert's suggestion with further evidence. Oil company exploration drilling and seismic reflection surveys on the present Australian shelf have provided much more evidence for the continuity of terrigenous Mesozoic facies between the present Australian shelf and Timor (Crostella and Powell, 1975). Analysis of gravity data in Timor, Timor Trough and the Australian shelf has been shown by Chamalaun et al. (1975) to indicate that Australian crust is continuous under the Timor Sea and extends to northern Timor. Of particular interest is the remarkable similarity in Triassic palynological sequences in Timor and Australian shelf (Crostella and Powell, 1975). As Umbgrove (1938) and Teichert (1939) have shown, Mesozoic facies in all the islands of the Outer Banda Arc and Sula Spur are so similar, that it must follow that if Timor formed part of the Australian continental margin during the Mesozoic then so did all the other islands of the Outer Banda Arc and the Sula Spur.

Smith et al. (1973) on the basis of palaeomagnetic data for eastern Australia have shown that Australia formed part of Gondwanaland throughout the Permian and Mesozoic, occupying throughout this time a much more southerly position than at present (Fig. 3). From their plot of the Mesozoic position of Australia it can be seen that the area of the present northwest shelf opposite Timor was about 35°S throughout the Mesozoic.

The Mesozoic reconstruction for the Indonesian islands which contain Mesozoic Australian facies (shown in Fig. 7) is based on the view that during the Late Cainozoic the islands of the Banda Arc acquired their sinuosity, so that during the Mesozoic they were closer together, having been pulled apart by extension around the Cainozoic arc. The suggestion that the Banda Arc acquired its sinuosity is based on two main factors: the presence in Timor and Seram, now occupying opposite sides of the Banda Arc, of large overthrust sheets of metamorphic rocks whose direction of movement indicates they were derived from a continent that must have lain on what is now the Banda Sea side of these islands (Van Bemmelen, 1949; Barber and Audley-Charles, 1976). The second line of evidence, for the sinuosity of the Banda Arc having been acquired after the Mesozoic, is that the presence of considerable thicknesses of terrigenous sediments of Australian facies in the islands of eastern Sulawesi, Buru, Seram, Misool as well as in the Tanimbar–Timor group, implies an important source of clastic sediment such as could only be provided by the erosion of a continent. The source of this Mesozoic sediment is thought to have been northeast Australia for the Tanimbar–Timor group. For the islands of Sula, Obi and Misool, which were called the Sula Spur by Klompé (1961), much of the Mesozoic sediment is thought to have been derived locally from within what is called here the Sula Peninsula (which incorporates the Sula Spur).

The details of the reconstruction of the Mesozoic Sula Peninsula are speculative. There is insufficient evidence of any kind for controlling the geographical details of the inter-island relationships. The recon-

Fig. 7. Diagrammatic Mesozoic reconstruction of Indonesia and the Philippines. Palaeolatitudes are taken from McElhinny et al. (1974) and Smith et al. (1973). The details of the Mesozoic geography of the Sundaland Peninsula of Asian Indonesia and of the Sula Peninsula of Australian Indonesia are speculative. Note the Mesozoic position of the present allochthonous Asian elements (shown in black) now found in the islands of the Outer Banda Arc. The reconstruction of the Vogelkop (Tjendrawasih) of New Guinea follows the suggestions of Hermes (1968). The reconstruction of the Asian and Australian parts of Indonesia is modified after Carter et al. (1976a). Present-day outlines are used for reference only and have no palaeogeographical significance.

struction shown in Fig. 7 is schematic in this Peninsula and a great deal of field work will be necessary before such geographical details can be reconstructed with confidence.

The Mesozoic position of the islands from Tanimbar to Timor has probably not moved significantly relative to the Australian shelf as far as can be judged by the facies pattern and isopachytes of the Mesozoic and Permian strata (Lofting et al., 1975).

The Mesozoic position of Sumba is very difficult to interpret because so little is known about the pre-Cainozoic rocks of the island (Van Bemmelen, 1949). It has recently been interpreted (Audley-Charles, 1975) as a part of the Australian continental margin that became partly detached during the Late Jurassic as a result of sea-floor spreading in the Wharton Basin (Falvey, 1972; Veevers and Heirtzler, 1974). The interpretation of Sumba as part of Mesozoic Australia was based on facies pattern and sediment source direction (Warris, 1973); there is no available diagnostic faunal evidence.

It is very important to recognise that all the Mesozoic Australian facies in Timor are para-autochthonous. It seems very likely from the available literature that the Mesozoic Australian facies have this structural condition throughout all the islands of the Outer Banda Arc, although it is possible that locally they might be autochthonous. In the islands of the Sula Spur these rocks are autochthonous (Van Bemmelen, 1949), and in Irian Jaya they appear to be autochthonous (Visser and Hermes, 1962).

Palaeogeographical reconstruction of the Asian (Laurasian) part of Mesozoic Indonesia

Palaeomagnetic evidence from the Malay Peninsula (McElhinny et al., 1974) has shown that it occupied a position about 15°N during the Late Palaeozoic and throughout the Mesozoic. The presence in Sumatra, Borneo (De Neve, 1961), Malay Peninsula (Gobbett and Hutchison, 1973) and the Philippines (Gervasio, 1973) of abundant Permian marine faunas in dominantly shallow-water carbonate sediments with a rich calcareous algal flora, large foraminifera, waagenophyllid corals and oldhaminid brachiopods associated with a high diversity of other invertebrate forms and reef facies, establishes beyond doubt that these islands must have occupied tropical latitudes during the Permian (Stauffer and Gobbett, 1972).

These rich faunas and micro-floras in a shallow marine carbonate facies ranging from Early to Late Permian in age are found in the exclusively allochthonous Maubisse Formation of Timor (Audley-Charles, 1968; Wanner, 1914–1929) and in Roti, Leti and Babar islands of the Outer Banda Arc (Van Bemmelen, 1949). The similarity in lithofacies and faunas seems to be overwhelming evidence indicating these allochthonous Maubisse Permian rocks in the Banda Arc were originally deposited in the Permian tropics. Several other lines of evidence have been adduced (Carter et al., 1976a) in support of the suggestion that the Maubisse Formation probably accumulated close to or along the strike of the very similar facies, which are also associated with alkaline volcanics, at the margins of Permian Sundaland in Sumatra. Similarly warm-water Asian facies of Triassic and Jurassic age occur in thrust sheets and as olistoliths on the islands of the Outer Banda Arc and possibly on Misool (see correlation chart). Their tectonic detachment from Asia and their subsequent mode of emplacement on Timor and other islands of the Outer Banda Arc during the Late Cainozoic has been discussed by Carter et al. (1976a).

It is difficult to find direct evidence from which to reconstruct the detailed geography of any large area of the Asian part of Mesozoic Indonesia. The details of the reconstruction of the Mesozoic Asian allochthonous elements of the Outer Banda Arc islands of Sawu, Roti, Timor, Tanimbar, Seram and Buru are only schematic (see Figs. 10A, 11A, 12A). Their geographical position during the Mesozoic is not known precisely, there being no available evidence to control the details of the reconstruction.

Various lines of evidence indicate the broad geographical outline of Mesozoic Asian Indonesia. The pre-Cainozoic rocks of Sumatra (Adinegoro and Hartoyo, 1975; De Costa, 1975; Van Bemmelen, 1949) and Java (Patmosukismo and Yahya, 1975; Van Bemmelen, 1949) are igneous and metamorphic rocks with limestones of Palaeozoic and Mesozoic age. The metamorphosed sedimentary rocks contain much greywacke, quartzite and other clastic rocks that imply an important source of terrigenous sediment that must have been on the Asian mainland side of these islands. There is a strong pattern discernible in the Mesozoic rocks of Sumatra that indicate increasing oceanic influence to the west towards the present Indian Ocean (see Figs. 10A, 11A, 12A). Exploration drilling by oil companies in the Sunda shelf and Java Sea (Adinegoro and Hartoyo, 1975; De

Costa, 1975; Patmosukismo and Yahya, 1975; Katili, 1973b; Pupilli, 1973) has found similar pre-Cainozoic igneous and metamorphic rocks. These observations, together with the presence of the belt of Mesozoic granites of Triassic, Jurassic and Cretaceous age (Katili, 1973a; Haile and Bignell, 1971) that extend from the Malay Peninsula, Riau and Lingga archipelagos to Bangka and Belitung and Java Sea, as well as the presence of Mesozoic granites intruding central, north and south Sumatra (Katili, 1973a), and the other area of Mesozoic granites in Anambas, Natuna and Tembelan islands of the western part of the South China Sea around western Borneo, suggest this whole region of western Indonesia is underlain by continental crust that has existed since at least the Late Palaeozoic. This conclusion is supported by seismic refraction studies in the Sunda shelf area (Dash et al., 1972; Sander et al., 1975) where the crust is about 22 km thick. Recent tectonic reconstructions for this region have placed Mesozoic subduction systems outside this area (Fig. 8) of the Malay Peninsula, west Borneo and Sumatra (Haile, 1973; Hutchison, 1973), as there is no evidence of any important suture implying large displacements within this part of western Indonesia; on the contrary what evidence there is outlined above and shown in Figs. 10A, 11A and 12A suggests this zone has behaved as a continental unit since the Late Palaeozoic.

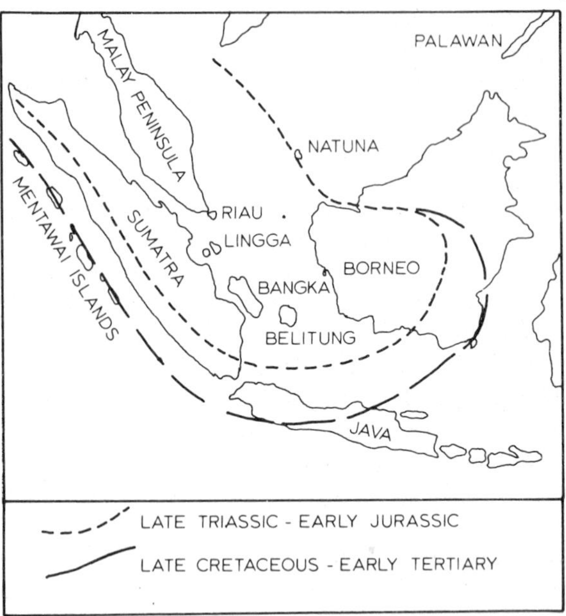

Fig. 8. Hutchison's (1973) view of the Mesozoic palaeographical evolution of western Indonesia shows the outward migration of the subduction zone. Note Hutchison's interpretation of the Meratus Mountains of southeast Borneo and the northwest part of Sarawak (Borneo) as part of the Mesozoic convergent plate margin.

The Mesozoic position of Java is less easy to interpret because there is so little available information about the pre-Cainozoic rocks. Hutchison placed Java on the oceanward side of his Late Triassic to Early Jurassic subduction system (Fig. 8) at the southern and western margin of Sumatra but he did not provide any direct evidence for this interpretation. His highly speculative interpretation was based upon his general view that there was a subduction system forming a tightly westward facing U-shaped arc extending through central Borneo to link up with the Serabang Line in northwest Borneo (Fig. 8), an interpretation supported by Haile (1973). However, Haile pointed out that there are four important characters of a subduction zone lacking in this area. For these reasons and because Hutchison's (1973) proposal, that this subduction system migrated eastwards in Borneo during the Mesozoic, where he interpreted the Meratus Mountains of southeast Borneo as part of a Late Cretaceous subduction system, seems to be contradicted by the available evidence from the Java Sea; this reconstruction is not followed in the maps (Figs. 10A, 11A, 12A) for this chapter.

Another totally different interpretation for the Mesozoic position of Borneo by Ben-Avraham and Uyeda (1973) suggested that Borneo moved southwards during the Late Jurassic along a north–south transform at the eastern margin of Indochina, from a position between what is now Hainan and Taiwan, at the margin of continental China (Fig. 9). Some major weaknesses of this interpretation are that it requires Late Jurassic–Cretaceous destruction of oceanic lithosphere between southern Borneo and northern Java, and it implies that during the Mesozoic oceanic crust must have occupied the area of the north Java Sea. This appears to conflict with an absence of any indication of a Cretaceous subduction trench in the Java Sea and with the presence there of Cretaceous granites and Cretaceous metamorphism of terrigenous clastic sediments (Katili, 1973b).

In conclusion, the available evidence appears to indicate that there have not been any major post-Permian displacements between the islands of Sumatra, Java and Borneo and the western part of the South China Sea, although this does not exclude

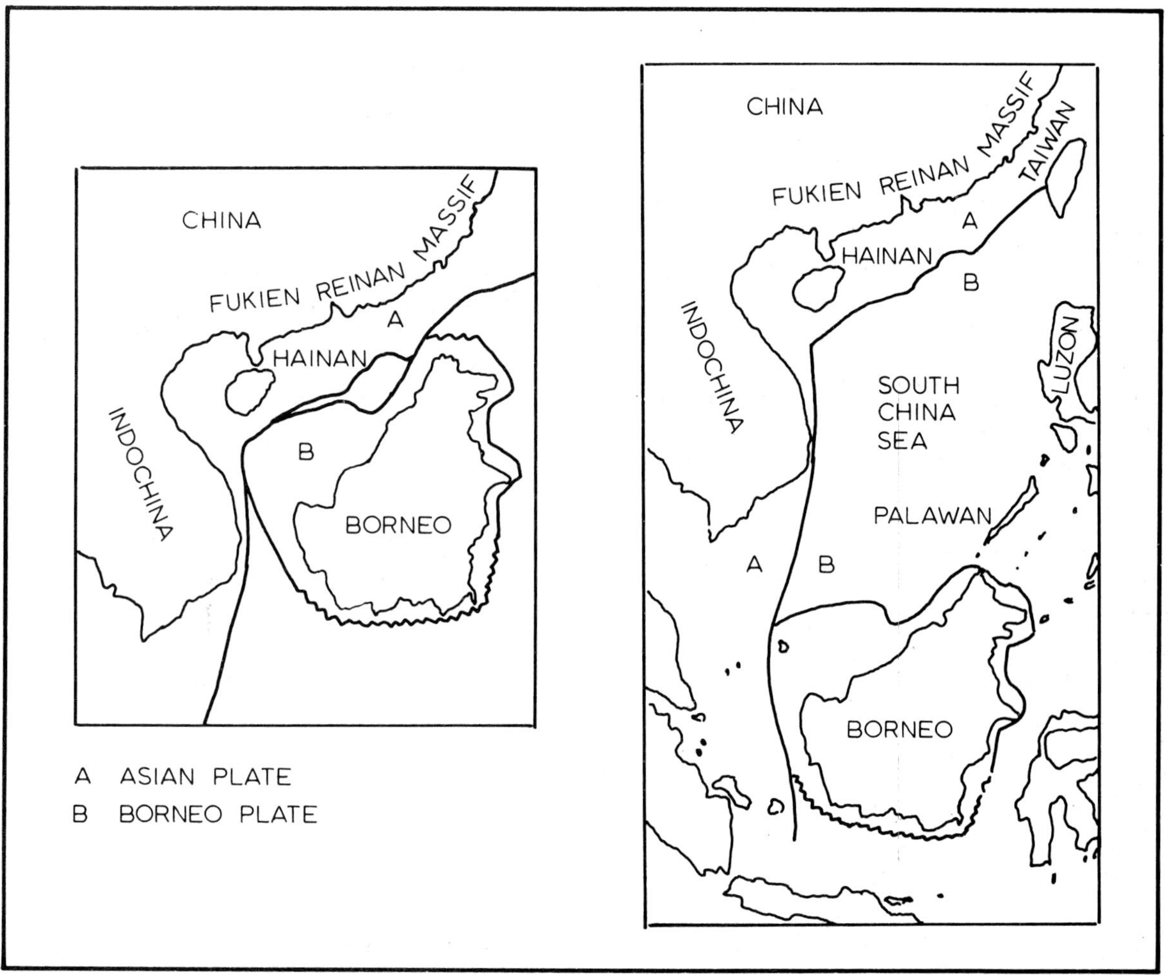

Fig. 9. Mesozoic plate-tectonic evolution of the South China Sea and Borneo region according to Ben-Avraham and Uyeda (1973).

the possibility of some transcurrent movements on large faults (Katili, 1970; Tjia, 1973). It does not seem possible to diagnose Mesozoic movements on such faults with the available information.

The Mesozoic position of western Sulawesi is difficult to determine precisely, although the similar Cainozoic successions in southeastern Borneo and in southwestern Sulawesi suggest the two areas have not changed much in relative position. However, the presence of the 2 km deep Makassar Strait separating Sulawesi from eastern Borneo has been considered to indicate some limited Cainozoic extension with possibly oceanic crust forming the floor of part of the Strait (Sander et al., 1975). Curray et al. (1977) report crustal structure intermediate between oceanic and continental between the Makassar Strait and north Bali. This implies that during the Mesozoic the present Makassar Strait did not exist and that western Sulawesi was contiguous with eastern Borneo as shown on the palaeogeographical maps (Figs. 10A, 11A, 12A). As nothing is known of the Triassic and Jurassic rocks of western Sulawesi it is difficult to comment on its Mesozoic evolution. The Cretaceous rocks are a fully marine facies with indications of both deep-water and marginal conditions with volcanism.

Palaeogeographical reconstruction of the Mesozoic Philippines

The position of the Philippines during the Permian was tropical and marine as indicated by the faunas, including fusulinids (Gervasio, 1973) mentioned in an earlier section. The widespread thick successions of immature sediments together with cherts and spilites indicate a fully marine position close to an eroding landmass where calcalkali submarine lavas and intrusions accumulated during the Mesozoic. These conclusions have led to the interpretation shown in this chapter (Figs. 10A, 11A, 12A), where the Philippines are regarded as having developed at the margin of mainland China during the Triassic and Jurassic, where terrigenous detritus from the Fukien Reinan Massif of southern China accumulated (Fig. 9). During the Late Jurassic the interpreted spreading of the floor of the South China Sea (Ben-Avraham and Uyeda, 1973) moved the Philippines away from the China mainland towards the Pacific. The Philippine region is regarded in this chapter as having formed part of a volcanic arc at the margin of China throughout the Mesozoic and to have been related to the Asian mainland in much the same way as was Japan (Lee, 1974; Uyeda and Miyashiro, 1974). According to Murphy (1973) the Philippine crust is of intermediate thickness (20–25 km) and built entirely of oceanic and island arc materials. This was interpreted by Murphy as suggesting the Philippines developed during the Oligocene from the coalescence of several arc–trench complexes. The view taken of these facts in this chapter is that the widespread extent of the pre-Cainozoic basement in the Philippines and the presence of various Mesozoic flysch sequences including about 6000 m of ammonite-bearing Jurassic flysch from southern Mindoro (Teves, 1953) is more suggestive of close proximity to a continental margin. It is unlikely that island-based palaeomagnetic data will resolve this difference in interpretation because of the small shifts in latitude involved in these models. As can be said of many parts of Indonesia the better understanding of the Mesozoic history of the Philippines probably depends more on establishing details of the Mesozoic lithologies and faunas and the stratigraphy of the marginal basins, than on any other type of study. It is probably widely accepted that during the Mesozoic the Philippines formed part of the arc–trench system marginal to Asia extending from Japan, through Ryukyu and Taiwan (Murphy, 1973; Lee, 1974; Uyeda and Miyashiro, 1974). It is the exact position of the Philippine region that is uncertain and also whether it formed at the continental margin, as suggested here, or whether it was represented by a series of separate arcs with marginal seas as Murphy (1973) suggested.

General summary of the palaeogeographical reconstructions

Although no direct palaeomagnetic measurements are available for Indonesian or Philippine Late Palaeozoic or Mesozoic rocks, the measurements of Permian and Mesozoic rocks of the Malay Peninsula (McElhinny et al., 1974) and eastern Australia (Smith et al., 1973) are very useful in controlling the palaeogeographical position of the Indonesian and Philippine regions located between these defined latitudes of Australia and the Malay Peninsula.

There is abundant faunal and lithofacies evidence for concluding that the Philippine Archipelago and most of Indonesia (including the allochthonous elements of the Outer Banda Arc) formed part of the Asian (Laurasian) continental margin in the Mesozoic tropics, while the para-autochthonous elements of the Outer Banda Arc of eastern Indonesia, together with the islands of the Sula Spur and New Guinea, formed part of the Australian continental margin, which was part of Gondwanaland located in high southern latitudes. The available palaeomagnetic evidence indicates that the Malay Peninsula was about 15°N and that part of the Australian margin represented by the basement of Timor was about 35°S. These two continental margins were separated by the Tethys Ocean, which had probably formed during the southward movement of Australia in the Late Palaeozoic. During the Mesozoic this ocean seems to have been about 2500 km wide between the two parts of Indonesia (Fig. 3).

Much of the geographical detail within the two reconstructed continental margins (shown in Figs. 7, 10–15) is highly speculative; but only those departures from present geography for which there is some evidence are included. In other words the Mesozoic geographical assembly of the areas of Indonesia and the Philippines almost certainly involved more changes than are shown in these maps. The very widespread thick cover of sediments and volcanics developed during the Cainozoic obscures much of the Mesozoic, and the complex movements, brought about by Cainozoic tectonic development of the arcs and marginal basins, has altered the original Mesozoic geography in ways that cannot yet be unravelled. For these reasons, as well as the lack of detailed information about the Mesozoic rocks themselves, the account of Mesozoic evolution that forms the remainder of this chapter is highly simplified.

PALAEOGEOGRAPHICAL EVOLUTION OF ASIAN (LAURASIAN) INDONESIA AND THE PHILIPPINES

The Mesozoic evolution of this region will be described in terms of conditions at the end of the three Mesozoic periods and illustrated by three maps (Figs. 10A, 11A and 12A). The geography will be described in very simple terms related to the principal facies of the rocks summarised in the correlation charts of this chapter. In some areas, such as western Sulawesi, where no rocks of Early Mesozoic age are known, large-scale geographical features such as ocean margins, arcs and trenches are extrapolated because to indicate a break would seem to be more misleading than to speculate on their continuance. The evidence for these palaeogeographical interpretations is indicated on the maps.

Triassic

Three principal sedimentary facies may be interpreted in this region: (1) paralic and shallow marine, (2) flysch indicating proximal continental margin marine deposits, and (3) bathyal facies of various lithologies that may be regarded as distal continental margin deep marine deposits. Four principal morphotectonic geographical units can be interpreted: (1) a volcanic arc with intrusive granites, (2) subduction trench, (3) Tethys Ocean, and (4) Asian landmass with paralic and back-arc basins. In these terms the geographical units are poorly supported with evidence. Only the volcanic arc with intrusive granites seems well established, however, there is considerable difficulty understanding the significance of the Serian Volcanics in the west Kalimantan and west Sarawak districts of western Borneo. These rocks were interpreted as a volcanic arc (Fig. 8) by Hutchison (1973), but the apparently conflicting evidence pointed out by Haile (1973) leads to the suspicion that these Serabang Line volcanics may be the result of back-arc extensional tectonics. However, this is too speculative to be included on the palaeogeographical

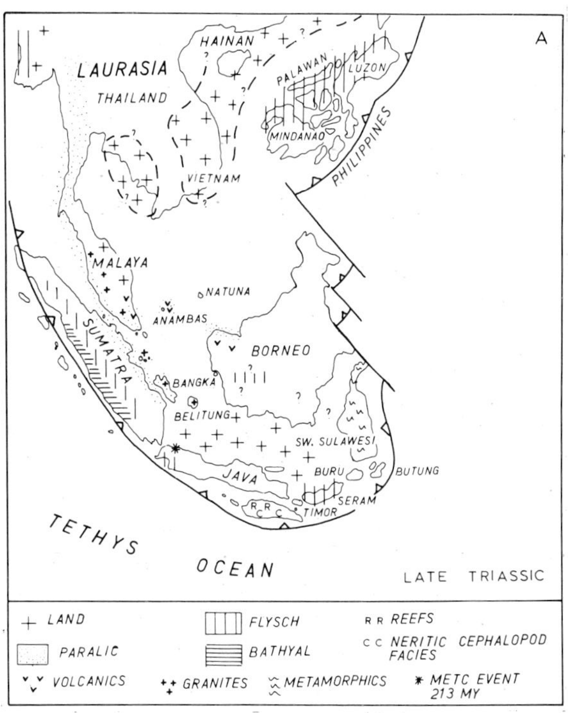

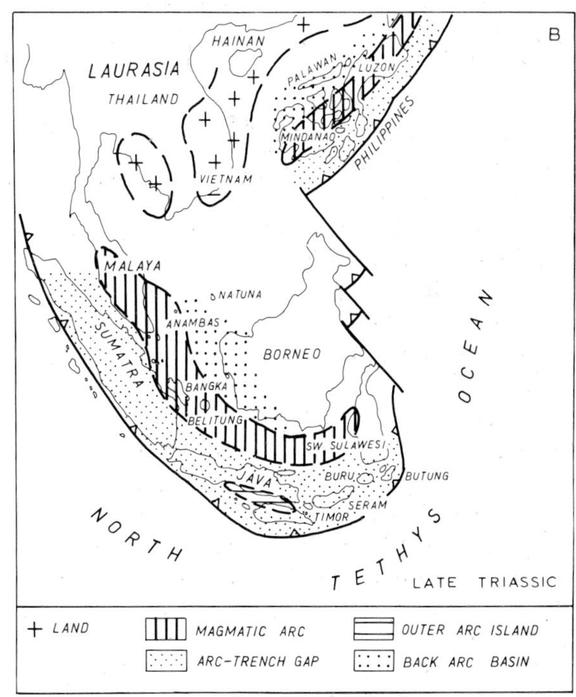

Fig. 10. A. Late Triassic reconstruction of Asian Indonesia and the Philippines. The zone of granitic plutons appears to be related to the land areas of the Sundaland Peninsula. Note the position of the present allochthonous elements now found in the islands of the Outer Banda Arc (see areas shown in black in Fig. 7). These elements were emplaced as thrust sheets on the Australian facies of these islands shown in Fig. 4 during the Late Cainozoic collision of the Asian and Australian parts of Indonesia according to Carter et al. (1976b). Present-day outlines are used for reference only and have no palaeogeographical significance. B. Late Triassic reconstruction of Asian Indonesia and the Philippines showing the palaeogeography of the northern Tethys margin. Present-day outlines imply no palaeogeographical significance.

map (Fig. 10A) where the presence of the volcanics is recorded without interpretation.

The presence of flysch facies along the western and southern margins of Sundaland in Sumatra and in the greywacke and phyllite formations of Seram maybe related to the uplift and erosion of the volcanic arc with Late Triassic granite intrusions. The strong Permian volcanic activity in Sumatra died out during the Triassic according to Katili (1974). The extensive belt of flysch in the Palawan–Luzon part of the Philippines may be related to the uplift and erosion of the Fukien Reinan Massif of mainland southern China associated with the underthrusting of Tethys along the Japan–Ryukyu–Philippine arc (Lee, 1974).

The siliceous shales and radiolarian cherts associated with *Halobia* and *Daonella*-bearing shales of Sumatra are interpreted as a deep-water marine facies.

The neritic facies of Timor is interpreted from what are now known to be olistoliths of Triassic limestone in the Bobonaro Scaly Clay olistostrome emplaced in Timor during the Late Cainozoic (Audley-Charles, 1968). These olistoliths include two types of allochthonous Triassic rocks in Timor, where the whole of the Early Triassic was claimed to be represented by cephalopod limestones only 2 m thick (Wanner, 1931b; Umbgrove, 1938). This facies was regarded as analogous to the Hallstatter facies of Europe deposited in the deeper parts of the neritic zone. Triassic coral reef limestones facies in Timor only occur as olistoliths in the Bobonaro Scaly Clay. It is this that arouses suspicion that the Norian coral reef limestones of Misool may be allochthonous and derived from the Asian margin, because they seem so out of character with the other Mesozoic facies of Misool.

Jurassic

Two main trends may be recognised from the Late Triassic to Late Jurassic: (1) The absence of a volcanic arc from Sundaland associated with a great reduction in emplacement of granitic plutons. This may be interpreted as indicating a lack of an active subduction system at the Sundaland margin for part of the Late Jurassic at least. (2) A general regression

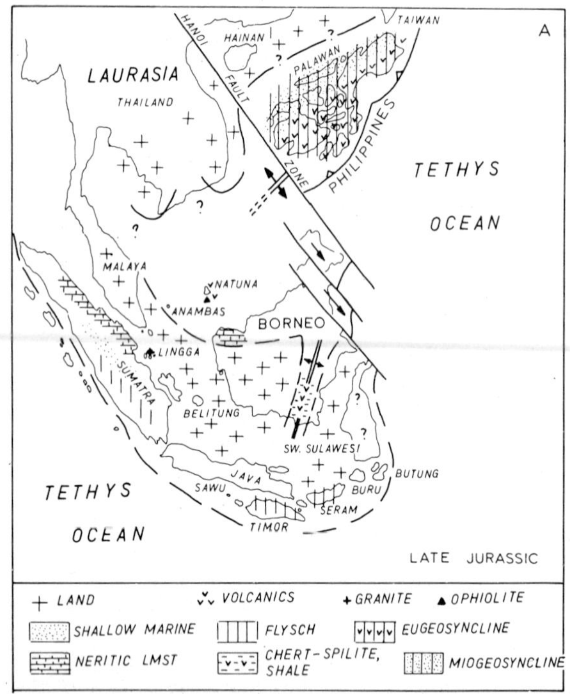

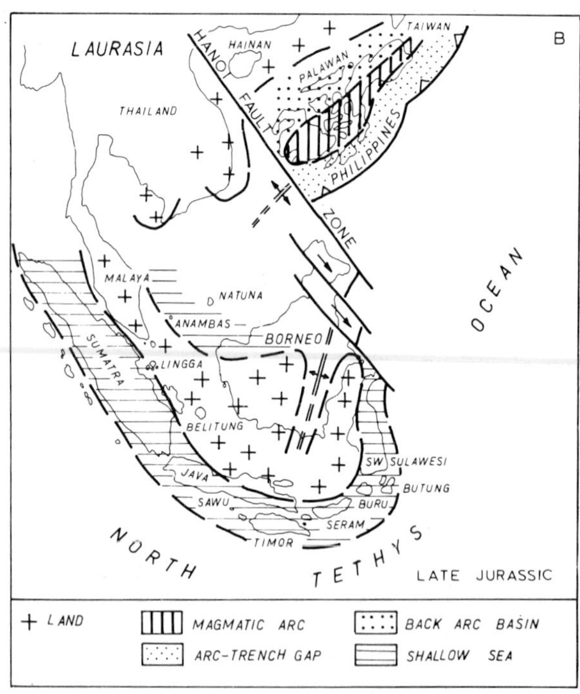

Fig. 11. A. Late Jurassic reconstruction of Asian Indonesia and the Philippines. The interpretation of sea-floor spreading in the South China Sea is based upon evidence in Ben-Avraham and Uyeda (1973). The volcanic and radiolarian chert facies of the Meratus Mountains of southeast Borneo is interpreted as developing from local submergence consequent upon extensional tectonics in this region. Present-day outlines are used for reference only and have no palaeogeographical significance. B. Late Jurassic reconstruction of Asian Indonesia and the Philippines showing the palaeogeography of the northern Tethys margin. Present-day outlines imply no palaeogeographical significance.

throughout the region associated with the uplift of a wide land area in Indonesia. (3) Initiation of the South China Sea marginal basin with associated extensional tectonics (Ben-Avraham and Uyeda, 1973). The Jurassic rocks of the Meratus Mountains of southeastern Kalimantan are difficult to understand or to fit into the overall geographical picture of the Late Jurassic. The presence of radiolarian cherts and basic volcanics there is interpreted in this chapter as the result of pull apart tectonics associated with widespread crustal updoming of extensive land areas and the formation of a graben feature in the Meratus Mountains. These extensional tectonics are correlated with those of the South China Sea at that time (Fig. 11A). This is highly speculative, but it appears to be more in line with the occurrence of diamond-bearing breccia pipes during the Cretaceous in this Meratus region than with the convergent plate margin with subduction zone (Fig. 8) as postulated by Hutchison (1973) and others.

The Philippine region during the Late Jurassic appears to be characterised by miogeosynclinal-type sedimentation along the margin of mainland China, while the oceanward side of the Philippines at this time was more typically eugeosynclinal. This model is shown in Fig. 11A where the oceanward eugeosynclinal part of the eastern Philippines are interpreted as forming part of a Jurassic volcanic arc–trench system continuous with the developing Japan–Ryukyu arc (Lee, 1974; Uyeda and Miyashiro, 1974).

Cretaceous

There is a great deal more information available about the Cretaceous system than any other part of the Mesozoic. Consequently it is possible to recognise more facies and to offer more detailed interpretations of their distribution. There appear to be two main differences from the Jurassic geography. Two different lines of evidence suggest that a subduction trench was active at the margin of Sundaland from northwestern Sumatra to a region south of east Java. This arc–trench system appears to have been active along the

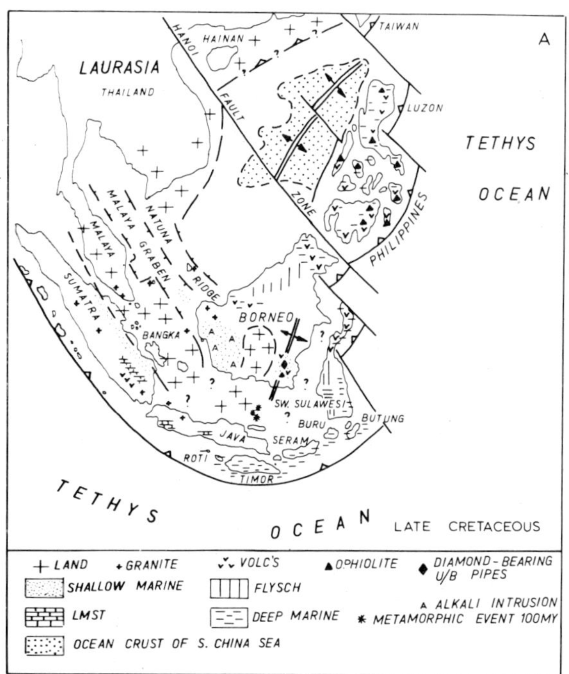

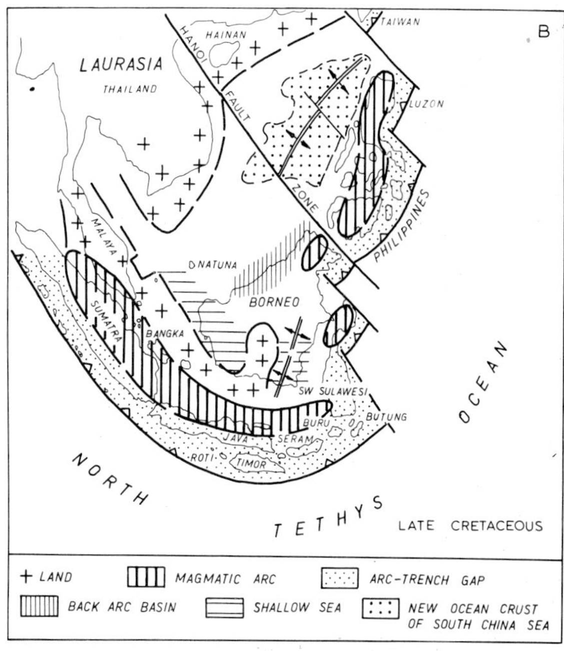

Fig. 12. A. Late Cretaceous reconstruction of Asian Indonesia and the Philippines. The interpretation of sea-floor spreading in the South China Sea is based on evidence in Ben-Avraham and Uyeda (1973). Diamond-bearing ultrabasic pipes in the Meratus Mountains of southeast Borneo are interpreted as having developed under extensional crustal conditions. Present-day outlines are used for reference only and have no palaeogeographical significance. B. Late Cretaceous reconstruction of Asian Indonesia and the Philippines showing the palaeogeography of the northern Tethys margin. Present-day outlines imply no palaeogeographical significance.

eastern margin of the Philippines so that a continuous system marginal to this part of Southeast Asia is postulated to be continuous with the Ryukyu–Japan system. The other major development in this region is the evolution of the marginal South China Sea with the migration of the Philippine region eastward and south towards the Pacific Ocean. An uplifted volcanic arc associated with Late Cretaceous intrusion of granite plutons is indicated for the region between northern Sumatra and western Borneo although the granites, alkali plutons and volcanics are also known on either flank of this trend in central and southern Sumatra and western Borneo. Late Cretaceous metamorphic events have been dated in rocks encountered in wells drilled in the east Java Sea (Katili, 1973a). The reported ophiolites and diamond-bearing ultrabasic breccia pipes in the Meratus Mountains are interpreted here as the result of deep fractures of an extensional tectonic system. Hutchison (1973) on the other hand interpreted these rocks as associated with a convergent plate margin. Hutchison's model appears to be unsupported by other evidence. The extensional model put forward here seems more in keeping with the development of the South China Sea and the extensional faulting reported from Sundaland (De Costa, 1975).

Some of the allochthonous Cretaceous rocks of the Outer Banda Arc islands such as the Cretaceous components of the Seical Formation of Timor and Cretaceous olistoliths in the Bobonaro Scaly Clay olistostrome (Audley-Charles, 1968) show evidence of having accumulated within reach of detritus from the Lolotoi Complex, others show evidence of having been associated with a trench in which oceanic deposits of red clay with manganese nodules and floods of shark teeth accumulated tectonically (Audley-Charles, 1968). This Cretaceous trench and the Lolotoi Complex are interpreted as having been marginal to Sundaland (Carter et al., 1976a) as shown in Fig. 12A.

PALAEOGEOGRAPHICAL EVOLUTION OF AUSTRALIAN (GONDWANAN) INDONESIA

The Mesozoic evolution of this region will be described in terms of conditions existing at the end of each period of the Mesozoic (Figs. 13A, 14A and 15A). The geography will be described in very simple terms related to the principal facies. One of the main difficulties in trying to understand the Gondwanan geology of the Outer Banda Arc is that throughout these islands the Australian facies are para-autochthonous and form the imbricate zone of the Late Cainozoic orogenic belt (Lofting et al., 1975), developed at the margin of Australia during the Mio-Pliocene collision between northward-drifting Australia and southward-migrating Banda Arc moving away from the Asian margin (Carter et al., 1976a). A consequence of this is that the Gondwanan Mesozoic facies in these islands of the Banda Arc is very strongly folded and imbricated with a strong vergence towards Australia. Locally some important thrusting is involved, such as that at the base of the Cretaceous bathyal deposits that characterise these islands. The shortening effects on these Mesozoic cover rocks above the Australian basement make it very difficult to unravel the various facies palinspastically because their stratigraphy and faunas are not known in detail. All that can be done in the present state of knowledge is to interpret the various facies as zones of increasing depth of deposition with decreasing proximal characteristics with reference to the Australian margin.

These difficulties are not encountered in the islands of the Sula Spur or in Irian Jaya because there the Mesozoic rocks appear to be nearly all autochthonous and much less deformed than in the Banda Arc islands.

Triassic (Fig. 13)

Essentially three major geographical units can be distinguished in this region during the Triassic. The land area of the Australian mainland has been interpreted by Lofting et al. (1975) and by Harrison (1969). The land area in Sumba and its position at the Australian margin have been interpreted by Audley-Charles (1975) from data provided by Warris (1973), Falvey (1972) and Van Bemmelen (1949). The land area in the Sula Spur islands of the Sula Peninsula was suggested by Van Bemmelen (1949). The Triassic sediments of this region indicate three main geographical units: (1) Shallow marine and fluvio-deltaic deposits that can be grouped together as shallow marine and paralic (coals are common locally). (2) The deeper-water zone of the outer shelf and perhaps delta-slope fan forms the other main unit. (3) The radiolarian calcilutites with *Halobia* and *Daonella* and thin shales present a problem of

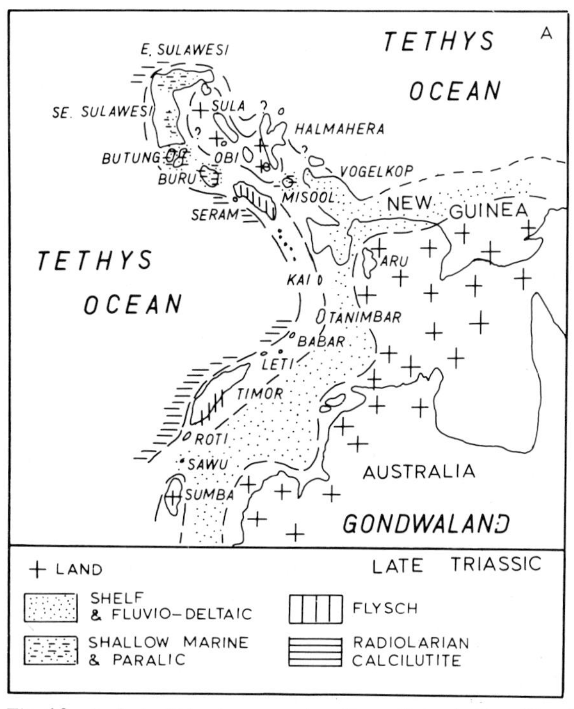

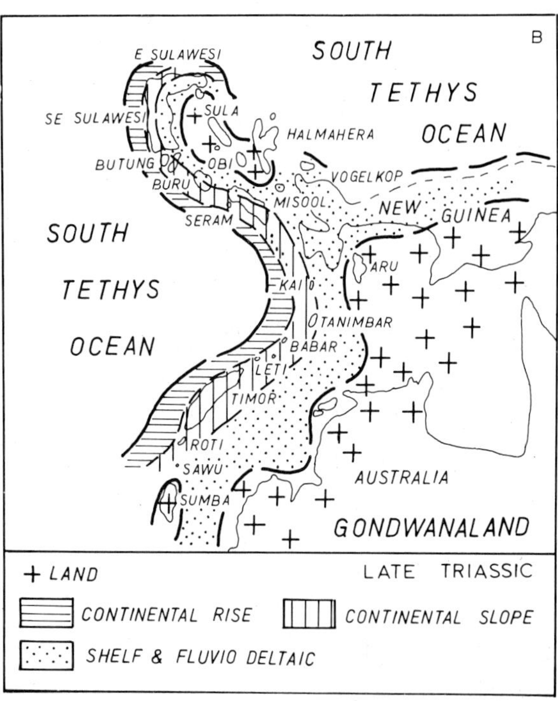

Fig. 13. A. Late Triassic reconstruction of Australian (Gondwanan) Indonesia. Note that the eastern Outer Banda Arc islands surround part of the Sula Spur of Klompé (1961) to form the much larger feature of the Sula Peninsula interpreted here. The reconstruction of the Vogelkop (Tjendrawasih) follows the suggestion of Hermes (1968). Present-day outlines are used for reference only and have no palaeogeographical significance. B. Late Triassic reconstruction of Gondwanan Indonesia showing the palaeogeography of the southern Tethys margin. Present-day outlines imply no palaeogeographical significance.

defining precisely their environment of deposition. This was discussed by Audley-Charles (1968), but all that can be said with confidence is that they were deposited below wave base in a region where very little terrigenous detritus entered. This may not have been in very deep water although it must have been isolated from the terrigenous material deposited at the same time in southern Timor (Babulu Member) and on the present Australian shelf. Perhaps it formed carbonate platforms forming fault blocks on the outer shelf, but the absence of a shelly benthos can only be understood at the moment in terms of deeper water beyond the Late Triassic shelf. The interpretation of these deeper-water deposits as having accumulated north of Timor is based on the palinspastic reconstruction that regards these rocks as having been moved towards the Australian margin by the Late Cainozoic collision.

Jurassic

Two main developments occurred in this region during the Late Jurassic. One was the initiation of a spreading centre in the Wharton Basin of the eastern Indian Ocean (Falvey, 1972; Veevers and Heirtzler, 1974). This was associated with the development of about 300 m of basaltic lavas in the Ashmore–Sahul block (which now forms part of the Australian shelf region), which appears to have been elevated above sea level for much of the Jurassic (Laws and Kraus, 1974). Sumba contains Jurassic tuffs, agglomerates and calcalkaline plutons of possible Late Jurassic age, that could be correlated with both the volcanism in the adjacent Sahul–Ashmore area and the spreading Indian Ocean in the Wharton Basin. The Mesozoic "western landmass", which Audley-Charles (1975) has equated with Sumba, moved northwards along a major right lateral fault zone according to Warris (1973). This movement can also be correlated with the development of the Wharton Basin as shown in Fig. 14A.

There is a general increase in the marine influence on the north Australian shelf during the Late Jurassic with a reduction of areas that were characterised by paralic deposits. The great reduction of carbonate

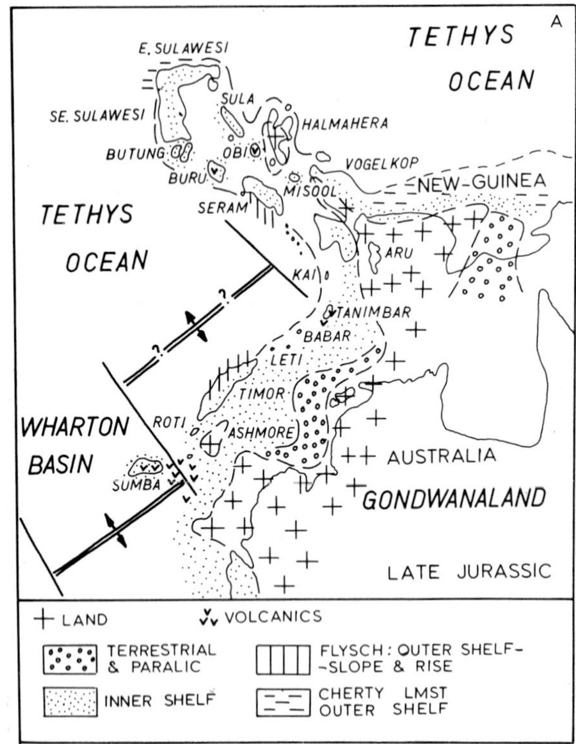

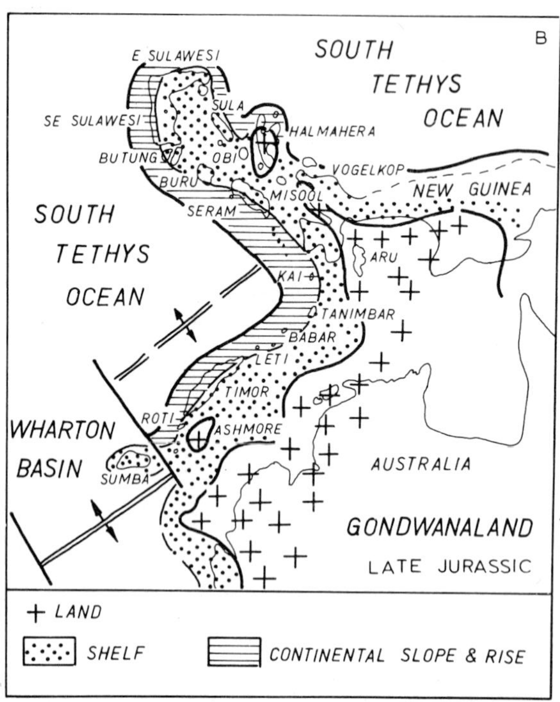

Fig. 14. A. Late Jurassic reconstruction of Australian (Gondwanan) Indonesia. The spreading ocean floor in the Wharton Basin of the Indian Ocean is based on the interpretation by Veevers and Heirtzler (1974). Present-day outlines are used for reference only and have no palaeogeographical significance. B. Late Jurassic reconstruction of Gondwanan Indonesia showing the palaeogeography of the southern Tethys margin. Present-day outlines imply no palaeogeographical significance.

lutites and radiolarian-rich lutites with the increase in sand and silt in the Jurassic deposits of the Outer Banda Arc islands is interpreted as a reflection of the erosion and movement of sediment stripped from the areas of the north Australian shelf during the Middle and Early Jurassic (Laws and Kraus, 1974). However, changes in the depth of the sea floor in this region may have been another important factor associated with block faulting related to Late Jurassic sea-floor spreading in the Wharton Basin and possibly in the region west of the Sula Peninsula (Fig. 14A). Thermal doming associated with the development of the spreading centre could have uplifted the Australian continental margin. The presence of Jurassic volcanics in the Tanimbar and Buru islands may be associated with this spreading.

Cretaceous

During the Late Cretaceous the dominating overall development was the marine transgression onto the margins of the Australian continent. The only part of Gondwanan Indonesia that may have been above sea level during the Late Cretaceous was the small region between the Vogelkop (Tjendrawasih) of Irian Jaya and Misool (Fig. 15A). Four major geographical zones may be distinguished: (1) The inner and outer shelf around the Australian landmass which extended into New Guinea. (2) The outer part of the shelf and possibly the slope where deep-water limestones accumulated in some areas while deep-water clastics and flysch were deposited in other parts of this zone. (3) In the northern province of New Guinea deep-water limestones accumulated with spilites (Harrison, 1969). (4) The calcilutites and marls with planktonic foraminifera and radiolaria associated with radiolarites and radiolarian shales and cherts in which the limestones and marls show evidence of extensive reworking and slumping throughout the Cretaceous (Audley-Charles and Carter, 1972) are interpreted as having been deposited on the Australian continental rise. It is this facies that characterises the islands of the

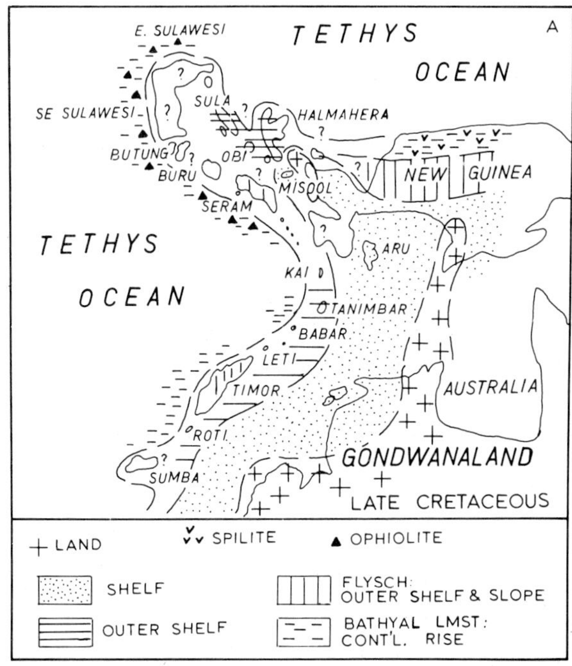

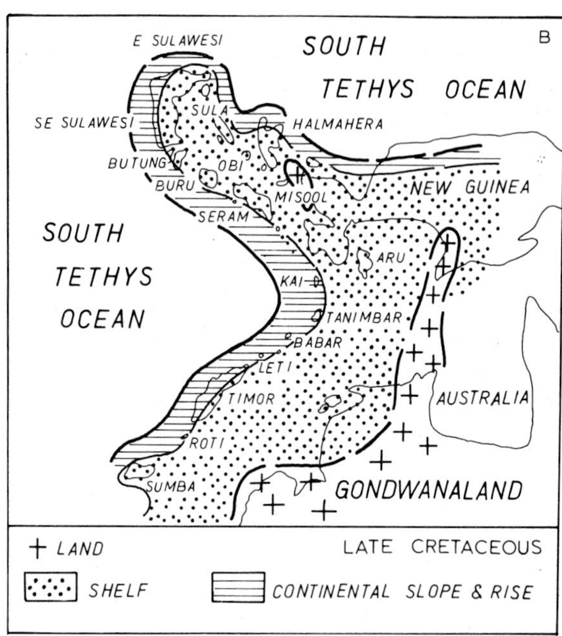

Fig. 15. A. Late Cretaceous reconstruction of Australian (Gondwanan) Indonesia. Note the palinspastic interpretation of the allochthonous bathyal lutites that characterise the Outer Banda Arc islands and the apparent restriction of the associated ophiolites to the eastern margin of the Sula Peninsula. B. Late Cretaceous reconstruction of Gondwanan Indonesia showing the palaeogeography of the southern Tethys margin. Present-day outlines imply no palaeogeographical significance.

Outer Banda Arc where it is allochthonous, forming strongly imbricated thrust sheets. Typically this facies continues through the Palaeogene and in Timor it has been found to continue into the Early Neogene (Carter et al., 1976a). With the exception of Buru in all the Outer Banda Arc islands of the Sula Peninsula (Fig. 15) this facies is associated with basic and ultrabasic igneous rocks (often called ophiolites in the literature). Kundig (1956) showed that in Sulawesi this facies is imbricated with the "ophiolites". In contrast the Banda Arc islands from Tanimbar to Sawu do not have "ophiolites" associated with this imbricated overthrust facies. As these rocks were emplaced on the islands of the Outer Banda Arc during the Mio-Pliocene collision between Australia and the Asian arc (Carter et al., 1976a), the presence of associated "ophiolites" in islands of the Sula Peninsula may be the result of differences in Late Cainozoic tectonic behaviour. It might be expected that similar facies would have accumulated on both the continental rise of northern Australia and on the ocean floor south of Sundaland. The Late Cainozoic collision of these two margins forming the present Banda Arc leads to difficulties in separating bathyal sediments thought to have accumulated at the Australian margin from those deposited on the ocean in front of the Asian margin (Carter et al., 1976b). The principal criterion seems to be the presence of derived particles characteristic of one margin or the other.

MESOZOIC PLATE MARGINS, STABLE ZONES AND EXTENSIONAL TECTONICS IN THE INDONESIAN AND PHILIPPINE ARCHIPELAGOS

The remarkable similarities between the Australian Mesozoic facies in the islands of the Outer Banda Arc, the Sula Spur and on the Australian northwest shelf may indicate that these areas were on the same plate throughout the Mesozoic.

During the Mesozoic the Gondwanan part of Indonesia, which formed part of the Australian continental shelf, slope and rise appears to have formed part of the same plate as the Tethys Ocean until the Late Jurassic, when a spreading system in the Wharton Basin of the eastern Indian Ocean divided a large part of Tethys from the Gondwanaland plate. Since the Late Jurassic Australia has remained part of the same plate as the new Indian Ocean and Wharton

Basin. In contrast to this the Asian part of Indonesia with the Philippines have been separated from the Tethys–Australian–Gondwana plate by a subduction trench at the Asian margin of Tethys throughout the Mesozoic, except possibly during the Late Jurassic when Asian Indonesia appears to have been part of the Tethys plate as the subduction system became inactive.

The Philippines appear to have been separated from the Tethys plate throughout their Mesozoic history. During the Early Mesozoic until the Late Jurassic, when the South China Sea developed, the Philippines and Asian Indonesia appear to have been part of the Asian plate. Since the Late Jurassic the Philippines have formed a separate plate from Asian Indonesia. Asian Indonesia seems to have been part of a much larger Asian plate (Laurasia) throughout the Mesozoic.

In terms of plate behaviour the Late Jurassic seems to have been the occasion for three apparently quite separate but important events. New ocean or marginal basin spreading began near the margins of the two opposite-facing continents of Asia (Laurasia) and Australia (Gondwanaland), in the South China Sea and Wharton Basins, respectively. These two spreading centres being about 3000 km apart (Fig. 7). At the same time the subduction system at the margin of Asian Indonesia (i.e., the Sundaland Peninsula) seems to have become inactive.

The Australian part of Indonesia at the margin of the Gondwanaland continent remained a passive continental margin throughout the Mesozoic, as it had since this margin separated from Asia during the Late Palaeozoic, when it began to drift south as a trailing Atlantic-type coast. This part of the continental margin did not become active until it collided with Asian Indonesia, in the form of the southward-migrating Banda Arc during the Mio-Pliocene.

In contrast to this the Asian margin of Indonesia appears to have been an active continental margin throughout most of the Mesozoic (Fig. 3) with perhaps a short passive phase during the Jurassic.

In general terms Australia seems to have separated from Asia about 300 m.y. ago. The Tethys Ocean, that had formed between these continents at their Late Paleozoic separation, finally subsided and underthrust Asia when this ocean crust was about 200 m.y. old during the Cretaceous, although some of it had been subducted earlier during the Permian, Triassic and Jurassic.

Extensional tectonics appear to have been an important factor in the development of both the Asian and Australian parts of Indonesia. Australian Indonesia was probably influenced by extensional features during the Jurassic (Lofting et al., 1975) when important block faulting occurred in the north-west Australian shelf. In Asian Indonesia there seems to have been a widespread phase of extensional tectonics during the Late Cretaceous (De Costa, 1975), which might have been associated with the initial separation of the Banda Arc allochthon from the margin of Sundaland by the development of the Banda Sea (Carter et al., 1976a). Another important phase of extension tectonics is interpreted as having developed in association with the Late Jurassic to Cretaceous formation of the South China Sea, the migration of the Philippines and the deep-seated events in the Meratus mountains.

The tendency since the Late Jurassic for Sundaland and Southeast Asia in general to split off narrow slivers of the continental margin, to form migrating island arcs with marginal seas separating them from the Asian continent, may be associated with the long history of oceanic lithosphere underthrusting this continental margin.

ACKNOWLEDGEMENTS

The author wishes to record his debt to D.J. Carter for stimulating debate extending over many years on the problems discussed in this chapter.

REFERENCES

Adinegoro, U. and Hartoyo, P., 1975. Palaeogeography of N.E. Sumatra. *Proc. Indones. Pet. Assoc., 3rd Annu. Conv.*, pp. 45–61.

Adiwidjaja, P. and De Costa, G.L., 1973. Pre-Tertiary palaeotopography and related sedimentation in South Sumatra. *Proc. Indones. Pet. Assoc., 2nd Annu. Conv.*, pp. 89–103.

Aleva, G.J.J., 1960. The plutonic igneous rocks from Billiton, Indonesia. *Geol. Mijnbouw,* 22: 427–436.

Amato, F.L., 1965. Stratigraphic palaeontology in the Philippines. *Philipp. Geol.*, 19: 1–24.

Andal, D.R., Esquerra, J.S., Hashimoto, W., Reyes, B.P. and Sato, T., 1968. The Jurassic Mansalay formation, southern Mindoro, Philippines. In: T. Kobayashi and R. Toriyama (Editors), *Geology and Palaeontology of Southeast Asia, 4.* Univ. Tokyo Press, Tokyo, pp. 179–197.

Audley-Charles, M.G., 1965. Permian palaeogeography of the northern Australia–Timor region. *Palaeogeogr., Palaeoclimatol., Palaeoecol.*, 1: 297–305.

Audley-Charles, M.G., 1968. The geology of Portuguese Timor. *Mem. Geol. Soc. Lond.*, 4: 1–76.

Audley-Charles, M.G., 1974. Sulawesi. In: *Mesozoic–Cainozoic Orogenic Belts. Geol. Soc. Lond., Spec. Publ.*, 4: 365–378.

Audley-Charles, M.G., 1975. The Sumba fracture: A major discontinuity between eastern and western Indonesia. *Tectonophysics*, 26: 213–228.

Audley-Charles, M.G., 1976. Mesozoic evolution of the margins of Tethys in Indonesia and the Philippines. *Proc. Indones. Pet. Assoc., 5th Annu. Conv., Jakarta, 1976*, pp. 179–198.

Audley-Charles, M.G. and Carter, D.J., 1972. Palaeogeographical significance of some aspects of Palaeogene and Early Neogene stratigraphy and tectonics of the Timor Sea region. *Palaeogeogr., Palaeoclimatol., Palaeoecol.*, 11: 247–264.

Audley-Charles, M.G. and Carter, D.J., 1974. Petroleum prospects of the southern part of the Banda Arcs, eastern Indonesia. *United Nations ESCAP, CCOP Tech. Bull.*, 8: 55–70.

Audley-Charles, M.G., Carter, D.J. and Barber, A.J., 1975. Stratigraphic basis for tectonic interpretation of the Outer Banda Arc, eastern Indonesia. *Proc. Indones. Pet. Assoc., 3rd Annu. Conv.*, pp. 25–44.

Audley-Charles, M.G., Barber, A.J., Carter, D.J. and Norvick, M.S., in preparation. Reconnaissance stratigraphy and structure of Seram and its relationship to the Banda Arc.

Barber, A.J. and Audley-Charles, M.G., 1976. The significance of the metamorphic rocks of Timor in the development of the Banda Arc, Indonesia. *Tectonophysics*, 30: 119–128.

Barber, A.J., Audley-Charles, M.G. and Carter, D.J., 1977. Thrust tectonics in Timor. *J. Aust. Geol. Soc.*, 24: 51–62.

Baumann, P., Genevraye, P. de, Samuel, L., Mudjito and Sajekti, S., 1973. Contribution to the geological knowledge of south west Java. *Proc. Indones. Pet. Assoc., 2nd Annu. Conv.*, pp. 105–108.

Bell, R.M. and Jessop, R.G.C., 1974. Exploration and geology of the west Sulu Basin, Philippines. *APEA J.*, 14: 21–28.

Ben-Avraham, Z. and Emery, K.O., 1973. Structural framework of Sunda Shelf. *Am. Assoc. Pet. Geol. Bull.*, 57: 2323–2366.

Ben-Avraham, Z. and Uyeda, S., 1973. The evolution of the China Basin and the Mesozoic palaeogeography of Borneo. *Earth Planet. Sci. Lett.*, 18: 365–376.

Boehm, G., 1905. Ueber Brachiopoden aus einem alteren Kalkstein der Insel Ambon. *Jaarb. Mijnw. Ned. Oost-Indië, Wet. Ged.*, pp. 88–93.

Bothe, A.Ch.D., 1927. Voorloopige mededeeling betreffende de geologie van Zuidoost Celebes. *Mijningenieur*, 8: 97–103.

Bothe, A.Ch.D., 1928a. Brief outline of the geology of the Rhio Archipelago and the Anambas islands. *Jaarb. Mijnw. Ned. Oost-Indië, 1925, Verh. II*, pp. 97–100.

Bothe, A.Ch.D., 1928b. Geologische verkenningen in den Riouw-Lingga Archipel en de eilandengroep der Poelau Toedjoeh (Anambas en Natoena eilanden). *Jaarb. Mijnw. Ned. Oost-Indië, 1925, Verh. II*, pp. 101–152.

Brondijk, J.F., 1964. The Danau Formation in northwest Borneo. *Br. Borneo Geol. Surv. Annu. Rep.*, 1963: 167–177.

Brouwer, H.A., 1919. Geologische onderzoekingen in Oost-Ceram. *Tijdschr. K. Ned. Aardrijkskd. Gen.*, 2(36): 715.

Brouwer, H.A., 1921. Geologische onderzoekingen op de Soela Eilanden. *Jaarb. Mijnw. Ned. Oost-Indië*, 1920: 2.

Brouwer, H.A., 1922. Geologische onderzoekingen op het eiland Roti. *Jaarb. Mijnw. Ned. Oost-Indië*, 1920(3): 33–106.

Brouwer, H.A., 1923. Geologische onderzoekingen op het eiland Halmaheira. *Jaarb. Mijnw. Ned. Oost-Indië*, 1921: 73–105.

Brouwer, H.A., 1924. Bijdrage tot de geologie der Obi-eilanden. *Jaarb. Mijnw. Ned. Oost-Indië*, 1923: 63–136.

Brouwer, H.A., 1927. Over mesozoische afzettingen en eenige vulkanische gesteenten van het eiland Ambon. *Jaarb. Mijnw. Ned. Oost-Indië*, 1923: 3.

Brouwer, H.A., 1942. Summary of the results of the expedition. *Geol. Exped. Lesser Sunda Islands*, 4: 345–402.

Brouwer, H.A., 1947. *Geological Explorations in the Islands of Celebes.* North-Holland, Amsterdam, 64 pp.

Brouwer, H.A., Hetzel, W.H. and Straeter, H.E.G., 1934. Geologische onderzoekingen op het eiland Celebes. *Verh. K. Ned. Geol-Mijnbouwkd. Gen.*, 10: 39–171.

Burton, C.K., 1974. Peninsular Thailand. In: *Mesozoic–Cainozoic Orogenic Belts. Geol. Soc. Lond. Spec. Publ.*, 4: 301–315.

Carter, D.J., Audley-Charles, M.G. and Barber, A.J., 1976a. Stratigraphical analysis of island arc–continental margin collision in eastern Indonesia. *J. Geol. Soc. Lond.*, 132: 179–198.

Carter, D.J., Audley-Charles, M.G. and Barber, A.J., 1976b. Discussion of stratigraphical analysis of island arc–continental margin collision in eastern Indonesia. *J. Geol. Soc. Lond.*, 132: 358–361.

Chamalaun, F.H., Lockwood, K. and White, A., 1976. The Bouguer gravity field and crustal structure of eastern Timor. *Tectonophysics*, 30: 241–259.

Crostella, A.A. and Powell, D.E., 1975. Geology and hydrocarbon prospects of the Timor area. *Proc. Indones. Pet. Assoc., 4th Annu. Conv.*, pp. 149–171.

Curray, J.R., Shor, G.G., Raitt, R.W. and Henry, M., 1977. Seismic refraction and reflection studies of crustal structure of the Banda and eastern Sunda arcs. *J. Geophys. Res.*, in press.

Dash, B.P., Shepstone, C.M., Dayal, S., Guru, S., Hains, B.L.A., King, G.A. and Ricketts, G.A., 1972. Seismic investigations on the northern part of the Sunda Shelf, south and east of Great Natuna Island. *United Nations ECAFE, CCOP Tech. Bull.*, 6: 179–196.

De Costa, G.G., 1975. The geology of the central and south Sumatra basins. *Proc. Indones. Pet. Assoc., 3rd Annu. Conv.*, pp. 77–110.

De Marez Oyens, F.A.H., 1913. De geologie van het eiland Babar. *Nat. Geneeskd. Congr.*, 1913, pp. 463–468.

De Neve, G.A., 1961. Correlation of fusulinid rocks from southern Sumatra, Bangka and Borneo, with similar rocks from Malaya, Thailand and Burma. *Proc. 9th Pac. Sci. Congr., Bangkok, 1957,* 12: 249.

Deninger, K., 1918. Zur Geologie von Mittel-Seran (Ceram). *Palaeontographica,* 4: 3.

De Roever, W.P., 1940. Geological investigations in the southwestern Moetis region (Netherlands Timor). *Geol. Exped. Lesser Sunda Islands,* 2: 97–344.

De Waard, D., 1954a. Geological research in Timor, an introduction. *Indones. J. Nat. Sci.,* 110: 1–8.

De Waard, D., 1954b. The orogenic main phase in Timor. *Indones. J. Nat. Sci.,* 110: 9–20.

De Waard, D., 1954c. Structural development of the crystalline schists in Timor, tectonics of the Lalan Asu massif. *Indones. J. Nat. Sci.,* 110: 143–153.

De Waard, D., 1954d. The second geological Timor expedition, preliminary results. *Indones. J. Nat. Sci.,* 110: 154–160.

De Waard, D., 1955a. On the tectonics of the Ofu Series. *Indones. J. Nat. Sci.,* 111: 137–143.

De Waard, D., 1955b. Tectonics of the Sonnebait overthrust unit near Nikiniki and Basleo. *Indones. J. Nat. Sci.,* 111: 144–150.

De Waard, D., 1956. Geology of a N–S section across western Timor. *Indones. J. Nat. Sci.,* 112: 101–114.

De Waard, D., 1957. Contributions to the geology of Timor, XII. The third Timor geological expedition, preliminary results. *Indones. J. Nat. Sci.,* 113: 7–42.

Dickinson, W.R., 1973. Reconstruction of past arc–trench systems from petrotectonic assemblages in the island arcs of the western Pacific. In: P.J. Coleman (Editor), *The Western Pacific.* University of Western Australia Press, Nedlands, W.A., pp. 569–601.

Dieckmann, W. and Julius, M.W., 1925. Algemeene geologie en ertsafzettingen van Zuidoost Celebes. *Jaarb. Mijnw. Ned. Oost-Indië, 1924, Verh.,* pp. 11–65.

Egeler, C.G., 1946. *Contribution to the Petrology of the Metamorphic Rocks of Western Celebes.* Thesis University of Amsterdam; North-Holland, Amsterdam.

Falvey, D.A., 1972. Sea-floor spreading in the Wharton Basin (Northeast Indian Ocean) and the breakup of eastern Gondwanaland. *APEA J.,* 12: 86–88.

Fitch, F.H., 1963. Geological relationship between the Philippines and Borneo. *Philipp. Geol.,* 17: 41–47.

Fitch, T.J. and Hamilton, W., 1974. Reply to the comment by M.G. Audley-Charles and J.S. Milsom on 'Plate convergence, transcurrent faults and internal deformation adjacent to southeast Asia and the western Pacific.' by T.J. Fitch. *J. Geophys. Res.,* 79: 4982–4985.

Froidevaux, C.M., 1975. Geology of Misool island (Irian Jaya). *Proc. Indones. Pet. Assoc., 3rd Annu. Conv.,* pp. 189–196.

Gageonnet, R. and Lemoine, M., 1958. Contribution à la connaissance de la géologie de la province portugaise de Timor. *Est. Ensaios Docum. Jta. Invest. Ultramar,* 48: 1–138.

Geiger, M.E., 1963. Palaeogeography of the Late Cretaceous–Eocene geosyncline in northwest Borneo. *Malaysia Geol. Surv. Annu. Rep.,* pp. 179–187.

Germeraad, J.H., 1946. *Geology of Central Seran.* De Bussy, Amsterdam.

Gervasio, F.C., 1968. Age and nature of orogenesis of the Philippines. *United Nations ECAFE, CCOP Tech. Bull.,* 1: 113–128.

Gervasio, F.C., 1973. Geotectonic development of the Philippines. In: P.J. Coleman (Editor), *The Western Pacific.* University of Western Australia Press., Nedlands, W.A., pp. 307–324.

Giani, L., 1971. *The Geology of the Belu District of Indonesian Timor.* Thesis, University of London, 122 pp.

Gobbett, D.J. and Hutchison, C.S., 1973. *Geology of the Malay Peninsula.* Wiley, New York, N.Y., 438 pp.

Grunau, H.R., 1953. Geologie von Portugiesisch Ost-Timor. Eine kurze Übersicht. *Eclog. Geol. Helv.,* 46: 29–37.

Grunau, H.R., 1956. Zur Geologie von Portugiesisch Ost-Timor. *Mitt. Naturforsch. Ges. Bern,* 13: 11–18.

Grunau, H.R., 1957. Neue Daten zur Geologie von Portugiesisch Ost-Timor. *Eclog. Geol. Helv.,* 50: 69–98.

Haile, N.S., 1969. Geosynclinal theory and the organizational pattern of the north-west Borneo geosyncline. *Q.J. Geol. Soc. Lond.,* 124: 171–194.

Haile, N.S., 1970. Notes on the geology of the Tambelan, Annambas and Bunguran (Natuna) islands, Sunda shelf, Indonesia, including radiometric age determinations. *United Nations ECAFE, CCOP Tech. Bull.,* 3: 55–89.

Haile, N.S., 1973. The recognition of former subduction zones in southeast Asia. In D.H. Tarling and S.K. Runcorn (Editors), *Implications of Continental Drift to the Earth Sciences, 2,* Academic Press, London, pp. 885–892.

Haile, N.S., 1974. Borneo. In: *Mesozoic–Cainozoic Orogenic Belts. Geol. Soc. Lond. Spec. Publ.,* 4: 334–347.

Haile, N.S. and Bignell, J.D., 1971. Late Cretaceous age based on K/Ar dates of granitic rock from Tambelan and Bunguran islands, Sunda Shelf, Indonesia. *Geol. Mijnbouw,* 50: 687–690.

Haile, N.S., McElhinny, M.W. and McDougall, I., 1977. Palaeomagnetic data and radiometric ages from the Cretaceous of West Kalimantan (Borneo), and their significance in interpreting regional structure. *J. Geol. Soc. Lond.,* 133: 133–144.

Hamilton, W., 1972. *Preliminary Tectonic Map of the Indonesian Region.* Scale 1 : 5,000,000 open file report. U.S. Geol. Surv., Washington, D.C.

Hamilton, W., 1973. Tectonics of the Indonesian region. *Geol. Soc. Malaysia Bull.,* 6: 3–10.

Hamilton, W., 1974. Earthquake map of the Indonesian region. *U.S. Geol. Surv.,* Map 1-875 C.

Hamilton, W., 1976. Subduction in the Indonesian region. *Proc. Indones. Pet. Assoc., 5th Annu. Conv., Jakarta, 1976,* 13 pp.

Harloff, Ch.E.A., 1933. *Toelichting bij blad 67 (bandjarnegara). Geol. kaart van Java 1 : 100,000.* Bandung.

Harrison, J., 1969. A review of the sedimentary history of the island of New Guinea. *APEA J.,* 9: 41–48.

Hartono, H.M.S., Tjokrosaputro, S., Suwitodirdjo and Rosidi, H.M.D., 1977. Some notes on the geologic map of Timor – Indonesia. *Conf. Geol. Min. Res. SE Asia. Jakarta, 1975,* in press.

Hashimoto, W. and Sato, T., 1968. Contribution to the geology of Mindoro and neighbouring islands, the Philippines. In T. Kobayashi and R. Toriyama (Editors), *Geolo-*

gy and Palaeontology of Southeast Asia, 5. Univ. Tokyo Press, Tokyo, pp. 192–210.

Hermes, J.J., 1968. The Papuan geosyncline and the concept of geosynclines. *Geol. Mijnbouw,* 47: 81–97.

Hetzel, W.H., 1936. Verslag van het onderzoek naar het voorkomen van asfaltgesteenten op het eiland Boeton. *Wet. Meded. Dienst Mijnb. Ned Oost-Indië,* 21.

Hutchison, C.S., 1973. Tectonic evolution of Sundaland: A Phanerozoic synthesis. *Geol. Soc. Malaysia Bull.,* 6: 61–86.

Hutchison, C.S., 1975. Ophiolite in Southeast Asia. *Geol. Soc. Am. Bull.,* 86: 797–806.

Irving, E.M., 1952. Geological history and petroleum possibilities of the Philippines. *Bull. Am. Assoc. Pet. Geol.,* 36: 437–476.

Jaworski, E., 1915. Die Fauna der Obertriadischen *Nucula*-Mergel von Misol. In: J. Wanner (Editor), *Palaeontologie von Timor,* 2: 73–174.

Jaworski, E., 1927. Obertriadische Brachiopoden von Ambon (Molukken). *Jaarb. Mijnb. Ned. Oost-Indië,* 1926: 55.

Kanmera, K. and Nakazawa, K., 1973. Permian–Triassic relationship and faunal changes in the eastern Tethys. *Can. Soc. Pet. Geol. Mem.,* 2: 100–119.

Katili, J.A., 1970. Large transcurrent faults in southeast Asia with special reference to Indonesia. *Geol. Rundsch.,* 59: 581–600.

Katili, J.A., 1971. A review of the geotectonic theories and tectonic maps of Indonesia. *Earth-Sci. Rev.,* 7: 143–163.

Katili, J.A., 1973a. Geochronology of west Indonesia and its implication on plate tectonics. *Tectonophysics,* 19: 195–212.

Katili, J.A., 1973b. On fitting certain geological and geophysical features of the Indonesian island arc to the new global tectonics. In: P.J. Coleman (Editor), *The Western Pacific.* University of Western Australia Press, Nedlands, W.A., pp. 287–305.

Katili, J.A., 1974. Sumatra. In: *Mesozoic–Cainozoic Orogenic Belts. Geol. Soc. Lond. Spec. Publ.,* 4: 317–331.

Keyzer, F.G., 1945. Upper Cretaceous smaller Foraminifera from Buton. *Verh. Ned. Akad. Wet.,* 48: 338–339.

Kho, C.H., 1972. Regional Geology: Sarawak and Sabah. *Malaysia Geol. Surv. Annu. Rep.,* pp. 67–69.

Kirk, H.J.C., 1968. The igneous rocks of Sarawak and Sabah. *Geol. Surv. Borneo Reg. Malaysia Bull.,* 5: 210 pp.

Klompé, Th.H.F., 1961. Pacific and Variscan orogeny in Indonesia. A structural synthesis. *Proc. 9th Pac. Sci. Congr., Bangkok, 1957,* 12: 76–115.

Klompé, Th.H.F., Johannas and Soekendar, 1961. Late Palaeozoic–early Mesozoic volcanic activity in the Sunda Land area. *Proc. 9th Pac. Sci. Congr., Bangkok, 1957,* 12: 204–217.

Koolhoven, W.C.B., 1930. Verslag over eene verkenningstocht in den Oostarm van Celebes en den Banggai-archipel. *Jaarb. Mijnw. Ned. Oost-Indië,* 1929: 187–228.

Koolhoven, W.C.B., 1932. De Geologie van het Maliliterrein (Midden Celebes). *Jaarb. Mijnw. Ned. Oost-Indië,* 1930(3): 127–153.

Koolhoven, W.C.B., 1933. Het primaire diamant voorkomen in Z-Borneo. *Mijningenieur,* 14: 138–144.

Koolhoven, W.C.B., 1935. Het primaire voorkomen van den Zuid-Borneo diamant. *Verh. K. Ned. Geol.-Mijnbouwkd. Gen.,* 11: 189–232.

Koperberg, M., 1929–1930. Bouwstoffen voor de geologie van de Residentie Menado. *Jaarb. Mijnw. Ned. Oost-Indië,* 1928: 1 and 2.

Krol, L.H., 1955. The Mesozoic folding in Borneo, Netherlands Indies, and surrounding territories, and its value for mapping unexplored non-fossiliferous areas. *Bull. Geol. Surv. Dep. Br. Terr. Borneo,* 2: 17–38.

Krumbeck, L., 1922. Zur Kenntnis des Juras der Insel Rotti. *Jaarb. Mijnw. Ned. Oost-Indië,* 1920: 3.

Kuenen, Ph.H., 1942. Obilatoe, Kisar and Siboetoe. *Geol. Mijnbouw,* 4: 81–90.

Kundig, E., 1956. Geology and ophiolite problems of east-Celebes. *Verh. K. Ned. Geol.-Mijnbouwkd. Gen.,* 16: 210–235.

Laws, R.A. and Kraus, G.P., 1974. The regional geology of the Bonaparte Gulf Timor Sea area. *APEA J.,* 14: 77–84.

Lee, S.M., 1974. The tectonic setting of Korea with relation to plate tectonics. *United Nations ESCAP, CCOP Tech. Bull.,* 8: 39–53.

Leichti, P., Roe, F.W. and Haile, N.S., 1960. The geology of Sarawak, Brunei and the western part of North Borneo. *Bull. Geol. Surv. Dep. Br. Terr. Borneo,* 3.

Leme, J. de A., and Coelho, A.V.P., 1962. Geologia do encrave de Ocussi. (provincia de Timor). *Garcia Orta (Lisboa),* 10: 553–566.

Lemoine, M., 1959. Un example de tectonic chaotique: Timor, essai de coordination et d'interpretation. *Rev. Géogr. Phys. Géol. Dyn.,* 2: 205–230.

Leong, K.M., 1972. Regional geology: eastern Sabah. *Malaysia Geol. Surv. Annu. Rep.,* pp. 73–77.

Lofting, M.J.W., Crostella, A.A. and Halse, J.W., 1975. Exploration results and future prospects in the northern Australasian region. *World Pet. Congr., Tokyo, 9th, Panel Disc.,* 7(3),

Loth, J.E. and Zwierzycki, J., 1926. De Kristallijne schisten op Java ouder dan Krijt. *Mijningenieur,* 2: 22–25.

McElhinny, M.W., Haile, N.S. and Crawford, A.R., 1974. Palaeomagnetic evidence shows Malay Peninsula was not a part of Gondwanaland. *Nature,* 252: 641–645.

Margolis, S.V., Ku, T.L., Glasby, G.P., Fein, C.D. and Audley-Charles, M.G., 1977. Fossil manganese nodules from Timor: geochemical and radiochemical evidence for deep-sea origin. *Chem. Geol.,* 20, in press.

Marks, P., 1956. Malayan archipelago. Asia. Lexique Stratigraphique International. *Congr. Géol. Int.,* Paris, 7: 3–241.

Molengraaff, G.A.F., 1920. Mangaanknollen in mesozoische diepzee-afzettingen van Nederl. Timor. *Versl. K. Akad. Wet.,* 29: 688–689.

Mollan, R.G., Craig, R.W. and Lofting, M.J.W., 1970. Geologic framework of continental shelf off northwest Australia. *Bull. Am. Assoc. Pet. Geol.,* 54: 583–600.

Murphy, R.W., 1973. Diversity of island arcs: Japan, Philippines, northern Moluccas. *APEA J.,* 13: 19–25.

Murphy, R.W., 1975. Tertiary basins of southeast Asia. *SEAPEX Proc.,* 2: 1–36.

Musper, K.A.F.R., 1928. Indragiri en Pelalawan. *Jaarb. Mijnw. Ned. Oost-Indië,* 1927(1): 1–247.

Musper, K.A.F.R., 1930. Beknopt verslag over uitkomsten van nieuwe geologische onderzoekingen in the Padangse Bovenlanden. *Jaarb. Mijnw. Ned. Oost-Indië,* 1930: 261–331.

Nogami, Y., 1968. Trias-Conodonten von Timor, Malaysien und Japan. *Mem. Fac. Sci. Kyoto Univ., Ser. Geol. Min.,* 34(2): 115–136.

Parke, M.L., Emery, K.O., Szymankiewicz, R. and Reynolds, L.M., 1971. Structural framework of continental margin in South China Sea. *Bull. Am. Assoc. Pet. Geol.,* 55: 723–751.

Patmosukismo, S. and Yahya, I., 1975. The basement configuration of the northwest Java area. *Proc. Indones. Pet. Assoc., 3rd Annu. Conv.,* pp. 129–152.

Powell, D.E., 1976. The geological evolution of the continental margin off northwest Australia. *APEA J.,* 16: 13–23.

Pupilli, M., 1973. Geological evolution of South China Sea area: Tentative reconstruction from borderland geology and well data. *Proc. Indones. Pet. Assoc., 2nd Annu. Conv.,* pp. 223–241.

Roothaan, H.Ph., 1928. Geologische en petrografische schets der Talaud en Nanoesa eilanden. *Jaarb. Mijnw. Ned. Oost-Indië,* 1925(2): 174–220.

Rosidi, H.H.D., Tjokrosapoetro, S. and Pendowo, B., 1976. Geologic map of the Painan and northeastern part of the Muarasiberut quadrangles, Sumatra. *Geol. Surv. Indones.,* 5/VIII.

Rutten, L.M.R., 1927. *Voordrachten over de geologie van Nederlandsch Oost-Indië.* Wolters, Groningen–Den Haag.

Sander, N.J., Humphrey, W.E. and Mason, J.F., 1975. Tectonic framework of southeast Asia and Australasia: Its significance in the occurrence of petroleum. *World Pet. Congr., Tokyo, 9th, Panel Disc.,* 7(4).

Schubert, R.J., 1913. Beitrag zur fossilen Foraminifera-fauna von Celebes. *Jahrb. K.K. Geol. Reichsanst.,* 1913: 127–150.

Schubert, R.J., 1915. Ueber Foraminiferengesteine der Insel Letti. *Jaarb. Mijnw. Ned. Oost-Indië,* 1914(1): 169–184.

Simons, A.L., 1940. Geological investigations in north-east Netherlands Timor. *Geol. Exped. Lesser Sunda Islands,* 1: 107–213.

Smith, A.G., Briden, J.C. and Drewry, G.E., 1973. Phanerozoic World Maps. *Spec. Pap. Palaeontol.,* 12: 1–42 (Palaeontol. Assoc. Lond.).

Snelling, N.J., Bignell, J.D. and Harding, R.R., 1968. Ages of Malayan granites. *Geol. Mijnbouw,* 47: 358–359.

Stauffer, P.H., 1974. Malaya and Southeast Asia in the pattern of continental drift. *Geol. Soc. Malaysia Bull.,* 7: 89–138.

Stauffer, P.H. and Gobbett, D.J., 1972. Southeast Asia a part of Gondwanaland? *Nature, Phys. Sci.,* 240: 139–140.

Stoneley, R., 1974. Evolution of the continental margins bounding a former Tethys Ocean. In: C.A. Burk and C.L. Drake (Editors), *The Geology of Continental Margins.* Springer-Verlag, Berlin, pp. 889–903.

Sukamto, R., 1975. *Geologic Map of Indonesia – Unjung Pandang Sheet VIII.* Geological Survey of Indonesia, Jakarta.

Tamesis, E.V., Manalac, C.A., Reyes, C.A. and Ote, L.M., 1973. Late Tertiary geologic history of the continental shelf off northwestern Palawan, Philippines. *Bull. Geol. Soc. Malaysia,* 6: 165–176.

Tappenbeck, D., 1940. Geologie des Mollogebirges und einiger benachbarter Gebiete. *Geol. Exped. Lesser Sunda Islands,* 1: 1–105.

Teichert, C., 1939. The Mesozoic transgressions in western Australia. *Aust. J. Sci.,* 2: 84–86.

Teichert, C., 1941. Upper Palaeozoic of western Australia: Correlation and palaeogeography. *Bull. Am. Assoc.Pet. Geol.,* 25: 371–415.

Teves, J.S., 1953. The pre-Tertiary geology of southern oriental Mindoro. *Philipp. Bur. Mines Rep. Invest.,* 10: 1–27.

Teves, J.S., 1956. Philippines: Asia. Lexique Stratigraphique International. *Congr. Géol. Int., Paris,* 3, 5: 1–167.

't Hoen, C. and Ziegler, K., 1917. Verslag over de resultaten van geol. mijnb. verkenningen in ZW Celebes. *Jaarb. Mijnw. Ned. Oost-Indië,* 1915(2): 253–363.

Tjia, H.D., 1973. Displacement patterns of strike-slip faults in Malaysia–Indonesia–Philippines. *Geol. Mijnbouw,* 52: 21–30.

Umbgrove, J.H.F., 1935. De Pretertiaire Historie van den Indischen Archipel. *Leidsche Geol. Meded.,* 7: 119–155.

Umbgrove, J.H.F., 1938. Geological history of the East Indies. *Bull. Am. Assoc. Pet. Geol.,* 22: 1–70.

United States Geological Survey, 1965. Geologic Map of Indonesia. *U.S. Geol. Surv. Misc. Publ.,* Map 1-414.

Uyeda, S. and Miyashiro, A., 1974. Plate tectonics and the Japanese islands: A synthesis. *Geol. Soc. Am. Bull.,* 85: 1159–1170.

Valk, W., 1945. *Contributions to the Geology of West Seran.* De Bussy, Amsterdam.

Van Bemmelen, R.W., 1949. *The Geology of Indonesia.* Government Printing Office, The Hague.

Van der Sluis, J.P., 1950. *Geology of East Seran.* De Bussy, Amsterdam.

Van West, F.P., 1941. Geological investigations in the Miomaffo region (Netherlands Timor). *Geol. Exped. Lesser Sunda Islands,* 3: 1–131.

Veevers, J.J. and Heirtzler, J.R., 1974. Tectonics and palaeogeographic synthesis of Leg 27. In: J.J. Veevers and J.R. Heirtzler et al., *Initial Reports of the Deep Sea Drilling Project, Leg. 27.* U.S. Government Printing Office, Washington, D.C., pp. 1049–1054.

Verbeek, R.D.M., 1905. Geologische beschrijving van Ambon. *Jaarb. Mijnw. Ned. Oost-Indië,* 1905: 1–323.

Verbeek, R.D.M., 1908. Molukken Verslag. *Jaarb. Mijnw. Ned. Oost-Indië,* 37: 1–835.

Visser, S.W. and Hermes, J.J., 1962. Geological results of the explorations for oil in Netherlands New Guinea. *Verh. K. Ned. Geol.-Mijnbouwk. Gen.,* 20: 1–265.

Von Loczy, L., 1933–4. Geologie van Noord Boengkoe en het Bongka gebied tusschen de Golf van Tomini en de Golf van Tolo in Oost Celebes. *Verh. K. Ned. Geol. Mijnbouwkd. Gen.,* 10: 219–227.

Wandel, J., 1936. Beitrag zur Kenntnis der jurassischen Molluskenfauna von Misool. *Neues Jahrb. Min. Pet.,* 75B: 447.

Wanner, J., 1907. Triaspetrefakten der Molukken und des

Timor Archipels. *Neues Jahrb. Min. Pet.,* 24: 161–220.

Wanner, J., 1914–1929. *Palaeontologie von Timor.* Stuttgart, 16 volumes.

Wanner, J., 1931a. Echinodermata from the Netherlands Indies. *Leidsche Geol. Meded.*, 5: 436–460.

Wanner, J., 1931b. Mesozoicum. Feestbundel Martin. *Leidsche Geol. Meded.,* 5: 567–610.

Wanner, J., 1940. Gesteinsbildende Foraminiferen aus Malm und Unterkreide des ostlichen Ostindischen Archipels. *Palaeontol. Z.,* 22: 75–99.

Warris, B.J., 1973. Plate tectonics and the evolution of the Timor Sea, Northwest Australia. *APEA J.,* 13, 13–18.

Westerveld, J., 1941. Three geological sections across South Sumatra. *Verh. K. Ned. Akad. Wet.,* 44: 1131–1139.

Wichmann, A., 1925. Geologische Ergebnisse der Siboga Expedition. *Siboga Monogr.,* 66,

Witkamp, H., 1912–1913. Een verkenningstocht over het eiland Soemba. *Tijdschr. K. Ned. Aardrijkskd. Gen.,* 29: 744–775; 30: 8–27, 484–505, 619–637.

Wolfenden, E.B. and Haile, N.S., 1963. Sematan and Lundu areas, west Sarawak. *Rep. Geol. Surv. Dep. Br. Terr. Borneo,* 1.

Wong, P.Y., 1972. Regional Geology: Northwest Borneo geosyncline. *Malaysia Geol. Surv. Annu. Rep.,* pp. 70–73.

Yamagiwa, N., 1963. Some Triassic corals from Portuguese Timor: Palaeontological Study of Portuguese Timor (I). *Mem. Osaka Univ. Lib. Arts Educ., Nat. Sci.,* 12: 83–87.

Zwierzycki, J., 1930. *Toelichting bij blad VIII van den geol. overzichtskaart van den Ned. Ind. Archipel. 1 : 1,000,000.* Dienst Mijnw. Ned. Oost-Indië.

Zwierzycki, J., 1931. *Toelichting bij blad 1 (Teloek Betoeng). Blad 1 Geol. Kaart van Sumatra, 1 : 200,000.* Dienst Mijnw. Ned. Oost-Indië.

Zwierzycki, J., 1932. *Toelichting bij blad 2 (Kotaagoeng). Geol. kaart Z. Sumatra. 1 : 200,000.* Dienst Mijnw. Ned. Oost-Indië.

Chapter 7

AUSTRALIA

N.H. LUDBROOK

INTRODUCTION

The Mesozoic geology of Australia is set within the framework of events associated with the dispersal of Gondwanaland and the evolution of the island continent. Differences in history, structural setting and tectonic style of these events can be distinguished in three main regions:

(1) the western margin, mostly in the state of Western Australia but including the northwest of the Northern Territory, which presents a history of sea-floor spreading from rifting to the development of the mature ocean (Veevers and Johnstone, 1974);

(2) eastern Australia, embracing the states of Queensland, New South Wales and eastern Tasmania, the northern part of which contains the geological record of tectonic evolution from transitional domain to craton (Day et al., 1974);

(3) southern Australia which includes areas marginal to the present southern coasts of Western Australia, South Australia and Victoria, the sedimentary and tectonic history of which also gives evidence of rift development and sea-floor spreading associated with the separation of the Australian plate from the Antarctic plate (Griffiths, 1971).

A similar division but into four sectors was made by Veevers and Evans (1973).

Sedimentary basins in which Mesozoic strata occur are shown in Fig. 1, adapted from the Tectonic Map of Australia and New Guinea (Geological Society of Australia, 1971). This figure also demonstrates the overlap of post-Triassic basins, formed during or after the fragmentation of Gondwanaland, upon pre-Jurassic basins containing Permian–Triassic or Permian sediments.

Although there is evidence of incipient rifting in the Permian of the Perth Basin (Johnstone et al., 1973; Veevers and Johnstone, 1974), and the East Australian Orogenic Province was subjected to intense volcanism and marine sedimentation during the Palaeozoic (Day et al., 1974), the Late Permian was on the whole a period of continental stability. Australia still remained part of Gondwanaland, the final breakup of which had not commenced.

The continent was affected by Mesozoic global events such as the marine transgression in the Early Scythian, which Balme and Helby (1973) attributed to eustatism, and the pulse of sea-floor spreading in the Atlantic and Pacific Oceans manifested in Late Jurassic and Early Cretaceous volcanism (see Fig. 7) and an epicontinental marine transgression in the Aptian (see Fig. 8) (Veevers and Evans, 1973).

The following summary of the Mesozoic geology of the continent as a whole relies heavily on the papers of Veevers and Johnstone (1974) and Day et al. (1974).

Early Triassic

In the Early Triassic (Fig. 2), the northwest coast was open to Tethys and marine transgressions occurred in the Fitzroy Trough and northern Perth Basin, and also in the Gympie Basin in Queensland. Elsewhere it was a period of continental sedimentation during which red beds were deposited over a large elongated area extending from the Sydney to the Bowen Basins and westerly to the Galilee and Cooper Basins. The Tasmania Basin received similar sediments. Granitic intrusions represent the extension of the late orogenic transitional phase in south-eastern Queensland to the Early Triassic, and during this tensional regime the formation of graben-like depressions such as the Esk and Abercorn Troughs was accompanied by the piling of thick andesitic volcanics.

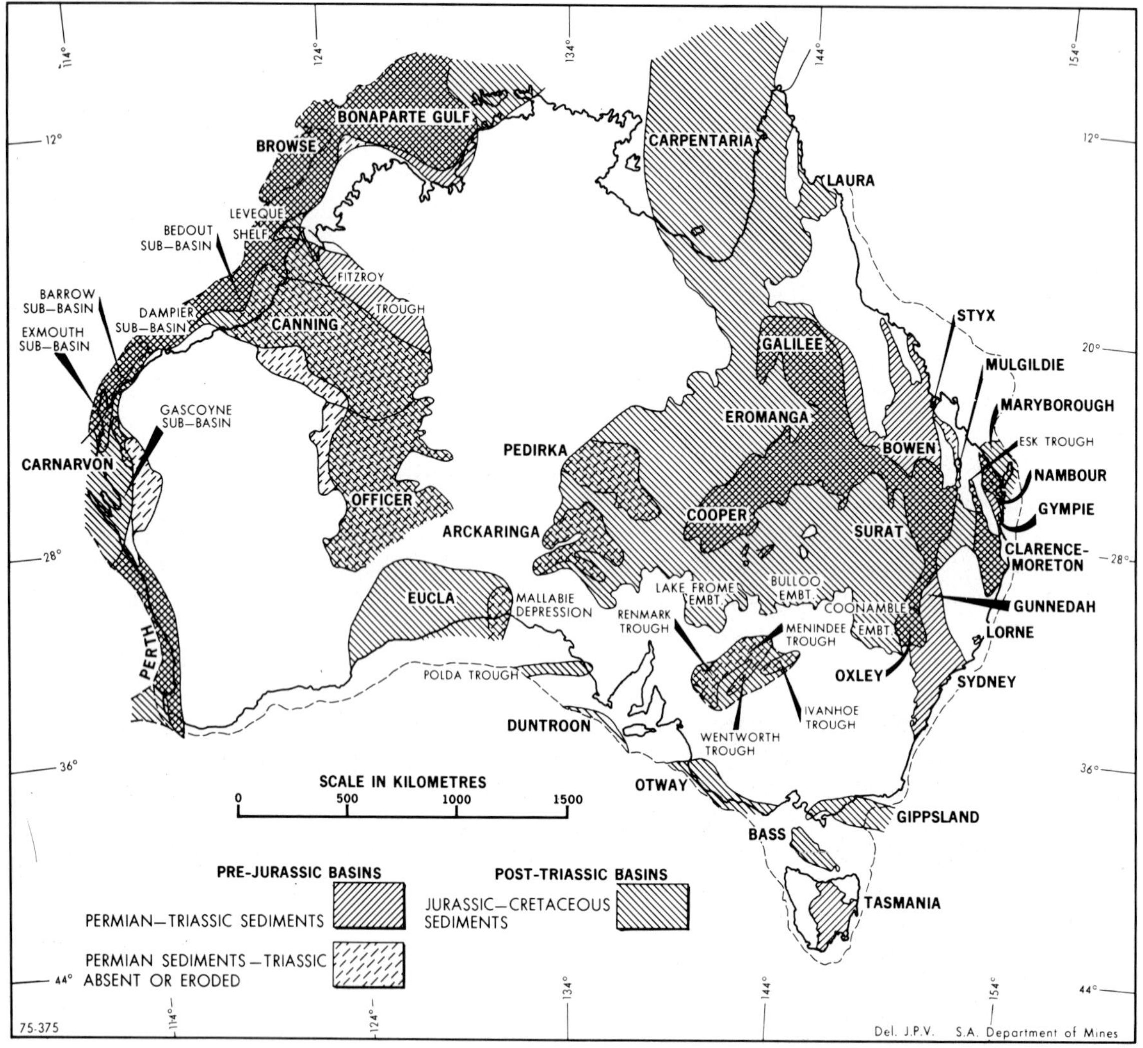

Fig. 1. Sedimentary basins containing Mesozoic rocks.

Middle Triassic

The withdrawal of the sea and the deposition of fluvial, paludal and deltaic sandstones in most areas of sedimentation characterized the Middle Triassic (Fig. 3). Crustal tension was accompanied by faulting and the continuation of andesitic and trachytic flows in southeastern Queensland.

Late Triassic

Except for marine carbonates on Ashmore and Scott Reefs, deposition in the western sector during the Late Triassic (Fig. 4) was almost entirely non-marine offshore from the present continental margin. Red beds were deposited near the margin of the Bonaparte Gulf Basin. In the east, tectonism led to the development of discrete intermontane basins containing important economic reserves of coal, particularly in Queensland, South Australia and Tasmania.

The Triassic closed with the deformation of the sediments in the Clarence–Moreton and Ipswich Basins, the termination of the post-orogenic transitional phase and the evolution of the craton in the area (Day et al., 1974).

Early Jurassic

The western sector was tectonically quiet in the Early Jurassic (Fig. 5) and over surfaces of high struc-

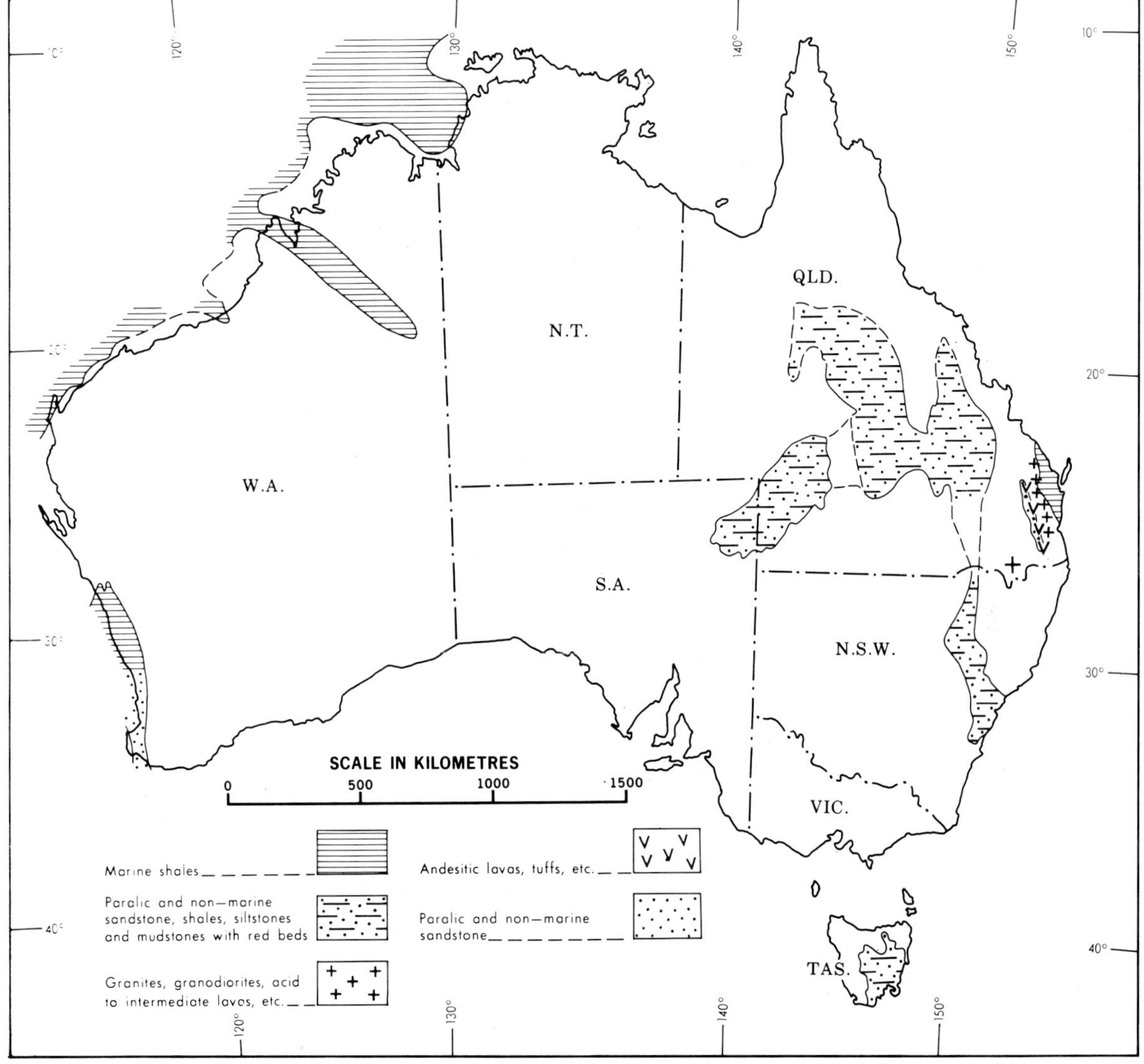

Fig. 2. Distribution of Early Triassic rocks.

tural relief non-marine sediments were deposited in the Perth Basin and non-marine to deltaic and marginal marine sediments along the northwest shelf. In the eastern sector deformation was followed by the shedding from uplifted areas of great quantities of quartzose sand which blanketed a wide area of central eastern Australia.

Middle Jurassic

The northern part of the Perth Basin was again entered by the sea in the Bajocian, and over most of the northwest shelf deposition continued under deltaic or marginal marine conditions. The Middle to earliest Late Jurassic experienced a further pulse of rifting with the initiation of a ridge off the northwest shelf. Deposition in the Bonaparte Gulf Basin was non-marine.

In the eastern sector the predominantly fluvial sands were replaced by fluvial, lacustrine and paludal deposits with coal. Volcanism was minor and local.

In the southern sector, however, the initiation of the rift between Australia and Antarctica is indicated by the extrusion of dolerite and tholeiitic basalt over much of eastern Tasmania as well as in the Otway Basin and on Kangaroo Island (Fig. 6).

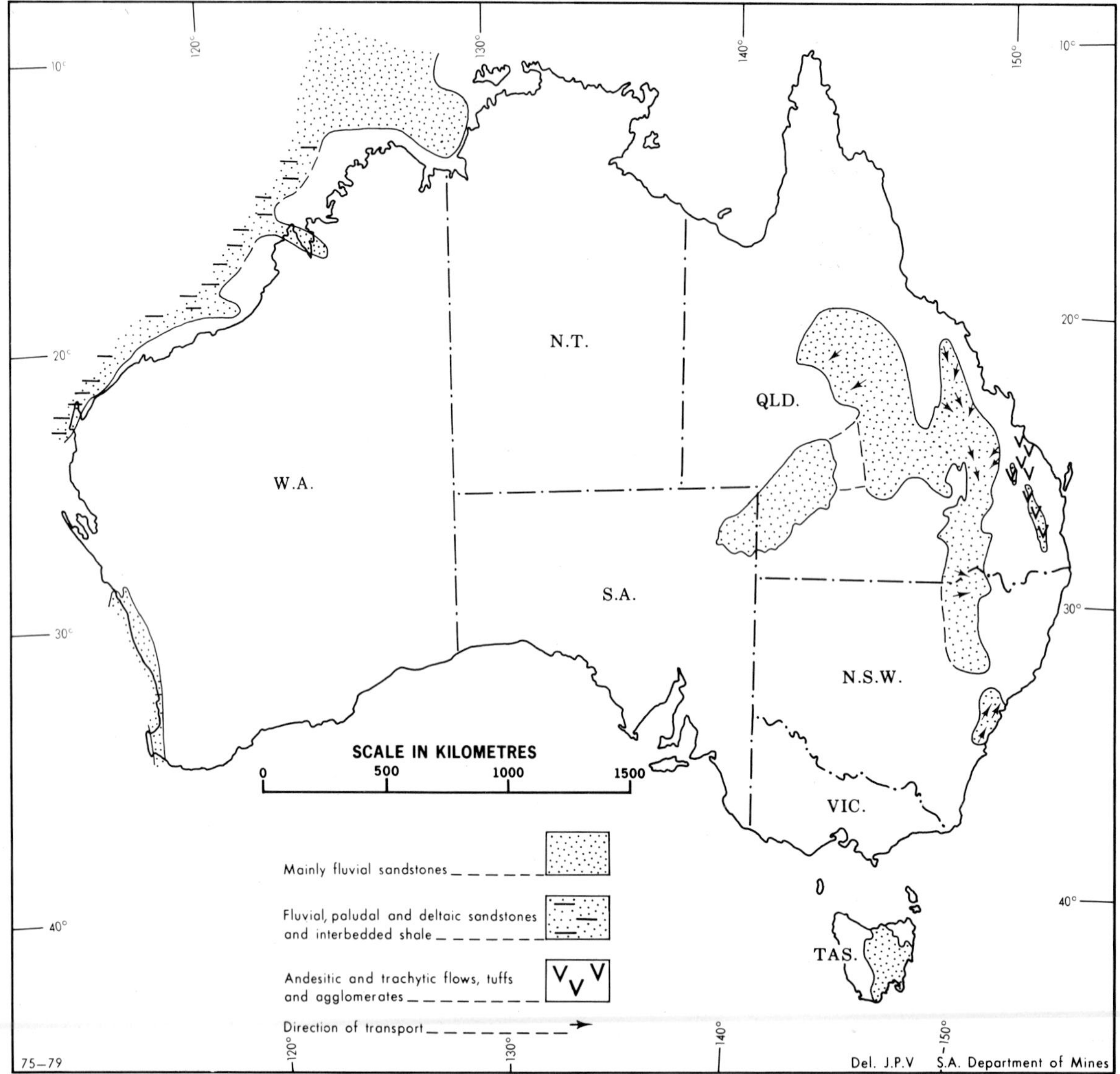

Fig. 3. Distribution of Middle Triassic rocks.

Late Jurassic to Neocomian

The Late Jurassic (Fig. 7) was a period of rupture and initiation of the juvenile ocean on the northwest margin. Oceanic basalt was emplaced on the Argo Abyssal Plain (DSDP 261) and on a spreading ridge which was probably a zone of Jurassic uplift forming the seaward margin of Mesozoic basins such as the Browse Basin (Laws and Kraus, 1974). Rupture in the southwest occurred somewhat later, in the Late Jurassic to Neocomian, and was possibly contemporaneous with the development of the rift along the southern margin of the present continent.

The disruption of Gondwanaland with the inception of sea-floor spreading in the west and of rifting in the south was reflected on the northeastern margin by intense volcanic activity from the Maryborough Basin northward to the Whitsunday Islands (Figs. 7, 8). There was progressive expansion of fresh-water to deltaic sedimentation over the wide area of the Great Artesian Basin as a result of uplift on ridges marginal to the Surat (Power and Devine, 1970), Eromanga and Carpentaria Basins, and subsidence following a very long period of peneplanation in the west of the

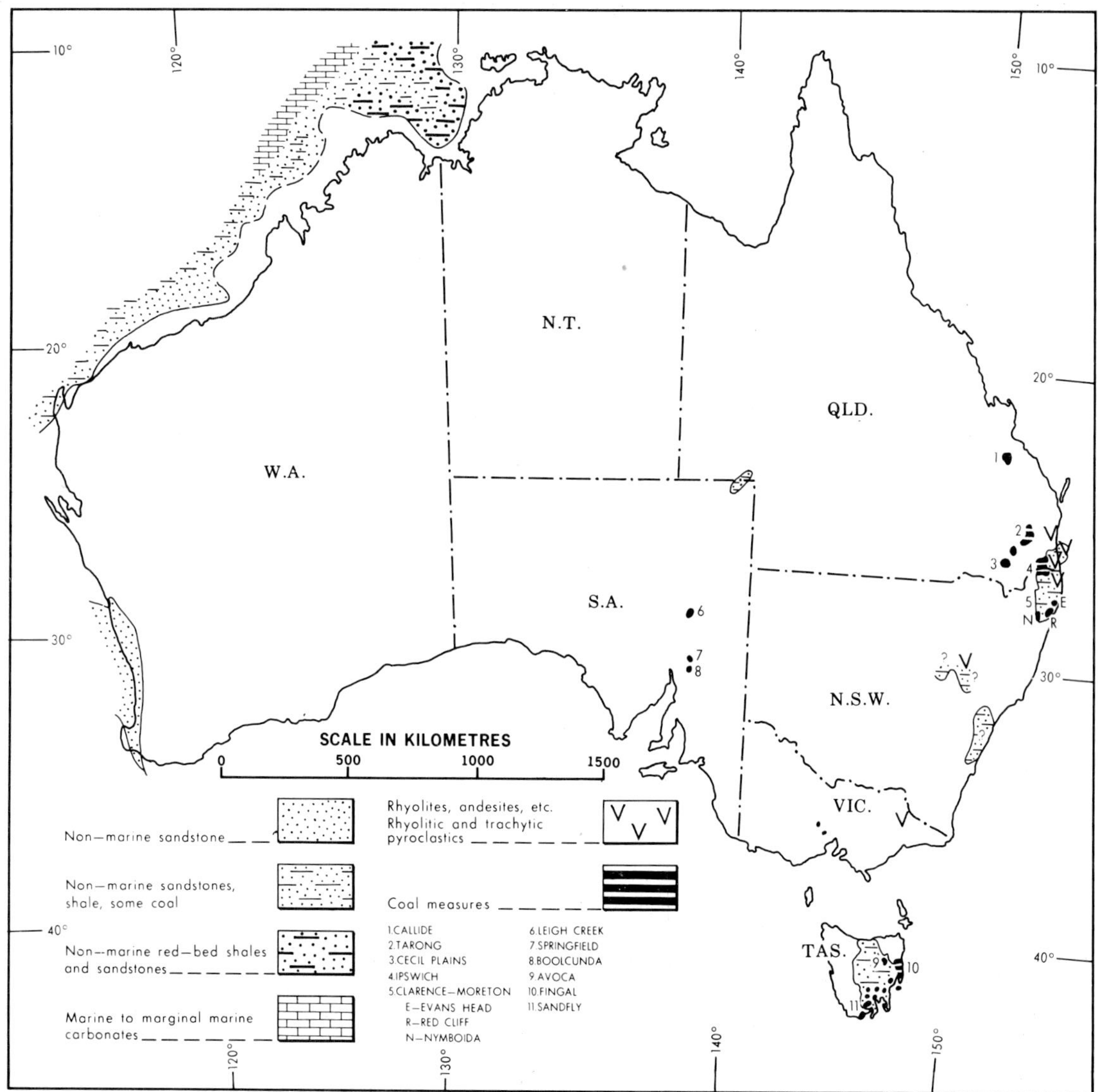

Fig. 4. Distribution of Late Triassic rocks.

basin (Wopfner, 1964). By the Late Neocomian, the sea had begun to transgress the area.

Early Cretaceous (Aptian–Albian)

In the Early Cretaceous (Aptian–Albian; Fig. 8), most of Australia was covered by an epicontinental sea which reached its maximum in the Aptian following the initiation of sea-floor spreading in the west and the magmatism in the east (Veevers and Evans, 1973). From the identification of a rich Neocomian molluscan fauna in the coastal belt of the Northern Territory (Skwarko, 1966), it may be assumed that the transgression came from the north. It is likely that there was connection between the southeast Canning Basin, Eucla Basin and Great Artesian Basin at this time, but the evidence for connection between the Canning Basin and the offshore Bedout Sub-Basin is negative, since no Aptian–Albian sediments are known from the northwest Canning Basin (M.H. Johnstone, pers. comm., 1975).

On the western margin the juvenile ocean phase continued to the Santonian. Evidence of this was provided by DSDP 263, drilled on the Cuvier Abyssal Plain in 5065 m of oceanic water. Between 100 m and the total depth of 746 m the drillhole inter-

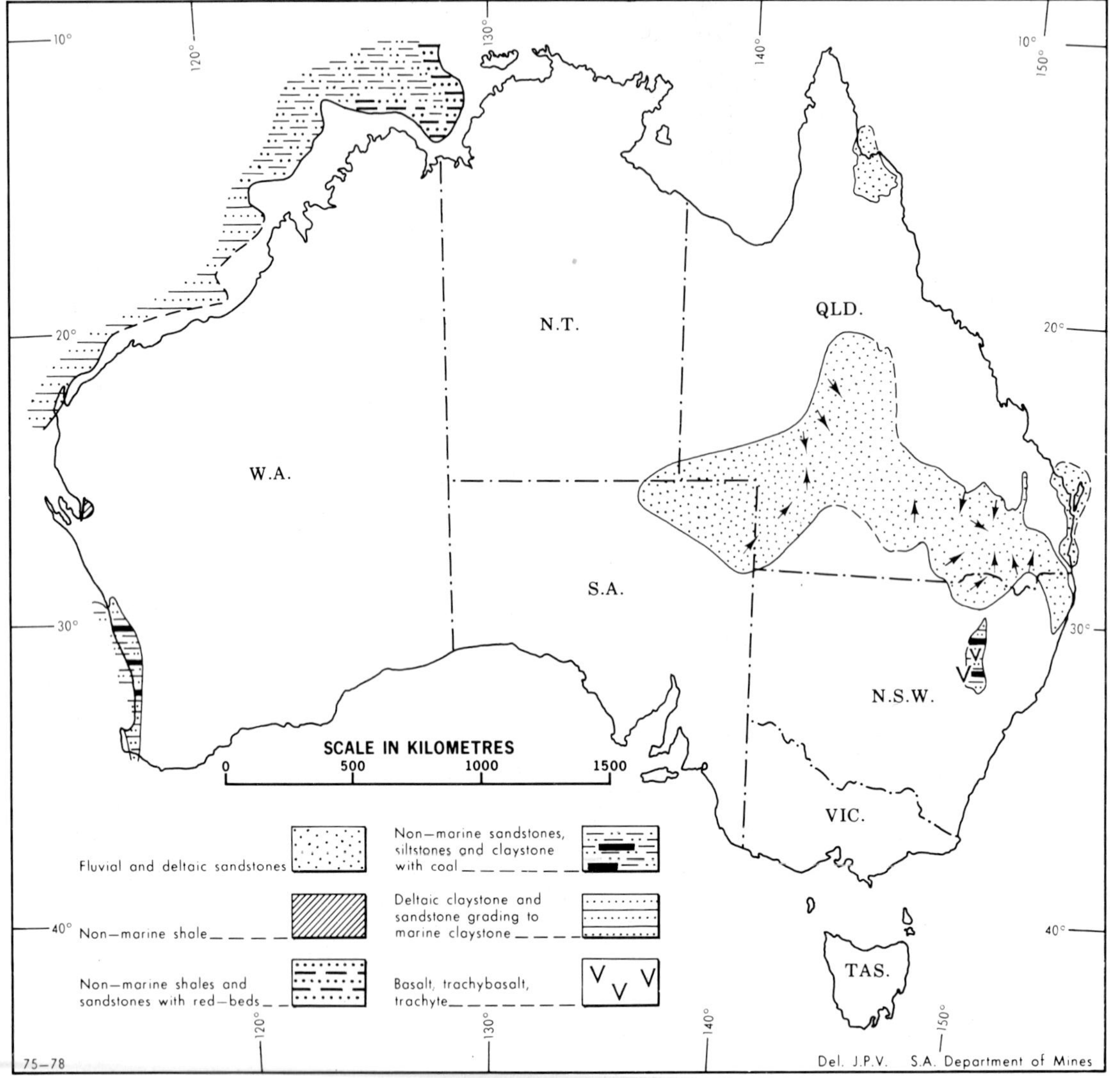

Fig. 5. Distribution of Early Jurassic rocks.

sected Albian or Aptian–Albian black clays deposited in shallow water overlain by Albian clayey nannoplankton ooze indicative of a deeper-water environment (Veevers and Johnstone, 1974).

The margin of the Eucla Basin lay on the continental shelf where it was probably separated from the rift valley forming along the southern margin by the uparching of the northern flank of the rift (Johnstone et al., 1973). The sea had not yet entered the rift on the southern margin; sediments of Aptian–Albian age in the Duntroon, Otway, Bass and Gippsland Basins are non-marine.

Late Cretaceous

The Late Cretaceous (Fig. 9) is distinguished by a change in regime. On the western margin deposition of detrital sediments continued till the Turonian, when there was a regional hiatus covering part of the Turonian and the Coniacian. The sediments following the hiatus belong to a carbonate mature ocean sequence beginning in the Santonian and generally persisting to the present day. Exceptions are on the western margin of the Bedout and Browse Basins, where the hiatus occurred in the Cenomanian and on

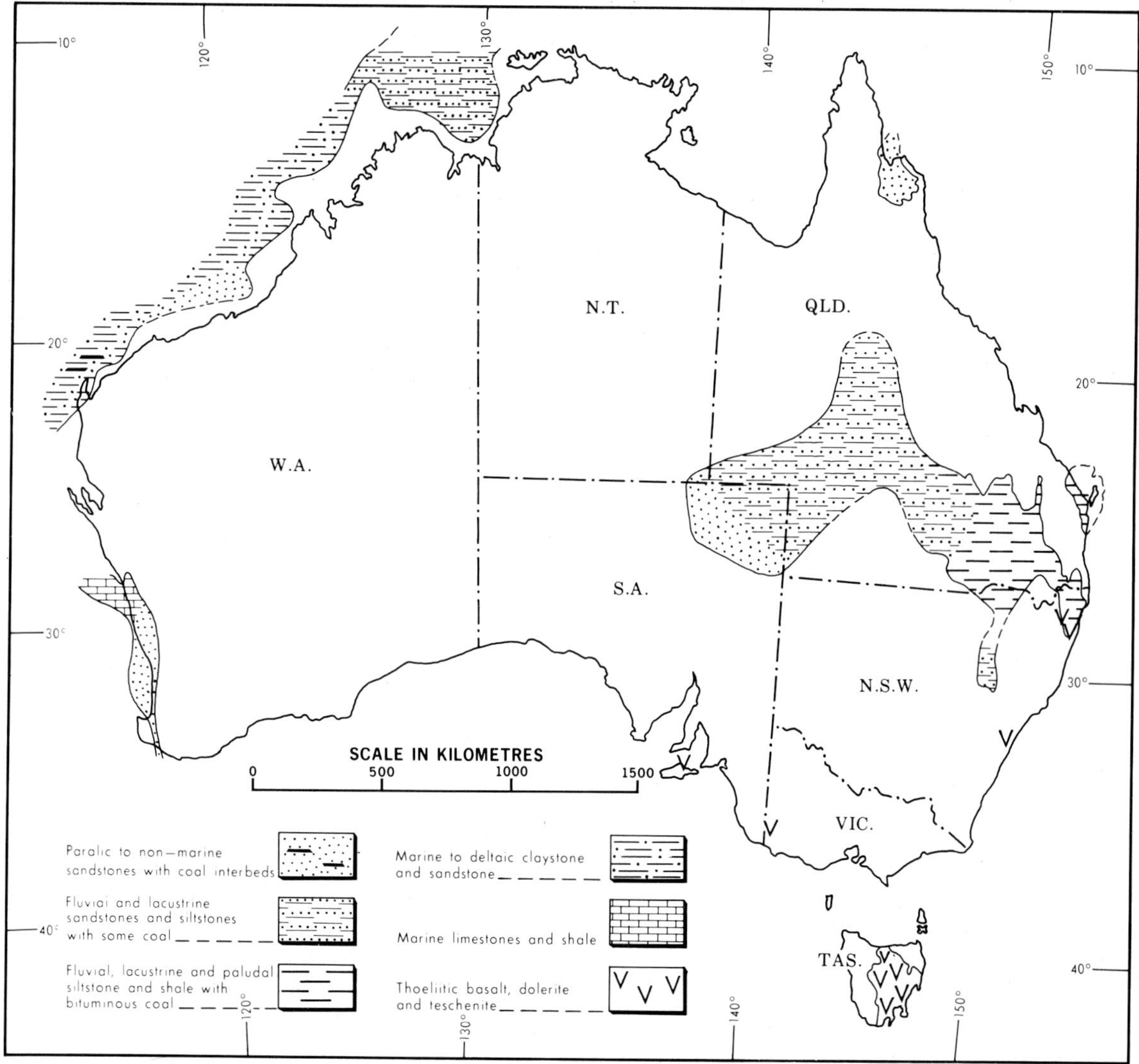

Fig. 6. Distribution of Middle Jurassic rocks.

the Naturaliste Plateau (DSDP 258 and 264) where carbonate sedimentation began in the Cenomanian–Turonian.

At this time the sea gained temporary entry into the southern rift and deposited Late Cretaceous marine to paralic sediments in the western part of the Eucla Basin, in the Duntroon Basin, and in the Otway Basin. It was not until the Middle Eocene that final opening of the rift with the deposition of marine carbonates occurred.

In the eastern sector the sea had retreated northwards. Cenomanian–Turonian ammonites have been described from Bathurst Island and Cenomanian ammonites from Melville Island (Wright, 1963). Only a lacustrine remnant of the former Great Artesian Basin remained, of somewhat lesser extent than the total drainage area of the Lake Eyre Basin today. Subsequent Late Cretaceous deposition was confined to river channels.

The Tasman Basin (under the Tasman Sea) between Australia and New Zealand opened in the Late Cretaceous (Griffiths, 1971; Elliott, 1972; Vogt and Connolly, 1971). The oldest sediments intersected in DSDP 207 and 208 of Leg 21, drilled on the Lord Howe Rise and to the north in the Coral Sea, were Maastrichtian. Site 207 on the summit of the southern end of the Lord Howe Rise bottomed in rhyolitic rocks below Maastrichtian silty claystone.

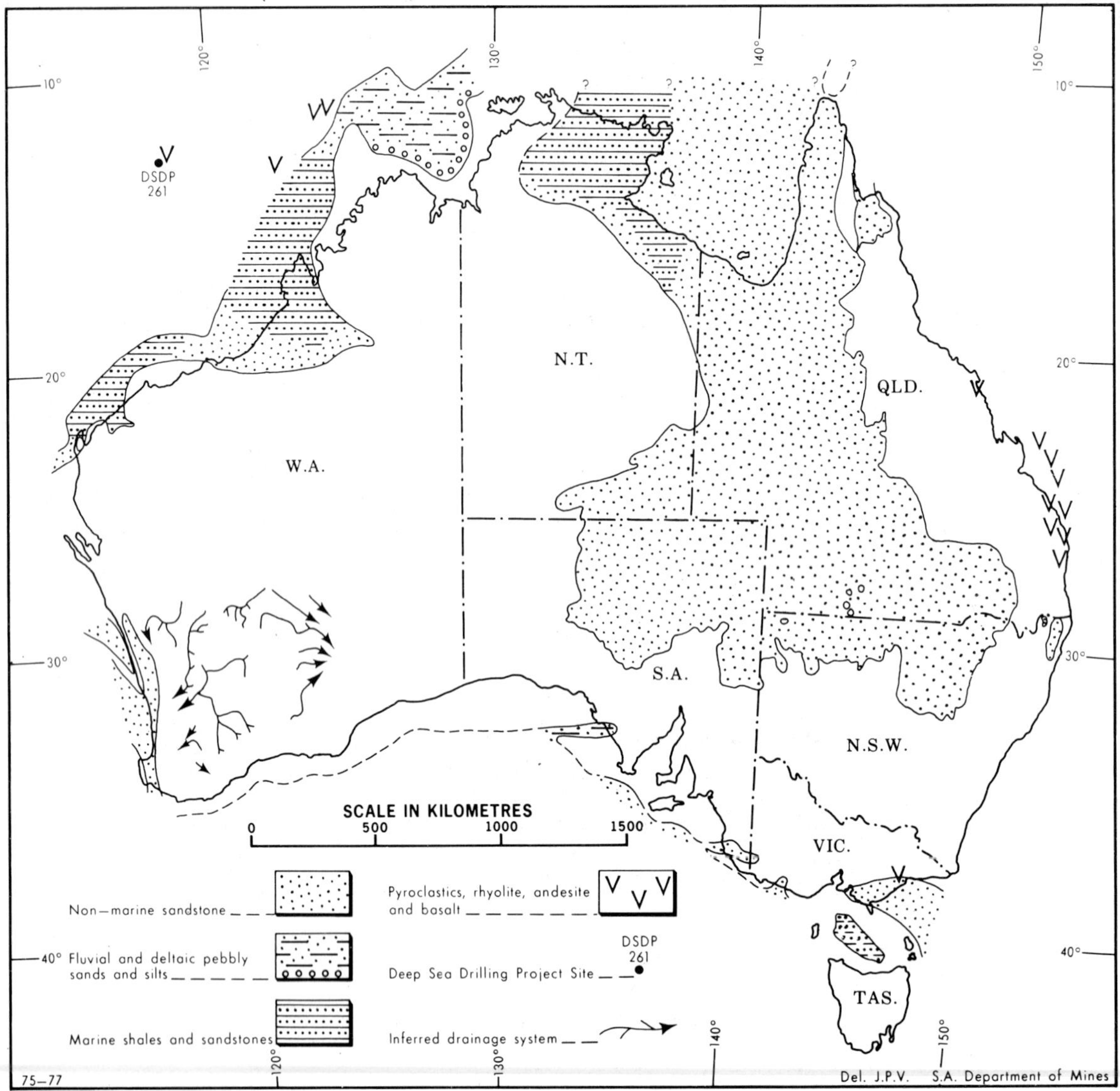

Fig. 7. Distribution of Late Jurassic to Early Cretaceous rocks.

Sanidines from the rhyolites gave a mean $^{40}Ar/^{39}Ar$ age of 93.5 ± 1.8 m.y. (McDougall and Van der Lingen, 1974). McDougall and Van der Lingen interpret these data as indicating that the separation of the Lord Howe Rise from the Australian–Antarctic continent began in the Late Cretaceous between 94 and 70 m.y. ago. This separation completed the formation of the eastern continental margin and as a result of the subsidence of the Tasman Basin sea-floor spreading commenced. The eastern margin of the Australian plate and its junction with the Pacific plate lies in the highly complex tectonic zone running from the Solomon Islands southward through New Zealand to south of the Macquarie Ridge where it joins the Antarctic plate.

WESTERN AUSTRALIA

Basins of the western continental margin

The western continental margin is considered by Veevers and Johnstone (1974) to have developed in three phases: (1) rifting on the southwest margin

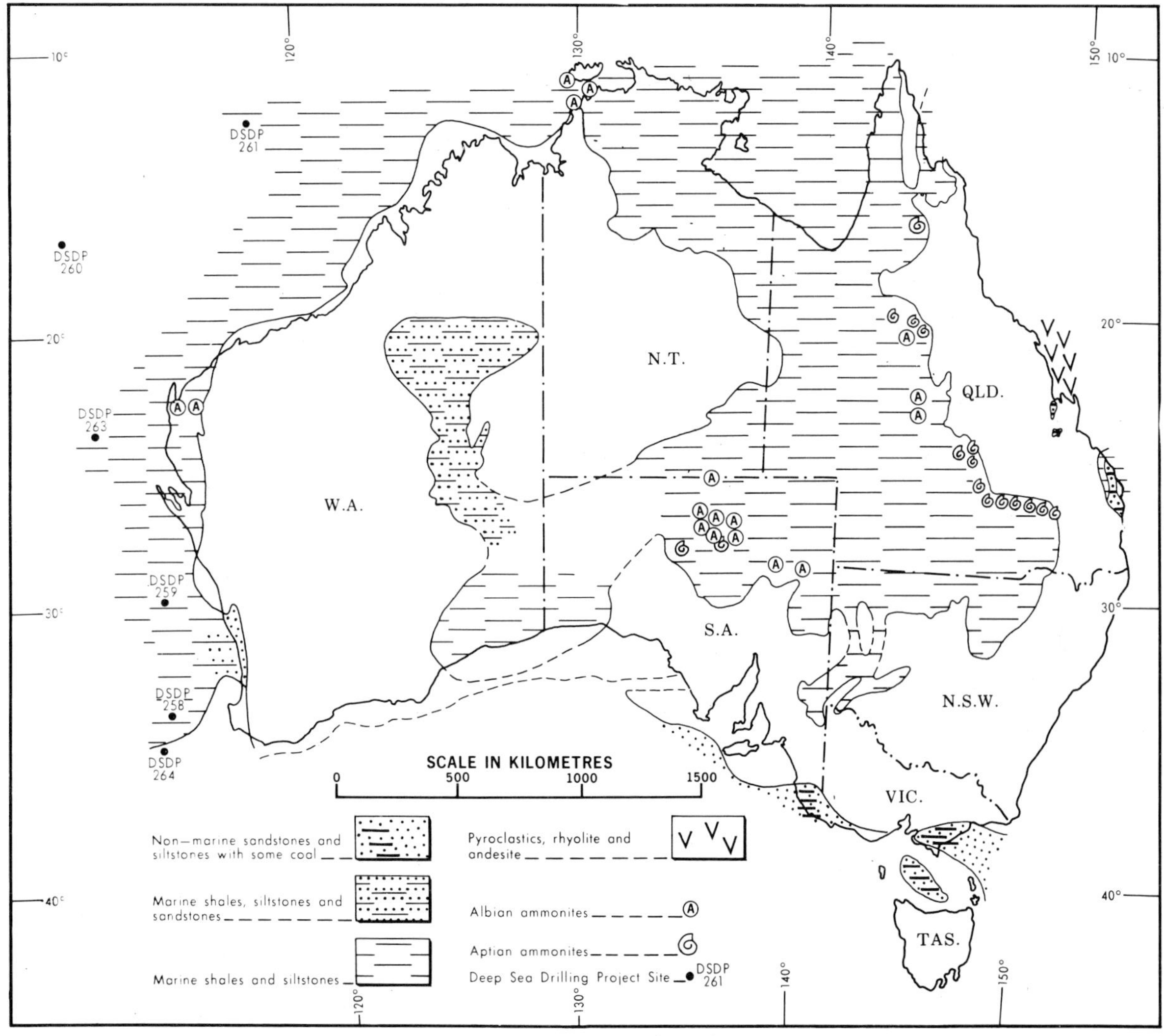

Fig. 8. Distribution of Early Cretaceous (mostly Aptian–Albian) rocks.

which commenced in the Early Permian and extended to the middle of the Neocomian, with main pulses in the Early Permian, Kazanian, Middle Triassic, and Middle to earliest Late Jurassic; (2) juvenile ocean development with eruption of basalt and marine transgression commencing in the Middle to Late Jurassic and lasting until the Santonian; and (3) development of the mature Indian Ocean in the Late Cretaceous with accumulation of shallow-water carbonates.

Knowledge of the structure and stratigraphy of offshore and onshore basins of Western Australia has increased dramatically over the last two decades. It has been gained from numerous wells drilled on the continental shelf during intensive and successful exploration for oil and gas in Mesozoic sediments, leading to the development of the Early Cretaceous Barrow Island oil field in the Carnarvon Basin and commercial gas discoveries in the Early Triassic and Early Jurassic of the Dandaragan Trough in the northern Perth Basin. It has been further supplemented by data from the DSDP sites in the eastern and southern Indian Ocean.

The northern basins – the Bonaparte Gulf Basin, Fitzroy Trough, and the onshore Canning Basin – trend generally northwest–southeast. They differ from those in the south in that they are filled with predominantly marine sediments deposited through-

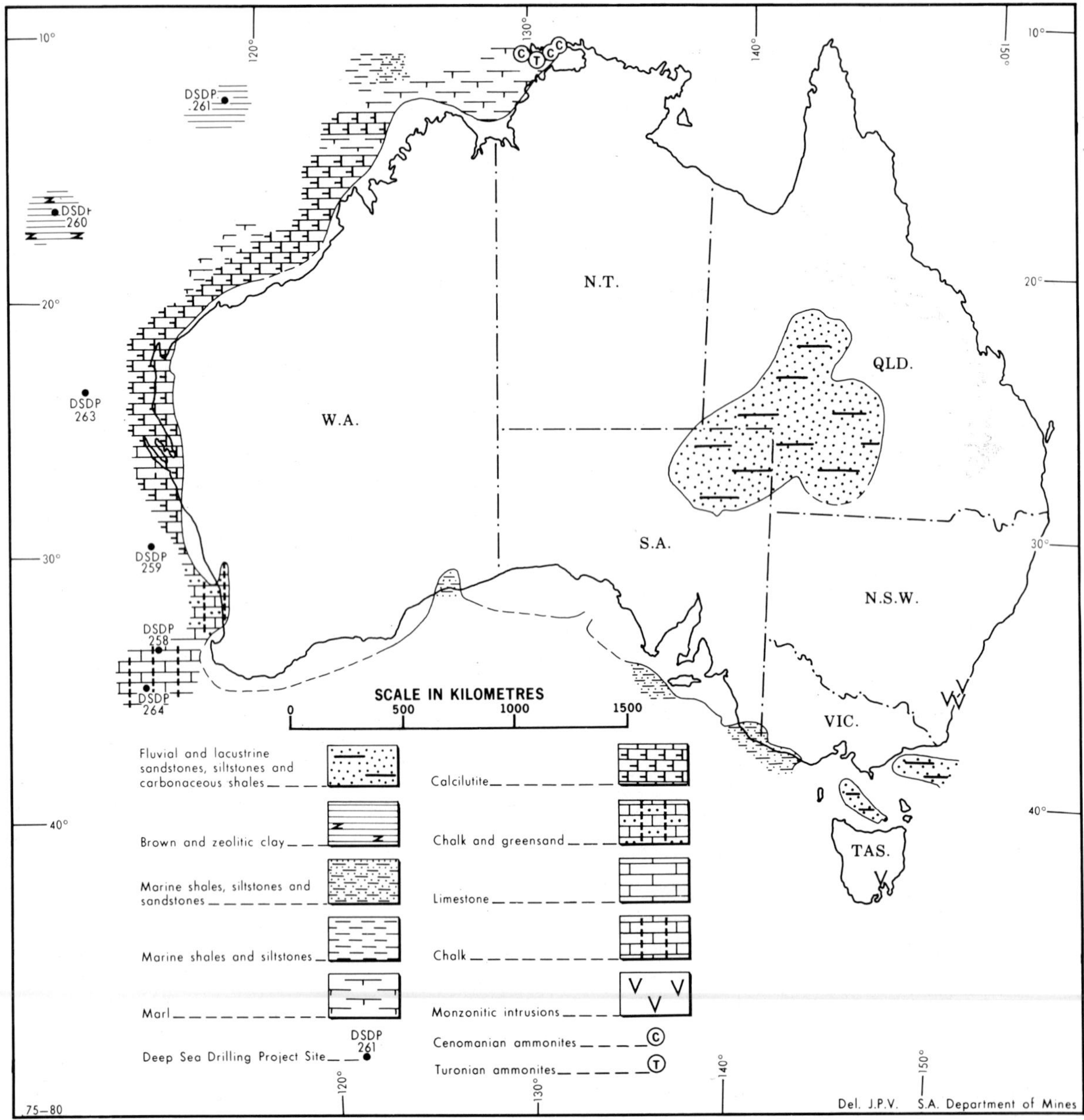

Fig. 9. Distribution of Late Cretaceous rocks.

out the Phanerozoic on broad platforms (Brunnschweiler, 1954) as a result of major crustal downwarp of or between continental blocks. The southern Carnarvon and Perth Basins are north–south-trending and the result of rifting or "pull-apart" tectonics (Veevers, 1972; Johnstone, in discussion of Veevers, 1972), the highly faulted north–south rift of the Perth Basin (Fig. 11) having a high proportion of nonmarine Permian and Mesozoic sediments. Veevers (1971, 1972) regards the northerly basins as lying on the continental margin fronting Tethys, while the Perth Basin lay within Gondwanaland until Australia moved away from the land to the west in the Cretaceous. The sequences and principal events of these basins are:

Bonaparte Gulf Basin

Onshore the Bonaparte Gulf Basin (Mollan et al.,

1969; Roberts and Veevers, 1973; Laws and Kraus, 1974) is almost exclusively Palaeozoic, with some Triassic and Cretaceous south of Port Keats in the Northern Territory. Offshore the basin and its components are well known from seismic studies and well sections. The basin formed between the stable Kimberley, Sturt and Darwin Precambrian blocks as the result of tensional block faulting, the main grain in the Petrel Sub-basin between them being northwest–southeast. Late Jurassic to Holocene trends on the northwest margin bordering the Timor Trough are predominantly northeast–southwest (Laws and Kraus, 1974).

Browse Basin

The Browse Basin (Halse and Hayes, 1971) is an offshore shelf area produced by a major crustal downwarp of the Kimberley Block.

Canning Basin and Fitzroy Trough

These are the results of crustal downwarp between the Precambrian shield of the Pilbara and Kimberley Blocks. They contain great thicknesses of Phanerozoic sediments each sequence of which is characteristic of a particular sub-basin: the South Canning Basin with Early Palaeozoic sediments, the Fitzroy Trough with a Late Palaeozoic and a Permian–Triassic sequence and the offshore Bedout Sub-basin with a complex of Palaeozoic and Mesozoic sediments (Brunnschweiler, 1954; Veevers and Wells, 1961; Balme, 1969a; Challinor, 1970).

Carnarvon Basin

This basin (Parry, 1967; Mollan et al., 1969; Challinor, 1970; Thomas and Smith, 1974) is intermediate in form between the northern basins and the Perth Basin. Its northern province is a major Mesozoic downwarp containing Triassic, Jurassic and Neocomian sediments (Thomas and Smith, 1974) in the offshore Dampier and Barrow Sub-basins, the Dampier Sub-basin having a similar sequence to that of the Bedout Sub-basin and thus being one of the connections between the offshore Carnarvon Basin and the offshore Canning Basin (Challinor, 1970). The Barrow Sub-basin (Parry, 1967; Burdett et al., 1970; Crank, 1973) has a known section of Late Jurassic to Miocene sediments. It contains the Barrow Island oil field, about 98% of the production coming from the Windalia Sand Member of Aptian–Albian age.

The southern Carnarvon Basin (Geary, 1970) is divided by a number of north–south-trending basement ridges and includes the Exmouth, North Gascoyne and South Gascoyne Sub-basins. The thick Permian section of the eastern part of the basin and the Triassic and Jurassic sequences of the Exmouth Sub-basin are missing in the offshore North Gascoyne Sub-basin where control is limited to one well, Pendock No. 1, which intersected Cretaceous sediments unconformably overlying Carboniferous dolomite.

Perth Basin

The Perth Basin (Jones and Pearson, 1972; Veevers, 1972; Johnstone et al., 1973) is an elongated north–south-trending highly faulted trough or half graben marginal to the Precambrian shield against which it is downfaulted by the Darling Fault. Less than half is onshore. It has four sub-basins: the Dandaragan and Bunbury Troughs and the Abrolhos and Vlaming Sub-basins, separated by basement highs. It was initiated in the Palaeozoic by broad down-warping accompanied by widespread marine sedimentation connected with the Carnarvon Basin during the Early Permian. It is filled with more than 15,000 m of Late Palaeozoic and Mesozoic sediments on its eastern margin against the Darling Fault and shallows to the northwest and south onto basement highs of which the Northampton Block is the northern and the Leeuwin–Naturaliste Block the southwestern. Mild tectonism during the Late Permian initiated a series of NNW–SSE-trending faults which progressively truncate the Permian and Mesozoic sequences as they lap onto the Northhampton Block (Hosemann, 1971).

The chief structural elements of the Perth and Carnarvon basins are shown in Figs. 10 and 11.

Stratigraphy of the western sector

The stratigraphic relationships of the Mesozoic rocks of the western sector are shown in Table I.

Triassic

Rifting is assumed to have begun in the southwest in the Permian and extended to the Neocomian, with the main pulses in the Early Permian, Kazanian, Middle Triassic and Middle to earliest Late Jurassic (Veevers and Johnstone, 1974). During the widespread Early Triassic transgression (Fig. 2), the sea temporarily gained entrance to the Fitzroy Trough and the northern part of the Perth Basin. The marine

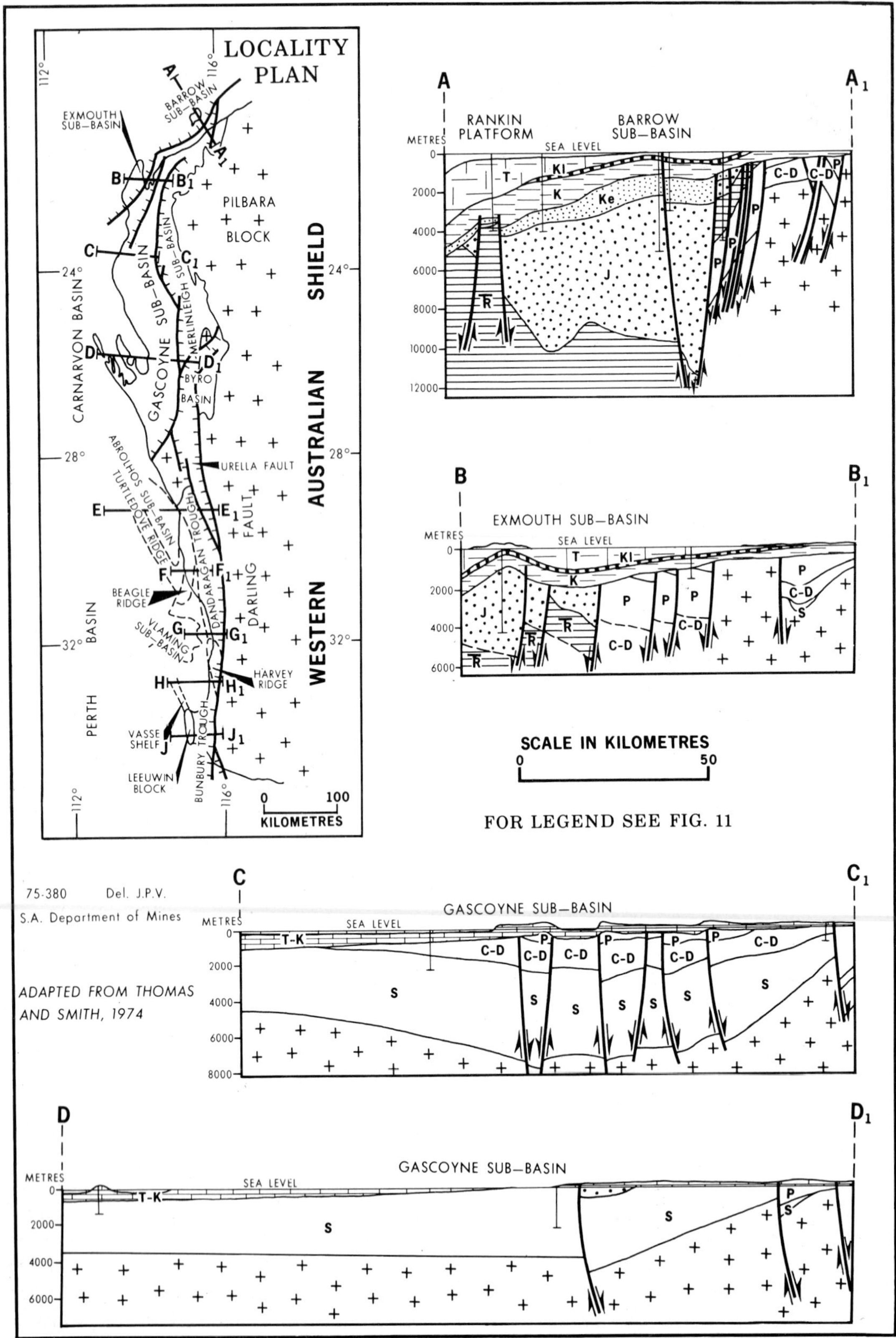

Fig. 10. Structural cross-sections Carnarvon Basin.

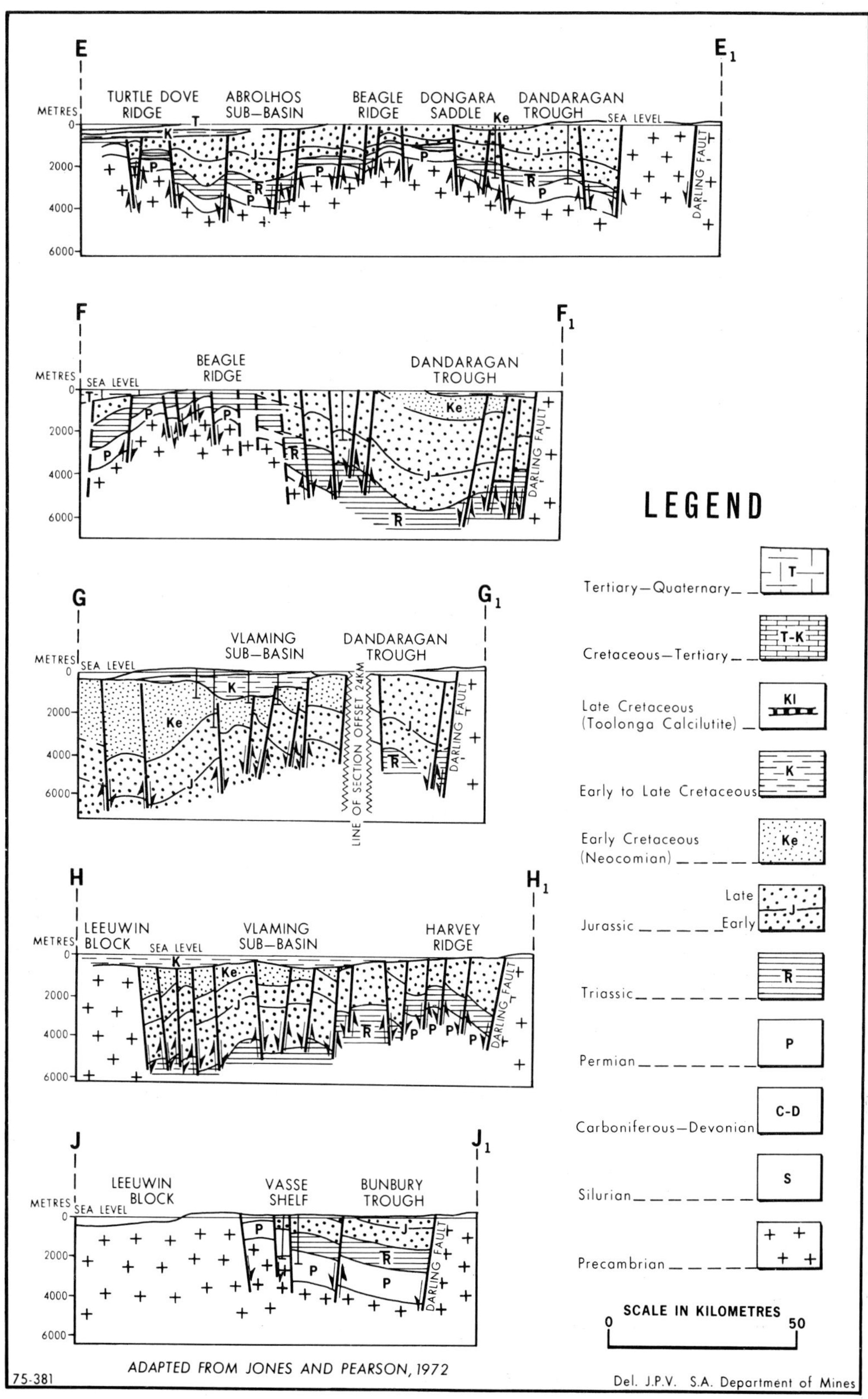

Fig. 11. Structural cross-sections Perth Basin.

transgression in the Perth Basin occurred after mild tectonic uplift and tilting of the Precambrian Northampton Block in the Late Permian, and composite nearshore marine and littoral sediments were deposited on progressively truncated strata and Precambrian basement in an area of low topographic relief (Hosemann, 1971).

Although exposures are limited, Triassic sediments have been intersected in many wells throughout the Perth Basin (Balme, 1969a). In the north, marine sedimentation commenced with the deposition of the Kockatea Shale at the base of which is a diachronous and discontinuous sandstone which has been a major hydrocarbon objective in the area following the successful development of commercial gas in the Dongara gas field.

The Kockatea Shale overlying the Dongara Sandstone consists of fine-grained shale and siltstone with sandstone interbeds deposited in a gently subsiding trough. It is of considerable chronostratigraphic importance since its marine fauna is consistent with impoverished Scythian faunas throughout the world, carrying the important ammonoid genus *Ophiceras* and the cosmopolitan bivalve *Claraia* (Kummel, 1973). Two of the ammonites, *Ophiceras (Discophiceras)* sp. cf. *O.(D.) subkyokticum* (Spath) and *Subinyoites kashmiricus* (Diener) give the age of the Kockatea Shale as Early Scythian (Otoceratan Zone).

The Kockatea Shale is succeeded by the Woodada Formation of fine-grained thinly bedded and cross-bedded sands and interbedded dark-grey siltstones with a microflora indicating a Late Scythian to Middle Triassic age. Leiospheres and spinose acritarchs occur, and the deposits are considered to be those of a fluvial environment representing the last relatively stable phase of the Scythian marine transgression.

The Locker Shale of the Carnarvon Basin and the Blina Shale of the Fitzroy Trough and northeast Canning Basin are lithologically similar to the Kockatea Shale. Neither is adequately represented in outcrop. The Locker Shale passes gradationally upwards into the Middle or Late Triassic Mungaroo Beds which show considerable lateral and vertical facies variation representing deltaic, fluvial, and paludal depositional environments. They carry spectacular microfloral assemblages representative of highly diversified vegetation. *Samarosporites speciosus* Goubin, *Rimaesporites aquilonalis* Goubin and *Camerosporites* sp., which occur in the Late Triassic of Madagascar, are present (Balme, 1969a).

In the Fitzroy Trough the Blina Shale occupies structural lows and becomes rapidly thinner towards the southeast. It contains coquinas of the conchostracan *Isaura,* the bivalve *Pseudomonotis,* lingulid brachiopods, fish teeth, reptilian remains, and near the base, a 3.3 m bone bed (Veevers and Wells, 1961). Correlation with the Kockatea Shale is supported by the occurrence in both formations of a labyrinthodont *Deltasaurus,* allied to *Rhytidosteus* of the *Cynognathus* Zone of South Africa and *Peltostega* of the *Posidonomya* beds of Spitsbergen and a brachyopid of comparable affinities (Cosgriff, 1965).

The Blina Shale is considered to have been deposited in a shallow bay or marine delta. It is conformably overlain by the continental Erskine Sandstone with poorly preserved plant fossils, and may be partly equivalent to the Culvida Sandstone, a thick-bedded cross-bedded quartz sandstone containing an *Isaura*-bearing shale: diverse plants such as *Dicroidium odontopteroides* (Morris) Gothan and *D. feistmanteli* (Johnston) Gothan, *Taeniopteris* sp., *Danaeopsis hughesi* Feistmantel, *Equisetites woodsi* Jones & de Jersey have been recorded (White, in Veevers and Wells, 1961). The precise relationships of the Culvida Sandstone with the Blina Shale and Erskine Sandstone are not clear, and much of the sequence has been removed by erosion (Veevers and Wells, 1961).

Grey laminated micaceous sandy siltstone and shale with lingulid brachiopods, vertebrate remains and estheriid conchostracans, and correlated with the Blina Shale, overlie Late Permian marine rocks with Tethyan affinities in the Port Keats area of the Bonaparte Gulf Basin (Thomas, 1957; Dickins et al., 1972).

The Late Triassic was a period of diastrophism during which the Fitzroy Trough was folded and faulted and intruded in the Rhaetian by dolerite with K/Ar age of 196 m.y. (Harding, 1969; Veevers, 1971). Triassic deposition was terminated, and considerable erosion took place in the Fitzroy Trough and Canning Basin.

The Perth Basin experienced the onset of graben tectonics with movement on the Darling Fault adjacent to which over 2,000 m of the Lesueur Sandstone were deposited (Jones and Pearson, 1972). The intense normal faulting and uplift of the basin margins during its deposition is considered evidence of active rifting and the initiation of a new cycle of sedimentation in the Jurassic. The poorly sorted continental sandstones are succeeded by 1,000–3,000 m

TABLE I

Correlation table for the Mesozoic rocks of western Australia

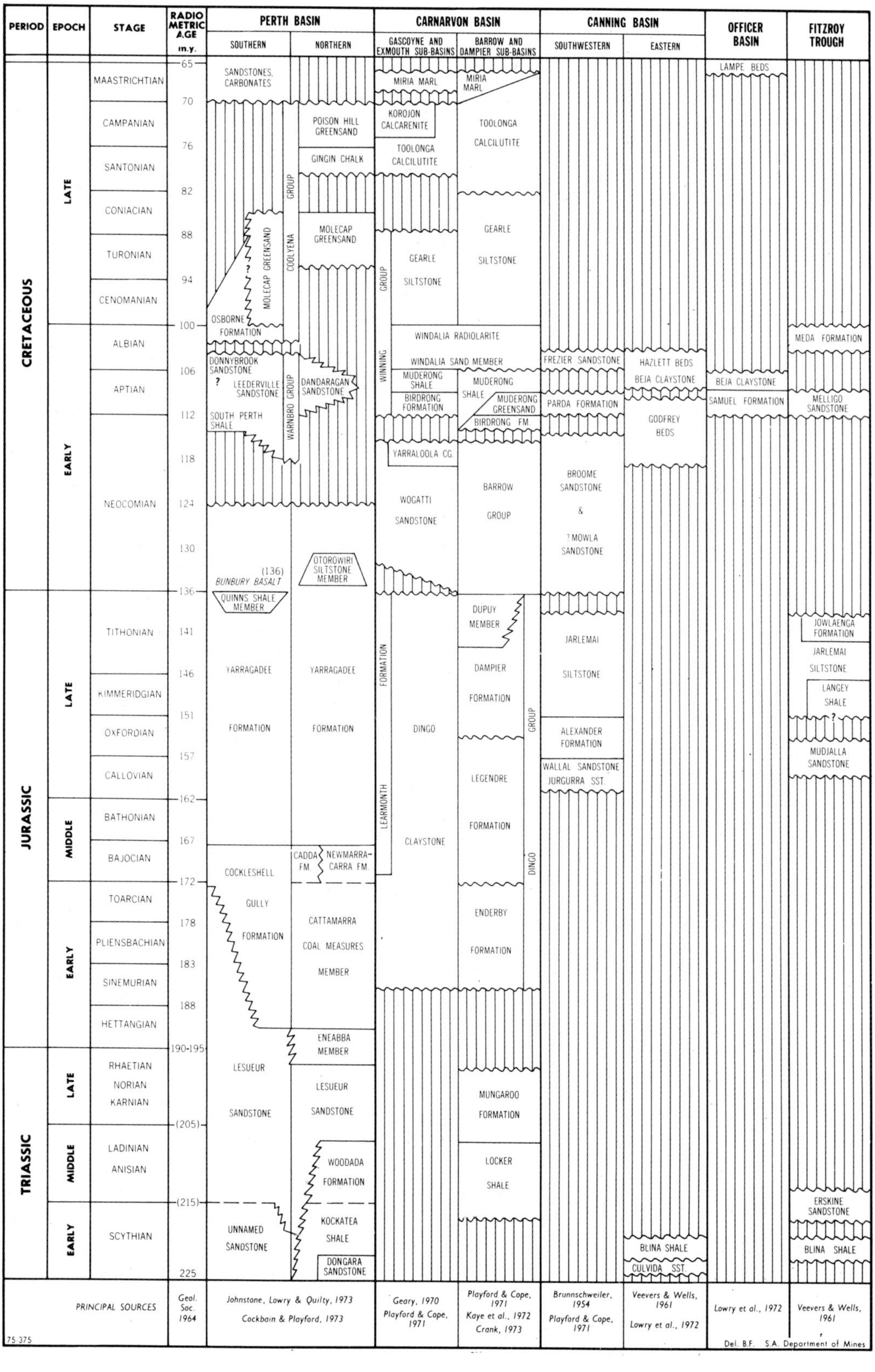

PERIOD	EPOCH	STAGE	RADIOMETRIC AGE m.y.	PERTH BASIN: SOUTHERN	PERTH BASIN: NORTHERN	CARNARVON BASIN: GASCOYNE AND EXMOUTH SUB-BASINS	CARNARVON BASIN: BARROW AND DAMPIER SUB-BASINS	CANNING BASIN: SOUTHWESTERN	CANNING BASIN: EASTERN	OFFICER BASIN	FITZROY TROUGH
CRETACEOUS	LATE	MAASTRICHTIAN	65	SANDSTONES, CARBONATES		MIRIA MARL	MIRIA MARL			LAMPE BEDS	
		CAMPANIAN	70		POISON HILL GREENSAND	KOROJON CALCARENITE	TOOLONGA CALCILUTITE				
		SANTONIAN	76		GINGIN CHALK	TOOLONGA CALCILUTITE					
		CONIACIAN	82		COOLYENA GROUP						
		TURONIAN	88	? MOLECAP GREENSAND	MOLECAP GREENSAND	GEARLE SILTSTONE (WINNING GROUP)	GEARLE SILTSTONE				
		CENOMANIAN	94	OSBORNE FORMATION							
	EARLY	ALBIAN	100	DONNYBROOK SANDSTONE		WINDALIA RADIOLARITE; WINDALIA SAND MEMBER	WINDALIA RADIOLARITE	FREZIER SANDSTONE	HAZLETT BEDS		MEDA FORMATION
		APTIAN	106	? LEEDERVILLE SANDSTONE	DANDARAGAN SANDSTONE (WARNBRO GROUP)	MUDERONG SHALE; BIRDRONG FORMATION	MUDERONG SHALE; MUDERONG GREENSAND; BIRDRONG FM.	PARDA FORMATION	BEJA CLAYSTONE	BEJA CLAYSTONE; SAMUEL FORMATION	MELLIGO SANDSTONE
		NEOCOMIAN	112, 118, 124, 130	SOUTH PERTH SHALE; (136) BUNBURY BASALT	OTOROWIRI SILTSTONE MEMBER	YARRALOOLA CG.; WOGATTI SANDSTONE	BARROW GROUP	BROOME SANDSTONE & ? MOWLA SANDSTONE	GODFREY BEDS		
JURASSIC	LATE	TITHONIAN	136	QUINNS SHALE MEMBER			DUPUY MEMBER				JOWLAENGA FORMATION
		KIMMERIDGIAN	141, 146	YARRAGADEE FORMATION	YARRAGADEE FORMATION	DINGO CLAYSTONE (LEARMONTH FORMATION)	DAMPIER FORMATION (DINGO GROUP)	JARLEMAI SILTSTONE			JARLEMAI SILTSTONE; LANGEY SHALE ?
		OXFORDIAN	151					ALEXANDER FORMATION			
		CALLOVIAN	157				LEGENDRE FORMATION	WALLAL SANDSTONE; JURGURRA SST.			MUDJALLA SANDSTONE
	MIDDLE	BATHONIAN	162								
		BAJOCIAN	167	COCKLESHELL GULLY FORMATION	CADDA FM.; NEWMARRA-CARRA FM.						
	EARLY	TOARCIAN	172		CATTAMARRA COAL MEASURES MEMBER		ENDERBY FORMATION				
		PLIENSBACHIAN	178								
		SINEMURIAN	183								
		HETTANGIAN	188		ENEABBA MEMBER						
TRIASSIC	LATE	RHAETIAN, NORIAN, KARNIAN	190-195	LESUEUR SANDSTONE	LESUEUR SANDSTONE		MUNGAROO FORMATION				
	MIDDLE	LADINIAN, ANISIAN	(205)		WOODADA FORMATION		LOCKER SHALE				ERSKINE SANDSTONE
	EARLY	SCYTHIAN	(215)	UNNAMED SANDSTONE	KOCKATEA SHALE; DONGARA SANDSTONE				BLINA SHALE; CULVIDA SST.		BLINA SHALE
			225								
PRINCIPAL SOURCES			Geol. Soc. 1964	Johnstone, Lowry & Quilty, 1973; Cockbain & Playford, 1973		Geary, 1970; Playford & Cope, 1971	Playford & Cope, 1971; Kaye et al., 1972; Crank, 1973	Brunnschweiler, 1954; Playford & Cope, 1971	Veevers & Wells, 1961; Lowry et al., 1972	Lowry et al., 1972	Veevers & Wells, 1961

75-375

Del. B.F. S.A. Department of Mines

of Early Jurassic continental coal-bearing sandstones, siltstones and shales of the Cockleshell Gully Formation (Johnstone et al., 1973).

In the southern Perth Basin deposition of coarse predominantly fluvial sandstones, mainly Lesueur Sandstone, continued throughout the Triassic and Early Jurassic (Figs. 2–5).

Jurassic

There was a minor marine incursion in the Perth Basin in the Middle Jurassic when the Newmarracarra Limestone and Cadda Formation were deposited. The Newmarracarra Limestone carries a rich molluscan fauna including *Trigonia moorei* Lycett, and an important ammonite assemblage with species of *Sonninia, Fontannesia, Otoites, Pseudotoites, Zemistephanus* and *Stephanoceras,* indicative of a Middle Bajocian age (Sowerbyi Zone) (Arkell and Playford, 1954).

Tectonic movements in the basin and on its margins continued during the Jurassic, with intense faulting in the Late Jurassic (Fig. 11). These movements accompanied the deposition from the Middle Jurassic to Early Cretaceous of the Yarragadee Formation, a thick sequence of alternating sandstone and micaceous siltstone with clay, shale, and conglomerate interbeds. They are mainly continental onshore and partly marine offshore. A major unconformity in the Neocomian separates the Yarragadee Formation from the overlying Early Cretaceous Warnbro Group (Cockbain and Playford, 1973) (Fig. 11).

In the Carnarvon Basin, sinking of the outer margin of the basin in the Barrow and Exmouth Sub-basins (Fig. 10, section A–A_1, B–B_1) during the Early Jurassic initiated deposition of a monotonous sequence of dark-grey micaceous siltstones, shales and minor sandstones, the Dingo Claystone, in part of which the ammonites *Perisphinctes, Kossmatia, Macrocephalites,* the belemnites *Belemnopsis alfurica* (Boehm) and *B. calloviensis* (Oppel) and bivalves *Inoceramus galoi* Boehm, *Meleagrinella* and *Quenstedtia* have been identified (McWhae et al., 1958).

Deposition continued to the Early Cretaceous, but while over most of the Carnarvon Basin there is an unconformity in the Neocomian, in the Barrow Sub-basin there is no hiatus until the Late Turonian. The Barrow Sub-basin (Parry, 1967; Burdett et al., 1970; Crank, 1973) has a number of hydrocarbon accumulations in the Late Jurassic–Early Cretaceous, 98% of the production from the Barrow oil field coming from the Aptian–Albian Windalia Sand Member in the Barrow Island anticline, a north–south closed anticline believed to have been initiated by basement uplift in the Late Triassic–Early Jurassic (Parry, 1967; Crank, 1973).

Although the onshore Canning Basin was relatively stable in the Early Jurassic (Challinor, 1970), there was continued subsidence offshore and movement on old fault trends. This widespread uplift and erosion in the Late Triassic and Early Jurassic is believed to be associated with the breakup of Gondwanaland.

A new sedimentary cycle began in the Late Jurassic in the Canning Basin with the deposition of conglomeratic sandstones and massive quartz sandstones and shales. The younger transgressive Alexander Formation contains the ammonoids *Virgatosphinctes* and *Kossmatia,* the bivalves *Inoceramus* and *Meleagrinella,* and abundant small Ophiuroidea. The sandstones are overlain by Oxfordian to Tithonian siltstones, sandy siltstones and glauconitic siltstones mainly of the Jarlemai Siltstone, which contains species of *Buchia* and *Meleagrinella,* and the Langey Shale, a thin bed less than 2 m thick, of glauconitic siltstone with moulds of *Belemnopsis* spp., *Kossmatia, Buchia* sp. cf. *B. subspitiensis* (Krumbeck), *Calpionella schneebergeri* Brunnschweiler and *Malayomaorica malayomaorica* (Krumbeck) (Brunnschweiler, 1960). C.A. Fleming (1958) and Jeletzky (1963) regard *Malayomaorica malayomaorica* as Middle Kimmeridgian in New Zealand and Indonesia.

Transgression continued into the Neocomian, with the sea advancing in a southeasterly direction (Veevers and Wells, 1961).

Cretaceous

Following the break-up of eastern Gondwanaland, sea-floor spreading persisted throughout the Cretaceous in the eastern Indian Ocean. No rocks dated older than Neocomian were recovered from DSDP Sites 258, 259, 263 and 264 of Legs 26, 27 and 28 (Luyendyk et al., 1973; Veevers et al., 1973; Hayes et al., 1973).

Cretaceous sediments crop out over a wide area of the Perth Basin although exposures are generally poor (Cockbain and Playford, 1973). The Neocomian was a period of intense tectonism accompanied by local arching and faulting and the extrusion of the Bunbury Basalt in the south. Subsidence of the offshore Vlaming Sub-basin took place, with the deposition of as much as 6,000 m of fluvial sand and estuarine shale and sand (Jones and Pearson, 1972; Johnstone et al., 1973) (Fig. 11, sections $G–G_1$, $H–H_1$). During the transgression following subsidence of the continental margin the Warnbro Group of marine and continental clastics was deposited. The group is best known from subsurface sections in the Perth area, and correlation is based on foraminiferal and microplankton assemblages.

The Late Cretaceous Coolyeana Group includes the marine Molecap Greensand and the Gingin Chalk. The richly fossiliferous Gingin Chalk carries the Santonian crinoids *Marsupites* and *Uintacrinus* (Withers, 1924, 1926), abundant bivalves including *Inoceramus, Gryphaea, Pycnodonte* and *Exogyra* (Etheridge, 1913; Feldtmann, 1963) and a rich foraminiferal assemblage in which *Rugoglobigerina* spp., *Globotruncana lapparenti* Brotzen, *G. marginata* (Reuss), *Hedbergella cretacea* (d'Orbigny), *Globigerinelloides asperus* (Ehrenberg) and *Bolivinoides strigillatus* (Chapman) are common (Belford, 1958, 1960).

In the Carnarvon Basin the best developed and thickest Cretaceous sections are in the Exmouth (Geary, 1970) and Barrow (Crank, 1973) Sub-basins. Neocomian sandstones comprise the lowermost units in the Exmouth Sub-basin (Fig. 10, section $B–B_1$). The Aptian Birdrong Sandstone at the base of the Winning Group is transgressive and partly glauconitic. It is followed by an Aptian to Turonian sequence of marine shales and greensand: the Muderong Shale, Windalia Radiolarite and Gearle Siltstone. After a hiatus in the Coniacian the group is overlain by Santonian to Campanian carbonate sediments of the Toolonga Calcilutite which carries a microfauna similar to that of the Gingin Chalk of the Perth Basin, and the Korojon Calcarenite in which *Bostrychoceras indicum* (Stoliczka) and abundant *Inoceramus* fragments occur with a Campanian microfauna including *Globotruncana arca* (Cushman), *Bolivina incrassata* Reuss and *Cibicides voltziana* (d'Orbigny). An erosional interval follows before the deposition of the Maastrichtian Miria Marl (Table I, Fig. 10) (Belford, 1958), from which an ammonite fauna containing species of *Baculites, Eubaculites, Giralites, Eubaculiceras, Cardabites, Nostoceras, Glyptoxoceras* and *Neohamites* has been described (Brunnschweiler, 1966).

A similar sequence occurs in the Barrow Sub-basin (Fig. 10, section $A–A_1$), where deltaic sedimentation from the Tithonian to the Neocomian resulted in the deposition of the sandstone complex of the Barrow Group. The base of the Aptian transgression is marked by the Muderong Greensand Member of the Muderong Shale, the upper member of which, the Windalia Sand Member, is the main source of production of the Barrow Island oil field.

Although a major unconformity separates the Santonian–Campanian Toolonga Calcilutite from the overlying Tertiary limestones, they are lithologically similar, and the carbonate sedimentation which began with the Toolonga Calcilutite has persisted to the present day (Crank, 1973).

The Canning Basin experienced changes of coastline in the Late Jurassic, with a southerly shift in both the area of glauconite deposition and a major transgression in the Neocomian resulting in the deposition of the partly continental and partly shallow marine Broome Sandstone. There may have been a narrow opening on the present continental margin in the Aptian (Veevers and Wells, 1961, p. 187), but as no Aptian–Albian sediments are known from the northwest Canning Basin (M.H. Johnstone, pers. comm., 1975), evidence is lacking. There is, however, a clear indication from the Aptian molluscan faunas described by Skwarko (1967), mainly from the Bejah Claystone of the Gibson Desert, that during the epicontinental marine inundation the southeast Canning Basin was connected with the Great Artesian Basin or with the Eucla Basin – probably by way of the Officer Basin, a possibility recognised also by Skwarko (Fig. 8).

Offshore in the Dampier Sub-basin a sequence similar to that of the Barrow Sub-basin has been intersected in drilling (Challinor, 1970).

EASTERN AUSTRALIA

Mesozoic sedimentary basins

The sedimentary basins of eastern Australia containing Mesozoic rocks are intracratonic basins which may be divided into two main groups: (1) pre-Jurassic basins formed during the later stages of an orogenic cycle; (2) post-Triassic basins formed after stabilisation of the area as a craton. They differ structurally from and are frequently superimposed upon those containing Permian–Triassic sequences.

Pre-Jurassic basins

The pre-Jurassic basins are mainly aligned in a

TABLE II

Correlation table for the Mesozoic rocks of eastern Australia

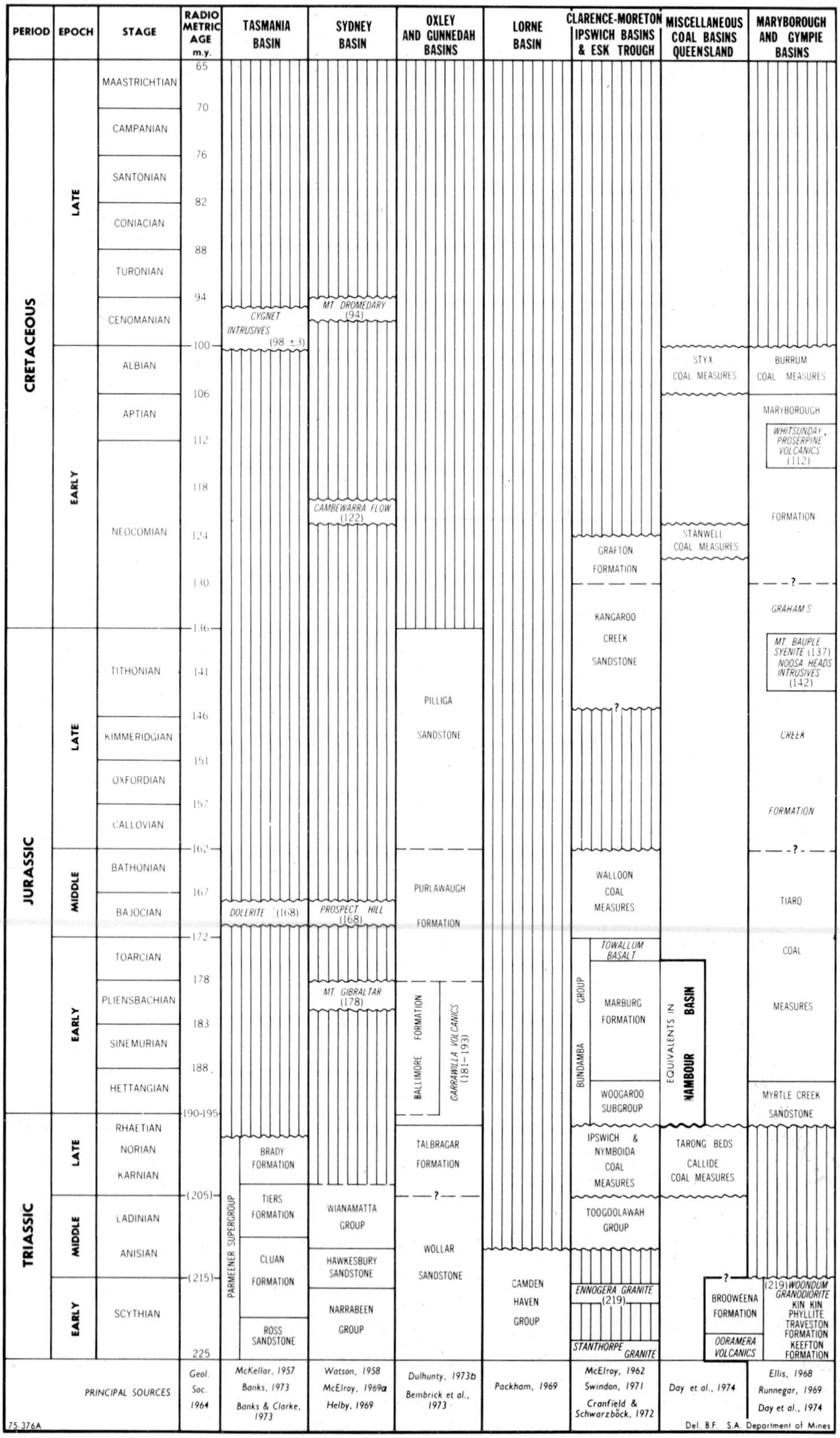

PERIOD	EPOCH	STAGE	RADIOMETRIC AGE m.y.	TASMANIA BASIN	SYDNEY BASIN	OXLEY AND GUNNEDAH BASINS	LORNE BASIN	CLARENCE-MORETON IPSWICH BASINS & ESK TROUGH	MISCELLANEOUS COAL BASINS QUEENSLAND	MARYBOROUGH AND GYMPIE BASINS
CRETACEOUS	LATE	MAASTRICHTIAN	65, 70							
		CAMPANIAN	76							
		SANTONIAN	82							
		CONIACIAN	88							
		TURONIAN	94							
		CENOMANIAN	100	*CYGNET INTRUSIVES* (98 ±3)	*MT DROMEDARY* (94)					
	EARLY	ALBIAN	106						STYX COAL MEASURES	BURRUM COAL MEASURES
		APTIAN	112							MARYBOROUGH; *WHITSUNDAY, PROSERPINE VOLCANICS* (112)
		NEOCOMIAN	118, 124, 130, 136		*CAMBEWARRA FLOW* (122)			GRAFTON FORMATION	STANWELL COAL MEASURES	FORMATION; ?; *GRAHAM'S*
JURASSIC	LATE	TITHONIAN	141			PILLIGA SANDSTONE		KANGAROO CREEK SANDSTONE; ?		*MT BAUPLE SYENITE* (137); *NOOSA HEADS INTRUSIVES* (142)
		KIMMERIDGIAN	146							*CREEK*
		OXFORDIAN	151							
		CALLOVIAN	157							*FORMATION*
	MIDDLE	BATHONIAN	162			PURLAWAUGH FORMATION		WALLOON COAL MEASURES		?; TIARO
		BAJOCIAN	167	*DOLERITE* (168)	*PROSPECT HILL* (168)					
	EARLY	TOARCIAN	172					*TOWALLUM BASALT*		COAL
		PLIENSBACHIAN	178		*MT GIBRALTAR* (178)	BALLIMORE FORMATION; *GARRAWILLA VOLCANICS* (181–193)		BUNDAMBA GROUP: MARBURG FORMATION	EQUIVALENTS IN NAMBOUR BASIN	MEASURES
		SINEMURIAN	183							
		HETTANGIAN	188					WOOGAROO SUBGROUP		MYRTLE CREEK SANDSTONE
TRIASSIC	LATE	RHAETIAN, NORIAN, KARNIAN	190-195	PARMEENER SUPERGROUP: BRADY FORMATION		TALBRAGAR FORMATION		IPSWICH & NYMBOIDA COAL MEASURES	TARONG BEDS; CALLIDE COAL MEASURES	
	MIDDLE	LADINIAN	(205)	TIERS FORMATION	WIANAMATTA GROUP	?; WOLLAR SANDSTONE		TOOGOOLAWAH GROUP		
		ANISIAN		CLUAN FORMATION	HAWKESBURY SANDSTONE					
	EARLY	SCYTHIAN	(215), 225	ROSS SANDSTONE	NARRABEEN GROUP		CAMDEN HAVEN GROUP	*ENNOGERA GRANITE* (219); *STANTHORPE GRANITE*	?; BROOWEENA FORMATION; *OORAMERA VOLCANICS*	(219) *WOONDUM GRANODIORITE*; KIN KIN PHYLLITE; TRAVESTON FORMATION; KEEFTON FORMATION
PRINCIPAL SOURCES			*Geol. Soc. 1964*	*McKellar, 1957; Banks, 1973; Banks & Clarke, 1973*	*Watson, 1958; McElroy, 1969a; Helby, 1969*	*Dulhunty, 1973b; Bembrick et al., 1973*	*Packham, 1969*	*McElroy, 1962; Swindon, 1971; Cranfield & Schwarzböck, 1972*	*Day et al., 1974*	*Ellis, 1968; Runnegar, 1969; Day et al., 1974*

75-376A

Del. B.F. S.A. Department of Mines

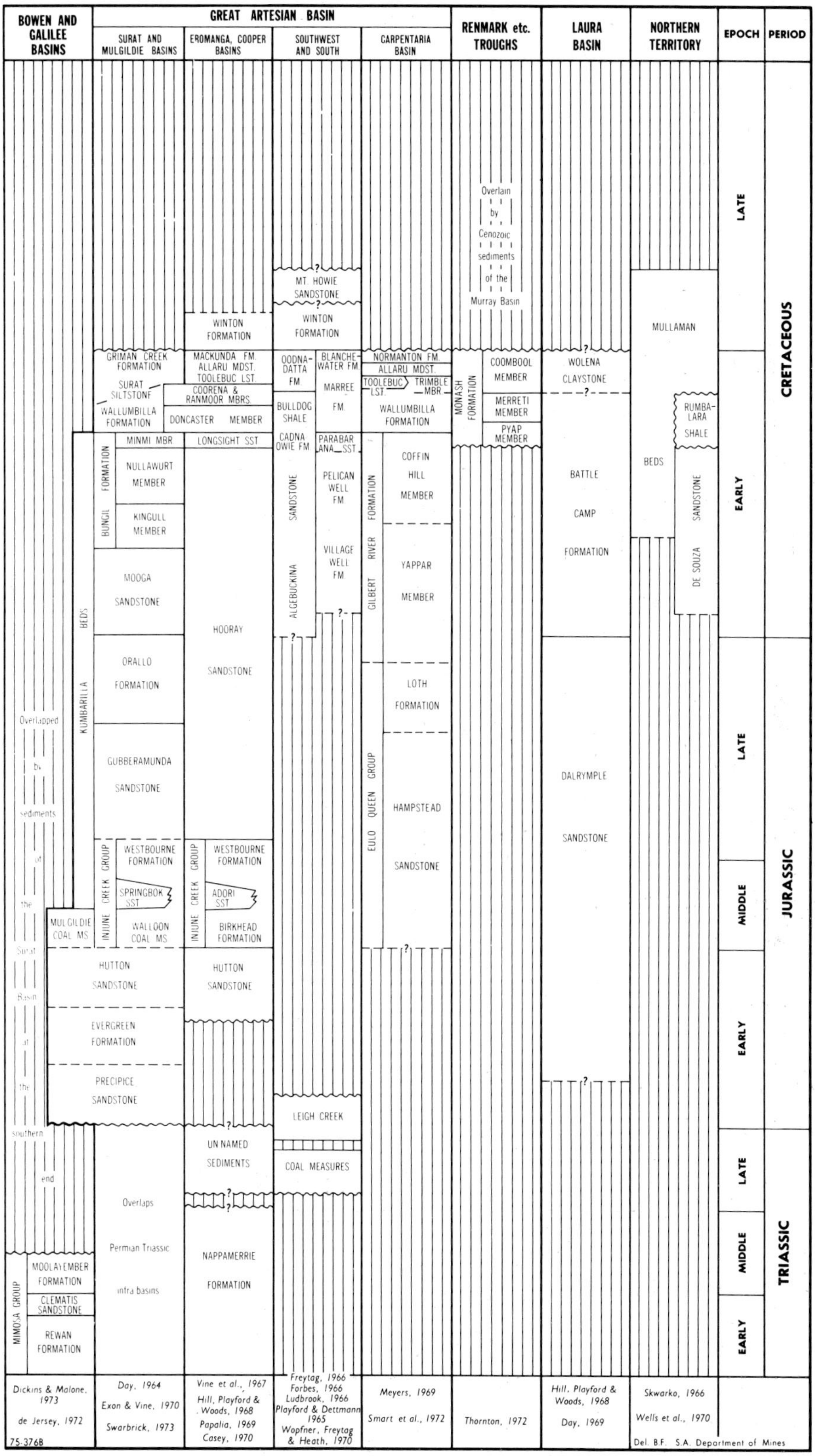

Basin	Units (youngest to oldest)	Sources
BOWEN AND GALILEE BASINS	Overlapped by sediments of the Surat Basin at the southern end; MULGILDIE COAL MS; MIMOSA GROUP: MOOLAYEMBER FORMATION, CLEMATIS SANDSTONE, REWAN FORMATION	Dickins & Malone, 1973; de Jersey, 1972
GREAT ARTESIAN BASIN – SURAT AND MULGILDIE BASINS	GRIMAN CREEK FORMATION; SURAT SILTSTONE; WALLUMBILLA FORMATION; KUMBARILLA BEDS: BUNGIL FORMATION (MINMI MBR, NULLAWURT MEMBER, KINGULL MEMBER), MOOGA SANDSTONE, ORALLO FORMATION, GUBBERAMUNDA SANDSTONE, INJUNE CREEK GROUP (WESTBOURNE FORMATION, SPRINGBOK SST, WALLOON COAL MS); HUTTON SANDSTONE; EVERGREEN FORMATION; PRECIPICE SANDSTONE; Overlaps Permian Triassic intra basins	Day, 1964; Exon & Vine, 1970; Swarbrick, 1973
GREAT ARTESIAN BASIN – EROMANGA, COOPER BASINS	WINTON FORMATION; MACKUNDA FM; ALLARU MDST; TOOLEBUC LST; COORENA & RANMOOR MBRS; DONCASTER MEMBER; LONGSIGHT SST; HOORAY SANDSTONE; INJUNE CREEK GROUP: WESTBOURNE FORMATION, ADORI SST, BIRKHEAD FORMATION; HUTTON SANDSTONE; UNNAMED SEDIMENTS; NAPPAMERRIE FORMATION	Vine et al., 1967; Hill, Playford & Woods, 1968; Papalia, 1969; Casey, 1970
GREAT ARTESIAN BASIN – SOUTHWEST AND SOUTH	MT. HOWIE SANDSTONE; WINTON FORMATION; OODNADATTA FM; BLANCHEWATER FM; MARREE FM; BULLDOG SHALE; CADNA-OWIE FM; PARABARANA SST; PELICAN WELL FM; VILLAGE WELL FM; ALGEBUCKINA SANDSTONE; LEIGH CREEK COAL MEASURES	Freytag, 1966; Forbes, 1966; Ludbrook, 1966; Playford & Dettmann 1965; Wopfner, Freytag & Heath, 1970
GREAT ARTESIAN BASIN – CARPENTARIA BASIN	NORMANTON FM; ALLARU MDST; TOOLEBUC LST; TRIMBLE MBR; WALLUMBILLA FORMATION; GILBERT RIVER FORMATION: COFFIN HILL MEMBER, YAPPAR MEMBER; EULO QUEEN GROUP: LOTH FORMATION, HAMPSTEAD SANDSTONE	Meyers, 1969; Smart et al., 1972
RENMARK etc. TROUGHS	Overlain by Cenozoic sediments of the Murray Basin; MONASH FORMATION: COOMBOOL MEMBER, MERRETI MEMBER, PYAP MEMBER	Thornton, 1972
LAURA BASIN	WOLENA CLAYSTONE; BATTLE CAMP FORMATION; DALRYMPLE SANDSTONE	Hill, Playford & Woods, 1968; Day, 1969
NORTHERN TERRITORY	MULLAMAN BEDS; RUMBALARA SHALE; DE SOUZA SANDSTONE	Skwarko, 1966; Wells et al., 1970

PERIOD	EPOCH
CRETACEOUS	LATE
	EARLY
JURASSIC	LATE
	MIDDLE
	EARLY
TRIASSIC	LATE
	MIDDLE
	EARLY

75-376B

Del. B.F. S.A. Department of Mines

general north–south direction in the East Australian Orogenic Province or occur as infrabasins below the Jurassic–Cretaceous basins constituting the Great Artesian Basin. Most of the Permian–Triassic Sydney, Ipswich, Gympie and Bowen Basins contain important coal deposits. The Gunnedah, Cooper and Galilee Basins are infrabasins covered by sediments of the Great Artesian Basin complex.

Post-Triassic basins

The Great Artesian Basin is essentially a composite basin with a hydrological connotation (Bourke et al., 1974), made up of a number of sub-basins and embayments shown on Fig. 1 and Table II.

The interrelationships of the Mesozoic basins in southeast Queensland and northeast New South Wales, including the Surat, Clarence–Moreton, Maryborough and Nambour Basins are shown in Fig. 12 (Day et al., 1974).

The Laura Basin is an open syncline plunging offshore to the north-northwest (Allen and Hogetoorn, 1970).

Stratigraphy of the eastern sector

The stratigraphic relationships of Mesozoic rocks in the eastern sector are shown in Table II.

Triassic

Triassic sediments are widespread in eastern Australia, where they occur principally in two subparallel elongated north–south areas of accumulation: the westerly on the western side of the Hunter Thrust System and New England Fold Belt, including the major structure of the Sydney, Gunnedah and Bowen Basins; the easterly including the small isolated Lorne Basin in the south and the Clarence–Moreton, Ipswich and Gympie Basins in the north.

Gympie Basin. Triassic marine sedimentation is confined to the Gympie Basin (Ellis, 1968), concealed except on the western side by younger sediments of the Maryborough and Nambour Basins. An assemblage of bivalves containing *Bakewellia, Nuculopsis* and *Neoschizodus* occurs in the Brooweena Formation, which consists of dark-grey siltstone and subgreywacke with interbeds of quartz sandstone. From the apparent absence of *Claraia* the fauna is regarded as Early Triassic, not older than Flemingitan (P.J.G. Fleming, 1966) or Dienerian in the correlation used by McTavish and Dickins (1974).

A significant small marine fauna containing the ammonoids *Latisageceras, Flemingites, Anaflemingites, Paranorites* and *Pseudohedenstroemia* occurs in massively bedded olive-green and grey-green shales with thin friable sandstone interbeds of the Traveston Formation. The age has been determined as Early Triassic *Meekoceras gracilitatus* Zone (Runnegar, 1969a). The occurrence may belong to a single marine transgression in the Gympie–Maryborough area in the Smithian when it is suggested that Early Triassic seas were most extensive (McTavish and Dickins, 1974).

Both the Brooweena Formation and the lowest Early Triassic unit, the Keefton Formation below the Traveston, are coarse-grained or conglomeratic sandstones overlying Early Permian rocks.

The Early Triassic was also a period of volcanism in the basin, and granites, granodiorites and tonalites radiometrically dated at from 215 ± 5 to 222 ± 5 m.y. were intruded over a wide area into the Early Permian Biggenden Beds. Copper, gold, silver and bismuth mineralization in the Mount Perry–Biggenden area is associated with the granite intrusions (Ellis, 1968).

Sydney Basin. Apart from the northerly marine incursions, Early Triassic sedimentation in eastern Australia, particularly in the Sydney–Bowen Basins, followed almost without perceptible break after the deposition of extensive coal measures in the Late Permian, and the environment throughout most of the Triassic and Jurassic was fluvial, lacustrine or paludal.

Prior to the recognition of the Scythian (Otoceratan) datum in the Kockatea Shale of the Perth Basin, the Permian–Triassic boundary was determined at the marked change from coal measures to quartz sandstones and polymictic conglomerates. Tradition also accepted the base of the Triassic at the replacement of the *Glossopteris* flora and its associated *Striatites* microflora (Balme, 1964) by floras with *Dicroidium.* However, Evans' (1966) correlation, on palynological evidence, of the Collaroy Claystone in the Narrabeen Group of the Sydney Basin (Table III) with the Kockatea Shale, and palynological studies of Balme (1969b), Balme and Helby (1973) and Helby (1973) indicate that the extinction of the *Glossopteris* flora took place in the Late Permian, and that rapid floral changes occurred before the *Dicroidium* flora became established in the Late Scythian. The subsequent entry of species of *Aratri-*

TABLE III

Triassic stratigraphy of the Sydney Basin

LATE TO MIDDLE TRIASSIC	WIANNA-MATTA GROUP	CAMDEN SUBGROUP	PRUDHOE SHALE: shales with lithic sandstone lenses (37 m)
			PICTON FORMATION: alternating lithic sandstones and shales (30 m)
			RAZORBACK SANDSTONE: massive lithic sandstone with shales (21 m)
			ANNAN SHALES: dark green and black shales with plant remains and iron oxide nodules (12 m)
			POTTS HILL SANDSTONE: lithic sandstone with shale lenses (12 m)
		LIVERPOOL SUBGROUP	BRINGELLY SHALE: dark green and black shales with plant fragments and iron oxide nodules (60 m)
			MINCHINBURY SANDSTONE: calcareous lithic sandstone with shale lenses and siderite nodules (60 m)
			ASHFIELD SHALE: black mudstones and silty shales with sideritic mudstone bands (60 m)
			MITTAGONG FORMATION: alternating bands of black shale and fine quartz sandstone (0–15 m)
	HAWKESBURY SANDSTONE: massive quartz-rich sandstone with occasional dark grey shale and siltstone lenses (30–240 m)		

Age	Group	Subgroup	north	south	west	Lithology
EARLY TRIASSIC			GOSFORD FORMATION		BURRALOW FORMATION (137 m)	shales, shaly sandstones and sandstones
	NARRA-BEEN GROUP	CLIFTON SUBGROUP	COLLAROY CLAYSTONE (137 m)	BALD HILL CLAYSTONE (150 m)		green and red shales and clay-stones red beds
			TUGGERAH FORMATION (98 m)	BULGO SANDSTONE (119 m)	GROSE SANDSTONE (213 m)	sandstones shales and fine conglomerates
				STANWELL PARK CLAYTON (37 m)		
?			MUNMORRAH CONGLOMERATE (156 m)	SCARBOROUGH SANDSTONE (26 m)	CALEY FORMATION (46 m)	conglomerates, sandstones red, green and blue shales
?PERMIAN				WOMBARRA SHALE (37 m)		
				COAL CLIFF SANDSTONE (9 m)		

	conformable on coal measures of Permian age, usually without discernible break

Compiled from McElroy (1969a), Standard (1969), and Lovering and McElroy (1969).

sporites in the Collaroy Claystone is regarded as an important stratigraphic event which took place in the Late Scythian (Helby, 1967). The marine transgression in the Early Triassic is reflected in the appearance of vast numbers of spinose acritarchs.

The Early Triassic rocks of the Sydney Basin are gently dipping and conformable on but less deformed than the underlying sediments. The sequence is shown in Table III. It commences with the Narrabeen Group (Hanlon et al., 1953) (McElroy, 1969a) which covers most of the northern part of the basin and passes to the north and west under Jurassic sediments of the Oxley and Coonamble Basins. The great cliff escarpments and plateaux of the Blue Mountains west of Sydney are formed in massive sandstones, chiefly the Grose Sandstone, of the Narrabeen Group.

Red beds are common, particularly in the Bald Hill Claystone and the Stanwell Park Claystone. They are considered by Loughnan et al. (1964) to have been deposited in a piedmont environment.

In the central part of the basin the Narrabeen Group is overlain by the Hawkesbury Sandstone consisting of highly lenticular cross-bedded mature orthoquartzites often forming coastal cliffs and cliff outcrops in and around the city of Sydney. The main source area lay to the south (Standard, 1969) with secondary contributions from tributary rivers to the west and north.

Fossils are not common in the Narrabeen Group. The Gosford Formation contains a diverse flora including species of *Phyllotheca, Dicroidium, Taeniopteris, Rhipidopsis,* a fish fauna including *Palaeoniscus, Myriolepsis, Belonorhynchus,* and an amphibian. Although rare, fossils in the Hawkesbury Sandstone include numerous insects and fish, plant macrofossils and microfossils.

The Hawkesbury Sandstone is overlain by a conformable sequence of interbedded dark shales and mudstones and lithic sandstones forming the Wianamatta Group of Middle to Late Triassic age. The Ashfield Shale exposed in brick quarries in the Sydney district typically includes black shales, siltstones and sideritic mudstones with ironstone nodules. The basal beds have yielded fossil plants such as *Phyllotheca, Cladophlebis, Dicroidium, Macrotaeniopteris* and *Cycadopteris,* fresh-water bivalves, insects and a complete skeleton of the large labyrinthodont *Paracyclotosaurus davidi* related to *Paracyclotosaurus hemprichi* of the Late Keuper of Germany and regarded by Watson (1958) as of Late Triassic age, and a brachyopid, *Notobrachyops picketti* Cosgriff. Plant microfossils are abundant throughout the group and acanthomorphic acritarchs occur intermittently (Helby, 1969).

Gunnedah Basin. The small structural Gunnedah Basin (Bembrick et al., 1973) also contains a Triassic sequence of massive quartz sandstones and conglomerates, flaggy sandstones and sandy shales related to the sediments of the Sydney Basin. Volcanism commenced in the Late Triassic with the extrusion on sediments of the Talbragar Formation of trachyte and trachybasalts, the Garrawilla Volcanics, which have a K/Ar dating of 171 to 201 m.y. (Dulhunty and McDougall, 1966). The sequence is overlain by Jurassic sediments of the Oxley Basin (Dulhunty, 1973b).

Bowen Basin. The Bowen Basin (Dickins and Malone, 1973) is the most northerly extension of the Sydney Basin. It is essentially a large elongated Permian to Triassic synclinal or miogeosynclinal basin with a steeper eastern flank and overlain at its southern end by Jurassic and Cretaceous sediments of the Surat Basin. The Triassic sequence is of basin-wide extent, following conformably on the Permian and covering the same area, but the Mimosa Group of which it is composed is distinguishable from the Permian sequence by a marked difference in gross lithology, the rarity of carbonaceous matter and a major change in tectonic regime. The Mimosa Group consists of three non-marine formations: the Rewan Formation, Clematis Sandstone and Moolayember Formation.

The Rewan Formation was deposited as a result of renewed tectonism with vigorous uplift of provenance areas in the south and southeast of the basin, volcanism and intrusion. It contains massive beds of red-brown mudstone with desiccation cracks. Similar red beds occur widely in the Triassic of eastern Australia. The Rewan microflora includes distinctive elements of the Kockatea Shale microflora, such as *Densoisporites playfordi* (Balme) Dettmann, *Kraeuselosporites cuspidus* Balme, *K. saeptatus* Balme, *Lundbladispora brevicula* Balme, *L. willmotti* Balme, *Osmundacidites senectus* Balme, and also of the Ross Formation of Tasmania (De Jersey, 1970a).

The Clematis Sandstone and Moolayember Formation contain macrofloras dominated by *Dicroidium* and abundant microfloras indicating an Early to Middle Triassic age for the Clematis Sandstone and

Middle to (?)Late Triassic for the Moolayember (Evans, 1966; De Jersey, 1968).

Tectonic activity was temporarily reduced during the deposition of the Clematis Sandstone in braided channels and on flood plains. The formation is characterized by massive cliff-forming cross-bedded quartz sandstone interbedded with grey and white flaggy siltstone; near the provenance area in the south, it contains a significant proportion of volcanic detritus.

During the deposition of the overlying Moolayember Formation, which is present only in synclinal areas (Allen et al., 1960), tectonism was renewed, the formation containing thick beds of polymictic mainly volcanolithic pebble and cobble conglomerates, some of andesitic provenance, deposited in the southern part of the basin during the final stages of subsidence of the Mimosa Syncline in which some 5,400 m of Triassic sediments were deposited. It was subsequently truncated by erosion before the deposition in the Surat Basin of the Jurassic Precipice Sandstone.

The Galilee Basin, an intracratonic basin to the west of and continuous with the Bowen Basin, contains red beds similar to those of the Bowen Basin. It possibly also had connection with the Cooper Basin to the southwest.

Cooper Basin. Triassic rocks are known only from subsurface sections in the Cooper Basin where an Early to Middle Triassic sequence with a maximum thickness of 800 m of greenish-grey to green interbedded dolomitic siltstone and sandstone and red and green mottled shale and siltstone, the Nappamerrie Formation (Papalia, 1969; Wopfner, 1969), forms the seal for the Permian gas reservoirs of the Gidgealpa, Moomba, Daralingie and Toolachie gas fields.

Lorne Basin. The small Lorne Basin is an important link between the Sydney and Clarence–Moreton Basins. Here basal conglomerates of the Camden Haven Group rest with marked unconformity on Early Permian sediments and are overlain by sandstones and red and purple shales correlated with the Narrabeen Group (Packham, 1969).

Esk and Abercorn Troughs. Crustal tension established in the Permian and intensifying in the Early Triassic caused downfaulting which produced the graben of the Esk Trough (Figs. 1, 12) and the fault-bounded Abercorn Trough beneath the Mulgildie Basin, in both of which thick sequences of andesitic boulder beds and pyroclastics and terrestrial sediments accumulated. The Esk Trough contains as much as 900 m of andesitic boulder beds in the Neara Volcanics and a gabbro-granophyre intrusive dated at 210 m.y. (Webb and McDougall, 1967). The Cynthia Beds of the Abercorn Trough consist of mudstone, feldspathic, lithic and tuffaceous sandstone, conglomerate and volcanics (Day et al., 1974).

Ipswich Basin. The Ipswich Coal Measures were formed in a partly fault-controlled intermontane depression, underlying the northern part of the Clarence–Moreton Basin (Allen and Hogetoorn, 1970). The coal measures consist of 1,400 m of lenticular shale and sandstone, with conglomerate, breccia, tuff and basalt. The lower relatively thin (320 m) part of the succession forming the Kholo Group overlying the Carboniferous Brisbane Metamorphics consists of locally derived detrital breccia, spilitic basalt (Houston, 1965) and polymictic conglomerates which are mostly composite alluvial fans of piedmont origin formed by rapid deposition during sustained subsidence of the basin. The productive coal seams with a total thickness of 1,070 m occur in the lenticular shales and sandstones of the Tivoli, Cooneana and Blackstone Formations. The Ipswich Coal Measures have an abundant *Dicroidium* macroflora, dominated by *D. odontopteroides,* together with species of *Cladophlebis, Doratophyllum, Stenopteris, Czekanowskia, Yabiella, Ginkgoites, Fraxinopteris, Neocalamites, Rienitsia* and *Sphenopteris* (Jones and De Jersey, 1947), and microflora dominated by *Alisporites,* some of which are believed to be the pollen of *Dicroidium,* of Middle to Late Triassic age (De Jersey, 1962, 1970b).

Fossil insects, mainly Coleoptera, Orthoptera, Homoptera and Blattoidea occur in the Mount Crosby Formation (Tindale, 1945) and Blackstone Formation (Tillyard and Dunstan, 1924).

Clarence–Moreton Basin. The lower part of the Triassic sequence with the massive sandstones and red beds is missing in the more easterly Clarence–Moreton Basin where productive coal measures formed in the Middle to Late Triassic. Widespread volcanism in the form of tuffs, basalt and rhyolitic tuffs in the coal measures is considered to have an effusive and local origin (McElroy, 1969b). The Chillingham Volcanics, with a maximum thickness of 1,500 m, crop out as a

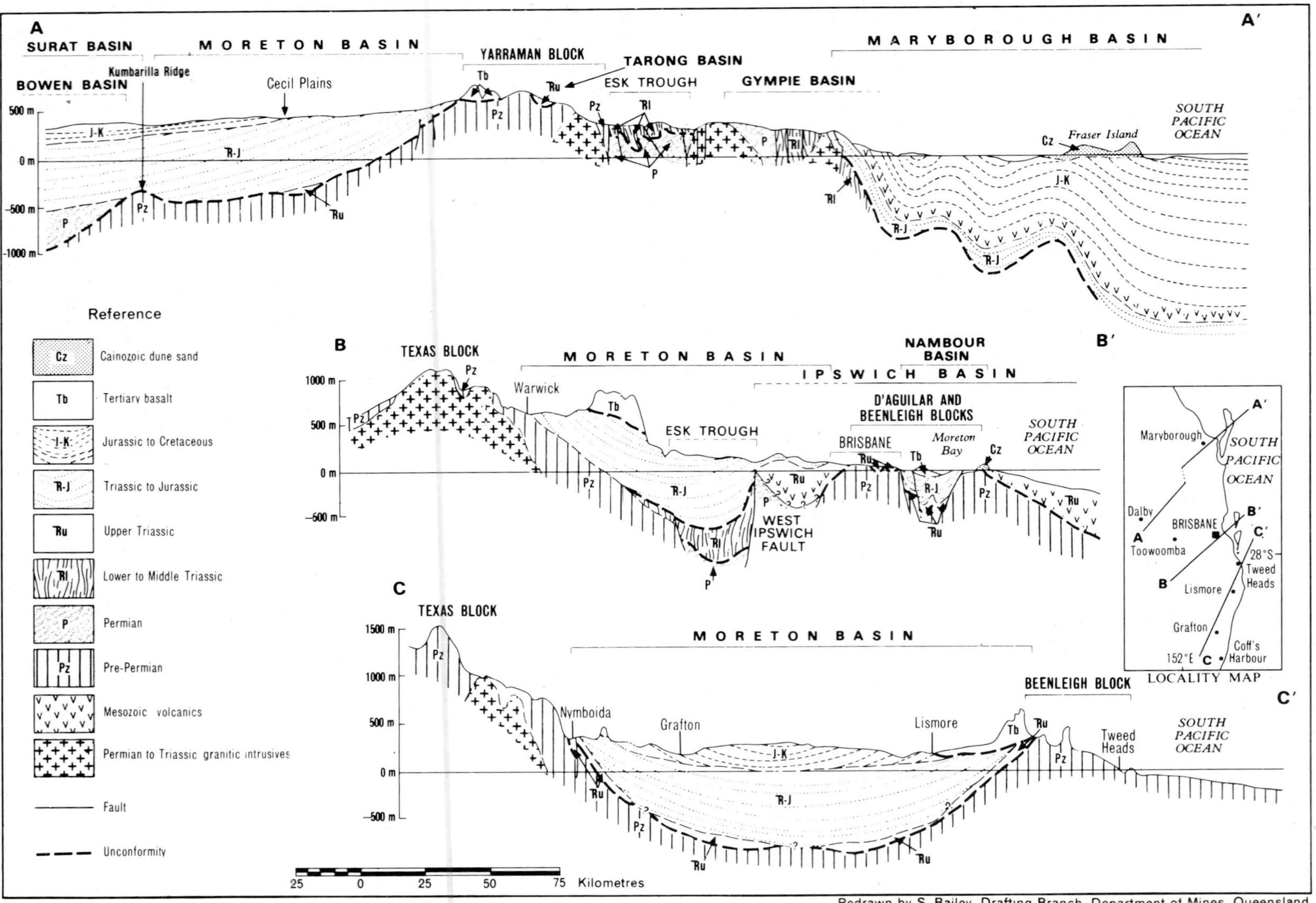

Fig. 12. Diagrammatic geological cross-sections showing interrelationships of Mesozoic basins.

belt of massive rhyolites and lithic tuffs over some 35 km in the middle of the basin.

The Nymboida, Evans Head and Red Cliff Coal Measures in the southern (Clarence) part of the basin (McElroy, 1962, 1969b) are equivalent to the Ipswich Coal Measures (Table IV). The Nymboida Coal Measures are a sequence of sandstones, shale, cobble conglomerate, tuff and coal. An abundant macroflora with species of *Dicroidium, Phoenicopsis, Taeniopteris, Pterophyllum, Cladophlebis* and *Ginkgo* occurs in shales of the Basin Creek Formation, the exposed section of which comprises 750 m of sandstone, shale and conglomerate.

The Clarence–Moreton Basin sediments unconformably overlying the Ipswich and Nymboida Coal Measures are related to those of the Surat Basin, the southeast sub-basin of the Great Artesian Basin. They form the Bundamba Group, predominantly psammitic, or cobble and pebble conglomerates, sandstones and siltstone with some thin carbonaceous shale. The flora, dominated by *Dicroidium odontopteroides* (Morris) Gothan, and the microflora, with abundant *Alisporites* (of the *Pteruchus* type) and containing in the lower part *Partitisporites novimundanus* Leschik and *Duplicisporites granulatus* (Leschik) Klaus restricted to the Carnian of Austria, are

TABLE IV

Detailed correlation table for the Clarence–Moreton and Ipswich Basins and Esk Trough.

Period	Epoch	Clarence-Moreton Basin					Ipswich Basin			Esk Trough	
CRET.		Grafton Formation (266 m)									
JURASSIC	LATE	Kangaroo Creek Sandstone (150 m)									
JURASSIC	MIDDLE	Walloon Coal Measures (600 m)		Walloon Coal Measures (244 m)							
JURASSIC	MIDDLE	*Towallum Basalt* (15 m)									
JURASSIC	EARLY	Marburg Formation (600 m)	Koukandowie Sandstone Member (120 m)	Bundamba Group	Marburg Formation	Heifer Creek Sandstone Member (213 m)					
JURASSIC	EARLY					Ma Ma Creek Sandstone Member (76 m)					
JURASSIC	EARLY		Blaxland Fossil Wood Conglomerate Member (45 m)			Winwill Conglomerate Member (46 m)					
JURASSIC	EARLY					Gatton Sandstone Member (46 m)					
TRIASSIC	LATE	"Bundamba Group"	Mill Creek Siltstone (60 m)		Woogaroo Sub-Group	Ripley Road Sandstone (198 m)					
TRIASSIC	LATE		Layton's Range = Corindi Conglomerate			Raceview Formation (122 m)					
TRIASSIC	LATE					Aberdare Conglomerate (12 m)					
TRIASSIC	LATE	Nymboida Coal Measures (1000 m)	Basin Creek Formation				Ipswich Coal Measures	Brassall Sub-Group	Blackstone Formation (244 m)		
TRIASSIC	LATE		Georges Knob Conglomerate Member						Cooneana Formation (244 m)		
TRIASSIC	LATE								Tivoli Formation (488 m)		
TRIASSIC	LATE		Farquhars Creek Seam	*Chillingham Volcanics*		*Brisbane Tuff*		Kholo Sub-Group	Cribb Conglomerate		
TRIASSIC	LATE		*Copes Creek Tuff*						*Hector Tuff*		
TRIASSIC	LATE		Bardool Conglomerate						Colleges Conglomerate		
TRIASSIC	LATE		Goolang Siltstone Member						Mt. Crosby Formation		
TRIASSIC	LATE		Cloughers Creek Formation						*Weir and Sugars Basalts*		
TRIASSIC	LATE								Blackwall Breccia		
TRIASSIC	MIDDLE TO EARLY									Toogoolawah Group	Esk Formation
TRIASSIC	MIDDLE TO EARLY										*Neara Volcanics*
TRIASSIC	MIDDLE TO EARLY										Bryden Formation

75-377 Del. B.F. S.A. Department of Mines

regarded as of Late Triassic (Carnian to Norian) age (De Jersey, 1964).

Tasmania Basin. The Triassic System of the Tasmania Basin is entirely non-marine, of fluvial and lacustrine origin, and composed of quartzite sandstones, lithic arenites and shales with minor fine-grained conglomerates and coal beds (Hale, 1962), remnants of a former extensive cover over more than half of eastern Tasmania. It is represented in the upper part of the Parmeener Super Group (Banks, 1973; Banks and Clarke, 1973), beginning with the cliff-forming Ross Sandstone. Pale greyish-buff shaly siltstone 93 m topographically above the base of the Ross Sandstone contains a microfloral assemblage of Early Triassic (Scythian) age (Playford, 1965). The overlying Cluan Formation consisting of quartz arenite and carbonaceous siltstones carries *Deltasaurus kimberleyensis,* described from the Blina Shale, and other amphibians permitting determination of the age as Scythian, probably Otoceratan (Banks and Clarke, 1973) (Fig. 2).

The Tiers Formation is composed of siltstones and carbonaceous siltstones with minor quartz and lithic arenites (Fig. 3). It contains *Aratrisporites* and *Tigrisporites* resembling forms from the Austrian Carnian, *Osmundacidites fissus* (Leschik) Playford from the Middle Keuper of Switzerland and *Punctatosporites walkomi* de Jersey, common in the Australian Middle and Late Triassic (Playford, 1965).

The youngest formation in the succession, the Brady Formation contains subbituminous coal with lithic quartz-arenite siltstone and shale (Fig. 4). The macroflora includes *Dicroidium* spp., *Cladophlebis* spp., *Linguifolium* and *Phoenicopsis elongatus* (Morris) Seward. The microflora indicates a Late Triassic, probably Rhaetian age.

South Australia. Late Triassic to Early Jurassic freshwater lacustrine sediments occur in isolated intramontane basins in the north of South Australia. The Leigh Creek Basin contains productive coal measures in a sequence some 700 m thick with a basal conglomerate, carbonaceous shales and coal seams with red ferruginous siltstone and lenticular sandstone interbeds (Parkin, 1953). On microfloral evidence the Leigh Creek Coal Measures are dated as Late Triassic (Rhaetic) to Early Jurassic (Liassic), the lower part having elements in common with the Ipswich Coal Measures and the upper part containing a younger assemblage with *Classopollis classoides* (Pflug) Pocock and Jansonius, common in the Early Jurassic of Australia and elsewhere (Playford and Dettmann, 1965). *Dicroidium* dominates the macroflora and fresh-water molluscs (Ludbrook, 1961b) and a fish (Wade, 1953) have been described.

Similar sediments occur in the Springfield Basin (Johnston, 1960), the Boolcunda Basin and in structurally low areas of the Cooper Basin (Wopfner, 1969) (Fig. 4).

Victoria. A Triassic dating is tentatively retained (Douglas, 1969) for isolated occurrences of mudstones and sandstones containing poorly preserved plant remains in a fault zone at Yandoit Hill, 105 km northwest of Melbourne and overlying Permian tillite conglomerate, sandstone and mudstone at Bald Hill near Bacchus Marsh, 50 km northwest of Melbourne (Singleton, 1973).

In the extreme southeast of the Australian mainland near Benambra syenites and granite porphyries were intruded in the Late Triassic, determined by K/Ar dating, in a complex area deformed and intruded during an orogeny in the Late Devonian (Talent, 1969).

Jurassic

Great Artesian Basin. In the eastern basins, Jurassic deposits are predominantly non-marine, although the presence of acritarchs and arenaceous foraminifera (Evans, 1966) in the Early Jurassic of the Surat Basin indicates that some are probably deltaic. No open-sea faunas have been discovered. As in Western Australia, uplift, folding and volcanism occurred in the Late Triassic and in the Early Jurassic the eastern lobes and sub-basins of the Great Artesian Basin began to form (Fig. 5). These events, also believed to be associated with the Jurassic dispersal of Gondwanaland, were followed by widespread deposition of fluvial sandstones transgressive across the erosion surface of pre-Jurassic infra-basins. The sandstones are of great economic importance as they form the aquifers of the Great Artesian Basin. Their sequences and lithology in the Surat, Clarence–Moreton, Maryborough and Nambour Basins show a similarity which suggests that during the Early Jurassic they were deposited in a single basin (Power and Devine, 1970) (Fig. 5, 13). The Precipice Sandstone at the base of the sequence in the Surat Basin is a diachronous, strongly current-bedded, quartzose sandstone of fluvial origin; it forms the reservoir for oil and gas on the Roma Shelf and for oil at Moonie and Alton. Its Jurassic age and that of its equivalents is determined from its microflora, with an abundance of *Classopollis* and distinctive pteridophytic spores including species of *Cadargasporites* (De Jersey and Paten, 1964).

Fluvial, deltaic and lacustrine sedimentation continued in the basins during the Early Jurassic, and acritarch swarms and chamositic oolites occurring in the upper part of the Evergreen Formation of the Surat Basin and Esk Trough have proved useful in stratigraphic correlation. After the deposition of the Hutton Sandstone in the regressive phase of the cycle, progression towards more stable and quiescent conditions produced paludal, floodplain and lacustrine environments in which the Walloon and Tiaro Coal Measures were formed.

The Middle to Late Jurassic Walloon Coal Measures and their lithological equivalents, the Mulgildie Coal Measures of the lower Injune Creek Group, are widespread units of the Great Artesian Basin. They contain rich macrofloras with *Cladophlebis australis* Morris, *Taeniopteris spatulata* McClelland, *Otozamites feistmanteli* Zigno, *Brachyphyllum crassum* Tenison Woods, *Phyllopteris feistmanteli* Etheridge Jr., and microfloras with *Regulatisporites ramosus* de Jersey, *Verrucosisporites walloonensis* de Jersey, *Lycopodiumsporites rosewoodensis* (de Jersey) de Jersey and *Contignisporites cooksoni* (Balme) Dettmann.

They are followed by a succession of transgressive quartz sandstones deposited during uplift which commenced in the Surat Basin in the Middle Jurassic and extended in a westerly direction across the Eromanga Basin to the western margin of the Great Artesian Basin in the Late Jurassic and Early Cretaceous. These sandstones constitute the main aquifers of the Great Artesian Basin. Their relationships in the Surat Basin are shown in Fig. 13. The most prominent crop out in conformable sequence in the Roma–Wallumbilla area (Day, 1964) and dip into the basin at a low angle: the Gubberamunda Sandstone, predominantly a coarse-grained quartz sandstone 15–60 m thick, with large-scale cross-bedding, deposited in a high-energy fluvial environment; the Orallo Formation 105–135 m thick of medium to coarse-grained subgreywackes with fragmentary fossil wood and some coal, deposited in a lacustrine environment; and the Mooga Sandstone 150–170 m thick in oil wells in the Surat Basin.

Stratigraphic equivalents of the Bungil Formation (Exon and Vine, 1970, redefined from the previously widely used Blythesdale Formation) lap onto Precambrian basement rocks on the western and southwestern margins of the Great Artesian Basin some 1,500 km to the west (Forbes, 1966; Wopfner et al., 1970). The outcropping rocks are probably for the most part of Early Cretaceous age, but a likely Late Jurassic age has been determined for a microflora occurring in subsurface carbonaceous clay interbedded with sands in a possible lateral facies development of the Algebuckina Sandstone overlying rocks of the Arckaringa Basin (Harris, 1970).

Cretaceous

The Early Cretaceous transgression. During the Early Cretaceous a vast area of Australia was invaded by the sea which not only inundated the Great Artesian Basin but transgressed over the old erosion surface developed on Precambrian and Palaeozoic rocks.

The transgression over the craton coincided with volcanism (see Figs. 7, 8) off the northwest margin, and on the southwest margin of the continent and in an offshore belt progressing northwards from the Maryborough Basin. The Maryborough Basin volcanics comprise a thick sequence of pyroclastics and andesitic and trachytic flows of the Grahams Creek Formation of Late Jurassic or Neocomian age; the Whitsunday Volcanics (waterlaid pyroclastics) and the Proserpine Volcanics (mainly rhyolite flows) are mostly Aptian with a K/Ar isotopic age of 96–115 m.y. (Clarke et al., 1971). Veevers and Evans (1973) have interpreted these events as being due to eustatic rise in sea level associated with sea-floor spreading, and transgression over continental interiors adjacent to subduction zones with pulsating orogenic movements on the margins in the manner of the Haug Effect described by Johnson (1971, 1972).

Evidence of the transgression commencing in the Late Neocomian or Early Aptian is given by the appearance of marine molluscs in quartz sandstones such as the lower units of the Mullaman Beds from the second unit of which Skwarko (1966) described a rich fauna of endemic molluscs including species of *Pterotrigonia (Rinetrigonia)* and *Iotrigonia (Zaletrigonia),* the Wrotham Park Sandstone of the Carpentaria Basin in which marine fossils occur rarely (Woods, 1961), the Minmi Member of the Bungil (formerly Blythesdale) Formation, from which Day

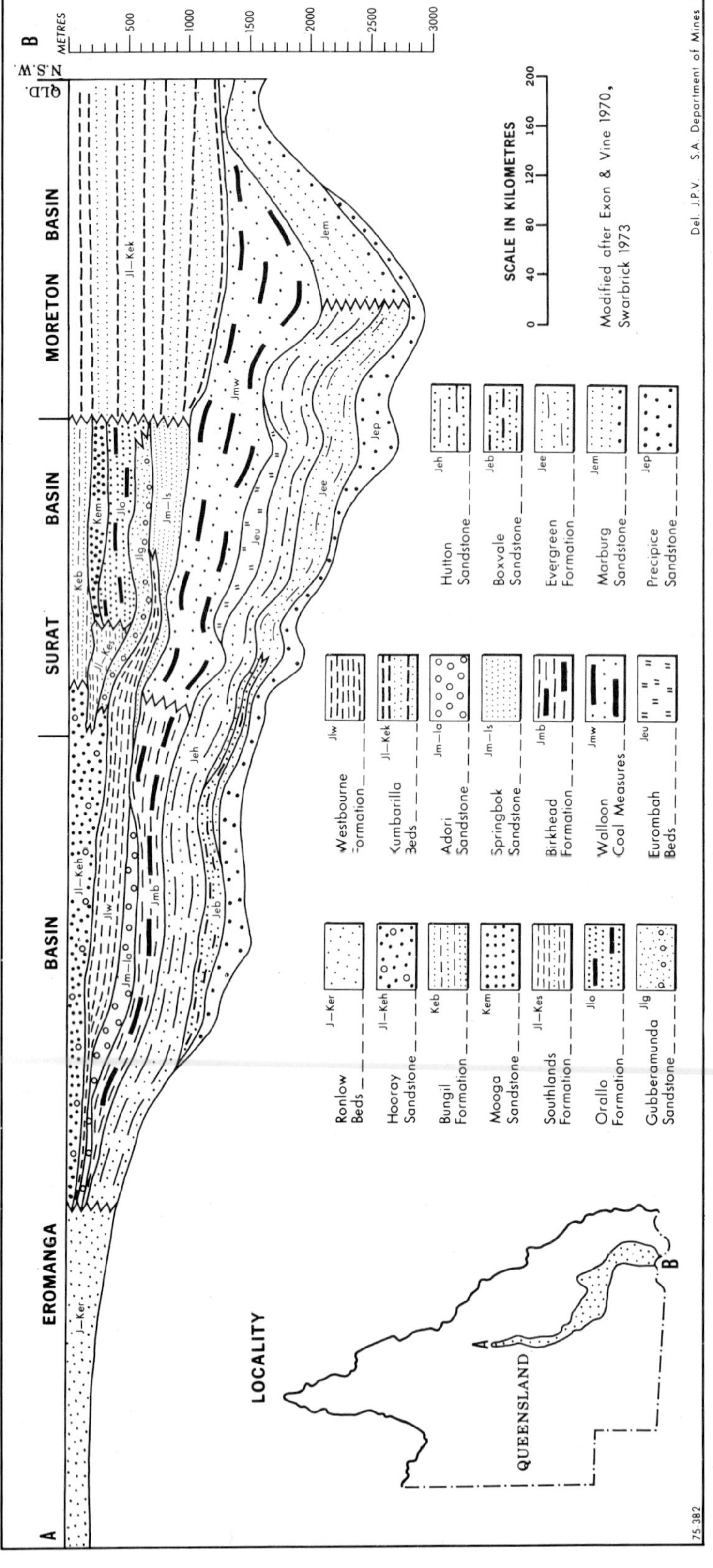

Fig. 13. Relationships of Jurassic–Early Cretaceous units outcropping on the eastern margin of the Great Artesian Basin.

(1964) recorded some thirty species about two-thirds of which continue into the overlying Aptian shales, and the Parabarana Sandstone with sparse molluscs associated with plant remains (Ludbrook, 1966).

The maximum transgression occurred during the Late Aptian when the Carpentaria, Eromanga and Surat Basins were united and connection existed between the Great Artesian Basin and the Eucla Basin to the southwest and troughs underlying the Murray Basin to the south (Fig. 8).

Great Artesian Basin (Surat, Eromanga and Carpentaria Basins). Under stable basin conditions in the Aptian–Albian, glauconitic and carbonaceous clays, silts and fine sands were deposited over a very extensive area. Virtually undisturbed since the Early Cretaceous these strata are generally poorly exposed and are known from flat surface exposures, dissected plateaux, rolling downs and stream bed sections and from numerous groundwater and oil exploration wells. Their maximum thickness exceeds 1,500 m in the Eromanga Basin. Together they form the Rolling Downs Group.

Invertebrate faunas of the Great Artesian Basin have been studied for over a century and for a comprehensive summary of the work of previous authors and documentation of the molluscan assemblages in Queensland and tentative correlation of the Early Cretaceous formations, the reader is referred to Day (1969).

Provincialism and a high degree of endemism in the molluscan and foraminiferal faunas places a restraint on firm and detailed correlation with areas outside Australia. Since the significant advance of Whitehouse (1926), who recognised the Aptian age of the "Roma" fauna and the Albian age of the "Tambo" fauna, effective correlation has been by means of ammonites.

The Neocomian age assigned to the earliest molluscan faunas depends on the presence of *Hatchericeras,* otherwise known only from Patagonia, stratigraphically below Aptian faunas in the Battle Camp Formation of the Laura Basin.

Direct correlation with the type Aptian zones of Europe cannot be achieved because hoplitoids are unknown from Australia, but both *Australiceras* and *Tropaeum,* regarded by R. Casey (1960) as restricted to the Aptian are represented. Australian species of *Australiceras* are related to Aptian forms found in Madagascar, the Caucasus, India, Colombia and California where they are associated with Late Aptian hoplitoids, while species of *Tropaeum* are close to those from Arctic Canada, Spitsbergen and southernmost South America, and also to those from Madagascar and northern Europe where they have a time range of *bowerbanki* to *nutfieldensis* Zones (Day, 1969). Associated large bivalves which occur in the Doncaster Member of the Wallumbilla Formation, the lower part of the Marree Formation and in the Bulldog Shale are *Maccoyella barklyi* (Moore), *Fissilunula clarkei* (Moore), *Eyrena linguloides* (Hudleston) and *Tancretella plana* (Moore). With the exception of *Maccoyella,* which occurs also in New Zealand and Patagonia, these are restricted to Australia.

Albian faunas include in the Allaru Mudstone species of *Labeceras, Appurdiceras* and *Myloceras* known only from New Guinea, Moçambique, Zululand and Madagascar, and *Falciferella,* described from the lower Gault Middle Albian, *Hoplites dentatus* Zone (R. Casey, 1954) and occurring in a somewhat modified form (Casey in Ludbrook, 1966) in the Wooldridge Limestone Member of the Oodnadatta Formation. The desmoceratids *Beudanticeras* and *Brewericeras* occur in the Ranmoor Member of the Wallumbilla Formation. Albian strata contain species of *Aucellina, Inoceramus* and *Dimitobelus* and planktonic foraminifera with *Hedbergella infracretacea* (Glaessner).

Micropalaeontological correlation is beset with problems similar to those experienced in Canada and elsewhere (Chamney, 1973), and interbasin biostratigraphic correlation depends largely upon the use of provincial microfaunas and palynological zonations. Most of the agglutinating foraminifera were described by Crespin (1963) from established rock units; the first attempt to coordinate foraminiferal and molluscan assemblages from surface and subsurface sections was made by Ludbrook (1966) who proposed a scheme of broad foraminiferal zones, dated where possible from available though sparse ammonite occurrences in South Australia. Haig (1973) related these zones to a numerical foraminiferal sequence used in subsurface correlation in the Surat Basin. Scheibnerova (1971) had also proposed a zonal sequence for two bores at Bourke near the southern margin of the Cunnamulla Shelf, southwest-west of the Surat Basin. Palynological zonations have been similarly recognised by Dettmann and Playford (1969).

Withdrawal of the epicontinental sea began in the Albian, but the survival of tolerant widely distributed Aptian–Albian foraminifera into the Blanchewater Formation (Ludbrook, 1966) suggests that even in the southern part of the Great Artesian Basin marine influence may have continued to the Cenomanian. Withdrawal was to the north. From the Mullaman Beds of Bathurst Island north of Darwin, Wright (1963) described a sequence of Late Cenomanian–Turonian ammonite assemblages, the stratigraphically lowest of which contains *Euomphaloceras lonsdalei* (Adkins), linking it with the Eagle Ford of Texas; the next includes *Hamites (Stomohamites) simplex* (d'Orbigny), *Turrilites costatus* Lamarck and species of *Sciponoceras* and *Acanthoceras*; and the uppermost contains *Collignoniceras* sp. cf. *C. woolgari* (Mantell), a widely distributed species in the Turonian. These faunas indicate an open sea connection with Tethys to the northwest.

With the regression of the sea, fluvial and lacustrine conditions returned to a large part of the Great Artesian Basin and the Winton Formation was deposited. This is an entirely non-marine unit characterised by interbedded fine-grained sandstone, siltstone and carbonaceous shale with coal beds. Green and white "salt and pepper" sands and concretionary sandy limestones have a high percentage of ferro-magnesian minerals, plant fragments and calcified fossil tree trunks.

The age of the Winton Formation is regarded as Cenomanian; its top is marked by an erosion surface and post-Cenomanian deposition in the Great Artesian Basin was limited to fluvial deposits such as the Mount Howie Sandstone (Wopfner, 1963) and kaolinitic sandstone equivalents disconformably overlying the Winton Formation near the southwestern margin (Forbes, 1972).

Renmark–Tararra–Menindee, Wentworth and Ivanhoe Troughs. During the Aptian transgression the sea gained access to a more-or-less parallel series of northeast–southwest-trending troughs related to the Darling Lineament and underlying the northern part of the Murray Basin. The Renmark–Tararra–Menindee Troughs, the Wentworth Trough and the Ivanhoe Trough (Thornton, 1974) contain up to 1,000 m of Early Permian mostly marine sediments on which Aptian–Albian shales, sandstones and siltstones were deposited. These are unknown at the surface but a thickness of as much as 440 m has been intersected in a number of oil exploration wells in the area (Ludbrook, 1958a, 1961a, 1969, fig. 86; and others in a series of unpublished reports of the South Australian Department of Mines) and named by Thornton (1972) the Monash Formation. The Aptian Pyap and Merreti Members are marine and contain typical Aptian foraminifera of the Great Artesian Basin. The uppermost Coombool Member is non-marine and distinguished by abundant megaspores, *Arcellites reticulatus,* and an associated Albian microflora.

The northerly withdrawal of the sea began in these troughs in the Late Aptian. Cretaceous sedimentation was related to that of the Great Artesian Basin and not to that of the Otway Basin to the south. The Monash Formation is overlain by Tertiary strata of the Murray Basin which is essentially a Cenozoic basin invaded by the sea from the south after the final separation of Australia from Antarctica in the Early Tertiary.

SOUTHERN AUSTRALIA

The Mesozoic and Cenozoic history of southern Australia is linked to rifting in the Jurassic and Cretaceous and the final separation of the Australian and Antarctic plates in the Early Tertiary (Table V).

Basins along the southern rift

During the break-up of Gondwanaland a rift developed along the southern continental margin (Griffiths, 1971) extending from west to east and giving rise to the Polda Trough and offshore Duntroon Basin, and cutting across the predominantly north–south structural trend of eastern Australia tc form the Otway Basin and the Strzelecki and Gippsland Basins (Weeks and Hopkins, 1967). The Polda Trough and Robe-Penola Trough of the Otway Basin are en echelon structures ("splays" of Griffiths, 1971) associated with the "tear apart" tectonics of the west–east rifting.

The onset of rifting was accompanied by the extrusion over most of eastern Tasmania of dolerites with a K/Ar age of 162 m.y., extremely similar to the Ferrar Dolerite of Antarctica (Edwards, 1942) dated at 160–165 m.y. (Evernden and Richards, 1962). Basaltic lavas and pyroclastic rocks with a Geochron K/Ar dating of 153 ± 5 m.y. quoted by Harding

TABLE V

Correlation table for the Mesozoic rocks of southern Australia

PERIOD	EPOCH	STAGE	RADIOMETRIC AGE m.y.	EUCLA BASIN	POLDA TROUGH	DUNTROON BASIN & KANGAROO ISLAND	OTWAY BASIN	BASS BASIN	GIPPSLAND & STRZELECKI BASINS
CRETACEOUS	LATE	MAASTRICHTIAN	65, 70				TIMBOON SAND; CURDIES FM.		
		CAMPANIAN	76	?			SHERBROOK GROUP: PAARATTE FORMATION		
		SANTONIAN	82					EASTERN VIEW GROUP	LATROBE GROUP
		CONIACIAN	88			NEOCOMIAN TO MAASTRICHTIAN SEDIMENTS	BELFAST MDS. MBR.		
		TURONIAN	94	MADURA FORMATION			FLAXMANS FM.		
		CENOMANIAN	100				WAARRE SST.		
	EARLY	ALBIAN	106						
		APTIAN	112				OTWAY GROUP	EQUIVALENTS OF OTWAY AND STRZELECKI GROUPS	
		NEOCOMIAN	118, 124, 130, 136	LOONGANA SANDSTONE ?			? VOLCANICS (b)		STRZELECKI GROUP
JURASSIC	LATE	TITHONIAN	141, 146		?		?		?
		KIMMERIDGIAN	151		?				
		OXFORDIAN	157		"POLDA FORMATION"		VOLCANICS (a)		
		CALLOVIAN	162		?				
	MIDDLE	BATHONIAN	167			BASALT KANGAROO ISLAND			
		BAJOCIAN	172						
	EARLY	TOARCIAN	178						
		PLIENSBACHIAN	183						
		SINEMURIAN	188						
		HETTANGIAN	190-195						
PRINCIPAL SOURCES			*Geol. Soc. 1964*	*Ludbrook, 1958b; Ingram, 1968; Lowry, 1970*	*Harris, 1964; Harris & Foster, 1974*	*Boeuf & Doust, 1975; McDougall & Wellman, 1976*	*Dettmann, 1963; Kenley, 1971; Harding, 1969; (a) Geochron dating; (b) A.N.U. dating*	*Robinson, 1974*	*Dettmann, 1963; James & Evans, 1971*

75-378

Del. W.J.E. S.A. Department of Mines

(1969) and Cundill (in Wopfner and Douglas, 1971) occur within the Casterton Beds intersected in several wells in the Gambier Embayment of the Otway Basin, and similar basalts and tuffs are recorded from the northern margin of the Strzelecki Basin (Hocking, 1972). An apparently anomalous "reliable minimum" K/Ar age of 120 ± 10 m.y. was obtained at Australian National University for olivine basalt from Casterton also quoted by Harding (1969). If this dating is correct, the lavas in the Casterton Beds are of Neocomian age, contemporaneous with the Cambewarra Flow (Table II). The most westerly occurrence is that of tholeiitic basalt on Kangaroo Island having a whole-rock age of 170 ± 5 m.y. or Middle Jurassic (McDougall and Wellman, 1976).

The earliest sedimentary evidence of rifting is contained in the Polda Trough where 100 m of Late Jurassic fine sands, silts and lignites underlie Middle Eocene to Early Oligocene sapropelic clays and sands (Harris, 1964; Harris and Foster, 1974). Seismic data indicate that this fault-controlled trough extends westwards offshore and crosses the continental shelf before merging with the continental margin (Smith and Kamerling, 1969). On aeromagnetic and seismic evidence, the Duntroon Basin (also named the Duntroon Embayment of the Great Australian Bight Basin) is considered, like the Otway Basin, to contain up to 10 km of sediments. Offshore drilling penetrated an almost continuous Early Cretaceous to Middle Miocene sequence beginning with Neocomian to Albian fluvial–lacustrine sediments, followed by Cenomanian to Danian deltaic deposits, Late Paleocene to Middle Eocene clastics and Late Eocene to Middle Miocene carbonates (Boeuf and Doust, 1975).

The easterly progression of the rift is reflected in the opening of the Otway and Strzelecki Basins in the Late Jurassic to Early Cretaceous and the deposition of a monotonous sequence of more than 3,000 m of current-bedded alternating greywacke-type arkoses and mudstones. These Neocomian to Albian sediments form the Otway Group of the Otway Basin and the Strzelecki Group of the Strzelecki Basin; they are of fluvial and paludal or lacustrine origin, derived from igneous rocks and uplifted Palaeozoic sediments on the northern margins. The remarkably fresh feldspars and unworn grains attest to their rapid deposition.

Otway Basin

The Otway Basin (Wopfner and Douglas, 1971) is a long deep trough with onshore embayments containing a great thickness of Cretaceous terrigenous clastic sediments. The northern structural margin of the Gambier Embayment on the western side of the basin is a complex hinge structure marked by differential block faulting in the Late Jurassic and Cretaceous, during the period of deposition, and steepened dips with associated truncation of the younger sediments of the Otway Group (Rochow, 1971). The southern offshore margin is a basement ridge on the edge of the continental shelf below which the Mesozoic sediments dip basinwards to the north (Weeks and Hopkins, 1967; Griffiths, 1971).

The Early Cretaceous Otway Group is a monotonous, lithologically uniform sequence of more than 5,000 m of feldspathic and lithic greywackes, siltstones and carbonaceous mudstones with some coal, unconformably overlying Palaeozoic rocks on the northern onshore boundary of the basin. It is of non-marine origin and from study of the microfloras (Dettman, 1963) and megaplants (Douglas, 1969, 1971) a Neocomian to Albian age is indicated. No microplankton has been reported.

The further opening of the rift and continued subsidence resulted in the entry of the sea during the Late Cretaceous and the deposition of the Sherbrook Group, a thick sequence of paralic sediments rapidly thickening to the south (Fig. 14). The new sedimentary cycle began with the deposition of the Waarre Sandstone, predominantly quartz sandstone with minor mudstone and containing microplankton (Bock and Glenie, 1965). The overlying Flaxmans and Paaratte Formations, particularly the Belfast Mudstone Member, contain abundant foraminifera of Turonian and Santonian age which exhibit an alternation of arenaceous with arenaceous/calcareous assemblages (Taylor, 1964). Planktonic foraminifera and marine megafaunas including ammonites and species of *Inoceramus* occur in the Belfast Mudstone.

The marine incursion was probably synchronous with that in the western part of the Eucla Basin, and it was followed by a regressive phase in which the environment became progressively non-marine (Taylor, 1971) or sporadic, with incursions in the Middle and Late Paleocene, until the Middle Eocene when, with the separation of Australia and Antarctica, the Eucla, St. Vincent and Murray Basins were opened from the south and marine sedimentation in the Gambier Embayment became continuous with that of the Murray Basin (Fig. 14).

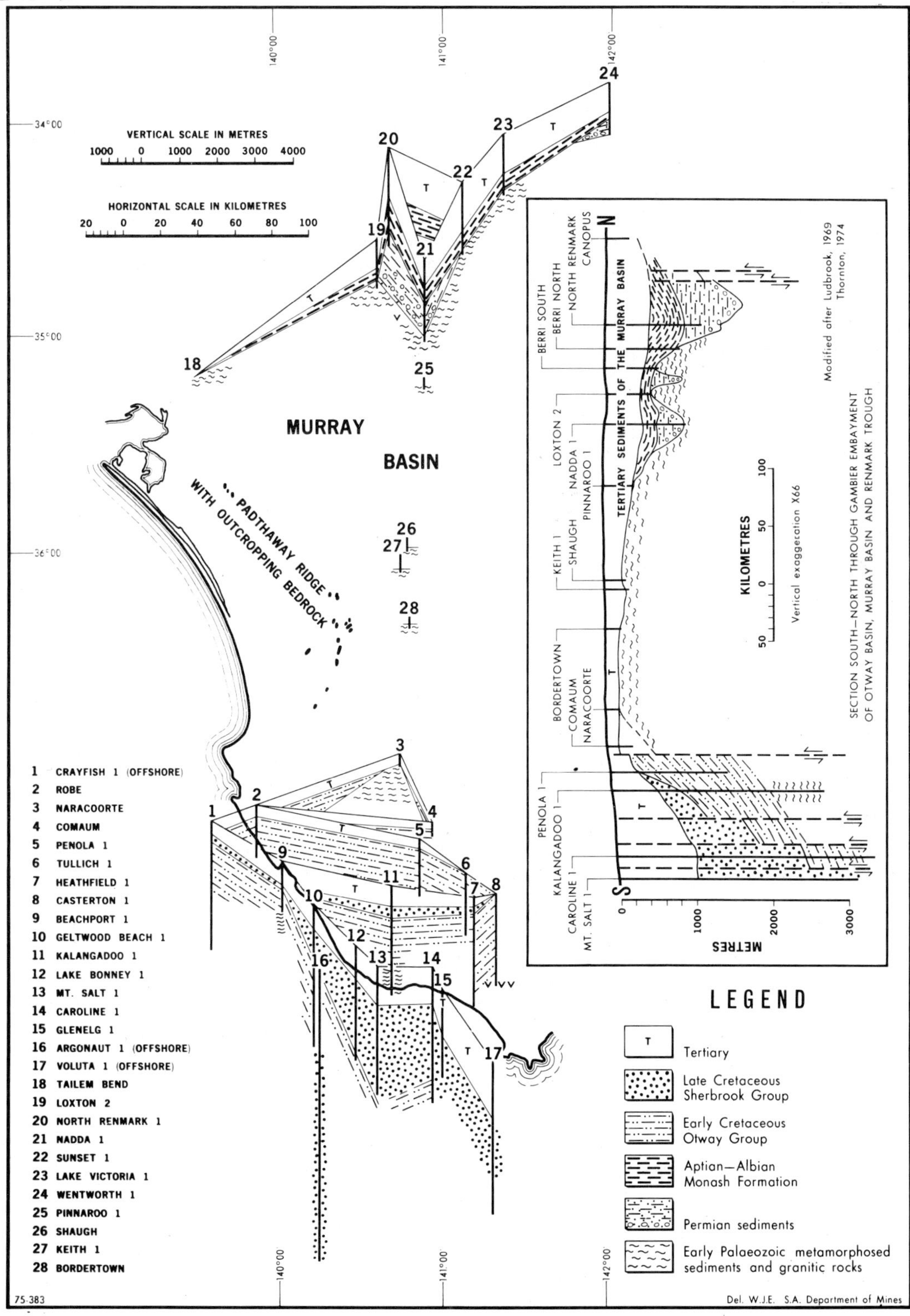

Fig. 14. Fence diagram and section showing Cretaceous sedimentation in the Gambier Embayment (Otway Basin) and Renmark and Wentworth Troughs.

Bass Basin

The Bass Basin (Robinson, 1974) commenced to form in the Late Jurassic and was subjected to tensional stress from the early stages of the separation of Australia from Antarctica until the Late Tertiary. Extensive block faulting occurred, and folding associated with intrusive and extrusive volcanism, particularly in the Late Cretaceous and Late Miocene. The oldest basin sediments are non-marine Early Cretaceous siltstone and sandstone related to those of the Otway and Strzelecki Groups. They are overlain by non-marine Late Cretaceous and Early Tertiary sandstone, shale and coal of the Eastern View Group.

Strzelecki Basin

The Strzelecki Basin is the onshore half-graben ancestor of the Gippsland Basin, four-fifths of which formed offshore as the result of taphrogenic and epeirogenic movements (Hocking, 1972) associated with the opening of the Tasman Sea and separation of Australia and New Zealand. Greywacke-type arkoses, mudstones, shales and feldspathic grits of the Strzelecki Group are similar to those of the Otway Group and it is likely that during the initial period of rifting the Otway and Strzelecki Basins were continuous (Hocking, 1972).

The depositional environment was relatively stable on the western side of the Strzelecki Basin where bituminous coal has been mined at several fields and the important Koonwarra fishbed occurs. Waldman's (1971) study of the 7.1 m of laminated mudstone and its biota has provided clear evidence of the palaeoenvironment. The laminations are regarded as varves representing seasonal floodings in a shallow lake which was ice-covered in winter. The mass mortality of the fish resulted from anoxia under the ice – the phenomenon known as "winter kill". The Koonwarra fishbed has a rich insect fauna which must have inhabited the margins of a cold lake (Riek, in Waldman, 1971).

Gippsland Basin

The Gippsland Basin formed a centre for the deposition of more than 8,000 m of Early Cretaceous to Holocene sediments (James and Evans, 1971). Both Strzelecki Group greywackes and Late Cretaceous sediments of the Latrobe Group are confined to the deep basin. The Latrobe Group is the source and reservoir for the oil and gas of the Gippsland Basin. It consists of 5,000 m of Late Cretaceous to Holocene fluvio-deltaic quartzose sandstone, siltstones, shales and coals, constituting two-thirds of the basin section. The sediments are more mature than those of the Strzelecki Group with less abundant feldspar and volcanolithic grains and the grain size is more diverse (Hocking, 1972). The late stages of Latrobe Group deposition are marked by complex channel or submarine canyon erosion in a general northwest–southeast direction and infilling with Late Eocene to Early Oligocene restricted and shallow marine clastics. The canyon configuration, erosion and fill are mainly responsible for the hydrocarbon traps at the eroded topographic surface of the Latrobe Group (Richards and Hopkins, 1969; James and Evans, 1971).

Eucla Basin

The Eucla Basin is an intracratonic basin marginal to the present coastline. It is essentially a Cretaceous–Tertiary basin initiated by gentle downwarping in the Neocomian. A northwest–southeast-trending infra-basin or trough, the "Denman Basin" (Whitten, 1972), associated in the south with the Mallabie Depression (Thomson, 1970), underlies it on the eastern side. This trough is filled with Late Proterozoic and Palaeozoic rocks and, like the Renmark and other troughs underlying the northern part of the Murray Basin, was entered by the sea in the Early Permian and again during the Aptian epicontinental transgression.

The palaeogeographic implications of Cretaceous sedimentation in the Eucla Basin are not yet fully understood. Cretaceous strata are known only in the subsurface, and wells are widely separated. Neocomian to Santonian sediments have been proved by drilling to a depth of 640 m in the western part of the basin (Ludbrook, 1958b; Ingram, 1968) but to the east and north the youngest are of Albian age. Early Cretaceous sediments are related to those of the Great Artesian Basin. They transgress over an uneven floor of crystalline basement, Proterozoic and Palaeozoic rocks including the Permian of the Mallabie Depression (Harris and Ludbrook, 1966) (Figs. 1, 15). The oldest are feldspathic, micaceous and calcareous sandstones with fontainebleau textures (Lowry, 1970) which are lithologically and stratigraphically equivalent to the Cadna-owie Formation (Wopfner et al., 1970) of the western Great Artesian Basin. They are overlain by the imperfectly known Aptian–Santonian Madura Formation in

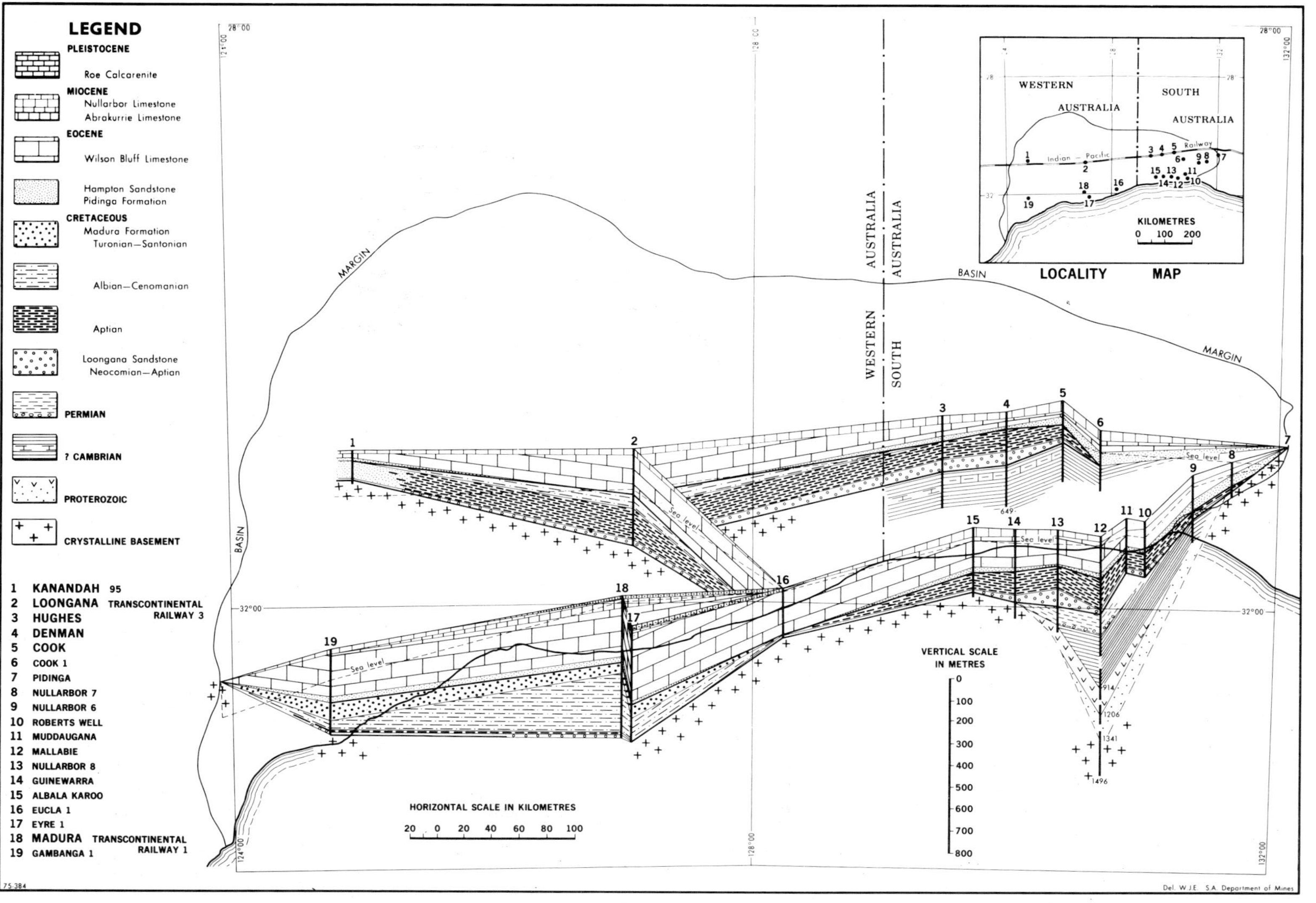

Fig. 15. Fence diagram of the subsurface of the Eucla Basin.

which there is evidence of two periods of deposition (Lowry, 1970).

The lowest strata of the Madura Formation are mudstones, glauconitic, carbonaceous and pyritic siltstones which are lithologically and faunally similar to those of the Great Artesian Basin and not to the non-marine greywacke-type sediments of the Otway Basin rapidly deposited during the west–east rifting. Agglutinating foraminifera typical of Aptian marginal sediments throughout the Great Artesian Basin and its sub-basins are common.

The overlying highly glauconitic mudstone and siltstone are generally devoid of foraminifera but have radiolaria in varying abundance. Radiolaria-bearing shales and radiolarites are extremely widespread in Australia (Brunnschweiler, 1959) in the Canning, Great Artesian, Carnarvon and Eucla Basins. They are mostly of Albian age (Crespin, in Veevers and Wells, 1961) at the top of the Early Cretaceous sequence before the Cenomanian regression.

Late Cretaceous sediments have been intersected only in the western part of the Eucla Basin and have no counterparts in the Canning or southern part of the Great Artesian Basin. They are restricted to a northerly-trending trough extending from the continental shelf to Madura (Fig. 9) (Tectonic Map of Australia, 1971; Lowry, 1970) and are related to greensands and glauconitic sandstones, siltstones and shales of the Osborne Formation, Molecap Greensand and Gingin Chalk of the Perth Basin (Ludbrook, 1958b; Ingram, 1968). The trough is apparently an offshoot of the west–east rift by which the sea gained entry to the western part of the Eucla Basin possibly as early as the Albian and continuing to the Santonian (Fig. 15). Marine connection with the Perth Basin around the Naturaliste Block was probably established by the Late Neocomian–Aptian, although drilling in DSDP Site 258 on the Naturaliste Block ceased in an unusually thick sequence of pre-Late Albian sediments beneath Cenomanian to Santonian carbonate sediments with abundant foraminifera, coccoliths and radiolaria (Luyendyk et al., 1973).

Marine carbonate sedimentation did not begin in the Eucla Basin until the Middle Eocene, when separation of the Antarctic and Australian plates was completed.

ACKNOWLEDGEMENTS

I am greatly indebted to the Director of Mines and to the Chief Draftsman for allowing the figures and tables to be drawn in the Department of Mines, South Australia; the Australian Petroleum Exploration Association Limited, West Australian Petroleum Proprietary Limited and the authors for permission to use in Figs. 10 and 11 the sections published by D.K. Jones and G.R. Pearson and B.M. Thomas and D.N. Smith in the *APEA Journal,* to M.H. Johnstone, Chief Geologist, West Australian Petroleum Proprietary Limited, for valuable help in the preparation of the maps in Figs. 2–9; to the Chief Government Geologist, Geological Survey of Queensland and the Queensland Division, Geological Society of Australia for permission to use the sections shown in Figs. 12 and 13; and to Shell Development (Australia) Proprietary Limited for supplying unpublished information on Mesozoic sediments in the Duntroon Basin.

REFERENCES

Allen, R.J. and Hogetoorn, D.J., 1970. Petroleum resources of Queensland. *Rep. Geol. Surv. Qd.,* 43: 1–42.

Allen, R.J., Denmead, A.K., Phillips, K. and Tweedale, G.W., 1960. The Bowen Basin. In D. Hill and A.K. Denmead (Editors), *The Geology of Queensland. J. Geol. Soc. Aust.,* 7: 280–283.

Arkell, W.J. and Playford, P.E., 1954. The Bajocian ammonites of Western Australia. *Philos. Trans. R. Soc.,* 237: 547–604.

Balme, B.E., 1964. The palynological record of Australian pre-Tertiary floras. In: L.M. Cranwell (Editor), *Ancient Pacific Floras.* University of Hawaii Press, Honolulu, pp. 49–80.

Balme, B.E., 1969a. The Triassic System in Western Australia. *J. Aust. Pet. Explor. Assoc. (APEA),* 9(2): 67–78.

Balme, B.E., 1969b. The Permian–Triassic boundary in Australia. *Spec. Publ. Geol. Soc. Aust.,* 2: 99–112.

Balme, B.E. and Helby, R.J., 1973. The Permian and Triassic Systems and their mutual boundary. *Mem. Can. Soc. Pet. Geol.,* 2: 433–444.

Banks, M.R., 1973. General Geology. In: M.R. Banks (Editor), *Symposium on the Lake Country of Tasmania. R. Soc. Tas.,* pp. 25–34.

Banks, M.R. and Clarke, M.J., 1973. Tasmania. Parmeener Super-group. In: *Field Trip No. 1. Upper Carboniferous to Triassic Rocks in Southeastern Australia. Third International Gondwana Symposium, Canberra, Australia, August, 1973,* pp. 23–47.

Belford, D.J., 1958. Stratigraphy and micropalaeontology of the Upper Cretaceous of Western Australia. *Sonderdr. Geol. Rundsch.,* 47(2): 629–647.

Belford, D.J., 1960. Upper Cretaceous Foraminifera from the Toolonga Calcilutite and Gingin Chalk, Western Australia. *Bull. Bur. Miner. Resour. Geol. Geophys. Aust.*, 57.

Bembrick, C.S., Herbert, C., Scheibner, E. and Stuntz, J., 1973. Structural subdivision of the New South Wales portion of the Sydney–Bowen Basin. *Q. Notes Geol. Surv. N.S.W.*, 11: 1–13.

Bock, P.E. and Glenie, R.C., 1965. Late Cretaceous and Tertiary depositional cycles in south-western Victoria. *Proc. R. Soc. Vict.*, 79(1): 153–163.

Boeuf, M.G. and Doust, H., 1975. Structure and development of the southern margin of Australia. *J. Aust. Pet. Explor. Assoc. (APEA)*, 15(1): 33–43.

Bourke, D.J., Hawke, J.M. and Scheibner, E., 1974. Structural subdivision of the Great Australian Basin in New South Wales. *Q. Notes Geol. Surv. N.S.W.*, 16: 10–16.

Brunnschweiler, R.O., 1954. Mesozoic stratigraphy of the Canning Desert and Fitzroy Valley, Western Australia. *J. Geol. Soc. Aust.*, 1: 35–54.

Brunnschweiler, R.O., 1959. New Aconeceratinae (Ammonoidea) from the Albian and Aptian of Australia. *Bull. Bur. Miner. Resour. Geol. Geophys. Aust.*, 54.

Brunnschweiler, R.O., 1960. Marine fossils from the Upper Jurassic and Lower Cretaceous of the Dampier Peninsula, north-western Australia. *Bull. Bur. Miner. Resour. Geol. Geophys. Aust.*, 59.

Brunnschweiler, R.O., 1966. Upper Cretaceous ammonites from the Carnarvon Basin of Western Australia. *Bull. Bur. Miner. Resour. Geol. Geophys. Aust.*, 58.

Burdett, J.W., Parry, J.C. and Willmott, S.P., 1970. Evaluation of the Windalia Sand, Barrow Island. An empirical approach. *J. Aust. Pet. Explor. Assoc. (APEA)*, 10(2): 91–96.

Casey, D.J., 1970. Northern Eromanga Basin. *Rep. Geol. Surv. Qd.*, 41.

Casey, R., 1954. *Falciferella*, a new genus of Gault ammonites, with a review of the family Aconeceratidae in the British Cretaceous. *Proc. Geol. Assoc.*, 65(3): 262–277.

Casey, R., 1960. A monograph of the Ammonoidea of the Lower Greensand. Part 1. *Palaeontogr. Soc. (Monogr.)*, pp. i–xxxvi, 1–44, pls. I–X.

Challinor, A., 1970. The geology of the offshore Canning Basin, Western Australia. *J. Aust. Pet. Explor. Assoc. (APEA)*, 10(2): 78–90.

Chamney, T.P., 1973. Micropalaeontological correlation of the Canadian Boreal Lower Cretaceous. In: R. Casey and P.F. Rawson (Editors), *The Boreal Lower Cretaceous. Geol. J., Spec. Iss.*, No. 5: 19–40.

Clarke, D.E., Paine, A.G.L. and Jensen, A.R., 1971. Geology of the Proserpine 1 : 250,000 Sheet Area, Queensland. *Rep. Bur. Miner. Resour. Geol. Geophys. Aust.*, 144.

Cockbain, A.E. and Playford, P.E., 1973. Stratigraphic nomenclature of Cretaceous rocks in the Perth Basin. *Annu. Rep. Geol. Surv. W. Aust.*, 1972: 26–31.

Cosgriff, J.W., 1965. A new genus of Temnospondyli from the Triassic of Western Australia. *J.R. Soc. W. Aust.*, 48(3): 65–90.

Cosgriff, J.W., 1973. *Notobrachyops picketti*, a brachyopid from the Ashfield Shale, Wiannamatta Group, New South Wales. *J. Paleontol.*, 47(6): 1094–1101.

Cranfield, L.C. and Schwarzböck, H., 1972. Nomenclature of some Mesozoic rocks in the Brisbane and Ipswich areas. *Qd. Gov. Min. J.*, 73: 414–416.

Crank, K., 1973. Geology of Barrow Island Oil Field. *J. Aust. Pet. Explor. Assoc. (APEA)*, 13(1): 49–57.

Crespin, I., 1963. Lower Cretaceous arenaceous foraminifera of Australia. *Bull. Bur. Miner. Resour. Geol. Geophys. Aust.*, 66.

Day, R.W., 1964. Stratigraphy of the Roma–Wallumbilla area. *Publ. Geol. Surv. Qd.*, 318.

Day, R.W., 1969. The Lower Cretaceous of the Great Artesian Basin. In: K.S.W. Campbell (Editor), *Stratigraphy and Palaeontology. Essays in Honour of Dorothy Hill.* Australian National University Press, Canberra, A.C.T., pp. 140–173.

Day, R.W., Cranfield, L.C. and Schwarzböck, H., 1974. Stratigraphy and structural setting of Mesozoic basins in southeastern Queensland and northeastern New South Wales. In A.K. Denmead, G.W. Tweedale and A.F. Wilson (Editors), *The Tasman Geosyncline – A Symposium. Geol. Soc. Aust. Qd. Div.*, pp. 319–363.

De Jersey, N.J., 1962. Triassic spores and pollen grains from the Ipswich Coalfield. *Publ. Geol. Surv. Qd.*, 307.

De Jersey, N.J., 1964. Triassic spores and pollen grains from the Bundamba Group. *Publ. Geol. Surv. Qd.*, 321.

De Jersey, N.J., 1968. Triassic spores and pollen grains from the Clematis Sandstone. *Publ. Geol. Surv. Qd.*, 338.

De Jersey, N.J., 1970a. Early Triassic miospores from the Rewan Formation. *Publ. Geol. Surv. Qd.*, 345.

De Jersey, N.J., 1970b. Triassic miospores from the Blackstone Formation, Aberdare Conglomerate and Raceview Formation. *Publ. Geol. Surv. Qd.*, 348.

De Jersey, N.J., 1972. Triassic miospores from the Esk Beds. *Publ. Geol. Surv. Qd.*, 357.

De Jersey, N.J. and Paten, R.J., 1964. Jurassic spores and pollen grains from the Surat Basin. *Publ. Geol. Surv. Qd.*, 322.

Dettmann, M.E., 1963. Upper Mesozoic microfloras from southeastern Australia. *Proc. R. Soc. Vict.*, 77(1): 1–148 (pls. 1–27).

Dettmann, M.E. and Playford, G., 1969. Palynology of the Australian Cretaceous. A review. In K.S.W. Campbell (Editor), *Stratigraphy and Palaeontology. Essays in Honour of Dorothy Hill.* Australian National University Press, Canberra A.C.T., pp. 174–210.

Dickins, J.M. and Malone, E.J., 1973. Geology of the Bowen Basin, Queensland. *Bull. Bur. Miner. Resour. Geol. Geophys. Aust.*, 130.

Dickins, J.M., Roberts, J. and Veevers, J.J., 1972. Permian and Mesozoic geology of the northeastern part of the Bonaparte Gulf Basin. *Bull. Bur. Miner. Resour. Geol. Geophys. Aust.*, 125: 75–102 (pls. 11, 12).

Douglas, J.G., 1969. The Mesozoic floras of Victoria. *Mem. Geol. Surv. Vict.*, 28.

Douglas, J.G., 1971. Biostratigraphical subdivision of Otway Basin Lower Cretaceous sediments. In: H. Wopfner and J.G. Douglas (Editors), *The Otway Basin of southeastern Australia. Spec. Bull. Geol. Surv. S. Aust. Vict.*, pp. 187–191.

Dulhunty, J.A., 1973a. Potassium–argon dating and occurrence of Tertiary and Mesozoic basalts in the Binnaway District. *J. Proc. R. Soc. N.S.W.*, 106: 104–110.

Dulhunty, J.A., 1973b. Mesozoic stratigraphy in central New South Wales. *J. Geol. Soc. Aust.*, 20(3): 319–328.

Dulhunty, J.A. and McDougall, I., 1966. Potassium–Argon dating of basalts in the Coonabarabran–Gunnedah District, New South Wales. *Aust. J. Sci.*, 28(10): 393–394.

Edwards, A.B., 1942. Differentiation of the dolerites of Tasmania. *J. Geol.*, 50: 451–480, 579–610.

Elliott, J.L., 1972. Continental drift and basin development in southeastern Australia. *J. Aust. Pet. Explor. Assoc. (APEA)*, 12(2): 46–51.

Ellis, P.L., 1968. Geology of the Maryborough 1 : 250,000 Sheet area. *Rep. Geol. Surv. Qd.*, 26.

Etheridge, Jr., R., 1913. The Cretaceous fossils of the Gingin Chalk. *Bull. Geol. Surv. W. Aust.*, 55.

Evans, P.R., 1966. Mesozoic stratigraphic palynology in Australia. *Australas. Oil Gas J.*, 12: 58–63.

Evernden, J.F. and Richards, J.R., 1962. Potassium–argon ages in eastern Australia. *J. Geol. Soc. Aust.*, 9(1): 1–49.

Exon, N.F. and Vine, R.R., 1970. Revised nomenclature of the "Blythesdale" sequence. *Qd. Gov. Min. J.*, 71: 3–7.

Feldtmann, F.R., 1963. Some pelecypods from the Cretaceous Gingin Chalk, Western Australia, together with descriptions of the principal chalk exposures. *J.R. Soc. W. Aust.*, 46(4): 101–125.

Fleming, C.A., 1958. Upper Jurassic fossils and hydrocarbon traces from the Cheviot Hills, North Canterbury. *N.Z. J. Geol. Geophys.*, 1(2): 375–394.

Fleming, P.J.G., 1966. Eotriassic marine bivalves from the Maryborough Basin, south-east Queensland. *Publ. Geol. Surv. Qd.*, 333 (Palaeontol. Pap. 8): 17–29 (pls. 7–9).

Forbes, B.G., 1966. The geology of the Marree 1 : 250,000 Map Area. *Rep. Invest. Geol. Surv. S. Aust.*, 28: 1–47.

Forbes, B.G., 1972. Possible post-Winton Mesozoic rocks northeast of Marree, South Australia. *Q. Geol. Notes, Geol. Surv. S. Aust.*, 41: 1–3.

Freytag, I.B., 1966. Proposed rock units for marine Lower Cretaceous sediments in the Oodnadatta region of the Great Artesian Basin. *Q. Geol. Notes, Geol. Surv. S. Aust.*, 1: 3–7.

Geary, J.K., 1970. Offshore exploration of the southern Carnarvon Basin. *J. Aust. Pet. Explor. Assoc. (APEA)*, 10(2): 9–15.

Geological Society, 1964. Phanerozoic Time-Scale. *Q.J. Geol. Soc. Lond.*, 120s: 260–262.

Geological Society of Australia, 1971. *Tectonic map of Australia and New Guinea. 1 : 5,000,000.* Sydney, N.S.W.

Gould, R.E., 1968. The Walloon Coal Measures. *Qd. Gov. Min. J.*, 69: 509–515.

Griffiths, J.R., 1971. Continental margin tectonics and the evolution of southeast Australia. *J. Aust. Pet. Explor. Assoc. (APEA)*, 11(1): 75–80.

Haig, D.W., 1973. Lower Cretaceous Foraminiferida, Surat Basin, southern Queensland: A preliminary stratigraphic appraisal. *Qd. Gov. Min. J.*, 74: 3–11.

Hale, G.E., 1962. Triassic System. In: A. Spry and M.R. Banks (Editors), *The Geology of Tasmania. J. Geol. Soc. Aust.*, 9(2): 217–231.

Halse, J.W. and Hayes, J.D., 1971. The geological and structural framework of the offshore Kimberley Block (Browse Basin) Area, Western Australia. *J. Aust. Pet. Explor. Assoc. (APEA)*, 11(1): 64–70.

Hanlon, F.N., Osborne, G.D. and Raggatt, H.G., 1953. Narrabeen Group: Its subdivisions and correlations between the South Coast and Narrabeen–Wyong Districts. *J.R. Soc. N.S.W.*, 87: 106–120.

Harding, R.R., 1969. Catalogue of age determinations on Australian rocks, 1962–1965. *Rep. Bur. Miner. Resour. Geol. Geophys. Aust.*, 117: 1–117.

Harris, W.K., 1964. Mesozoic sediments of the Polda Basin, Eyre Peninsula. *Q. Geol. Notes, Geol. Surv. S. Aust.*, 12: 6–7.

Harris, W.K., 1970. An Upper Jurassic microflora from the western margin of the Great Artesian Basin, South Australia. *Q. Geol. Notes, Geol. Surv. S. Aust.*, 35: 3–8.

Harris, W.K. and Foster, C.B., 1974. Stratigraphy and palynology of the Polda Basin, Eyre Peninsula. *Miner. Resour. Rev. S. Aust. Dep. Mines*, 136: 56–78.

Harris, W.K. and Ludbrook, N.H., 1966. Occurrence of Permian sediments in the Eucla Basin, South Australia. *Q. Geol. Notes, Geol. Surv. S. Aust.*, 17: 11–14.

Hayes, D.E., Frakes, L.A., Barrett, P., Burns, D.A., Chen, Pei-Hsin, Ford, A.B., Kaneps, A.G., Kemp, E.M., McCollum, D.W., Piper, D.J.W., Wall, R.E. and Webb, P.N., 1973. Leg 28, Deep-sea drilling in the southern ocean. *Geotimes*, 18(6): 19–24.

Helby, R.J., 1967. Triassic microfossils from a shale within the Wollar Sandstone, N.S.W. *J. Proc. R. Soc. N.S.W.*, 100: 61–73.

Helby, R.J., 1969. Age of the Narrabeen Group as implied by the microfloras. In: G.H. Packham (Editor), *The Geology of New South Wales. J. Geol. Soc. Aust.*, 16(1): 405.

Helby, R.J., 1973. Review of late Permian and Triassic palynology of New South Wales. *Spec. Publ. Geol. Soc. Aust.*, 4: 141–155 (pls. 1–3).

Hill, D., Playford, G. and Woods, J.T., 1968. Cretaceous fossils of Queensland. *Qd. Palaeontogr. Soc.*, pp. K1–K35.

Hocking, J.B., 1972. Geologic evolution and hydrocarbon habitat, Gippsland Basin. *J. Aust. Pet. Explor. Assoc. (APEA)*, 12(1): 132–137.

Hosemann, P., 1971. The stratigraphy of the basal Triassic sandstone, North Perth Basin, Western Australia. *J. Aust. Pet. Explor. Assoc. (APEA)*, 11(1): 59–63.

Houston, B.R., 1965. Triassic volcanics from the base of the Ipswich Coal Measures, southeast Queensland. *Publ. Geol. Surv. Qd.*, 327: 1–18.

Ingram, B.S., 1968. Stratigraphic palynology of Cretaceous rocks from bores in the Eucla Basin, Western Australia. *Annu. Rep. Geol. Surv. W. Aust.*, 1967: 64–67.

James, E.A. and Evans, P.R., 1971. The stratigraphy of the offshore Gippsland Basin. *J. Aust. Pet. Explor. Assoc. (APEA)*, 11(1): 71–74.

Jeletzky, J.A., 1963. *Malayomaorica* gen. nov. (Family Aviculopectinidae) from the Indo-Pacific Upper Jurassic, with comments on related forms. *Palaeontology*, 6(1): 148–160 (pl. 21).

Johnson, J.G., 1971. Timing and coordination of orogenic, epeirogenic and eustatic events. *Geol. Soc. Am. Bull.*, 82: 3263–3298.

Johnson, J.G., 1972. Antler Effect equals Haug Effect. *Geol. Soc. Am. Bull.*, 83: 2497–2498.

Johnston, W., 1960. Exploration for coal, Springfield Basin in the hundred of Cudlamudla, Gordon–Cradock District. *Rep. Invest. Geol. Surv. S. Aust.*, 16: 1–62.

Johnstone, M.H., 1972. In discussion, Veevers, J.J. Evolution of the Perth and Carnarvon Basins. *J. Aust. Pet. Explor. Assoc. (APEA)*, 12(2): 54.

Johnstone, M.H., Lowry, D.C. and Quilty, P.G., 1973. The geology of southwestern Australia – A review. *J.R. Soc. W. Aust.*, 56(1–2): 5–15.

Jones, D.K. and Pearson, G.R., 1972. The tectonic elements of Perth Basin. *J. Aust. Pet. Explor. Assoc. (APEA)*, 12(1): 17–22.

Jones, O.A. and De Jersey, N.J., 1947. The flora of the Ipswich Coal Measures – Morphology and floral succession. *Pap. Univ. Qd. Dep. Geol.*, 3 (N.S.) (3): 1–88 (pls. I–X).

Kaye, P., Edmond, G.M. and Challinor, A., 1972. The Rankin Trend, Northwest Shelf, Western Australia. *J. Aust. Pet. Explor. Assoc. (APEA)*, 12(1): 3–8.

Kenley, P.R., 1971. Cainozoic geology of the eastern part of the Gambier Embayment, southwestern Victoria. In: H. Wopfner and J.G. Douglas (Editors), *The Otway Basin of Southeastern Australia. Spec. Bull. Geol. Surv. S. Aust. Vict.*, pp. 89–153 (Enclosures 5-1,2,3,4).

Kennett, J.P., Houtz, R.E., Andrews, P.B., Edwards, A.R., Gostin, V.A., Hajos, M., Hampton, M., Jenkins, D.G., Margolis, S.V., Ovenshine, A.T. and Perch-Nielsen, K., 1973. Deep-sea drilling in the roaring forties. Leg 29. *Geotimes*, 18(7): 14–17.

Kummel, B., 1973. Lower Triassic (Scythian) molluscs. In: A. Hallam (Editor), *Atlas of Palaeobiogeography*. Elsevier, Amsterdam, pp. 225–233.

Laws, R.A. and Kraus, G.P., 1974. The regional geology of the Bonaparte Gulf Timor Sea area. *J. Aust. Pet. Explor. Assoc. (APEA)*, 14(1): 77–84.

Loughnan, F.C., Ko Ko, M. and Bayliss, P., 1964. The red beds of the Triassic Narrabeen Group. *J. Geol. Soc. Aust.*, 11(1): 65–77.

Lovering, J.F. and McElroy, C.T., 1969. Wianamatta Group. In: G.H. Packham (Editor), *The Geology of New South Wales. J. Geol. Soc. Aust.*, 16(1): 417–426.

Lowry, D.C., 1970. Geology of the Western Australian part of the Eucla Basin. *Bull. Geol. Surv. W. Aust.*, 122.

Lowry, D.C., Jackson, M.J., Van de Graaff, W.J.E. and Kennewell, P.J., 1972. Preliminary results of geological mapping in the Officer Basin, Western Australia, 1971. *Annu. Rep. Geol. Surv. W. Aust.*, 1971: 50–56.

Ludbrook, N.H., 1958a. The Murray Basin in South Australia. In: M.F. Glaessner and L.W Parkin (Editors), *The Geology of South Australia. J. Geol. Soc. Aust.*, 5(2): 102–114.

Ludbrook, N.H., 1958b. Stratigraphic sequence in the western portion of the Eucla Basin. *J.R. Soc. W. Aust.*, 41(4): 108–114.

Ludbrook, N.H., 1961a. Stratigraphy of the Murray Basin in South Australia. *Bull. Geol. Surv. S. Aust.*, 36.

Ludbrook, N.H., 1961b. Mesozoic non-marine Mollusca (Pelecypoda: Unionidae) from the north of South Australia. *Trans. R. Soc. S. Aust.*, 84: 139–147.

Ludbrook, N.H., 1966. Cretaceous biostratigraphy of the Great Artesian Basin in South Australia. *Bull. Geol. Surv. S. Aust.*, 40.

Ludbrook, N.H., 1969. Tertiary Period. In: L.W. Parkin (Editor), *Handbook of South Australian Geology. Geol. Surv. S. Aust.*, Adelaide, S.A., pp. 172–203.

Luyendyk, B.P., Davies, T.A., Rodolfo, K.S., Kempe, D.R.C., McKelvey, B.C., Leidy, R.D., Horvath, G.J., Hyndman, R.D., Thierstein, H.R., Boltovskoy, E. and Doyle, P., 1973. Across the southern Indian Ocean aboard Glomar Challenger. Leg 26, Deep-sea Drilling Project. *Geotimes*, 18(3): 16–19.

McDougall, I. and Van der Lingen, G.J., 1974. Age of the rhyolites of the Lord Howe Rise and the evolution of the southwest Pacific Ocean. *Earth Planet. Sci. Lett.*, 21: 117–126.

McDougall, I. and Wellman, P., 1976. Potassium–argon ages for some Australian Mesozoic igneous rocks. *J. Geol. Soc. Aust.*, 23(1): 1–9.

McElroy, C.T., 1962. The geology of the Clarence–Moreton Basin. *Mem. Geol. Surv. N.S.W.*, 9.

McElroy, C.T., 1969a. Triassic System. Narrabeen Group. In: G.H. Packham (Editor), *The Geology of New South Wales. J. Geol. Soc. Aust.*, 16(1): 388–407.

McElroy, C.T., 1969b. The Clarence–Moreton Basin in New South Wales. In: G.H. Packham (Editor), *The Geology of New South Wales. J. Geol. Soc. Aust.*, 16(1): 457–479.

McKellar, J.B.A., 1957. Geology of portion of the Western Tiers. *Rec. Queen Vict. Mus. Launceston*, N.S., 7: 1–13.

McTavish, R.A. and Dickins, J.M., 1974. The age of the Kockatea Shale (Lower Triassic), Perth Basin – A reassessment. *J. Geol. Soc. Aust.*, 21(2): 195–201.

McWhae, J.R.H., Playford, P.E., Lindner, A.W., Glenister, B.F. and Balme, B.E., 1958. The stratigraphy of Western Australia. *J. Geol. Soc. Aust.*, 4(2): 1–161.

Meyers, N.A., 1969. Carpentaria Basin. *Rep. Geol. Surv. Qd.*, 34.

Mollan, R.G., Craig, R.W. and Lofting, M.J.W., 1969. Geological framework of the Continental Shelf off northwest Australia. *J. Aust. Pet. Explor. Assoc. (APEA)*, 9(2): 49–59.

Nicol, G.N., 1970. Exploration and geology of the Arafura Sea. *J. Aust. Pet. Explor. Assoc. (APEA)*, 10(2): 56–61.

Packham, G.H. (Editor), 1969. The geology of New South Wales. *J. Geol. Soc. Aust.*, 16(1).

Papalia, N., 1969. The Nappamerri Formation. *J. Aust. Pet. Explor. Assoc. (APEA)*, 9(2): 108–110.

Parkin, L.W., 1953. The Leigh Creek Coalfield. *Bull. Geol. Surv. S. Aust.*, 31.

Parry, J.C., 1967. The Barrow Island Oilfield. *J. Aust. Pet. Explor. Assoc. (APEA)*, 7(2): 130–133.

Playford, G., 1965. Plant microfossils from Triassic sediments near Poatina, Tasmania. *J. Geol. Soc. Aust.*, 12(2): 173–210 (pls. 6–11).

Playford, G. and Dettmann, M.E., 1965. Rhaeto-Liassic plant microfossils from the Leigh Creek Coal Measures, South Australia. *Senckenberg. Leth.*, 46(2/3): 127–181 (pls. 12–17).

Playford, P.E. and Cope, R.N., 1971. The Phanerozoic stratigraphy of Western Australia: A correlation chart in two parts. *Annu. Rep. Geol. Surv. W. Aust.*, 1970: 32–33.

Power, P.E. and Devine, S.B., 1970. Surat Basin, Australia – Subsurface stratigraphy, history and petroleum. *Bull. Am. Assoc. Pet. Geol.*, 54(12): 2410–2437.

Richards, K.A. and Hopkins, B.M., 1969. Exploration in the Gippsland, Bass and Otway Basins, Australia. *E.C.A.F.E. Nat. Res. Symp., Canberra.*

Roberts, J. and Veevers, J.J., 1973. Summary of BMR studies of the onshore Bonaparte Gulf Basin 1963–1971. In: *Geological Papers, 1970–1971. Bull. Bur. Miner. Resour. Geol. Geophys. Aust.*, 139: 29–58.

Robinson, V.A., 1974. Geologic history of the Bass Basin. *J. Aust. Pet. Explor. Assoc. (APEA)*, 14(1): 45–49.

Rochow, K.A., 1971. Geological interpretation of seismic time sections in the Gambier Embayment. In: H. Wopfner and J.G. Douglas (Editors), *The Otway Basin of Southeastern Australia. Spec. Bull. Geol. Surv. S. Aust. Vict.*, pp. 285–315.

Runnegar, B., 1969. A Lower Triassic ammonoid fauna from southeast Queensland. *J. Paleontol.*, 43(3): 818–828.

Scheibnerova, V., 1971. Palaeoecology and palaeogeography of Cretaceous deposits of the Great Artesian Basin (Australia). *Rec. Geol. Surv. N.S.W.*, 13(1): 1–48 (charts 1–3).

Sclater, J.G., Von der Borch, C.C., Gartner, S., Hekinian, R., Johnson, D.A., McGowran, B., Pimm, A.C., Thompson, R.W. and Veevers, J.J., 1972. Deep Sea Drilling Project. Leg 22. *Geotimes*, 17(6): 15–17.

Singleton, O.P., 1973. Geology of the Bacchus Marsh District. In: J. McAndrew and M.A.H. Marsden (Editors), *Regional Guide to Victorian Geology.* School of Geology, University of Melbourne, Publ., 2nd ed., 1: 59–64.

Skwarko, S.K., 1966. Cretaceous stratigraphy and palaeontology of the Northern Territory. *Bull. Bur. Miner. Resour. Geol. Geophys. Aust.*, 73.

Skwarko, S.K., 1967. Mesozoic Mollusca from Australia and New Guinea. *Bull. Bur. Miner. Resour. Geol. Geophys. Aust.*, 75.

Smart, J., Grimes, K.G. and Doutch, H.F., 1972. New and revised stratigraphic names – Carpentaria Basin. *Qd Gov. Min. J.*, 73: 190–201.

Smith, R. and Kamerling, P., 1969. Geological framework of the Great Australian Bight. *J. Aust. Pet. Explor. Assoc. (APEA)*, 9(2): 60–66.

Standard, J.C., 1969. Hawkesbury Sandstone. In: G.H. Packham (Editor), *The Geology of New South Wales. J. Geol. Soc. Aust.*, 16(1): 407–415.

Stevens, N.C., 1969. The volcanism of southern Queensland. *Spec. Publ. Geol. Soc. Aust.*, 2: 193–202.

Swarbrick, C.F.J., 1973. Stratigraphy and economic potential of the Injune Creek Group in the Surat Basin. *Rep. Geol. Surv. Qd.*, 79.

Swindon, V.G., 1971. Moreton District. In: G. Playford (Editor), *Geological Excursions Handbook. Aust. N.Z. Assoc. Adv. Sci., 43rd Congr., Brisbane, Geol. Soc. Aust. Qd. Div.*, pp. 105–125.

Talent, J.A., 1969. Geology of East Gippsland. In: *East Gippsland Symposium. Proc. R. Soc. Vict.*, 82(1): 37–60.

Taylor, D.J., 1964. The depositional environment of the marine Cretaceous sediments of the Otway Basin. *J. Aust. Pet. Explor. Assoc. (APEA)*, 1964: 140–144.

Taylor, D.J., 1971. Foraminifera and the Cretaceous and Tertiary depositional history in the Otway Basin in Victoria. In: H. Wopfner and J.G. Douglas (Editors), *The Otway Basin of Southeastern Australia. Spec. Bull. Geol. Surv. S. Aust. Vict.*, pp. 217–233.

Thomas, B.M. and Smith, D.N., 1974. A summary of the petroleum geology of the Carnarvon Basin. *J. Aust. Pet. Explor. Assoc. (APEA)*, 14(1): 66–76.

Thomas, G.A., 1957. Oldhaminid brachiopods in the Permian of northern Australia. *J. Palaeontol. Soc. India*, 2: 174–182.

Thomson, B.P., 1970. A review of the Precambrian and Lower Palaeozoic tectonics of South Australia. *Trans. R. Soc. S. Aust.*, 94: 193–221.

Thornton, R.C.N., 1972. Lower Cretaceous sedimentary units beneath the western Murray Basin. *Q. Geol. Notes, Geol. Surv. S. Aust.*, 44: 5–11.

Thornton, R.C.N., 1974. Hydrocarbon potential of western Murray Basin and infrabasins. *Rep. Invest. Geol. Surv. S. Aust.*, 41.

Tillyard, R.J. and Dunstan, B., 1924. Mesozoic insects of Queensland. *Publ. Geol. Surv. Qd.*, 273.

Tindale, N.B., 1945. Triassic insects of Queensland. *Proc. R. Soc. Qd.*, 56: 37–46.

Veevers, J.J., 1971. Phanerozoic history of Western Australia related to continental drift. *J. Geol. Soc. Aust.*, 18(2): 87–96.

Veevers, J.J., 1972. Evolution of the Perth and Carnarvon Basins. *J. Aust. Pet. Explor. Assoc. (APEA)*, 12(2): 52–54.

Veevers, J.J. and Evans, P.R., 1973. Sedimentary and magnetic events in Australia and the mechanism of worldwide Cretaceous transgressions. *Nature Lond., Phys. Sci.*, 245(142): 33–36.

Veevers, J.J. and Johnstone, M.H., 1974. Comparative stratigraphy and structure of the Western Australian margin and the adjacent deep ocean floor. In: J.J. Veevers and J.R. Heirtzler et al., *Initial Reports of the Deep Sea Drilling Project XXVII.* U.S. Government Printing Office, Washington, D.C., pp. 571–585.

Veevers, J.J. and Wells, A.T., 1961. The geology of the Canning Basin, Western Australia. *Bull. Bur. Miner. Resour. Geol. Geophys. Aust.*, 60.

Veevers, J.J., Heirtzler, J.R., Bolli, H.M., Carter, A.N., Cook, P.J., Krasheninnikov, V.A., McKnight, B.K., Proto-Decima, F., Renz, G.W., Robinson, P.T., Rocker, K. and Thayer, P.A., 1973. Deep Sea Drilling Project, Leg 27, in the Eastern Indian Ocean. *Geotimes*, 18(4): 16–17.

Vine, R.R., Day, R.W., Milligan, E.N., Casey, D.J., Galloway, M.C. and Exon, N.F., 1967. Revision of the nomenclature of the Rolling Downs Group in the Eromanga and Surat Basins. *Qd. Gov. Min. J.*, 68: 144–151.

Vogt, P.C. and Connolly, J.R., 1971. Tasmanitid guyots, the age of the Tasman Basin and motion between the Australian Plate and the mantle. *Geol. Soc. Am. Bull.*, 82: 2577–2584.

Wade, R.T., 1953. Note on a Triassic fish fossil from Leigh Creek, South Australia. *Trans. R. Soc. S. Aust.*, 76: 80–81.

Waldman, M., 1971. Fish from the freshwater Lower Cretaceous of Victoria, Australia, with comments on the palaeoenvironment. *Spec. Pap. Palaeontol., Palaeontol. Assoc. Lond.*, 9.

Watson, D.M.S., 1958. A new labyrinthodont (*Paracyclotosaurus*) from the Upper Trias of New South Wales. *Bull. Br. Mus. (Nat. Hist.), Geol.*, 3(7): 235–263.

Webb, A.W. and McDougall, I., 1964. Granites of Lower Cretaceous age near Eungella, Queensland. *J. Geol. Soc. Aust.*, 11(1): 151–153.

Webb, A.W. and McDougall, I., 1967. Isotopic dating evidence on the age of the Upper Permian and Middle Triassic. *Earth Planet. Sci. Lett.*, 2: 483–488.

Weeks, L.G. and Hopkins, B.M., 1967. Geology and exploration of three Bass Strait Basins, Australia. *Bull. Am. Assoc. Pet. Geol.*, 51(5): 742–760.

Wells, A.T., Forman, D.J., Ranford, L.C. and Cook, P.J., 1970. Geology of the Amadeus Basin, Central Australia. *Bull. Bur. Miner. Resour. Geol. Geophys. Aust.*, 100.

Whitehouse, F.W., 1926. The Cretaceous Ammonoidea of eastern Australia. *Mem. Qd. Mus.*, 8(3): 195–242.

Whitten, G.F., 1972. *Annual Report of the Director of Mines and Government Geologist for the year ended 30th June, 1972.* Dep. Mines, South Australia, Adelaide, S.A.

Withers, T.H., 1924. The occurrence of the crinoid *Uintacrinus* in Australia. *J.R. Soc. W. Aust.*, 11: 15–18.

Withers, T.H., 1926. The crinoid *Marsupites* in the Upper Cretaceous of Western Australia. *J.R. Soc. W. Aust.*, 12: 97–100.

Woods, J.T., 1961. Mesozoic and Cainozoic sediments of the Wrotham Park area. *Publ. Geol. Surv. Qd.*, 304: 1–6.

Wopfner, H., 1963. Post-Winton sediments of probable Upper Cretaceous age in the central Great Artesian Basin. *Trans. R. Soc. S. Aust.*, 86: 247–253.

Wopfner, H., 1964. Permian–Jurassic history of the western Great Artesian Basin. *Trans. R. Soc. S. Aust.*, 88: 117–128 (pls. 1–2).

Wopfner, H., 1969. Mesozoic Era. In: L.W. Parkin (Editor), *Handbook of South Australian Geology. Geol. Surv. S. Aust.*, Adelaide, S.A., pp. 133–171.

Wopfner, H. and Douglas, J.G. (Editors), 1971. The Otway Basin of southeastern Australia. *Spec. Bull. Geol. Surv. S. Aust. Vict.*

Wopfner, H., Freytag, I.B. and Heath, G.R., 1970. Basal sediments (Jurassic–Cretaceous) of western Great Artesian Basin, South Australia: Stratigraphy and environment. *Bull. Am. Assoc. Pet. Geol.*, 54.

Wright, C.W., 1963. Cretaceous ammonites from Bathurst Island, Northern Australia. *Palaeontology*, 6(4): 597–614.

Chapter 8

NEW ZEALAND

G.R. STEVENS and I.G. SPEDEN

INTRODUCTION: STRUCTURAL FRAMEWORK AND TECTONIC EVENTS (G.R. STEVENS)

The Triassic and Jurassic rocks of New Zealand were deposited in the New Zealand Geosyncline (Wellman, 1956; Fleming, 1962, 1970), the filling of which probably started in the Carboniferous–Permian (Figs. 1, 2). The New Zealand Geosyncline extended northwest at least as far as New Caledonia, east to Chatham Islands and south to Auckland and Campbell Islands (Fig. 3). Facies belts have been recognized in the sediments that accumulated during this phase of geosynclinal deposition and for those of Mesozoic age the terms "Hokonui Facies" (= shelf and transitional) and "Alpine" (= redeposited) were introduced by Wellman (1952). The rock-stratigraphic names Torlesse Supergroup and Murihiku Supergroup have been applied to rocks of the Alpine and Hokonui Facies, respectively (Suggate, 1961; Campbell and Coombs, 1966) (Fig. 1).

Rocks of Torlesse Supergroup are complexly folded and faulted argillite and greywacke sandstone with minor conglomerate, limestone and submarine volcanic rocks. Fossils are rare, but Triassic and Jurassic fossils have been found and it has been postulated that these lie roughly in belts, with the Triassic rocks generally to the west and Jurassic to the east (Fig. 28). Torlesse Supergroup rocks and their presumed metamorphosed equivalents (Haast Schist Group) largely make up the axial ranges of both islands (see Figs. 26, 27).

The Murihiku Supergroup sediments have been folded into a broad regional synclinal structure (Wellman's "Marginal Syncline"), with Jurassic sediments in its axis in the west coast of the North Island and in Southland and southeast Otago (Figs. 1, 2). In the South Island the Marginal Syncline has been cut and displaced by the Alpine Fault (Suggate, 1963).

One interpretation of the New Zealand Geosyncline envisages a pair of elongate north- or northwest-

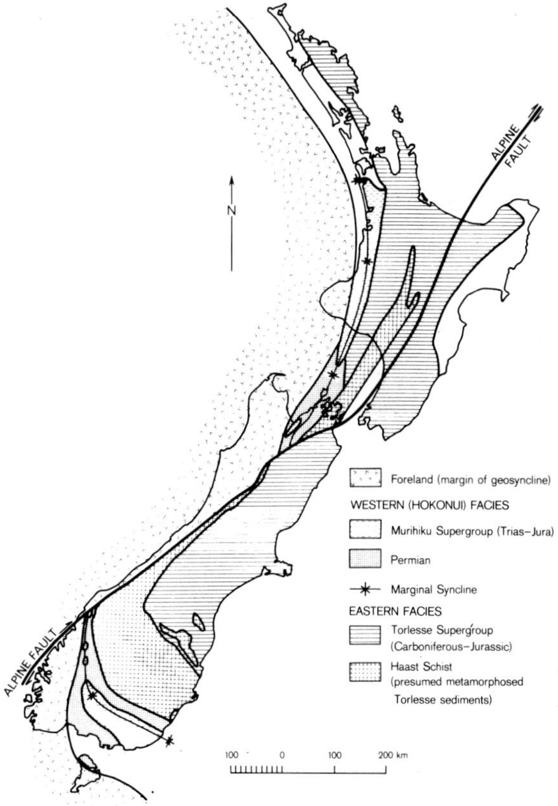

Fig. 1. The main elements of the New Zealand Geosyncline. In the preparation of this generalized diagram Cretaceous and Cenozoic covering strata have been ignored. Presumed subsurface extensions of the rock units shown have been included. Figs. 4, 5, 14, 15, 26, 27 should be consulted for surface outcrops of Murihiku and Torlesse Super Groups and Haast Schist. Modified from Fleming (1970).

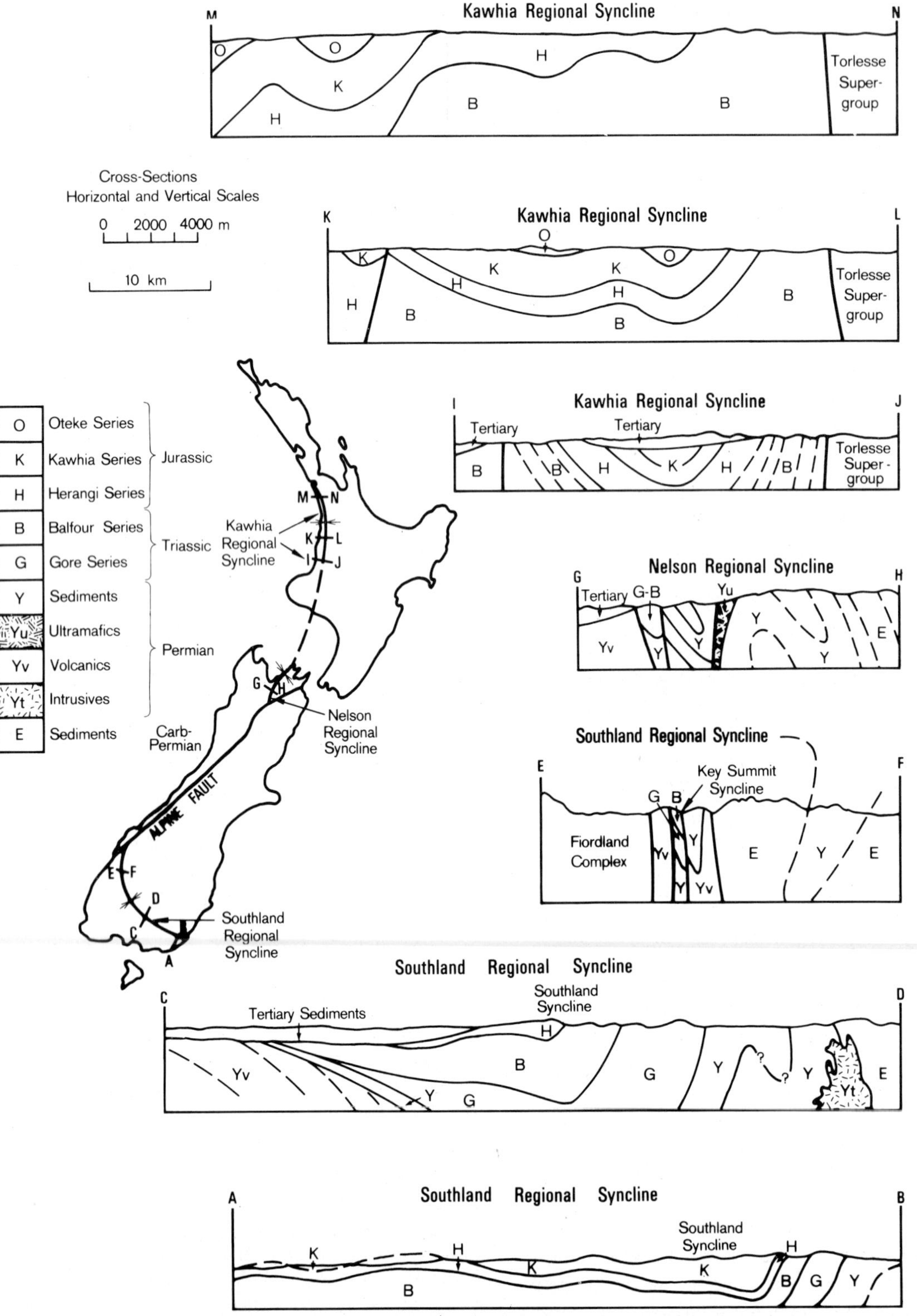

Fig. 2. Serial cross-sections across the Kawhia, Nelson and Southland Regional Synclines showing typical post-metamorphic folding and faulting and correlation of the major stratigraphic and tectonic units across the Alpine Fault. (After Suggate et al. 1977, fig. 5.6.)

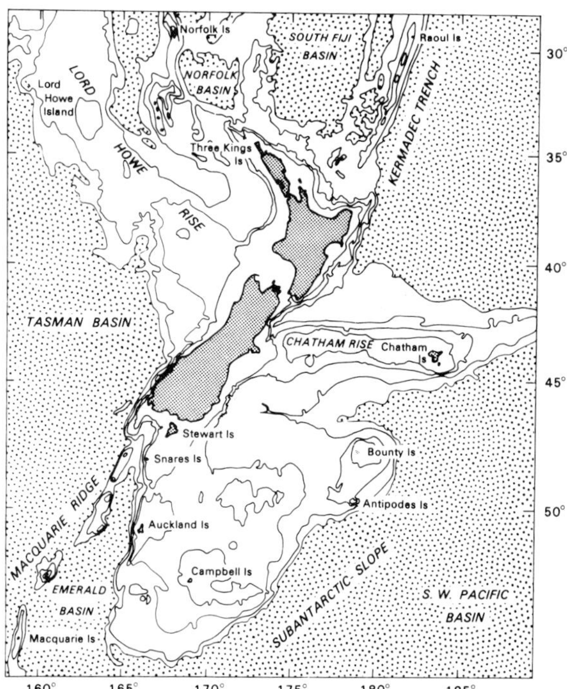

Fig. 3. Bathymetry of the New Zealand region. Isobaths are drawn at 500 m, 1000 m, 2000 m and 3000 m. Oceanic areas deeper than 3000 m are stippled. Modified from Lawrence (1967).

trending sedimentary realms or facies belts. The Murihiku belt flanked a stable cratonic foreland, presumably now forming parts of Antarctica, Australia, Lord Howe Rise and Norfolk Ridge, the sedimentary cover of which is represented by the Parapara assemblage of rocks in northwest Nelson (Clark et al., 1967; Carter et al., 1974, p. 9). The Torlesse belt, on the other hand, lay on the Pacific side of the Murihiku belt and, taking sea-floor spreading into account, may have been derived from Marie Byrd Land, West Antarctica (Bradshaw and Andrews, 1973).

The Murihiku belt is characterized by fossiliferous, structurally simple, marine and subsidiary non-marine sediments. Andesitic tuffs and volcanic-derived clastics are typical. The Torlesse belt, on the other hand, is characterized by deep-water "redeposited" or turbidite sediments and is generally quartzo-feldspathic in composition.

The Murihiku/Torlesse pattern is interpreted as reflecting a Mesozoic marginal sea and island-arc complex (Murihiku) and a deep-water, oceanic trench or subduction zone accumulation on Pacific oceanic crust (Torlesse) (Fleming, 1970; Landis and Bishop, 1972; Dickinson, 1971a, b; cf. Bradshaw and Andrews, 1973).

Although the Murihiku model summarized above has been generally accepted, recent sedimentological work in the South Island has cast doubt on a general deep-water origin for most of the Torlesse. Bradshaw and Andrews (1973), while recognizing the presence of graded turbidite type facies, particularly in the west of the Torlesse belt, have postulated an eastern source for much of the Torlesse in the South Island and derivation from submarine deltaic and other shallow marine, rather than deep-water environments. These authors and Blake et al. (1974) place the possible source area for Torlesse sediments in West Antarctica and suggest they were subsequently rafted by sea-floor spreading into juxtaposition with the Hokonui sediments.

Sedimentation in the New Zealand Geosyncline is generally thought to have been terminated, or at least severely restricted, by the Rangitata Orogeny in the latest Jurassic and earliest Cretaceous (Kingma, 1959; Grindley, 1961, 1975; Waterhouse, 1975). In the Middle Jurassic however, tectonic movements within the geosyncline, interpreted as precursor events of the main orogenic phase (Speden, 1961, 1971), displaced the major axis of sedimentation and led to cessation of deposition at different places.

The geological history of the New Zealand Mesozoic is dominated by events associated with the Rangitata Orogeny. Pre-Cretaceous rocks, in particular the very thick sedimentary fill of the New Zealand Geosyncline (e.g. Fleming, 1970), are in most places strongly indurated, tectonized and partly or completely metamorphosed.

Eversion and folding of the New Zealand Geosyncline undoubtedly produced widespread changes in the geography of the Southwest Pacific and accompanying changes in sedimentation and dispersal routes available for terrestrial and marine organisms (Fleming, 1967).

The continental landmass established after the Rangitata Orogeny extended far beyond the present-day New Zealand coastline. The blocks comprising the Challenger Plateau, Lord Howe Rise, Campbell Plateau and Chatham Rise were all part of it, and possibly also the Norfolk Ridge (Fig. 3). Although little is known of the early history of the deeply depressed basins and troughs between these blocks

(New Caledonia Basin, Bounty Trough, Hikurangi Trough), they too probably formed part of this ancient landmass.

No Neocomian marine fossils are known in New Zealand and it is thought that during the Neocomian land newly created by the Rangitata Orogeny occupied most of the area that is now modern New Zealand.

After Aptian times however, terrestrial and marine erosion had reduced the landmass of the Rangitata Orogeny to such an extent that the sea was beginning to encroach in at least two areas: Raukumara Peninsula and Northland. In the former area the relationship of the Aptian–Albian sediments to older formations, where this is known, is strongly unconformable (Speden, 1972, 1973, 1975a), while in the latter area they are likely to be part of large-scale olistostrome deposits of unknown origin (Kear and Waterhouse, 1967; Katz, 1968). Encroachment by the sea over the old landmass continued throughout the remainder of Cretaceous time, and in the Campanian–Maastrichtian extensive marine transgressions occurred both in the west (Lord Howe Rise) and east (e.g., Stevens and Suggate, in Suggate et al., 1977, figs. 11.17–11.19).

From Late Cretaceous through into Early Tertiary large areas of the old landmass became peneplaned. Subsequent tectonic events broke the landmass into large fragments which became detached and temporarily or semipermanently submerged. This caused formation of extensive epi-continental basins, thickest sedimentation occurring in the hinge area between the various blocks. Marine transgression reached a maximum in the Oligocene, when some three-quarters of present-day New Zealand was submerged.

The later Cenozoic history of New Zealand is characterized by movements relating to extension tectonics and development of the New Zealand Shear Belt (Suggate et al., 1977, p. 38) For convenience in treatment, the discussion of the marginal sea and island-arc complex, the Murihiku, is separated from the Torlesse Supergroup, without regard to the environment of deposition of the latter. Their general paleobiogeographical and tectonic relationships will be described in separate sections.

In the New Zealand Mesozoic, correlations with international divisions can be provided only at scattered isolated horizons in the stratigraphic column, and to provide for stability in stratigraphic nomenclature a system of local stages has been employed, broadly correlated with international divisions (Marwick, 1951, 1953a; Wellman, 1959).

MURIHIKU SUPERGROUP – TRIASSIC (G.R. STEVENS)

Triassic rocks of the Murihiku Supergroup occur in regional synclines in southwest Auckland, Nelson, Southland and south Otago, but are unrepresented in the central part of the South Island as a result of displacement of the synclines along the Alpine Fault (Figs. 1, 2, 4, 5). In the Kawhia Regional Syncline (southwest Auckland) Upper Triassic and Jurassic are exposed. Further south, in the Nelson Regional Syncline, the sequence includes Upper and Middle Triassic, but the top of the sequence is not preserved and the older Triassic rocks are faulted against Permian. Triassic remnants are preserved in the highly deformed Key Summit section of the Southland

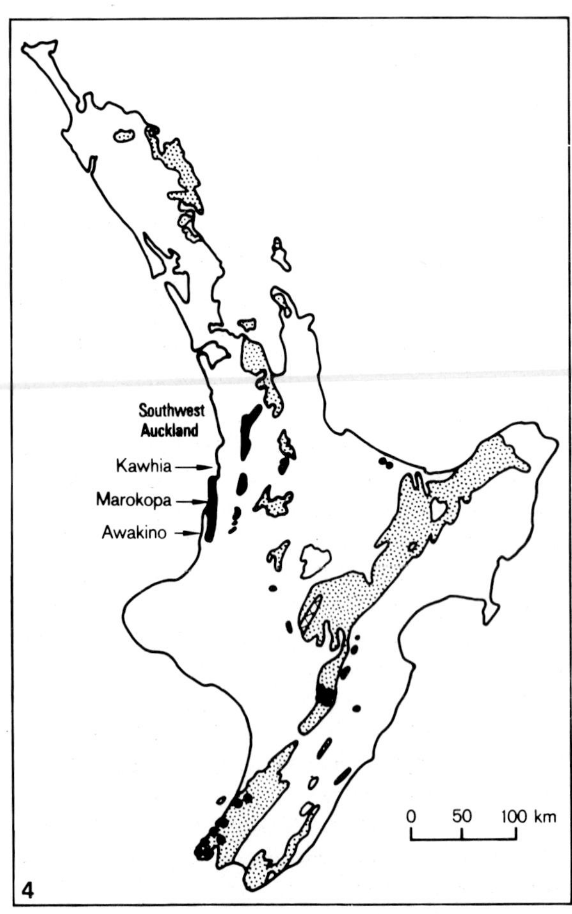

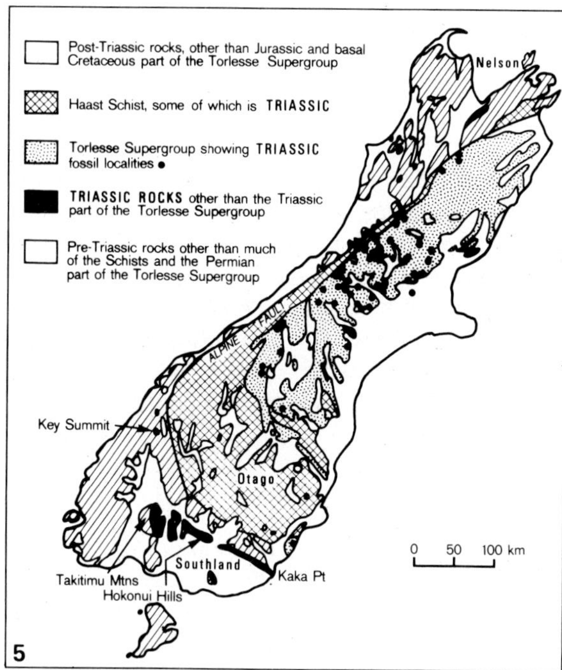

Figs. 4, 5. Maps of New Zealand showing distribution of Triassic rocks and Triassic fossil localities in the Torlesse Supergroup. (After Suggate et al., 1977, figs. 4.33, 4.34.)

Regional Syncline, but further south broader structures expose the complete sequence of Lower, Middle and Upper Triassic, underlain by Upper Permian and overlain by Lower Jurassic. The presence of unconformities and disconformities, however, commonly results in incomplete sections, and in the most southerly sections Middle Triassic is unconformable on Permian (cf. Force, 1973). Inferences from facies and thickness suggest that the southerly sequences were deposited close to the tectonically active margin of the New Zealand Geosyncline (Mutch, 1957).

The sequences comprise mainly moderately to strongly indurated sandstone and siltstone, with tuffaceous beds common and conglomerate at various horizons. The progressive alteration of tuffs by low-temperature metamorphism, resulting from depth of burial, is well developed in Southland and has provided a standard sequence for the study of such alteration (Coombs et al., 1959).

Much of the Triassic sequence is sufficiently fossiliferous to have encouraged the setting up of local stages and their use.

Biostratigraphy

Marine invertebrate fossils of Triassic age are comparatively abundant along the western and southern flanks of New Zealand, through southwest Auckland, Nelson, Southland and south Otago (Murihiku Supergroup), and are found sporadically in the sediments making up the axial ranges (Torlesse Supergroup). The subdivisions based on faunas are comparatively easy to recognize. The type areas are all in Southland. At present the lower two divisions are based chiefly on ammonoid genera. Brachiopods predominate in the Middle Triassic whilst the Upper Triassic is especially rich in bivalves (Table I).

TABLE I

Some representative fossils of the New Zealand Triassic (see Fig. 6 for tentative overseas correlations)

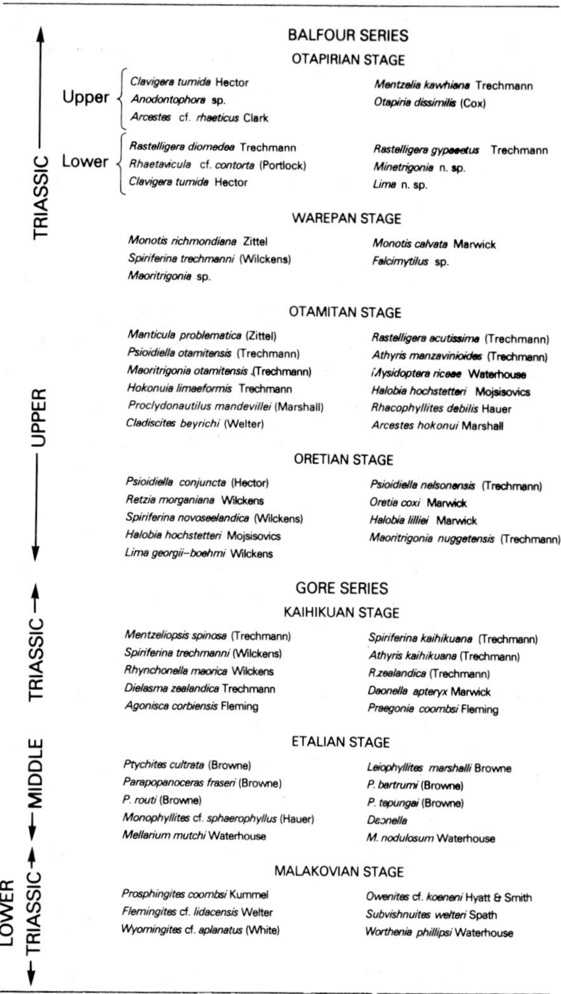

Triassic	Series / Stage	Fossils	Fossils
UPPER TRIASSIC	BALFOUR SERIES		
	OTAPIRIAN STAGE		
	Upper	*Clavigera tumida* Hector	*Mentzelia kawhiana* Trechmann
		Anodontophora sp.	*Otapiria dissimilis* (Cox)
		Arcestes cf. *rhaeticus* Clark	
	Lower	*Rastelligera diomedea* Trechmann	*Rastelligera gypaetus* Trechmann
		Rhaetavicula cf. *contorta* (Portlock)	*Minetrigonia* n. sp.
		Clavigera tumida Hector	*Lima* n. sp.
	WAREPAN STAGE		
		Monotis richmondiana Zittel	*Monotis calvata* Marwick
		Spiriferina trechmanni (Wilckens)	*Falcimytilus* sp.
		Maoritrigonia sp.	
	OTAMITAN STAGE		
		Manticula problematica (Zittel)	*Rastelligera acutissima* (Trechmann)
		Psioidiella otamitensis (Trechmann)	*Athyris manzavinioides* (Trechmann)
		Maoritrigonia otamitensis (Trechmann)	*Mysidoptera riceae* Waterhouse
		Hokonuia limaeformis Trechmann	*Halobia hochstetteri* Mojsisovics
		Proclydonautilus mandevillei (Marshall)	*Rhacophyllites debilis* Hauer
		Cladiscites beyrichi (Welter)	*Arcestes hokonui* Marshall
	ORETIAN STAGE		
		Psioidiella conjuncta (Hector)	*Psioidiella nelsonensis* (Trechmann)
		Retzia morganiana Wilckens	*Oretia coxi* Marwick
		Spiriferina novoseelandica (Wilckens)	*Halobia lilliei* Marwick
		Halobia hochstetteri Mojsisovics	*Maoritrigonia nuggetensis* (Trechmann)
		Lima georgii–boehmi Wilckens	
MIDDLE TRIASSIC	GORE SERIES		
	KAIHIKUAN STAGE		
		Mentzeliopsis spinosa (Trechmann)	*Spiriferina kaihikuana* (Trechmann)
		Spiriferina trechmanni (Wilckens)	*Athyris kaihikuana* (Trechmann)
		Rhynchonella maorica Wilckens	*R. zealandica* (Trechmann)
		Dielasma zealandica Trechmann	*Daonella apteryx* Marwick
		Agonisca corbiensis Fleming	*Praegonia coombsi* Fleming
	ETALIAN STAGE		
		Ptychites cultrata (Browne)	*Leiophyllites marshalli* Browne
		Parapopanoceras fraseri (Browne)	*P. bartrumi* (Browne)
		P. routi (Browne)	*P. tepungai* (Browne)
		Monophyllites cf. *sphaerophyllus* (Hauer)	*Daonella*
		Mellarium mutchi Waterhouse	*M. nodulosum* Waterhouse
LOWER TRIASSIC	MALAKOVIAN STAGE		
		Prosphingites coombsi Kummel	*Owenites* cf. *koeneni* Hyatt & Smith
		Flemingites cf. *lidacensis* Welter	*Subvishnuites welteri* Spath
		Wyomingites cf. *aplanatus* (White)	*Worthenia phillipsi* Waterhouse

Two series, Gore and Balfour, are recognized in New Zealand. The older Gore Series has three stages, and the Balfour Series four. Some of the stages correspond approximately with stages of the world standard sequence, but others are more restricted in time (Fig. 6).

Gore Series

Malakovian Stage. In the type area, Malakoff Hill, western Southland, the sediments of the Malakovian Stage rest unconformably on Upper Permian sediments (Mutch, 1966, 1972). Fossils are moderately common in the type area, and include *Owenites* cf. *koeneni, Flemingites* cf. *lidacensis, Subvishnuites welteri, Wyomingites* cf. *aplanatus, Worthenia phillipsi,* brachiopods and bivalves. The ammonoid *Prosphingites coombsi* has been described from near Kaka Point, south Otago (Kummel, 1965).

The ammonoids known from the Malakovian form two distinct zones. Those from Malakoff Hill are mid-Scythian in age, whereas *Prosphingites* is Late Scythian (Kummel, 1959, 1965). According to the subdivisions of the Scythian proposed by Tozer (1967) and Silberling and Tozer (1968) the Malakoff Hill ammonoids fall within the Smithian Stage, and the Kaka Point *Prosphingites coombsi* in the Spathian Stage (cf. Fig. 6).

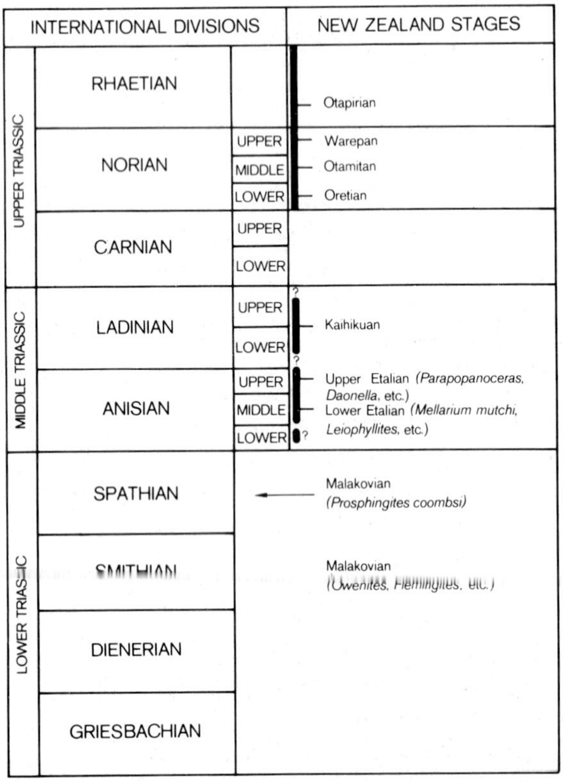

INTERNATIONAL DIVISIONS			NEW ZEALAND STAGES
UPPER TRIASSIC	RHAETIAN		Otapirian
	NORIAN	UPPER	Warepan
		MIDDLE	Otamitan
		LOWER	Oretian
	CARNIAN	UPPER	
		LOWER	
MIDDLE TRIASSIC	LADINIAN	UPPER	? Kaihikuan
		LOWER	?
	ANISIAN	UPPER	Upper Etalian (*Parapopanoceras, Daonella,* etc.)
		MIDDLE	Lower Etalian (*Mellarium mutchi, Leiophyllites,* etc.)
		LOWER	?
LOWER TRIASSIC	SPATHIAN		← Malakovian (*Prosphingites coombsi*)
	SMITHIAN		Malakovian (*Owenites, Flemingites,* etc.)
	DIENERIAN		
	GRIESBACHIAN		

Fig. 6. Correlation table of the New Zealand Triassic, showing levels in the New Zealand stage sequence at which correlations can be made with international divisions. Campbell (1974a and in Carter, 1974, p. 190) believes that a gap corresponding to Carnian time occurs in the New Zealand Triassic marine sequence and correlates Oretian, Otamitan and Warepan with Norian. The broken lines show levels in the New Zealand sequence at which, although strata of equivalent age are probably present, no direct correlations with international divisions are possible, owing to the absence of fossils with overseas affinities.

Etalian Stage. Lower and Upper Etalian faunas can be distinguished. Lower Etalian faunas contain the gastropod *Mellarium mutchi* and the ammonoids *Leiophyllites* and *Discoptychites.* Upper Etalian faunas contain the ammonoid *Parapopanoceras* and the first occurrence in New Zealand of the important bivalve *Daonella.*

Parapopanoceras is known from the Anisian of Spitsbergen, Greenland and British Columbia. *Discoptychites* is a genus typical of the Anisian in the Alps and Middle East, and is also recorded from the Himalayas. The presence together of *Parapopanoceras* and *Daonella* point to a Late Anisian age for the Upper Etalian.

Kaihikuan Stage. Kaihikuan faunas are rich in brachiopods, notably *Mentzeliopsis spinosa, "Spiriferina" kaihikuana* and *"Athyris" kaihikuana* (see Marwick, 1953a; Campbell and Force, 1973). The most widespread bivalve is *Daonella,* and other species as well as gastropods are known. The fauna occurs widely throughout the Southland and Nelson Regional Synclines. No Kaihikuan is known in the North Island.

An exact overseas correlation is not possible. The stage was assumed by Trechmann (1918) and Marwick (1953a) to be Ladinian, in view of the presence of *Daonella* and absence of *Parapopanoceras. S. kaihikuana* and *Mentzeliopsis spinosa* are found in New Caledonia above beds with basal Triassic ammonoids (Drot, in Avias, 1953, pp. 87, 88). Campbell (in Carter, 1974, p. 190) assigns a Ladinian age to the Kaihikuan.

Balfour Series

Oretian Stage. Oretian faunas are characterized by *Oretia coxi* and species of *Halobia.* Other taxa include *Retzia morganiana, Psioidiella conjuncta,* and *P. nelsonensis. Oretia coxi* is not overwhelmingly abundant, and several of the common fossils such as *Halobia,* persist into the overlying stage. Nevertheless *Halobia* is a useful marker, for it replaces *Daonella* of the Kaihikuan, and is widespread.

Oretian faunas are known from both limbs of the Southland Regional Syncline. In the North Island the Triassic of the Kawhia Regional Syncline, at Marokopa and Awakino, commences with Oretian faunas (Campbell, 1955; Grant-Mackie, 1959).

The Oretian has been provisionally correlated with Lower Carnian (Marwick, 1953a). Campbell (1974a; in Carter, 1974, p. 190) maintains however that Carnian correlatives are absent from the New Zealand Triassic sequence and assigns an Early Norian age to the Oretian (Fig. 6).

In New Caledonia probable Oretian faunas are represented by *Halobia,* found below *Manticula* and above brachiopods with Kaihikuan affinities (Avias, 1953).

Otamitan Stage. The Otamitan Stage is richly fossiliferous and is widespread throughout New Zealand and New Caledonia. The most typical species is *Manticula problematica,* a mytilid-shaped bivalve which occurs in extraordinary abundance. The incoming of this species has been used by Campbell and McKellar (1960) to mark the base of the stage. Apart from other bivalves such as *Halobia* and *Mysidioptera,* both brachiopods (*Athyris manzavinoides* and *Psioidiella otamitensis*) and gastropods are common.

The stage is widely distributed in the Southland and Kawhia Regional Synclines and is well represented at Nelson. *Manticula* occurs in abundance in New Caledonia (Avias, 1953).

The stage has been assigned to the Upper Carnian (Marwick, 1953a). Campbell (1974a; in Carter, 1974, p. 190) however assigns a Norian age to the Otamitan (Fig. 6). An overlap of Otamitan and Warepan fossils, indicating possible close time equivalence, is documented by Speden (1975b).

Warepan Stage. Warepan faunas are distinguished by the presence of the bivalve *Monotis richmondiana,* which replaces *Manticula problematica,* the characteristic bivalve of the preceding stage.

The stage is widely distributed in New Zealand. It is well represented at Nelson and in the Kawhia Regional Syncline, but is missing from several areas in the Southland Regional Syncline (Campbell, 1959; Watters, 1952; Wood, 1956). *Monotis richmondiana* has been widely reported from the Torlesse Supergroup of the South Island (Campbell and Warren, 1965; Andrews et al., 1976).

Marwick (1953a) had assumed *Monotis* to characterize the Norian stage of the world standard succession, but Tozer (1967) and Silberling and Tozer (1968) have assigned it chiefly to the Upper Norian. Tozer (1967, p. 41) specifically referred *M. richmondiana* to the Upper Norian lower *Suessi* zone (see also Campbell, 1974a).

Otapirian Stage. Campbell and McKellar (1956) (see also Campbell 1956, 1968) defined the base of the Otapirian as being marked by the incoming of *Rastelligera diomedea.*

Campbell (1956) differentiated Lower and Upper Otapirian faunas. Lower Otapirian faunas include *Spiriferina* aff. *trechmanni, Psioidea* aff. *nelsonensis, Rastelligera diomedea, Clavigera tumida, Myophoria* sp., *Rhaetavicula* cf. *contorta, Lima* sp. and *Torastarte bensoni.* Upper Otapirian faunas include *Mentzelia kawhiana, Clavigera tumida* and *Otapiria dissimilis.*

Martin (1975) has pointed out, however, that the use of *Clavigera* as an Otapirian index is doubtful as the genus is known to occur in *Monotis*-bearing strata in both New Zealand and New Caledonia. Also, work by Speden (in Kear, 1961) has shown that the range of *Torastarte bensoni* extends from Otapirian into Lower Jurassic (Ururoan).

Rastelligera and *Clavigera* are found widely in the Southland Regional Syncline. A small area of Otapirian is present at Nelson, and the faunas are also known along the western limb of the Kawhia Regional Syncline.

The presence in the Otapirian of *Arcestes* cf. *rhaeticus* suggests correlation with the Rhaetian, although Tozer (1967, p. 41) considered that the stage probably belonged to the uppermost Norian or upper *Suessi* zone. Nevertheless, no compelling evidence has been offered to support this view, and the absence of *Monotis* from the Otapirian, as from the Rhaetian, is probably significant (Stevens, 1970a).

Regional stratigraphy

South Auckland

The Murihiku Supergroup rocks of southwest Auckland are folded into the broad Kawhia Regional Syncline. The oldest beds in the syncline are Late Triassic, and these are exposed on both flanks (Figs. 7–9) but are limited eastwards by the Waipa Fault (Kear, 1960), which throws down Jurassic rocks of the Manaia Hill Group, and westwards by the Tasman Sea, or locally by down-faulted Jurassic rocks. Southwards they are concealed by younger beds, but reappear at Nelson in the South Island. Most common are well-jointed, well-bedded or massive indurated siltstone and sandstone ("argillite" and "greywacke"), with abundant tuffaceous material and with local vitric tuffs and shell beds. All rocks belong to the Zeolite Mineral Facies of Coombs (1960). Jointing is well developed and the joint-planes at many exposures are prominently coated with zeolites.

The distribution and lithology of the Triassic rocks of the Kawhia Regional Syncline are illustrated and described in Figs. 7 and 8. There are important differences between the rocks of the two limbs of the syncline. On the western limb the thicknesses of the various stages of the Balfour Series are remarkably constant: about 1200 m of Otapirian (Rhaetian), 225 m of Warepan (Norian), 225–380 m of Otamitan (Karnian), and 600–760 m of Oretian (Carnian), excluding nearly 1200 m of the very coarse granitic Moeatoa Conglomerate (at Marokopa), which may also be Oretian (Fig. 8). The western coastal sequences are generally well exposed and have yielded, in different localities, the typical fossils of all these stages. Dips generally range from 45° to 85°.

On the eastern limb of the syncline exposures are far fewer, and, apart from *Monotis* and *Manticula* shell beds, fossils are rare. Minor folding is common, and faults parallel to the strike repeat considerable thicknesses of rocks. It appears that the sequences are thinner in the eastern than in the western limb of the Syncline. The Otapirian Stage, for example, is between 600 and 300 m thick on the eastern limb, compared with 1200 m in the west. The beds are as indurated as those in the west, however, and their dip is as steep, or steeper.

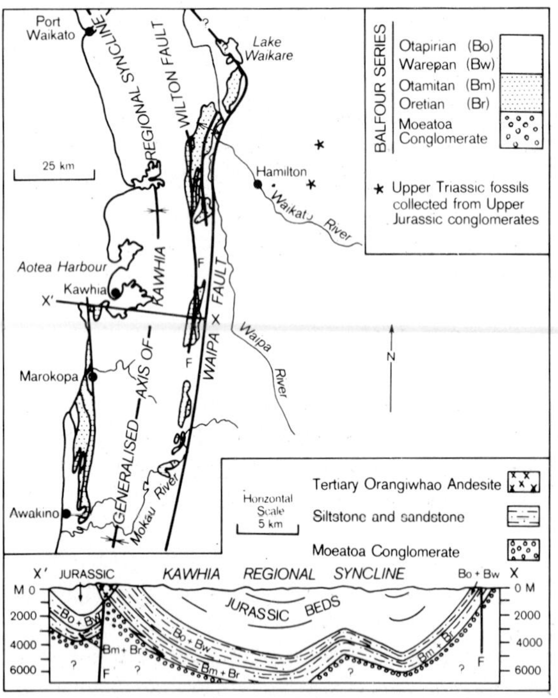

Fig. 7. Map of southwest Auckland region showing the distribution of Triassic rocks and a geological cross-section along line X^1–X. Stratigraphic columns for Awakino, Marokopa and Kawhia are shown in Fig. 8. (After Suggate et al., 1977, fig. 4.38.)

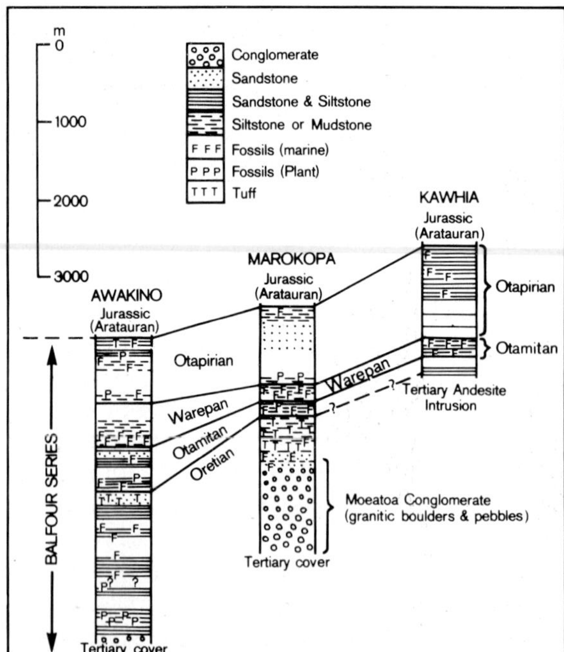

Fig. 8. Generalized stratigraphic columns for the Triassic of southwest Auckland (cf. Fig. 7). (After Suggate et al., 1977, Fig. 4.39.)

Fig. 9. Aerial view of the coastal Triassic section at Kiritehere, 25 km south of Kawhia Harbour (see Fig. 7). Sediments of the Otapirian Stage outcrop on the coast at the left (northern) edge of the photo; Warepan sediments in the middle; Otamitan sediments to the right (Kear, 1960). (Photo: D.L. Homer, New Zealand Geological Survey.)

Nelson

The Triassic beds at Nelson are preserved in a narrow fault-bounded strip 32 km long and not more than 2.5 km wide. They form a belt of low hills between the Waimea Plain to the west (filled with Tertiary and Quaternary) and the ranges of Permian rocks to the east. Within the strip the Triassic rocks are preserved in a disjointed series of synclines (Fig. 10).

The dominant rock types are moderately hard, well-jointed sandstone and siltstone, with conglomerate bands occurring throughout. Fossils are abundant (Campbell, 1955, 1974a; Force and Campbell, 1974) locally in shell beds, and plant fragments are common in the upper part of the sequence.

Canterbury

Isolated blocks of fossiliferous Triassic strata occur in the foot-hills of the Southern Alps, in-faulted into Torlesse Supergroup rocks (Campbell and Warren, 1965; Campbell and Force, 1972).

Southland and southeast Otago

In Otago and Southland Triassic rocks of the Murihiku Supergroup extend for some 270 km south from Key Summit (on the Alpine divide) to the Hokonui Hills, where they bifurcate into two south-east-trending belts, bounding a core of Jurassic rocks (Figs. 11–13). Dips in the Key Summit Syncline are high and the eastern limb is overturned. In the Southland Regional Syncline, on the other hand, although

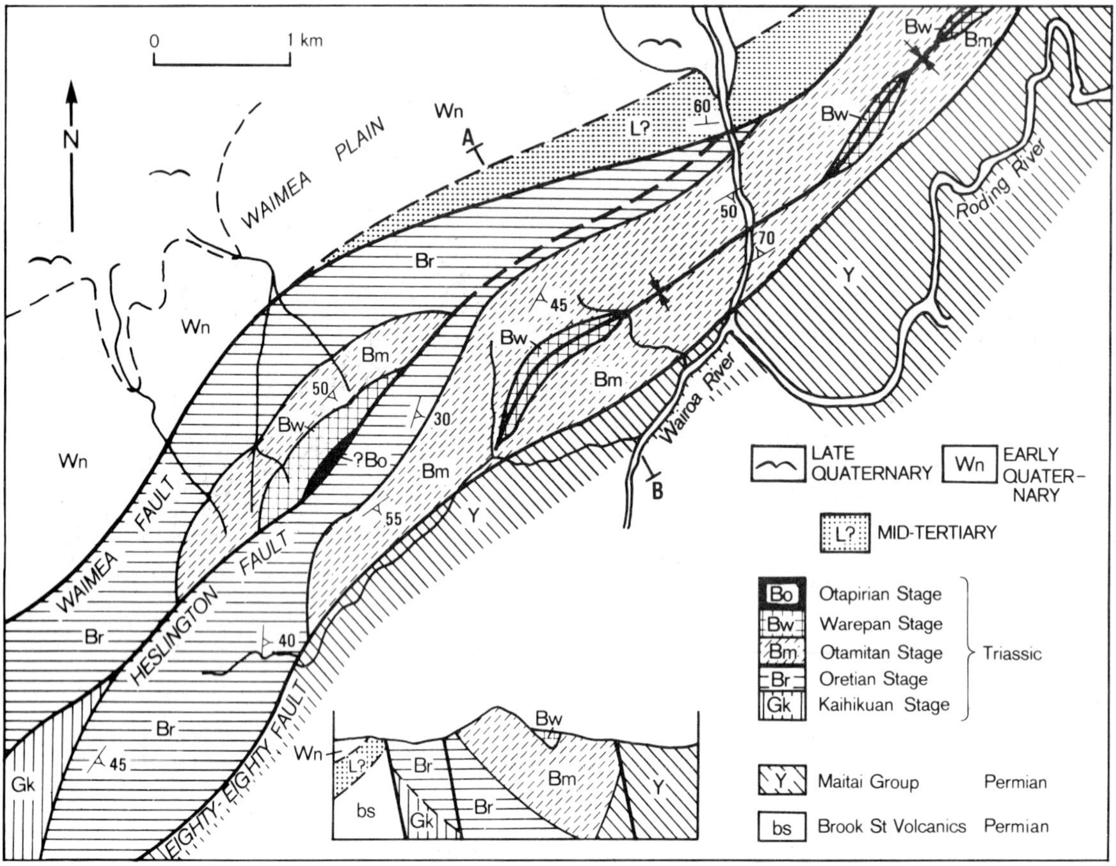

Fig. 10. Map and geological cross-section of the Triassic in the Nelson area. (After Suggate et al., 1977, fig. 4.46.)

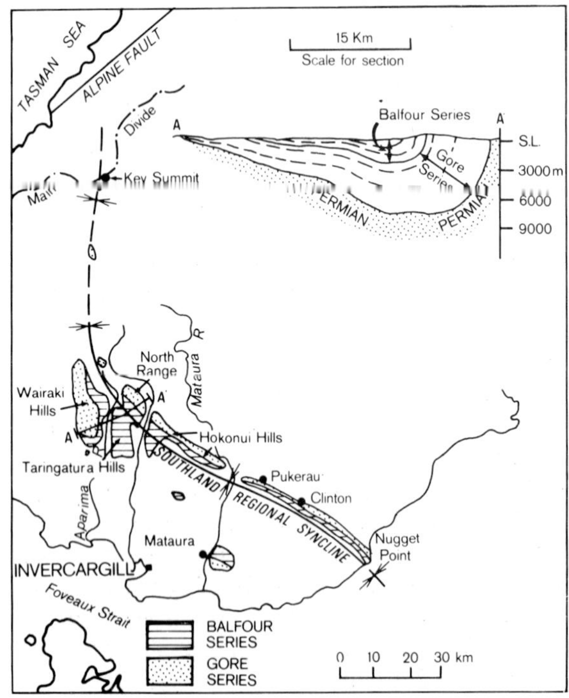

the northern limb is steep, dips in the undulating southern limb rarely exceed 20° and an extensive area of Triassic rocks is exposed. Although conglomerates lie between uppermost Permian and lowermost Triassic beds (Wood, 1956, Mutch, 1972), there is no evidence of angular unconformity.

Throughout the Southland Regional Syncline the sequences are mainly moderately hard, well-jointed blue-grey tuffaceous and volcanic greywacke, crystal, lithic and vitric tuff and dark blue-grey argillite, but conglomerate bands are found throughout (e.g. Boles, 1974). The tuffs become less common upwards, and more acid upwards. More tuff beds are present on the southern limb of the syncline than on the northern. Fossils are abundant, and shell beds and plant

Fig. 11. Map and geological cross-section showing the distribution of Triassic rocks in Otago and Southland (Southland Regional Syncline). (After Suggate et al., 1977, fig. 4.48 (cf. Fig. 12).)

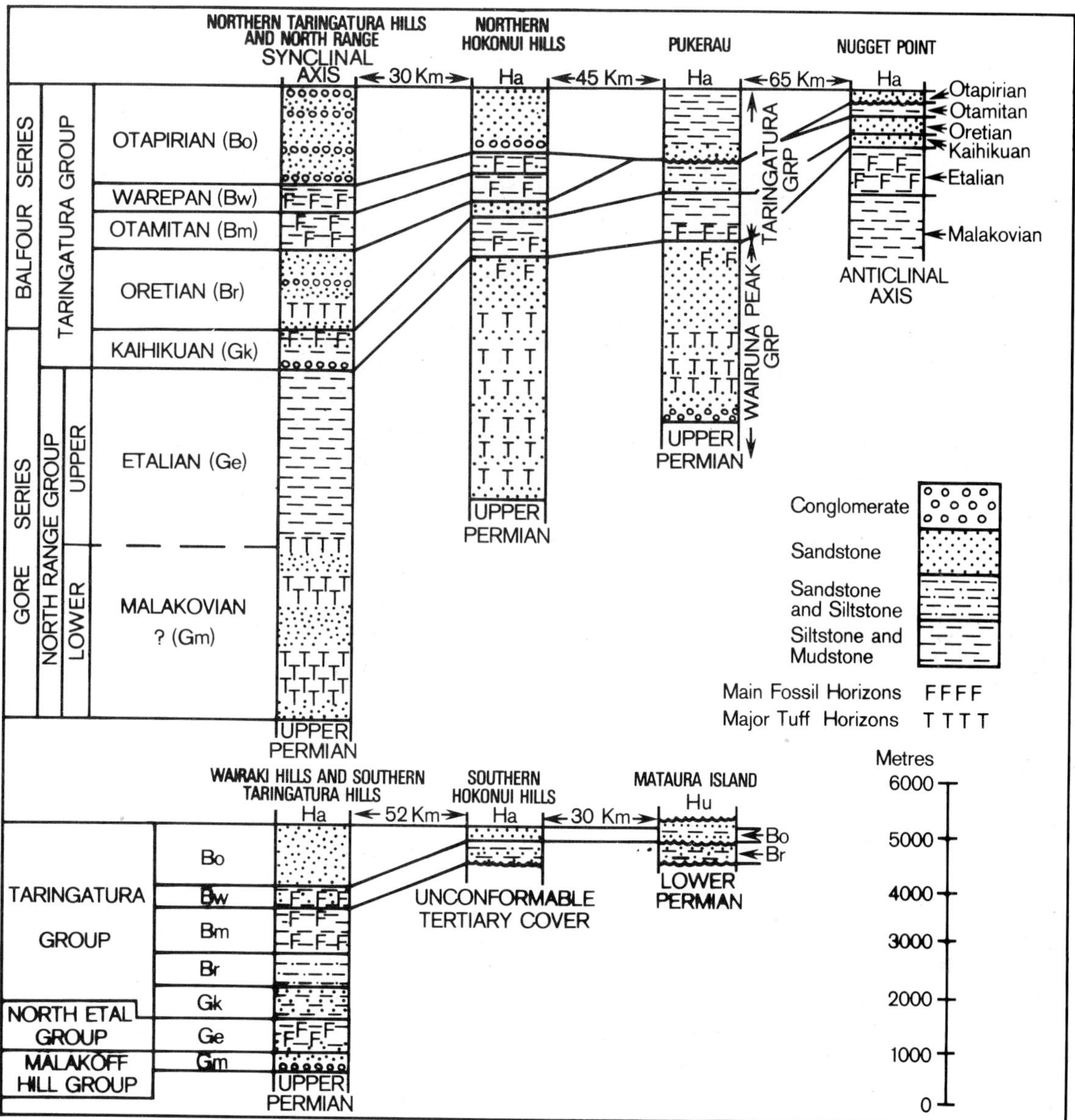

Fig. 12. Stratigraphic columns for the Triassic of the Southland Regional Syncline. (*Ha* = Aratauran Stage; *Hu* = Ururoan Stage; both Lower Jurassic). (After Suggate et al., 1977, fig. 4.49 (cf. Fig. 11).)

fragments are locally common in the lower and upper parts of the sequence.

MURIHIKU SUPERGROUP – JURASSIC (G.R. STEVENS)

Jurassic rocks of the Murihiku Supergroup in sequence with Triassic, form the cores of the Kawhia and Southland Regional Synclines (Figs. 1, 2, 14, 15). The Early Jurassic sequences are wholly marine but by the end of the Jurassic non-marine conditions apparently prevailed everywhere.

As a result of folding and erosion Jurassic rocks of the Murihiku Supergroup are nowhere in sequence with Cretaceous. The youngest are non-marine beds over 1200 m thick conformably following Tithonian marine beds (in the Port Waikato region). Despite their thickness, however, they are not usually thought to extend into the Cretaceous. The lack of Oxfordian

Fig. 13. Aerial view of Triassic–Jurassic sediments exposed in the axis of the Southland Syncline, Southland. Otapirian beds (uppermost Triassic) underlie the low hills in the foreground. On the flanks of the cuesta in the middle of the photo (called locally "The Bastion"), the Otapirian beds are overlain by Aratauran and Ururoan (Jurassic), and the cuesta itself is capped by a resistant sandstone of Ururoan–earliest Temaikan age. Faulting repeats the sequence again in the far right distance of the photo – the cuesta visible is Flag Hill, again capped by Ururoan–Temaikan sandstone (McKellar, 1969). (Photo: S.N. Beatus, New Zealand Geological Survey.)

fossils in a sequence in which the remainder of the Jurassic is represented might be taken as indicating a major interruption of sedimentation, even of precursor movements of the Cretaceous Rangitata Orogeny.

Lithologies range from conglomerate to mudstone, commonly in massive units, but in places including alternating sandstone–siltstone sequences. Tuffaceous horizons are found in both the Kawhia and Southland Regional Synclines, but do not make up as high a proportion of the sequences as in the Triassic rocks of the Murihiku Supergroup. Regional metamorphic changes are of very low grade and zeolites are developed in some rocks. Carbonaceous beds are common in the non-marine upper parts of the sequences, and the only pre-Cretaceous coal to be mined in New Zealand came from a poor seam in Southland. The marine sequences are fossiliferous and this has led to the establishment of local stages. Thicknesses range up to 7600 m in the Kawhia Regional Syncline, and at least 4600 m in the Southland Regional Syncline. Despite facies variations within the two regions, there is no clear indication of direction in the overall variations of thickness, or of systematic variations of the thicknesses of individual stages in relation to the margins of the geosyncline.

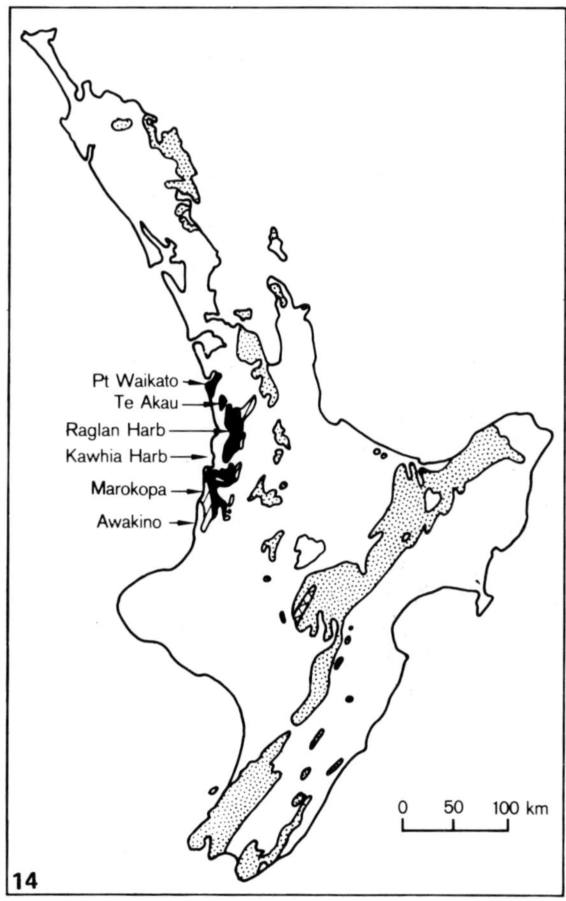

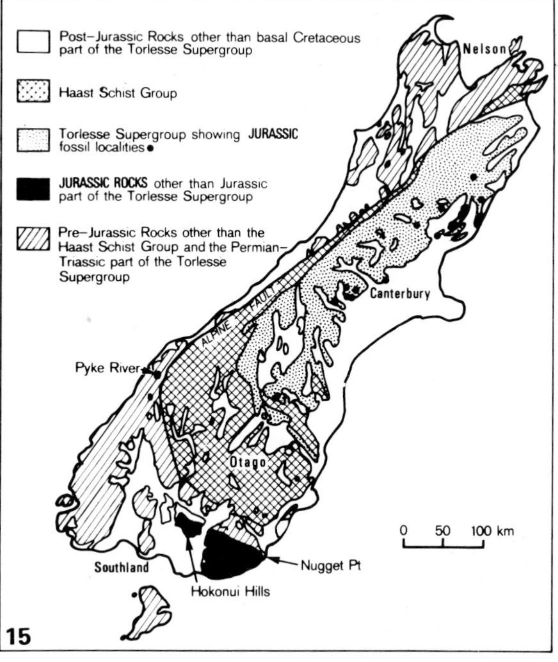

Figs. 14, 15. Maps of New Zealand showing distribution of Jurassic rocks and Jurassic fossil localities in the Torlesse Supergroup. (After Suggate et al., 1977, figs. 4.57, 4.58.)

Biostratigraphy

New Zealand Jurassic faunas are commonly rich in bivalves, but many of the ammonites and other fossils useful for external correlation are often sparsely distributed, particularly in the Lower and Middle Jurassic. Thus, as correlations with international divisions can be provided only at scattered, isolated horizons in the stratigraphic column, local stages have been recognized, broadly correlated with the international divisions (Fig. 16).

Modern subdivision of the Jurassic is based on Marwick (1951, 1953a) and revisions by Fleming (1958, 1960) and Fleming and Kear (1960). It is divided into three series: Herangi (oldest) Kawhia and Oteke (Table II). The Herangi–Kawhia junction approximately corresponds with the Lower and Middle Jurassic junction. Six stages are recognized within these series, and all were originally based by Marwick (1953a) on the sequences near Kawhia Harbour, on the west coast of the North Island. Later work has shown however, that the Aratauran, Temaikan, and to some extent the Ururoan, are incompletely represented at Kawhia and that more complete sequences are available in Southland (Hokonui Hills) and the south Otago coast.

The base of the Jurassic System is particularly well exposed in the south Otago coast and Hokonui Hills, Southland (see Figs. 23, 24). The base is drawn immediately below the incoming of psiloceratid ammonites and/or the pelecypod *Otapiria marshalli.*

The Jurassic–Cretaceous boundary is apparently nowhere clearly exposed in the New Zealand stratigraphic sequence as it is now known. The Huriwai Formation (Purser, 1961), the highest rock unit recognized in the Puaroan Stage (the highest Jurassic time-stratigraphic unit, see Fig. 16), is non-marine and although its plant fossils were originally assigned to the Neocomian, a Tithonian age is now favoured (Edwards, 1934; McQueen, 1955; Norris, 1968; Norris and Waterhouse, 1970; Waterhouse and Norris, 1972). Marine faunas immediately preceding the Huriwai Formation were formerly thought to be of Neocomian age (e.g. Arkell, 1956, p. 454) but later study has shown them more likely to be Tithonian (Fleming and Kear, 1960).

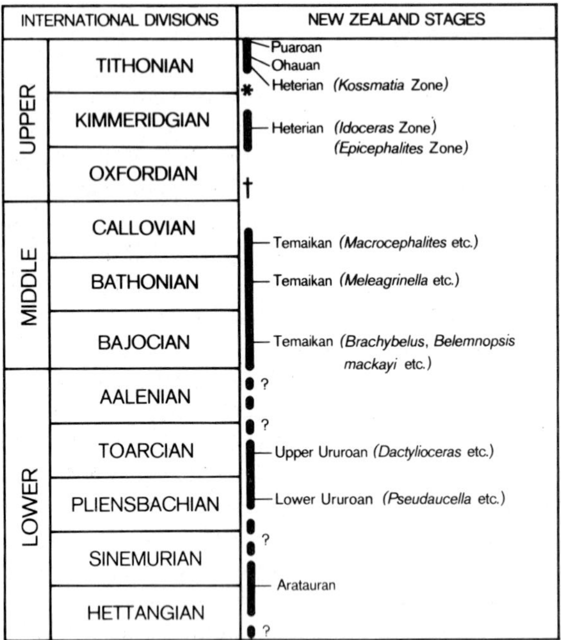

INTERNATIONAL DIVISIONS		NEW ZEALAND STAGES
UPPER	TITHONIAN	Puaroan Ohauan * Heterian *(Kossmatia Zone)*
	KIMMERIDGIAN	Heterian *(Idoceras Zone)* *(Epicephalites Zone)*
	OXFORDIAN	†
MIDDLE	CALLOVIAN	Temaikan *(Macrocephalites etc.)*
	BATHONIAN	Temaikan *(Meleagrinella etc.)*
	BAJOCIAN	Temaikan *(Brachybelus, Belemnopsis mackayi etc.)*
LOWER	AALENIAN	?
	TOARCIAN	? Upper Ururoan *(Dactylioceras etc.)*
	PLIENSBACHIAN	Lower Ururoan *(Pseudaucella etc.)*
	SINEMURIAN	? Aratauran
	HETTANGIAN	?

Fig. 16. Correlation table of the New Zealand Jurassic showing levels in the New Zealand stage sequence at which firm correlations can be made with international divisions. The broken lines show levels in the New Zealand sequence at which, although strata of equivalent age are probably present, no direct correlations with international divisions are possible, owing to the absence of fossils with overseas affinities. The asterisk indicates a level in the New Zealand sequence in which the exact extent of an apparent time gap has yet to be assessed (see details in accompanying text). The dagger indicates a time gap that is thought to represent an important hiatus in the New Zealand stratigraphic record.

TABLE II

Some representative fossils of the New Zealand Jurassic (see Fig. 16 for tentative overseas correlations)

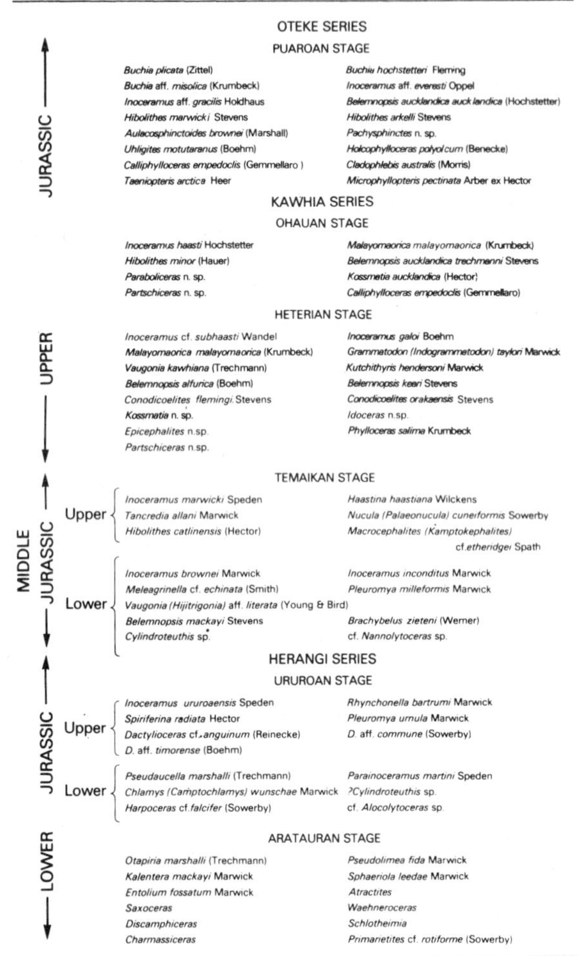

		OTEKE SERIES	
		PUAROAN STAGE	
UPPER JURASSIC		*Buchia plicata* (Zittel)	*Buchia hochstetteri* Fleming
		Buchia aff. *misolica* (Krumbeck)	*Inoceramus* aff. *everesti* Oppel
		Inoceramus aff. *gracilis* Holdhaus	*Belemnopsis aucklandica aucklandica* (Hochstetter)
		Hibolithes marwicki Stevens	*Hibolithes arkelli* Stevens
		Aulacosphinctoides brownei (Marshall)	*Pachysphinctes* n. sp.
		Uhligites motutaranus (Boehm)	*Holcophylloceras polyolcum* (Benecke)
		Calliphylloceras empedoclis (Gemmellaro)	*Cladophlebis australis* (Morris)
		Taeniopteris arctica Heer	*Microphyllopteris pectinata* Arber ex Hector
		KAWHIA SERIES	
		OHAUAN STAGE	
		Inoceramus haasti Hochstetter	*Malayomaorica malayomaorica* (Krumbeck)
		Hibolithes minor (Hauer)	*Belemnopsis aucklandica trechmanni* Stevens
		Paraboliceras n. sp.	*Kossmatia aucklandica* (Hector)
		Partschiceras n. sp.	*Calliphylloceras empedoclis* (Gemmellaro)
		HETERIAN STAGE	
		Inoceramus cf. *subhaasti* Wandel	*Inoceramus galoi* Boehm
		Malayomaorica malayomaorica (Krumbeck)	*Grammatodon (Indogrammatodon) taylori* Marwick
		Vaugonia kawhiana (Trechmann)	*Kutchithyris hendersoni* Marwick
		Belemnopsis alfurica (Boehm)	*Belemnopsis keari* Stevens
		Conodicoelites flemingi Stevens	*Conodicoelites orakaensis* Stevens
		Kossmatia n. sp.	*Idoceras* n.sp.
		Epicephalites n.sp.	*Phylloceras salima* Krumbeck
		Partschiceras n.sp.	
		TEMAIKAN STAGE	
MIDDLE JURASSIC	Upper	*Inoceramus marwicki* Speden	*Haastina haastiana* Wilckens
		Tancredia allani Marwick	*Nucula (Palaeonucula) cuneiformis* Sowerby
		Hibolithes catlinensis (Hector)	*Macrocephalites (Kamptokephalites)* cf. *etheridgei* Spath
	Lower	*Inoceramus brownei* Marwick	*Inoceramus inconditus* Marwick
		Meleagrinella cf. *echinata* (Smith)	*Pleuromya milleformis* Marwick
		Vaugonia (Hijitrigonia) aff. *literata* (Young & Bird)	
		Belemnopsis mackayi Stevens	*Brachybelus zieteni* (Werner)
		Cylindroteuthis sp.	cf. *Nannolytoceras* sp.
		HERANGI SERIES	
		URUROAN STAGE	
LOWER JURASSIC	Upper	*Inoceramus ururoaensis* Speden	*Rhynchonella bartrumi* Marwick
		Spiriferina radiata Hector	*Pleuromya urnula* Marwick
		Dactylioceras cf. *anguinum* (Reinecke)	*D.* aff. *commune* (Sowerby)
		D. aff. *timorense* (Boehm)	
	Lower	*Pseudaucella marshalli* (Trechmann)	*Parainoceramus martini* Speden
		Chlamys (Camptochlamys) wunschae Marwick	?*Cylindroteuthis* sp.
		Harpoceras cf. *falcifer* (Sowerby)	cf. *Alocolytoceras* sp.
		ARATAURAN STAGE	
		Otapiria marshalli (Trechmann)	*Pseudolimea fida* Marwick
		Kalentera mackayi Marwick	*Sphaeriola leedae* Marwick
		Entolium fossatum Marwick	*Atractites*
		Saxoceras	*Waehneroceras*
		Discamphiceras	*Schlotheimia*
		Charmassiceras	*Primarietites* cf. *rotiforme* (Sowerby)

Herangi Series

Aratauran Stage. The Aratauran as defined by Marwick (1953a), based on the Kawhia sequence, had as its lower boundary the incoming of *Otapiria marshalli* (Trechmann), replacing the Otapirian (Upper Triassic) *Otapiria dissimilis* (Cox), and as its upper boundary the incoming of *Pseudaucella marshalli* (Trechmann).

Successions in the coastal region of south Otago (Speden and McKellar, 1958; Speden, 1961, 1971) and Hokonui Hills, Southland (Campbell and McKellar, 1956; McKellar, 1969) provide good definition of the Otapirian–Aratauran boundary (Figs. 23, 24).

The Aratauran ammonite assemblages (Spath, 1923; Marwick, 1953a, pp. 116–7) have been used as the basis for external correlation, and the stage has been correlated by Marwick (1953a, p. 11) with the Hettangian–Sinemurian. This correlation has been confirmed by Arkell (1953, 1956, p. 456), and also by study of the ammonite sequences of Southland and Kawhia (Stevens, unpubl.) where Hettangian correlatives and the *bucklandi* zone of the Sinemurian are present.

Hettangian–Sinemurian ammonite assemblages similar to those of New Zealand have been described by Avias (1953) from New Caledonia. *Otapiria marshalli* has been recorded from New Caledonia by Routhier (1950) and Avias (1951, 1953, p. 149) and is associated with ammonites which, according to Avias, have affinities with those of the Upper Hettangian of Europe. *Otapiria* species similar to *O. marshalli* are known from northern Alaska and northeastern Siberia (Imlay, 1967), as well as from Austria

(Grant-Mackie and Zapfe, 1973), Equador (Geyer, 1974) and Argentina (S.E. Damborenea, pers. comm.).

Ururoan Stage. The base of the Ururoan has been taken as the incoming of dwarf *Pseudaucella marshalli* (Trechmann) (Speden and McKellar, 1958; Grant-Mackie, 1959; Kear, 1961; Martin, 1975).

Grant-Mackie (1959, pp. 776–9) recognized an Upper and Lower Ururoan, on the basis of two separate fossil assemblages. The Lower Ururoan is characterized by *Pseudaucella* and rare *Chlamys* and the Upper Ururoan by *Dactylioceras* and rhynchonelloid brachiopods. The Lower Ururoan is defined by Grant-Mackie as those beds deposited between the first appearance of *Dactylioceras* and that of *Inoceramus.*

The *Dactylioceras* assemblage (i.e., Grant-Mackie's Upper Ururoan) has been correlated with the Toarcian (Spath, 1923, p. 301; Marwick, 1953a, p. 22; Arkell, 1956, p. 456; Grant-Mackie, 1959, p. 777), and other Toarcian ammonites, cf. *Alocolytoceras* and *Harpoceras* cf. *falcifer* (J. Sowerby), have been identified (G.R. Stevens) in recent collections from the Upper Ururoan.

The age of the *Pseudaucella* assemblage (i.e., of the Lower Ururoan) is not well defined, but by inference is regarded as Pliensbachian (Arkell, 1956, p. 456; Grant-Mackie, 1959, p. 777). *P. marshalli* has been recorded from New Caledonia (Routhier, 1950; 1953, p. 26; Avias, 1951, 1953, pp. 154–5) but with no definite age connotation.

Kawhia Series

Temaikan Stage. Criteria for defining the base of the Temaikan are still being sought (Speden, 1970).

At Kawhia, the Rengarenga Group, mainly coarse-grained terrestrial beds, lies between marine Ururoan and Heterian rocks (see Fig. 20). Only the middle formation (Opapaka Sandstone, 23 m) contains Temaikan marine fossils (Fig. 20). *Meleagrinella* cf. *echinata* (Smith), which first appears at the base of the Opapaka Sandstone, was used by Marwick (1953a) and Fleming and Kear (1960) to recognize the base of the stage at Kawhia and by Purser (1961) in the Port Waikato region. But Speden (1971) has documented the occurrence of *Meleagrinella* below *Pseudaucella marshalli.*

In the Hokonui Hills the Ururoan–Temaikan boundary has been drawn below the first appearance of *Belemnopsis mackayi* Stevens, some 6 m stratigraphically above the highest *Pseudaucella marshalli* and some 225 m stratigraphically below the first appearance of *Meleagrinella echinata* and *Pleuromya milleformis* Marwick (McKellar, 1969). The same *Belemnopsis* is in the lower 5–10 m of the Opapaka Sandstone at Kawhia and at the base of the Temaikan in the Awakino region (Stevens, 1965, p. 72), but appears to be missing from the southeast Otago sequence (Speden, 1961, 1971) and here the base of the Temaikan has been defined by the incoming of *Inoceramus inconditus* and by the first occurrence of *M.* cf. *echinata* above *P. marshalli.*

Speden (1961, 1971) recognized two divisions of the Temaikan in the southeast Otago sequence. According to him the Lower Temaikan is characterized by the *Pleuromya milleformis* assemblage zone (with *Meleagrinella* cf. *echinata* and *Inoceramus inconditus*) and the Upper Temaikan by the *Haastina haastiana* assemblage zone (with *Tancredia allani* Marwick, *Inoceramus marwicki* Speden and macrocephalitid ammonites).

The base of the Temaikan can be correlated with the Bajocian, as *Brachybelus zieteni* (Werner) has a range of Middle Lias to Bajocian in Europe and *Belemnopsis mackayi,* which occurs stratigraphically below *Brachybelus* in New Zealand, has marked affinities with those of the Bajocian of Europe and Western Australia (Stevens, 1965). An ammonite from the Lower Temaikan of southeast Otago has been identified as cf. *Nannolytoceras,* of Bajocian–Bathonian age in Europe and North America.

Meleagrinella cf. *echinata,* appearing above *Belemnopsis,* has basically Bathonian affinities (Arkell, 1956, p. 456) but has also been recorded from the Lower Callovian (e.g., Cox, 1940, p. 92; *Macrocephalus* Beds of Kachh).

Beds occurring at the top of the Temaikan in southeast Otago can be correlated with the Lower and Middle Callovian by means of the macrocephalitid ammonites recorded by Speden (1958, 1961, 1971). The best preserved of the ammonites can be compared with *Macrocephalites* (*Kamptokephalites*) *etheridgei* Spath from the Lower Callovian of New Guinea. Ammonites from the Upper Temaikan of the Awakino region have been compared (Stevens, 1965, p. 32) with macrocephalitids described by Boehm (1912) from the Callovian of Taliabu, Sula Islands, Indonesia (Lower and Middle Callovian according to Arkell, 1956, pp. 438–9).

Heterian Stage. The base of the stage is defined as the lowest occurrence of *Inoceramus galoi* Boehm. *Malayomaorica malayomaorica* (Krumbeck) first appears in the upper part of the Heterian and ranges upwards into the Ohauan Stage (Fleming and Kear, 1960, pp. 42–3).

Inoceramus galoi, Malayomaorica malayomaorica and belemnites are valuable for internal correlation, but the Heterian ammonite assemblage has not yet been recorded outside Kawhia, except that the holotype of *Idoceras speighti* (Marshall) was found as a river boulder in north Canterbury (Arkell, 1953).

According to Arkell (see Fleming and Kear, 1960), an Early and Middle Kimmeridgian age is indicated by the Heterian ammonites (see below for discussion on Arkell's "Kimmeridgian"). Three distinctive ammonite zones may be distinguished in the Heterian – an *Epicephalites* zone in the Oraka Sandstone, an *Idoceras* zone in the lower part of the Ohineruru Formation and a *Kossmatia* zone in the remaining formations assigned to the Heterian at Kawhia (Fig. 20). The Early–Middle Kimmeridgian boundary appears to fall between the *Idoceras* and *Kossmatia* zones (Arkell, in Fleming and Kear, 1960, p. 44). The *Kossmatia* zone probably extends upwards into the lower part of the Ohauan Stage. Enay (1972a, b), however, assigns a Late Tithonian age to the New Zealand *Kossmatia* (see below). On the basis of this assessment a time gap separates the *Idoceras* (Kimmeridgian) and *Kossmatia* (U. Tithonian) zones.

Inoceramus galoi, Malayomaorica malayomaorica and the Heterian belemnites have all been recorded from Indonesia, where they occur in beds dated by previous workers as Oxfordian. These age determinations, however, have not been unreservedly accepted by New Zealand and Australian workers (see Fleming, 1958, 1960; Fleming and Kear, 1960; Brunnschweiler, 1960; Stevens, 1965), as some of the critical Indonesian collections (e.g., those from the Sula Islands; Boehm, 1907; Arkell, 1956, p. 437) are stream collections, derived from beds of several different ages.

No fossils of definite Late Callovian or Oxfordian age are known from New Zealand, and this gap falls between the Temaikan (Bajocian–Middle Callovian) and Heterian (Early and Middle Kimmeridgian). At Kawhia the Temaikan assemblage in the Opapaka Sandstone is separated from the lowest Heterian assemblage in the Oraka Sandstone by the non-marine Wharetanu Measures (Fig. 20). This formation, together with a possible unrecognized disconformity, may represent the missing Late Callovian and Oxfordian time, as well as the Early and Middle Callovian (Upper Temaikan) that Speden (1970) has shown to be missing in the Kawhia section.

In Southland, when the Upper Temaikan is present, it is succeeded by non-marine beds (Speden, 1961, 1971) that may represent the Upper Callovian and Oxfordian. In the Awakino region (Grant-Mackie, 1959) Upper Temaikan beds pass without apparent break into Heterian beds.

Ohauan Stage. The range zone of *Inoceramus haasti* Hochstetter appears to coincide approximately with the boundaries of the Ohauan. The precise boundaries cannot, however, be defined at present because, towards the lower boundary, in the lower part of the Kowhai Point Siltstone of the Kawhia sequence (Fig. 20) the *Inoceramus* are intermediate in morphology between *I. galoi* Boehm (Heterian Stage) and *I. haasti,* and have been identified as *I.* cf. *subhaasti* Wandel (Fleming and Kear, 1960, pp. 32, 43; Fleming, 1960).

Malayomaorica malayomaorica (shared with Upper Heterian) and *Inoceramus haasti* (restricted) are the most useful Ohauan fossils for internal correlation. *M. malayomaorica* is known from Indonesia and Australia (see Fleming, 1958; Brunnschweiler, 1960; Jeletzky, 1963); *I. haasti* and *I. subhaasti* from Indonesia (Wandel, 1936). Their ages in these areas are however not well defined, and ammonites are used as the basis for external correlation of the stage.

Two ammonite zones have been recognized in the Ohauan by Arkell (in Fleming and Kear, 1960, p. 44). The lower part of the stage (Kowhai Point Siltstone) apparently shares a *Kossmatia* zone with the upper part of the Heterian Stage. A *Paraboliceras* zone has been recognized in the Kinohaku Siltstone (Fig. 20).

Spath (1923), on the basis of the presence of *Berriasella novoseelandica* (Hochstetter), assigned the Ohauan to the uppermost zone of the Tithonian. However, as Marwick pointed out (1953a, p. 28), this age assignment had a number of problems, amongst which is the identity of *B. novoseelandica* (cf. Boehm, 1911, p. 21; Spath, 1923, p. 303).

Arkell (in Fleming and Kear, 1960, pp. 32–7) regarded *B. novoseelandica* as a squashed *Kossmatia* or *Paraboliceras,* and correlated the Ohauan with the early Middle Kimmeridgian. This age determination was based however on the concept of the Kimmeridgian then current in Great Britain (e.g. Arkell, 1956).

Recent international colloquia on subdivision of the Jurassic (Maubeuge et al., 1964, 1974; Gerasimov and Mikhailov et al., 1967; Geczy et al., 1971; Enay et al., 1975) have had a substantial bearing on correlation of the New Zealand Upper Jurassic. "Kimmeridgian" as referred to in the recent literature (e.g. Fleming and Kear, 1960; Stevens, 1965, 1967, 1968, 1974a) is used in the sense of Arkell (1956). Correlation between the various faunal realms and provinces that were developed at the end of the Jurassic (Boreal, Tethyan, etc.) has been a contentious issue for many years, so much so that separate biostratigraphic units have been used in the various realms or provinces, e.g. Volgian (Boreal), Portlandian (Anglo-Paris Basin), Tithonian (Tethys and Pacific) and Kimmeridgian *sensu anglico,* differed from Kimmeridgian *sensu gallico.* Recent studies have however largely clarified inter-provincial correlations and enabled Kimmeridgian *sensu anglico* to be correlated with European sequences (e.g. Cope and Zeiss, 1964; Cope, 1967; Casey, 1967, 1973; Enay, 1972a, b; 1973).

The currently accepted correlations equate Arkell's Middle and Upper Kimmeridgian with Lower Tithonian (e.g., Enay, 1972a, p. 385, footnote). Thus, Arkell's age for the Ohauan Stage of New Zealand (and the *Kossmatia* zone of the Heterian; cf. Arkell et al., 1957, p. L.323) is Early Tithonian in modern terms, using a two-fold division of the Tithonian (e.g. Enay, 1964, 1972a, b; 1973).

Enay (1972a, pp. 361, 380–1) maintains however that Early Tithonian is absent in the entire Western Pacific and regards the assemblages of *Kossmatia, Paraboliceras, Aulacosphinctoides,* etc., occurring in the Himalayan, Malayo-Pacific and Australasian regions as Late Tithonian. This opinion is also shared by Verma and Westermann (1973, pp. 209–211). Enay would therefore regard the Upper Heterian (i.e. *Kossmatia* zone), Ohauan and Puaroan of the New Zealand Jurassic as Upper Tithonian.

As the *Idoceras* zone of the Lower Heterian is presumably Kimmeridgian in the modern sense (e.g., Verma and Westermann, 1973) an apparent gap exists between the *Idoceras* and *Kossmatia* zones in New Zealand, spanning at least Early Tithonian time (and an as yet unknown portion of Kimmeridgian time). However, as overlap exists between the *Idoceras* and *Kossmatia* zones at Kawhia (Fleming and Kear, 1960, pp. 28–30) the physical presence of the gap may be questionable.

Clearly, the relationship of the Heterian, Ohauan and Puaroan stages to the international divisions requires further study.

Oteke Series

Puaroan Stage. Definition of the base of the Puaroan is problematic (see Stevens, 1974a, p. 744). Purser (1961), working in the Port Waikato region, distinguished two zones in the Puaroan; a lower zone, Puaroan A, comprising beds between the first appearance of *Buchia hochstetteri* Fleming and that of *Buchia plicata*, and Puaroan B, comprising the remainder of the marine sequence above the first appearance of *B. plicata.* However, beds that might be expected to contain *B. hochstetteri* are not exposed at Kawhia, and Challinor (1970, 1974, 1975, 1977) has redefined the base of the stage using a belemnite zonation.

The top of the Puaroan Stage has not yet been defined by fossils. At Kawhia Harbour the upper part of the Puti Siltstone is unfossiliferous, and no higher formation is preserved (Fig. 20). But to the north correlatives of it are known and these contain *Buchia hochstetteri,* and are overlain by conglomerate with *Buchia plicata* (Fleming and Kear, 1960, p. 44; Player, 1958). In this way correlation has been made with Purser's Puaroan A (with *Buchia hochstetteri*) and Puaroan B (with *Buchia plicata*).

In the Port Waikato region beds assigned to Puaroan B are overlain by non-marine plant-bearing measures (Huriwai Formation; Purser, 1961, p. 10) (Fig. 18). The plant fossils found in the Huriwai Formation, principally *Cladophlebis australis* (Morris) and *Taeniopteris arctica* Heer, were originally given a Neocomian (Early Cretaceous) age by Arber (1917) but a Late Jurassic (Edwards, 1934; McQueen, 1955; Purser, 1961) or Tithonian–Neocomian (Harris, 1962) age is favoured by later workers. Recent work by Norris (1964, 1968, pp. 341–343) on the spores and pollen of Puaroan B and Huriwai Formation has revealed the presence of two microfloral assemblage zones in the Huriwai Formation that are quite distinct from two other zones in the Puaroan B beds. Norris, considered the Huriwai microflora to be post-Early Tithonian, and pre-Berriasian, with the implication that they are of Middle and perhaps Late Tithonian age. Microfloras from the highest part of the formation have several species in common with microfloras from the Australian basal Cretaceous (Dettman, 1963), but also contain species which indicate that the microfloras are older than Cretaceous.

Recent field parties have found that marine beds with *Buchia plicata* are interbedded with plant beds at the base of the Huriwai Formation in at least three places in the Port Waikato area. Thus modern practice is to place the Huriwai Formation within the Puaroan Stage (cf. Marwick, 1953a, p. 30; Fleming, 1960, p. 267; Kear, 1966; Schofield, 1967). In the Port Waikato region, *Buchia* aff. *misolica* (Krumbeck) forms a very distinctive zone between *Malayomaorica malayomaorica* and *Buchia hochstetteri.*

The Puaroan ammonites, *Buchia* and belemnites are particularly useful for internal correlation and the overseas affinities of all three help with external correlation of the stage.

The ammonites from the Puti Siltstone have been accepted by all workers as Lower Tithonian since it was suggested by Boehm (1911) and Spath (1923, pp. 304–5). Arkell (in Fleming and Kear, 1960, pp. 44–5) showed that the Puti Siltstone is characterized by an *Aulacosphinctoides* zone and suggested that the Middle Kimmeridgian and Early Tithonian boundary is between this zone and the preceding *Paraboliceras* zone (Kinohaku Siltstone, Ohauan Stage). In terms of Enay's (1972a, b) understanding of the Tithonian Stage, however, the Puaroan would be regarded as entirely Late Tithonian.

The overseas relationships of New Zealand *Buchia* have been summarized by Fleming (1958, 1959). The Puaroan *Buchia* are closely related to those of Indonesia, Australia and India, although only in India is the stratigraphy well enough defined to serve as a check on the age of the New Zealand *Buchia.* There, the *hochstetteri–plicata* zones of New Zealand appear to be equivalent to part of the Chidamu Beds (Middle Spiti Shales) which contain species of *Aulacosphinctoides* and *Uhligites* similar to those of the Puaroan.

Within the Waiharekeke Conglomerate (Fig. 20) the belemnite assemblage changes from dominant *Belemnopsis spathi* in the lower part to *Hibolites arkelli* Stevens near the top. Fleming and Kear (1960, p. 38) have suggested that the base of the Tithonian may lie within this formation The *Hibolites* of the Puaroan are closely related to those of Indonesia and Madagascar (Stevens, 1965).

Some 700 m of non-marine beds (Huriwai Formation) overlie the highest horizon of marine fossils (*Buchia plicata*) at Port Waikato, and Norris (1964, 1968) has recognised two microfossil assemblage zones in the Huriwai that are of post-Early Tithonian, pre-Berriasian age. If the Puaroan, Ohauan and Upper Heterian Stages in the New Zealand sequence are all of Late Tithonian age, as Enay (1972a, b) believes, the Huriwai Formation must be approaching the Jurassic–Cretaceous boundary very closely (if not actually crossing) (cf. Norris and Waterhouse, 1970; Waterhouse and Norris, 1972).

Regional stratigraphy

Southwest Auckland

The Jurassic rocks of southwest Auckland, over 7600 m thick, are exposed in the Kawhia Regional Syncline, west of the Waipa Fault (Figs. 17–20) and extend some 145 km from Waikato Heads to Awakino. Thirty kilometres south of the outcropping Jurassic, siltstone with Jurassic Foraminifera was obtained from beneath Tertiary rocks at a depth of about 335 m in an oil prospecting well (Hornibrook, 1953).

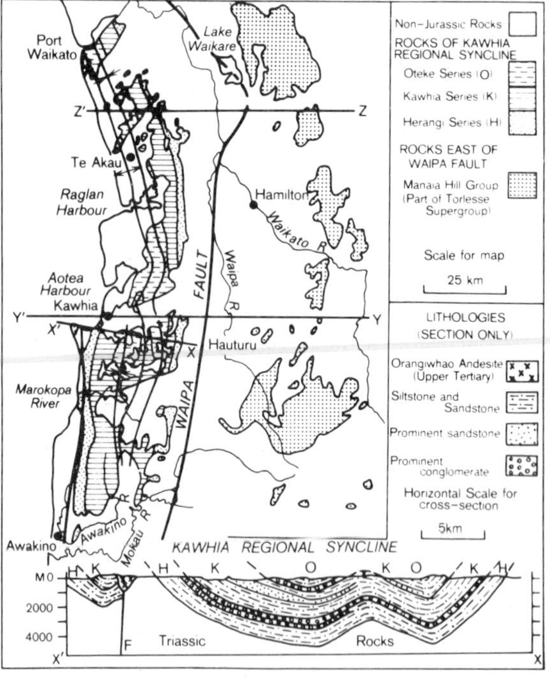

Fig. 17. Map showing distribution of Jurassic rocks in southwest Auckland and a geological cross-section along the line X^1–X. For cross-sections along the lines Y^1–Y and Z^1–Z see Fig. 21. Stratigraphic columns for Port Waikato, Te Akau, Kawhia, Hauturu and Awakino are shown in Fig. 18. The Kawhia Regional Syncline is the major structural feature, with minor medial folds. (After Suggate et al., 1977, fig. 4.64.)

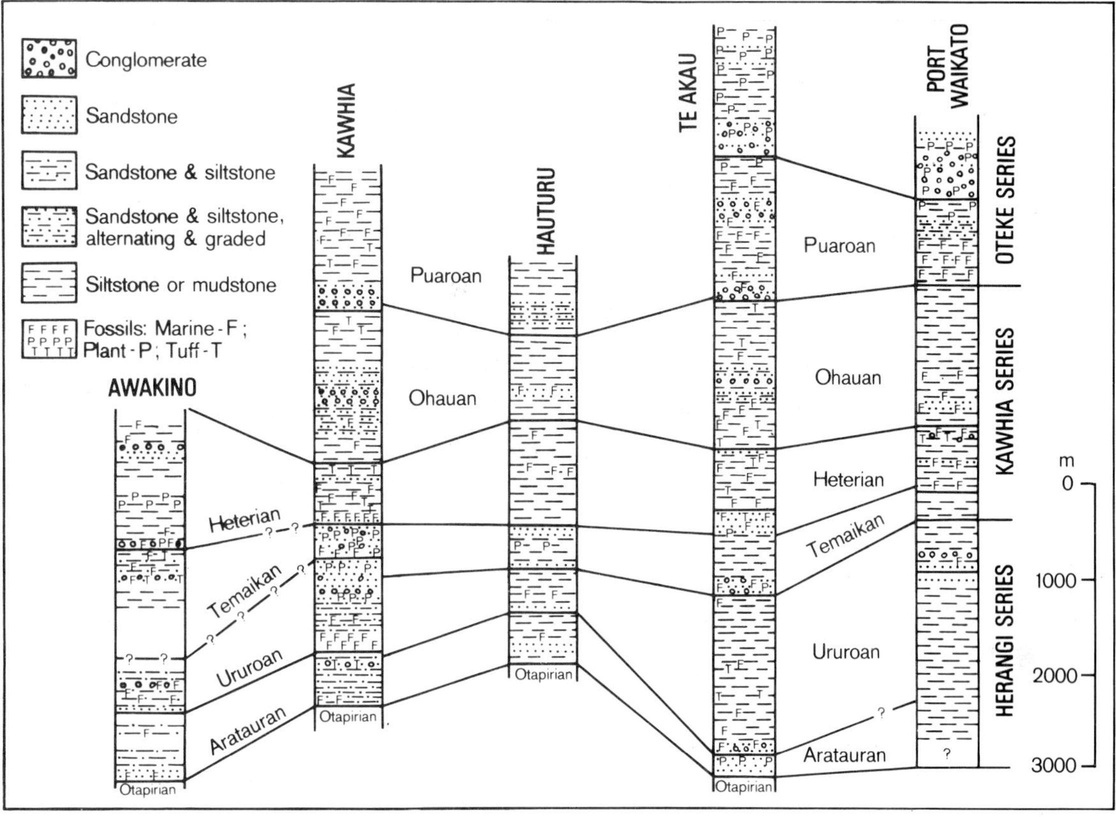

Fig. 18. Generalized stratigraphic columns for the Jurassic of the Kawhia Regional Syncline, southwest Auckland. (cf. Fig. 17). (After Suggate et al., 1977, fig. 4.65.)

The Jurassic rocks are mostly fossiliferous, tuffaceous, shelf-facies siltstone, but sandstone and conglomerate locally form prominent and continuous strike ridges. The latter facilitate formation mapping, and detailed sequences have been measured at Kawhia, Te Akau and Waikato Heads. Subdivision of the New Zealand Jurassic has been based on fossils from the sequence at Kawhia (Fig. 20), as discussed above; a detailed description of the section is given by Fleming and Kear (1960) and Kear and Fleming (1976). Martin (1975) has recently studied in detail the lower part of the section.

The section at Waikato Heads (Figs. 17–18) is important as it preserves details of the uppermost Jurassic, lacking at Kawhia (Purser, 1961; Challinor, 1974).

The Kawhia Regional Syncline has Jurassic beds in its core and Triassic beds on both flanks (Fleming and Kear, 1960, p. 16); but it is more than a simple syncline, since it contains important minor folds (Fig. 21). The beds of the main east limb dip more steeply than those of the west by 10°–15°, typical values for both being 45°–16°, but dips locally reach vertical. The central fold, the Kawaroa Anticline, has been mapped in detail from Waikato South Head to well south of the Marokopa River, as have the minor Toe and Kaimango Synclines on its east and west sides, respectively. Together they make up the Kawhia Regional Syncline.

The structures plunge north (where progressively younger formations appear) at about 2° from Marokopa to the Te Akau District (Kear, 1966), and then the plunge reverses to south, again at 2°, at Port Waikato (Purser, 1961, p. 21).

Facies and thickness changes across the Kawhia Regional Syncline are shown in Fig. 21, which also shows the relationship of these rocks of the Oparau Facies to the off-shore Manaia and Hunua Facies of the Jurassic rocks of the Torlesse Supergroup (Kear, 1971).

Fig. 19. Albatross Point, immediately to the south of the entrance to Kawhia Harbour. The exposures consist of sandstone, conglomerate, carbonaceous siltstone and shale of the Kaiate Formation, Rengarenga Group, of Temaikan age (Martin, 1975). (Photo: D.L. Homer, New Zealand Geological Survey.)

The Oparau Facies was deposited in an open-water continental shelf environment in shallow to moderate depths, with recurring acidic volcanic ash fall. It is separated from the Manaia Hill and Hunua Facies in the east by the Waipa Fault (Kear, 1960). The latter rocks are of the same age as those of the Oparau Facies, but are much thicker, less fossiliferous, significantly more sandy, and were apparently deposited beyond the foot of the continental slope (Kear, 1971). The present Waipa Fault contact is therefore thought to correspond to the Jurassic continental slope, probably steepened by recurrent fault movement, and was the site of serpentinite injection (Kear, 1960) (Fig. 22).

Mid-Canterbury

Carbonaceous conglomerate, sandstone and shale containing a Jurassic flora crop out at a number of localities in the eastern foothills of Canterbury between the Waimakariri and Rangitata Rivers (Fig. 15). These rocks differ markedly in lithology, induration and structure from the neighbouring Torlesse Supergroup greywacke and argillite, and are thought to overlie them unconformably. The floras are dominated by *Cladophlebis* and *Coniopteris* with *Taeniopteris, Elatocladus, Dicroidium,* (*Thinnfeldia* of Arber, 1917). Arber suggested a Rhaetic or Early Jurassic age for both collections; Edwards (1934) re-examined the Malvern Hills plants, and concluded that "the various Jurassic floras of New Zealand . . . are probably not earlier than Middle Jurassic".

In North Canterbury rocks containing Jurassic marine faunas and spore-bearing carbonaceous beds

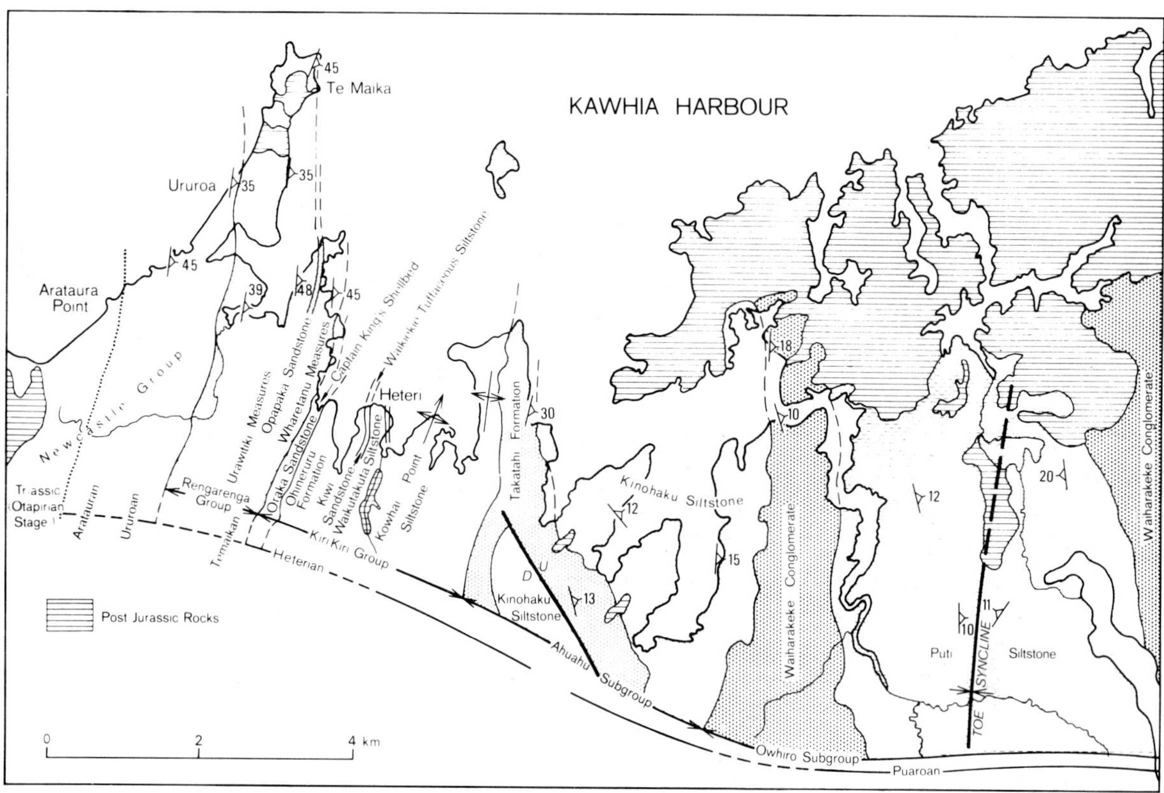

Fig. 20. Map of the Jurassic sediments on the south side of Kawhia Harbour, showing the boundaries of formations and stages. (After Suggate et al., 1977, fig. 4.66.)

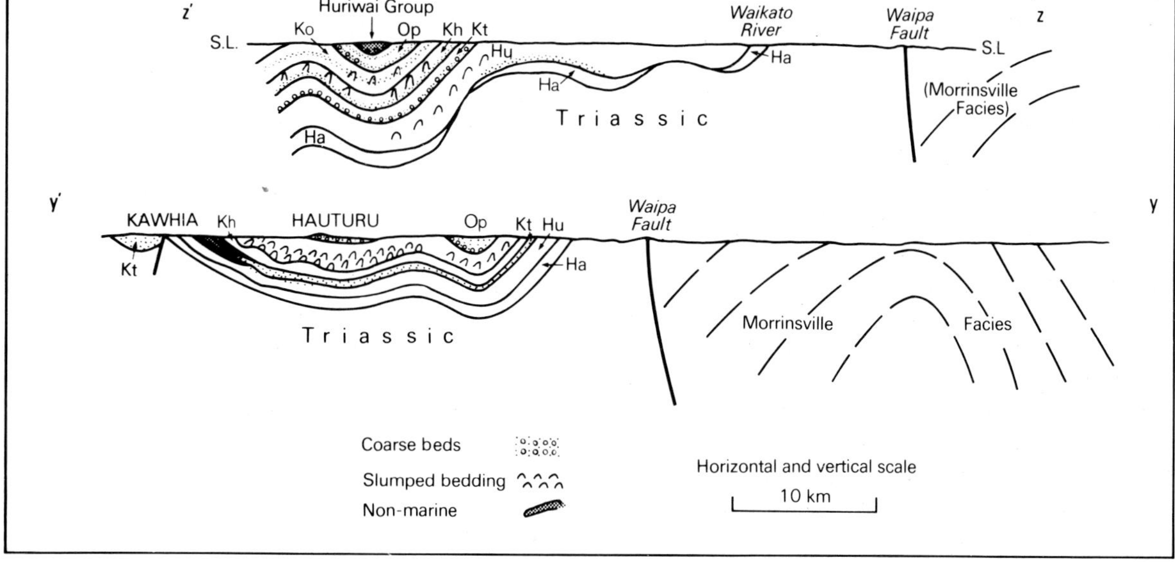

Fig. 21. Diagrammatic sections to illustrate facies and thicknesses in the Jurassic of southwest Auckland. An increase in coarseness towards the west has been slightly exaggerated. For location of section lines see Fig. 17. (After Suggate et al., 1977, fig. 4.78.)

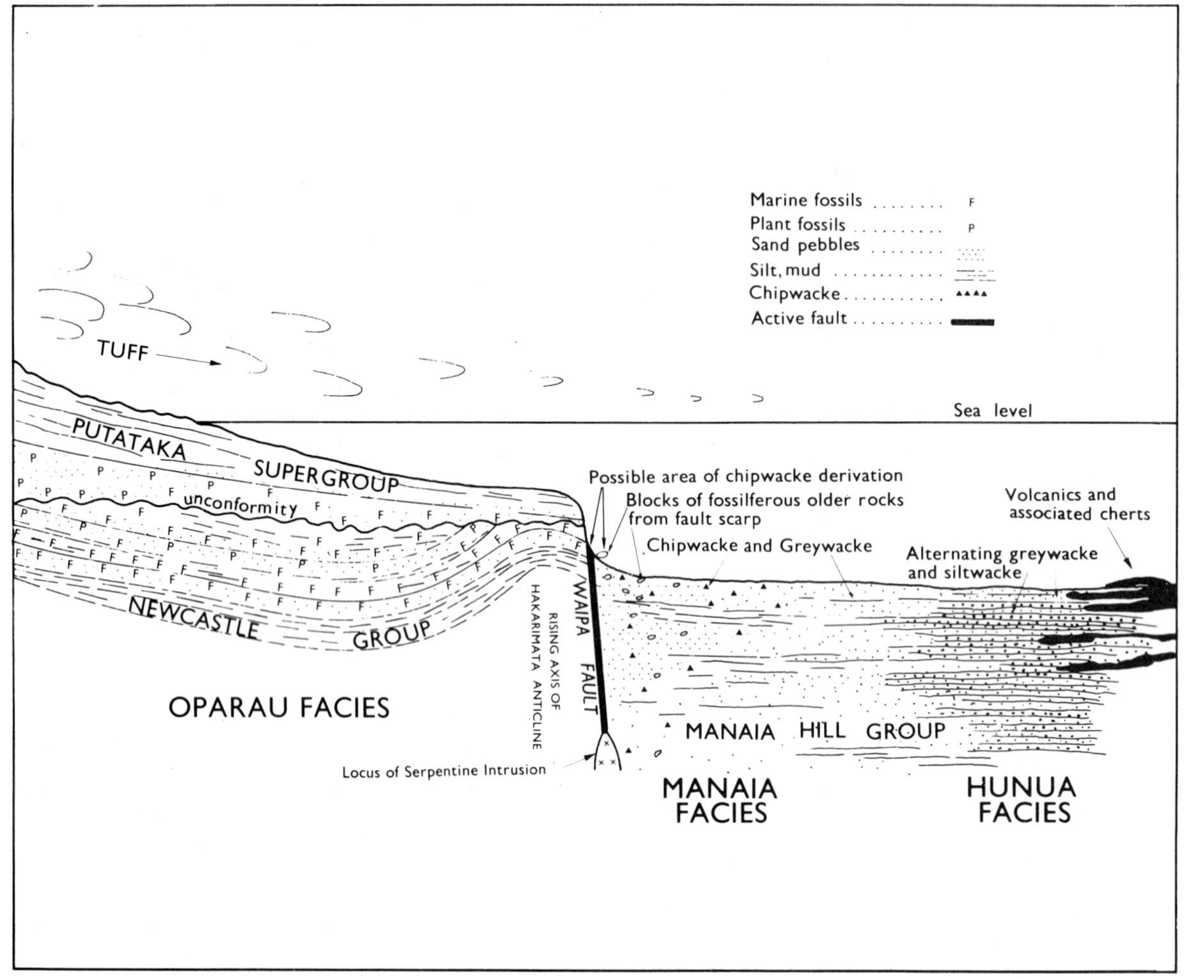

Fig. 22. Cross-section interpretation of Late Jurassic palaeogeography in southwest Auckland. (After Kear, 1967.)

are widely distributed in greywacke–argillite sequences of the Torlesse Supergroup (q.v.)

Northwest Otago

An important isolated area of Upper Jurassic rocks lies near the Pyke River, a few kilometres east of the Alpine Fault (McKellar et al., 1962; see Fig. 15). The nearest fossiliferous marine rocks of similar age are 420 km to the northeast (in north Canterbury), where similar belemnites have been found in the Torlesse Supergroup. In south Otago, 240 km to the southeast, no marine Upper Jurassic is preserved, the highest Jurassic beds apart from non-marine conglomerate and sandstone being Callovian (Upper Temaikan; Speden, 1961, 1971). The Pyke River occurrences may be interpreted to mean that Upper Jurassic rocks were more widespread in the past in the South Island.

Southland and southeast Otago

In southeast Otago and Southland Jurassic rocks of the Murihiku Supergroup occupy the axial zone of the Southland Regional Syncline for 120 km, from the coast at False Islet northwest to Bastion Hill near the Oreti River (Figs. 23–25). Small areas crop out west of the lower Mataura River and west of the Oreti River at the northwest end of the regional syncline. On the north limb of the regional syncline the Jurassic forms a narrow strip 3–5 km wide of steeply dipping strata, although small folds broaden the outcrop in the vicinity of the Mataura River and near Kuriwao (south of Clinton). The extensive area of Jurassic on the south limb is due to several double-pitching secondary folds, which rarely have dips exceeding 50° (Speden, 1959). From a "high" in the vicinity of the upper Waipahi River the axis of the

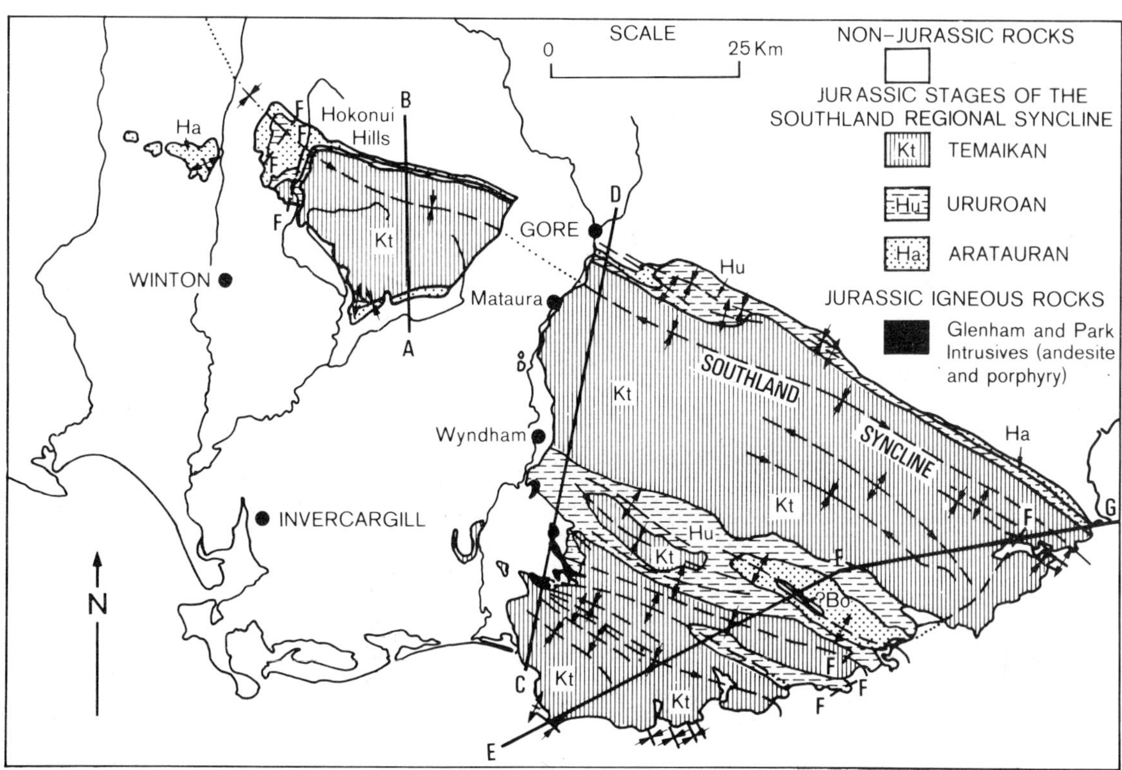

Fig. 23. Distribution of Jurassic stages in the Southland Regional Syncline. Major folds are shown. See Fig. 24 for cross-sections *A–B, C–D, E–F–G.* (After Suggate et al., 1977, fig. 4.80.)

regional syncline pitches gently southeast towards the coast and northwest to the Mataura River. The association of interbedded lithologies and structure results in prominent steep strike ridges on the steeply dipping north limb and extensive inclined flat surfaces and escarpments on the gently folded south limb of the regional syncline (Fig. 25).

On both limbs of the regional syncline the Jurassic usually conformably overlies Otapirian Stage (Triassic) sediments, although local unconformities sometimes cut out the Aratauran Stage.

The youngest Jurassic marine sediments are Late Temaikan (Callovian) in age. These are exposed near the coast, where they are conformably succeeded by non-marine beds of uncertain but probable latest Temaikan age.

Characteristic sediments of the regional syncline are sandstone, siltstone and conglomerate. Sequences of thin-bedded alternating sandstone and siltstone occur, but generally either sandstone or siltstone predominate. Conglomerates are of variable thickness and tend to be discontinuous laterally (Wood, 1956). Carbonaceous and plant beds are common, particularly in sandstone sequences characteristic of the Temaikan Stage, and in many cases indicate terrestrial deposition. Variations in grain size, sedimentary structures and the presence or absence of marine fossils and plant beds indicate a diversity of paleoenvironments, especially in the sandstone sequences, but these are poorly known. The sequence clearly represents shelf deposition, mostly inner and marginal shelf.

At the coast, on the north limb of the Southland Regional Syncline, the marine Lower and Middle Jurassic is some 4,000 m thick (Speden, 1961, 1971), compared with some 3,350 m about 65 km to the northwest at Gore (Wood, 1956), and 3,600 m 30 km further northwest in the Hokonui Hills (McKellar, 1969). On the south limb of the regional syncline, Speden (op. cit.) records 4,560 m at the coast, where the basal beds are not exposed. In the lower Mataura River, where the Jurassic rests on Permian, the thickness is about 3,600 m. The non-marine beds at the top of the sequence are at least 730 m thick in the Mataura

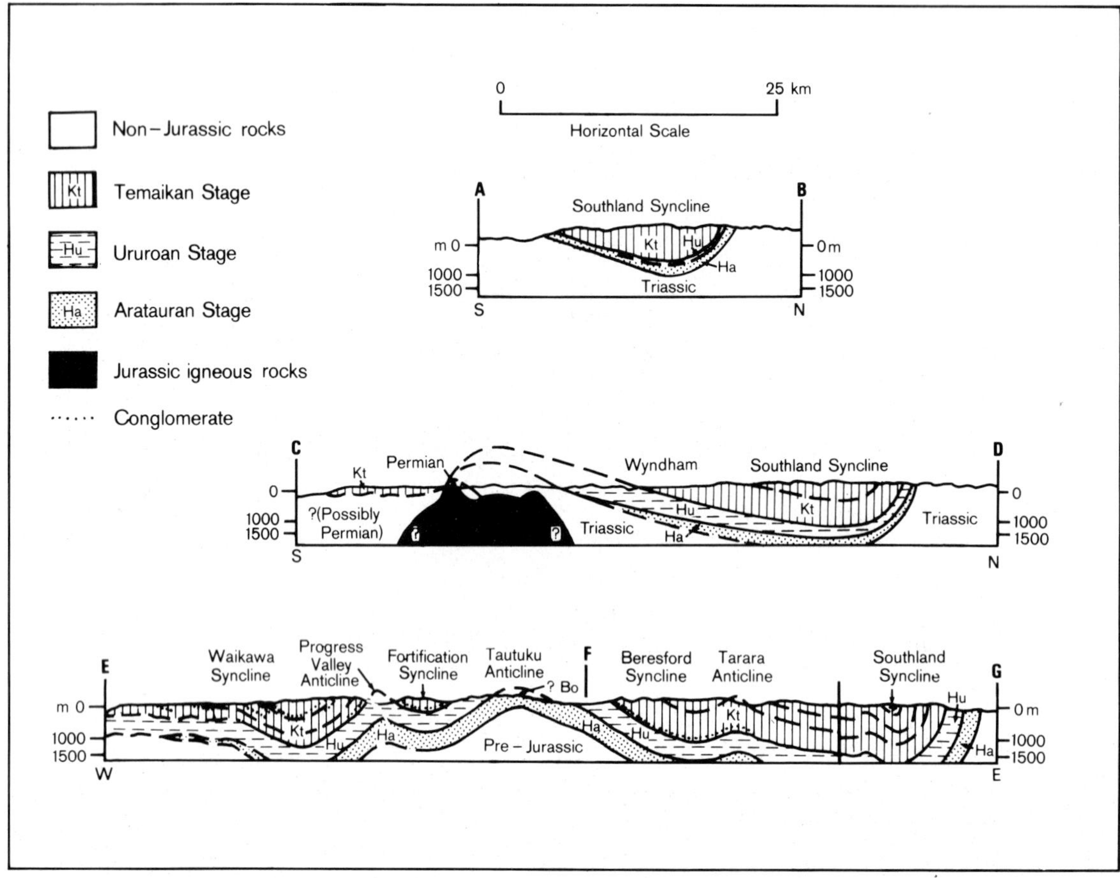

Fig. 24. Cross-sections through the Jurassic of the Southland Regional Syncline. See Fig. 23 for location of section lines. (After Suggate et al., 1977, fig. 4.81.)

region where the upper beds are not preserved. Thus, the thickest sequence is at the coast on the south limb of the regional syncline south of the major structural axis (Speden, 1961, 1971).

Because of relatively good exposures and common fossils the classic area for the Lower Jurassic of Southland is in the Hokonui Hills. Cox (1878) and McKay (1878) divided the Jurassic part of the Triassic–Jurassic sequence in the Hokonui Hills into four "Series" – Bastion, Flag Hill, Putataka and (at the top) Mataura, which is non-marine. Marwick (1953a) placed the faunas of the lowest three divisions into the standard sequence of New Zealand Jurassic stages, recognizing the Aratauran and Ururoan (Lower Jurassic) and Temaikan (Middle Jurassic). McKellar (1969) has remapped the Hokonui Hills and subdivided the sequence on the basis of Marwick's (1953a) stage units.

Non-marine beds, indicated by plant and carbonaceous beds, are present throughout the Jurassic of Southland and south Otago, but are particularly common in, and characteristic of, the Temaikan Stage. With a few exceptions they have been little studied or collected. Plant macrofossils from several localities were described by Arber (1913, 1917) and Edwards (1934). Edwards considered the floras to be "probably not earlier than Middle Jurassic" and gave ages generally younger than those allotted by Arber. All are now known to be probably Temaikan.

TORLESSE SUPERGROUP (G.R. STEVENS)

The name Torlesse for much of what are informally but commonly referred to by New Zealand geologists as "the greywackes" goes back to Haast

Fig. 25. Looking ENE along the axis of the Southland Regional Syncline in the vicinity of Dunvegan. Jurassic (Temaikan) strike ridges in foreground. Triassic in left background. (Photo: D.L. Homer, New Zealand Geological Survey.)

(1865) who was then mapping the largest continuous belt of these rocks, in Canterbury. After long disuse it was revived by Suggate (1961) and extended to cover all the "undifferentiated" Permian–Jurassic rocks of the South Island east of the main outcrop of the Haast Schist Group. Suggate gave it the status of a Group, but it embraces such a thickness of sediments, with significant local lithologic differences, that the higher-rank status of Supergroup (cf. Campbell and Coombs, 1966) has to follow greater subdivision and the inclusion of Groups within it.

The rocks comprise a "greywacke suite", predominantly of argillite, greywacke (used with a connotation of induration), spilitic and basaltic lava, tuff, jasper and chert. Limestone is rare and everywhere associated with volcanics. No country-wide study of Torlesse Supergroup sedimentation has been published, but local detailed structural (e.g., Lillie and Gunn, 1964; Mayer, 1968) and petrological (Reed, 1957; Skinner, 1972; Andrews, 1974) studies have been made.

A concept representing these rocks as a single sedimentary facies was embodied in the term "Alpine Facies", introduced by Wellman (1952), who noted "the structure is complex, fossils are rare . . .". Even with an age range of Late Paleozoic to Late Mesozoic, as documented by Campbell and Warren (1965) and Speden (1976b), lithologic distinctions are lacking or well enough known only to allow the broadest age subdivision on this basis. Wellman (1952) noted, with respect to the Mesozoic part, that "bedding is variable, massive sandstone bands . . . alternating with bands of

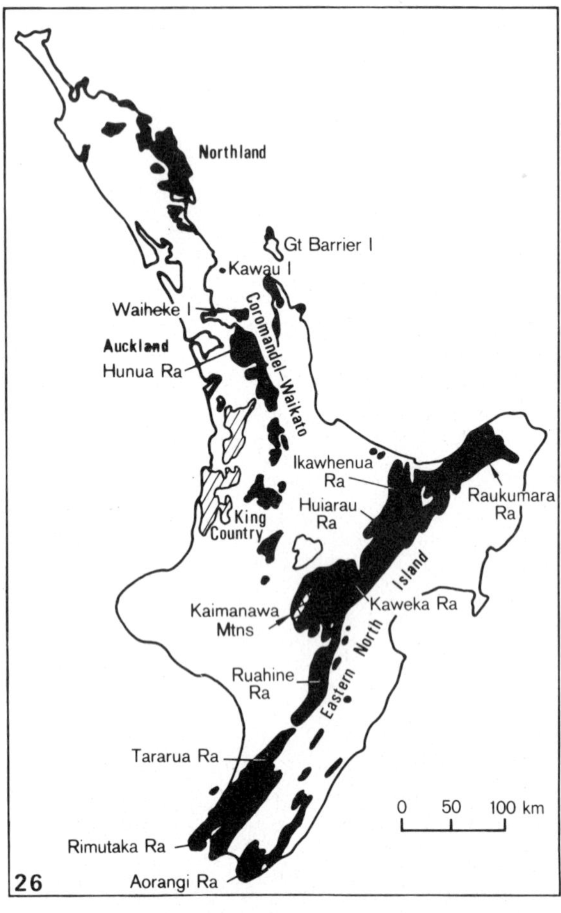

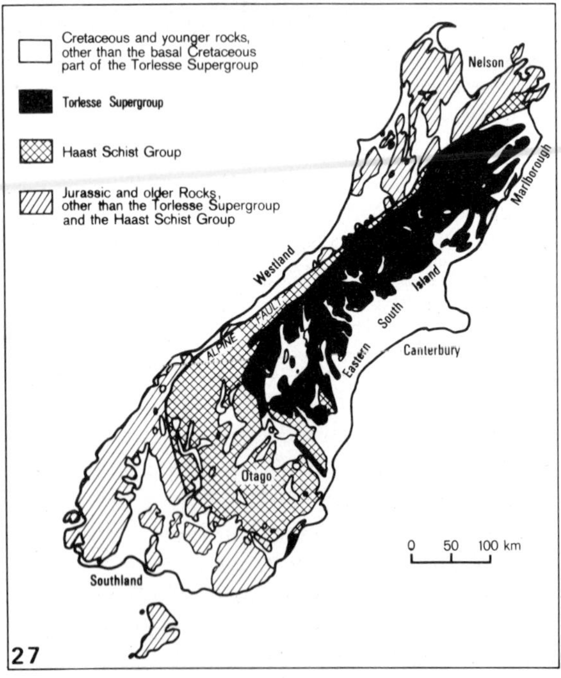

Figs. 26, 27. Maps of New Zealand showing distribution of Torlesse Supergroup rocks. (After Suggate et al., 1977, figs. 4.93, 4.94; see also Speden, 1972.)

sandstone and siltstone . . .. Thick beds of mudstone or siltstone are rare. Graded bedding is conspicuous in most of the alternating bands . . . Current bedding and ripple marks are absent. Many of the sandstone bands contain abundant fragments of darker argillite (and) often show small scale slump bedding". This description gives little impression of the overall induration and perhaps somewhat exaggerates the prevalence of graded bedding in alternating beds. Nevertheless it is apparent that there is much in common with deposits of supposed turbidity current deposition, and Wellman (1956) used the expression "Alpine Facies (greywacke and thin interbedded layers of spilitic basalt) of redeposited sediments". Fleming (1962), however, preferred the term "axial", but this implies knowledge of the position of the axis (presumably the most rapidly-subsiding part) of the New Zealand Geosyncline.

Although the Torlesse rocks are all well indurated, they are not uniformly so. On the 1 : 250,000 maps of the N.Z. Geological Survey the boundary between the Torlesse Supergroup and the Haast Schist Group is placed, following Suggate (1961), at the incoming of incipient schistosity in greywacke, so that the boundary has to be arbitrary within a transitional sequence. Close to the boundary the rocks are highly indurated, with argillite tending to have a slaty cleavage and greywacke containing abundant quartz veins. These are some of the most resistant rocks of the Southern Alps, forming the main divide for much of its length. With decrease of induration, jointing, especially of the finer beds, is a most conspicuous feature, and is characteristic through to the least indurated rocks, which are commonly described as sandstone and siltstone rather than greywacke and argillite.

Increasing induration is not a simple factor of age, although only Permian and Triassic fossils are known close to the schist boundary, and the least indurated rocks are probably all of Jurassic or Early Cretaceous age (Figs. 26–28). The upper limit of the Torlesse Supergroup is most commonly set by a major unconformity beneath Upper Cretaceous or Tertiary sediments laid down in a totally different paralic shelf environment. In a few places in Canterbury, however, non-marine beds of probable Jurassic age are inferred

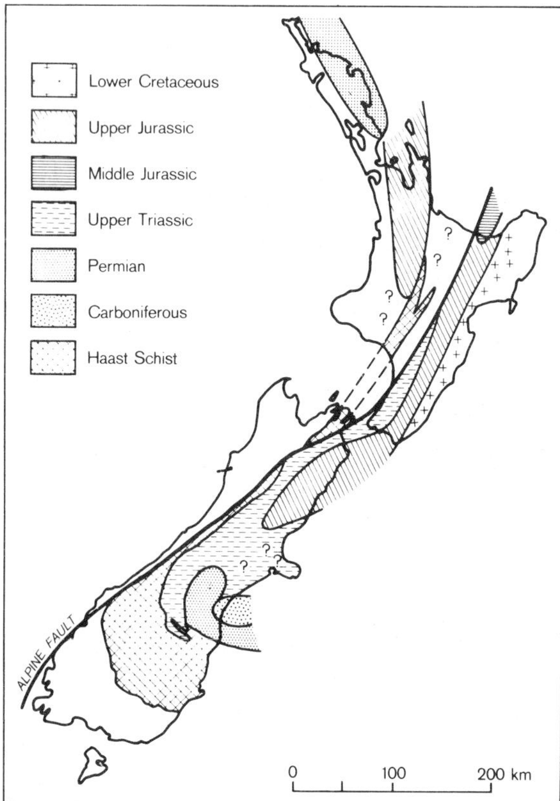

Fig. 28. Generalized distribution of dated fossils in the Torlesse Supergroup. (After Fleming, 1970, fig. 13; Suggate et al., 1977, fig. 4.95.)

to overlie unconformably (see p. 270), and have been excluded from the Torlesse Supergroup. The placing of the top in areas where continuity with marine Upper Cretaceous has been claimed, as in Marlborough (Wellman, 1955; Lensen, 1962) and the east coast of the North Island (cf. Wellman, 1959; Kingma, 1962, 1966, 1967) is more uncertain. The boundary in such areas is almost certainly diachronous, as is to be expected in an environment that was continuously marine but was changing in character from that of a geosyncline to that of a region of slow and intermittent subsidence. Although further work may result in changes, the oldest sediments in these areas to be excluded from the Torlesse Supergroup are of Korangan (Aptian) age. Even where the overall character of sedimentation did not change until later in the Cretaceous, for example in the Raukokere area of the Raukumara Peninsula and in Northland, fossils became more abundant and the strata are notably less deformed in the younger rocks. It is accordingly convenient not to describe rocks younger than Motuan as Torlesse Supergroup, with the Korangan, Urutawan and Motuan Stages represented in both Torlesse and other units.

Biostratigraphy

Most of the sediments of the Torlesse Supergroup, regardless of the mode of deposition within the New Zealand Geosyncline, accumulated in an environment generally unfavourable for bottom organisms, and so usually contain few traces of life. Rare incursions of shelf faunas are found, however, among the sparse assemblages that are preserved.

Six assemblages can be recognized. They are characterized by: (1) apparently deep-water benthic forms such as tube worms (*Terebellina, Titahia,* etc.); (2) planktonic forms (Radiolaria, microplankton, etc.) or forms with either long-lived planktonic larvae or an epiplanktonic or pseudoplanktonic habit (e.g., *Daonella, Halobia, Monotis, Buchia, Inoceramus*), which have fallen into the benthic zone; (3) the remains of free-swimming vertebrates that have fallen into the benthic zone; (4) rare transported fossils such as leaf and stem fragments and spores and fragments of the more robust invertebrate shells; (5) rare incursions of shelf faunas, perhaps related to local shallowing; and (6) reef faunas, inhabiting highs, often volcanic, on the sea floor.

Fossils from the South Island rocks of the Torlesse Supergroup are listed and discussed by Campbell and Warren (1965) and Speden (1975c). North Island localities are far fewer, but have yielded fossils of some ages not represented in the South Island (Fleming, 1970; Stevens, 1972; Speden, 1972, 1976b). The total age span of the Torlesse Supergroup is Carboniferous to Early Cretaceous.

A number of Mesozoic stages are not represented, or are poorly represented, by fossils in Torlesse rocks (Stevens, 1972), and although this may result from collection failure the possibility must be considered that genera with a planktonic or pseudoplanktonic habitat may have been absent or rare in the faunas of the times. This possibility is difficult to gauge as it presupposes knowledge of the ecology of the animals concerned. A number of pointers may be given, however. Genera are known from some stages (e.g., *Rhaetavicula* – Otapirian; *Meleagrinella* – Temaikan) that probably had some means of becoming widely dispersed (e.g., pelagic larvae, epiplanktonic or pseudoplanktonic habit), as they are cosmopolitan in distribution. But they have not as yet been found in Torlesse rocks. For other stages (e.g., Aratauran, Ururoan), however, the entire known fauna was probably neritic in habitat and therefore less likely to be preserved in Torlesse rocks.

Shallowing, possibly local, of a part of the New Zealand Geosyncline in the Middle Triassic is reflected in the Torlesse rocks of the Waitaki Valley, which contain a plentiful Kaihikuan fauna with bottom-dwelling forms. It is succeeded by a sequence with carbonaceous bands, and well-preserved plant remains, almost certainly not fully marine, forming the youngest beds locally preserved. A similar local shallowing of the geosyncline in the north Canterbury area in the Upper Jurassic may be reflected in the number and variety of fossils recorded from the Torlesse rocks of this area (Campbell and Warren, 1965).

The annelids *Terebellina* and *Titahia,* characteristic of the Torlesse rocks of a large part of eastern central Canterbury (Campbell and Warren, 1965) and Wellington Peninsula (Webby, 1967), are not known to occur in the western (?shelf) facies of the New Zealand Geosyncline and have been interpreted as deep-water benthic forms. Such fossils are known only from the Upper Triassic part of Torlesse Supergroup, and comparable deep-water benthic fossils are not known from the remainder.

Ammonoid and ammonite faunas are known from the western facies rocks of the Triassic and Jurassic but only a few fossils (*Anaptychus, Idoceras speighti* and phylloceratids) are known from the Torlesse Supergroup rocks of these areas; these have probably been transported. Thus it is likely that most of the New Zealand ammonites were probably necto-benthic in habitat (Allan, 1956, p. 376) with few truly planktonic types, and, furthermore, little oceanic drift of floating shells.

Few belemnites are found in Late Jurassic rocks of the Torlesse Supergroup, in contrast to their abundance in rocks of the same age in the western (shelf) facies. Few are known, also, from the Early Cretaceous rocks. While it is recognized that belemnites were probably capable of a free-swimming oceanic existence it is thought that as their prey was generally confined to the shelves, so they were confined. Some scattered occurrences in geosynclinal Torlesse rocks may result from transport of isolated guards.

Presumed pseudoplanktonic or epiplanktonic types (*Inoceramus, Aucellina,* etc.) are characteristic of the Cretaceous rocks of the Torlesse Group, but rare ammonites and belemnites appear in the late Early Cretaceous, perhaps reflecting the spread of shelf conditions. Sandstone with the type Korangan (Aptian) fauna which contains many undoubted neritic types (Fleming, in Wellman, 1959, p. 151) is excluded from the Torlesse Supergroup, but probable Korangan fossils are found within it in the Hawai River area of the Raukumara Peninsula. Comparably, some Urutawan and Motuan (Albian) fossils are found in Torlesse-like rocks in Marlborough.

Regional stratigraphy

Coromandel – Waikato – King Country

The Manaia Hill Group, a predominantly sandstone facies, embraces most of the Torlesse rocks of this region (Skinner, 1972). It forms the higher and more rugged country of a belt extending discontinuously from the Coromandel Peninsula (with the Manaia Hill type locality) for 320 km southwards to the Taumarunui District of the King Country (Fig. 26). Lithic–volcanic greywacke sandstone, commonly including dark siltwacke chips, constitutes the predominant lithology in western localities, but interbedded siltstone/argillite and conglomerate become increasingly prominent to the east. Bedding is typically poorly developed, except where argillite is interbedded, and large-scale spheroidal weathering is a characteristic feature. Roundstone conglomerate is perhaps the most striking rock of the group. The pebbles include acid, intermediate and basic volcanics, intermediate to acid plutonics (Great Barrier) and metamorphic and sedimentary rocks, some of which are fossiliferous: Jurassic fossils at Great Barrier Island and Kuaotunu (Puaroan) and at Manaia Hill (Heterian to Ohauan) (Stevens, 1970b); Jurassic (Heterian) and Triassic (Warepan and Otamitan–Oretian) fossils near Morrinsville (Kear, 1955, pp. 108–110; Kear and Schofield, 1965). Dark siltwacke (argillite) is less common than the coarser lithologies but is important locally, as at Torehape (western Hauraki Plains), where a 300 m thick member has been recognized. Prehnite, laumontite and analcime are commonly found near shear zones but the zeolites are more abundant in the east. The rocks range from the Zeolite to the lower Prehnite–Pumpellyite Metagreywacke Facies of Coombs (1960); pumpellyite itself has not been discovered.

The fossiliferous conglomerate boulders, the greywacke sandstone with siltwacke chips ("chipwacke") and other sedimentary features indicate that redeposition was of some importance in the formation of these beds, but that the site of deposition was on the continental slope (Kear, 1971; Skinner, 1972).

As well as the fossils found as pebbles, the Manaia Hill Group contains sparse but widely distributed *in situ* fossils, all being Late Jurassic.

Auckland – northwest Waikato

Torlesse rocks, in which a few Jurassic fossils have been found, occur on a number of islands near Auckland City – Waiheke, Motutapu, Motuihe, and Ponui – and in the Hunua and Hapuakohe Ranges, northwest Waikato (Fig. 26). The rocks have been subdivided by Schofield (1967, 1974) into the Waiheke and Manaia Hill Groups.

The Waiheke Group is characterized by a predominance of fine-grained rocks, argillite and siltstone, commonly black in colour; by the presence of the red-bed suite – radiolarian chert, spilite, red argillite and manganese ore; and by the absence of conglomerate and macrofossils (Schofield, 1974). Flysch-type sediments are not common, particularly in the upper part of the sequence. Although laumontite is relatively rare, prehnite is common and pumpellyite has been recorded (Reed, 1966).

The Manaia Hill Group in this region is characterized by thick unbedded greywacke sandstone, greywacke sandstone with siltstone chips ("chipwacke") and by thick sequences of flysch-type interbedded siltstone and sandstone (Schofield, 1974). The sandstones are highly cemented, but unlike those in the Waiheke Group, the finer-grained sediments crumble when exposed to air and many are lighter coloured. Macrofossils are generally uncommon and nowhere as common as within the rocks of the western facies of the Jurassic.

The significance of the Waipa Fault in separating the western and eastern facies of the New Zealand Geosyncline in the south Auckland region led Kear (1967, 1971) to propose that it marks the former position of the continental slope, and was indeed tectonically active in the Jurassic (Fig. 22).

Eastern North Island

The greywacke sandstone and argillite of the main North Island ranges are composed of mixed lithic–volcanic and granitic–feldspathic detritus, the former being dominant in the north and the latter in the south. Rhyolitic material is notable in the eastern facies. The ranges, which result from horizontal as well as vertical movements of the Late Cenozoic Kaikoura Orogeny, stretch from Wellington northeast almost to Cape Runaway, and separate the east coast region of predominantly Cretaceous and Tertiary rocks from the Taupo Volcanic Zone and Wanganui Depression to the west. Their total length is 510 km, and their width varies greatly, from 65 km in the Kaimanawa–Kaweka Mountains to as little as 8 km where the Ruahine and Tararua Ranges approach each other at the Manawatu Gorge.

The ranges fall into four groups: in the north the Huiarau–Ikawhenua–Raukumara Group, the central Ahimanawa–Kaweka–Kaimanawa Block, the southern Ruahine–Tararua–Rimutaka Ranges and to the southeast the Aorangi Mountains (Fig. 26). No sharp lines of demarcation separate the rocks within the groups.

The feldsphathic greywacke and argillite of the central and southern ranges is generally similar to that of Canterbury, whereas the lithic–volcanic greywacke of the northern ranges is similar to that typical of the Northland–Waikato region. There are notable differences in induration from place to place. Few fossils have been found, and all are of Mesozoic age (Speden, 1976b); it is possible, however, that the oldest rocks are as old as Permian.

Eastern South Island

About a quarter of the South Island is occupied by Torlesse rocks which predominate throughout much of southern Marlborough (Fig. 29) and inland Canterbury, and extend also into north Otago and Westland. The Torlesse Supergroup grades up from the least metamorphosed rocks ("semi-schists") of the Haast Schist Group, the change being arbitrarily placed at the transitional metamorphic boundary between Chlorite Subzone 1 and Chlorite Subzone 2. In many areas this boundary is parallel to bedding (and therefore time) planes, but this is certainly not so on a regional scale. In north Otago and south Canterbury the age of the Torlesse Supergroup extends back to at least the Permian, as judged by the presence in several places of *Atomodesma* fragments in non-schistose rocks, whereas in the Trent River area, northern Westland, rocks containing the Late

Fig. 29. View looking southwest along the Awatere Valley, Marlborough, northeastern South Island. The Awatere Fault, active in 1848, separates Cretaceous (left) from Torlesse Supergroup (right) (for other details see Stevens, 1974b, fig. 4.5).

Triassic bivalve *Monotis* are close to the local base of the Supergroup, while further south *Terebellina* is recorded from rocks of Chlorite Subzone 2 (Campbell and Warren, 1965).

Throughout north Otago and Canterbury (Fig. 27), the top of the Torlesse Supergroup is marked by a major unconformity between well-indurated geosynclinal sediments and markedly less indurated Cretaceous and Cenozoic rocks of near-shore or non-marine facies. Nowhere in these areas is there evidence that any of the Torlesse Supergroup rocks are younger than Jurassic, whereas the rocks above the unconformity are Late Cretaceous or younger, with the exception of small deposits of conglomerate and carbonaceous shale containing Jurassic plants in eastern Canterbury. In parts of Marlborough, however, no similar regional unconformity appears to be present. Southeast of the Awatere Fault there appears to be local transition from Torlesse Supergroup rocks dominated by poorly sorted, commonly thin-bedded greywacke and argillite into more massive beds of better sorted sandstone and siltstone, more or less calcareous and including relatively minor greywacke–argillite sequences.

Extensive tectonic slides have been revealed by recent detailed studies in Torlesse rocks of the South Island (Bradshaw, 1972, 1973; Andrews et al., 1974).

Source and depositional environment

The principal source of sediment was undoubtedly a granitic or granodioritic terrain, as is shown by the abundance of granitic material in the conglomerates, the predominance of quartz and feldspar grains in the finer sediments, and their bulk chemical composition (see, for example, Reed, 1959; and Mason, 1962). Except in some of the Permian rocks, the proportion of volcanic rock fragments is small, in marked contrast with that in the western shelf facies rocks of similar age. This seems to imply a difference in depositional situation between the rocks of the two facies of a more fundamental nature than mere distance from the margin of the geosyncline.

Many lines of evidence, but particularly the sparseness of fossils, the graded sediments, the penecontemporaneous conglomerates and the sedimentary structures evidently produced by strong spasmodic currents, combine to suggest that many of the Torlesse rocks were deposited by high density flows ("turbidity currents") from a postulated primary source elsewhere. It is not suggested that all sediment accumulated in this way, and there is little evidence of the depth of water. Some of the argillite beds and many of the roundstone conglomerate and carbonaceous bands may well be primary deposits, the latter having been formed during times when filling of the geosyncline had locally exceeded, for one or more of many possible causes, the rate of subsidence (Andrews, 1974).

Some Torlesse sediment may have been derived from a western crystalline source and transported across near-shore basinal areas to be deposited farther off-shore (Landis and Bishop, 1972). On the other hand, considerable evidence points to the existence of a landmass, lying east of the Torlesse terrain, which supplied sediment to the New Zealand Geosyncline during at least Late Triassic to Late Jurassic time (Bradshaw and Andrews, 1973). Bradshaw and Andrews (1973) and Blake et al. (1974) place the possible source area for the Torlesse sediment in the Marie Byrd Land–Jones Mountains area of West Antarctica and suggest they were subsequently rafted by sea-floor spreading into juxtaposition with the Murihiku sediments.

CRETACEOUS (I.G. SPEDEN)

Cretaceous rocks are distributed widely over New Zealand, and are known in offshore prospecting drill-holes and on the Chatham Islands (Figs. 30–31).

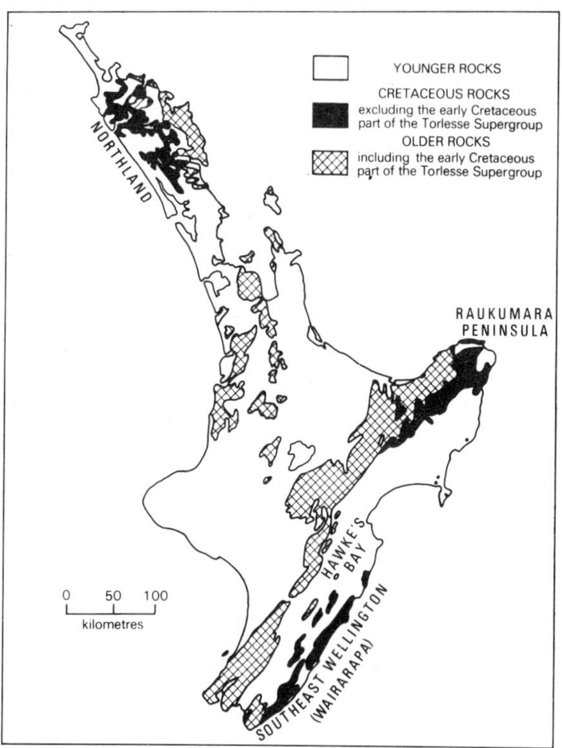

Fig. 30. Distribution of Cretaceous rocks in the North Island. (After Suggate et al., 1977, fig. 6.1.)

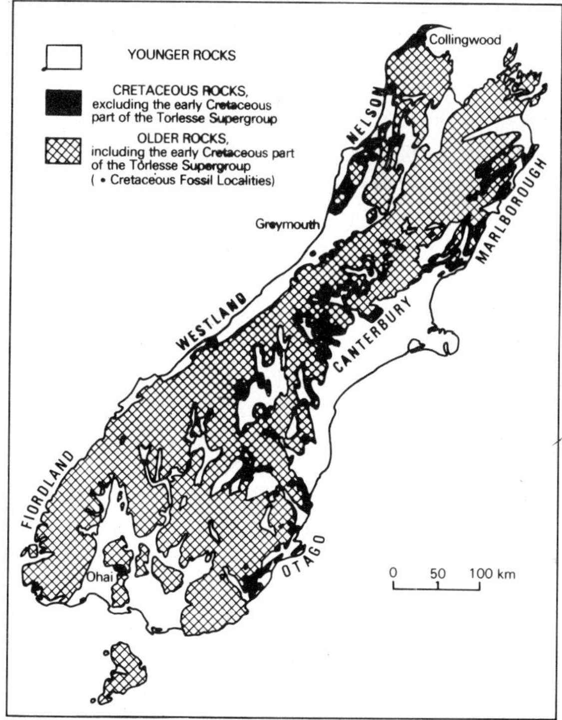

Fig. 31. Distribution of Cretaceous rocks in the South Island. (After Suggate et al., 1977, fig. 6.1.)

Three major types of sequences are known:

(1) Thick, non-marine breccia–conglomerate sandstone sequences, mainly derived from granitic and metamorphic terrains, were deposited in local fault-controlled basins in Nelson, Westland, Otago and Southland.

(2) Thick marine sequences dominated by sandstone, siltstone and thin-bedded sandstone–siltstone units are developed primarily in four regions – Northland, Raukumara Peninsula, Wairarapa and Marlborough.

(3) Thin non-marine and marine transgressive sequences of latest Cretaceous age are present, particularly on the east of the South Island, where steeply dipping and complexly folded rocks of the New Zealand Geosyncline had been exposed to subaerial erosion (Suggate, in Suggate et al., 1977).

Sequences of the third type overlap in space and time those characterized by thick marine and non-marine deposition, notably in southern Marlborough and east Otago. Geographic separation of areas of thick marine and non-marine deposition, with the latter being restricted to the west and south of the South Island, is one of the distinctive characteristics of the New Zealand Cretaceous.

The oldest known fossils are Late Aptian, which implies there is no proven stratigraphic record (but see p. 283) in New Zealand for some 25 m.y. (Neocomian–Early Aptian). Thus Early Cretaceous events are poorly known, and are generally attributed to the Rangitata Orogeny which had a profound effect on the geological development of the New Zealand region. Studies indicate local events, including metamorphism, several phases of folding (Brown, 1968; Andrews et al., 1974) and plutonism (Aronson, 1968; Harper and Landis, 1967; Waterhouse, 1975) were important. Radiometric dating (Sheppard et al., 1975) and geological studies (Suggate, 1963) suggest that sequences were uplifted into the zone of argon retention at varying rates and at different times. Much, if not all, of the metamorphism and plutonism may, therefore, have occurred at considerable depth.

Clearly Rangitata Orogeny involves a complex series of events including, in Westland, closely related tectonism, intrusion, volcanism and sedimentation near the end of the orogeny (Nathan, 1974b). These events have been explained recently by subduction along converging plate margins (Coombs et al., 1976). The influence of vertical and horizontal movement along major faults in latest and post-orogenic time is poorly understood, although undoubtedly widespread (Grindley in Suggate et al., 1977). Lack of major evidence for extensive planation and terrestrial sediments in the areas of deposition of thick marine sequences suggests that uplift above sea level did not occur in these areas. A possible explanation for the known stratigraphic relations and sediments (see p. 284) is their control by movement of major fault blocks, including tilting, with consequent changes in submarine topography and sedimentation as reported for the Cretaceous of east Greenland (Surlyk, 1975) and California (Bishop, 1970). Fault-controlled deposition and deformation may have been part of a gradational change from the tectonics of Rangitata Orogeny to the fault-dominated regime characteristic of the subsequent geologic history of the New Zealand region (Bishop, 1974; P.R. Moore, 1977).

The relation of marine fossiliferous Cretaceous sequences to underlying rocks in some areas is uncertain, particularly where the underlying rocks are unfossiliferous or sparsely fossiliferous sequences included in Torlesse Supergroup. In many parts of New Zealand a basal unconformity separates fossiliferous Cretaceous from older rocks. For example, Torlesse-like rocks of Late Jurassic–Early Cretaceous age are unconformably overlain by Koranga (Late Aptian) or Te Wera (Albian) formations in the Koranga–Motu District, Raukumara Peninsula (Speden, 1975a, 1977), and an unconformity is almost certainly present below the Whatarangi Formation (Albian) in the Aorangi Range, Wairarapa (Bates, 1969). In these areas the unconformity marks a significant event, or events, generally attributed to Rangitata Orogeny.

In Marlborough, Wairarapa and one area of Raukumara Peninsula where Cretaceous marine rocks are well developed there is a marked change within the Motuan from indurated, structurally complex, steeply dipping, sparsely fossiliferous Torlesse-like sequences to less indurated, structurally simpler, open-folded, concretionary fossiliferous sequences. Torlesse-like rocks which have Motuan fossils near the top are overlain unconformably by open-folded Motuan in the Waioeka Gorge, western Raukumara Peninsula (Speden, 1973). Recent mapping in the Wairarapa indicates that rocks previously mapped as Torlesse (Jurassic–?Early Cretaceous; Kingma, 1967; Johnston, 1975) are late Early Cretaceous (?Urutawan–Motuan), are separated from typical Torlesse by major faults, and overlain unconformably by less deformed Motuan which commences with breccia and conglomerate (Moore and Speden, in prep.; see p. 300). Similarly in Marlborough, Torlesse-like sequences with *Inoceramus* species of Urutawan–Motuan age are overlain unconformably by fossiliferous Motuan in the Awatere Valley (Allen, 1962; Speden, 1977), and probably also at Coverham where the presence of an unconformity was inferred by Thomson (1917) and Gair (1967), but Lensen (1962) and Hall (1963) considered the sequence conformable. In Northland the stratigraphic relationship of Torlesse-like rocks of Urutawan–Motuan age with younger Cretaceous rocks is unknown (see p. 293).

Two major conclusions can be drawn from the evidence. First, a major intra-Motuan event occurred over much of New Zealand. Its absence in some areas, notably in the Koranga–Motu district of western Raukumara Peninsula (Speden, 1976a), suggests complex depositional patterns and tectonic control. Second, environments of deposition similar to and perhaps the same as those of the New Zealand Geosyncline persisted at least locally into the Cretaceous, to Motuan time. Detailed mapping, sedimentary analysis and paleontological studies are required to determine whether deposition was continuous from Late Jurassic through the Early Cretaceous, and whether some of the Early Cretaceous sequences are best included in the Torlesse Supergroup of the New Zealand Geosyncline or should be classed in local units deposited in separate basins, as considered by Hay (1975) for Torlesse-like rocks in Northland.

The boundary between Cretaceous and Tertiary is placed between the Haumurian (Maastrichtian) and Teurian (Paleocene) Stages (Hornibrook, 1962; Webb, 1973a, b). The contact is best defined by foraminifera in Te Uri Stream, southern Hawkes Bay, where there is a disconformity between the stages. In other well-known sections, including Haumuri Bluff, the Cretaceous–Cenozoic boundary is unconformable, disconformable, faulted or difficult to locate because of poor microfaunas (Hornibrook, 1962; Webb, 1973a and b; Edwards, 1973). Continuity of deposition across the Cretaceous–Cenozoic boundary seems probable in some areas (e.g., Raukumara Peninsula and Wairarapa), but is yet to be demonstrated convincingly in the northeast of the South Island.

Other notable features of the Late Cretaceous in many areas were the latest Cretaceous–Early Tertiary peneplanation and complementary transgression (Suggate, 1950; Suggate in Suggate et al., 1977; Wilson, 1956). Neither peneplanation nor transgression was simple. Locally peneplanation may have occurred over a long time, and continued intermittently and concurrently with differential movements which resulted in minor unconformities and disconformities. It is possible that there was no period of overall tectonic calm or continuous subaerial erosion.

Associated with the Late Cretaceous transgression is the last major feature of the Cretaceous. In Haumurian (Maastrichtian) time began the widespread deposition over the east coast of the North and South Islands, and in Northland, of grey-black micaceous, spasmodically siliceous siltstone of the "Whangai facies". This distinctive facies locally contains glauconitic and quartz sandstones and sideritic concretions, and develops a jarositic efflorescence on weathering (Kingma, 1974). The facies is poor in macrofossils, except for thin tubes, possibly pogonophorid (M.R. Gregory, pers. comm.), and is characterized by arenaceous foraminiferal assemblages (Webb, 1966). Although siltstone beds resembling the Whangai appear as early as Late Teratan and occur in Piripauan–Haumurian alternating sequences, the widespread deposition of the facies generally started suddenly in the Haumurian and continued into the Cenozoic. Commencement of deposition was markedly diachronous regionally (see p. 301).

Introduction of the widespread Whangai facies reflects the culmination of a major change in the tectonic environment to relative stability associated with transgression and, possibly, regional planation (Fleming, 1970). This change appears to coincide with a diminution of fault movements which, nevertheless, continued throughout the latest Cretaceous and earliest Tertiary (Bowen, 1964b; Stoneley, 1968). The environment of deposition of the Whangai facies remains controversial (Webby, 1967). Deposition in deep water is preferred by some, whilst the intimate association with shallow shelf sediments in most areas favours deposition in specialized shelf environments, possibly behind largely submerged offshore barriers.

Not enough is known of the structure and stratigraphy of the Cretaceous in most areas to suggest reliable depositional or tectonic models. Locally, gross lithofacies patterns remained consistent through large segments of Cretaceous time. However, thickness of individual formations and stages varies rapidly and significantly, and is often associated with minor lithofacies and faunal changes. Penecontemporaneous movements on faults perhaps best explain observed marked differences, such as the sudden reduction in the thickness of the Motuan at Koranga (Speden, 1975a; P.R. Moore, 1977), and local coarse conglomerate and breccia. Sedimentation may have occurred in local troughs of deposition controlled by faulting (Austin et al., 1973), with renewal and reversal of movement and associated tilting affecting grain size, thickness and preservation of units (P.R. Moore, 1977).

Minor unconformities, patterns of grain size and thickness, and source directions indicate that much of the Cretaceous sediment was derived from a landmass to the west (Grindley, 1960; Speden, 1975a; Lensen in Suggate et al., 1977). Sources to the east and south are suggested locally as in Marlborough (Lensen, 1962, and in Suggate et al., 1977). Although facies patterns on a regional scale are at present inconclusive, preliminary studies of clay mineral and trace element distributions by geologists of New Zealand Aquitaine Petroleum Limited (Rumeau, 1965; Kingma, 1974), suggest that during the Late Cretaceous a component characterized by kaolinite and orthoclase was derived from an actively eroding landmass to the east. In contrast, sediment derived from the west is distinguished by illite, illite–montmorillonite and chlorite. A complex double-sided basin of deposition within the east coast fold and fault belt (Katz, 1974) is probable, as indicated on the paleogeographic maps of Stevens and Suggate (see Figs. 54–55). In view of the known instability of clay minerals during diagenesis, and after, further research is required to establish the proposed double-sided model.

A characteristic feature of the Cretaceous, especially in Raukumara Peninsula and Wairarapa, is the presence of conglomerates dominated by acid to intermediate igneous pebbles, mainly of volcanic composition (Johnston and Browne, 1973; Speden, 1975a, 1976a). The original source of these pebbles is obscure, but may have been from the presumed granitic terrain to the east.

The Cretaceous was a period of relatively intense igneous activity, especially in Northland and Marlborough. The ages of volcanic masses in Northland, generally included in the Tangihua Volcanic Group (p. 293), and the comparable Matakoa Volcanics at East Cape (p. 296), are uncertain. Elsewhere, two sets of igneous rocks are present; basaltic–spilitic flows, tuffs and cherts, in older Clarence (Albian) sequences, and younger doleritic sills, teschenites and basalt flows in Ngaterian to Haumurian (Late Albian–Maastrichtian) rocks. Challis (1968) classified the igneous rocks into a western belt characterized by high potash/soda ratios, and an eastern belt of low potash/soda ratios.

Biostratigraphy

Present subdivision of the Cretaceous is based on Wellman (1959) who recognized twelve stages grouped in four series (Fig. 32, Tables III and IV). Revisions of stages in the Clarence and Mata Series removed the Coverian from the sequence (Vella, 1961; Hall, 1963; Henderson, 1973) and transferred the Teurian Stage to the Cenozoic (Hornibrook,

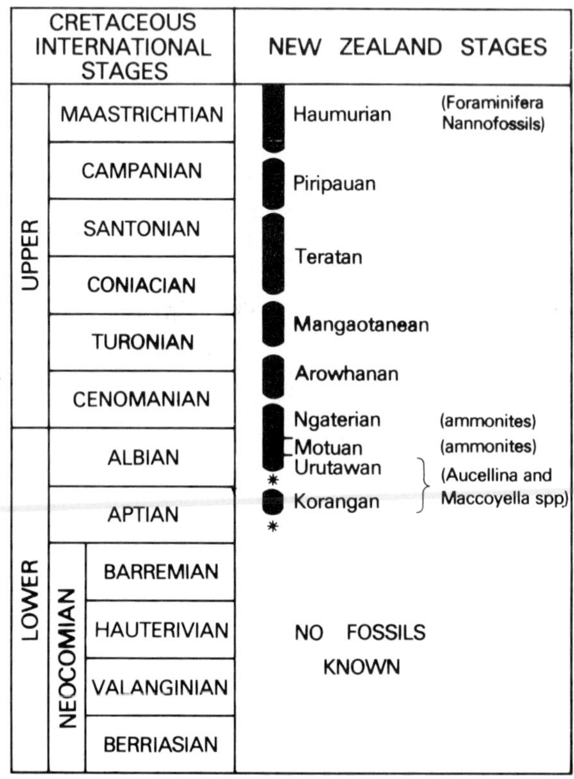

Fig. 32. Correlation chart of the New Zealand Cretaceous, showing levels at which firm correlations can be made with international divisions. The broken lines show levels in the New Zealand sequence at which, even though strata of equivalent age are present, no direct correlations with the international divisions are possible owing to the absence of fossils with overseas affinities. The asterisk indicates levels in the New Zealand sequence in which, due to unconformities, the exact extent of a time gap is uncertain.

TABLE III

History of classification and correlation of the New Zealand Cretaceous since Wellman (1959)

INTERNATIONAL STAGES	NEW ZEALAND SERIES AND STAGES: SERIES	STAGES: Wellman 1959	Vella 1961	Hall 1963	Henderson 1970, 1973	SERIES	STAGES: Speden (this paper)	Symbol	INTERNATIONAL STAGES
Maastrichtian	MATA	Haumurian		Haumurian	? Haumurian	MATA	Haumurian	Mh	Maastrichtian
Campanian	MATA	Piripauan		Piripauan	Haumurian / Piripauan	MATA	Piripauan	Mp	Campanian
Santonian	RAUKUMARA	Teratan / Mangaotanean		Teratan / Mangaotanean	Piripauan	RAUKUMARA	Teratan	Rt	Santonian
Coniacian	RAUKUMARA	Arowhanan		Mangaotanean / Arowhanan	Teratan	RAUKUMARA	Teratan	Rt	Coniacian
Turonian	CLARENCE	Ngaterian	Arowhanan	Arowhanan ?	Mangaotanean	RAUKUMARA	Mangaotanean	Rm	Turonian
Cenomanian	CLARENCE	Motuan / Urutawan	Ngaterian	Ngaterian	Arowhanan / Ngaterian	RAUKUMARA / CLARENCE	Arowhanan / Ngaterian	Ra / Cn	Cenomanian
Albian	CLARENCE	Coverian	Motuan / Urutawan	Motuan / Urutawan	Motuan / Urutawan	CLARENCE	Motuan / Urutawan ?- ?	Cm / Cu	Albian
Aptian	TAITAI	Korangan	Korangan	Korangan	Korangan	TAITAI	Korangan	Uk	Aptian
Neocomian	TAITAI	Mokoiwian	Mokoiwian	Mokoiwian	Mokoiwian	TAITAI	(no fossil evidence for older rocks)		Neocomian

TABLE IV

Characteristic species of the New Zealand Cretaceous stages

Series	Stage	Species	Species
UPPER CRETACEOUS	HAUMURIAN	*Inoceramus matotorus* *Dimitobelus hectori* *Anisomyon* sp. *Kossmaticeras bensoni* *Foraminifera*	*Pacitrigonia* aff. *hanetiana* *Pterotrigonia pseudocaudata* *Ostrea lapillicola* *Mixtipecten amuriensis* *Callistina wilckensi* *Lahillia* aff. *neozelanica* *Struthioptera novo-seelandica* *Rotularia ornata* *Baculites rectus* *Kossmaticeras* spp. and other ammonites
	PIRIPAUAN	*I. australis* *I. pacificus* *Dimitobelus lindsayi*	
	TERATAN	*Inoceramus nukeus* *I. opetius*	
	MANGAOTANEAN	*I. bicorrugatus*	
	AROWHANAN	*I. rangatira*	*Eselaevitrigonia meridiana* *Modiolus kaikourensis*
	NGATERIAN	*I. concentricus* *I. tawhanus* *I.* sp. ex gr. *fyfei– hakarius* *Scaphites equalis* *Worthoceras parvum* *Hypoturrilites varians* *Mariella (M.) thomsoni*	
LOWER CRETACEOUS	MOTUAN	*Aucellina euglypha* *Ptychoceras* sp.	*Inoceramus* sp. ex gr. *ipuanus–kapuus*
	URUTAWAN	*Aucellina* cf. *gryphaeoides* *Maccoyella* n.sp. A & B *Pseudolimea* cf. *echinata* *Parvamussium* sp.	
	KORANGAN	*Aucellina* cf. *radiatostriata* *Maccoyella* cf. *reflecta* *Pseudolimea* n.sp. A	*Spondylus* sp. *Camptonectes* sp.

1962). Speden (1976a, 1977) has shown the Mokoiwian Stage, previously considered to be the oldest Cretaceous stage, to be equivalent to some part of the Motuan and perhaps Late Urutawan (Albian). Thus, at present the New Zealand Cretaceous is divided into nine stages placed in four series.

The Jurassic–Cretaceous boundary is apparently nowhere recognizable, and the Cretaceous–Cenozoic boundary appears to be marked by stratigraphic breaks in well-known sections (p. 283). Faunal provincialism and a paucity of ammonoids make overseas correlation difficult and, at best, very approximate. Correlation with the international standard succession is reasonable at only four levels (see Fig. 32): Korangan Stage bivalves with the Late Aptian, Motuan ammonoids with the Late Albian, Upper Ngaterian ammonoids with the Early–Middle Cenomanian; and Haumurian foraminifera and nannofossils with the Maastrichtian (Henderson, 1973; Edwards, 1973; Speden, 1977). The Lower–Upper Cretaceous boundary falls within the Ngaterian Stage (Henderson, 1973). Otherwise, the position of the boundaries of the New Zealand stages relative to those of the international stages is uncertain; exact box-like equivalence is unlikely.

Macrofossils are generally sporadically distributed in Cretaceous rocks. When present, the fossils are usually epifaunal or endobyssate forms such as *Aucellina* and *Inoceramus*. Relatively diverse faunas from shallow-water sandstone facies are known locally from all stages except the Mangaotanean and Teratan which, however, have modest faunas of mostly undescribed gastropods in alternating sequences in southeastern Hawkes Bay (Kingma, 1971). Strong facies control of assemblages and species is demonstrated in some areas (Speden, 1975a; Warren and Speden, 1977). Ammonoids and belemnoids are rare except locally as at Coverham (Henderson, 1970, 1973). Only in the Late Cretaceous (Mata Series) are foraminifera abundant enough to be useful for subdivision (Webb, 1966, 1971, 1973a, b, and in Hornibrook, 1969). Good foraminiferal faunas are, however, found throughout the Cretaceous record (Stoneley, 1962) and, with pollen, spores and microplankton, are being actively studied. Early and Late Cretaceous reptiles are described by Fleming et al. (1971) and Welles and Gregg (1971), and annelids by Fleming (1971).

The general paucity of macrofossils, their frequent fragmentation, and the separation of zones by barren intervals prompted Wellman to define the Cretaceous stages in terms of the range zones of "key fossils", primarily species of *Inoceramus*. The "stages" are effectively biostratigraphic units, as recently emphasized by Carter (1974) who considered the stages were equivalent to Oppel-zones. Many of the New Zealand Cretaceous stages (e.g. Arowhanan) in fact represent the range-zone of single species, with the base being determined by the incoming of the index species and the top by the appearance in the same or a different section of the species diagnostic of the succeeding "stage". Other of the stages (e.g. Piripauan) are Oppel-zones.

Taitai Series (?Neocomian–Aptian)

Wellman (1959) used the name Taitai Series in a time-stratigraphic sense essentially for rocks of pre-Clarence Series age. He included in the series the Korangan Stage and the supposedly older rocks of the Mokoiwian Stage. The Mokoiwian was placed below the Korangan on the basis of a lithological correlation of the Taitai "sandstone", a formation conformably overlying Mokoiwi "siltstone" in Tapuaeroa Valley (Wellman, 1959), with the Koranga Sandstone some 80 km to the southwest, and equated by Wellman with Neocomian and Early Aptian, largely because of its presumed stratigraphic position.

Equivalence of the Mokoiwian Stage with part of the mid-Clarence Series (?Urutawan–Motuan; Speden, 1976a, 1977), leaves the series with only one formal stage, the Korangan. In spite of this, Speden (1977) has recommended that Taitai Series be retained for rocks of post-Oteke and pre-Clarence Series age. It must be emphasized however that there is at present no evidence for pre-Korangan (Neocomian) fossils in New Zealand, no convincing evidence for continuity of deposition, and no means of determining the lower limit of the Taitai Series or the upper limit of the underlying Oteke Series.

Korangan Stage (Late Aptian). The stage is based on fossils (Marwick, 1939) in the Koranga Sandstone which overlies with marked angular unconformity Torlesse-like rocks of Late Jurassic–Early Cretaceous age at Koranga, Raukumara Peninsula (Speden, 1975a). The stratotype is unconformably overlain by Te Wera Sandstone (Urutawan Stage). These unconformities prevent full definition of the stage.

A diverse faunal assemblage of at least 31 species, mostly epifaunal, occurs mainly in or adjacent to conglomerate-breccia bands. Characteristic fossils are *Aucellina* cf. *radiatostriata* Bonarelli, *Maccoyella* cf. *reflecta* (Moore), *Trigonia* (*T.*) n. sp. cf. *tenuis* Kitchin, *Pseudolimea* sp. A and *Camptonectes* sp. (Speden, 1976a). Absence of identifiable specimens of *Inoceramus* and belemnites prevents full integration of the local sequence of zones.

Correlation with the Late Aptian of Australia (Day, 1969) is based on the presence of apparently identical species of *Aucellina* and *Maccoyella* (Day and Speden, 1968; Speden, 1975a). Support for an Aptian age is given by the age of *A. radiatostriata* Bonarelli (Aptian: Cox, 1953) in South America and of closely similar Boreal species *A. pavlowi* Sokolov and *A. aptiensis* (d'Orbigny).

Clarence Series (Albian–Cenomanian)

The stratotypes of the three stages of the Clarence Series conformably follow each other in the section at Motu Falls (Fig. 33), Raukumara Peninsula (Wellman, 1959). A sequence of moderately indurated silt-

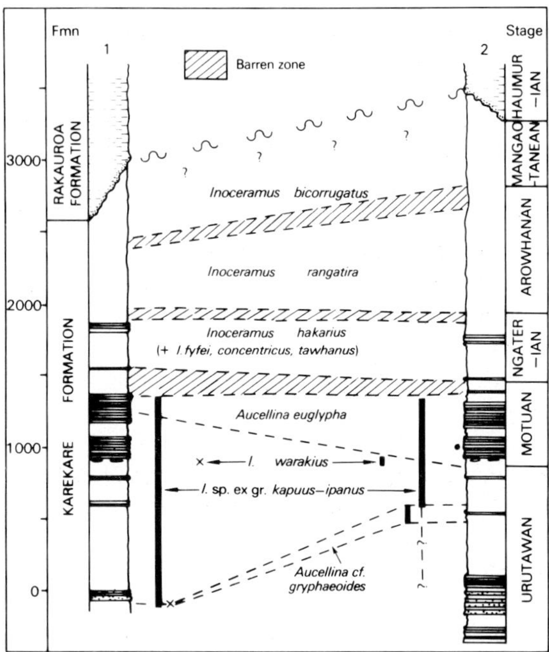

Fig. 33. Range-zones of index species of macrofossils in the Cretaceous at Motu Falls, Raukumara Peninsula. (After Speden, in prep.) *1* = Te Waka Stream section; *2* = Motu River–Waitangirua Stream–Koura Stream composite section. The lower part of this section is equivalent to the type section of the Clarence Series of Wellman (1959).

stone with concretionary layers and units of alternating sandstone and siltstone less than 200 m thick is exposed on the eastern overturned limb of an anticline. Conformably below the siltstone in the core of the anticline is a sandstone-dominated unit, considered to be Korangan by Wellman. Speden (in prep.) has shown the oldest rocks are Urutawan, and that the top of the Ngaterian is poorly exposed and disrupted by faults.

Recognition of the three stages of the Clarence Series is based largely on species of *Inoceramus* (Wellman, 1959; Henderson, 1973). Morphological variation of *I. kapuus* Wellman, the index species of the Urutawan, and *I. ipuanus* Wellman, the index species of the Motuan, is such that specimens typical of each species range through both stages. The species are part of a lineage which includes *I. urius* Wellman. Instead, species of *Aucellina* are used to distinguish the stages as the range-zone of *A.* cf. *gryphaeoides* (Sowerby) consistently underlies the range-zone of *A. euglypha* Woods in the type area and elsewhere (Speden, in prep.).

The Ngaterian index species are distinctive and fall into two groups which possibly represent lineages: (1) The species group of *I. concentricus* Parkinson and *I. tawhanus* Wellman, of moderately to strongly inflated, subequivalve to inequivalve specimens; and (2) the species group of *I. hakarius* Wellman and *I. fyfei* Wellman, of weakly to moderately inflated, equivalve to subequivalve specimens.

Urutawan Stage (Early–Middle Albian). As originally defined by Wellman (1959) the stage comprises the dominantly siltstone beds in the range zone of *I. kapuus* (Fig. 33). Underlying sparsely fossiliferous beds with *Oculina? nefrens* Squires and *Inoceramus* sp. indet. were considered to be possibly Coverian. Discovery of *Aucellina* cf. *gryphaeoides* in the sandstone sequence demonstrates that the basal limit of the stage in its type section is unknown (Speden, in prep.). The top of the stage is fixed by the incoming of the Motuan *A. euglypha.*

At Koranga the Urutawan is represented by the Te Wera Sandstone and the lower 30 m of the Karekare Siltstone (Speden, 1975a). As an unconformity is present below the Te Wera Formation (Speden, 1975a; P.R. Moore, 1977) no base is known to the stage at Koranga. Diverse assemblages dominated by epifaunal species, including two diagnostic species, *Aucellina* cf. *gryphaeoides* and *Maccoyella* n. sp. A, characteristic of moderately high-energy sandstone environments are present at many localities. These assemblages contrast with the low diversity, *Inoceramus*-dominated, deeper-shelf siltstone faunas at Motu Falls.

There is no positive record of Urutawan faunas elsewhere in New Zealand. However, the presence of the stage is suggested by the occurrence of *I.* sp. ex gr. *ipuanus–kapuus* in sections below Motuan at Coverham and the Awatere Valley, Marlborough, and perhaps in Northland (Speden, 1977; Hay, 1975).

Wellman (1959) correlated the Urutawan with the Cenomanian on the basis of the apparent similarity of *I. kapuus* with Cenomanian species. Henderson's (1973) correlation of the overlying Ngaterian and Motuan ammonoid faunas with latest Albian to Middle Cenomanian and Late Albian respectively, and the stratigraphic position of the Urutawan at Koranga, indicate an Albian age. The occurrence in the Uratawan of *Aucellina* cf. *gryphaeoides* and species of *Maccoyella* common in the Late Albian of the Great Artesian Basin (Day, 1969), and the absence of Early Albian species such as *A. hughendenensis*

(Etheridge Sr.), *I. constrictus* (Etheridge Jr.) and *Pseudavicula* spp., favour a Middle–Late Albian age. Thus the unconformity below Te Wera Formation at Koranga possibly represents some part of the Late Aptian to Middle Albian.

Motuan Stage (Middle–Late Albian). As originally defined by Wellman (1956, 1959) the Motuan is characterized by *I. urius* Wellman and *Aucellina euglypha* Woods in its lowest part, and *I. ipuanus* and *Aucellina* sp. in its upper part. Revision of the type section (Fig. 33) has shown that the range-zone of *A. euglypha* is the best indicator of the stage. Coinciding with lower 100 m of the range of *euglypha* is a large flat *Inoceramus* best identified as *I. warakius* (Speden, 1976a, 1977), the index species of the Mokoiwian Stage. This confirms evidence of a mostly Motuan age for the Mokoiwian Stage at Mt Taitai, Tapuaeroa Valley (Speden, 1977). The Motuan is the first widely developed and easily recognized stage in New Zealand Cretaceous marine sequences.

Diverse faunas, including *A. euglypha, Maccoyella incurvata* Waterhouse, *Maccoyella* n. sp. B, and *Inoceramus* cf. *neocomiensis* d'Orbigny, occur in sandstone facies unconformably overlying Torlesse-like rocks in the Waioeka Gorge (Speden, 1973), and in the lower Waimana and Waiotahi Valleys (Speden, 1975a).

Ammonites (Henderson, 1973) and belemnites (Stevens, 1965) are too sparsely distributed or long-ranging for the establishment of zones. Foraminifera recorded from the Motuan are described by Stoneley (1962) and listed by Webb (in Suggate et al., 1977). Palynomorphs are relatively common (e.g. Wilson, 1976). Indeterminable Ichthyosauria have been collected in the Wairarapa (Fleming et al., 1971).

The presence of the ammonoid *Ptychoceras* and the stratigraphic position of the stage below Ngaterian ammonoid zones, which give a firm correlation with Late Albian to Middle Cenomanian (Henderson, 1973), indicate a Middle–Late Albian age.

Ngaterian Stage (Late Albian–Middle Cenomanian). The Ngaterian was defined by Wellman (1959) as including those rocks with *Inoceramus fyfei* Wellman, *I. hakarius* Wellman, and *I. tawhanus* Wellman (= *I. concentricus* var. *porrectus* Woods). In the Motu Falls section (Fig. 33), the type Ngaterian section is poorly exposed and the upper part is disrupted by faults (Speden, in prep.). An adjacent section along Te Waka Stream gives a continuous sequence. The upper contact of the Ngaterian below the type Arowhanan in Mangaotane Valley is conformable.

In Northland, Raukumara Peninsula, Wairarapa and most of Marlborough, the stage is represented by siltstone or alternating sandstone and siltstone facies characterized by species of *Inoceramus.* Inner-shelf sandstone facies with diverse macrofossil assemblages are present in the middle Awatere Valley (Fleming in Wellman, 1959, p. 145; Challis, 1966) and in the Gridiron Formation, Clarence Valley (Fleming in Suggate, 1958).

Henderson (1973) recognized two distinctive ammonite assemblages within the Ngaterian; a lower *Worthoceras parvum* Assemblage Zone (known only at Coverham), and an upper *Scaphites equalis coverhamensis* Assemblage Zone. Belemnites, particularly *Dimitobelus superstes,* are relatively common. Foraminifera have been listed by Hornibrook (in Wellman, 1959, p. 145–6), Stoneley (1962) and Webb (in Suggate et al., 1977). Rich microfloras, dominated by *Trisaccites microsaccatus* (Couper) and other coniferous pollen species, contain rare dicotyledonous pollen (Couper, 1960). Plant macrofossils, including pteridophytes and cycads described by McQueen (1956), and non-marine molluscs, are present in coal measures at Seymour River, Clarence Valley (Suggate, 1958).

The Coverian Stage as defined by Wellman (1959) consisted of a lower zone with *Inoceramus* cf. *anglicus* Woods and an upper zone with *I. concentricus* Parkinson which, in the type area at Coverham, was overlain by a barren zone about 300 m thick. Hall (1963), however, showed the *anglicus* zone to be Motuan and the Coverian to be equivalent to a lower part of the Ngaterian (see also Henderson, 1973).

The ammonites of the *Worthoceras parvum* (Lower Ngaterian) and *Scaphites equalis coverhamensis* (Upper Ngaterian Assemblage Zones) provide the best evidence for overseas correlation (Henderson, 1973). *Hypoturrilites, Sciponoceras baculoides, Desmoceras (Pseudouhligella) ezoanum* Matsumoto and *D.* (*P.*) *poronaicum* Yabe, support firm correlation of the Ngaterian with the Lower to Middle Cenomanian. The ammonoids in the lower zone, and its stratigraphic position, favour correlation with latest Albian and, perhaps, earliest Cenomanian. A Late Albian to Middle Cenomanian correlation is compatible with the widespread presence in the

Ngaterian of the *Inoceramus concentricus* species group which, although reported from the Aptian of Patagonia (Bonarelli and Nagera, 1921; Cox, 1953), is typically Albian in Europe (Sornay, 1966), Albian–Cenomanian in Siberia (Pergament, 1965) and Cenomanian in Japan (Matsumoto, 1963). The incoming of angiosperm pollen in the Ngaterian is also compatible with a position near the Albian–Cenomanian boundary. Webb (in Hornibrook, 1969) considered the Foraminifera favoured correlation of the stage with the Aptian–Albian.

Raukumara Series (Cenomanian–?Santonian)

The three stages of this series are based on species of *Inoceramus* occurring in zones separated by barren interzones in a siltstone-dominated sequence at Mangaotane Valley (Wellman, 1959). The stages follow in continuous succession, with the lowest stage, the Arowhanan, overlying conformably Ngaterian alternating sandstone and siltstone (Fig. 37). Because of the limited faunas and the endemism of the inoceramids, overseas correlations are approximate. The limits are set by the stratigraphic position of the rocks of the series conformably above the Ngaterian (Late Albian–Middle Cenomanian) and below the Piripauan (?Late Santonian–?Maastrichtian).

Arowhanan Stage (Late Cenomanian–Turonian). The type Arowhanan as defined by Wellman (1959), some 150 m thick, comprises the range-zone of *Inoceramus rangatira* Wellman and an overlying thin barren zone. *Inoceramus rangatira* is the only widely distributed macrofossil in the siltstone or thin- to thick-bedded alternating sandstone and siltstone lithofacies characteristic of the stage in most areas. Sandstone lithofacies in the middle Clarence Valley (Suggate, 1958; Lensen in Suggate et al., 1977), Marlborough, and near Wairata, Raukumara Peninsula (Speden, 1973), have more diverse assemblages including *Eselaevitrigonia* n. sp. (Woods), *Nototrigonia* n. sp., *Iotrigonia glyptica* (Woods), and *Dimitobelus superstes* (Hector) (Wellman, 1959, p. 144). Arowhanan rocks locally overlie unconformably older rocks in Marlborough (Lensen in Suggate et al., 1977) and at Wairata.

McQueen (1956) and Couper (1960) recorded plant fossils from Raukumara Series rocks, probably including some from Arowhanan. Poor foraminifera, including *Rzehakina* (Scott, 1961), are reported (Webb in Suggate, et al., 1977).

Correlation of the Arowhanan is inferred largely from its stratigraphic position (Wellman, 1959; Hall, 1963). Correlation of part of the stage with the Late Cenomanian is supported by the record of *Desmoceras,* a genus that ranges no higher than Cenomanian (Henderson, 1973).

Mangaotanean Stage (Turonian–Coniacian). The type section of the stage in Mangaotane Valley, some 150 m thick, includes the range zone of *I. bicorrugatus* Marwick and an overlying thin barren zone. The index species is the only widely distributed macrofossil in the siltstone or thin- to thick-bedded alternating sandstone and siltstone lithofacies characteristic of the stage in most areas. An alternating sequence with grits and conglomerates, often with *I. bicorrugatus,* is present in Wairarapa and southeastern Hawkes Bay, where other species, particularly gastropods, are found (Kingma, 1971).

Dimitobelus superstes (Hector) occurs locally. No ammonites are known. Foraminifera from the stage are listed by Webb (in Suggate et al., 1977).

The Mangaotanean cannot be correlated accurately. Henderson's (1973) assignment of at least part of the underlying Arowhanan to Cenomanian, and the correlation of the overlying Teratan with the Coniacian–Santonian, provide the only reasonable limits to correlation. *I. bicorrugatus* grossly resembles several species from the Turonian–Coniacian of Europe and Japan. However, none of the overseas species have the distinctive juvenile phase of *I. bicorrugatus.*

Teratan Stage (Coniacian–Santonian). This stage, the uppermost in the Raukumara Series, includes 300 m of beds with *Inoceramus nukeus* Wellman and *I. opetius* Wellman and an overlying thin barren zone. The type section lies conformably above the type Mangaotanean and below Piripauan alternating sandstone and siltstone (Wellman, 1959). Siltstone locally with thin-bedded sandstone and siltstone, deposited in shelf environments and containing abundant shell beds of *Inoceramus,* characterizes the stage in most areas. Sandstone-dominated sequences occur locally, as in southeastern Hawkes Bay (Kingma, 1971), while at the Chatham Islands volcanic tuff and breccia of Kahuitara Tuff contain *I. opetius* and other macrofossils (Hay et al., 1970; Speden, 1976c).

Inoceramus nukeus and *I. opetius* are the only common macrofossils. *Nukeus* overlies *opetius* in some but not all sections (W.R. Moore, 1961, p. 61; P.R. Moore, 1977). More diverse assemblages have been collected in southeastern Hawkes Bay and Chatham Islands.

Three Teratan ammonite specimens, all associated with *I. nukeus* have been found (Henderson, 1973). *Dimitobelus? ongleyi* Stevens, now known to range into the overlying Piripauan and Haumurian stages, is present in four areas (Stevens in Suggate et al., 1977). Foraminifera from Teratan beds are listed by Hornibrook (in Wellman, 1955, pp. 111–2, and 1959, p. 143), Scott (1961) and Webb (in Suggate et al., 1977).

Wellman (1959) correlated the Teratan with Upper Santonian and Lower Campanian, a correlation supported by the resemblance of *I. opetius* to specimens of *I. (Platyceramus) rhomboides* Seitz (1961), and of *I. opetius* and *nukeus* to specimens of *I. lingua* Goldfuss and *I. lobatus* Goldfuss. Conversely, the similarity of *I. opetius* to specimens of *I. uwajimensis* (Yehara; Noda, 1975) and *I. nukeus* to specimens of *I. lamarcki* Parkinson (Woods, 1911) suggests a Turonian–Coniacian age. A Santonian–Campanian correlation is supported by the occurrence in the Teratan of *Globotruncana fornicata* Plummer (Webb in Stevens, in Suggate et al., 1977) and by the close affinities of *Dimitobelus? ongleyi* with a belemnite from the Toolonga Calcilutite, Western Australia (Stevens, 1970a). An upper limit of Santonian is suggested by the Santonian–Campanian affinities of the planktonic foraminifera in the overlying Piripauan (see below).

Recent studies by Mildenhall (in prep.) on pollen and spores suggest the Kahuitara Tuff, Chatham Islands, spans the Mangaotanean to Teratan (Turonian–Santonian). A radiometric date of 77.3 ± 1 m.y. (Grindley et al., 1977) on basalts overlying the Kahuitara Tuff (with *I. opetius*) is consistent with a Late Coniacian–Santonian age; but the stage would overlap the Campanian if the radiometric time scale of Obradovich and Cobban (1975) is adopted.

Mata Series (?Santonian–Maastrichtian)

The two stages of the Mata Series, the Piripauan and Haumurian, are based on faunas from the sequence at Haumuri (Amuri) Bluff, southern Marlborough (Fig. 34, 35). The sparsely fossiliferous nature of the marginal marine sediments, strong facies control of faunas, and the presence of a major unconformity at the base and a disconformity at the top of the section, implying that an unknown amount of Piripauan and Haumurian time is not represented at

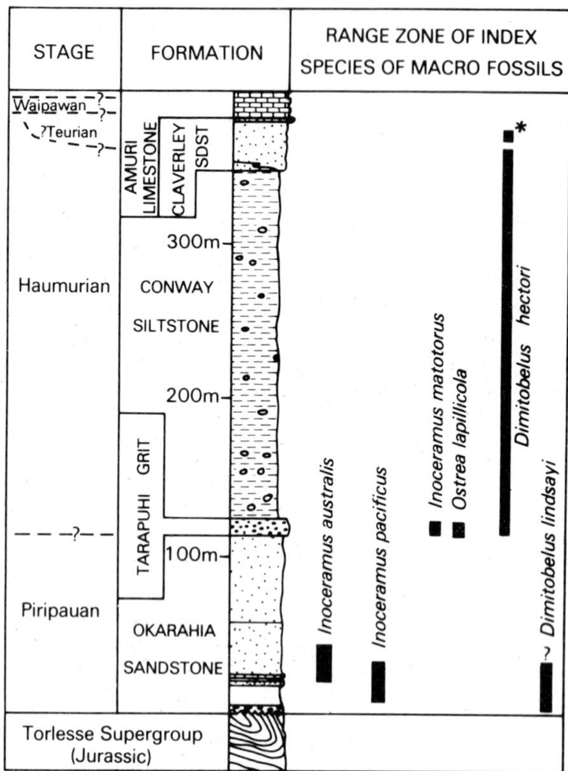

Fig. 34. The Haumuri Bluff section, showing the stratigraphic ranges of index species of macrofossils. (After Warren and Speden, 1977.) * Uppermost occurrences of *D. hectori* reworked.

Haumuri Bluff, seriously impair the utility of the type section for local and regional correlation (Warren and Speden, 1977).

Piripauan Stage (Santonian–Campanian). As defined by Wellman (1959) the type Piripauan consists of about 120 m of calcareous quartzose sandstone, concretionary bands and conglomerate with *Inoceramus australis* Woods and *I. pacificus* Woods, which unconformably overlies Torlesse Supergroup (?Late Jurassic) and grades up into Haumurian (Fig. 34). Rare shellbeds have a diverse macrofauna of more than 70 species, dominantly bivalves and gastropods (Fleming in Wellman, 1959; Warren and Speden, 1977). Only three species – *I. australis, I. pacificus* and *Dimitobelus lindsayi* (Hector) – are sufficiently common and widespread to function as index species (Warren and Speden, 1977).

Diverse Piripauan macrofaunas are known in Northland (Wellman, 1959), and in shallow marine sequences south of Lake Waikaremoana where there

Fig. 35. North face of Haumuri Bluff, Marlborough, showing slumped tree-covered slopes. The bare cliffs are of Okarahia Sandstone (Piripauan) and the far point (Piripaua) is in Amuri Limestone (Cenozoic). (Photo: G. Warren.)

are also beds with non-marine molluscs (Speden, 1973). Elsewhere on the east coast of the North Island and in Marlborough the Piripauan is represented mostly by alternating sandstone and siltstone, often with glauconitic beds, normally containing only the three index species. South of Haumuri Bluff the Piripauan is known definitely only in the valleys of the Conway and Leader Rivers. Non-marine beds are present in Mikonui Stream, north-west of Haumuri Bluff. Part of the non-marine sequences at the base of the Late Cretaceous transgression in Otago and Canterbury, and some in the coal measures of Otago–Southland and the west coast of the South Island, may be Piripauan.

Fossils described from the Piripauan are listed in Wellman (1959) and include bivalves (Woods, 1917), gastropods (Wilckens, 1922), ammonites (Henderson, 1970), belemnites (Stevens, 1965), annelids, corals (Squires, 1958), reptiles (Welles and Gregg, 1971) and fish remains (Chapman, 1918). Foraminifera are listed by Webb (in Suggate et al., 1977).

Wellman (1959, p. 121), on the basis of similarities between species of *Inoceramus,* correlated the Piripauan with Middle and Upper Campanian. Approximate correlation with Campanian and, possibly, Late Santonian and Early Maastrichtian, is suggested by the similar composition of macro- and microfossil assemblages, containing species of *Inoceramus* closely resembling *australis* and *pacificus,* in formations spanning Santonian–Maastrichtian in Chile and Patagonia (Katz, 1963; Scott, 1966; Charrier and Lahusen, 1969).

Henderson (1970) recognized two ammonite zones in the Mata Series; a lower *Kossmaticeras (Kossmaticeras)* "zone" in the Piripauan and an upper *Kossmaticeras (Natalites)* "zone" spanning the Late Piripauan and Haumurian. He concluded that the fauna of the *Kossmaticeras (Natalites)* "zone" was consistent with a Campanian–Maastrichtian age, while the older *Kossmaticeras (K.)* "zone" was probably Santonian because of its stratigraphic position and the upper limit of the range of the subgenus *Kossmaticeras* in the Santonian of Madagascar.

Recent sampling has demonstrated the presence in the type Piripauan of a meagre araneceous foraminiferal assemblage which is unsuitable for correlation. Microfaunas from other sections have planktonic taxa very similar to those in the underlying Teratan Stage. Webb (in Hornibrook, 1969, p. 128) considered these planktonics indicated correlation with Campanian, but later (in Stevens, in Suggate et al., 1977) favoured a Santonian–Campanian age.

The evidence is generally consistent with a Campanian age, probably with some overlap of Late Santonian and possible overlap of Early Maastrichtian.

Haumurian Stage (?Campanian–Maastrichtian). This stage, the youngest recognized in the New Zealand Cretaceous (Hornibrook, 1962), as defined by Wellman (1959, pp. 117, 134–5) is characterized by rocks with *I. matotorus* Wellman and *Ostrea lapillicola* Marwick. In the stratotype at Haumuri Bluff these species are present only in the Tarapuhi (= Black) Grit, the lowest of the three formations placed in the stage (Warren and Speden, 1977; Fig. 34). The top of the stage is marked by a disconformity within the Claverley Sandstone, the oldest beds above the disconformity being Waipawan. The lowest Tertiary stage, the Teurian, is apparently missing. Stevens (1965) has shown that *Dimitobelus hectori* Stevens, which ranges through the Tarapuhi Grit to above the disconformity where it is probably reworked, is also a useful index species for the stage.

In other well-known sections in the northeast of the South Island and on the east coast of the North Island the Cretaceous–Tertiary boundary is unconformable, disconformable, faulted, or uncertain or difficult to locate because of poor microfaunas (Hornibrook, 1962; Edwards, 1973; Webb, 1973a, b). Recent studies by Webb (1973a, b) confirm the post-Cretaceous age of the distinctive Wangaloan macrofaunas (Finlay and Marwick, 1937; Wellman, 1959).

Fleming (in Wellman, 1959, pp. 137–40) has listed the diverse macrofaunas collected from the Haumurian. Marine reptiles have been described recently (Welles and Gregg, 1971), while brachiopods, echinoids and sponges remain undescribed. Over 130 species are known, dominantly molluscs, including some 35 species of ammonites (Henderson, 1970).

The only macrofossils useful as index species for the stage are *Dimitobelus hectori* and *Inoceramus matotorus*; the utility of *Ostrea lapillicola* is uncertain (Speden, 1973; Warren and Speden, 1977). Examination of collections has not substantiated any of the records of *I. matotorus* with index species of older stages (Warren and Speden, 1977; see Stevens in Suggate et al., 1977). *Inoceramus matotorus* is restricted to Lower Haumurian in sections in northeast Raukumara Peninsula, northern Wairarapa (Kingma, 1962), and Marlborough (Lensen, 1963), a distribution probably controlled by facies as at Haumuri Bluff (Warren and Speden, 1977).

Foraminifera provide the best indices for the stage and are discussed by Hornibrook (1958, 1962, in Wellman, 1959) and Webb (1966, 1971, 1973a, b) who proposed three assemblage zones: (1) *Trochammina globigeriniformis* Zone, (2) *Rzehakina epigona* Zone, and (3) *Globotruncana circumnodifera* Zone.

The first two zones are characterized by arenaceous foraminifera, are believed to be facies controlled and are therefore of limited use in correlation. The zones are, in part, lateral equivalents (Webb, 1966, fig. 4).

Nannofossils are also useful for zonation and correlation, and Edwards (1971, 1973) recognizes a *Nephrolithus frequens* Zone of Late Haumurian age. Microfloras from marine Haumurian beds are listed by Couper (1960, pp. 7–9), and are dominated by *Dacrydiumites mawsonii* Cookson, *Podocarpidites otagoensis* Couper, *Microcachryidites antarcticus* Cookson, and other coniferous species, although *Nothofagus kaitangata* Te Punga is common.

Correlation of Haumurian with Maastrichtian is strongly indicated by foraminifera and nannofossils; the foraminifera include such world-wide Maastrichtian forms as *Rugoglobigerina, Globotruncana (Abathomphalus), Heterohelix, Bolivina incrassata* Reuss and *Bolivinoides draco dorreeni* Finlay. The last occurrences of *Inoceramus*, ammonites, belemnites and probably also of marine reptiles, also support a Maastrichtian correlation. Reliable correlation by macrofossils is at present not possible, although Henderson (1970) concluded that the ammonites of his *Kossmaticeras (Natalites)* zone, thought to span the Haumurian and Late Piripauan, are consistent with a Campanian–Maastrichtian age.

Regional stratigraphy

Northland

Cretaceous rocks are widespread in Northland

(Fig. 30, Table V). Complex structure and poor exposure means that stratigraphic relationships have been established only locally. Interpretation of Cretaceous stratigraphy and paleogeography is complicated by the Onerahi Chaos Breccia (Kear and Waterhouse, 1967; Katz, 1968), an olistostome and/or mélange-gravity slide complex probably emplaced in the Late Oligocene to Early Miocene, and thought to be derived from the west (Kear and Waterhouse, 1967) or east (Brothers, 1974), and by the considerable uncertainties over the age and stratigraphic relationships of the igneous masses so characteristic of Northland. These masses, considered to be coeval, originally submarine geosynclinal basic and acidic intrusives and extrusives (Quennell and Hay, 1964; Farquhar, 1969), but now allochthonous, range in age from Late Jurassic to perhaps Oligocene (Brothers, 1974; Hornibrook and Hay, 1977). If the stratigraphic relations apparent are primary, volcanism extended over a considerable period of time.

The oldest sedimentary rocks (Table V) are locally interbedded with altered basic and acidic volcanic rocks (Hay, 1975), and are generally considered to be a continuation of sedimentation characteristic of the New Zealand Geosyncline (Waipapa Group; Hay in Suggate et al., 1977). However, the stratigraphic relationships of units containing Clarence fossils with Waipapa Group rocks are unknown (Hay, 1975). The common volcanics and, especially the sediments of the younger Awapoko, Tupou and Whatuwhiwhi Formations, are not typical of Torlesse sequences; they are more characteristic of Clarence Series rocks deposited under conditions of crustal instability. Hay (1975) interprets the Clarence and Arowhanan formations as representing progressive upward change from flysch-like to shelf sediments filling a subsidiary geosyncline which formed during eversion of the New Zealand Geosyncline.

Late Cretaceous sediments represent a great range of facies from shallow-water to redeposited (Hay, 1960, 1975), and are possibly up to 2000 m thick. Mata formations unconformably overlie older rocks in many areas, and are the oldest cropping-out over much of Northland.

Raukumara Peninsula

The peninsula has a core of older rocks, included in Torlesse Supergroup or Taitai Series (Kingma, 1966), flanked to the northeast and south by bands of successively younger Cretaceous and Cenozoic rocks (Figs. 36, 37).

Structure is complex, particulary in Taitai and

TABLE V

Classification of Cretaceous sedimentary rocks in Northland

International unit	Stage	Formation	Key fossils
Maastrichtian	Haumurian	Ngatuturi Claystone Waiari Formation	Haumurian microfauna and macrofossils
		Punakitere Sandstone	ammonites in concretions *Ostrea lapillicola*, *Inoceramus* sp.
Campanian	Piripauan	not named	*I. australis* *I. pacificus*
Santonian	Teratan	Callaghan Siltstone	*I. opetius*
Coniacian	Mangaotanean	Waikaraka Siltstone	*I. bicorrugatus*
Turonian	Arowhanan	Whangape Sandstone Awapoko Formation	*I. rangatira*
Cenomanian	Ngaterian	Patutahi Formation	*I. tawhanus*
Albian	Motuan	Tupou Formation	*Aucellina euglypha*
	Motuan–Urutawan	Whatuwhiwhi Formation Tokerau Formation	*I.* sp. ex gr. *ipuanus–kapuus*

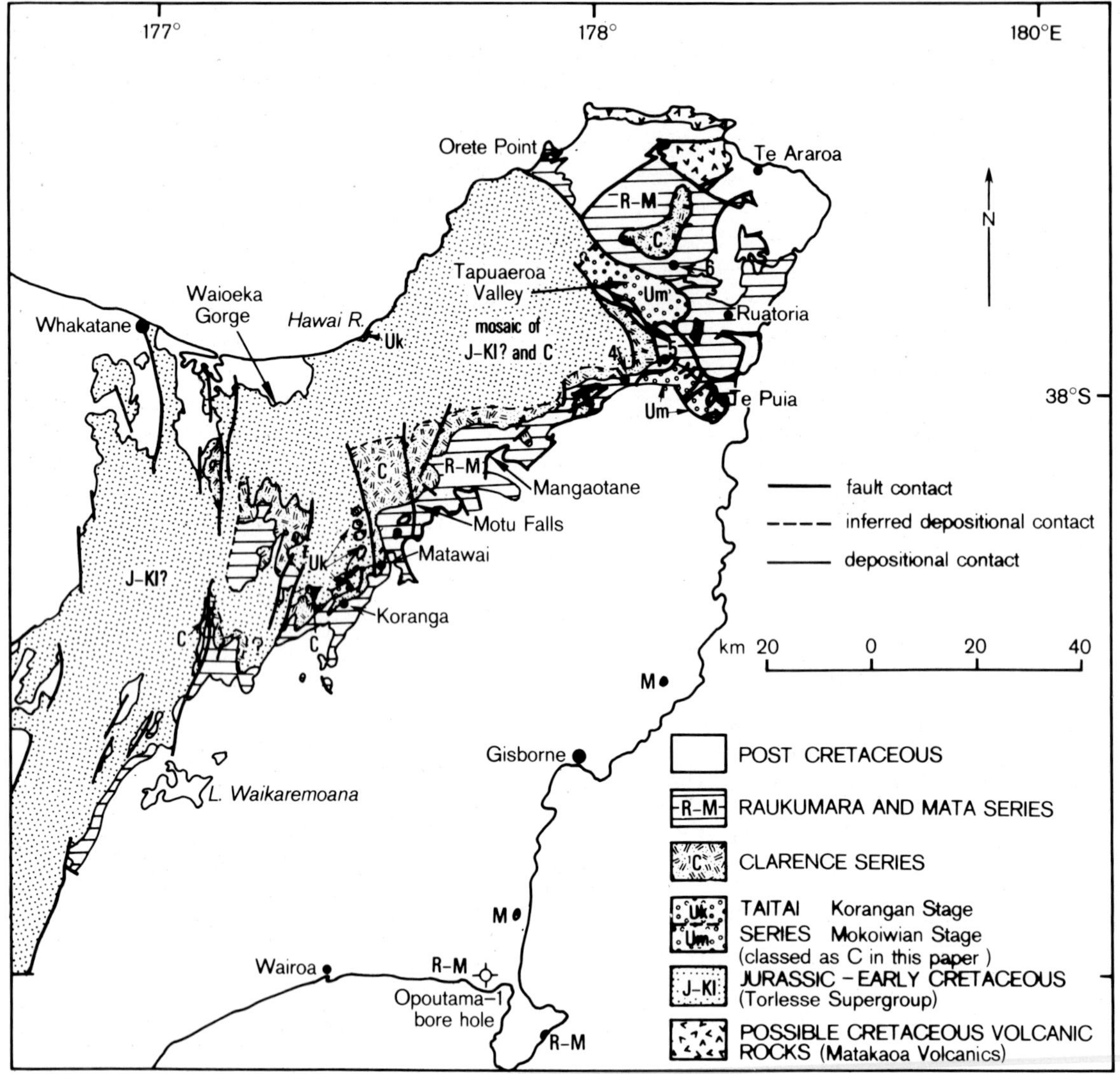

Fig. 36. Major time stratigraphic units, Raukumara Peninsula. (After Kingma and Speden in Suggate et al., 1977, Speden, 1975a, 1976a.) For sections, including numbers *4–6*, see Fig. 39.

older Clarence Series. Recent mapping by P.R. Moore and I.G. Speden suggests the presence of two areas of contrasting structure bounded by a line extending from the mouth of Raukokere River, west of Orete Point, south-southeast to Puketoro Stream (Fig. 39, loc. 4) and then southwest along and intersecting the southern margin of the belt of youngest Cretaceous rocks to just west of Mangaotane where the boundary disappears under the Late Tertiary cover. Northwest and west of this line the sequence is largely if not entirely autochthonous and dominated by north-northeast to northeast-trending faults, although east-west faults and lineaments tend to predominate in the eastern part of the region. Northeast and south of the line the succession is dominated by décollement slices emplaced in Late Oligocene–Miocene time (Stoneley, 1968) and is largely, if not completely, allochthonous. To the south of the main belt of Cretaceous rocks, Late Cretaceous occurs in diapiric and fault-controlled structures (Ridd, 1968, 1970). A section of Raukumara and younger sediments in the Opoutama-1 drill-hole near Mahia Peninsula has been described by Katz (1974).

A major change in intensity of deformation, degree of induration and environment of deposition

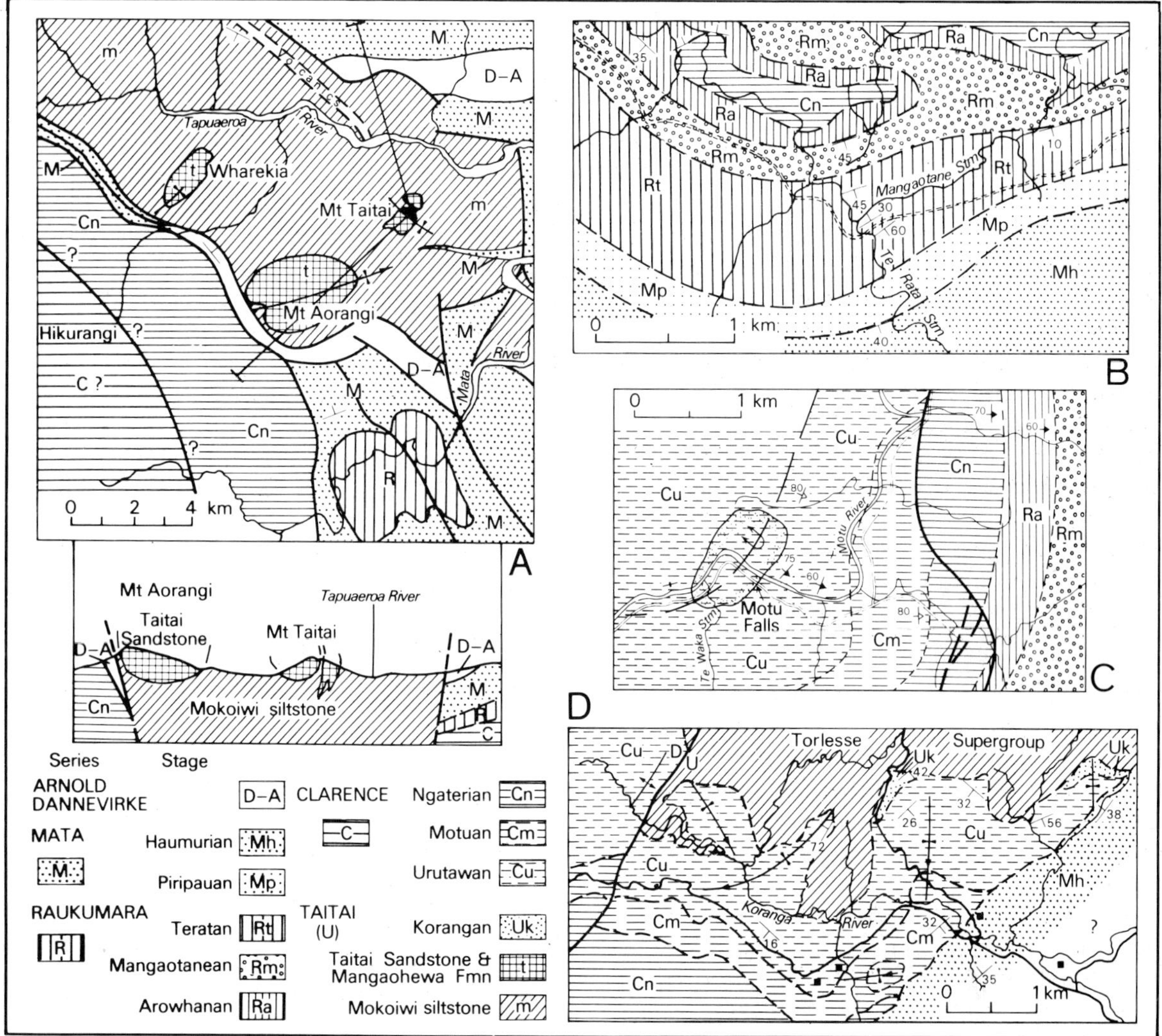

Fig. 37. Geology of key areas in Raukumara Peninsula. See Fig. 36 for location. (After Suggate et al., 1977, fig. 6.11.) A. Tapuaeroa Valley near Mt Taitai (Speden, 1976a). B. Mangaotane Valley (Wellman, 1959). C. Motu Falls (Speden, in prep.). D. Koranga (Speden, 1975a).

coincides with an unconformity which separates Torlesse-like from overlying shelf sediments. This change is diachronous as it occurs below the Korangan (Late Aptian) in the Koranga District (Speden, 1975a; P.R. Moore, 1977) but within the Motuan (mid–Late Albian) in Waioeka Gorge some 15–20 km to the northwest (Speden, 1973). A second major regional event involving deformation, erosion and strong angular unconformity locally, occurred over the sea of the autochthonous block prior to deposition of the Haumurian (?Late Campanian–Maastrichtian; Speden, 1975a; Stoneley, 1968). No unconformity is present over the area of the allochthonous block.

In the west, north of Lake Waikaremoana, the source of sediments appears to have been from the west and south, from a landmass which persisted during the Cretaceous. Continuing instability along the eastern margin of this landmass is suggested by facies and thickness patterns, and local unconformities (Grindley, 1960; Speden, 1973, 1975a; P.R. Moore, 1977). Although facies patterns are at present indefinite, clay mineral distributions suggest that a component of the Late Cretaceous sediments was derived from an actively eroding granitic terrain to the east (Rumeau, 1965; Kingma, 1974). Development of a complex double-sided basin of deposition within the East Coast fold and fault belt (Katz, 1974) during the Cretaceous has been proposed (see p. 284; Kingma, 1974; Suggate et al., 1977).

Basic spilitic volcanic rocks are known in place within the Mokoiwi Formation (Motuan), Tapuaeroa Valley (Wellman, 1959; Speden, 1976a). More extensive igneous activity is represented by the basic Matakaoa Volcanics, which form prominent ranges at the north end of the peninsula (Kingma, 1966). The volcanics have been generally considered to be Late Jurassic to Early Miocene in age (Chapman-Smith and Grant-Mackie, 1971; Speden, 1976b). However, Strong's (1976) study of foraminifera in apparently interbedded limestone favours a late Early Cretaceous (Taitai or Clarence Series) age. The probable allochthonous nature of the volcanic masses could partly explain the wide range of ages previously allotted.

Taitai Series. The Korangan Stage is represented by the Koranga Formation, a minimum of 150 m of predominantly sandstone, which unconformably overlies Torlesse-like rocks between Koranga and Motu (Speden, 1975a, 1977; Fig. 37), and by rocks of unknown stratigraphic relations containing *Aucellina* cf. *radiatostriata* Bonarelli at Hawai River (Wellman, 1959).

Much of the poorly known Torlesse-like rocks forming the main ranges may be Early Cretaceous and thus classed in the Taitai Series. Characteristic lithologies are essentially massive lithic and feldspathic sandstone units and well- to thin-bedded alternating sandstone and siltstone units which rarely exceed 500 m in thickness, dark blue-grey and dark grey siltstone, mostly lacking concretions, and subordinate conglomerate bands. Estimation of total thickness is impossible, atlhough several thousand metres of sediments are known within fault-bounded blocks. The abundance of plant material, coarse grain size of much of the thick sandstone units, and a paucity of features typical of turbidity-current deposits suggest deposition in deltaic to submarine fan environments close to source areas.

Clarence Series. In southwestern Raukumara Peninsula Torlesse-like rocks of the Koranga Formation are overlain unconformably by low-dipping, open-folded, fossiliferous, predominantly inner shelf sandstones of the Te Wera Formation (Urutawan) which grade laterally and vertically into the blue-gray shelf siltstone of the Karekare Formation (Urutawan to Teratan; Speden, 1975a, P.R. Moore, 1977; Figs. 37, 38). Locally within the Karekare Formation are thin units of alternating sandstone and siltstone. These are well-developed at Motu Falls where they reach a thickness of 200 m (Fig. 33).

Lateral variation in facies and thickness is marked (Fig. 39). The Urutawan consists largely of sandstone at Koranga, but mainly of siltstone at Motu Falls. Fossiliferous shelf siltstones characterize the Motuan except to the northwest, in the Waioeka Gorge and the lower Waimana–Waiotahi Valleys, where there are sandstone sequences with diverse faunas and, locally, apparently non-marine beds (Speden, 1973, 1975a). In both areas the sandstone facies is less than 150 m thick and grades rapidly up into siltstone resembling the Karekare Formation. The Ngaterian is represented by the Karekare Formation in all areas except Mangaotane Valley where there is thin- to well-bedded sandstone and siltstone (Wellman, 1959). Complex facies and depositional patterns are also indicated by the Motuan age of the top of Torlesse-like sequences unconformably below the shelf-Motuan in the Waioeka Gorge.

Little is known of the Cretaceous between Mangaotane Valley and the northeast part of the Raukumara Peninsula. In the Tapuaeroa region the Clarence lithologies are relatively uniform. The oldest fossiliferous rocks known are the probable shelf–upper slope siltstone and thin sandstone beds of the Mokoiwi Formation (?Late Urutawan–Motuan) with with its laterally equivalent, probable submarine channel-fill or fan-deposits of the Taitai Sandstone (Speden, 1976a). Although the Mokoiwi Formation occurs in klippen (Katz, 1974; Speden, 1976a), the formation passes up into well- to thin-bedded alternating sandstone and siltstone characteristic of the Ngaterian in other décollement slices. This Motuan to Ngaterian sequence, generally folded and faulted, forms part of the ranges north of and most of the ranges south and west of Tapuaeroa Valley for at least 10 km. Thicknesses are difficult to estimate because of the complex structure; neither the Mokoiwi Formation nor the Ngaterian sediments are likely to exceed 1000 m in thickness.

The stratigraphy and structure of the rocks forming the main Raukumara ranges are too poorly known for reliable interpretation.

Fig. 38. Koranga Valley, showing rugged forested ridges of the Torlesse Supergroup (Late Jurassic–Early Cretaceous) in background, with the lower ridges of folded Koranga Sandstone (Korangan Stage, Aptian) and Te Wera Sandstone (Urutawan, Albian) in the middle distance and low subdued ridges of Karekare Siltstone (Urutawan–Ngaterian, Albian–Cenomanian) in foreground. See Speden (1975a).

Raukumara Series. Rocks of the three stages are well developed in many parts of Raukumara Peninsula (Figs. 35, 36). Deposition is continuous above the Ngaterian Stage except in the west and, perhaps, at Puketoro Stream where marked differences in induration and structure between the Ngaterian and Arowhanan favour an unconformity (Wellman, 1959, p. 110; cf. Pick, 1962).

Three major geographically separated lithofacies are evident (Figs. 39, 40–42). To the west shallow shelf sandstones unconformably overlie Torlesse-like rocks (W.R. Moore, 1961; Speden, 1973). Between Koranga and Puketoro Stream a fossiliferous shelf siltstone facies (Karekare Formation, Speden, 1975a) is characteristic. Green and purplish siltstone and very fine-grained poorly sorted sandstone beds, due either to conditions of oxidation and reduction in the basin of deposition or to a volcanic component in the sediment, occur throughout the series in the type section at Mangaotane Valley. North of Puketoro Stream the Arowhanan and Mangaotanean Stages consist of well- to thin-bedded alternating sandstone and siltstone, the beds varying considerably in grain size and thickness geographically and stratigraphically, and containing abundant plant fragments, carbonaceous laminae, and scattered to common macrofossils. In the Teratan in most sections the alternating sequence grades up into fossiliferous shelf siltstone, frequently with purple and green bands, and with common shell beds of *Inoceramus opetius* and *nukeus*.

The thicknesses of stages varies significantly over short distances (Fig. 39). North of Puketoro Stream much of this variation is secondary and due to folding

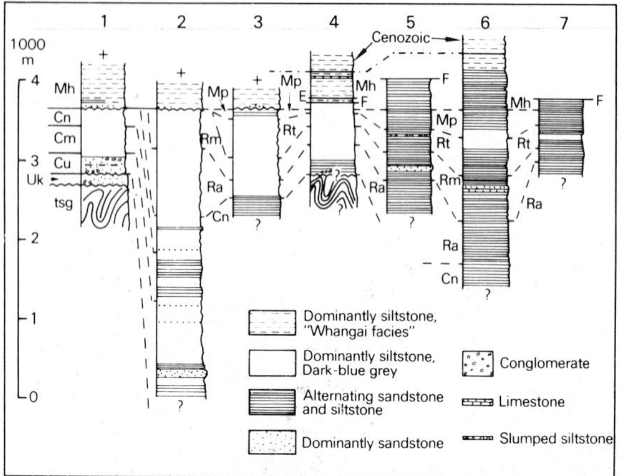

Fig. 39. Stratigraphic columns for important sections, Raukumara Peninsula. See Fig. 36 for location and Table III for symbols and stages. 1 = Koranga (N87-8); Speden (1975a). 2 = Motu Falls (N88); Speden (in prep.). 3 = Mangaotane River (N79); Wellman (1959), W.R. Moore (1961). 4 = Puketoro Stream (N80); Pick (1962) and observations of I.G. Speden. 5 = Mata River (N71); Pick (1962) and observations of I.G. Speden. 6 = Waiorongomai and Mangaoporo Valleys (N71); Pick (1962) and observations of I.G. Speden and P.R. Moore. 7 = Orete Point (N62); W.R. Moore (1957) and observations of I.G. Speden.

and faulting during Late Oligocene and Early Miocene (Stoneley, 1968). Intertonguing of the southwestern siltstone and the northeastern alternating facies (Fig. 39) suggests the latter represents part of a submarine fan or delta complex.

In western Raukumara Peninsula transgression of the Series appears to have been primarily westward and southward on to a nearby landmass, a source also suggested by sedimentary structures (Grindley, 1960; Speden, 1973, 1975a; P.R. Moore, 1977; W.R. Moore, 1961). The pattern of major facies and the spatial distribution of the thickness of sandstone beds favour derivation of the alternating facies in northern Raukumara Peninsula from the west.

Studies of clay mineral and feldspar distribution by geologists of New Zealand Aquitaine Petroleum Company (Rumeau, 1965) suggest that an actively eroding landmass providing orthoclase and kaolinite lay to the east-southeast during at least Late Raukumara time. The possibility of a northeast-trending ridge separating basins of deposition as proposed by Wellman (1959, fig. 1) was, however, disproved by the Opoutama-1 Well which penetrated 1140 m of monotonous flysch-like sandstone and siltstone of Raukumara age (Katz, 1974).

Mata Series. Rocks of this series are well represented, and are stratigraphically continuous with Early Tertiary sediments in most areas.

South of Lake Waikaremoana thin, highly fossiliferous Piripauan marine quartzose sandstone, locally with non-marine beds, unconformably overlies Torlesse Supergroup (Speden, 1973). Between Lake Waikaremoana and Mangaotane Valley, where no Piripauan is known in place, Haumurian rocks unconformably overlie older rocks (Stoneley, 1968; Speden, 1975a). At Mangaotane Valley the Piripauan reappears conformably above Teratan and is disconformably or unconformably overlain by Haumurian (Fig. 39). Over most of the region the Haumurian consists of dark grey micaceous siltstone, locally siliceous, typical of the Whangai facies and assigned to the Rakauroa Formation (Webb, 1971). Southwest of Motu a bedded quartzose and glauconitic sandstone basal phase, the Tahora "sandstone" (Webb, 1971), rarely exceeding 20 m, is generally present. The westernmost Haumurian outcrops consist dominantly of sandstone lithologically similar to Tahora "sandstone" and up to several hundred metres thick (Speden, 1973; P.R. Moore, 1977).

North of Mangaotane Valley the Mata, including Piripauan, conformably overlies Teratan. At Puketoro Stream the section is dominated by siltstone, as in the southwest. Elsewhere the Mata consists mostly of plant-rich, near-shore, alternating quartzose and glauconitic sandstone and siltstone closely similar to the Raukumara Series sequences, except north of Tapuaeroa River where a basal part of the Piripauan is locally formed of siltstone (Fig. 39). Thin conglomerates, grits and glauconitic bands are common in the Tapuaeroa region, particulary in the Haumurian where they frequently contain *Ostrea lapillicola* and *Inoceramus matotorus.* The Haumurian beds have been classed in Tapuaeroa Formation (Wellman, 1959), and underlying alternating sequences poor in conglomerate and grit in the Waiorongomai Sandstone (Pick, 1962; Laing, 1972).

Paleogeographically, the gross pattern is a continuation of that existing during Raukumara time; deposition of mainly shelf sediments between a landmass to the west and a possible landmass to the east. In western Raukumara Peninsula, however, deformation and erosion during Late Teratan to Piripauan time, probably due to movement on fault blocks and following similar but less extensive movement in the Late Mangaotanean to Teratan (P.R. Moore, 1977),

Fig. 40. Alternating thin-bedded sandstone and siltstone, part of a unit of Motuan age at Motu Falls, Raukumara Peninsula: overturned and dipping west (to right). Sequence is typical of units in shelf siltstones of Karekare Formation (Speden, 1975a) and the Clarence and Raukumara Series in western Raukumara Peninsula, although thickness of beds varies. These units have sedimentary features characteristic of turbidity current and non-turbidity current deposits, and may be depositional lobes or channel-fills (rare).

resulted in folding and rapid erosion of soft Clarence to Raukumara sediments (Speden, 1975a). This important regional event, and the subsequent westward and southward transgression of the Mata, introduces the significant change to sedimentation characteristic of the Whangai facies. Differences in the geographic and stratigraphic distribution of lithofacies (Fig. 39) suggest depositional patterns were not simple.

Southern Hawkes Bay – Wairarapa

In this region the Cretaceous rocks occur mostly as inliers surrounded by Cenozoic rocks (Fig. 43). The oldest known fossiliferous rocks are the sandstone and conglomerate of the Whatarangi Formation (?Urutawan), Aorangi Range (Bates, 1969; Speden, 1969). Motuan and younger rocks are widespread.

Evidence is lacking for Taitai Series rocks. Recent field work by P.R. Moore and Speden (in prep.) has confirmed (Speden, 1977) that most areas mapped as Torlesse Supergroup (Johnston, 1975) and Taitai Series (Eade, 1966; Johnston, 1975; Neef, 1974; Van den Heuvel, 1960) are Clarence Series. The only place where the oldest fossiliferous Cretaceous rocks are in juxtaposition with Torlesse rocks similar to those of the main ranges is in the Aorangi Range where the contacts are faulted.

Two distinct sequences are present in the Clarence Series in the Wairarapa: (1) a lower tectonized, sparsely fossiliferous sequence, Motuan and possibly older, of mostly alternating sandstone and siltstone, massive sandstone and minor siltstone units, which appear to be lateral facies equivalents; (2) an upper less indurated and deformed, structurally less com-

Fig. 41. Mangaotanean siltstone containing single valves of *Inoceramus bicorrugatus* Wellman, Mangaotane River, showing the typical lithology of the siltstone lithofacies of Clarence and Raukumara age, western Raukumara Peninsula. (Photo: G.W. Grindley.)

plex Motuan and younger fossiliferous shelf siltstone, sandstone and alternating sequence. In most areas the upper sequence has at the base a distinctive breccia-conglomerate. The succession is well developed in the Tinui–Awatoitoi District (Johnston, 1975) where the older Clarence rocks are equivalent to Waewaepa and Taipo Formations, the breccia-conglomerate to Gentle Annie Breccia and Bideford Member, and the overlying sequence to the Mangapurupuru Group. The paleogeographic significance of the intra-Motuan break and change in depositional environments is not fully understood.

In southern Hawkes Bay probable Motuan beds, including limestones, occur at Red Island (Marwick, 1966; Kingma, 1971), and Ngaterian siltstones are preserved in the cores of anticlines (Lillie, 1953). Siltstone (Springhill Formation, Johnston, 1975) is also characteristic of the Ngaterian at Tinui, although alternating sandstone, siltstone and conglomerate is typical of eastern Wairarapa (Van den Heuvel, 1960; Eade, 1966; Johnston, 1975).

A similar pattern of deposition of siltstone in western Tinui and conglomeratic alternating sequences elsewhere continued throughout the Raukumara Series into Piripauan. The proportion of conglomerate-breccia and sandstone beds varies significantly geographically, with beds of conglomerate attaining a thickness of 20 m at Ngahape. Black siltstone forms most of the Teratan at Waimarama, southern Hawkes Bay (Kingma, 1971).

Deposition of alternating beds with conglomerate and grit beds continued into the Haumurian in southern Hawkes Bay, where the sequence is very similar to the Tapuaeroa Formation of Raukumara Peninsula (Lillie, 1953, Kingma, 1971), and through the Haumurian at Tinui (Te Mai Formation, Johnston,

Fig. 42. Thick-bedded alternating sandstone and siltstone of Arowhanan age, upper Waikura River, Raukumara Peninsula. Units such as this make up a subordinate part of the alternating lithofacies typical of most of the Ngaterian to Lower Haumurian of northeastern Raukumara Peninsula. Typically, the alternating facies is thinner bedded, with beds rarely exceeding 50 cm thick, and contains a higher proportion of siltstone.

1975). In all areas the coarser sequences appear to grade into the Whangai siltstone facies, the entry of which seems to be markedly diachronous. It commences in the Haumurian at Ngahape, in the Middle Haumurian in southern Hawkes Bay, and in the latest Haumurian or earliest Tertiary in the Tinui District.

Facies and thickness patterns are poorly understood. The Clarence to Mata sequence at Tinui is some 4800 m thick. Reported thicknesses for the Raukumara and Mata Series range from 500 to 1400 and 400 to 1000 m, respectively. Major sources of sediment probably lay both to the west and east. Local facies changes, however, suggest land to the north (Waterhouse and Bradley, 1957). The alternating beds are generally interpreted as redeposited flysch sequences. Many features, including a large component of plant material and the size and shape of conglomerate clasts, possibly favour deposition as part of or near submarine deltas or fans close to sources of supply. A wide range of structures, including spectacular sedimentary dikes (Waterhouse and Bradley, 1957) in both Early and Late Cretaceous rocks, demonstrate the widespread and continuing occurrence of soft rock deformation, both penecontemporaneous and later. This diking was probably largely due to slumping caused by the elevation and erosion of submarine growing folds and fault blocks.

Two sets of low potash/soda igneous intrusions are present in the region: basaltic–spilitic flows and asso-

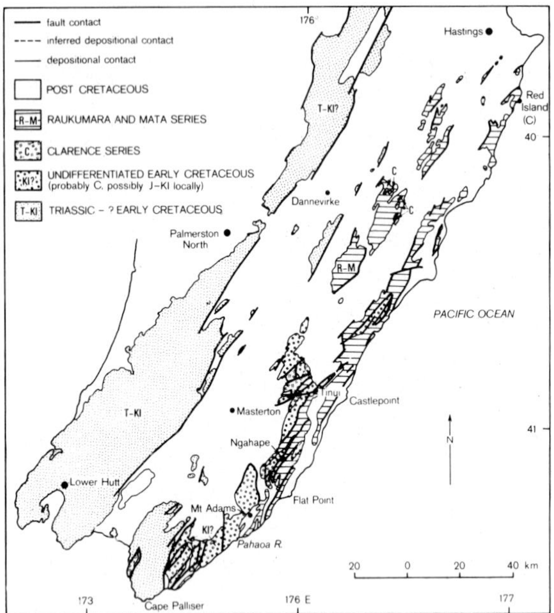

Fig. 43. Major time stratigraphic units in southern Hawkes Bay and southeast Wellington. (After Lillie, 1953; Kingma, 1966, 1967, 1971; and field observations of P.R. Moore and I.G. Speden.)

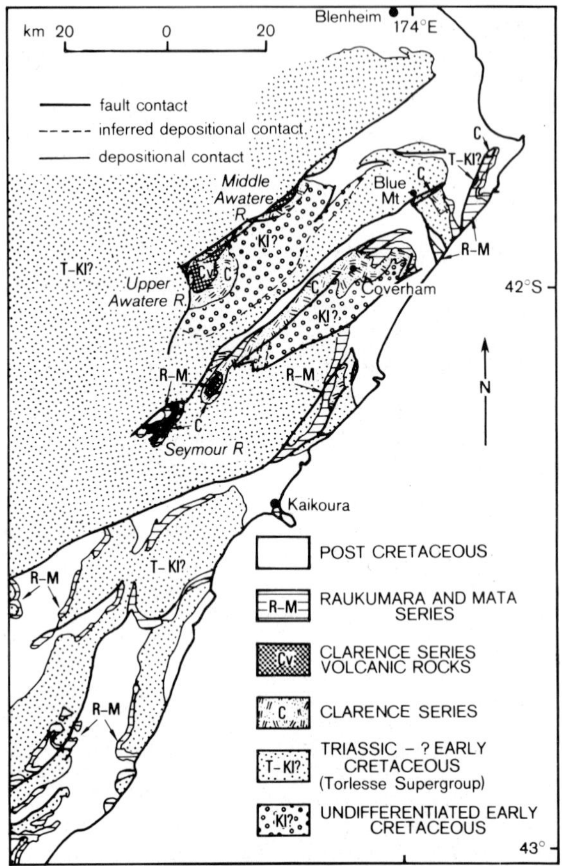

Fig. 44. Major time stratigraphic units in Marlborough and north Canterbury. (After Lensen, 1963; Lensen in Suggate et al., 1977.)

ciated tuffs and cherts in the older tectonized Clarence (= Mokoiwian) sequences, and younger doleritic sills and basalt flows intruded into Ngaterian to Haumurian rocks (Challis, 1968; Grapes, 1970).

Marlborough and Canterbury

Major Cretaceous sequences, up to 4100 m thick, occur in Marlborough, especially in the Awatere and Clarence Valleys and at Coverham, where they have been extensively studied (Thomson, 1919; Wellman, 1955, 1959; Suggate, 1958; Hall, 1963; Lensen, 1962; Challis, 1966). Lensen (in Suggate et al., 1977) recognizes three lithostratigraphic groups in Marlborough, each representing a different phase of sedimentation related to the eversion of the New Zealand Geosyncline and the change to the mobility characteristic of the succeeding Cretaceous and Cenozoic (Figs. 44–46, Table VI).

Sawtooth Group (Taitai Series–Motuan). Indurated, structurally complex siltstone, greywacke sandstone and minor conglomerate, classed in three formations generally included in Torlesse Supergroup and considered to be geosynclinal. Rocks of the group attain a maximum total thickness of some 2600 m. They

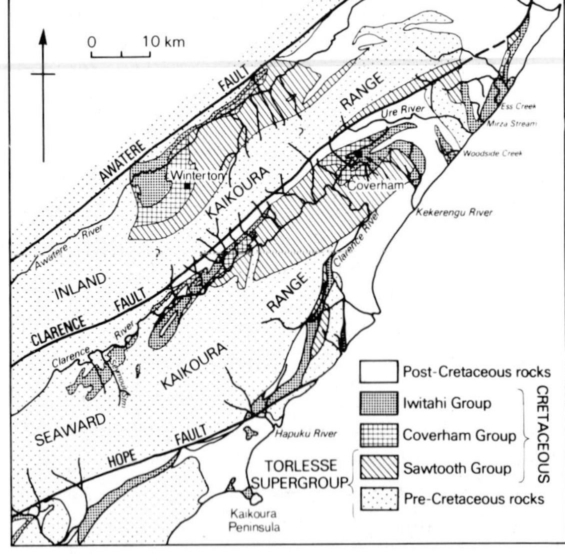

Fig. 45. Distribution of major lithological units in Marlborough. (After Lensen in Suggate et al., 1977, fig. 6.27.)

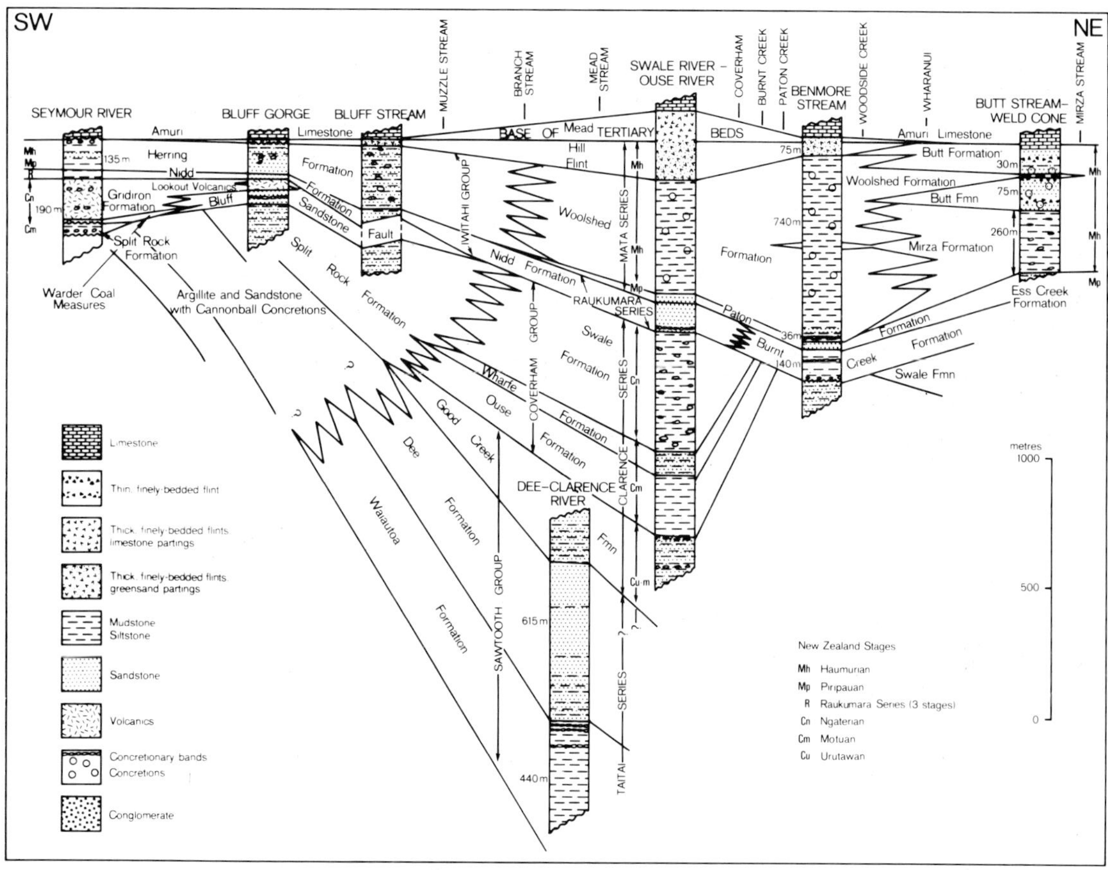

Fig. 46. Stratigraphic columns for important sections in Marlborough. (After Lensen in Suggate et al., 1977, fig. 6.28.)

crop out extensively in the Clarence and Awatere Valleys, being considered by Lensen to conformably overlie older Torlesse rocks in the former area and to be faulted against Torlesse in the latter (Allen, 1962; Challis, 1966). Specimens of *Inoceramus* sp. ex gr. *ipuanus–kapuus* Wellman suggest a Urutawan–Motuan age for the Good Creek Formation, and probable Cretaceous *Inoceramus* sp. favour a late Early Cretaceous age (Clarence Series) for the top of the Waiautoa Formation (Figs. 45–46; Speden, 1977).

Coverham Group (Motuan–Ngaterian). This group includes shelf sediments (Split Rock Formation) in south and west Marlborough which pass into deeper-water siltstone (Ouse and Swale Formations) and alternating sandstone and siltstone (Wharfe Sandstone) formations in north and east Marlborough. The formations represent the filling-in of the geosyncline. The units are highly fossiliferous, particularly at Coverham where ammonites (Henderson, 1973) and dicotyledonous pollen (Couper in Wellman, 1955; Couper, 1960) are common. Lateral equivalents are present in the Awatere Valley where the Winterton Formation has at its top coal measures (Challis, 1966; Speden, 1977). Thickness and facies vary widely (Figs. 45–46).

Split Rock Formation unconformably overlies Torlesse rocks in the Clarence Valley, and units of the group are faulted against Torlesse in Awatere Valley. However, the nature of the boundary between Sawtooth and Coverham Groups at Coverham is controversial; Thomson (1917) and Gair (1967) considered it to be unconformable, while Lensen (1962) and Hall (1963) mapped it as conformable. The boundary separates rocks differing markedly in degree of structural complexity, environment of deposition and, to a lesser degree, induration. It coincides with a major stratigraphic change of uncertain tectonic and time significance, but is similar to the

TABLE VI

Lithostratigraphic units of the Coverham and Iwitahi Groups, Marlborough (after Lensen in Suggate et al., 1977)

	South and West	Coverham	Northeast
(Tertiary)		(Amuri Limestone)	
	unconformity	(Mead Hill Flint)	Butt Formation
Haumurian	Herring Formation	Woolshed Formation	Mirza Formation
Piripauan		Paton Formation	Ess Creek Formation
Teratan	Nidd Formation		Burnt Creek Formation
Mangaotanean		Nidd Formation	
Arowhanan	Gridiron Formation		*unconformity* (Sawtooth Group)
Ngaterian	*unconformity*	Swale Formation	
Motuan	Split Rock Formation	Wharfe Sandstone	
		Ouse Siltstone	
	unconformity (Torlesse Supergroup)	*?unconformity* (Sawtooth Group)	

intra-Motuan (Albian) change in parts of Raukumara Peninsula and Wairarapa (Speden, 1977).

Iwitahi Group (Ngaterian–Haumurian). The group consists of intertonguing formations of fossiliferous sandstone, glauconitic sandstone, siltstone (both massive and bedded) and alternating sandstone and siltstone (Figs. 45–47), which were deposited in shallow water marginal to a landmass sited to the west and an impersistent land area to the southeast. To the southwest, where the group unconformably overlies Coverham Group, the lowest beds are non-marine (Warder Coal Measures), thick volcanics (Lookout Volcanic Member) are present, and there is an unconformity at the top. The volcanics are thickest in the Awatere Valley (1500 m, Challis, 1966), but in the Clarence Valley they do not exceed 360 m and thin rapidly to the northeast (Suggate, 1958). Plant macrofossils, including rare angiosperms (McQueen, 1956) and dicotyledonous pollen (Couper, 1960), and non-marine molluscs are present in Warder Coal Measures.

Overlying the Coverham Group in central and southern Marlborough is the glauconitic muddy sandstone of the Midd Formation which thins and becomes coarser-grained both to the northeast and southeast where there are shell beds with *Inoceramus rangatira, Megatrigonia glyptica* and *Eselaevitrigonia meridiana.* In northeast Marlborough the lowest formation of the Iwitahi Group, the micaceous siltstone of Burnt Creek Formation, unconformably overlies Sawtooth Group, due to local uplift during the Arowhanan, and the Cretaceous–Cenozoic boundary falls within Mead Hill Flint.

The Mata Series occurs extensively, and consists of intertonguing formations which are generally sandy

Fig. 47. Coverham, Marlborough, looking southwest down Wharf Stream. Low country in the foreground and middle distance is underlain by Cretaceous (Coverham and Iwitahi Groups), flanked to the west (right) by prominent ridges of Tertiary limestone and to the east (left) by ridges of Sawtooth Group. Mt Tapuaenuku (2885 m) forms the highest point of the range (Sawtooth Group) in the background. See Fig. 45 for geology. (Photo: S.N. Beatus.)

and coarser to the south (Fig. 46). South of Kaikoura only Mata rocks are present with the Piripauan restricted to the area north of the Conway River and Haumuri Bluff, the type locality of the Mata stages (Warren and Speden, 1977; Fig. 34). Elsewhere a generally thin Haumurian sequence is widely distributed, often with basal coal measures or conglomerate resting on a deeply leached surface cut in Torlesse Supergroup. Some 1500 m of Mata rocks have been recorded in the Malvern Hills (Speight, 1928).

Igneous rocks. The main volcanic activity occurred in the Ngaterian (Late Albian–Middle Cenomanian) when the Lookout Volcanics erupted, mostly as subaerial porphyritic alkaline–olivine basalt lava and agglomerate. The upper part of the Mt Tapuaenuku Massif consists of a layered pyroxenite–peridotite–anorthosite–gabbro–syenite complex (Challis in Suggate et al., 1977). Teschenite sills and variolitic basalt flows intrude Haumurian sediments in northeast Marlborough (Lensen, 1962). Acid and intermediate volcanics at Mt Somers, Malvern Hills and Rakaia Gorge, and possibly the gabbro–syenite intrusions at Mandamus, are poorly dated by stratigraphy, but may be early Late Cretaceous in age as suggested by a radiometric date for an andesite at Mt Somers (Hulston and McCabe, 1972, p. 419). Challis (1968) classified the Marlborough volcanics in terms of an island arc origin.

Nelson and Westland

The Cretaceous rocks of the western part of the

South Island are non-marine except in south Westland where a breccia 0–1500 m thick, correlated with the Hawks Crag Breccia (see below), is overlain by "coal measures" and marine Haumurian (Maastrichtian) sandstones reaching a maximum of 2000 m thick (Bowen in Suggate et al., 1977). Cretaceous non-marine rocks in north Westland and Nelson are classed in the Pororari Group (Lower Cretaceous), Paparoa Coal Measures (Raukumara–Mata Series; Nathan, 1974a), and the main part (Maastrichtian) of the Pakawau Group of northwest Nelson.

Pororari Group (Albian–?Cenomanian). The group (Bowen, 1964a; Nathan, 1974a) includes a fanglomerate as a middle formation (Hawks Crag Breccia) enclosed in finer-grained formations below (Ohika Formation) and above (Topfer Formation). Incomplete sequences occur in isolated areas; however, the Hawks Crag Breccia is present in most. The total thickness may exceed 4000 m.

The Ohika Formation (450 m) unconformably overlies the Greenland Group (Lower Paleozoic) in the lower Buller Gorge, where it grades laterally and vertically into Hawks Crag Breccia (500 m). The breccia is unconformably overlain by Brunner Coal Measures (Eocene). On the west side of the Paparoa Range, Hawks Crag Breccia rests unconformably on 600 m of sandstone and siltstone correlated with Ohika Formation, and grades up into fluviatile sandstone and conglomerate correlated with Topfer Formation, which is overlain by Paparoa Coal Measures. East of Paparoa Range southeast of Reefton, the Hawks Crag Breccia rests on granite and grades up into the Topfer Formation (Suggate, 1957; Beck et al., 1958).

The Pororari Group is thought to have been deposited during late locally intense tectonic activity with rapid terrestrial sedimentation taking place in fault-angle depressions and commencing in places with lacustrine sandstone and siltstone, followed by locally derived breccias and conglomerate representing fans, slump and flood plain deposits eroded from rapidly disintegrating fault scarps (Beck and Nathan in Suggate et al., 1977). As tectonic activity diminished, the coal measures were laid down.

Plant remains are common in the finer beds of the Pororari Group (Walkom in Wellman, 1950). Recent studies of microfloras (Norris and Waterhouse, 1970; Waterhouse and Norris, 1972) of samples from the Hawks Crag Breccia indicate an Albian age, probably Late Albian. The interbedding at the base of the Ohika Formation of igneous rocks and tuffs erupted from the Berlins Porphyry, and radiometric dates (S. Nathan, 1974b, pers. comm.) on the porphyry, support the age indicated by palynology. A Late Albian age for the Hawks Crag Breccia suggests the Topfer Formation is latest Albian and possibly Early Cenomanian in age.

Paparoa Coal Measures (Cenomanian–Maastrichtian). Rocks of this group crop out in the Greymouth Coalfield, south of Hokitika (Gage and Wellman, 1944), and were penetrated by the Arahura 1 oil-prospecting well (Nathan, 1974a). In the Greymouth Coalfield the fluviatile sandstone, conglomerate and siltstone have been subdivided into four coal-measure members separated by three siltstone–mudstone members (Nathan, 1974a; originally mapped as formations by Gage, 1952). Pillow lavas of olivine basalt are present in the third member above the base. Members and coal seams vary greatly in thickness. The greatest known thickness is 800 m, but to the northwest the group is less than 500 m thick. The rank of coal at any one stratigraphic horizon diminishes eastward and westward from the axis of sedimentation.

The overall pattern of sedimentation was controlled by a western source area and by a northeast-trending axis of subsidence towards which formations thicken, with a steep eastern flank between the axis and an intermittantly rising area to the east (Gage, 1952). Tectonic activity continued, with slight volcanism in the Greymouth area.

Most contacts between the Paparoa Coal Measures and the underlying Pororari Group are faulted in the Greymouth area (Gage, 1952), although there appears to be no unconformity above the Hawks Crag Breccia [= Jay (1) breccia]. An almost unbroken sequence may be present in the Pororari River area (M.G. Laird in Nathan, 1974a). A disconformity is probably present between the Paparoa Coal Measures and the Brunner Coal Measures on the western side of the Greymouth Coalfield, while on the eastern side the Brunner Coal Measures rest conformably on the middle part of the Paparoa Coal Measures. No physical break is apparent between the formations in the central Paparoa Range, in the central part of the Greymouth Coalfield or in the Arahura 1 drillhole.

Pakawau Group (Maastrichtian–Eocene?). In northwest Nelson rocks of the group unconformably overlie Paleozoic basement and are apparently overlain unconformably by younger units. The stratigraphy of the group is complex and little known, although Suggate (1956) established four formations in the principal coal-mining area. Thicknesses vary greatly; the group may exceed 1050 m in the east but be no more than 100 m thick in the southwest (Bishop, 1968). Most of the group is Haumurian (Bowen in Suggate et al., 1977). However, the oldest beds may range down into the Piripauan and an uppermost part of the youngest formation is probably Paleocene and Eocene in age.

Otago and Southland

Cretaceous sequences in southeastern New Zealand are dominantly non-marine, except for small areas of Maastrichtian marine beds (Table VII). In a few places there are two sets of non-marine beds separated by an unconformity; a lower set consisting mainly of breccia-conglomerate, and an upper set composed mostly of coal measures. The highest beds of the upper set overlap pre-Cretaceous basement rocks and, with the succeeding estuarine and marine deposits, document a westward transgressing sea during the latest Cretaceous.

Early to Mid-Cretaceous terrestrial and marginal marine beds (?Aptian–Early Senonian). The oldest Cretaceous rocks are thick sequences of schist and greywacke breccia or conglomerate and arkosic sandstone, which unconformably overlie Torlesse, metamorphic or igneous basement rocks in north and southeast Otago and Fiordland, and are thought to have accumulated in growing fault-angle depressions.

The Kyeburn Formation (Harrington, 1955; Bishop and Laird, 1976), north Otago, some 4000 m thick, has common red-stained beds and carbonaceous beds sporadically throughout. Microfloras with angiosperm pollen, and associated with poorly preserved angiosperm leaves, suggest an age not older than Late Albian (Mildenhall in Bishop and Laird, 1976) somewhat younger than an early Albian age indicated by a radiometric date of 104 m.y. obtained on biotite

TABLE VII

Cretaceous lithostratigraphic units in Otago and Southland

International unit	N.Z. Series and Stage	Oamaru District	Shag Point area	North Otago	East Otago Dunedin	Kaitangata	Ohai Coalfield
Paleocene					Abbotsford Mudstone	Wangaloa Formation	
Maastrichtian	Haumurian Stage Mata Series	Papakaio Formation	Katiki Formation Herbert Formation		Saddle Hill Siltstone Brighton Limestone	Taratu Formation	Morley Coal Measures New Brighton Conglomerate Wairio Coal Measures
		?	?				?
Senonian	Piripauan Stage		Shag Point Group		Taratu Formation		
			?	?	?	?	
Turonian	Raukumara Series					Henley Breccia	
Cenomanian				Kyeburn Formation			
Albian	Clarence Series					Tokomairiro Beds	
						?	
Aptian	Taitai Series (part)			?			

from a tuff band low in the formation (Bishop and Laird, 1976, p. 59).

The Henley Breccia (Mutch and Wilson, 1952; Harrington, 1958), some 900 m thick, extends as a narrow inland band south of Dunedin. Angiosperm leaves occur 150 m above the base; consequently, the entire formation may be younger than Late Albian. An upper limit of lower Senonian is set by the Late Senonian age of the unconformably overlying Taratu Formation (Harrington, 1958; Couper, 1960).

Two small areas of Early Cretaceous rocks are preserved in southwestern Fiordland. (1) The non-marine Puysegur Formation (Wood, 1960) is, locally at least, of late Clarence Series age by microfloras (I. Raine and G. Wilson, New Zealand Geological Survey, pers. comm.). In other areas the formation is latest Eocene or earliest Oligocene (Carter and Lindqvist, 1975). (2) 55 km to the east a small area of red-stained conglomerate and arkosic sandstone some 300 m thick, the Sand Hill Formation of Wood (1969), faulted into plutonic and metamorphic rocks of the Fiordland complex and probably unconformably overlain by the Eocene Hump Ridge Formation, is correlated with the Puysegur Formation.

Latest Cretaceous coal measures and marine beds (Senonian–Maastrichtian). In many places near the coast in eastern Otago, non-marine and estuarine quartzose conglomerates interbedded with siltstones and coal seams of variable thickness, unconformably overlie the leached surface of peneplained older rocks. From Otago Peninsula south the coal measures are called the Taratu Formation, the type area of which is in the Kaitangata Coalfield where the formation includes 17 named members (Harrington, 1958) having a total thickness of at least 600 m. Harrington (1958) followed Ongley (1939) in considering the Taratu Formation passed eastwards laterally and vertically into estuarine and fossiliferous marine beds of the Wangaloa Formation (Paleocene; Hornibrook and Harrington, 1957; Hornibrook, 1962; Webb, 1973a, b). Webb (1973a, b), although noting that the contact was obscure, favoured a probably unconformable relationship between the Taratu and Wangaloa Formations in the type area of the latter. The floras of the Taratu Formation have not been studied in detail. A sample from the upper part was given a Mata Series (Late Senonian–Maastrichtian) age by Couper (1960, p. 76). The age of the basal beds does not seem to be significantly older.

Near Dunedin Haumurian (Brighton Limestone, Saddle Hill Siltstone) sediments unconformably overlie the Taratu Formation (Henderson, 1970; Webb, 1973a, b). In southern North Otago, breccia-conglomerate and coal measures of the Shag Point Group (ca. 900 m thick) unconformably overlie schist and grade up into the non-marine and marine Herbert Formation (30 m) with Haumurian macro- and microfaunas. Overlying the Herbert Formation is the Katiki Formation, some 100 m of grey to black concretionary argillaceous sandstone containing Haumurian foraminifera (Webb, 1971). Webb (1973b) considers that the Herbert Formation unconformably overlies the Shag Point Group.

Further to the north, in the Oamaru District, basal quartzose coal measures of the Papakaio Formation (Gage, 1957) rest unconformably on a peneplained surface, and are overlain by marine beds of latest Cretaceous or Early Tertiary age. Inland the upper part of the Papakaio Formation may be Early Tertiary (Mutch, 1963). Gage (1957) favoured a history of repeated marine transgressions with reworking and erosion of quartz conglomerate during Papakaio time.

In Southland, Late Cretaceous non-marine sediments (Ohai Group) are restricted to the Ohai Coalfield, some 160 m west from the nearest Maastrichtian sediments at Kaitangata, where they are preserved in a major syncline faulted along the south limb and locally faulted on the north limb and unconformably overlie Paleozoic–Mesozoic basement rocks. The group was divided by Bowen (1964b) into three formations (Table VII). The lenticular nature and irregular repetition of lithologies, the relationship of secondary folds to the intervals between coal horizons, and non-sequences in sections crossing faults, indicate that sedimentation was accompanied by faulting and folding. Deformation also occurred prior to the deposition of the unconformably overlying Tertiary rocks. Couper (in Bowen, 1964b) concluded that microfloras indicated a Late Cretaceous (Piripauan–Haumurian) age.

Chatham Islands

These islands are situated near the edge of the continental shelf some 900 km east of Christchurch. They are notable for a relatively stable geological history from at least late Early Cretaceous to the present day. Cretaceous rocks are known on The Sisters Group, north of the main islands, where a siltstone sample provided a latest Albian–Turonian (Ngaterian–Arowhanan) microplankton assemblage (Mildenhall and Wilson, 1976), and on Pitt Island, the southernmost of the main islands.

Grindley et al. (1977) have reclassified the lithological units established by Hay et al. (1970) into a lower Waihere Bay Group, consisting of the non-marine to marginal marine Headland Conglomerate (10 m) and the overlying Tupuangi Sandstone [ca. 600 m; including the Rauceby Sandstone of Hay et al. (1970)], and an upper probably unconformably overlying Pitt Island Group consisting of marine palagonite tuff and breccia (Kahuitara Tuff, 225 m) and overlying basaltic lavas (Southern Volcanics). Although no basal contact is exposed the occurrence in Headland Conglomerate of schist fragments and quartz pebbles derived from the Chatham Schist imply a major unconformity. The Cretaceous sequence is overlain unconformably by Eocene or Upper Miocene to Pliocene tuffs and interbedded marine sediments.

Microfloras indicate a Ngaterian age for the base of the Tupuangi Sandstone and a Teratan age for a lower part of the Kahuitara Tuff, with the interval between being of uncertain age (Mildenhall, in prep.). Thus the top of the Tupuangi might be Arowhanan, and the base of the Kahuitara tuff Mangaotanean or even Arowhanan. The stratigraphy and microfloras favour a Urutawan–Motuan age for the Headland Conglomerate, even though the assemblages obtained from the formation give indefinite Korangan to Motuan (Late Aptian–Albian) ages. The Teratan age for the top of the Kahuitara Tuff is compatible with the presence of *Inoceramus opetius* Wellman (Speden, 1976c), and a radiometric data of 77.3 m.y. obtained from the base of the Southern Volcanics (Grindley et al., 1977).

FAUNAL HISTORY AND PALAEOBIOGEOGRAPHY (G.R. STEVENS)

Throughout Palaeozoic and Early Mesozoic time New Zealand was an integral part of Gondwanaland. In Late Precambrian–Devonian New Zealand occupied a position between Victoria, Australia and Victoria Land, Antarctica (Cooper, 1975). Except for some possible differential movements of Tasmania and New Zealand in the Silurian–Devonian (Cooper, 1975, fig. 6) it is thought that New Zealand stayed at the edge of Gondwanaland for about the next 200 m.y., until Gondwanaland started to break-up in the Middle Mesozoic.

In the Carboniferous and Permian, while ice sheets scoured the Gondwanaland continents, the site of modern New Zealand was occupied by the developing New Zealand Geosyncline. Little land existed in the New Zealand area – at most, scattered volcanic archipelagos (Suggate et al., 1977, fig. 11.14).

The New Zealand Permo-Carboniferous sequence is entirely marine, and Waterhouse (1967, 1970, 1976) has described from the Permian alternating sequences of warm- and cool-water tolerant organisms that he has interpreted as resulting from the effects of interglacial periods in Gondwanaland. At this time close faunal relationships existed between New Zealand and eastern Australia, indicating the existence of adequate routes for the interchange of benthic faunas.

For the first time in New Zealand's faunal history distinctive "Austral" (or "anti-Boreal") faunas appear in the Permian. These Austral faunas are interpreted as cool-temperate stenothermal populations developed peripheral to the main Gondwanaland mass straddling the South Pole of the time (e.g. Stevens, 1975, fig. 13).

Close ties with eastern Australia are lost, however, in the Triassic when a distinct "Maorian Province" (Diener, 1916; Wilckens, 1927) developed in the New Zealand Geosyncline, indicating a disruption of the formerly available benthic migration routes.

Triassic

A gap occurs in the New Zealand faunal sequence during the transition from Late Paleozoic to Mesozoic. This gap may be a result of unconformity, unsuitable facies, or both.

The New Zealand Triassic faunal record (Figs. 48, 49) commences with ammonoids of Malakovian age (Smithian and Spathian Stage). These ammonoids are members of widely distributed Tethyan and Circum-Pacific groups and as presumably they had nektonic or planktonic stages in their life histories they were able to freely enter New Zealand seas, otherwise barred to less vagile forms.

Etalian (Anisian) faunas, like those of the Malakovian, are characterized by ammonoids of almost cosmopolitan distribution. They are joined however in the Etalian by *Daonella,* one of many pterioid

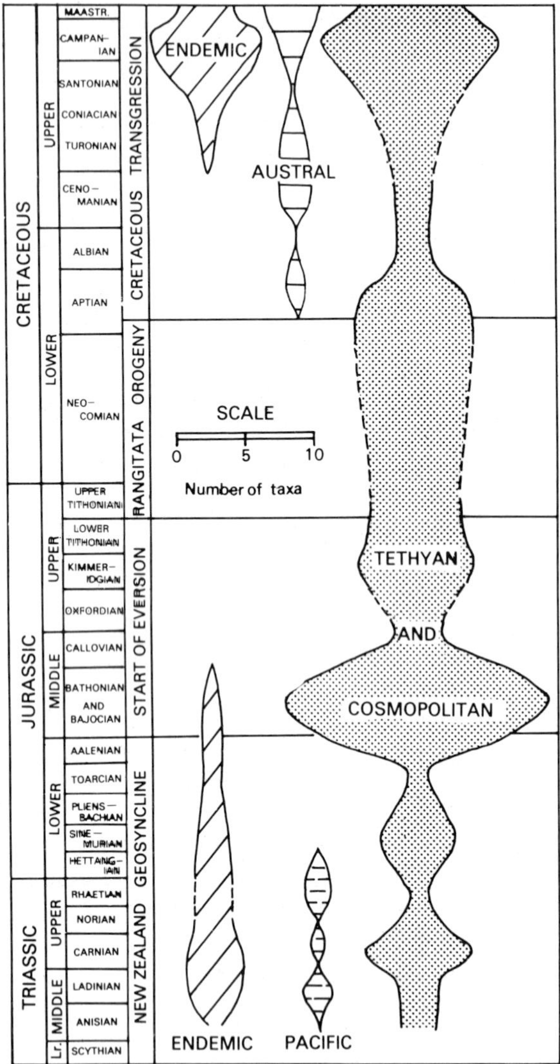

Fig. 48. Diagram showing the biogeographic elements represented among incoming taxa of Bivalvia and Gastropoda in the Mesozoic of New Zealand. (After Fleming, 1967.)

bivalves that acquired world-wide distribution in the Triassic and Jurassic, presumably as a result of a planktonic or pseudoplanktonic habit, or a long-lived planktonic larval stage.

On the other hand, other Etalian Mollusca and Brachiopoda, although few in number, show a degree of endemism, indicating isolation of the New Zealand region and heralding development of the Maorian Province.

Not until the Kaihikuan (Ladinian) is the Middle Triassic fauna represented by a rich assemblage of benthic invertebrates. A brachiopod assemblage characterized by large massive *"Spiriferina"* (representing endemic lineages), *Athyris* and *Mentzeliopsis* (endemic genus) is not closely related to any described from elsewhere, and was thought to be archaic by Trechmann (1918) and Wilckens (1927). Some characteristic elements extend to New Caledonia, which apparently shared with New Zealand not only a geosynclinal structure but also a benthic faunal history characterized by a certain degree of isolation to allow the development of such Maorian endemic elements.

The Late Triassic faunas show a continuation of the earlier faunal trends (Figs. 48, 49). Easily distributed cephalopods and pterioids (e.g., *Halobia* in the Oretian and Otamitan, *Monotis* in the Warepan, *Rhaetavicula* in the Otapirian) show Tethyan or cosmopolitan affinities and presumably origin. Additional cosmopolitan or Tethyan forms appeared in limited numbers and a Circum-Pacific faunal element is also discernible. Endemic genera and subgenera, some of which range to New Caledonia, continued to develop among Brachiopoda and benthic Mollusca, indicating a continuation of the conditions isolating

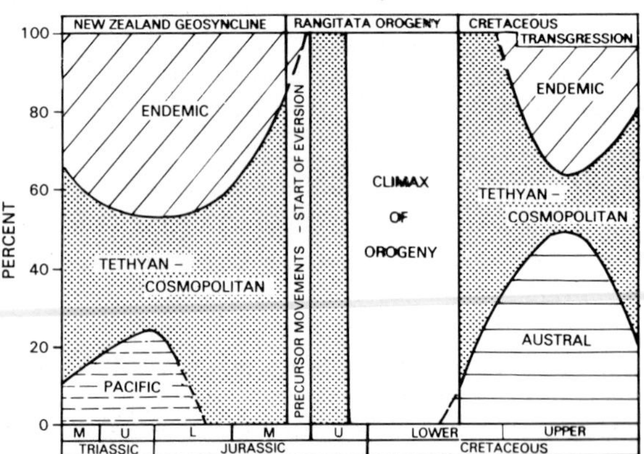

Fig. 49. Faunal changes related to orogenic events in the Mesozoic of New Zealand. The diagram shows percentages of the various biogeographical elements among new bivalve and gastropod arrivals in each stage, somewhat generalized. The increase in Tethyan–cosmopolitan elements in the Middle to Upper Jurassic corresponds with the climax of the Rangitata Orogeny, that temporarily ended isolation of the Southwest Pacific by establishing shallow-water routes to the northwest. The renewed development of endemic elements in the Upper Cretaceous reflects the increasing isolation of New Zealand. Appearance of Austral elements in the Cretaceous suggests that cool-temperate migration paths developed as Australasia, Antarctica and South America moved closer to the South Pole (cf. Fig. 51). (After Fleming, 1967.)

the New Zealand–New Caledonia segment of the New Zealand Geosyncline.

Jurassic

Early Jurassic faunal relationships resemble those of the Late Triassic (Figs. 48, 49). Some of the Maorian and Circum-Pacific elements continue (*Clavigera, Mentzelia, Otapiria*), and new Maorian endemic genera appear (*Kalentera, Sphaeriola, Pseudaucella*). Only a trickle of cosmopolitan or Tethyan incomers are recorded – *Oxytoma, Entolium, Lopha, Pseudolimea, Placunopsis* and *Camptochlamys.*

Isolation is suggested by the complete absence of Early Jurassic belemnites and trigoniids, abundant elsewhere in the world, but probably excluded from New Zealand seas by lack of suitable shallow-water migration routes. In contrast with restricted access for benthic forms, Early Jurassic seas, like those of the Triassic, were open to permit the immigration of world-wide ammonite genera (*Psiloceras, Schlotheimia, Dactylioceras, Harpoceras*), presumably with planktonic or nektonic stages in their life cycles.

In the Middle Jurassic (Temaikan Stage) there is a marked change in the proportions of the several faunal elements. A flood of Tethyan–cosmopolitan immigrants include benthic forms such as *Lithophaga, Indogrammatodon, Pholadomya, Camptonectes, Variamussium, Myophorella, Orthotrigonia, Hijitrigonia* and *Ostrea*, in addition to the pterioids *Meleagrinella, Inoceramus, Isognomon,* and Cephalopoda of presumably high vagility. During the Temaikan only one of the score or so of genera appearing for the first time is endemic (*Haastina*). This trend to fewer endemics and more widespread genera is accentuated in the Late Jurassic (Heterian to Puaroan), when no endemic genera are recorded as incomers, whereas a succession of ammonite and pterioid invaders (*Malayomaorica, Inoceramus, Buchia*), nektonic belemnites (*Belemnopsis, Conodicoelites, Hibolites*) and even typically benthic groups (several Trigoniidae, Limidae, *Astarte, Kutchithyris*) appear as immigrants, mainly from the Tethyan and Pacific border areas (Figs. 48–51).

Trias–Jura biogeographic trends may be generalized as follows. At no time during the Triassic and Jurassic was access denied to easily distributed groups such as ammonoids and pterioid bivalves, although the relatively small number of such organisms that colonized may suggest the existence of barriers. The Middle Triassic to late Early Jurassic interval, on the other hand, was characterized by high endemism

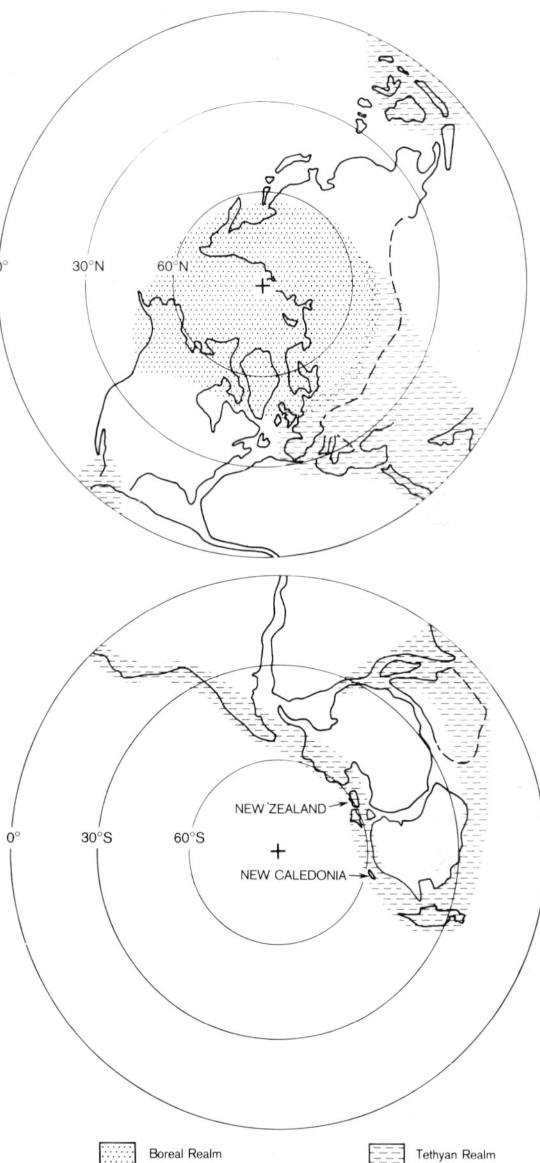

Fig. 50. Late Jurassic (Kimmeridgian–Tithonian) biotic provinces as delineated by belemnites. Grouping of the Laurasia landmasses around the North Pole at this time is reflected in the differentiation of Boreal belemnites, adapted to cool-temperate waters. On the other hand, grouping of the Gondwana countries away from the South Pole provided Tethyan belemnites with tropical and warm-temperate dispersal routes. (Continental reconstructions after Creer, 1973, fig. 4; faunal data from Stevens, 1963, 1965, 1967, 1971, 1973a.)

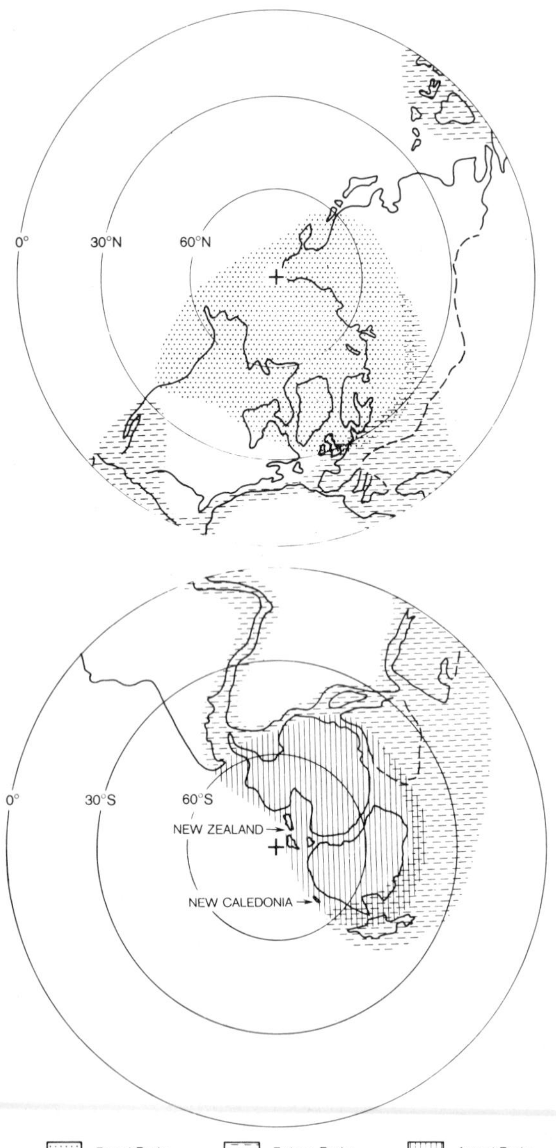

Fig. 51. Early Cretaceous (Aptian–Albian) biotic provinces as delineated by belemnites. As in the Jurassic (Fig. 50) Boreal belemnites, adapted to life in cool-temperate seas populated the northern fringes of the Laurasia landmasses, grouped around the North Pole. Fragmentation of Gondwanaland had been in progress since Middle and Late Jurassic time and by Aptian–Albian South America and Africa had drifted apart to form the South Atlantic Ocean. Tethyan belemnites, adapted to life in tropical and warm-temperate seas, migrated along the dispersal routes that became available at this time. Movement of Australasia and Antarctica had brought them closer to the South Pole (cf. Fig. 50) and these countries were populated by Austral belemnites, like the Boreal belemnites, adapted to cool-temperate seas. (Continental reconstructions after Creer, 1973, fig. 5; faunal data from Stevens, 1963, 1965, 1967, 1971, 1973b.)

(50–60% of fauna) among benthic organisms, so much so that Diener (1916) and Wilckens (1927, p. 59) recognized New Zealand and New Caledonia as constituting a "Maorian Province". The high endemism suggests difficulty of access for marine organisms at this time of extensive geosynclinal development in the Southwest Pacific. There is evidence, however, that access may have been better to the Torlesse or eastern basin of the New Zealand Geosyncline, as at various levels in the New Zealand Triassic–Jurassic distinctive Torlesse taxa appear that have marked Indo-Pacific affinities at times when equivalent Murihiku taxa show a high level of endemism (Campbell, 1974c).

This difference in faunal affinity between Murihiku and Torlesse fossils lends considerable support to the thesis that the two facies accumulated in separate basins, perhaps linked by a filter barrier of fluctuating effectiveness, and have been subsequently rafted together by sea-floor spreading (cf. Bradshaw and Andrews, 1973; Blake et al., 1974).

In both Torlesse and Murihiku basins a major change in access took place in the Middle Jurassic, putting an end to the period of high endemism, and initiating a flood of invaders from the Tethyan and Indo-Pacific realms (Fleming, 1967, 1970, 1975). During the Middle and Late Jurassic the New Zealand and New Caledonian region received numerous shallow-water immigrants, all having very close affinities with Indonesia, the Himalayas and Iran (Marwick, 1953a, b: Fleming, 1967; Avias, 1953).

The Middle and Late Jurassic spread of Tethyan and Indo-Pacific shallow-water faunas extended beyond New Zealand to West Antarctica and southern South America (Stevens, 1967, 1971, 1973a, b, c). Oxygen isotope studies of these migrant faunas suggest that they were warm-temperate forms (Stevens, 1971), and we look in vain in Australasia and Antarctica for a marked cool-temperate "Austral"

Fig. 52. Late Triassic (Oretian–Otapirian) palaeogeography of New Zealand. (After Suggate et al., 1977, fig. 11.15.) For ease of reference the distribution of rocks and positions of shorelines have been plotted on the modern outline of New Zealand. These patterns however have since been disrupted by 450 km of horizontal movement along the Alpine Fault. Although timing of this movement is the subject of debate (see Suggate et al., 1977, pp. 320–322, 678), few geologists doubt that movement has affected the distribution of Jurassic and older rocks and the inserts in Figs. 52–55 show the reconstructed shape of New Zealand, together with the probable distribution of land and sea.

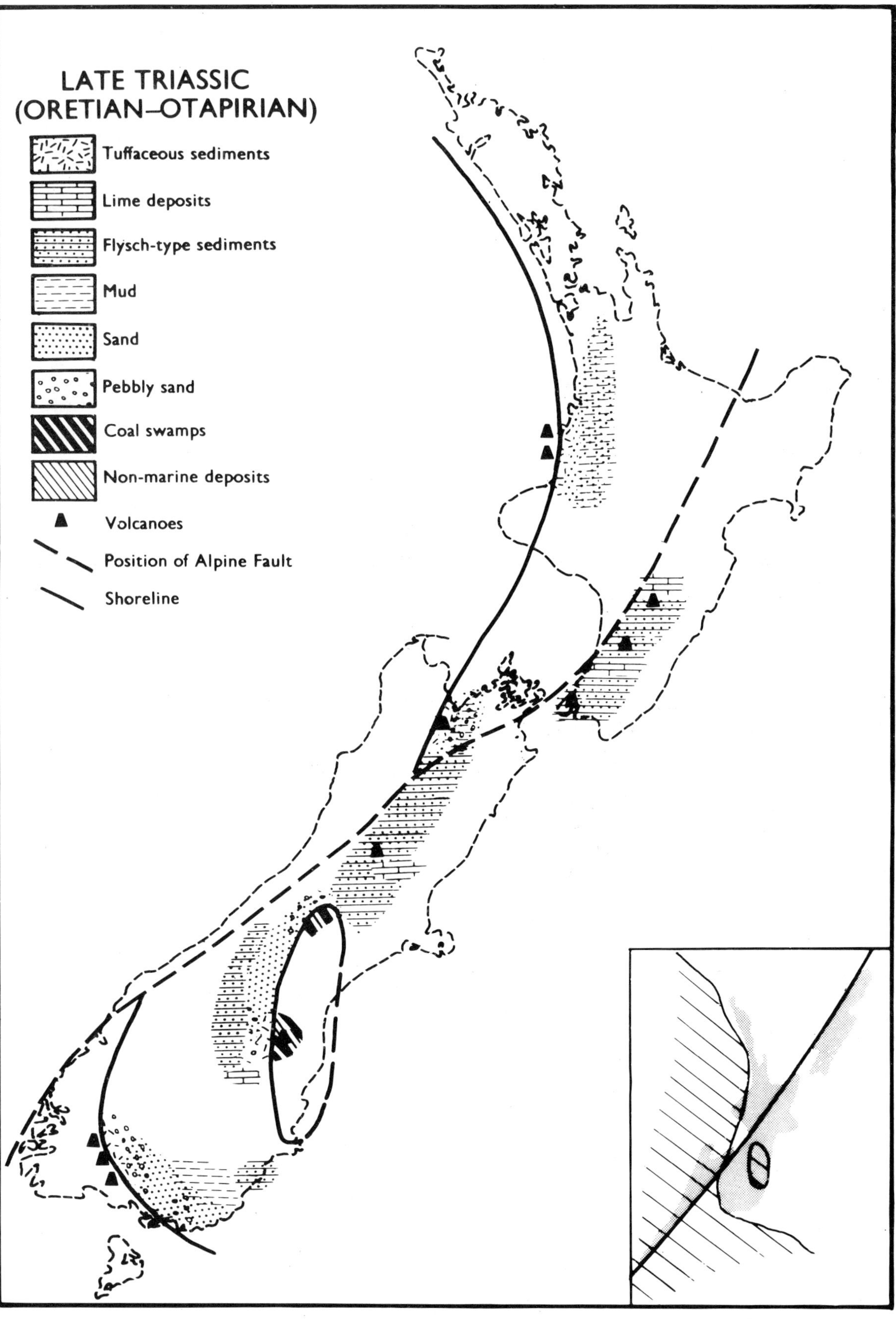

LATE TRIASSIC
(ORETIAN–OTAPIRIAN)
Tuffaceous sediments
Lime deposits
Flysch-type sediments
Mud
Sand
Pebbly sand
Coal swamps
Non-marine deposits
Volcanoes
Position of Alpine Fault
Shoreline

realm, in contrast to the situation in the Permian and Cretaceous–Paleocene.

The major biogeographic change initiated in the Middle Jurassic appears to have coincided with the beginning of geosynclinal eversion in the New Zealand Geosyncline, the first signs of which occurred in the Callovian (Speden, 1971; ca. 160 m.y.). Tectonism accompanying the eversion movements probably opened up shallow-water immigration routes into the Southwest Pacific that were used by colonizing benthic organisms.

Fig. 52 depicts the New Zealand palaeogeographic situation in the Late Triassic, when development of the New Zealand Geosyncline was at a maximum. Fig. 53 on the other hand shows the geosyncline at its waning phase, in Late Jurassic, as movements of the Rangitata Orogeny progressively restrict the areas of marine influence.

Cretaceous

No marine or non-marine fossils of Neocomian and Early Aptian age have been recognized in New Zealand and New Caledonia. Presumably, eversion of the New Zealand Geosyncline, initiated in Middle Jurassic, reached a climax in this interval ("Rangitata Orogeny"). This period is assumed to represent an interval of extended land and long fingers of land may have reached out from New Zealand into the Southwest Pacific.

The marine fossil record is taken up once more in the Late Aptian (Fig. 54). During the Late Cretaceous and Early Tertiary topographic relief became less and the land area diminished (Fig. 55), and by Middle Oligocene almost two thirds of the area of modern New Zealand had been submerged (e.g., Suggate et al., 1977, figs. 11.18–11.22).

Furthermore, sea-floor spreading to the west and south of New Zealand commenced in Late Cretaceous times (ca. 80 m.y.; Hayes and Ringis, 1973; Christoffel and Falconer, 1972). It is thought therefore that there is very little likelihood of land links to New Zealand subsequent to Late Mesozoic.

The sparse marine fauna of the Late Aptian (Korangan Stage) of New Zealand includes a number of Tethyan or cosmopolitan incomers among benthic Bivalvia (e.g., *Cucullaea, Spondylus*), at least two species of the widely distributed pterioid genus *Aucellina*, and two species of *Maccoyella*, otherwise restricted to Australia and South America (Fleming in Wellman, 1959; Fleming, 1975). Thus the Tethyan influence that dominated the Late Jurassic remains strong, but *Maccoyella* represents a southern element (i.e. "Austral"), not previously discernible in the Mesozoic record (Figs. 48, 49).

During the succeeding periods of the Early and Late Cretaceous the immigrant mollusca can all be classified as Tethyan–cosmopolitan, Austral or endemic. The Tethyan–cosmopolitan element, relatively strong in the Aptian, declines in the middle of Late Cretaceous but again assumes importance at the end of the period so that it dominates Campanian–Maastrichtian (Piripauan–Haumurian) molluscan immigrants. The Austral element, heralded by *Maccoyella* in the Aptian, had a fluctuating influence throughout the Cretaceous, but is quite strongly represented by taxa such as the belemnite *Dimitobelus*, southern forms of *Kossmaticeras* and other ammonites, *Nototrigonia, Struthioptera, Pacitrigonia, Acanthocardia acuticostata, Neilo, Nordenskjoldia*, and *Lahillia.* Such Austral elements, common to New Zealand and Australia or to New Zealand and Chile or Seymour Island, give a distinctive aspect to Cretaceous faunas that was lacking in the Jurassic and Triassic. No Cretaceous endemic genera are recognized before the Cenomanian (in which *Costacolpus* appears), but they become important in the Campanian and Maastrichtian with such forms as *Cucullastis, Mixtipecten, Tikia, Conchothyra* and *Cyclorismina.* Successive forms of *Inoceramus* in New Zealand, widely used in correlation, illustrate the fluctuating influence of immigration and isolation leading to endemism. They shared the repeated colonizing success of their predecessors among pterioid bivalves, so that *I.* cf. *neocomiensis, I. concentricus* and *I. ipuanus* appear to be the result of recurrent invasion from the north, while *I. warakius, I. rangatira, I. bicorrugatus, I. pacificus, I. australis* and *I. matatorus* show various degrees of endemism as a result of isolation.

Isolation is also indicated by the persistence in the New Zealand Cretaceous of several genera of Trigoniidae (*Pterotrigonia, Oistotrigonia*) after their disappearance from other lands, so that in the Late Mesozoic the New Zealand fauna already contained relict organisms that had suffered extinction in other parts of the world.

Among the angiosperms that arrived in the Late Cretaceous, *Nothofagus* and the proteaceous plants so characteristic of southern lands have a good fossil record. The angiosperms probably required con-

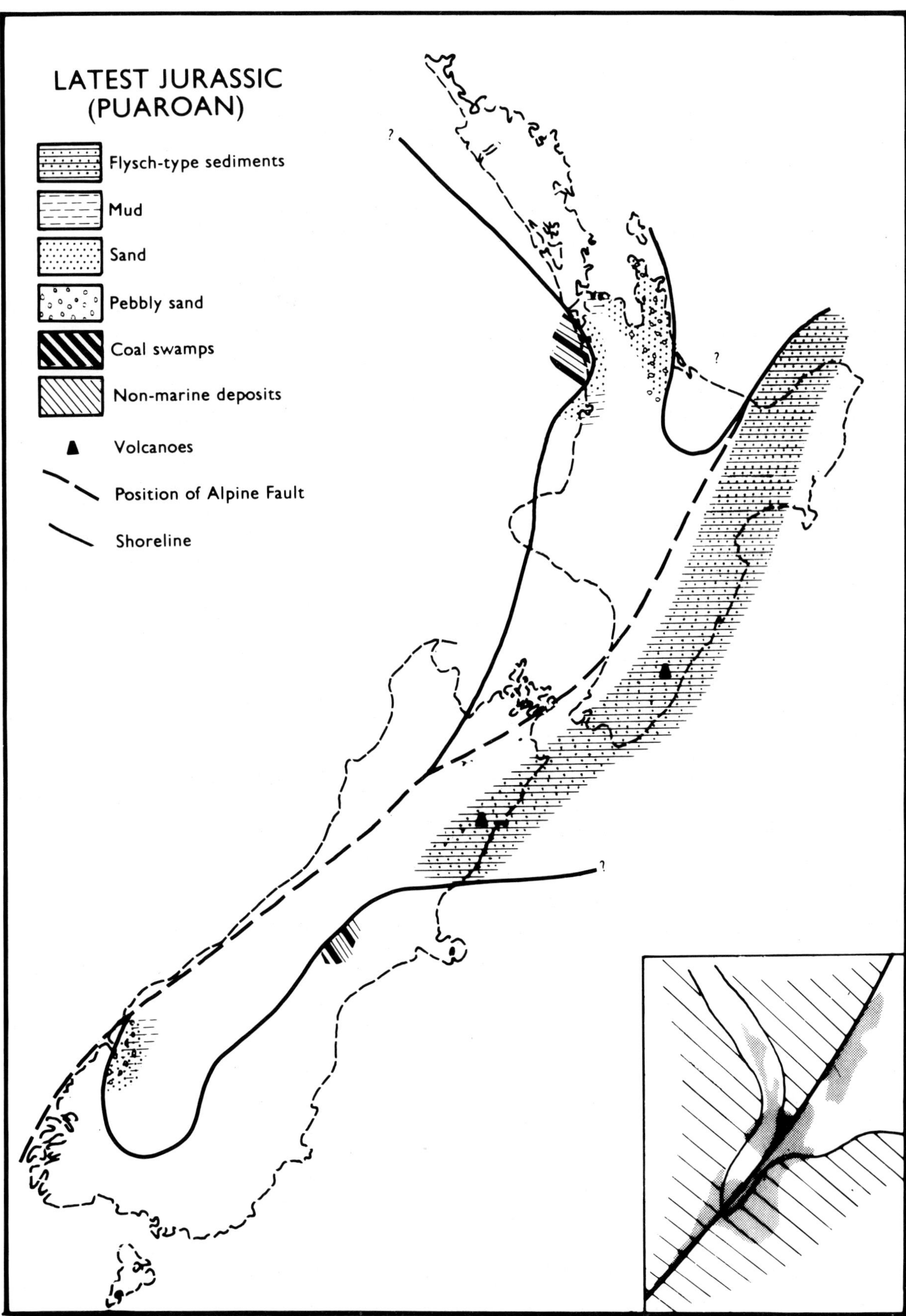

Fig. 53. Latest Jurassic (Puaroan) palaeogeography. (After Suggate et al., 1977, fig. 11.16.)

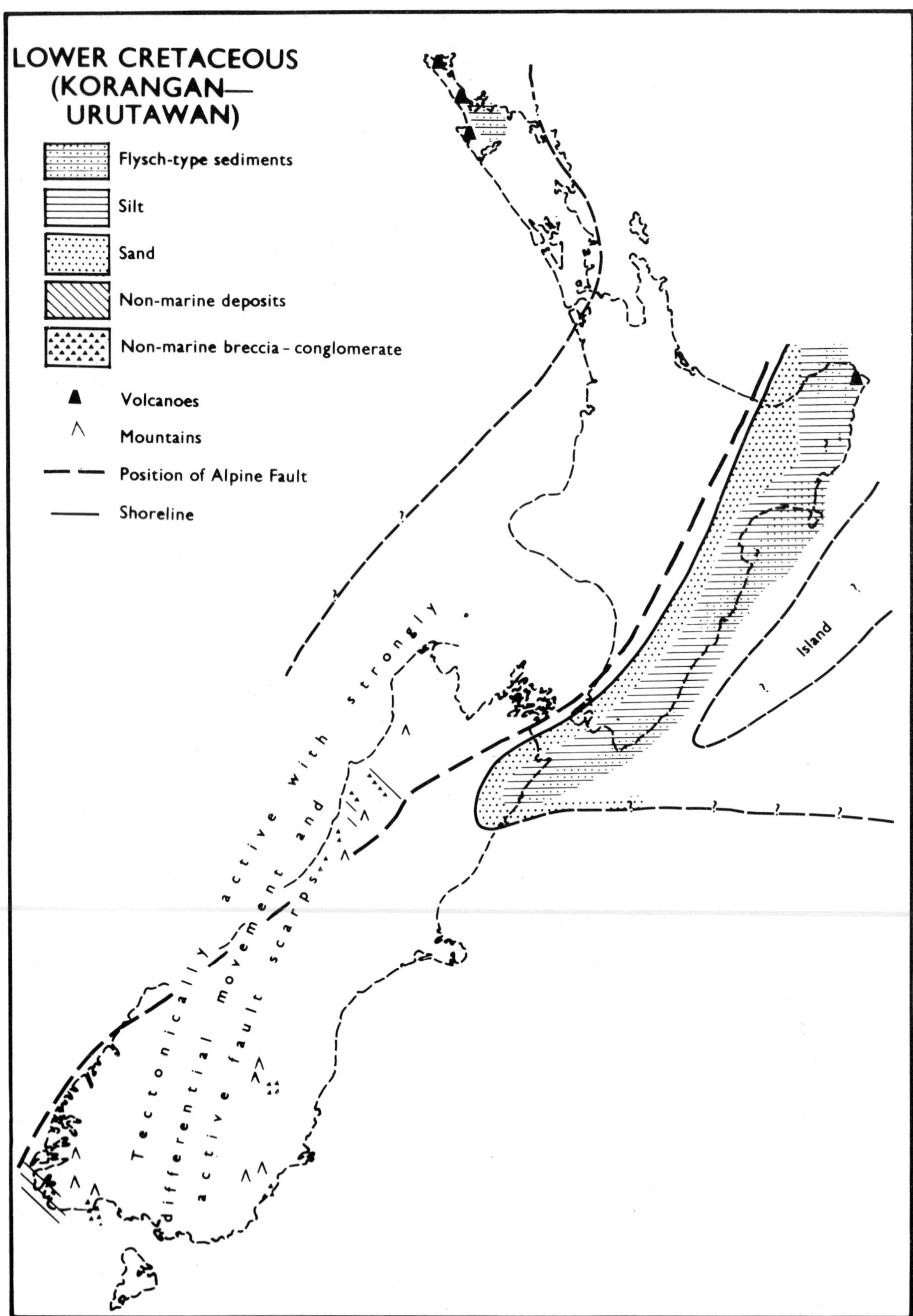

Fig. 54. Early Cretaceous (Korangan–Urutawan) palaeogeography. (After Suggate et al., 1977, fig. 11.17.)

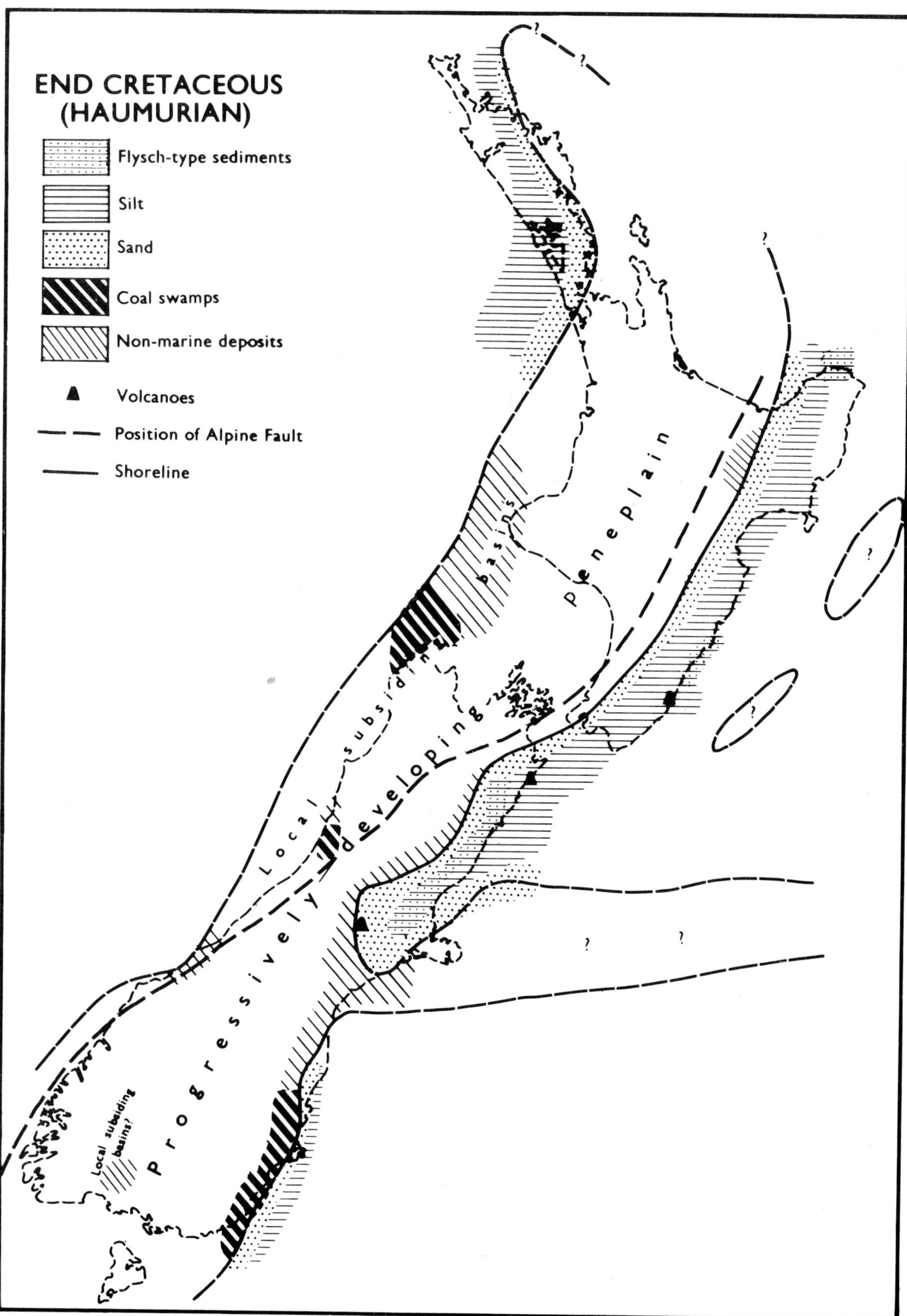

Fig. 55. End Cretaceous (Haumurian) palaeogeography. (After Suggate et al., 1977, fig. 11.18.)

tinuous land for their dispersal (Gressitt, 1963), as did the ratite moas and kiwis, unless they lost the power of flight after their arrival (Fleming, 1970). Isolation of New Zealand was, however, complete before marsupials and snakes reached Australia and New Guinea.

THE RELATIONSHIP OF NEW ZEALAND MESOZOIC FAUNAS TO PLATE TECTONICS

The changing affinities of New Zealand Mesozoic faunas may be interpreted as reflecting stages in the rotation and subsequent fragmentation of the Southern Hemisphere Gondwanaland landmasses. In the Carboniferous and Permian New Zealand was peripheral to Gondwanaland, lying adjacent to the Gondwana continental ice sheet, and palaeomagnetic and other data indicate a situation in high latitudes (e.g. Stevens, 1974, fig. 1.8; Waterhouse and Bonham-Carter, 1975). The Permian marine faunal succession in New Zealand has been interpreted as indicating alternations of dominance by cool (Austral) and warm (Tethyan) elements, compatible with a sequence of glacial and interglacial climates on the adjacent Gondwana lands (Waterhouse, 1967, 1970, 1976; Waterhouse and Bonham-Carter, 1975).

In the Triassic and Early Jurassic New Zealand probably remained peripheral to Gondwanaland, but by this time Gondwanaland had moved northwards away from the South Pole and glaciation had ceased (Creer, 1973). During this time many marine animals achieved a cosmopolitan distribution; climate was largely uniform and extensive shallow-water migration routes were available throughout many regions of the world. In the Southwest Pacific, however, although highly vagile marine invertebrates had apparently relatively free access, migration of benthic forms of lesser vagility was largely impeded (Stevens, 1965; Fleming, 1967), and New Zealand and New Caledonia formed a distinctive "Maorian Province" (Diener, 1916; Wilckens, 1927). At this time there is no biological evidence suggesting that New Zealand and New Caledonia were intimately associated with other lands.

The flood of Tethyan and cosmopolitan benthic invertebrates in Middle and Late Jurassic time coincides with the start of fragmentation of Gondwanaland (Heirtzler, 1968; Larson and Pitman, 1972) and with the start of the Rangitata Orogeny, although actual sea-floor spreading to west and south of New Zealand did not apparently commence until Late Cretaceous (ca. 80 m.y.), according to magnetic-stripe chronology (Hayes and Ringis, 1973; Christoffel and Falconer, 1972).

Movements associated with the Rangitata Orogeny probably produced extensive land masses fingering both north and south of New Zealand and New Caledonia. These movements, together with others linked with the start of fragmentation of Gondwanaland, ended isolation of the Southwest Pacific (albeit temporarily), by establishing shallow-water migration routes along which moved waves of Tethyan and Indo-Pacific immigrants.

Ancestors of the "Archaic" elements of New Zealand's biota, the podocarps, ferns and araucarians so conspicuous in New Zealand forests, and of the Tuatara (*Sphenodon*), New Zealand frog (*Leiopelma*), and ratite birds such as moas and kiwis (Fleming, 1963; Kuschel et al., 1975), probably arrived in New Zealand at about the time (Late Jurassic and Early Cretaceous) when the Rangitata Orogeny was at its peak, and presumably land extension was at its maximum. Whether these organisms travelled to New Zealand via continuous fingers of land, or across strings of archipelagos, separated by narrow stretches of shallow sea, is not known. This time, however, was the only period in New Zealand's geological history in which existed the strong possibility of land links between New Zealand and other Southwest Pacific lands.

Although some authors (e.g. Cracraft, 1975; Raven and Axelrod, 1972, 1974) favour a southern origin for the archaic elements of New Zealand's biota, the strong Indonesian–Himalayan affinities of the marine benthic immigrants that flooded into New Zealand at the same time suggest the availability of a terrestrial route from the north. Such a northern route may have allowed terrestrial organisms of tropical, subtropical and warm-temperate character (probably including the moas, kiwis, etc.) access to New Zealand from the north, whereas those with cool-temperate characteristics migrated via Antarctica (Stevens, 1977). Such links, however, were evidently broken by the time land dinosaurs, early mammals (marsupials) and snakes appeared in Australia.

Some Gondwanaland reconstructions for the Jurassic place West Antarctica, Australia and New Zealand in high southern latitudes (e.g. Smith et al., 1973, fig. 16A), in which case by analogy with the

Northern Hemisphere one would expect to find anti-Boreal (i.e. Austral) rather than Tethyan faunas in these countries (Stevens, 1967, 1971).

Oxygen isotopic evidence also suggests that these countries were populated by warm-temperate marine faunas (Stevens, 1971), and this, and the presence of strong Tethyan faunal affinities, have been used to support the idea that, contrary to many reconstructions for the Middle and Late Jurassic, the Gondwana lands were grouped in temperate climates some distance away from the Jurassic South Pole (Fig. 50) (Stevens, 1967, 1971, 1973a–c; Filatoff, 1975; Vakhrameev, 1972, 1975).

The situation changed markedly in the Cretaceous, however, and development of an Austral element, the first to appear since the Permian, is a notable feature of the New Zealand and Australian Cretaceous (Fig. 51). The presence of Austral elements in Madagascar, southern India, Australasia, West Antarctica and Patagonia may be interpreted as indicating that southward movement of the Gondwana landmasses, the first faunal evidence for which was evident in the Kimmeridgian–Tithonian (Stevens, 1973a) continued throughout the Cretaceous (Creer, 1973). Whereas these countries had shared Tethyan macrofaunas in the Jurassic, they shared Austral macrofaunas in the Cretaceous (Stevens, 1973b).

The change in affinity of the New Zealand marine invertebrates from Tethyan in the Jurassic to Austral in the Cretaceous was, however, not accompanied by a major difference in isotopic temperatures obtained from belemnites and it has been suggested that the climatic change involved in the New Zealand area was minimal – from marginal Tethyan to marginal Austral (perhaps equivalent to a change from warm-temperate to cool-temperate) (Stevens, 1971; Stevens and Clayton, 1971). Such a change might result from development of new oceanic current patterns in the Southwest Pacific due to palaeogeographic changes, following the Rangitata Orogeny, as well as to movements of Gondwanaland (cf. Creer, 1973).

In the Late Cretaceous the Austral affinities of the New Zealand marine macrofaunas are greatly reinforced (Fleming, 1975). At this time Austral affinities between the southern continents were so strong that some authors maintain that extensive land links were established (e.g. Gressitt, 1963). During this time New Zealand received early angiosperms as migrants – plants such as the southern beech (*Nothofagus*) and southern Proteaceae, that had travelled from southern South America via a cool-temperate route around the margins of Antarctica (Raven and Axelrod, 1972, 1974).

Both northern and southern land routes to New Zealand were however broken in the late Late Cretaceous (ca. 80 m.y.), when sea-floor magnetic information indicates that opening started of the Tasman Sea between New Zealand and Australia, and the Southwest Pacific Ocean between Campbell Plateau and West Antarctica (Hayes and Ringis, 1973; Christoffel and Falconer, 1972).

Exchange of cool-temperate ("Austral") marine faunas between southern South America, West Antarctica and New Zealand persisted however until the Paleocene (Fleming, 1975), indicating that the sea opening up south of New Zealand still had shallow-water migration routes across it that could be traversed by benthic marine organisms. Such shallow-water Austral links disappear or are lost after the Paleocene, and do not reappear in New Zealand's geological record.

In the Early Tertiary New Zealand moved northwards, as part of the Indian-Australian plate (Molnar et al., 1975). As it did so marine immigrants of Australian–Indonesian–Malaysian affinity appeared in the New Zealand fossil record, but it is generally agreed that the migrant taxa were all capable of being distributed across open ocean, and there is no suggestion of shallow-water links to these regions persisting into the Tertiary (Fleming, 1975).

As Australia and Antarctica started to separate in the Late Paleocene (ca. 55 m.y.; Weissel and Hayes, 1972), and faunal dispersal under the influence of the West Wind Drift was initiated (Fell, 1962), New Zealand began to receive southern immigrants again ("Neoaustral"), but like those from Australia and other northern countries these also were all of types with the capability of being widely distributed across open ocean (by means of epiplanktonic, planktonic or nektonic mode of life in larval and/or adult stages) (Fleming, 1975).

Regardless of whether the early mammals (marsupials) and snakes used northern or southern routes to reach Australia [and Owen's reconstructions (1976) show that both were geographically possible] land links to New Zealand had disappeared by the time they arrived on the scene (probably in Late Cretaceous or Paleocene). It is to this isolation of New Zealand from mammals, lasting until the coming of Polynesian man some 1000 years ago, that we owe the persistence of Mesozoic relics, like the Tuatara, and the adaptive radiation of flightless birds, like the moas and kiwis, to fill niches normally occupied by grazing and browsing animals.

ACKNOWLEDGEMENTS

The above contribution has been largely simplified and condensed from *The Geology of New Zealand* (Suggate et al., 1977), drawing on the contributions to that work of A.C. Beck, F.E. Bowen, C.A. Fleming, D.R. Gregg, G.W. Grindley, R.F. Hay, D. Kear, J.T. Kingma, G.J. Lensen, I.C. McKellar, A.R. Mutch, S. Nathan, J.J. Reed, J.C. Schofield, D.N.B. Skinner, I.G. Speden, G.R. Stevens, R.P. Suggate, G. Warren, J.B. Waterhouse, W.A. Watters and P.N. Webb. The present authors have up-dated the information in *The Geology of New Zealand* where appropriate. I.G. Speden has been responsible for the compilation of the Cretaceous section and G.R. Stevens for the remainder.

The authors are grateful to G.W. Grindley, P.R. Moore, J.I. Raine, C.P. Strong and R.P. Suggate (all of New Zealand Geological Survey) for reviewing the manuscript. P.R. Moore kindly provided additional information.

REFERENCES

Allan, R.S., 1956. Report of the Standing Committee on datum-planes in the geological history of the Pacific region. *Proc. 8th Pac. Sci. Congr.*, 2: 325–423.

Allen, A.D., 1962. *The Stratigraphy and Structure of the Middle Awatere Valley (S35)*. Thesis, lodged in the Library, Victoria University of Wellington, Wellington (unpublished).

Andrews, P.B., 1974. Deltaic sediments, Upper Triassic Torlesse Supergroup, Broken River, North Canterbury, New Zealand. *N.Z. J. Geol. Geophys.*, 17: 881–905.

Andrews, P.B., Bishop, D.G., Bradshaw, J.D. and Warren, G., 1974. Geology of the Lord Range, Central Southern Alps, New Zealand. *N.Z. J. Geol. Geophys.*, 17: 271–99.

Andrews, P.B., Bradshaw, J.D. and Speden, I.G., 1976. Lithological and palaeontological content of the Carboniferous–Jurassic Canterbury Suite, South Island, New Zealand. *N.Z. J. Geol. Geophys.*, 19: 791–819.

Arber, E.A.N., 1913. On the earlier Mesozoic flora of New Zealand. *Proc. Cambridge Philos. Soc.*, 17: 122–31.

Arber, E.A.N., 1917. The earlier Mesozoic floras of New Zealand. *N.Z. Geol. Surv. Paleontol. Bull.*, 6.

Arkell, W.J., 1953. Two Jurassic ammonites from South Island, New Zealand, and a note on the Pacific Ocean in the Jurassic. *N.Z. J. Sci. Technol.*, B35(3): 259–64.

Arkell, W.J., 1956. *Jurassic Geology of the World*. Oliver and Boyd, London and Edinburgh, 806 pp.

Arkell, W.J. et al., 1957. *Treatise on Invertebrate Palaeontology. Part L (Cephalopoda, Ammonoidea)*. Geol. Soc. Am., University of Kansas Press, Lawrence, Kansas, 490 pp.

Aronson, L., 1968. Regional geochronology of New Zealand. *Geochim. Cosmochim. Acta*, 32: 669–97.

Austin, P.M., Sprigg, R.C. and Braithwaite, J.C., 1973. Structure and petroleum potential of Eastern Chatham Rise, New Zealand. *Am. Assoc. Pet. Geol. Bull.*, 57: 477–97.

Avias, J., 1951. Note préliminaire sur la présence du Lias inférieur (Hettangien et Sinemurien) et moyen en Nouvelle-Calédonie. *C.R. Acad. Sci.*, Paris, 232: 172–4.

Avias, J., 1953. Contribution à l'étude stratigraphique et paléontologique des formations antécrétacées de la Nouvelle-Calédonie Centrale. *Sci. Terre*, 1: 1–276.

Bates, T.E., 1969. The Whatarangi Formation (Lower Cretaceous), Aorangi Range, Wairarapa, New Zealand. *Trans. R. Soc. N.Z., Geol.*, 6(11): 139–142.

Beck, A.C., Reed, J.J. and Willett, R.W., 1958. Uranium mineralization in the Hawks Crag Breccia of the lower Buller Gorge Region, South Island, New Zealand. *N.Z. J. Geol. Geophys.*, 1: 432–450.

Bishop, C.C., 1970. Upper Cretaceous stratigraphy on the west side of the Northern San Joaquin Valley, Stanislaus and San Joaquin Counties, California. *Spec. Rep. Calif. Div. Mines Geol.*, 104.

Bishop, D.G., 1968. *Sheet S2 – Kahurangi. Geological Map of New Zealand 1 : 63,360*. Department of Scientific and Industrial Research, Wellington, 1st edition.

Bishop, D.G., 1974. Stratigraphic, structural and metamorphic relationships in the Dansey Pass area, Otago, New Zealand. *N.Z. J. Geol. Geophys.*, 17: 301–336.

Bishop, D.G. and Laird, M.G., 1976. The stratigraphy and depositional environment of the Kyeburn Formation, Central Otago. (With an appendix by D.C. Mildenhall.) *J. R. Soc. N.Z.*, 6(1): 55–71.

Blake, M.C., Jones, D.L. and Landis, C.A., 1974. Active continental margins: Contrasts between California and New Zealand. In: C.A. Burk and C.L. Drake (Editors), *The Geology of Continental Margins*. Springer-Verlag, Berlin, pp. 853–872.

Boehm, G., 1907. Beiträge zur Geologie von Niederlandisch Indien. 1: Die Südküsten der Sula-Inseln Taliabu und Mangoli 3: Oxford des Wai Galo. *Palaeontographica, Suppl.*, 4 (Abt. 1): 3.

Boehm, G., 1911. Grenzschichten zwischen Jura und Kreide von Kawhia (Nordinsel Neuseelands). *Neues Jahrb. Min. Geol. Paläontol.*, 1911(1): 1–24.

Boehm, G., 1912. Beiträge zur Geologie von Niederländisch Indien. 1: Die Südküsten der Sula-Inseln Taliabu und Mangoli 4: Unteres Callovien. *Palaeontographica, Suppl.*, 4 (Abt. 1) Abs. 4.

Boles, J.R., 1974. Structure, Stratigraphy and Petrology of mainly Triassic rocks, Hokonui Hills, Southland, New Zealand. *N.Z. J. Geol. Geophys.*, 17: 337–374.

Bonarelli, G. and Nagera, J.J., 1921. Observaciones geologicas en las immediaciones del Lago San Martin (Territoria de Santa Cruz). *Boll. Min. Agric., Buenos Aires,* 27B (Geol.).

Bowen, F.E., 1964a. *Buller, Sheet 15. Geological Map of New Zealand 1 : 250,000.* Department of Scientific and Industrial Research, Wellington, 1st edition.

Bowen, F.E., 1964b. Geology of the Ohai Coalfield. *N.Z. Geol. Surv. Bull.,* 51.

Bradshaw, J.D., 1972. Stratigraphy and structure of the Torlesse Supergroup (Triassic–Jurassic) in the foothills of the Southern Alps near Hawarden (S60-61), Canterbury. *N.Z. J. Geol. Geophys.,* 15: 71–87.

Bradshaw, J.D., 1973. Allochthonous Mesozoic fossil localities in melange within the Torlesse Rocks of North Canterbury. *J. R. Soc. N.Z.,* 3: 161–167.

Bradshaw, J.D. and Andrews, P.F., 1973. Geotectonics and the New Zealand Geosyncline. *Nature, Phys. Sci.,* 241: 14–16.

Brothers, R.N., 1974. Kaikoura Orogeny in Northland, New Zealand. *N.Z. J. Geol. Geophys.,* 17: 1–18.

Brown, E.H., 1968. Metamorphic structures in part of the eastern Otago Schists. *N.Z. J. Geol. Geophys.,* 11: 41–65.

Brunnschweiler, R.O., 1960. Marine fossils from the Upper Jurassic and the Lower Cretaceous of Dampier Peninsula, Western Australia. *Bull. Bur. Min. Resour., Geol. Geophys., Aust.,* 59.

Campbell, J.D., 1955. The Oretian Stage of the New Zealand Triassic System. *Trans. R. Soc. N.Z.,* 82: 1033–1047.

Campbell, J.D., 1956. The Otapirian Stage of the Triassic System of New Zealand. Part 2. *Trans. R. Soc. N.Z.,* 84: 45–50.

Campbell, J.D., 1959. The Warepan Stage (Triassic): Definition and correlation. *N.Z. J. Geol. Geophys.,* 2: 198–207.

Campbell, J.D., 1968. *Rastelligera* (Brachiopoda) of the Upper Triassic of New Zealand. *Trans. R. Soc. N.Z., Geol.,* 6(3): 23–37.

Campbell, J.D., 1974a. *Heterastridium* (Hydrozoa) from Norian sequences in New Caledonia and New Zealand. *J. R. Soc. N.Z.,* 4: 447–453.

Campbell, J.D., 1974b. Biostratigraphy and structure of Richmond Group Rocks in Wairoa River–Mount Heslington Area, Nelson. *N.Z. J. Geol. Geophys.,* 17: 41–62.

Campbell, J.D., 1974c. The Indo-Pacific element in faunas of the Torlesse Supergroup. *Abstr., Geol. Soc. N.Z. Conf., 1974.*

Campbell, J.D. and Coombs, D.S., 1966. Murihiku Supergroup (Triassic–Jurassic) of Southland and south Otago. *N.Z. J. Geol. Geophys.,* 9: 393–398.

Campbell, J.D. and Force, E.R., 1972. Stratigraphy of the Mount Potts Group at Rocky Gully, Rangitata Valley, Canterbury. *N.Z. J. Geol. Geophys.,* 15: 157–167.

Campbell, J.D. and Force, E.R., 1973. Kaihikuan Stage (Middle Triassic): definition and type locality. *N.Z. J. Geol. Geophys.,* 16: 209–220.

Campbell, J.D. and McKellar, I.C., 1956. The Otapirian Stage of the Triassic System of New Zealand. Part 1. *Trans. R. Soc. N.Z.,* 83: 695–704.

Campbell, J.D. and McKellar, I.C., 1960. The Otamitan Stage (Triassic): definition and type locality. *N.Z. J. Geol. Geophys.,* 3: 643–659.

Campbell, J.D. and Warren, G., 1965. Fossil localities of the Torlesse Group in the South Island. *Trans. R. Soc. N.Z., Geol.,* 3(8): 99–137.

Carter, R.M., 1974. A New Zealand case-study of the need for local time scales. *Lethaia,* 7: 181–202.

Carter, R.M. and Lindquist, J.K., 1975. Sealers Bay submarine fan complex, Oligocene, southern New Zealand. *Sedimentology,* 22: 465–83.

Carter, R.M., Landis, C.A., Norris, R.J. and Bishop, D.G., 1974. Suggestions towards a high-level nomenclature for New Zealand rocks. *J. R. Soc. N.Z.,* 4: 5–18.

Casey, R., 1967. The position of the Middle Volgian in the English Jurassic. *Proc. Geol. Soc. Lond.,* 1640: 128–133.

Casey, R., 1973. The Ammonite succession at the Jurassic–Cretaceous boundary in eastern England. In: R. Casey and P.F. Rawson (Editors), *The Boreal Lower Cretaceous. Geol. J. Spec. Iss.,* 5: 193–266.

Challinor, A.B., 1970. *Uhligi*-complex Belemnites of the Puaroan (Lower–?Middle Tithonian) stage in the Port Waikato Region of New Zealand. *Earth Sci. J., Hamilton,* 4: 66–105.

Challinor, A.B., 1974. Biostratigraphy of the Ohauan and Lower Puaroan stages (Middle Kimmeridgian and Lower Tithonian), Port Waikato region, New Zealand, with a description of a new *Belemnopsis. N.Z. J. Geol. Geophys.,* 17: 235–269.

Challinor, A.B., 1975. New Upper Jurassic Belemnite from southwest Auckland, New Zealand. *N.Z. J. Geol. Geophys.,* 18: 361–71.

Challinor, A.B., 1977. Proposal to re-define the Puaroan stage of the New Zealand Jurassic System. *N.Z. J. Geol. Geophys.,* 20: 17–46.

Challis, G.A., 1966. Cretaceous stratigraphy and structure of the Lookout Area, Awatere Valley. *Trans. R. Soc. N.Z., Geol.,* 4(5): 119–137.

Challis, G.A., 1968. The K_2O : Na_2O ratios of ancient volcanic arcs in New Zealand. *N.Z. J. Geol. Geophys.,* 11: 200–211.

Chapman, F., 1918. Descriptions and revisions of the Cretaceous and Tertiary fish remains of New Zealand. *N.Z. Geol. Surv. Bull.,* 7.

Chapman-Smith, M. and Grant-Mackie, J.A., 1971. Geology of the Whangaparaoa area, eastern Bay of Plenty. *N.Z. J. Geol. Geophys.,* 14: 3–38.

Charrier, R. and Lahusen, A., 1969. Stratigraphy of the late Cretaceous–early Eocene, Seno Skyring–Straits of Magellan area, Magallanes Province, Chile. *Am. Assoc. Pet. Geol. Bull.,* 53: 568–590.

Christoffel, D. and Falconer, R.K.H., 1972. Marine magnetic measurements in the Southwest Pacific Ocean and the identification of new tectonic features. In: D.E. Hayes (Editor), *Antarctic Oceanology II. The Australian–New Zealand Sector. Am. Geophys. Union Antarc. Res. Ser.,* 19: 197–209.

Clark, R.H., Vella, P. and Waterhouse, J.B., 1967. The Permian at Parapara Peak, north-west Nelson. *N.Z. J. Geol. Geophys.,* 10: 232–246.

Coombs, D.S., 1960. Lower grade mineral facies in New Zealand. *Rep. 21st Int. Geol. Congr.,* 13: 339–351.

Coombs, D.S., Ellis, A.J., Fyfe, W.S. and Taylor, A.M., 1959. The zeolite facies, with comments on the interpretation of hydrothermal syntheses. *Geochim. Cosmochim. Acta,* 17: 53–107.

Coombs, D.S., Landis, C.A., Norris, R.J., Sinton, J.N., Borns, D.J. and Craw, D., 1976. The Dun Mountain Ophiolite Belt, New Zealand, its tectonic setting, constitution and origin, with special reference to the southern portion. *Am. J. Sci.,* 276(5): 561–603.

Cooper, R.A., 1975. New Zealand and Southeast Australia in the Early Paleozoic. *N.Z. J. Geol. Geophys.,* 18: 1–20.

Cope, J.C.W., 1967. The palaeontology and stratigraphy of the lower part of the Upper Kimmeridge Clay of Dorset. *Bull. Br. Mus. (Nat. Hist.), Geol.,* 15(1).

Cope, J.C.W. and Zeiss, A., 1964. Zur Parallelisiering des englischen Oberkimmeridge mit dem frankischen Untertithon (Malm Zeta). *Geol. Bl. Nordost Bayern,* 14: 5–14.

Cope, R.N. and Reed, J.J., 1967. The Cretaceous Paleogeology of the Taranaki–Cook Strait Area. *Proc. Australas. Inst. Min. Metall.,* 222: 63–72.

Couper, R.A., 1960. New Zealand Mesozoic and Cenozoic plant microfossils. *N.Z. Geol. Surv. Paleontol. Bull.,* 32.

Cox, L.R., 1940. The Jurassic lamellibranch fauna of Kuchh. *Mem. Geol. Surv. India, Paleontol. Indica, Ser. 9,* 3(3).

Cox, L.R., 1953. Lower Cretaceous Gastropoda, Lamellibranchia and Annelida from Alexander I Land (Falkland Island Dependencies). *Falkland Isl. Dep. Surv. Sci. Rep.,* 4.

Cox, S.H., 1878. Report on the geology of the Hokonui Ranges, Southland. *N.Z. Geol. Surv. Rep. Geol. Explor.,* 1877–8 (11): 25–48.

Cracraft, J., 1975. Mesozoic dispersal of terrestrial faunas around the southern end of the world. *Mém. Mus. Nat. Hist. Nat., Sér. A,* 88: 29–54.

Creer, K.M., 1973. A discussion of the arrangement of Paleomagnetic Poles on the map of Pangaea for epochs in the Phanerozoic. In: D.H. Tarling and S.K. Runcorn (Editors), *Implications of Continental Drift to the Earth Sciences, 1.* Academic Press, London, pp. 47–76.

Day, R.W., 1969. The Lower Cretaceous of the Great Artesian Basin. In: K.S.W. Campbell (Editor), *Stratigraphy and Paleontology. Essays in Honour of Dorothy Hill.* Australian National University Press, Canberra, A.C.T., pp. 140–173.

Day, R.W. and Speden, I.G., 1970. Probable synonymy of the Lower Cretaceous bivalves *Maccoyella reflecta* (Moore) and *Maccoyella magnata* Marwick. *N.Z. J. Geol. Geophys.,* 13: 647–654.

Dettmann, M.E., 1963. Upper Mesozoic microfloras from southeastern Australia. *Proc. R. Soc. Vict.,* 77: 1–148.

Dickinson, W.R., 1971a. Detrital modes of New Zealand greywackes. *Sediment. Geol.,* 5: 37–56.

Dickinson, W.R., 1971b. Clastic sedimentary sequences deposited in shelf, slope and trough settings between magmatic arcs and associated trenches. *Pac. Geol.,* 3: 15–30.

Diener, C., 1916. Die Marinen Reiche der Triasperiode. *Denkschr. K. Akad. Wiss. Wien, Math.-Naturwiss. Kl.,* 92: 405–550.

Drot, J., 1953. Description des Brachiopodes du Trias et de l'infralias de Nouvelle Calédonie. In: J. Avias, *Contribution à l'étude stratigraphique et paléontologique des formations antécrétacées de la Nouvelle Calédonie Centrale. Sci. Terre,* 1: 87–104.

Eade, J.V., 1966. Stratigraphy and structure of the Mt Adams Area, eastern Wairarapa. *Trans. R. Soc. N.Z., Geol.,* 4(4): 103–117.

Edwards, A.R., 1971. A calcareous nannoplankton zonation of the New Zealand Paleogene. *Proc. 2nd Planktonic Conf. Roma, 1970,* pp. 381–419.

Edwards, A.R., 1973. Calcareous nannofossils from the Southwest Pacific, Deep Sea Drilling Project, Leg 21. In: R.E. Burns, J.E. Andrews et al., *Initial Reports of the Deep Sea Drilling Project, 21.* U.S. Government Printing Office, Washington, D.C., pp. 641–691.

Edwards, W.N., 1934. Jurassic plants from New Zealand. *Ann. Mag. Nat. Hist., Ser. 10,* 13: 81–109.

Enay, R., 1964. L'étage Tithonique. In: P.L. Maubeuge et al., *Colloque du Jurassique à Luxembourg 1962.* St Paul, Luxembourg, pp. 355–379.

Enay, R., 1972a. Paléobiogéographie des Ammonites du Jurassique terminal (Tithonique/Volgien/Portlandien *s.l.*) et mobilité continentale. *C.R. Somm. Séances Soc. Geol. Fr.,* 1972, Fasc. 4: 163–167.

Enay, R., 1972b. Paléobiogéographie des Ammonites du Jurassique Terminal (Tithonique/Volgien/Portlandien *s.l.*) et mobilité continentale. *Geobios,* 5: 355–407.

Enay, R., 1973. Upper Jurassic (Tithonian) Ammonites. In: A. Hallam (Editor), *Atlas of Palaeobiogeography.* Elsevier, Amsterdam, pp. 297–307.

Enay, R. et al., 1975. Colloque sur la limite Jurassique–Cretacé (Lyon–Neuchatel, Septembre, 1973). *Mém. Bur. Rech. Géol. Min.,* 86.

Farquhar, O.C., 1969. Former seamounts in New Zealand and the volcanoes of modern oceans. *Oceanogr. Mar. Biol. Ann. Rev.,* 7: 101–172.

Fell, H.B., 1962. West-wind-drift dispersal of echinoderms in the Southern Hemisphere. *Nature,* 193: 759–761.

Filatoff, J., 1975. Jurassic palynology of the Perth Basin, Western Australia. *Palaeontographica,* 154B: 1–113.

Finlay, H.J. and Marwick, J., 1937. The Wangaloan and associated molluscan faunas of Kaitangata–Green Island Subdivision. *N.Z. Geol. Surv. Palaontol. Bull.,* 15: 140 pp.

Fleming, C.A., 1958. Upper Jurassic fossils and hydrocarbon traces from the Cheviot Hills, north Canterbury. *N.Z. J. Geol. Geophys.,* 1: 375–394.

Fleming, C.A., 1959. *Buchia plicata* (Zittel) and its allies, with a description of a new species, *Buchia hochstetteri. N.Z. J. Geol. Geophys.,* 2: 889–904.

Fleming, C.A., 1960. The Upper Jurassic sequence at Kawhia, New Zealand with reference to the ages of some Tethyan guide fossils. *Rep. 21st Int. Geol. Congr.,* Part 21: 264–269.

Fleming, C.A., 1962. New Zealand biogeography: A palaeontologist's approach. *Tuatara,* 10: 53–108.

Fleming, C.A., 1963. The nomenclature of biogeographic elements in the New Zealand biota. *Trans. R. Soc. N.Z. Gen.,* 1(2): 13–22.

Fleming, C.A., 1967. Biogeographic change related to Mesozoic orogenic history in the SW Pacific. *Tectonophysics,* 4: 419–427.

Fleming, C.A., 1970. The Mesozoic of New Zealand: Chapters

in the history of the circum-Pacific mobile belt. *Q. J. Geol. Soc. Lond.*, 125: 125–170.

Fleming, C.A., 1971. A preliminary list of New Zealand fossil polychaetes. *N.Z. J. Geol. Geophys.*, 14: 742–756.

Fleming, C.A., 1975. The Geological History of New Zealand and its Biota. In: G. Kuschel (Editor), *Biogeography and Ecology in New Zealand.* W. Junk, The Hague, pp. 1–86 (Monographiae Biologicae Vol. 27).

Fleming, C.A. and Kear, D., 1960. The Jurassic sequence at Kawhia Harbour, New Zealand. *N.Z. Geol. Surv. Bull.*, 67.

Fleming, C.A., Gregg, D.R. and Welles, S.P., 1971. New Zealand ichthyosaurs – a summary, including new records from the Cretaceous. *N.Z. J. Geol. Geophys.*, 14: 734–741.

Force, E.R., 1973. Permian–Triassic contact relations in circum-Pacific geosynclines. *Pac. Geol.*, 6: 19–23.

Force, E.R. and Campbell, J.D., 1974. Some faunas and formations of Kaihikuan age (middle Triassic) in the Murihiku Supergroup. *N.Z. J. Geol. Geophys.*, 17: 389–402.

Gage, M., 1952. The Greymouth Coalfield. *N.Z. Geol. Surv. Bull.*, 32.

Gage, M., 1957. Geology of Waitaki Subdivision. *N.Z. Geol. Surv. Bull.*, 55.

Gage, M. and Wellman, H.W., 1944. The geology of Koiterangi Hill, Westland. *Trans. R. Soc. N.Z.*, 73: 351–64.

Gair, H.S., 1967. *The Question of Post-Rangitata Peneplanation in New Zealand. An Investigation of Cretaceous Peneplanation in Relation to Unconformities in Upper Jurassic and Cretaceous Sequences.* Thesis, University of Canterbury Library, Christchurch, 339 pp.

Geczy, B. et al., 1971. Colloque du Jurassique Méditerranéen à Budapest, 1969. *Ann. Inst. Geol. Publ. Hung.*, 54(2).

Gerasimov, P.A. and Mikhailov, N.P. et al., 1967. *International Symposium on Upper Jurassic Stratigraphy on the U.S.S.R. June 6–18, 1967. Guidebook and Programme.* Moscow, 50 pp.

Geyer, O.F., 1974. Der Unterjura (Santiago Formation) von Ekuador. *Neues Jahrb. Geol. Paläontol. Monatsh.*, 1974(9): 525–541.

Grant-Mackie, J.A., 1959. Hokonui stratigraphy of the Awakino–Mahoenui area, southwest Auckland. *N.Z. J. Geol. Geophys.*, 2: 755–787.

Grant-Mackie, J.A. and Zapfe, H., 1973. *Otapiria* (Monotidae, Bivalvia) aus den Zlambach-Schichten des Salzkammergutes Oberösterreich. *Österr. Akad. Wiss. Anz. Math.-Naturwiss. Kl.*, 1973(7): 45–49.

Grapes, R.H., 1970. A petrological study of the Kaiwhata Sill, Ngahape, East Wairarapa, New Zealand. *Trans. R. Soc. N.Z., Earth Sci.*, 7(10): 177–196.

Gressitt, J.L. (Editor), 1963. *Pacific Basin Biogeography – A Symposium.* Bishop Museum Press, Honolulu, Hawaii, 563 pp.

Grindley, G.W., 1960. *Sheet 8, Taupo. Geological Map of New Zealand 1 : 250,000.* Department of Scientific and Industrial Research, Wellington.

Grindley, G.W., 1961. Mesozoic orogenies in New Zealand. *Proc. 9th Pac. Sci. Congr.*, 12: 71–75.

Grindley, G.W., 1974. New Zealand. In: A.M. Spencer (Editor), *Mesozoic–Cenozoic Orogenic Belts: Data for Orogenic Studies. Geol. Soc. Lond., Spec. Publ.*, 4: 387–416.

Grindley, G.W., Adams, C.J.D., Lumb, J.T. and Watters, W.A., 1977. Paleomagnetism, K–Ar dating and tectonic interpretation of Upper Cretaceous and Cenozoic Volcanic Rocks of the Chatham Islands, New Zealand, *N.Z. J. Geol. Geophys.*, 20(3).

Haast, J., 1865. Report on geological exploration of the West Coast. *Proc. Prov. Counc., Canterbury, Sess.*, 23: 13–21.

Hall, W.D.M., 1963. The Clarence Series at Coverham, Clarence Valley. *N.Z. J. Geol. Geophys.*, 6: 28–37.

Harper, C.T. and Landis, C.A., 1967. K–Ar ages from regionally metamorphosed rocks, South Island, New Zealand, and some tectonic implications. *Earth Planet. Sci. Lett.*, 2: 419–29.

Harrington, H.J., 1955. Geology of the Naseby District, central Otago. *N.Z. J. Sci. Technol.*, 36B: 581–599.

Harrington, H.J., 1958. Geology of the Kaitangata Coalfield. *N.Z. Geol. Surv. Bull.*, 59.

Harris, T.M., 1962. The occurrence of the fructification *Carnoconites* in New Zealand. *Trans. R. Soc. N.Z., Geol.*, 1(4): 17–27.

Hay, R.F., 1960. The geology of Mangakahia Subdivision. *N.Z. Geol. Surv. Bull.*, 61.

Hay, R.F., 1975. *Sheet N7, Doubtless Bay. Geological Map of New Zealand 1 : 63,360.* Department of Scientific and Industrial Research, Wellington, New Zealand, 1st edition.

Hay, R.F., Mutch, A.R. and Watters, W.A., 1970. Geology of the Chatham Islands. *N.Z. Geol. Surv. Bull.*, 61.

Hayes, D.E. and Ringis, J., 1973. Sea floor spreading in the Tasman Sea. *Nature,* 243: 454–458.

Hector, J., 1874. On the fossil Reptilia of New Zealand. *Trans. N.Z. Inst.*, 6: 333–358.

Heirtzler, J.R., 1968. Sea-floor spreading. *Sci. Am.*, December 1968.

Henderson, R.A., 1970. Ammonoidea from the Mata Series (Santonian–Maastrichtian) of New Zealand. *Palaeontol. Assoc., Spec. Pap. Palaeontol.*, No. 6.

Henderson, R.A., 1973. Clarence and Raukumara Series (Albian–?Santonian) Ammonoidea from New Zealand. *J. R. Soc. N.Z.*, 3: 71–123.

Hornibrook, N. de B., 1953. Jurassic foraminifera from New Zealand. *Trans. R. Soc. N.Z.*, 81: 375–378.

Hornibrook, N. de B., 1958. New Zealand Upper Cretaceous and Tertiary foraminiferal zones and some overseas correlations. *Micropaleontology,* 4: 25–38.

Hornibrook, N. de B., 1962. The Cretaceous–Tertiary boundary in New Zealand. *N.Z. J. Geol. Geophys.*, 5: 295–303.

Hornibrook, N. de B., 1969. News reports. New Zealand. *Micropaleontology,* 15: 128–130.

Hornibrook, N. de B. and Harrington, H.J., 1957: The status of the Wangaloan Stage. *N.Z. J. Sci. Technol.*, 38B: 655–670.

Hornibrook, N. de B. and Hay, R.F., 1977. Late Cretaceous agglutinating Foraminifera from sediments interbedded with the Tangihua Volcanics, Northland, New Zealand. *Bull. Aust. Bur. Min. Resour., Geol. Geophys.*, in press.

Hulston, J.R. and McCabe, W.J., 1972. New Zealand potassium–argon age list – 1. *N.Z. J. Geol. Geophys.*, 15: 406–432.

Imlay, R.W., 1967. The Mesozoic pelecypods *Otapiria* Marwick and *Lupherella* Imlay, new genus in the United States. *Prof. Pap. U.S. Geol. Surv.*, 573-B.

Jeletzky, J.A., 1963. *Malayomaorica* gen. nov. (Family Aviculopectinidae) from the Indo-Pacific Upper Jurassic; with comments on related forms. *Paleontology,* 6: 148–60.

Johnston, M.R., 1975. *Sheet N159 and part N158. Tinui–Awatoitoi Geological Map of New Zealand 1 : 63,360.* Department of Scientific and Industrial Research, Wellington, 1st edition.

Johnston, M.R. and Browne, P.R.L., 1973. Upper Jurassic and Cretaceous conglomerates in the Tinui–Awatoitoi district, eastern Wairarapa. *N.Z. J. Geol. Geophys.,* 16: 1055–1060.

Katz, H.R., 1963. Revision of Cretaceous stratigraphy in Patagonian cordillera of Ultima Esperanza, Magallanes Province, Chile. *Bull. Am. Assoc. Pet. Geol.,* 47: 506–524.

Katz, H.R., 1968. Potential oil formations in New Zealand, and their stratigraphic position as related to basin evolution. *N.Z. J. Geol. Geophys.,* 11: 1077–1133.

Katz, H.R., 1974. Recent Exploration for oil and gas. In: *Economic Geology of New Zealand* Monograph 4. Australasian Institute of Mining and Metallurgy, Parkville, Vic., pp. 463–480.

Kear, D., 1955. Mesozoic and Lower Tertiary stratigraphy and limestone deposits, Torehina, Coromandel. *N.Z. J. Sci. Technol.,* B37: 107–114.

Kear, D., 1960. *Sheet 4, Hamilton. Geological Map of New Zealand 1 : 250,000.* Department of Scientific and Industrial Research, Wellington, 1st edition.

Kear, D., 1961. Ururoan beds of the Hetherington inlier, west Auckland. *N.Z. J. Geol. Geophys.,* 4: 231–238.

Kear, D., 1966. *Sheet N55, Te Akau. Geological Map of New Zealand 1 : 63,360* Department of Scientific and Industrial Research, Wellington, New Zealand, 1st edition.

Kear, D., 1967. Economic geology of the Waikato. *Earth Sci. J., Hamilton,* 1: 1–18.

Kear, D., 1971. Basement rock facies – northern North Island. *N.Z. J. Geol. Geophys.,* 14: 275–283.

Kear, D. and Fleming, C.A., 1976. Detail of Kawhia Jurassic Type Section. *N.Z. Geol. Surv. Rep.,* 58.

Kear, D. and Schofield, J.C., 1965. *Sheet N65, Hamilton. Geological Map of New Zealand 1 : 63,360.* Department of Scientific and Industrial Research, Wellington, New Zealand, 1st edition.

Kear, D. and Waterhouse, B.C., 1967. Onerahi Chaos Breccia of Northland. *N.Z. J. Geol. Geophys.,* 10: 629–646.

Kingma, J.T., 1959. The tectonic history of New Zealand. *N.Z. J. Geol. Geophys.,* 2: 1–55.

Kingma, J.T., 1962. *Sheet 11, Dannevirke. Geological Map of New Zealand 1 : 250,000.* Department of Scientific and Industrial Research, Wellington, New Zealand, 1st edition.

Kingma, J.T., 1966. *Sheet 6, East Cape. Geological Map of New Zealand 1 : 250,000.* Department of Scientific and Industrial Research, Wellington, New Zealand, 1st edition.

Kingma, J.T., 1967. *Sheet 12, Wellington. Geological Map of New Zealand 1 : 250,000.* Department of Scientific and Industrial Research, Wellington, New Zealand, 1st edition.

Kingma, J.T., 1971. Geology of Te Aute Subdivision. *N.Z. Geol. Surv. Bull.,* 70.

Kingma, J.T., 1974. *The Geological Structure of New Zealand.* Wiley, New York, N.Y., 407 pp.

Kummel, B., 1959. Lower Triassic ammonoids from western Southland, New Zealand. *N.Z. J. Geol. Geophys.,* 2: 429–447.

Kummel, B., 1965. New Lower Triassic ammonoids from New Zealand. *N.Z. J. Geol. Geophys.,* 8: 537–547.

Kuschel, G. et al., 1975. *Biogeography and Ecology in New Zealand.* W. Junk, The Hague, 689 pp.

Laing, A.C.M., 1972. Geology and petroleum prospects of Ruatoria area, east coast North Island, New Zealand. *APEA J.,* 12: 45–52.

Landis, C.A. and Bishop, D.G., 1972. Plate tectonics and regional stratigraphic–metamorphic relationships in the southern parts of the New Zealand Geosyncline. *Geol. Soc. Am. Bull.,* 83: 2267–2284.

Larson, R.L. and Pitman, W.C., 1972. World-wide correlation of Mesozoic magnetic anomalies and its implications. *Geol. Soc. Am. Bull.,* 83: 3645–3662.

Lawrence, P., 1967. *New Zealand region bathymetry 1 : 6,000,000.* N.Z. Oceanographic Institute Chart. Miscellaneous Ser. 15, Department of Scientific and Industrial Research, Wellington, New Zealand.

Lensen, G.J., 1963. *Sheet 16, Kaikoura. Geological Map of New Zealand 1 : 250,000.* Department of Scientific and Industrial Research, Wellington, New Zealand, 1st edition.

Lillie, A.R., 1953. The Geology of the Dannevirke Subdivision. *N.Z. Geol. Surv. Bull.,* 46.

Lillie, A.R. and Gunn, B.M., 1964. Steeply plunging folds in the Sealy Range, Southern Alps. *N.Z. J. Geol. Geophys.,* 7: 403–423.

McKay, A., 1877. Report on Kaikoura Peninsula and Amuri Bluff. *N.Z. Geol. Surv. Rep. Geol. Explor.* 1874–6, No. 9: 172–184.

McKay, A., 1878. Notes on the sections and collections of fossils obtained in the Hokonui district. *N.Z. Geol. Surv. Rep. Geol. Explor.* 1877–8, No. 11: 49–90.

McKellar, I.C., 1969. *Sheet 169, Winton. Geological Map of New Zealand 1 : 63,360.* Department of Scientific and Industrial Research, Wellington, New Zealand, 1st edition.

McKellar, I.C., Mutch, A.R. and Stevens, G.R., 1962. An Upper Jurassic outlier in the Pyke Valley, north-west Otago, and a note on Upper Jurassic belemnites in the South Island. *N.Z. J. Geol. Geophys.,* 5: 487–492.

MacPherson, E.O., 1946. An outline of Late Cretaceous and Tertiary diastrophism in New Zealand. *N.Z. Dep. Sci. Ind. Res., Geol. Mem.* 6.

McQueen, D.R., 1955. Revision of supposed Jurassic angiosperms from New Zealand. *Nature,* 175: 177.

McQueen, D.R., 1956. Leaves of Middle and Upper Cretaceous pteridophytes and cycads from New Zealand. *Trans. R. Soc. N.Z.,* 83: 673–685.

Martin, K.R., 1975. Upper Triassic to Middle Jurassic stratigraphy of southwest Kawhia, New Zealand. *N.Z. J. Geol. Geophys.,* 18: 909–938.

Marwick, J., 1939. *Maccoyella* and *Aucellina* in the Taitai Series. *Trans. R. Soc. N.Z.,* 68: 462–465.

Marwick, J., 1951. Series and stage divisions of New Zealand

Triassic and Jurassic rocks. *N.Z. J. Sci. Technol.*, B32(3): 8–10.

Marwick, J., 1953a. Divisions and Faunas of the Hokonui System (Triassic–Jurassic). *N.Z. Geol. Surv. Paleontol. Bull.*, 21.

Marwick, J., 1953b. Faunal migrations in New Zealand seas during the Triassic and Jurassic. *N.Z. J. Sci. Technol.*, 34B: 317–321.

Marwick, J., 1966. An aberrant aucellinoid (Bivalvia, Pteriacea) from Red Island, Hawkes Bay. *N.Z. J. Geol. Geophys.*, 9: 495–503.

Mason, B.H., 1962. Metamorphism in the Southern Alps of New Zealand. *Bull. Am. Mus. Nat. Hist.*, 123(4): 211–248.

Matsumoto, T., 1963. The Cretaceous. In: F. Takai, T. Matsumoto and R. Toriyama (Editors), *Geology of Japan*. The University of Tokyo Press, Tokyo, pp. 99–128.

Maubeuge, P.L. et al., 1964. *Colloque du Jurassique à Luxembourg 1962.* St Paul, Luxembourg, 948 pp.

Maubeuge, P.L. et al., 1974. Colloque du Jurassique à Luxembourg 1967. *Mem. Bur. Rech. Géol. Min.*, 75.

Mayer, W., 1968. The stratigraphy and structure of the Waipapa Group on the islands of Motutapu, Rakino, and the Noisies Group, near Auckland, New Zealand. *Trans. R. Soc. N.Z.*, 5: 215–233.

Mildenhall, D.C. and Wilson, G.J., 1976. Cretaceous palynomorphs from the Sisters Islets, Chatham Islands, New Zealand. *N.Z. J. Geol. Geophys.*, 19: 121–126.

Molnar, P., Atwater, T., Mammerickx, J. and Smith, S.M., 1975. Magnetic anomalies, bathymetry and the tectonic evolution of the South Pacific since the Late Cretaceous. *Geophys. J. R. Astron. Soc.*, 40: 383–420.

Moore, P.R., 1977. Geology of Western Koranga Valley, Raukumara Peninsula. *N.Z. J. Geol. Geophys.*, 20(6).

Moore, W.R., 1957. *Geology of the Raukokere Area, Raukumara Peninsula, North Island.* Thesis lodged in the Library, Victoria University of Wellington, Wellington (unpublished).

Moore, W.R., 1961. *Ruatahuna–Motu Cretaceous Survey.* British Petroleum, Shell and Todd Petroleum Development Limited, Geological Report 12. Petroleum Report 314, lodged in the Library, N.Z. Geological Survey.

Mutch, A.R., 1957. Facies and thickness of the Upper Paleozoic and Triassic sediments of Southland. *Trans. R. Soc. N.Z.*, 84: 499–511.

Mutch, A.R., 1963. *Sheet 23, Oamaru. Geological Map of New Zealand 1 : 250,000.* Department of Scientific and Industrial Research, Wellington.

Mutch, A.R., 1966. Sheet S159, Morley. *Geological Map of New Zealand 1 : 63,360.* Department of Scientific and Industrial Research, Wellington, 1st edition.

Mutch, A.R., 1972. Geology of Morley Subdivision, Sheet S159. *N.Z. Geol. Surv. Bull.*, 78.

Mutch, A.R. and Wilson, D.D., 1952. Reversal of movement on the Titri Fault. *N.Z. J. Sci. Technol.*, B33: 398–403.

Nathan, S., 1974a. Stratigraphic nomenclature for the Cretaceous–Lower Quaternary rocks of Buller and north Westland, West Coast, South Island, New Zealand. *N.Z. J. Geol. Geophys.*, 17: 423–445.

Nathan, S., 1974b. Petrology of the Berlins Porphyry: a study of the crystallization of granitic magma. *J. R. Soc. N.Z.*, 4: 463–483.

Neef, G., 1974. *Sheet N153, Eketahuna. Geological Map of New Zealand 1 : 63,360.* Department of Scientific and Industrial Research, Wellington, 1st edition.

New Zealand Geological Survey, 1972. *North Island. Geological Map of New Zealand 1 : 1,000,000.* Department of Scientific and Industrial Research, Wellington, 1st edition.

Noda, M., 1975. Succession of *Inoceramus* in the Upper Cretaceous of Southwest Japan. *Mem. Fac. Sci., Kyushu Univ., Ser. D, Geol.*, 23(2): 211–261.

Norris, G., 1964. *Report on Spore and Pollen Assemblages From the Puaroan B and Huriwai Formation of Port Waikato, S.W. Auckland.* Report on File N51, N.Z. Geological Survey, Wellington, New Zealand.

Norris, G., 1968. Plant microfossils from the Hawks Crag Breccia, S.W. Nelson, New Zealand. *N.Z. J. Geol. Geophys.*, 11: 312–44.

Norris, G. and Waterhouse, J.B., 1970. Age of the Hawks Crag Breccia. *Trans. R. Soc. N.Z., Earth Sci.*, 7: 241–250.

Obradovich, J.D. and Cobban, W.A., 1975. A time scale for the Late Cretaceous of the Western Interior of North America. *Geol. Assoc. Can. Spec. Pap.*, No. 13: 31–54.

Ongley, M., 1939. The geology of the Kaitangata–Green Island Subdivision, Eastern and Central Otago Divisions. *N.Z. Geol. Surv. Bull.*, 38.

Owen, H.G., 1976. Continental displacement and expansion of the Earth during the Mesozoic and Cenozoic. *Philos. Trans. R. Soc., Lond.*, A281: 223–291.

Pergament, M.A., 1965. *Inoceramus* and Cretaceous stratigraphy of the Pacific Region. *Acad. Sci. U.S.S.R. Geol. Inst. Trans.*, 118.

Pick, M.C., 1962. *The Stratigraphy, Structure and Economic Geology of the Cretaceo-Tertiary rocks of the Waiapu District, New Zealand.* Dissertation, University of Bristol Library, Bristol. 532 pp. (lodged in the N.Z. Geological Survey Library, Lower Hutt).

Player, R.A., 1958. *The Geology of North Kawhia.* Thesis, University of Auckland, Auckland (unpublished).

Purser, B.H., 1961. Geology of the Port Waikato region (Onewhero Sheet N51). *N.Z. Geol. Surv. Bull.*, 69.

Quennell, A.M. and Hay, R.F., 1964. Origin of the Tangihua Group of North Auckland (letter). *N.Z. J. Geol. Geophys.*, 7: 638–640.

Raven, P.H. and Axelrod, D.I., 1972. Plate tectonics and Australasian paleobiogeography. *Science*, 176: 1379–1386.

Raven, P.H. and Axelrod, D.I., 1974. Angiosperm biogeography and past continental movements. *Ann. Mo. Bot. Gard.*, 61: 539–673.

Reed, J.J., 1957. Petrology of the Lower Mesozoic Rocks of the Wellington District. *N.Z. Geol. Surv. Bull.*, 57.

Reed, J.J., 1959. Chemical and modal composition of dunite from Dun Mountain, Nelson. *N.Z. J. Geol. Geophys.*, 2: 916–919.

Reed, J.J., 1965. Mineralogy and petrology in the N.Z. Geological Survey 1865–1965. *N.Z. J. Geol. Geophys.*, 8: 999–1087.

Reed, J.J., 1966. Geological and petrological investigations of wacke aggregates used in Auckland–Hamilton Motorway,

Redoubt and Takanini Sections. *N.Z. Geol. Surv. Rep.*, 20.

Ridd, M.F., 1968. Gravity sliding on Raukumara Peninsula. *N.Z. J. Geol. Geophys.*, 11: 547–548.

Ridd, M.F., 1970. Mud volcanoes in New Zealand. *Am. Assoc. Pet. Geol. Bull.*, 54: 601–616.

Routhier, P., 1950. Sur la présence de formations liasiques en Nouvelle-Calédonie. *Cah. Géol. Thoiry*, 3: 30.

Routhier, P., 1953. Etude géologique du versant occidental de la Nouvelle Calédonie entre le Col de Boghen et la pointe d'Arama. *Mém. Soc. Géol. Fr., N.S.*, 67: 1–271.

Rumeau, J.L., 1965. Note préliminaire sur le Crétacé du Nord de l'East Coast, Nouvelle Zélande. *N.Z. Aquitaine Pet. Limited, Pet. Rep.*, No. 477 (lodged in the Library, N.Z. Geological Survey, Lower Hutt).

Schofield, J.C., 1967. *Sheet 3, Auckland. Geological Map of New Zealand 1 : 250,000.* Department of Scientific and Industrial Research, Wellington, 1st edition.

Schofield, J.C., 1974. Stratigraphy, facies, structure and setting of the Waiheke and Manaia Hill Groups, East Auckland. *N.Z. J. Geol. Geophys.*, 17: 807–838.

Scott, G.H., 1961. Contribution to the knowledge of *Rzehakina* Cushman (Foraminifera) in New Zealand. *N.Z. J. Geol. Geophys.*, 4: 3–43.

Scott, K.M., 1966. Sedimentology and dispersal pattern of a Cretaceous flysch sequence, Patagonian Andes, southern Chile. *Bull. Am. Assoc. Pet. Geol.*, 50: 72–107.

Seitz, O., 1961. Die Inoceramen des Santon von Nordwestdeutschland, 1. Teil (Die Untergattungen *Platyceramus, Cladoceramus,* und *Cordiceramus*). *Beih. Geol. Jahrb.*, 46.

Sheppard, D.S., Adams, C.J. and Bird, G.W., 1975. Age of metamorphism and uplift in the Alpine Schists, New Zealand. *Geol. Soc. Am. Bull.*, 86: 1147–1153.

Silberling, N.J. and Tozer, E.T., 1968. Biostratigraphic classification of the marine Triassic in North America. *Spec. Pap. Geol. Soc. Am.*, 110.

Skinner, D.N.B., 1972. Subdivision and petrology of the Mesozoic rocks of Coromandel (Manaia Hill Group). *N.Z. J. Geol. Geophys.*, 15: 203–227.

Smith, A.G., Briden, J.C. and Drewry, G.E., 1973. Phanerozoic World maps. In: N.F. Hughes (Editor), *Organisms and Continents through Time. Palaeontol. Assoc. Spec. Pap. Palaeontol.*, 12: 1–42.

Sornay, J., 1966. Idées actuelles sur les Inocérames d'après divers travaux récents. *Ann. Paléontol.*, 52: 59–92.

Spath, L.F., 1923. On ammonites from New Zealand. *Q. J. Geol. Soc. Lond.*, 79: 286–308.

Speden, I.G., 1958. A note on the age of the Jurassic flora of Owaka Creek, south-east Otago, New Zealand. *N.Z. J. Geol. Geophys.*, 1: 530–532.

Speden, I.G., 1959. The alignment of fold axes in the Jurassic of S.E. Otago and southern Southland. *N.Z. J. Geol. Geophys.*, 2: 448–460.

Speden, I.G., 1961. *Sheet S184, Papatowai. Geological Map of New Zealand 1 : 63,360.* Department of Scientific and Industrial Research, Wellington, 1st edition.

Speden, I.G., 1969. Lower Cretaceous marine fossils, including *Maccoyella* sp., from the Whatarangi Formation, east side of Palliser Bay, New Zealand. *Trans. R. Soc. N.Z., Geol. Ser.*, 6: 143–153.

Speden, I.G., 1970. Three new inoceramid species from the Jurassic of New Zealand. *N.Z. J. Geol. Geophys.*, 13: 825–851.

Speden, I.G., 1971. Geology of the Papatowai Subdivision, southeast Otago. *N.Z. Geol. Surv. Bull.*, 81.

Speden, I.G., 1972. New fossil localities in the Torlesse Supergroup, western Raukumara Peninsula, New Zealand. *N.Z. J. Geol. Geophys.*, 15: 433–445.

Speden, I.G., 1973. Distribution, stratigraphy and stratigraphic relationships of Cretaceous sediments, western Raukumara Peninsula, New Zealand. *N.Z. J. Geol. Geophys.*, 16: 243–268.

Speden, I.G., 1975a. Cretaceous stratigraphy of Raukumara Peninsula Part 1. Koranga (Parts N87 and N88). Part 2. Lower Waimana and Waiotahi Valleys (Part N78) *N.Z. Geol. Surv. Bull.*, 91.

Speden, I.G., 1975b. An association of the Late Triassic *Bivalves Halobia, Manticula* and *Monotis* in the Torlesse rocks of North Canterbury. *N.Z. J. Geol. Geophys.*, 18: 279–284.

Speden, I.G., 1975c. Additional fossil localities of the Torlesse Rocks of the South Island. *N.Z. Geol. Surv. Rep.*, 69.

Speden, I.G., 1976a. Geology of Mt Taitai, Tapuaeroa Valley, Raukumara Peninsula. *N.Z. J. Geol. Geophys.*, 19: 71–119.

Speden, I.G., 1976b. Fossil localities in Torlesse Rocks of the North Island, New Zealand. *J. R. Soc. N.Z.*, 6: 73–91.

Speden, I.G., 1976c. *Inoceramus opetius* in the Kahuitara Tuff, Chatham Islands. *N.Z. J. Geol. Geophys.*, 19: 385–387.

Speden, I.G., 1977. The Taitai Series (Early Cretaceous) and the elimination of the Mokoiwian Stage. *N.Z. J. Geol. Geophys.*, in press.

Speden, I.G., in prep. Geology of the Cretaceous at Motu Falls, Raukumara Peninsula. *N.Z. J. Geol. Geophys.*

Speden, I.G. and McKellar, I.C., 1958. The occurrence of Aratauran beds south of Nugget Point, south Otago, New Zealand. *N.Z. J. Geol. Geophys.*, 1: 647–652.

Speight, R., 1928. Geology of the Malvern Hills. *N.Z. Dep. Sci. Ind. Res. Geol. Mem.*, 1.

Squires, D.S., 1958. The Cretaceous and Tertiary corals of New Zealand. *N.Z. Geol. Surv. Paleontol. Bull.*, 29.

Stevens, G.R., 1963. Faunal realms in Jurassic and Cretaceous belemnites. *Geol. Mag.*, 100: 481–497.

Stevens, G.R., 1965. The Jurassic and Cretaceous belemnites of New Zealand and a review of the Jurassic and Cretaceous belemnites of the Indo-Pacific region. *N.Z. Geol. Surv. Paleontol. Bull.*, 36.

Stevens, G.R., 1967. Upper Jurassic fossils from Ellsworth Land, West Antarctica, and notes on Upper Jurassic biogeography of the South Pacific region. *N.Z. J. Geol. Geophys.*, 10: 345–395.

Stevens, G.R., 1968. The Jurassic System in New Zealand. *N.Z. Geol. Surv. Rep.*, 35.

Stevens, G.R., 1970a. New Zealand Triassic and Cretaceous correlations. *N.Z. J. Geol. Geophys.*, 13: 718–721.

Stevens, G.R., 1970b. Upper Jurassic belemnites from Kuaotunu (Coromandel Peninsula) and northeastern Great Barrier Island. *N.Z. J. Geol. Geophys.*, 13: 721–725.

Stevens, G.R., 1971. Relationship of isotopic temperatures

and faunal realms to Jurassic–Cretaceous palaeogeography, particularly of the Southwest Pacific. *J. R. Soc. N.Z.*, 1: 145–158.
Stevens, G.R., 1972. Palaeontology of the Torlesse Supergroup. *N.Z. Geol. Surv. Rep.*, 54.
Stevens, G.R., 1973a. Jurassic belemnites. In: A. Hallam (Editor), *Atlas of Palaeobiogeography*. Elsevier, Amsterdam, pp. 259–274.
Stevens, G.R., 1973b. Cretaceous belemnites. In: A. Hallam (Editor), *Atlas of Palaeobiogeography*. Elsevier, Amsterdam, pp. 385–401.
Stevens, G.R., 1973c. The palaeogeographic history of New Zealand. *N.Z. Entomol.*, 5: 230–239.
Stevens, G.R., 1974a. The Jurassic System in New Zealand. *Mem. Bur. Rech. Géol. Min.*, 75: 739–751.
Stevens, G.R., 1974b. *Rugged Landscape: The Geology of Central New Zealand.* A.H. & A.W. Reed Limited, Wellington, 286 pp.
Stevens, G.R., 1975. Drifting continents. *N.Z. Cartogr. J.*, 5(3): 2–20.
Stevens, G.R., 1977. Mesozoic biogeography of the Southwest Pacific and its relationship to plate tectonics. *Proc. Int. Symp. on the Geodynamics of the Southwest Pacific, Noumea, New Caledonia.* Editions Technip, Paris.
Stevens, G.R. and Clayton, R.N., 1971. Oxygen isotope studies on Jurassic and Cretaceous belemnites from New Zealand and their biogeographic significance. *N.Z. J. Geol. Geophys.*, 14: 829–97.
Stoneley, H.M.M., 1962. New Foraminifera from the Clarence Series (Lower Cretaceous) of New Zealand. *N.Z. J. Geol. Geophys.*, 5: 592–616.
Stoneley, R., 1968. A Lower Tertiary decollement on the East Coast, North Island, New Zealand. *N.Z. J. Geol. Geophys.*, 11: 128–156.
Strong, C.P., 1976. Cretaceous Foraminifera from the Matakaoa Volcanic Group. *N.Z. J. Geol. Geophys.*, 19: 140–143.
Suggate, R.P., 1950. Quartzose coal measures of west Nelson and north Westland. *N.Z. J. Sci. Technol.*, B31: 1–14.
Suggate, R.P., 1956. Puponga Coalfield. *N.Z. J. Sci. Technol.*, B35: 539–559.
Suggate, R.P., 1957. Geology of the Reefton Subdivision. *N.Z. Geol. Surv. Bull.*, 56.
Suggate, R.P., 1958. The geology of the Clarence Valley from Gore Stream to Bluff Hill. *Trans. R. Soc. N.Z.*, 85: 397–408.
Suggate, R.P., 1961. Rock-stratigraphic names for the South Island schists and undifferentiated sediments of the New Zealand geosyncline. *N.Z.J. Geol. Geophys.*, 4: 392–399.
Suggate, R.P., 1963. The Alpine Fault. *Trans. R. Soc. N.Z. Geol.*, 2: 105–129.
Suggate, R.P. et al., 1977. *The Geology of New Zealand.* New Zealand Government Printer, Wellington, 2 vols.
Surlyk, F., 1975. Fault controlled marine fan-delta sedimentation at the Jurassic–Cretaceous boundary East Greenland. *IXth Int. Congr. Sedimentol., Nice*, pp. 305–311.
Thomson, J.A., 1917. Diastrophic and other considerations in classification and correlation and the existence of minor diastrophic districts in the Notocene. *Trans. N.Z. Inst.*, 49: 397–413.
Thomson, J.A., 1919. Geology of the middle Clarence and Ure Valleys, East Marlborough, New Zealand. *Trans. N.Z. Inst.*, 51: 289–349.
Tozer, E.T., 1967. A standard for Triassic time. *Geol. Surv. Can. Bull.*, 156: 1–103.
Trechmann, C.T., 1918. The Trias of New Zealand. *Q. J. Geol. Soc. Lond.*, 73: 165–246.
Vakhrameev, V.A., 1972. Mesozoic floras of the Southern Hemisphere and their relationship to the floras of northern continents. *Paleontol. Zh.*, 1972(3): 146–161 (in Russian).
Vakhrameev, V.A., 1975. Main features of phytogeography of the globe in Jurassic and Early Cretaceous time. *Paleontol. Zh.*, 1975(2): 123–132 (in Russian).
Van den Heuvel, 1960. The geology of the Flat Point area, eastern Wairarapa. *N.Z. J. Geol. Geophys.*, 3: 309–320.
Vella, P., 1961. An occurrence of an Albian ammonite in the Motuan Stage in the Upper Awatere Valley, *Trans. R. Soc. N.Z., Geol.*, 1: 1–4.
Verma, H.M. and Westermann, G.E.G., 1973. The Tithonian (Jurassic) ammonite fauna and stratigraphy of Sierra Catorce, San Luis Potosi, Mexico. *Bull. Am. Paleontol.*, 63(277): 107–320.
Wandel, G., 1936. Beiträge zur Paläontologie des Ostindischen Archipels; 13: Beiträge zur Kenntnis der jurassischen Mollusken-fauna von Misol, Ost-Celebes, Buton, Seran und Jamdena. *Neues Jahrb. Min. Geol. Paläontol., Abh. B*, 75(B-B): 447–526.
Warren, G. and Speden, I.G., 1977. The Piripauan and Haumurian stratotypes (Mata Series, Upper Cretaceous) and correlative sequences in the Haumuri Bluff District, South Marlborough. *N.Z. Geol. Surv. Bull.*, 92, in press.
Waterhouse, J.B., 1967. Proposal of series and stages for the Permian in New Zealand. *Trans. R. Soc. N.Z., Geol.*, 5(6): 161–180.
Waterhouse, J.B., 1970. The World significance of New Zealand Permian stages. *Trans. R. Soc. N.Z., Earth Sci.*, 7(7): 97–109.
Waterhouse, J.B., 1975. The Rangitata Orogen. *Pac. Geol.*, 9: 35–73.
Waterhouse, J.B., 1976. Permian palaeontology. Late Paleozoic biogeography. In: R.P. Suggate et al., *The Geology of New Zealand.* N.Z. Government Printer, Wellington, pp. 150–157; 706–710.
Waterhouse, J.B. and Bonham-Carter, G.F., 1975. Global distribution and character of Permian biomes based on brachiopod assemblages. *Can. J. Earth Sci.*, 12: 1085–1146.
Waterhouse, J.B. and Bradley, J., 1957. Redeposition and slumping in the Cretaceo-Tertiary Strata of S.E. Wellington. *Trans. R. Soc. N.Z.*, 84: 519–548.
Waterhouse, J.B. and Norris, G., 1972. Paleobotanical solution to a granite conundrum: Hawks Crag Breccia of New Zealand and the tectonic evolution of the Southwest Pacific. *Geosci. Man*, 4: 1–15.
Watters, W.A., 1952. The geology of the eastern Hokonui Hills, Southland, New Zealand. *Trans. R. Soc. N.Z.*, 79: 467–484.
Webb, P.N., 1966. *New Zealand Late Cretaceous Foraminifera and Stratigraphy.* Schotanus and Jens, Utrecht, 19 pp.
Webb, P.N., 1971. New Zealand Late Cretaceous (Haumurian)

Foraminifera and stratigraphy: a summary. *N.Z. J. Geol. Geophys.*, 14: 795–828.

Webb, P.N., 1973a. Paleocene Foraminifera from Wangaloa and Dunedin. *N.Z. J. Geol. Geophys.*, 16: 109–157.

Webb, P.N., 1973b. A re-examination of the Wangaloan problem. *N.Z. J. Geol. Geophys.*, 16: 158–169.

Webby, B.D., 1967. Tube fossils from the Triassic of southwest Wellington, New Zealand. *Trans. R. Soc. N.Z., Geol.*, 5: 181–191.

Weissel, J.K. and Hayes, D.E., 1972. Magnetic anomalies in the south east Indian Ocean. In: D.E. Hayes (Editor), *Antarctic Oceanology II: The Australian–New Zealand Sector. Am. Geophys. Union Antarc. Res. Ser.*, 19: 165–196.

Welles, S.P. and Gregg, D.R., 1971. Late Cretaceous marine reptiles of New Zealand. *Rec. Canterbury Mus.*, 9: 11–111.

Wellman, H.W., 1950. Ohika Beds and the Post-Hokonui Orogeny. *N.Z. J. Sci. Technol.*, 32B(3): 11–38.

Wellman, H.W., 1952. The Permian–Jurassic stratified rocks. In: C. Teichert (Editor), *Symposium sur les Series de Gondwana. Proc. 19th Int. Geol. Congr.*, pp. 13–24.

Wellman, H.W., 1955. A revision of the type Clarentian section at Coverham, Clarence Valley. *Trans. R. Soc. N.Z.*, 83: 93–118.

Wellman, H.W., 1956. Structural Outline of New Zealand. *N.Z. Dep. Sci. Ind. Res. Bull.*, 121.

Wellman, H.W., 1959. Divisions of the New Zealand Cretaceous. *Trans. R. Soc. N.Z.*, 87: 99–163.

Wellman, P., 1970. Geology of the Ngahape Area, Eastern Wairarapa. *Trans. R. Soc. N.Z., Earth Sci.*, 8(11): 151–171.

Wilckens, O., 1922. The Upper Cretaceous gastropods of New Zealand. *N.Z. Geol. Surv. Palaeontol. Bull.*, 9.

Wilckens, O., 1927. Contribution to the palaeontology of the New Zealand Trias. *N.Z. Geol. Surv. Palaeontol. Bull.*, 12.

Wilson, D.D., 1956. The Late Cretaceous and Early Tertiary transgression in South Island, New Zealand. *N.Z. J. Sci. Technol.*, 37B: 610–622.

Wilson, D.D., 1963. Geology of Waipara Subdivision. *N.Z. Geol. Surv. Bull.*, 64.

Wilson, G.J., 1976. An Albian–Cenomanian dinoflagellate assemblage from the Mokoiwi siltstone, Mt Taitai, Raukumara Range, New Zealand. *N.Z. J. Geol. Geophys.*, 19: 132–136.

Wilson, G.J., 1977. Late Cretaceous (Senonian) dinoflagellate cysts from the Kahuitara Tuff, Chatham Islands. *N.Z. J. Geol. Geophys.*, 19: 127–129.

Wood, B.L., 1956. The geology of the Gore Subdivision (S170). *N.Z. Geol. Surv. Bull.*, 53.

Wood, B.L., 1960. *Sheet 27, Fiord. Geological Map of New Zealand 1 : 250,000.* Department of Scientific and Industrial Research, Wellington, 1st edition.

Wood, B.L., 1969. Geology of Tuatapere Subdivision, western Southland. *N.Z. Geol. Surv. Bull.*, 79.

Woods, H., 1911. A monograph of the Cretaceous Lamellibranchia of England. *Palaeontol. Soc. Lond.*, 2(7).

Woods, H., 1917. The Cretaceous faunas of the northeastern part of the South Island of New Zealand. *N.Z. Geol. Surv. Bull.*, 4.

Chapter 9

NORTHERN AND EASTERN AFRICA

A.E.M. NAIRN

INTRODUCTION

Since the end of the Pan-African thermo-tectonic event near the beginning of the Phanerozoic (±500–600 m.y.), Africa has been relatively unaffected by major orogenic episodes. Only at its extremities are fold belts found, the Paleozoic Cape folds in the south, the Mauretanide Hercynian chain in the northwest and the northern Riff–Tellian Cenozoic chains of Maghreb and the Syrian arcs of Egypt. During this time, and in particular the Mesozoic with which we are here concerned, there have been, on the other hand, significant vertical movements. The magnitude of these movements is only just becoming apparent largely as the result of offshore petroleum exploration of the last two or three decades.

The vertical movements produced a number of coastal basins which, while bearing little relation to the Paleozoic depositional basins, appear in their distribution to reflect structural controls established during the Pan-African event (Kennedy, 1965). It is generally claimed that throughout most of the Mesozoic, Africa remained above sea level with sea invading only the marginal areas and Fig. 1 illustrates one such interpretation (Kennedy, 1965). The generalization is doubtless correct, but in the absence of detailed geological knowledge of so much of Africa and the paucity of boreholes, such generalizations are not without their dangers as the presence of marine horizons of Late Jurassic and Early Cretaceous age in the Congo implies (Cahen, 1954).

Igneous activity following the end of the intense volcanicity which terminated Karroo Supergroup times, was relatively light but widespread, consisting mainly of volcanic activity with the formation of intrusions during Cretaceous times. Other Cretaceous igneous volcanic activity from its distribution, is apparently closely related to the pattern of fractures which led to the formation of grabens, in particular that of the Red Sea. The main phase of Karroo volcanicity centred around Late Triassic–Early Jurassic times, with the waning phase in Mozambique dating as mid-Cretaceous (Fitch and Miller, 1971; Cox, 1972).

The area discussed in the following article comprises the Saharan and East African regions. The Alpine chains of Maghreb, because of their involvement in Alpine tectonics will be considered in the chapter devoted to Western Europe (see Volume B). Southern Africa between the Walvis Ridge and

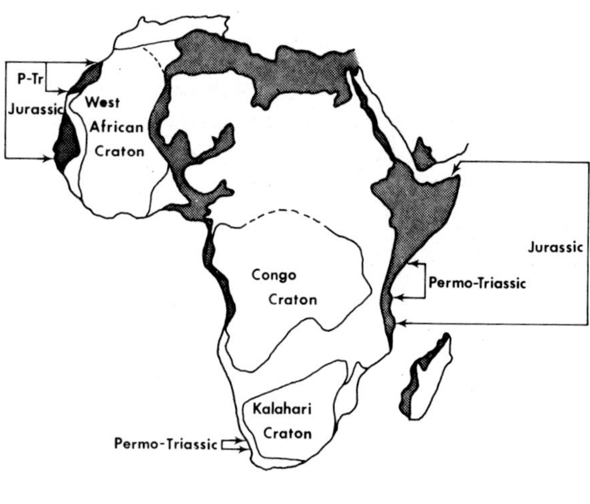

Fig. 1. The maximum extent of Mesozoic and Tertiary marine incursions over the African margin after Kennedy (1965). Note this does not include the marine horizons recognized by Cahen (1954) in the Congo.

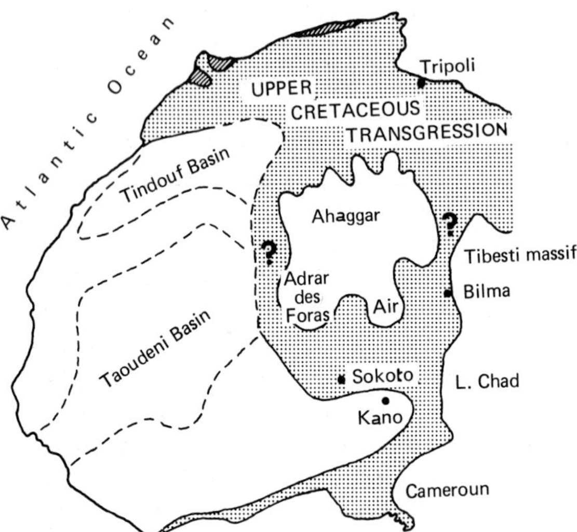

Fig. 2. The Upper Cretaceous transgression over the West African Craton after Furon (1963). There is still some uncertainty over the linkage between the Mediterranean and the Gulf of Guinea.

Mozambique is discussed by Dingle (this volume, Ch. 11), while the West African margin is handled by De Klasz (this volume, Ch. 10). In the region so defined the marine beds, with one principal exception lie to the north and east. That exception (see Fig. 1) is the result of the Cenomanian transgression which established a short-lived link between the Tethys and naissant South Atlantic around the western margin of the Ahaggar Massif (Fig. 2).

It is convenient to consider the geology of northern and eastern Africa in two segments, the Saharan realm and the East African. Much of the article has had to be drawn from the literature, and suffers from the author's lack of familiarity with parts of the area covered.

THE SAHARAN REALM

The continental facies and the Nubian problem

During the Mesozoic the Saharan realm appears to have formed a low-lying continental area bordered by ocean to the north and east. Over this continental mass thin arenaceous deposits form the lowest member of the Mesozoic sequence, and these deposits by common usage are termed the Nubian or Nubia Sandstone over most of the area. The inordinate stratigraphic complexities which arise are due to the fact that the sandstone is not everywhere the same age, there may be more than one sandstone of similar lithology and sandstones of different lithologies have been assigned the same name. In the continental area the age is at best poorly defined by plant remains, and the Nubian and other sandstones represent typical continental facies with conditions of deposition showing marked similarities over wide areas. The major marine transgressions of the Mesozoic provide marker horizons in some areas helping to delimit the age. Towards the Tethyan margin the Nubian Sandstone passes to fluvio-marine or a truly marine character, and is dated by associated marine sediments.

The term Nubian Sandstone was introduced by Russegger in 1837 who applied it to sandstones on either side of the Nile Valley near the Sudanese–Egyptian border. On the map they were shown as Early Cretaceous and in the text they were specified as not younger than Early Cretaceous. Since then the term has been applied to beds from Sinai to Niger and Chad, and ages assigned on the basis of similarities of sequences hundreds of miles apart. It has been shown to have been applied to beds of differing age (Weissbrod, 1970; Said, 1962) and lithology. In addition it has been regarded as the time equivalent of the "continental intercalaire" (see p. 331). Thus the term Nubian Sandstone may be "correctly" applied to continental Carboniferous in Sinai, marine Upper Cretaceous in Egypt or continental Lower Cretaceous in the Sudan. To add to the problem there are other Mesozoic beds of similar lithology to which other names have been assigned (such as the Gedaref Sandstone).

None of the numerous attempts to resolve the terminological confusion has proved successful, and Whiteman's (1970) proposal to erect a Nubian Group given the age uncertainties in the region where Nubian was originally defined seems to possess little advantage. Pomeyrol (1968) gave an extensive review and suggested cutting the Gordonion knot by abandonning the term. There is little doubt that eventually this suggestion will be adopted, but it can only come about by the establishment of precise chronostratigraphy and environmental descriptions which require extensive field work. Beginnings in this direction have already been made (Conant and Goudzari, 1967; Goudzari, 1970; Whiteman, 1971; Weissbrod, 1970) but in the meantime the term persists.

The solution adopted here is to use the term Nubian Sandstone in the area of original definition, and italicize every other record of it, giving age and location. The reader may be able to comprehend the stratigraphical confusion by recalling the basic paleogeography of sedimentation in a continental area which passes into marine conditions around the margins with the repetition of similar facies at different times and preservation reflecting local conditions. Thus throughout the Mesozoic similar environmental conditions persisted but these are only recorded in a few places, Triassic and low Jurassic (northern Ethiopia: Beyth, 1973), probable Jurassic (Egypt: El Shazly, 1977) and Cretaceous (Whiteman, 1971).

The Nubian Sandstone Formation of the Sudan and Egypt

The Nubian Sandstone covers much of north-central Sudan extending northwards into Egypt to Aswan close to the margin of the Tethys at its maximum extent, and westwards into Libya and Chad. Despite the geographic extent of the outcrops the lithological variations are relatively slight. For the most part of terrestrial origin (for some shallow marine horizons in Egypt have been termed *Nubian*) it implies widespread terrestrial sedimentary environments. Enough is known of the age of the Nubian for it to be established as diachronous, and it is thus best to regard it as the onshore Mesozoic facies in Africa north of the Sahara, although this runs into nomenclatural problems in regions formerly under French control where the same lithologies are assigned to "continental intercalaire" and "continental terminal" as for example in neighbouring Chad.

The age of the Nubian Sandstone is hard to define for only fossil plants of little chronological value have been found. These have been described as Early Cretaceous or Wealden (Neocomian, Barremian and Early Cenomanian ages have also been assigned). In some places the Nubian Sandstone is overlain by the Hudi Chert Formation to which an Early Tertiary (Oligocene) age is assigned. The lower limit is even less certain, for commonly the Nubian Sandstone rests upon Precambrian. Attempts to be more precise involve long-range correlation with the Egyptian section, a dangerous procedure.

It is evident that the present distribution of the Nubian Sandstone in the Sudan is controlled by a pattern of pre-Paleozoic synclinal and anticlinal warps (see Fig. 3 and Whiteman, 1971, fig. 4), but it is not clear whether this warping is post-depositional or whether the warping represents an old basement configuration. It is clear that the greatest thicknesses, in excess of 500 m, lie in the downwarped area but it cannot be seen whether thinning of the limbs is an erosional or an original, depositional effect.

In the vicinity of Khartoum four different lithological types of Nubian Sandstone have been described (Whiteman, 1971). The commonest facies is the Merkhiyat Sandstone type, a medium- to coarse-grained, yellow brown sandstone with a characteristically high clay-silt fraction. Pebbles are sometimes common consisting mostly of vein quartz with some phyllite. There is a considerable range in the degree of sorting found. The sandstone contains white flecks interpreted as kaolinized feldspar, no fresh feldspar occurs. The heavy mineral suite is restricted to stable varieties. Graded bedding sometimes occurs and cross-bedding, usually of tabular planar type but occasionally trough planar type, is common. The transport direction indicated is from the south to southeast. Plant remains are relatively abundant and seldom show signs of distant transport and *in situ* stumps are not uncommon.

The second variety, the Shendi type, is a medium- to fine-grained, clear, well-sorted and bedded, occasionally micaceous sandstone. It too has a heavy mineral suite restricted to the stable varieties. It has rib and furrow structures and ripple marks from which a current from the southeast has been deduced. Tabular–planar cross-bedding is common, sometimes the foreslope laminae are crumpled suggesting contemporaneous subaqueous slumping.

Pebble conglomerates and intraformational conglomerates form the third and fourth facies. The pebbles in the pebble conglomerates are dominantly quartz and phyllite though locally other components, e.g. rhyolite, may be important. The pebbles are usually well-rounded and smooth with disc and ellipsoidal shapes. Packing is poor and there are large amounts of sand and silt. The cement is principally ferruginous. Whiteman (1971) argues that the occurrence of relatively "soft phyllite pebbles" argues against reworking, but if primary, the absence of pebbles of metamorphic rocks is curious in view of the presence of staurolite and kyanite in the heavy mineral suite.

The clasts of the intraformational conglomerates are platey to rounded or pellet-shaped mudstone or siltstone fragments in a coarse sandy matrix. Quartz

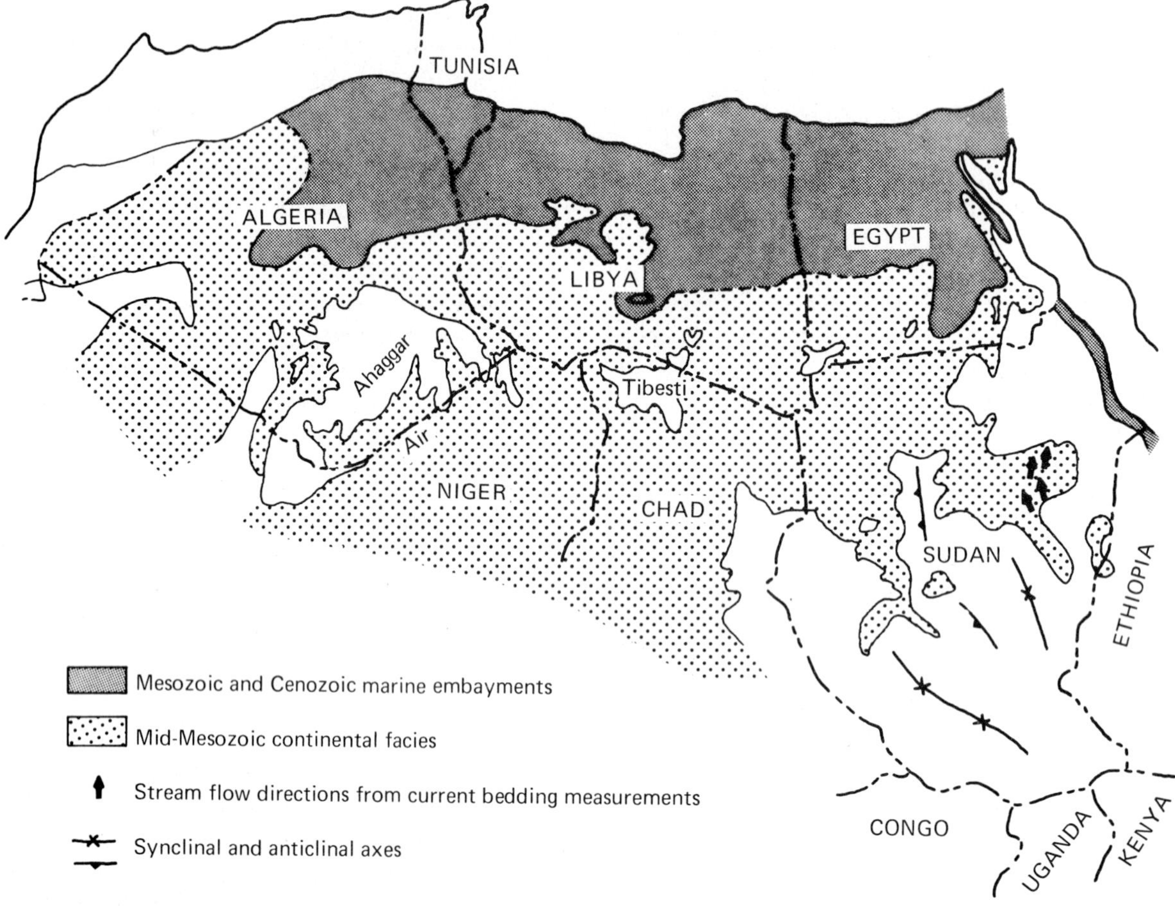

Fig. 3. The distribution of mid-Mesozoic continental facies north of the Sahara (after Whiteman, 1971 and others).

and phyllite pebbles also occur but are rare. Such conglomerates are commonly interbedded in sandstones of the Merkhiyat type.

Interbedded with these sandstones are mudstones, composed of clays of the kaolinitic group, with some micas, quartz and sericite. Occasionally they may be carbonaceous or pyritous. While these are widely distributed, with thicknesses in excess of 100 m having been recorded, they are seldom exposed due to their low resistivity to weathering.

Formational names have been proposed for the two sandstone types, Omdurman Formation for the Merkhiyat Sandstone type, and Shendi Formation for the fine-grained variety. However, only if clear stratigraphic relationships can be established do these names have value. Over much of the Sudan detailed descriptions such as those for the Khartoum region are not available, so that at the present stage not much is gained by using them.

From the northern Sudan the Nubian Sandstone may be traced northwards into Egypt, where it still remains thick, of the order of 200 m at Aswan and 635 m at Kharga oasis. According to Shukri and Said (1944) however the character of the sandstone changes to fluviomarine. The age of the Nubian Sandstone in Egypt is assigned from the overlying beds, and it is apparent from Said (1962) that two different formations are involved. One appears to the north of Suez in the Wadi Qiseib (the Malha Formation) and also in Wadi Qena below Cenomanian or Campanian beds. The other, succeeded by the Variegated Shales passes up into the Maastrichtian phosphate horizon near Quseir and is therefore assigned a Campanian age. In the extreme south of the Sudan a succession of sandstone and mudstone outliers has also been assigned to the *Nubian Sandstone.*

The Gedaref Sandstone Formation

The Gedaref Sandstone Formation crops out along the Ethiopian border southeast of Khartoum and is said to pass laterally into the Adigrat Sandstone Formation. As the youngest age of the latter is Oxfordian it is considerably older than the Nubian Sandstone of northern Sudan. Neither the size nor the shape of the Gedaref Basin is known. The maximum thickness recorded in the Gedaref bore is in excess of 140 m. Two outliers of it are known at different altitudes suggesting to Whiteman (1971) differential warping rather than faulting or original basin topography.

Lithologically it is made up of conglomerates, sandstones and sandy mudstone. Brecciated limestone was recovered from the Gedaref bore but in the absence of fossils its marine or continental character is in doubt. Since the Gedaref Sandstone passes laterally into the Adigrat Sandstone, which is the marine littoral facies of the transgressing Late Triassic–Jurassic sea, inferentially the Gedaref may have a similar association despite its stated lithological similarity to the Nubian Sandstone. The Adigrat Sandstone will be considered in the following section (p. 351).

The continental beds in Libya, Chad and Niger

Arenaceous deposits of continental origin may be traced westwards from the Sudan and Egypt into Chad and Libya where they sweep in a broad arc across southern Cyrenaica, around the periphery of the Murzuk Basin to the Fezzan and Algeria. The continental environment in the Fezzan and southern Cyrenaica however lasted from Carboniferous times until at least mid-Cretaceous, and deposits of several hundred meters accumulated. These beds were correlated in part at least with the "continental intercalaire" of the Sahara as well as with *Nubian Sandstone.*

They were initially separated by unconformity into an earlier post-Tassilian group ranging from Permian to Early Cretaceous and a Lower Cretaceous *Nubian Sandstone* (see Conant and Goudzari, 1967). In the Fezzan this was considerably refined by Lefranc (1959a, b), De Lapparent and Lelubre (1948) and Lelubre (1952). Formational names were introduced and the term Nubian Sandstone dropped (Table I), however the term *Nubian* is still in common usage, if generally restricted to the uppermost unit to which an Early Cretaceous or even Late Jurassic age is assigned [Murzuk Basin (Klitzsch, 1963), for which the name Messak Sandstone was also proposed].

In the Niger Republic, east of the Air Massif, the Cenomanian marine transgression (Fig. 2) provides a marker horizon at the top of the "continental intercalaire" The succession given by Furon (1963) is:

U. Cenomanian–L. Turonian	10 m limestone
L. Cenomanian Olandara Fm.	200 m sandstone with dinosaurs and several marine horizons with oysters and gastropods
"continental intercalaire" (Tefidet Group)	Tanguerat Sandstone: 200 m (= Dibella Sandstone of Chad)
	Tagrezou Sandstone: 100 m with silicified wood and dinosaurs
	Angornakauer Sandstone: 200 m with silicified wood

Once again the "continental intercalaire" renamed the Tefidet Group is split up into three sandstone units to which no precise time limits can be assigned.

The Chad Cuvette east and southeast of the Air Massif also has a cover of continental deposits, and once again the Cenomanian marine transgression provides the marker horizon. Overlying this horizon and extending to Senonian are the further continental beds of the terminal Cretaceous or Early Eocene and although thin here, more than 700 m (Chad Group) overlying the marine Cretaceous has been found in the Maiduguri bore in the southern part of the Chad Basin which lies within Nigeria.

Depositional environment of the continental beds

In the absence of precise dating for much of the *Nubian Sandstone,* and the interpretation problems where ages have been determined, it seems better to combine the available information on the Nubian, Gedaref and other sandstones to gain some impression of the Mesozoic continental environment. Fortunately the environment appears to have been reasonably stable so that the errors of such a broad generalization are not excessive.

Whiteman (1971) considered that the Nubian Sandstone was laid down in a system of fans and lakes separated by low isolated hills. Rigassi (1970) questions the extent of the lacustrine environment, favouring playa lakes, with generally drier conditions.

TABLE I

Stratigraphy of the continental beds in the Fezzan, Libya

Lefranc (1959)	Goudzari (1970), De Lapparent and Lelubre (1948)	Series	Lithology
Continental Intercalaire	"Nubian"	Djoua Series Wealden–Early Cenomanian	limestone, clay and marl
	post-Tassilian Group	Taouratine Series Kimmeridgian–Late Dogger	red to dark red claystone grading up to thick coarse sandstone, age by analogy with Algeria where fauna examined by De Lapparent
		Zarzaitine Series Triassic–Liassic	∿60 m basal red clay overlain by thick interbedded sandstone and marls; calcareous marl at top regarded by some as Muschelkalk (Late Triassic age by analogy with Algeria–De Lapparent)
post-Tassilian		Tiguentourine Series? Early Permian	50 m red-brown clay, locally dolomitic; rests unconformably upon Carboniferous; Base of clay is gypsiferous; a basal conglomerate developed in the southeast in Jebel Ben Ghenema

The sedimentological data suggest a relatively fine line can be drawn, the well-sorted sandstones suggest an aqueous environment, the poorly sorted Merkhiyat Sandstones with pebbly horizons are more consistent with the flash flood-type deposits and the presence of concretions favours a dry environment. The conglomerates may then be regarded as channel fill and the intraformational conglomerates as the rip up and redeposition of dried-out playa flats sediments. The presence of plants with xerophytic adaptations, tree rings and of *in situ* stumps suggest the climate was seasonally arid. The water-table remained high enough to support a plant cover and wet enough to flush out the soil preventing a build up of saline lakes, and in some regions it was moist enough to permit the development of swamps (at Dongola on the Nile where coal is reported).

Thus while there was probably no large network of lakes, some presumably did exist. The area was subject to a seasonal climate with flash floods forming temporary playas, with some semi-permanent water courses supporting tree growth. The climate differed from the present only in the increased rainfall (or lower temperature) implied.

Northward into Egypt these conditions give way to the fluvio-marine conditions of the Tethys margin. It would not be surprising to find a development of a zone of coastal dunes and sebkhas close to the sea margin, though these have not been positively identified.

Westwards in Libya, Hea (1971) described the petrography of the Mesozoic continental sandstones from the southern Sirte Basin in terms of two compositional end members, quartz arkoses and impure arkoses of the Guilat facies, typical of the *Nubian Sandstone* of the Merkhiyat type and orthoquartzites and feldspathic quartzites of the Amrha facies.

Sandstones of the Guilat facies show the more varied composition often with minimal alteration. The quartz arkoses provide evidence of some weathering and mixing with recycled quartz. The Amrha facies of quartz orthoquartzites may have subordinate feldspathic quartzites and is thus similar to the Shendi type of the Sudan. Given the lithological similarities it is not surprising that Hea describes essentially the same climate and environment of deposition. He does however stress the degree of reworking and recycling of sedimentary material, the ultimate source of which was the upwarped African craton regions to the south.

The Mediterranean margin and Gulf of Suez

The continuity of the sea around the northern and eastern margin of the African craton can be established by tracing the varying position of the shoreline through Libya, Egypt and Saudi Arabia to Somalia on the east coast of Africa. The Saudi Arabian segment of this tract considered by Saint-Marc (this volume, Ch. 12) provides a convenient break point permitting separate treatment of the Mediterranean and Indian Ocean shores. The region further to the west, from Tunisia to Algeria is obscured by the young fold mountains thrust over the edge of the craton during Cenozoic times. The history of this region will be discussed by Moullade.

During the Mesozoic the location of the ancient Tethyan shoreline varied as a function of the advance and retreat of the sea. The details of these shoreline migrations and the concomitant movement of the facies belts is known largely as a result of petroleum exploration, for over most of the area Mesozoic beds are generally concealed below a younger cover. Transgressions occurred during each of the Mesozoic periods. The Triassic transgression was more important in western Libya than in the east. Although formerly regarded as representing the classical transgression of the Muschelkalk present opinion (Busson, 1967) assigns it to the Carnian. The Jurassic transgression was particularly important in the east and is usually linked with the fragmentation of Gondwana. The most extensive was the Late Cretaceous transgression which began in the Cenomanian and reached its maximum in the east during the Maastrichtian–Paleocene, although in the west the link across the West African craton joining the Mediterranean and the naissant South Atlantic Ocean (see Fig. 3) appears to have occurred earlier. (Cenomanian–Turonian)

Unlike the Indian Ocean margin, the Mediterranean margin is not characterized by the development of fault grabens parallel to the coast, but basement control appears to be responsible for a sequence of NE–SW-trending ridges and basins, to which the NNW–SSE-Suez Basin clearly seen in lithological maps (Fig. 8, El Shazly, 1977) or taphrogeosyncline of Said (1962) may be added. Not all of the basins were active throughout the time, for example the subsidence of the Sirte Basin began only in the Mid-Cretaceous and continued into the Tertiary, whereas the Suez Basin was a zone of subsidence from the Trias to the Cretaceous, continuing a history dating back to the Carboniferous at least.

In Egypt the transition from the continent into the offshore can be followed in the gradual facies change, the replacement of sand by shale and eventually by limestone, seen in data from exploration wells. The nearshore facies can be studied along the Red Sea (Quseir-Safaga) and from outcrops in Dakhla, Kharga and other oases. In Tripolitania on the other hand, at certain epochs at least, the old shoreline can be clearly documented close to the present coast and the principal marine facies lie to the west and in the present offshore to the north, i.e., under the Pelagian shelf (see Burollet et al., 1976). The true continental–oceanic margin lay far to the north. The exact location of that boundary is obscured by the tectonic events of the Alpine Orogeny, however arguments based upon paleomagnetism by Gregor et al. (1975), Barbieri et al. (1974) and Schult (1973) suggest that part at least of Sicily belongs to the African craton. D'Argenio et al. (1975) have indicated that the Sicilian portion represented a region of block-faulting with the development of carbonate platforms and basins. This is consistent with the interpretation of Caire (1977) who defines an African promontory to include much of Italy and the Adriatic Sea over which during Tertiary times sediments were thrust. The region thus represents an offshore shallowly flooded fractured shelf. Further to the east the margin of the African craton may parallel the Hellenic arc under which it is presently descending (McKenzie, 1972; Papazachos and Cominakis, 1971). This general interpretation is inconsistent however with current ideas of the rotation of Italy. It may be that this rotation involves only the Mesozoic–Cenozoic sequence and no basement involvement.

Apart from a few small outcrops on or near the Cyrenaican coast, Mesozoic outcrops along the Mediterranean margin are confined to the west, to western Tripolitania and Tunisia in particular. The quality of the available subsurface data is variable; a complex pattern of horsts and grabens in the Sirte Basin makes a close delineation of the Mesozoic sedimentary pattern impossible at present. Somewhat fewer holes have been drilled in northern Egypt, and this may account for the relatively simple scheme appearing in the synthesis of El Shazly (1977).

The Mesozoic of Libya

The principal outcrops of Mesozoic rocks in western Libya are in the Gefara (Djeffara) escarpment which extends west-southwestwards from Homs on the coast to Nalut near the Tunisian border 300 km distant. At that point the coastal plane is 140 km wide. The escarpment which rises in parts to over

TABLE II

Composite stratigraphic section for western Libya

Period	Stage	Formation/Group	Unit	Description
UPPER CRETACEOUS	Maastrichtian		Lower Tar Marl Member	lowermost member of Zmam Fm., found in the Hamada el Hamra; est. thickness 250 m; a gray-green shale, often gypsiferous
	Campanian	Mizda Formation	Thala Member	soft marls with limestone and dolomite bands gypsiferous [min. 90 m exposed at Mizda oasis 200 km south of Tripoli]
	Santonian Coniacian		Mazuza Limestone Member	hard thick bedded limestone with some chert
			Tigrinna Marl Member	shales with interbedded limestone and dolomite; fossiliferous, shallow marine environment
	Turonian		Gharian Limestone Fm.	light-gray-white hard massive limestone and dolomite with chert fairly common in upper part deposited in very shallow marine environment; many fossils destroyed by diagenetic alteration; type section 60 m principal exposures in the Gefara escarpment
	Cenomanian		Jefren Marl Fm.	soft brown to yellow-gray marl with thin limestone bands; 85 m in type section; dated by microfauna
			Ain Tobi Limestone	light gray to yellow-gray thick-bedded dolomitic limestone often oolitic at base; although a transgressive unit as Cenomanian sea spread over non-marine Mesozoic, maintains a relatively constant thickness, 150–190 m; no unconformity apparent as top of Chicla Fm. reworked; very shallow marine environment
LOWER CRETACEOUS			Chicla Sandstone Fm.	non-marine light-coloured friable sandstone conglomerate with some brown, yellow, red silts and clays; widely distributed along Gefara scarp as far as Tunisia; has occasional thin lignitic bands; braided river deposit; thickness variable from a few metres to excess of 80 m to Tunisia; usually regarded as "Lower Cretaceous", but Late Jurassic age also proposed; plant material suggests Albian age
			Cabao Sandstone Fm.	white to greenish gray sandstone, red at base; it is friable, cross-bedded, medium to coarse-grained; clay lens near base contain plant fragments; top more massive weathering gray; age could be Late Jurassic (Portlandian)
JURASSIC	Kimmeridgian–Oxfordian		Sciucsciuc Limestone Fm.	alternating limestones and clays with occasional sandy beds; highly fossiliferous; 30 m at type section; passes up gradually to Cabao section
	Callovian	Tiji Formation of Burollet	Chameau Mort Sandstone Fm.	Continental sandstone with clay intercalations; the sandstone is fine-grained, subangular and cross-bedded; Jurassic age assigned on basis of included flora
			Giosc Shale Fm.	lacustrine clays with minor sandstone interbeds
	Bathonian		Tacbal Limestone Fm.	marly or sandy limestone alternating with green or yellow-green clays; occasional gypsiferous bands in lower part about 25 m in type section
	Bajocian	Bir el Ghnem Group	Abreghs Fm.	two lithologic units; lower thick gypsum anhydrite succession with dolomite and argillaceous interbeds overlain by green to yellow-green clays with minor limey and gypsiferous bands; clays occassionally carbonaceous 77.6 m thick in the type locality
			Bu en Niran Fm.	basal limestone overlain by succession of green and red clays with gypsum bands; relation to gypsum members suggests environment oscillating between open marine through shallow to lagoonal; thickness about 20 m
			(Bir el Ghnem Fm.)	thick white to gray gypsum with frequent dolomite interbeds which are sometimes fossiliferous but not diagnostic; assigned a Late Triassic age

TABLE II (*continued*)

	Lower Jurassic	Bu Gheilan Fm.	upper unit: about 30 m massive regularly bedded, stromatolitic dolomites, finely laminated and pelletal dolomites with subordinate flat pebble conglomerates and clays and secondary limestone (sulphate replacement) in lower part middle unit: 15–20 m oolitic and intraclastic calcarenites in thick cross-stratified beds, dolomicrites and subordinate stromatolitic dolomites lower unit: fine and coarse dolomitic breccias and edgewise conglomerates, finely laminated dolomicrites, stromatolitic and pelletal dolomicrites and laminated microgranular gypsum
	Norian–Lias	Bu Sceba Fm.	upper member: poorly cemented white or reddish sandstones alternating with clay and siltstones; clays in part bentonitic; deposited in predominantly continental environment from low sinuosity streams, flood plains, probably also transitional or marine conditions; aeolian dunes in flood plain lower member: red siltstones with carbonate cement sandy dolomites and fine sandstones; shallow-water neritic environment passing to transitional and non-marine, that is intertidal or shallow subtidal passing to alluvial deltaic in uppermost beds; at top channels eroded and filled with conglomerate; maximum thickness 173 m
TRIASSIC	Carnian	Azizia Limestone Fm.	thick sequence of well-bedded dark gray carbonates, limestones, dolomitic limestones and dolomicrites, pelmicrites, algal biomicrites, oosparites described; environment of deposition low-energy inner platform shallow subtidal to intertidal; max. thickness about 160 m
	Ladinian	Currusc Fm.	succession of yellow to green clays and pale red to brown micaceous sandstone; minor calcareous intercalations near top; max. exposed thickness + 40 m; contains a Ladinian faunal assemblage; exposed in cores of domal structures in Gefara escarpment
		Ouled Chebbi Fm.	known in subsurface only

This table with its short lithological description should be used in conjunction with Fig. 4 and the text. The ages assigned to some beds are by stratigraphic position, and where facies changes occur various interpretations are possible. Certain of the beds are confined to the westernmost part of the Nefusa escarpment.

800 m is the result of a major flexure of Miocene age. During the Mesozoic the region formed the margin of a gently subsiding platform which included the Pelagian shelf, Sicily, eastern and southern Tunisia and the Algerian salt basin (Assereto and Benelli, 1971). Subsidence was interrupted on a number of occasions resulting in the influx of sand, and on a smaller scale in the complex variations seen in detailed analyses of the limestone facies. The eastward–westward see-sawing of the facies belts provides a complex history over which there is still no general accord. The Tripolitanian section has been discussed by Christie (1955), Desio et al. (1963), Burollet 1963, 1976), Assereto and Benelli (1971), Hammuda (1971) and Hinnawy and Cheshilev (1975). For the Tunisian and Algerian region it would be difficult to improve on the monographic synthesis of Busson (1967b, 1970), although there are more general accounts of the Tunisian region by Salaj (1977) and Burollet (1967).

The positive movements which affected the Gefara region were stronger in the east, and in particular as a result of movement during the Early Cretaceous the section there is less complete exposing principally the Cenomanian, Lower Jurassic and Triassic rocks. The intervening horizons appear further to the west although simple interpretation is affected by facies change. The higher Cretaceous beds appear in the Mizda oasis section 90 km south of Garian. East of the Gefara lies the Sirte Basin where Eocene and younger rocks form the surface outcrop. Information on the subsurface crop draws heavily on the work of

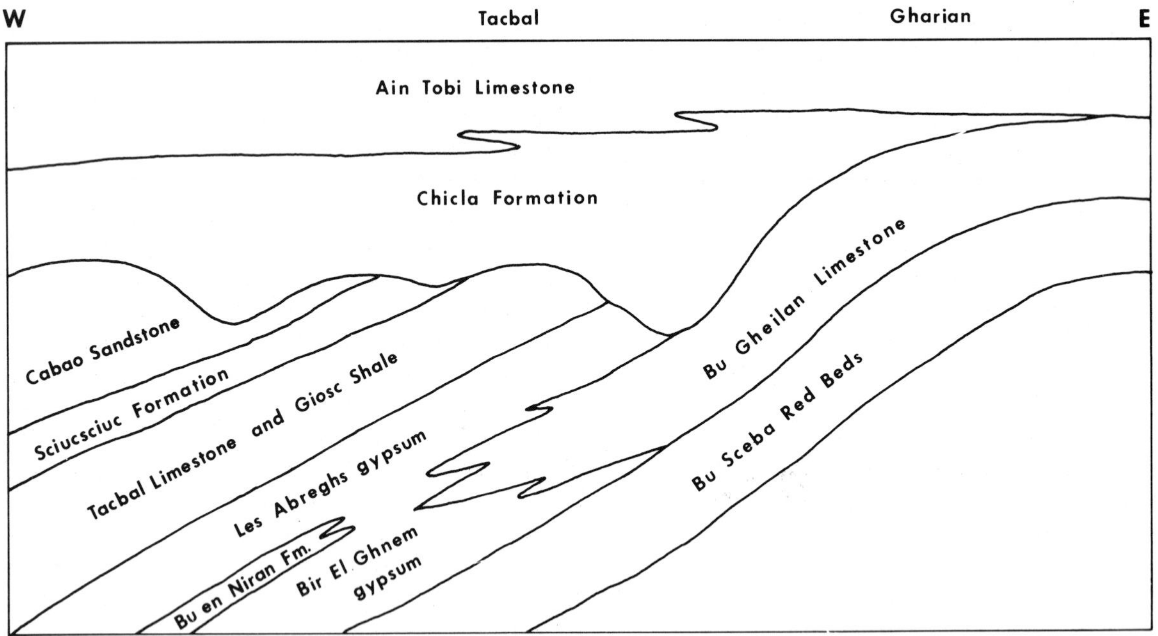

Fig. 4. East-west section illustrating the facies relationship of the Jurassic beds along the Gharian scarp, south of Tripoli (after Assereto and Benelli, 1971; Burollet, 1963, and others).

Barr and Weegar (1972), a source also used by Salah (1975). The separation between it and the Sirte Basin, which began to form in Late Cretaceous times is taken at the Hon graben.

A composite stratigraphic succession, with brief descriptions of the units found in the Gefara section is given in Table II. It must be realized that the section is a generalization for because of facies change not all units are found in all areas, for example the Bu Gheilan Limestone Formation is the lateral equivalent of some part of the Bir El Ghnem Group (see Fig. 4). Erosion too, following uplift, has taken its toll, in particular the Chicla Sandstone Formation rests on all formations from the Cabao Sandstone (Lower Cretaceous) to the Bu Gheilan Limestone (Lower Jurassic) (Hammuda, 1971). In the Sirte Basin, which initially had the form of a series of NW–SE-striking horsts and grabens, the most complete sections are derived from wells drilled into the grabens for these received continuous sedimentation from Cenomanian times onward. The principal formations are described in Table III and should be considered in conjunction with Fig. 5.

As a result of detailed sedimentological studies, the depositional environment in the Gefara is well known for certain time intervals. Unfortunately the same detail does not extend to all formations, thus the detailed description of the Late Cretaceous environments is lacking. The following discussion, drawn largely from the work of Busson (1967) and Assereto and Benelli (1971) applies in particular to the time interval Late Triassic to Early Cretaceous. The Cenomanian brought a flooding of the Gefara, and for the first time, of the more low lying, i.e. downfaulted, zones of the Sirte Basin. The sea was apparently shallow, and the region remained marine through the rest of the Cretaceous. The southward extent of the sea in the region of the present Hamada el Hamra is uncertain. This was the period during which a link of the Tethys with the South Atlantic is postulated (Fig. 2), the sea penetrating west of the Gefara through a region which had persisted as a depression through much of the Mesozoic and Late Paleozoic.

Only four of the Mesozoic formations have received detailed sedimentological treatment, the Upper Ladinian–Carnian Azizia Formation and the Lower Jurassic Bu Gheilan Formation, and two primary sandstone units, the Lower Jurassic–Upper Triassic Bu Sceba Formation and the Albian Chicla Sandstone Formation, which have been studied in the region south of Tripoli. However, as the history of the region during the Mesozoic involved only vertical movements of general small extent, the pattern revealed from the study of the four units can be regarded as typifying the Mesozoic, until the Cenomanian transgression. The Cenomanian limestones appear more uniform and are certainly more widespread but have not received detailed treatment while younger beds are poorly exposed.

In the Lower Jurassic Bu Gheilan limestones, Assereto and Benelli (1971) describe six facies: laminated dolomicrites, pellet dolomicrites, laminated algal biolithites (stromatolites), edgewise conglomerates, oolitic and intraclastic dolomitized calcarenites. The lithofacies of the lower of the three units into which the Bu Gheilan Limestone may be divided, are analogous to low-energy supratidal to shallow subtidal carbonate environments. The association of mud-cracked stromatolitic and laminated dolomites, evaporites and edgewise conglomerates suggest an environment similar to the recent sebkha and algal deposits in the Persian Gulf, an interpretation consistent with minor features such as dessication polygons, carbonate pseudomorphs of gypsum, planar or domal biolithites, etc. The upper division is similar, differing only in having fewer evaporites, and comparison with humid, rainy, flood-inundation flats of Andros Island and Florida is suggested rather than the arid supratidal evaporitic flats of the Trucial Coast.

The middle unit contains oolitic and fossiliferous calcarenites which are well sorted and current bedded, suggesting the strong current action associated with oolitic sand shoals and bars or tidal channels. This facies alternates with thin dolomitic micrites indicating low-energy conditions such as are found in shallow lagoons and shallow subtidal environments.

The upper part of the Bu Sceba Formation passes vertically and laterally to the Bu Gheilan facies, i.e. forming part of an onlap marine sequence passing from red alluvial mudstones and sandstones, green lacustrine clays to supratidal flat dolomicrites and evaporites to low supratidal stromatolitic dolomites and laminated pelmicrites, to shoal and tidal channel sediments, with lagoonal or coastal flat pelmicrites all represented (see Fig. 6).

The sequence is similar to the Persian Gulf. The lower member of the Bu Sceba Formation has the lithological and paleontological features of a shallow marine neritic environment passing upwards to a non-marine environment. Current action is indicated by sorting, absence of clay, ripple marks and parallel lamination. Silty dolomites and micrites, clays and fossils indicate alternation with quiet-water conditions. The sandstones at higher levels in the member are poorly sorted, which together with mudcracked marks and collapse breccias are regarded as indicating low intertidal to supratidal conditions while at the top deposits of an alluvial delta occur. These latter may also pass laterally to intertidal deposits.

The four facies recognized in the upper member suggest inland dune, shallow, low sinuosity, river channels in a large flood plain with overbank muds on the floodplain and restricted arid lagoon flats. Sediment was transported across this continental flood plain from an easterly or southeasterly source. The presence of feldspar suggests a basically arid environment while the maturity of the quartz grains suggest polycyclic origin for some of the sand. There is also some evidence of material from granitic and gneissic terrain hence a source such as the Tibesti is suggested, with basement and Paleozoic sediments undergoing erosion.

Westward lie the gypsum deposits of the Bir El Ghnem Group, raising the question of the origin of carbonate deposits on the landward margin of a restricted-circulation saline environment. The suggestion adopted by Assereto and Benelli (1971) of dilution by meteoric water from the continent, is not easily reconciled with the arid environment proposed for the greater part of the time involved, although it can also account for the dolomitization of the oolite shoals and bars.

The environment suggested by the facies of the Azizia Formation is a low-energy inner platform shallow subtidal–intertidal, comparable to the west coast of Andros Island and Florida. Thus while still shallow water, the water depth was apparently somewhat greater than that in which the Bu Gheilan Limestone formed (Fig. 7).

Reconstruction of the geological history of the region suggests that in western Tripolitania transitional to brackish and shallow marine environments existed throughout most of the later Triassic, and there is evidence of the existence of a salt basin in the offshore region. Major uplift affecting eastern Tripolitania caused a westward migration of these facies. The Triassic closed with a regional transgression, the transitional facies migrating eastwards from the region of the Tunisian border with the development of a typical onlap facies. Major uplift again affecting the east occurred in Late Jurassic–Early

TABLE III

Stratigraphic section for the Sirte Basin

Maastrichtian	Kalash Limestone Fm.	white, tan or gray, argillaceous, calcilutitic limestone with dark gray calcareous shale interbeds; occasionally upper part becomes calcarenitic; occurs over much of Sirte Basin 30–100 m thick but may exceed 160 m; over Dahra–Hofra high replaced by Lower Satal Fm. which is in part dolomitic and contains anhydrite horizons, and rests unconformably on clastic beds; Whah Limestone, a lateral equivalent is a very shallow marine limestone also resting on non-marine sandstone
Campanian	Sirte Shale Fm. *	Predominantly a dark brown waxy shale, carbonaceous or calcareous, with thin limestone interbeds; only occasional silty beds in type section but increasingly silty in adjacent areas; depending on location basal contact may be gradational or unconformable; in latter case base is sandy, pyritic and glauconitic; in eastern Sirte Basin, the lower part is calcareous (e.g., Tagrifet Limestone Fm, Maragh Limestone Fm.) and the name is restricted to upper shaly horizon; both limestone horizons transgress over former highs
Santonian–Coniacian	Rachmat Fm. *	Typically a shale sequence with minor sandstone and limestone and occasional dolomite interbeds; widely distributed in troughs, it may exceed 400 m in thickness, though absent on regional highs; top contact with Sirte Shale Fm. often marked by sandy, glauconitic zone with phosphate pebbles suggesting time break; may rest conformably upon Argub carbonate, Etel, Bahi or Maragh Formations or rest unconformably upon Lower Paleozoic or granitic rocks; in the eastern Sirte Basin passes laterally to Maragh Fm. and rests directly upon Amal high
Turonian	Etel Fm. *	a sequence of thin-bedded carbonates, shales, siltstones with some fine sandstone and anhydrite; carbonates predominantly dolomitic, subordinate silty calcilutites and less common, calcarenites; the presence of anhydrite is a characteristic feature; shales gray-green to brown, occasional red calcareous, in part pyritic; fine calcereous glauconitic sandstones, and siltstones occur but are not common; widely distributed in Sirte Basin (central and south) but absent over regional highs (Dahra–Hofra, Waha, Amal); in troughs the formation is over 1,000 m thick; equivalent of Argub carbonate in northwest Sirte Basin, latter presumed to be a more basinward facies
Cenomanian	Lidam Fm.	light-brown-gray saccharoidal dolomite sometimes angular, and light gray and brown pelletal calcarenites; lower part sandy indicating mixing with underlying sand, quartz becoming rare upwards; glauconite and oolites also rare; main occurrence in main Sirte trough where it may reach 200 m; unconformably overlies various units including "Nubian", Paleozoic and Basement; Etel Fm. usually conformable above but not always; shallow marine, in part lagoonal and intertidal environment inferred
	Bahi Fm.	Interbedded sandstone, siltstone, shale and conglomerate; glauconite common in top 3–6 m but absent at lower horizons; usually a 3–6 m basal conglomerate; variable in thickness from few metres to in excess of 130 m; unconformably overlain with an abrupt contact by Lidam Formation; unfossiliferous, no age assigned and is a facies which may occur at many horizons

The sequence given is typical of the northwest and southwest Sirte Basin. The irregular basin topography is reflected in many abrupt facies changes and thickness variations (cf. Fig. 5). Principal sources; Barr and Weegar (1972), Salah (1975).

* The Sirte, Rachmat and Etel Formations form part of the Rakb Group.

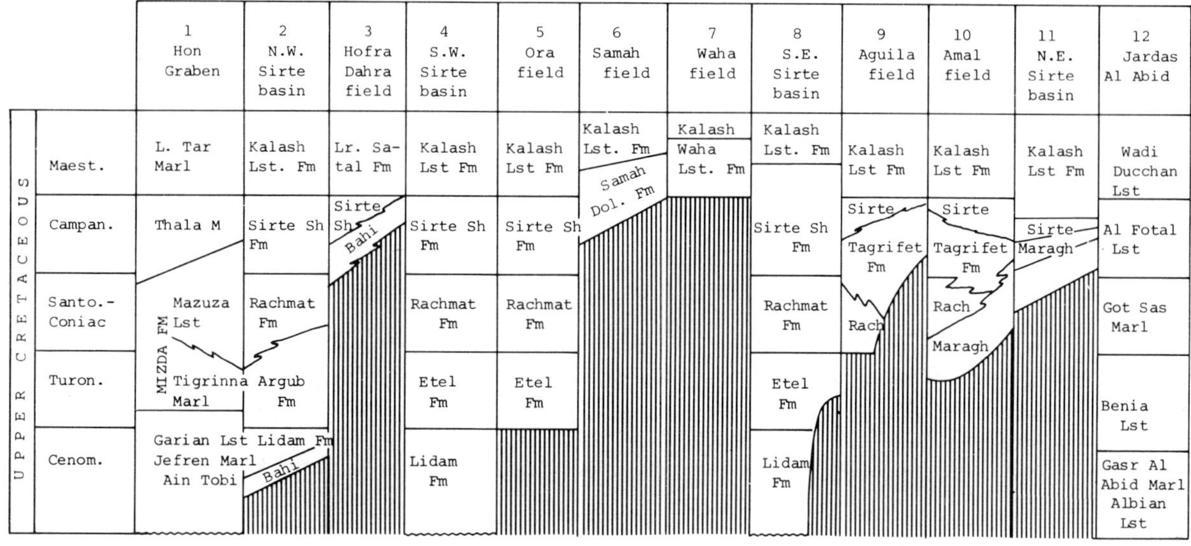

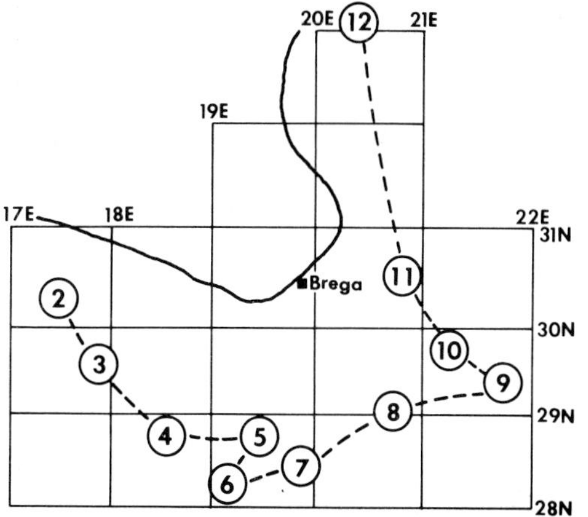

Fig. 5. Representative Mesozoic stratigraphic columns at various locations in the Sirte Basin (after Barr and Weegar, 1972). Locations of the sections shown on the inset map are from Barr and Weegar (1972). The precise location of *2, 4, 8,* and *11* is in some doubt.

Cretaceous times resulting in deep erosion and base levelling. Much of the Late Jurassic section was removed from the region south of Tripoli (see Fig. 4). The exact age of the movement or movements depends upon the interpretation of the Cabao and Chicla Sandstones. The Chicla Sandstone represents a cover of braided river deposits, the pattern of which is given by Hammuda (1971). Over this the Ain Tobi Limestone, the lowest member of the Nefusa Group, transgressed.

The later movements of which the Chicla Sandstone is one result, were presumably the same which resulted in the inception of the Sirte Basin, in which case some at least of the undated arenaceous beds at the base of the section may be time equivalents in the Sirte Basin of the Chicla Sandstone. It also puts a limit to the Early Cretaceous uplift of the eastern Jebel Nefusa.

The absence of Cretaceous in some areas of the Sirte Basin, or of pre-Maastrichtian in others while in yet others there is a more or less continuous section, and abrupt facies changes are evidence of contemporaneous movement during Late Cretaceous. This evidence of such movement is difficult to see in the Jebel Nefusa, because of poor exposure. It is however clear that at its inception the Sirte Basin consisted of a series of NW–SE-oriented horst and graben structures, of which the Hon graben, the only

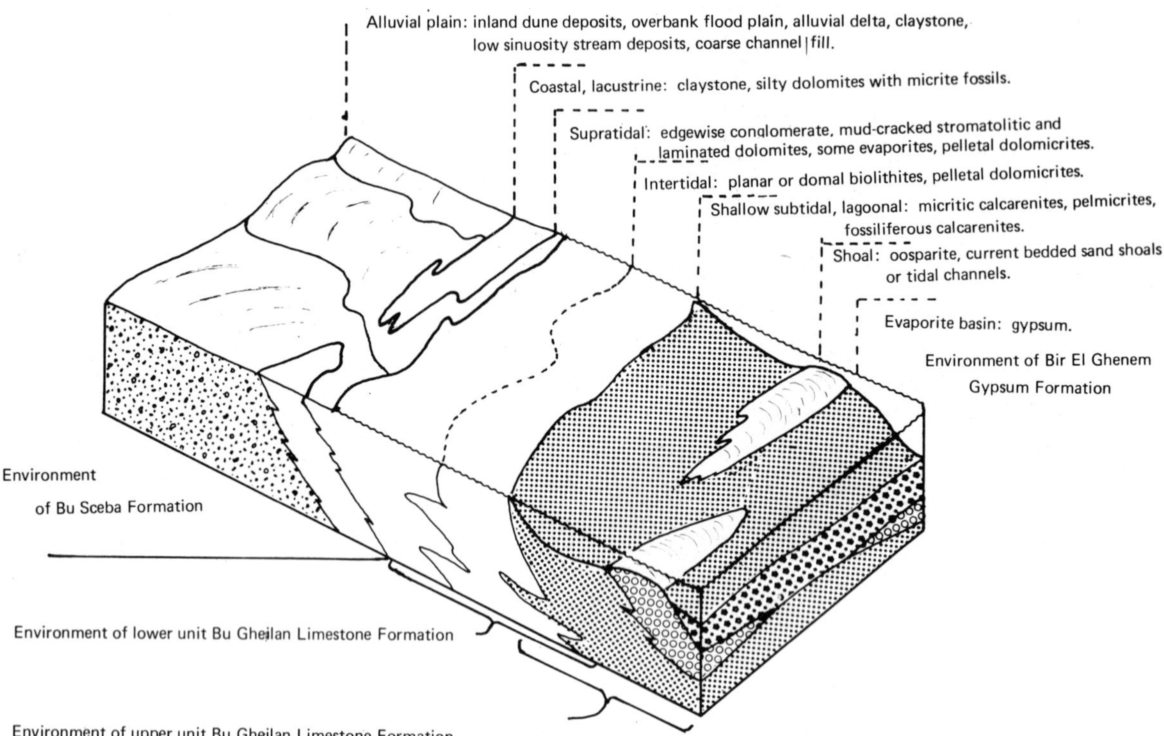

Fig. 6. Block diagram illustrating the environmental and facies relationships of the different lithologies of the Lower Jurassic Bu Gheilan limestone Formation (after Assereto and Benelli, 1971).

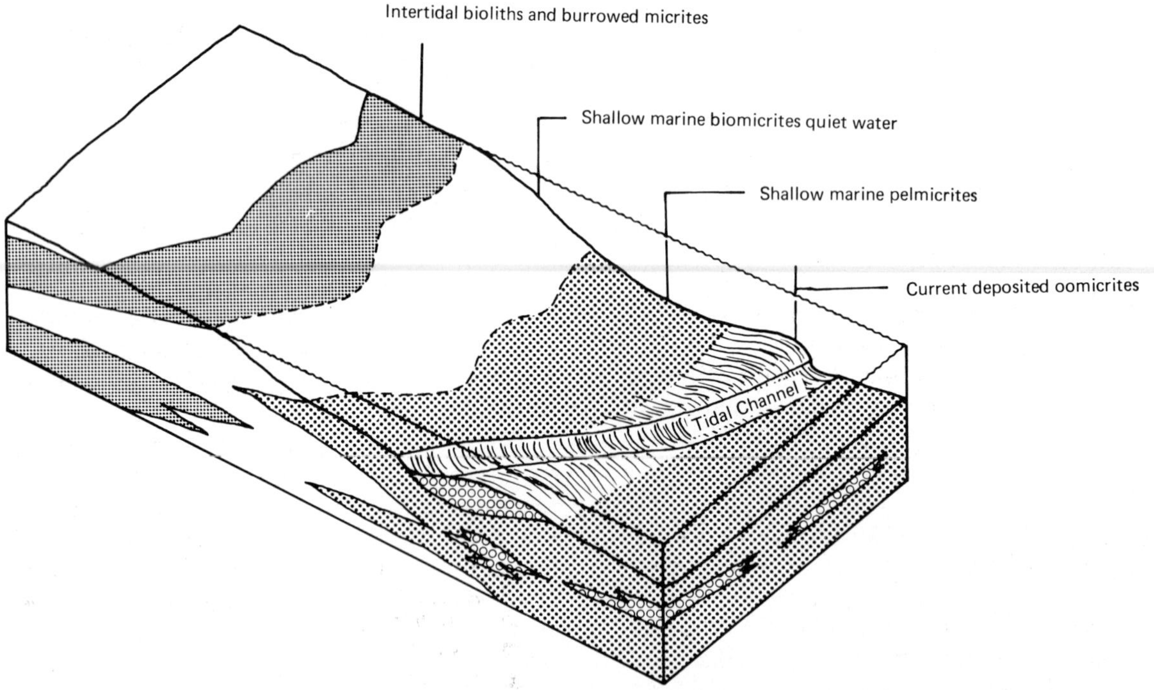

Fig. 7. Block diagram illustrating the environmental and facies relationships of the different lithologies found in the Azizia Limestone Formation (after Assereto and Benelli, 1971).

one exposed at the surface, forms the western margin. In the Jebel Nefusa the Gefara arch, which follows the same line as the Paleozoic Tripoli–Tibesti uplift (Klitzch, 1971), has the same NW–SE trend.

More uniform conditions in the Sirte Basin were only established in Maastrichtian times, with limestone deposition general both on the horsts and in the grabens.

The Mesozoic in Cyrenaica is still comparatively poorly known; thick sections but of limited outcrop consisting primarily of limestone are described in the Upper Cretaceous from Jardas Al Abid and the northeast Jebel Akhbar by Barr and Weegar (1972). The former is listed for comparison in Fig. 5.

The Mesozoic of Egypt

As in the case of the Sirte Basin the Mesozoic of the Western Desert of Egypt is deeply buried, the depth of the top of the Late Cretaceous chalk is of the order of 1 km in the coastal region (see El Shazly, 1977). The principal outcrops in the Western Desert, in the oases of Kharga, Dakhla and El Bahariya, expose near shore facies of Late Cretaceous rocks. The best exposures of marine Mesozoic rocks are in Sinai and along the Red Sea margin, approximately from Safaga to Quseir, but important outcrops occur in the Wadi Qena and in a small inlier (Abu Roash) immediately north of Cairo. With the exception of Sinai however there is little surface information available for pre-Cenomanian rocks.

The information resulting from oil exploration in the Western Desert has provided the basis for the interpretation of El Shazly (1977) and El Gezeery et al. (1975). The information, based on Jurassic and Cretaceous rocks, is summarized by them in terms of lithological diagrams and isopach maps (see Fig. 8A, B). The thickness of the Jurassic beds increases from about 0.5 km near 29°30′N to more than 2 km near the Mediterranean coast. A thickness of about 2 km is quite normal for Lower Cretaceous beds, but may expand to in excess of 3 km (near Mersa Matruh).

While the ancient shoreline positions for most of the Mesozoic are not known except by extrapolation, a sand : shale ratio of 8 : 1 suggests a nearshore environment. Falling clastic ratios indicate the increasing proportion of carbonates and hence increasingly marine environments. In general there is a progressive diminution of clastics northwards upon which the effects of basement control, in terms of basins and intervening relatively positive areas, result in the sinusoidal pattern seen in Fig. 8A, B. The variation in facies, at least in the Upper Cretaceous shows that these were syn-sedimentary features with the basins generally containing more carbonates, and the highs more arenaceous sediments. The sequence at El Bahariya oasis even suggests that the highs were exposed at least on some occasions (see El Shazly, 1977).

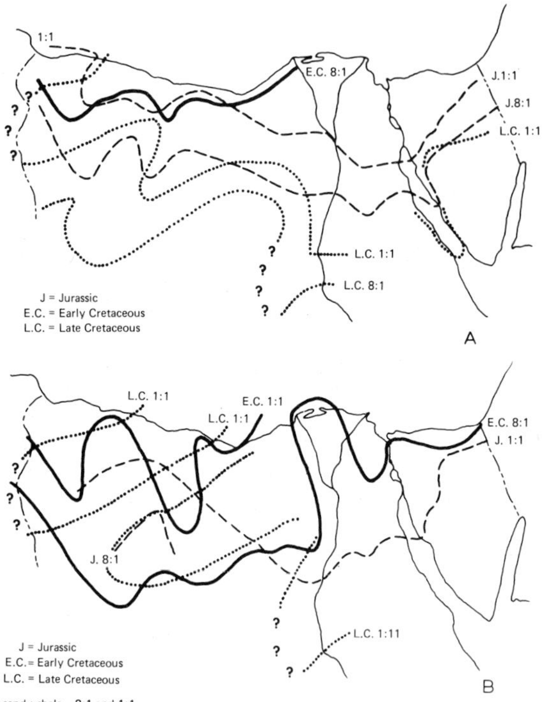

Fig. 8. A. Clastic ratio in northern Egypt (modified after El Gezeery et al., 1975). B. Sand : shale ratio in northern Egypt (modified after El Gezeery et al., 1975; from El Shazly, 1977).

Rocks of Triassic age are not known west of the immediate environment of Suez (Wadi Quseib-Abu Darag area; see Abdallah et al., 1963). A Muschelkalk trangression with a shoreline lying about 30°N is suggested by Said (1962). A northward transition from sandstones to limestones, that is to an offshore environment free of clastic sediments is suggested by outcrops near the Israeli border. Subsurface records in the Sinai and Suez reveal a thin clastic sequence in the south and limestones in the north. El Shazly (1977) suggests that a Triassic basin may have extended southwards in the Gulf of Suez region, an interpretation consistent with the established distribution of Carboniferous, Jurassic and Cretaceous rocks.

Marine Jurassic outcrops are more extensive than

those of the underlying Triassic and occur in northern Sinai in the Maghara structure, and on the western side of the Gulf of Suez in the El Galala–El Bahariya plateau. Continental Jurassic sediments are intercalated with marine in Maghara, and are known in the Western Desert, but since, however, they are poorly fossiliferous their extent may be greater than realized. Generally the sediments extend to about 30°N.

The principal feature of the Jurassic was a transgression from the Tethys which began in Callovian times and reached its maximum during the Kimmeridgian (Fig. 9). The transgression was more extensive in the east than in the west (El Shazly, 1977). By the end of the Jurassic, following uplift, the sea regressed and erosion of the younger Jurassic beds resulted.

The Jebel Maghara succession (Table IV), suggests that the area was near the coast, with the alternation of marine and fluviomarine horizons. Coal in the succession has been mined (see Abdel-Khalek, 1973).

A similar early Middle Jurassic transition from lagoonal to marginal marine clastics and carbonates is known in the Western Desert.

TABLE IV

The Jebel Maghara succession

Kimmeridgian–Late Bathonian	El Masajid Fm. (576 m) marine limestone, (maximum transgression)
"Bathonian"	Safa Fm. (215 m) fluviomarine with coal
Late Bajocian–Late Lias Bir Maghara Fm. (444 m)	massive limestone with some clay
Pliensbachian Shusha Fm. (272 m)	fluviomarine shales and sandstones with coaly material

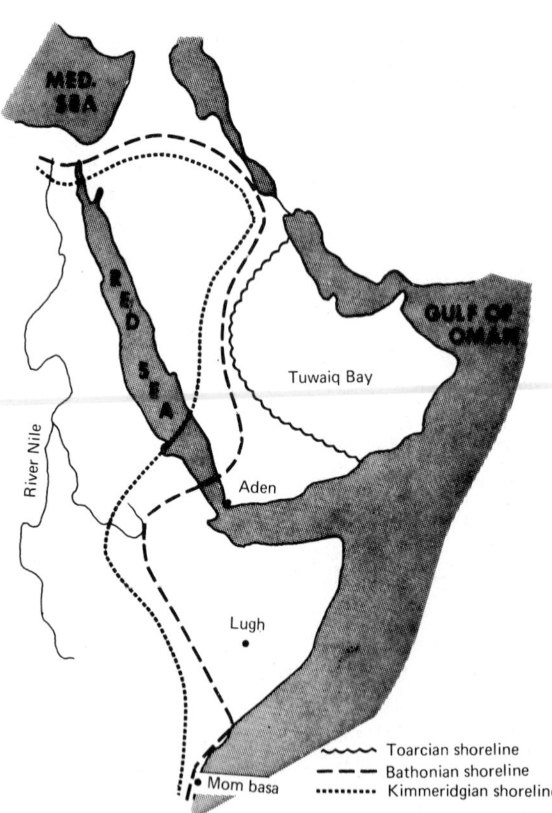

Fig. 9. Jurassic shorelines in northeastern Egypt, Arabia, and the Horn of Africa (from Abdallah et al., 1963).

Below marine Bathonian at the entrance to the Gulf of Suez a sandstone section rich in fossils of Rhaetic–Early Liassic age occurs. The Bathonian age assigned to the marine beds is based upon the brachiopod-lamellibranchs found, for the presence of ammonites has not yet been recorded. The flora of the sandstone is similar to that of the Kohlan Sandstone of Yemen. The Bathonian is overlain by Nubian here capped by Cenomanian marls.

In Egypt as in Libya the effects of the positive movements which took place at the end of the Jurassic are reflected in the primarily detrital sediments found in the subsurface in northern Egypt. However, while the principal Cretaceous trangression was clearly in progress during Cenomanian times, for example in the Wadi Qena and the Bahariya oasis, fossiliferous marine Cenomanian sediments overlie Nubian Sandstone, oil exploration in the Western Desert has revealed widespread Albian carbonates (e.g., El Alamein Formation), and in both outcrop and subsurface in the Gulf of Suez region marine horizons of Urgo-Aptian age are intercalated in continental, predominantly arenaceous beds, of the Malha Formation (formerly loosely termed *Nubian Sandstone*).

The southern progress of the transgression, which reached its peak in the Paleocene, is indicated by the progressive change in the age of the oldest marine fossiliferous horizon. In the Red Sea region between Quseir and Safaga, and in the Dakhla and Kharga oases the continental *Nubian* facies is followed without obvious break by the phosphate horizon to which a Campanian age can be assigned. This difference in age assigned to the *Nubian Sandstone* facies within a relatively short distance (cf. Wadi Qena) has created stratigraphical problems discussed in the previous section. In particular in Egypt the *Nubian Sandstone* is not purely continental but provides a transition to shallow, nearshore marine environments.

As a consequence of the Late Cretaceous transgression the sediments of this age in northern Egypt found in the subsurface are rich in carbonates, with a 1 : 4 clastic ratio. The facies of the exposed Upper Cretaceous rocks indicate that they formed under nearshore conditions. The only exposure of the carbonate facies in northern Egypt occurs in the Abu Roash inlier (Table V). In northern Egypt the Upper Cretaceous section reaches a thickness of 1–1.5 km, but thins southwards to less than 0.5 km. The Gulf of Suez Basin at this time was also receiving a thick sedimentary sequence and a structure map of the base of the Chalk unit shows a thickness of more than 2 km at the centre of the gulf which decreases outwards. There is also evidence of a basin or depression paralleling the western flank of the Eastern Desert, which between 28° and 29°N coincides roughly with the modern Nile Valley (Fig. 10).

In the Sudan, rocks of Maastrichtian age to which the name Mukawar Formation has been given were found in the Maghersum No. 1. borehole (Whiteman, 1971). These grey to red-brown silty shales with grey sandstone and occasional limestone bands are possibly the equivalent of marine marls found on the coastal plain near Jeddah on the opposite shore of the present Red Sea. They suggest that in Late Cretaceous times the Tethyan Gulf occupying the present Gulf of Suez region reached as far south as 23°.

Most of the outcrop information concerns the littoral facies of the Upper Cretaceous, of which the phosphate facies extending from Safaga–Quseir on the Red Sea coast to Qena–Idfu in the Nile Valley and westwards to Kharga and Dahkla oases is the most interesting. Over this region the succession is sufficiently similar that the Quseir section may be taken as typical (Table VI).

El Shazly (1977) interprets the facies pattern of the Campanian as passing from near shore marine and continental sands south of 24° (the *Nubian Sandstone* facies) into offshore marine sands and clays within which occur most of the exploitable phosphate beds, to more argillaceous marine beds between 27° and 29°N within which occasional phosphate beds occur and finally into predominantly marine carbonates north of 29°N. There is however some uncertainty about the correct environmental significance of the phosphates. According to Said (1962) they formed in a "... bay connected to the open sea with strong currents abrading the bottom and reworking sediments".

The phosphate beds, which show considerable lateral variation, are interbedded with shales, marls and silicified limestones. They contain large bones, oyster shells, irregularly distributed fossil fragments and pebbles showing evidence of rolling in a matrix of finer material. Collophanitic oolites in a variety of colours also occur. Most of the phosphate horizons show little evidence of bedding. Although some of the oyster shells show evidence of secondary phos-

TABLE V

Succession in the Abu Roash inlier

Maastrichtian–Campanian	Chalk 57 m poor microfauna
Santonian	*Plicatula* Series 34 m of limestones and marls Flint Series 34 m a white chalky limestone
Turonian	*Actaeonella* Series 22 m of limestone shale and marl Limestone Series 54 m poorly fossiliferous Rudistae Series 26 m of marl and shale with oyster bands
Cenomanian	Sandstone Series 33 m with some shale and oyster bands, base not seen.

TABLE VI

The Upper Cretaceous section at Quseir (after Said, 1962)

Paleocene	chalk
Early Paleocene–Maastrichtian	Dakhla Shale 165 m
Maastrichtian	Duwi Fm. siliceous limestones, marls and lenticular phosphate beds
Campanian	variegated shale 70 m (unfossiliferous) "*Nubian Sandstone*"

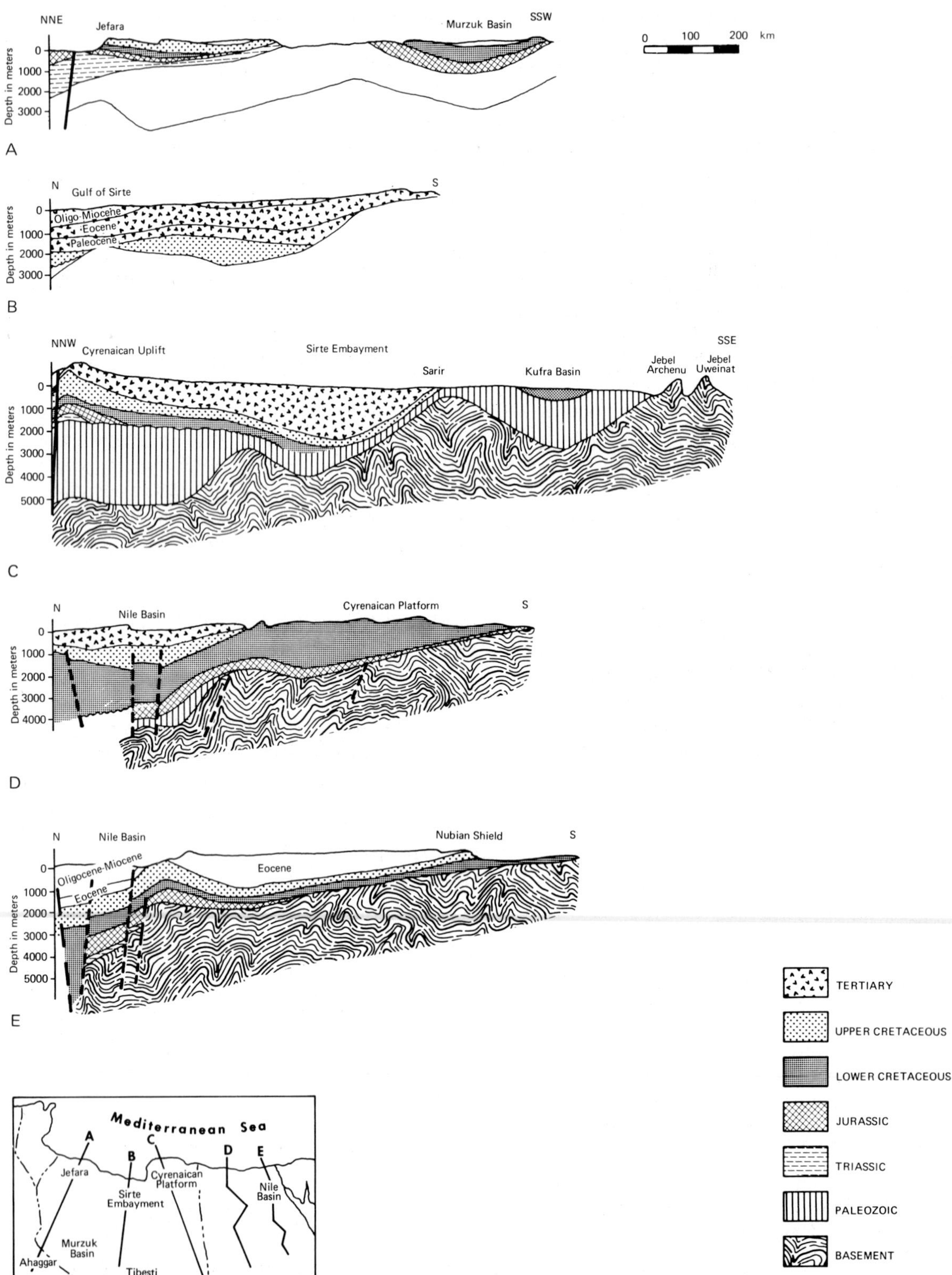

Fig. 10. Simplified cross-sections across the Mediterranean margin of the African craton.

phatization, the phosphate is generally regarded as of primary origin, derived from ammonium phosphate resulting from the decay of organic matter in sea water.

The conditions for the concentration of phosphate exist in isolated basins, and current action in such basins would not dissipate the phosphate concentration. However, the fauna suggests that the region was not isolated. Paleogeographic considerations rule out an isolated land-locked basin of the Black Sea type. The remaining possibility is of relatively shallow coastal basins behind barrier shoals or bars with the coarse deposits, essentially lag deposits, from storm disturbance rather than regular current action.

While the principal phosphate deposits are Campanian in age some are found in other Senonian rocks. This is consistent with a southward migration of facies, also seen in the ages assigned to the Chalk which is Campanian–Maastrichtian in the Abu Roash inlier north of Cairo and Paleocene in the south near Quseir (Said, 1962).

THE EAST AFRICAN REALM

As in the case of the Saharan realm, it is convenient to present the geology of the East African realm in terms of the continental and marine facies. The former essentially consists of the beds terminating the Karroo cycle. The Karroo System (Supergroup) was initially established in South Africa and a more refined description is found in Dingle (Ch. 11). The term has been commonly, if somewhat unwisely, applied to a variety of beds which have in common with the type area only the tendency for the lowest beds to be gray and locally coal-bearing, the middle group to be sandstone and shale and the upper group to consist of coarse clastics (Kent, 1974). These beds tend to be distributed in a series of separate basins whose form in some cases was fault-controlled. Although regarded as continental, short-lived marine phases are known, as in Kenya, Madagascar and Tanzania. Where the basins have subsequently become the site of later Mesozoic marine sedimentation they are considered with the latter; the inland basins where sedimentation virtually ended at the termination of the Karroo cycle can be examined separately.

The geology of the marine Mesozoic is most easily comprehended in terms of two basins, the Somali Basin to the north now connected by the Mozambique Channel to the Natal Basin (Fig. 11) in the south, and the working hypothesis that the true edge of the African continent lies east of Madagascar (and the Seychelles) as proposed by Baker and Miller (1963). This hypothesis explicitly rules out any significant continental displacement of Madagascar; more specifically it denies the placing of Madagascar close to either the Tanzanian or Mozambique shores despite tentative paleomagnetic results of the author (Nairn, 1964). Within the limits so defined areas of oceanic crust are not extensive (Francis et al., 1966), and while discussion of its origin go beyond the limits of this work, it may represent limited spreading or crustal thinning. Thus, the progressive stages of marine invasion regarded by Wegener (1922) and Du Toit (1954) as evidence of the progressive disruption of Gondwana, are strictly the record of the development of the Somali Basin as it flooded its margins. Continental drift is an unnecessary hypothesis in this context.

Fig. 11. The Jurassic–Cretaceous coastal basins of eastern Africa and the "Karroo" continental basins (after Blant, 1973). *L* = Lamu; *M* = Mombasa; *D* = Dar es Salaam; *B* = Beira; *L.M.* = Lourenço Marques; *Du* = Durban, *S* = Cape St. Pierre.

The continental facies

As previously indicated in Africa south of the Sahara, with minor exceptions, the continental beds represented belong to the Karroo cycle and are usually regarded as extending to Early Jurassic. Subsequent to the end of the Karroo cycle little deposition has taken place over the craton with the exception of the young Kalahari sands. The region covered by the Karroo beds may be conveniently, if somewhat arbitrarily, divided into a northern and southern area. The dividing line, or zone, extends from Windhoek to Belfast, east of Pretoria (Du Toit, 1954), separating the northern region with its restricted succession and slightly different facies from the Karroo Basin to the south. That basin has been described by Dingle (Ch. 11); it includes Botswana, where as Green (1966) pointed out the rocks which cover the eastern half of the country are relatively uniform and broadly similar to those in South Africa. With increasing distance from the southern basin, the Karroo succession becomes increasingly restricted or diverges more from the type. The section in southern Rhodesia is the most complete in the northern area.

Although not capped by Stormberg (Karroo) basalts at this point, the most complete section in southern Rhodesia is the Mafungabusi escarpment section near Gokwe (McGregor, 1941, 1947) where the following Triassic section is found:

	Approximate thickness
Nyamandhlovu Sandstone	20 m
Forest Sandstone	65 m
Pebbly arkose	85 m
Fine red marly sandstone	
(minor unconformity)	75 m
Escarpment Grit	20 m

The total thickness, about 265 m of essentially arenaceous beds, is small compared to the thicknesses found in South Africa. The beds are poorly fossiliferous and the correlation of the lower beds with the Molteno–Red Beds sequence of South Africa is based upon plant remains, plus the fact that the Forest Sandstone and Nyamandhlovu Sandstone are of aeolian origin and thus parallel the Cave Sandstone. Near Bulawayo, the Forest Sandstone rests directly upon basement and is capped by a thin sequence of weathered lavas. The thickness of lava, locally called the Batoka basalts, reaches 650 m at Victoria Falls. South of the falls in the Wankie Coalfield the Mafungabusi section can be recognized with the Escarpment Grit resting unconformably upon a shale sequence assigned to the Beaufort (Du Toit, 1954).

Recently a discovery near Gokwe of dinosaur bones in a sequence of sandstones with occasional thin impure pisolitic limestones was made by Bond and Bromley (1970). The importance of the discovery of these beds, to which the name Gokwe Formation has been assigned, is the rarity of post-Karroo cycle deposits. Although the dinosaurian remains are unidentified, and the single lamellibranch can only be tentatively identified, the beds are regarded as latest Jurassic to Early Cretaceous in age. They are regarded as having formed as the result of intermittent floods in a shallow lake depression comparable with the present Makarikari Lake Basin in Botswana.

In East Africa, both in Tanzania and Kenya, beds loosely referred to the Karroo System (Supergroup) were deposited in contemporaneously developing fault grabens. In some cases marine horizons make a brief appearance (Permian: Kidodi Basin, Tanzania; Lamu Embayment, Kenya, etc.), only in the Mandawa Basin does the presence of Permo-Triassic evaporites indicate a continued if restricted marine influx. These basins which lie close to the present coast, will be considered in the next section along with the later Mesozoic marine sequence.

Further inland, Jurassic beds have not been definitely identified although sequences assigned to the Karroo might pass up into lowermost Jurassic. Continental Cretaceous deposits, identified by vertebrate remains, have been found and it seems possible that more detailed field studies will reveal these to be more widely distributed than previously thought.

The principal outcrops in southwestern Tanzania are found in the Ruhuhu depression on the eastern side of Lake Malawi (Nyasa) and the Rukwa depression running northwest of the lake (Fig. 12). The depressions existed at the time of deposition (graben) and subsequently were reactivated by later faulting which served to preserve the soft Karroo sediments. The beds, originally more extensive, have been removed by erosion from the adjoining areas. Within each depression small coalfields are recognized, the Songwe–Kiwira, Galula and Muassa in the Rukwa trough, the Ngaka and Katewaka–Mchuchuma in the Ruhuhu depression.

In southwestern Tanzania the early Karroo beds overlap on to an irregular basement which had eleva-

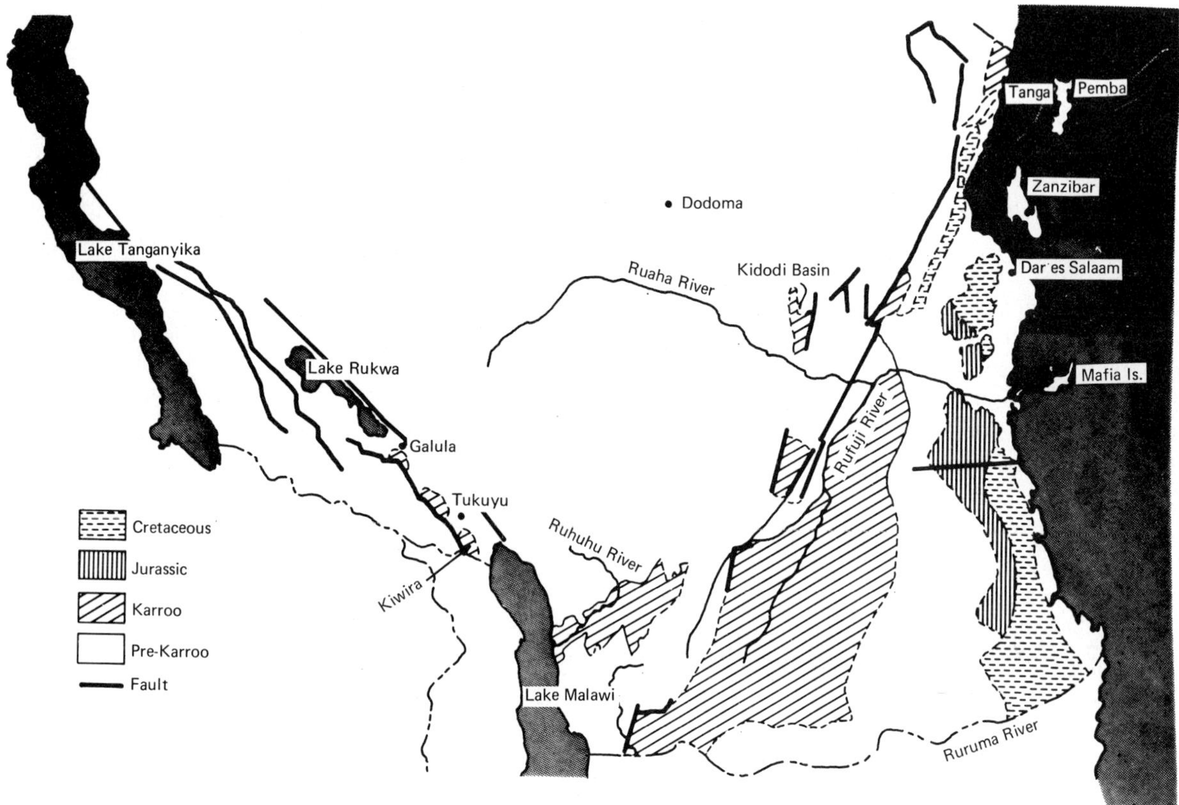

Fig. 12. Karroo basins of southwest Tanzania and Malawi.

tions of up to 300 m; these were covered by Late or Middle Ecca (Permian) deposits (Quennell et al., 1956). The name Songea Series was applied to the Karroo sequence which has an estimated thickness of the order of 3,000 m. This sequence in the Ruhuhu Basin is regarded as the type for southwest Tanzania. The earlier term (see Stockley and Oates, 1931) has been replaced by a number of defined units.(K1–K8, see McKinlay, 1954), of which only the upper two (K7 Kingori Sandstone, K8 Manda Beds and Upper Bone Bed) are regarded as Triassic.

Both units are essentially massive gritty feldspathic cross-bedded sandstones, the lower unit, the Kingori Sandstone, passes laterally into lithologies similar to the overlying Manda beds. Within the Manda beds are intercalated green and purple marls, and several bone-bearing horizons (consequently the name Upper Bone-bearing series has been proposed). The fauna compares to the Lower Beaufort–Stormberg (uppermost Permian to Upper Triassic) or top Lower Triassic to Upper Triassic according to various authors.

Continental Cretaceous beds have been identified in northern Malawi from dinosaur remains. Across the border in the Songwe Valley, Red Beds in which reptilian eggs and unidentified bone fragments are found, have also been assigned to the Cretaceous (Spence, 1954). Further north in the Songwe Valley near Mbeya unfossiliferous friable medium-grained subangular Red Beds, unconformably overlying Karroo beds have also been assigned to the Cretaceous.

The marine facies

The Somali Basin

The Somali Basin as here defined is open (or truncated) to the northeast. The southeastern margin is formed by the Seychelles Bank and northwestern Madagascar to about the latitude of Mozambique, with the western margin including Tanzania, Kenya, Somalia and part of Ethiopia, extending beyond into the Arabian Peninsula (see Saint-Marc, Ch. 12). Although commonly concealed by Quaternary sediments along much of the African margin, the evidence suggests that continued large-scale continental downwarping occurred with normal faulting playing an important role (Fig. 13). Where the contact is exposed faulting is common and the interpretation of geophysical data is consistent with faulting or step faulting (e.g. Walters and Linton, 1973).

The orientation of the basin is such that only a narrow coastal strip of Mesozoic sediments is found in Tanzania and Kenya south of Malindi. The Mesozoic cover widens over the Lamu Embayment and extends in a broad swath across Somalia and Ethiopia.

The geophysical basis for considering the Somali Basin as a unit was suggested by the demonstration of the extent of the continental crust of the Seychelles (Davies and Francis, 1964) whose presence was first indicated by the occurrence of Precambrian granites on Mahé and Praslin (Baker and Miller, 1963). Subsequently from a series of profiles Francis et al. (1966) produced a seismic profile from Lamu on the Kenya coast to the Seychelles (Fig. 14). This shows the stepped approach of the Moho discontinuity from typical continental depths under the mainland to a zone of intermediate crustal thickness and finally to shallow depths (8.5 km) over a relatively short distance close to the Seychelles Bank.

Within the basin compressional structures are only seen related to halokinetic processes (as in southern Tanzania) or to the injection of igneous rocks (Kent,

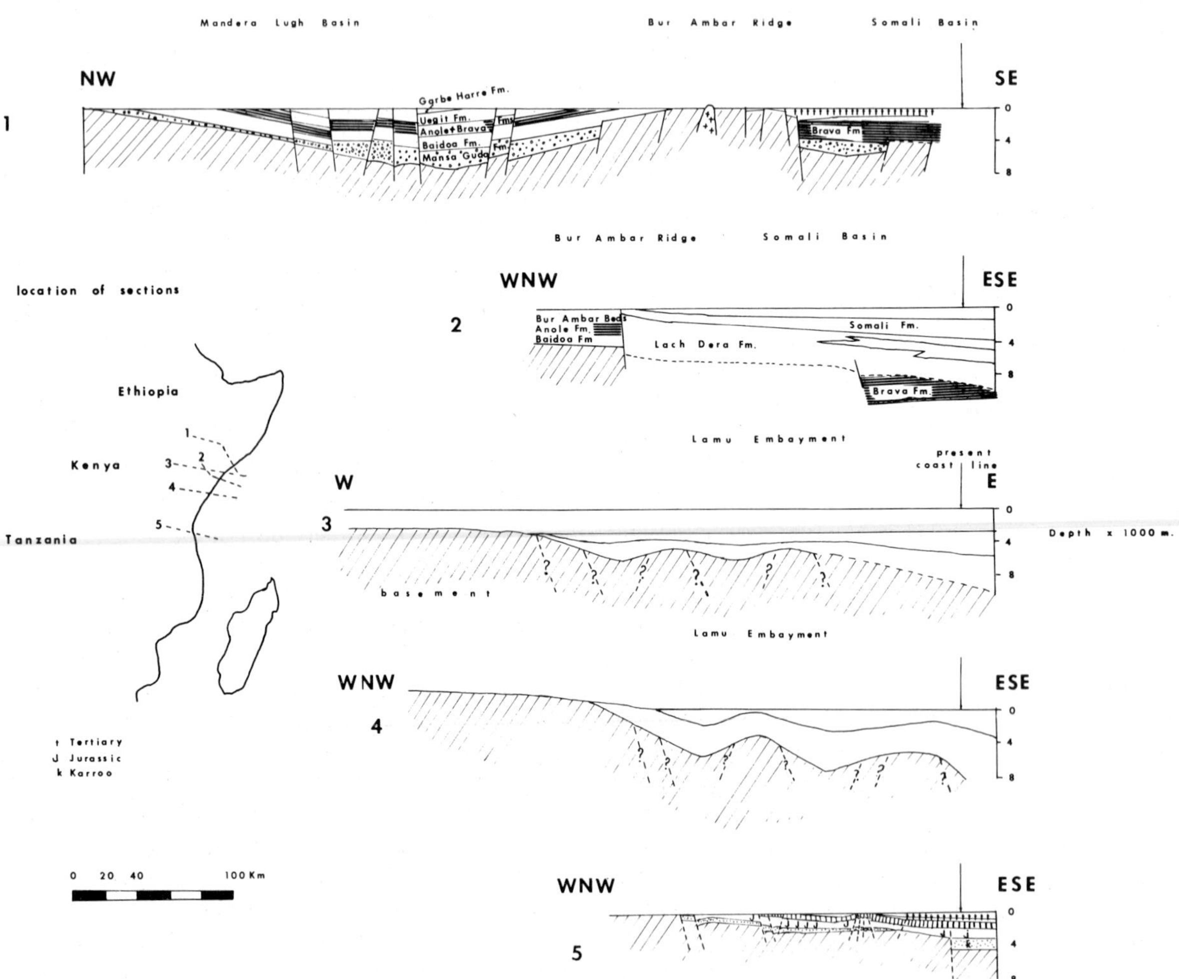

Fig. 13. Cross-sections of the western margin of the Somali Basin (from Beltrandi and Prye, 1973; Walters and Linton, 1973; Kent and Perry, 1973). The formations in section 1, excluding the Mansa Guda Formation, are Jurassic in age.

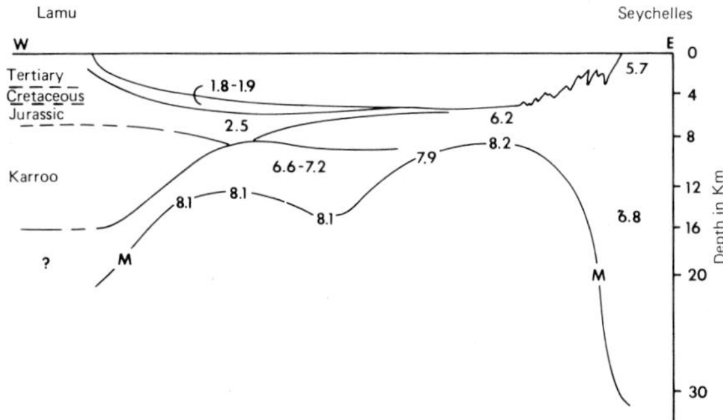

Fig. 14. Seismic cross-section from Lamu to the Seychelles (from Davies and Francis, 1964).

1974). These effects are minor compared to the extent of vertical movement both in space and time. Horizontal displacement of the axis of the Carlsberg Ridge along the Owens Fracture Zone was proposed by Matthews (1963). Faulting may have begun in Permian times but only became important during Late Triassic (the Somalia rift system), Early and Late Jurassic; Middle and end-Cretaceous times. The effects of these vertical movements is apparent in the pattern of clastic sediment deposition within the basin.

An interesting parallel may be drawn between the Somali Basin and the basin centred on the Pelagian Sea if allowance be made for the Alpine tectonics which affect the surrounds of the latter. Thus Madagascar parallels Sicily and the Lamu Embayment–Ethiopian region bears much the same relationship to the Congo as the Tunisian–Algerian–Libyan area does to the Nigerian Cretaceous.

The western margin of the Somali Basin as already indicated approximates to a line of flexing which subsequently fractured and which is oriented north–south. The limits of this line appear to be marked by grabens, the Mandawa graben of Tanzania in the south and a graben whose existence has been proposed, if not proven, running north from near Mombasa to the Red Sea (Beltrandi and Pyre, 1973). Sediments in the Mandawa graben which go back to Permian suggest that the earliest movements may date from that time. The presence of the evaporites in the graben also points to restricted access of sea water, during Permian and Triassic times. However, although there is a brief marine interlude near the base of the Triassic in the Lamu Embayment of Kenya, there is little sign of widespread marine conditions becoming established much before Middle Jurassic except in the northern area, i.e. the Horn of Africa and north-eastern Kenya.

The Horn of Africa (Somalia, Ethiopia, Eritrea). The marine transgression over the Horn of Africa, from southeast to northwest, the first major Phanerozoic transgression of which record remains, began in Late Triassic times. It is also the only region where transgression was not immediately bounded by a faulted margin. There is a large onshore section in the Lamu Embayment, however geophysical information suggests that the margin is fault bounded.

The surface over which the sea advanced in the Horn of Africa was still diversified, and some regions such as the Danakil Alps and the Socotra (Beydoun and Bichan, 1970) remained above water for a considerable period. Even when finally submerged, the thinner sequence suggests that they still acted as relatively positive areas. Mohr (1961) gives some estimates of the elevations based upon the non-represented thickness of deposits taken from nearby regions until deposition actually occurs, i.e. an elevation of 850 m for the Socotra Hills, while in central Somaliland Cretaceous sandstones resting upon basement suggest and elevation of 1,100 m.

The trangressing sea deposited littoral sands often over a basal conglomerate (Adigrat Sandstone) while in the quieter clear waters in the wake of the advancing sea richly fossiliferous limestones (Antalo limestones and equivalents) were deposited (Table VII). Between the two, in lagoonal conditions some gypsum was deposited (e.g. Abbai beds). The sea initially advanced rapidly across the Somali Republic and Ogaden during the Lias. The rate of transgression slowed down during Callovian and Oxfordian coming to a halt in Kimmeridgian times. Uplift in the Late Jurassic resulted in regression, and as the sea retreated a littoral sandstone facies followed it (the Upper Sandstone Series). The paleogeographic conditions are summarized in Fig. 15, while the facies relationships are shown in Fig. 16. Detailed descriptions of local areas and local formational names are given in Mohr (1961) (Table VII).

Fig. 15 is important for it clearly establishes the facies relationships and indicates how the temporal relations vary according to location. Marine limestone deposition held sway much longer in the southeast, while the Adigrat Sandstone, the nearshore marine sands, ranges in age from Late Triassic in Ogaden

TABLE VII

Summary of the lithologies of the Ethiopian–Somalia succession from Mohr (1961) (the many detailed local descriptions and names have here been ignored to present a generalized pattern of the sequence of lithologies.)

Upper Sandstone Series 100–500 m	massive white sandstone, occasionally cross-bedded, age range from Oxfordian–Kimmeridgian to Portlandian; equivalent formations Mandera Series (Danissa Beds) and Marehan Sandstone (a lateral equivalent of the upper part of the Mandera Series)
Antalo Limestone Formation 0–800 m	broadly concordant upon the Adigrat Sandstone; locally may be subdivided (e.g., in southwest Somalia the name Abbai beds is applied to basal limestones and dolomitic limestone with gypsum); many limestone varieties, and sand and silt are particularly common as limestone traced towards the shorelines to the north and west; generally formed in a clear shallow-water neritic environment with a fauna rich in ornamented pelecypods; reef corals occur and ammonites are found in presumably deeper-water deposits; age range Bajocian–Portlandian.
Adigrat Sandstone ~500 m	fairly constant thickness, range is from a few metres to 1,000 m, due mostly to irregular deposition surface; a deltaic or lacustrine origin proposed by Beyth (1973); may have a ferruginous basal conglomerate and sporadic levels or lenses of conglomerate or variegated shales. Typically a white, massive, angular, quartzose sandstone with sparse silica or kaolinitic, occasionally limonitic cement. Some fossil wood fragments otherwise unfossiliferous. In Mekele outlier sand is mature and may have passed through several cycles. Equivalent formations Mansa Guda (SW Somalia) Kholan Series (Yemen)

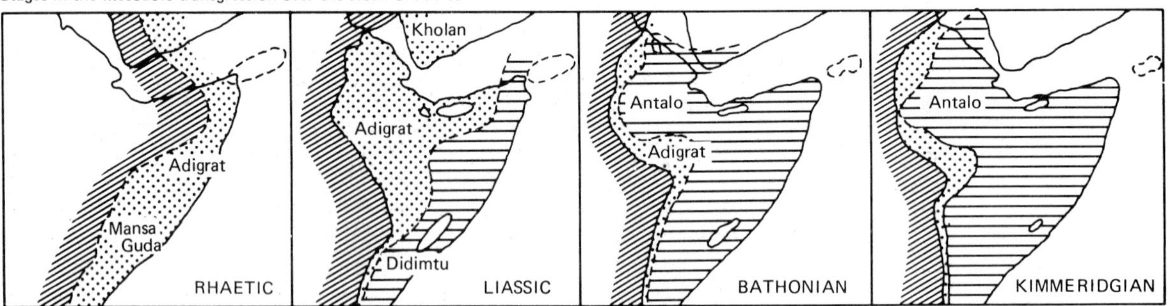

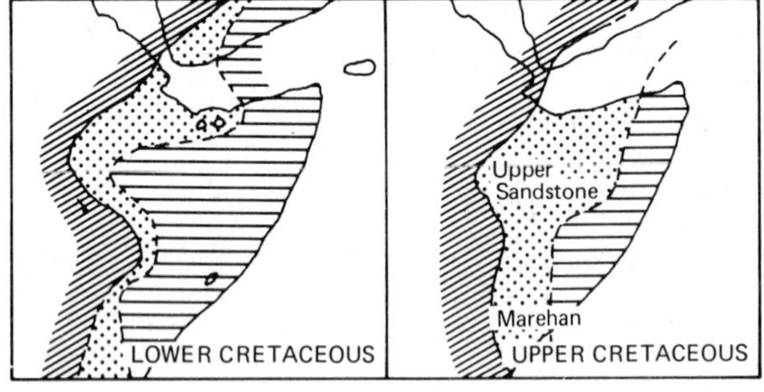

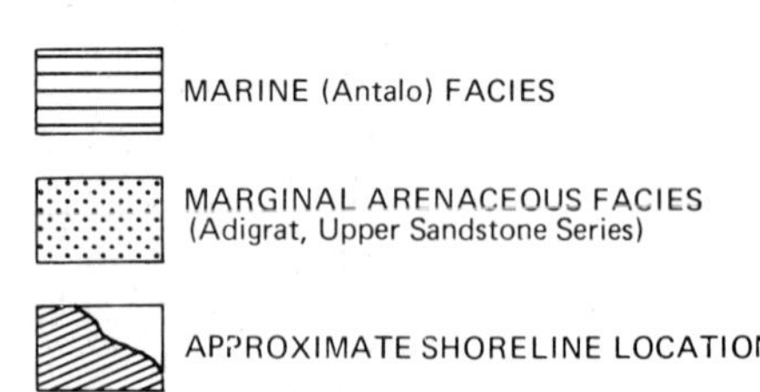

Fig. 15. The spatial distribution of marine and marginal arenaceous facies during the Jurassic and Cretaceous over the Horn of Africa (after Mohr, 1961).

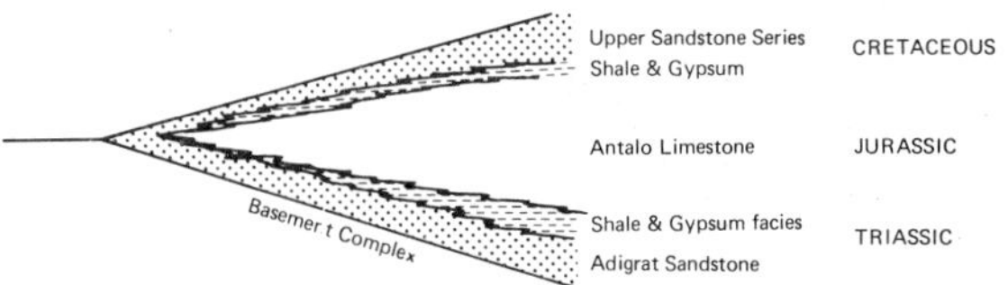

Fig. 16. The facies relationship of the Adigrat Sandstone, Antalo Limestone and Upper Sandstone Series and the intervening Shale and Gypsum facies in Ethiopia (after Mohr, 1961).

to Early or possibly Middle Jurassic in Eritrea (Table VIII). This facies meets the regressive sandstone facies, the Upper Sandstone facies, at the margin of the basin where its age is Oxfordian. The latter youngs to the southeast, the direction of marine retreat, and may be as young as Portlandian at Antalo on the Ethiopian Platform.

The maximum extent of the transgression, reached in Kimmeridgian times, is not known. According to Beyth (1973) it probably never extended further west than 37°E, no more than 150 km west of the Mekele outlier (Beyth, 1973), but Mohr (1961) considers a much greater westward extent into the present-day Sudan is possible.

While the transgression appears to have been smooth and regular, the same is not true of the regression. The regression was caused by the uplift of the Horn of Africa beginning in the latest Jurassic and continuing into the Cretaceous until by the end of that period the western part of the area was land and marine conditions were restricted to the region of Socotra. Only here is continuous sedimentation up into the Cenozoic found.

Oscillatory movements during the regression are suggested by the varied facies occurring, however their magnitude is still disputed. The succession is best developed in Somaliland where a sequence of Neocomian limestones and sandstones rest disconformably upon, and progressively overlap, Jurassic

TABLE VIII

Correlation chart for the Mesozoic of the Horn of Africa, Ethiopia and the Sudan (after Whiteman, 1971)

		Ogaden	Eritrea	E Sudan	S and C Sudan
CRETACEOUS	Maastrichtian	Jesomma Sandstone (reg.)			
	Senonian				
	Turonian	Belet Wen Limestone			Fareig Basalt
	Cenomanian	Ferfer Gypsum			Nubian Sandstone facies
	"Lower Cretaceous"				
JURASSIC	Portlandian	Mustahill Lst. U. Sst. Fm.			
	Kimmeridgian	Main Gypsum		Gedaref Sst.	Khor Shinab basalt 148 ± 30 m.y.
	Oxfordian	Antalo Lst. facies	Eritrean facies		
	Callovian				in grits, conglomerate and sandstone
	Bathonian				
	Bajocian	Adigrat Sst. facies			
	Lias				
TRIASSIC			period of erosion		period of erosion

beds and at the Ethiopian border even step over on to rocks of the basement complex. They are in turn unconformably overlain by limestone. Thus a minor transgression initiated in the Barremian reached its maximum in the Aptian to Albian before regression took over once again.

In Socotra, the absence of Neocomian and Jurassic with Cenomanian resting upon the basement complex indicates the length of time some of the positive areas persisted. On the island rudists and corals pass laterally to a marly facies. Deeper-water contemporary facies are found on nearby Semha.

As might be anticipated apart from endemic species the fauna shows its closest affinites with Eastern Europe, and secondarily with India. The presumed migration route was around Arabia.

Over the southern part of the basin, marine conditions became generally established in Middle Jurassic times. This was associated with movements along the marginal faults, with the upstanding blocks acquiring a westerly tilt, explaining according to Walters and Linton (1973), the relative paucity of coarse detritus. The absence of relict basins to the west however suggests that most of the erosional products eventually wound up in the Somali Basin, and these may be at least partially represented in the detritals recorded in the Lamu Embayment.

One feature of the faulting is that north of Lamu the trend changes to a more northeasterly direction, approximately parallel to the trend of the Owen Fracture. The Bur Ambar region, initially a positive area, was uplifted and, near the end of the Jurassic, separated the Mandera–Lugh Basin from coastal Somalia.

The Mandera–Lugh Basin appears to occupy a pre-Jurassic regional trough or graben whose existence is postulated back in Late Paleozoic. In southwest Somalia however there are no sedimentary rocks older than Jurassic exposed. The neritic carbonates of the Toarcian Didimtu Beds rest directly upon basement rocks of the Bur–Ambar uplift to the east. Along the western margin of the basin in Kenya a thick (650 m) sandstone and conglomerate sequence, the Mansa Guda Formation (Ayers, 1952), is interposed between the Didimtu Beds and the basement. This sandstone therefore is the stratigraphic analogue of the Adigrat Sandstone of northern Ethiopia and possibly of the Duruma Sandstone of Kenya. Geophysical evidence suggests that a still greater thickness may lie at depth in the central part of the Mandera–Lugh Basin or graben. The Jurassic and Lower Cretaceous section in the Mandera–Lugh Basin (lithological descriptions given in Table IX) bears a close resemblance to that of Ethiopia–Somalia, even to the Late Jurassic–Early Cretaceous formation of local evaporitic basins.

Drilling in the coastal zone of Somalia has penetrated as low as rocks of Middle to Late Jurassic age. Two formations have been recognized, the Brava Formation of Middle to Late Jurassic and Early Cretaceous age, and the Lach Dera Formation. Both show general lithological similarities to the sequence in Kenya, with essentially a lower, predominantly shale, succession in excess of 3,000 m and the upper, dominantly arenaceous succession. The top of the latter is Cenozoic, for a limestone and shale intercalation about the middle of the sequence contains nummulites. The lower part cannot be dated in the absence of fossils. The presence of so much sandstone, and the presence of lignitic fragments in the upper part of the sequence implies rejuvenation, i.e. uplift in the source area (which is presumably Late Cretaceous–Early Cenozoic).

Kenya and Tanzania. South of Ethiopia–Somalia, the widest onshore extent of Mesozoic beds is in the Lamu Embayment, but as in the coastal area, knowledge of the sequence has depended to a great extent upon drilling. The Lamu Embayment is limited to the north by the Wajir basement high, a high with which the Bur–Ambar positive zone may be related. The western margin of the Mesozoic basin, although obscured by recent Quaternary cover, appears to be bounded by step faults (Walters and Linton, 1973). In cross-sections Walters and Linton (1973) show smoothed variations in depth, but these may well be a buried horst graben topography similar to that found in the Sirte Basin (Barr and Weegar, 1972) given the existence of grabens further south in Tanzania.

The Lamu Embayment is of particular interest for it is one of the few places where the Late Paleozoic and Triassic ("Karroo") beds are known. These essentially arenaceous lithologies represented by the Taru Grits, Maji-ya-Chumvi Beds and Mariakani and Mazeras Sandstone, contain a single Eo-Triassic marine horizon. They also represent at least two coarse to fine-grained sedimentary cycles, Taru Grit to Maji-ya-Chumvi Beds and Marakani and Mazeras Sandstones. These may represent phases in the downwarping which eventually brought marine conditions

TABLE IX

Composite stratigraphic section for the southwestern part of the Mandera–Lugh Basin (modified from Beltrandi and Pyre, 1973)

UPPER JURASSIC–? LOWER CRETACEOUS	Garba Harre Formation 730 m upper member: dolomitic limestone, siltstones with gypsum lenses, green shales and massive cross-bedded sandstones; several intraformational unconformities lower member: alternating dolomitic limestone and red cross-bedded, ripple marked sandstones and fossiliferous calcarenites; base marked by thick massive sandstone unit; depositional environment ranging from littoral to neritic
KIMMERIDGIAN	Uegit Formation 740 m basal cross-bedded sandstones interbedded with coquinas, followed by thick grey oolites and bioclastic calcarenites with corals and oysters; 40 m shale member near middle; upper part more bioclastic with frequent sandstone intercalations; Bur Ambar beds 200 m of sandstone represent a possible nearshore facies against the Bur Ambar hills
OXFORDIAN	Anole Formation 520 m grey calcareous shales, marly limestones and occasional cross-bedded calcisiltites and coquinas; a dense argillaceous limestone at the base; formed in fairly quiet deep-water neritic environment
CALLOVIAN	Baidoa Formation Galado Member: detrital carbonates Baidoa Member: 900 m yellow brown limestones, sometimes detrital, generally oolitic with abundant shell fragments, shale interbeds up to 30 m in middle; limestones more fossiliferous near the top with detrital, calcarenite and algal limestones grading to coquinas
BATHONIAN	Deleb Sandstone 13–20 m coarse-grained sandstone grading laterally and vertically to Uarei Shale; ? Bathonian in age
TOARCIAN	Didimtu beds in southwest Somalia: limestones and calcareous mudstones Mansa Guda Formation 650 m: cross-bedded sandstones and conglomerates

to the Lamu Embayment as well as coastal Kenya and Tanzania to the south in Middle Jurassic times. Subsequent movements then appear in fluctuations in the marine facies with a major coarse to fine cycle beginning in pre-Aptian Early Cretaceous times. A summary of the principal lithologies and some, but by no means all, of the local formational names are given in Table X.

Although the Jurassic and Cretaceous sections are generally similar, there are some regional differences between coastal Tanzania–Kenya and the Lamu Embayment; for example in the embayment Coniacian is said to be absent whereas in Tanzania it is Cenomanian and Turonian which are locally not recorded. There is no indication whether these are local events but the probability is that the beds are present within the deeper parts of the basin. In the coastal regions, there is no information on "Karroo" rocks which are presumed to be present (Kent and Perry, 1973) although beyond the reach of the drill. Nor can it be neglected that although normal marine conditions in the Mandawa graben date from Bathonian (Kent,

TABLE X

Composite stratigraphic section based on the region west of Mombasa, with additions for the Lamu Embayment (modified from Walters and Linton, 1973)

System	Stage	Description
CRETACEOUS	Cenomanian–Campanian	shales in Lamu Embayment (boreholes) all stages excluding Coniacian, formed in a relatively quiet deep-water environment
	Albian–Cenomanian Aptian–Neocomian Middle Kimmeridgian	30 m blocky shale with Albian to Cenomanian foraminiferal assemblage; in Lamu Emment about 1000 m Freretown Limestone shallow-water shelf environment overlain by 20 m Aptian shale; the limestone is detrital, partly silty and occasionally bioclastic; underlain by shale; total about 60 m; in Lamu Embayment at least 1500 m Neocomian and 250 m Aptian
JURASSIC	Upper Oxfordian	1700 m of shale; upper part has lenticular limestones suggesting shallow marine environment, lower part false-bedded and probably deltaic; only difference from Neocomian shales is presence of nodules
	Lower Callovian	Late Callovian–Early Oxfordian apparently absent, either regressive phase with non deposition or poor outcrop
		Lamu Embayment 100 m; oolitic limestone grading up into 30 m siltstone and quartzite; has a ? Oxfordian fauna
		180 m Kibiongi Beds: sandy micaceous shale and thin fine-grained sandstones; shales show current bedding slump structures, rain pittings suggesting shallow possibly marine environment; has been proposed as a lateral equivalent of the Kambe Limestone
	Bathonian Upper Bajocian	150 m Kambe Limestone (but up to 600 m known): base where seen has a 12–15 m conglomerate with Mazeras Sandstone boulders in a pisolitic and coral limestone; mostly dark grey oolitic limestone with shale interbeds occasionally coralliferous "reefal" limestone; represents various shallow-water marine environments
	Lower Jurassic	Lamu Embayment has no definite Middle Jurassic but a 200 m sequence may include Middle Jurassic; Didimtu beds in Mandera Basin in north contain limestones and represents first marine transgression; no deposits known in Lamu Embayment or in southern Kenya (Mombasa region)
TRIASSIC	?Liassic Upper Triassic	450 m Mazeras Sandstone: massive, coarse, cross-bedded sandstones and grits with interbedded siltstones and silicified wood; probably deposits of extensive deltas, but upper beds aeolian (coastal dunes, or flood plain–sebkha? suggested); Late Triassic in age but may pass into Early Jurassic
	Triassic	2900–3300 m Mariakani Sandstone: fine-grained flaggy sandstones, micaceous siltstones and silty shales with indeterminate plant remains; regarded as Triassic; represents a refuse to coarser, more arenaceous conditions than the preceeding formation
	Lower Triassic	1200 m Maji-ya-Chumvi Beds: upper part (650 m) sequence as a whole represents shallow lagoonal and swamp environments, with finely laminated silty shales, siltstones and flaggy sandstones, these may be cross-bedded, sun-cracked and rippled
PERMIAN	Permian	marine incursion with fish fossils, suggesting an Eo-Triassic age
		Maji-ya-Chumvi Beds: lower part characterized by plant remains with *Voltzia* and *Ullmania*; sequence in part is Permian; division to upper and lower based upon fish bed
	Permo-Carboniferous	1500 m Upper Taru Grits: fluviatile feldspathic grits and sandstones derived from a westerly source

1965) restricted ingress of sea water occurred back to Permian at least as reflected in the development of more than 2,500 m of halite resting upon coarse arkoses in which there are lignite traces. This implies a marine link whose location is not clear.

Drilling on the offshore islands of Mafia and Zanzibar terminated in Upper Cretaceous black shales suggesting a neritic–pelagic environment.

The Congo. The Mesozoic beds of the Congo, regarded in the earliest days of exploration as Rhaetic and Triassic by analogy with South Africa and assigned to the Lualaba–Lubilash System are now known to be much younger. The recognition within the sequence of two marine horizons, one Jurassic and the other Cretaceous poses interesting paleogeographic problems for uplift and erosion has destroyed whatever traces of marine connections which may have formerly existed.

With the existence of Jurassic and Cretaceous beds established it seems appropriate to use a local sequence rather than force foreign terms. Cahen (1954) summarises the Mesozoic geology as follows:

Kwango Series	Nsele Stage	Late Cretaceous, Cenomanian
	Inzia Stage	Turonian almost certainly represented by marine horizon (top Inzia Stage)
Kamina Series		pre-end Cretaceous, age not known more specifically
Lualaba Series	Loia Stage	not dated with precision Late Jurassic–Early Cretaceous
	Stanleyville Stage	Certainly Late Jurassic
Red Beds		probably Triassic
Lukuga Series		Late Carboniferous–Permian

Although most of the beds of both the Lualaba and Kwango Series, the principal Mesozoic formations, are continental, the presence of marine horizons gives an interest out of all proportion to the magnitude of the incursions since obvious links with the ocean are absent. In the particular case of the Late Jurassic marine horizon a link to the northeast across Ethiopia with the Indian Ocean is presumed, while the Late Cretaceous marine link proposed by Cahen (1954) is with the Atlantic via the Gulf of Benue.

In northern Katanga near Lake Tanganyika and on the Lukuga south of Albertville and at Makunga north of lat. 5° isolated red-bed sequences are found. Lithologically they are similar to the Triassic–Rhaetic beds which overlie the coal sequence in southwestern Tanganyika. Fossils have been found only at Makunga, and these, while not definitive, suggest affinities with Late Triassic–Rhaetic forms.

The Lualaba Series crops out along the northeastern margin and in the central part of the Congo Basin. In the southern Congo beds of similar lithology have been assigned to the Lualaba Series, but in the absence of faunal control such long distance correlations are dubious. Two stages are recognized: the lower and more studied Stanleyville Stage, which can be dated as Late Jurassic, and the upper, Loia Stage, occupying the more central parts of the basin, to which Late Jurassic–Early Cretaceous age is assigned.

The beds of the Stanleyville Stage consist primarily of argillites and sandy feldspathic beds above a basal conglomerate (see Table XI). It is in beds of unit 2 (Table XI) that fossils ". . . d'origine presque certainement marine . . ." have been found. These beds referred to as the "Complexe de lime fine" contain a score of fish genera, lamellibranchs, ostracods and phyllopods. The affinities of *Lepidotes congolensis* Hussakof are with Oxfordian and Bathonian forms. Higher in the sequence occurs a fresh-water fish fauna whose aspect is Late Liassic and Cahen (1954) suggests that this is an "archaic" fauna. Beds of the Loia Stage have also yielded fossils, but at the present time they do not yield any conclusive evidence of age. Since they follow the beds of the Stanleyville Stage without break they are regarded as Late Jurassic, but may well extend into the Early Cretaceous.

The Kamina Series consists of cross-bedded pebbly concretionary sandstones with argillaceous sandstone horizons. Although there are numerous outcrops with this lithology, at only one, 10 km east of Kamina, can some approach be made to its stratigraphic position. There it is overlain by polymorphic sandstones separated by an erosion surface which is dated as terminal Cretaceous. No lower limit can be set. Lithologically it is similar to the dinosaur beds of Nyasaland. It could be correlated with the Kwango Series, although the nearest outcrop of the latter is 400 km distant, or the Lualaba Series, or represent a new unit.

The Kwango Series covers wide areas of the western part of the southern Congo and probably extends to the north. The beds rest either directly upon basement or upon rocks which have been attributed to the Lualaba Series. Brief lithological descriptions are given in Table XI. At the top of the Lower Inzia Stage, a marine fauna has been described (Saint Seine, 1953). Among the fish remains is a form only known in Cenomanian–Turonian beds elsewhere. This is the only marine horizon so far recognized in the Kwango Series.

The beds of the Lualaba and Kwango Series were

TABLE XI

Summary of the Mesozoic lithologies in the Congo basin (after Cahen 1954)

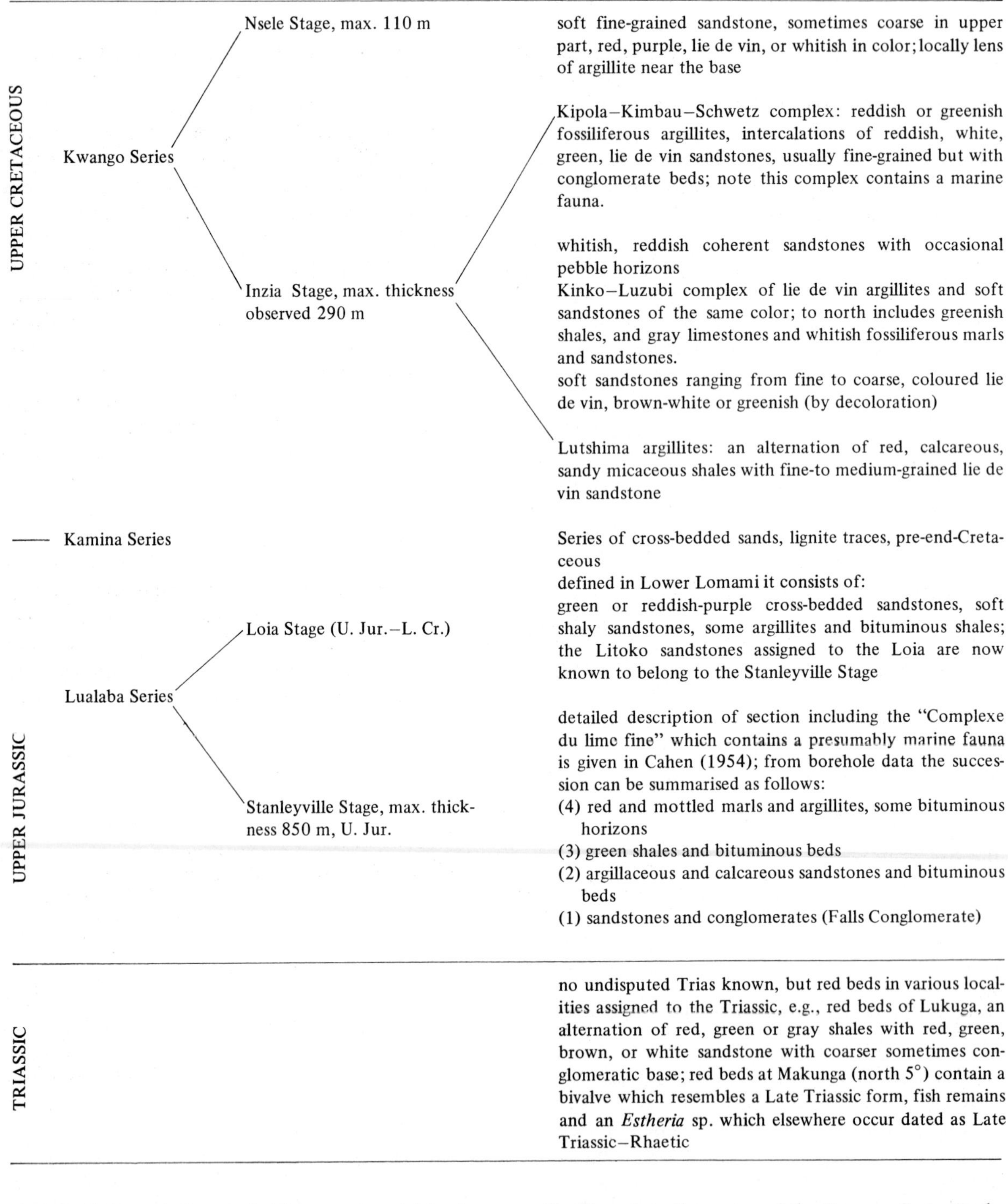

UPPER CRETACEOUS	Kwango Series	Nsele Stage, max. 110 m	soft fine-grained sandstone, sometimes coarse in upper part, red, purple, lie de vin, or whitish in color; locally lens of argillite near the base
		Inzia Stage, max. thickness observed 290 m	Kipola–Kimbau–Schwetz complex: reddish or greenish fossiliferous argillites, intercalations of reddish, white, green, lie de vin sandstones, usually fine-grained but with conglomerate beds; note this complex contains a marine fauna.
			whitish, reddish coherent sandstones with occasional pebble horizons Kinko–Luzubi complex of lie de vin argillites and soft sandstones of the same color; to north includes greenish shales, and gray limestones and whitish fossiliferous marls and sandstones. soft sandstones ranging from fine to coarse, coloured lie de vin, brown-white or greenish (by decoloration)
			Lutshima argillites: an alternation of red, calcareous, sandy micaceous shales with fine-to medium-grained lie de vin sandstone
——	Kamina Series		Series of cross-bedded sands, lignite traces, pre-end-Cretaceous
UPPER JURASSIC	Lualaba Series	Loia Stage (U. Jur.–L. Cr.)	defined in Lower Lomami it consists of: green or reddish-purple cross-bedded sandstones, soft shaly sandstones, some argillites and bituminous shales; the Litoko sandstones assigned to the Loia are now known to belong to the Stanleyville Stage
		Stanleyville Stage, max. thickness 850 m, U. Jur.	detailed description of section including the "Complexe du lime fine" which contains a presumably marine fauna is given in Cahen (1954); from borehole data the succession can be summarised as follows: (4) red and mottled marls and argillites, some bituminous horizons (3) green shales and bituminous beds (2) argillaceous and calcareous sandstones and bituminous beds (1) sandstones and conglomerates (Falls Conglomerate)
TRIASSIC			no undisputed Trias known, but red beds in various localities assigned to the Triassic, e.g., red beds of Lukuga, an alternation of red, green or gray shales with red, green, brown, or white sandstone with coarser sometimes conglomeratic base; red beds at Makunga (north 5°) contain a bivalve which resembles a Late Triassic form, fish remains and an *Estheria* sp. which elsewhere occur dated as Late Triassic–Rhaetic

deposited in a shallow, subsiding continental basin, the depocentre of which appears to have migrated westwards, for the principal outcrop of the Stanleyville Stage is in the east, and the Kwango Series in the west. Basin marginal facies indicate the margin of the Kwango lay to the south, southeast and southwest, and grain size diminishes northwards within the basin. Beds assigned tentatively to the Lualaba Series in the southern Congo include fluviatile conglomerates which apparently represent valley fill.

The beds are generally waterlain; subaerial exposure is indicated only during the Nsele Stage of the Kwango. Cahen (1954) suggests deposition in brackish water or lagoons. The common occurrence of bituminous horizons, particularly in the lower part of the section, implies a considerable organic content in the original sediments.

The presence of marine horizons implies open contact with the sea on at least two occasions (Leriche, 1938). During the Late Jurassic the marine link cannot be with the South Atlantic, and the obvious source is the Indian Ocean. An easterly direction is indicated sedimentologically for the "Complexe de lime fine" passes westwards into the sandy and argillaceous beds. It is tempting to regard this as the maximum of the marine incursion which covered southern Ethiopia. If true however it seems to imply an earlier maximum than the Kimmeridgian. All traces in the intervening region must have been removed by erosion following uplift of the region between the two. If this uplift is assigned to the Early Cretaceous, then Cahen's suggestion of a northerly link with the Atlantic through the Gulf of Benue for the Cenomanian–Turonian marine fauna of the Inzia Stage is a reasonable one. Shoreline facies both of the Kwango Basin and of the Cretaceous of the Atlantic margin rule out a direct link across Angola.

The Mozambique Basin

The term as used here essentially covers the southern, Mesozoic–Tertiary, basin of Mozambique, the southwest shores of Madagascar with a Karroo–Mesozoic succession and the Mozambique Channel which broadens southwards into the Natal Basin. The Mozambique Channel may be thought of as a culmination which separates the Somali and Mozambique Basins.

Dixey (1956) initially proposed that the Mozambique Channel was geosynclinal in nature, implying no relative motion of Madagascar. Some modification of this hypothesis seems necessary for it is inconsistent with the evidence from the Mozambique shelf of a sedimentary wedge thinning eastwards at the continental shelf edge (Fig. 17) and with Cliquet's (1957) observation that the "Karroo" of Madagascar was deposited in a deep trough adjoining the eastern margin of the basin. The beds appear to thin westwards, and the sinking of the basin is associated with later post-Karroo deposition. There may have been continuity of sedimentation across the Mozambique Channel in later times (Isalo Formation, Triassic–Middle Jurassic), but whether other independent Karroo fault troughs exist beneath deep water is unknown. On the basis of a small number of gravity

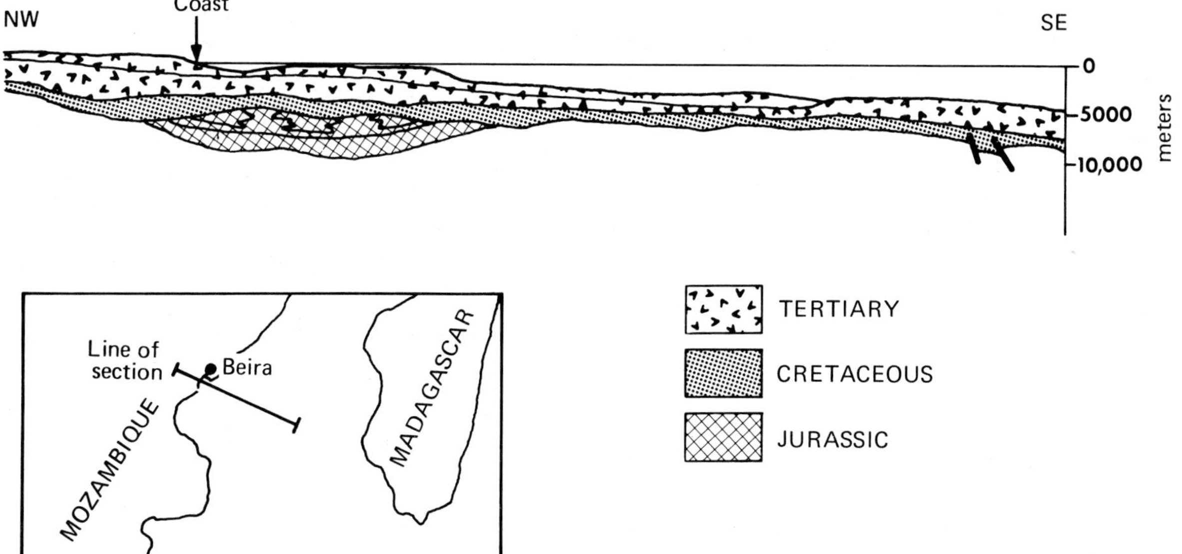

Fig. 17. Section across Mozambique into the Mozambique Channel (simplified from Kent, 1974).

stations roughly along the axis of the Mozambique Channel, Talwani (1962) concluded that the channel is isostatically compensated and structurally neither typically oceanic nor continental.

Tectonically, as in the case of so many of the Karroo basins, there was faulting on a large scale contemporaneous with sedimentation. It was active in Late Paleozoic times and essentially completed by Early Jurassic times. Thus Jurassic and Cretaceous sediments were deposited on open shelves in the absence of major tectonic dislocations. Later faulting which did occur was on a much smaller scale.

Mozambique. Short episodes of marine transgression progressively covered more and more extensive areas, from the Middle Permian transgression affecting only southern Madagascar to western Madagascar in the Bajocian, southern Mozambique in the Barremian to central Mozambique in the Campanian. The western channel was fully inundated by mid-Cretaceous times. This transgression from the south to southeast matches the southward penetration from the Somali Basin with marine episodes in the Permo-Triassic affecting the Kidodi and Mandawa Basins in Tanzania and northern Madagascar, and northwestern Madagascar and northern Mozambique in Late Jurassic (Förster, 1975).

The southern basin of Mozambique extends about 160 km in a north–south direction and at its widest reaches 440 km east to west although the exact eastern edge is conjectural. Pre-Pleistocene rocks in this area are exposed along the western margin of the basin and in a mosaic of horst structures, east of the Mabate–Funhalouro rift, an extension of the East African graben system. The latter occupies the eastward bulge of the coast between Beira and Lourenço Marques. Two distinctive features of the basin are the extensive volcanics which terminated the Karroo cycle and the absence of any record of Jurassic sediments in any of the onshore wells so far drilled. Kent (1974) however indicates their presence in the off-shore section (Fig. 17).

Most of the "Karroo" rocks are exposed in a belt parallel to and adjacent to the Zambezi River graben or trending north–south along the extension of the Lebombo Ridge. Below the igneous rocks occur representatives of the Ecca and Beaufort Series of South Africa. The beds are continental sandstones and shales. Flores (1973) indicates the presence of limestone in the Lunho Series northeast of Lake Malawi without establishing whether it is marine or not; if the former it may be associated with the Tanzanian Permo-Triassic marine horizons. The age of the series is no more closely defined than Beaufort and Ecca.

Although volcanic activity began in southern Africa in Late Triassic times, it shifted northeastwards during the Jurassic and Cretaceous. Mesozoic basalts however are not found north of Mozambique and Madagascar although carbonatites and kimberlite pipes do occur, and some subsurface basalts are found in the Somali Basin. The volcanics of Mozambique probably do not extend as far back as Triassic. Radiometric ages have established that rocks of Late Jurassic and Early Cretaceous ages are present. Plant remains associated with the intercalated sediments have suggested a Liassic age. Volcanic rocks have been penetrated in subsurface at –3,000 m, and six of eight DSDP boreholes have penetrated tholeiitic basalt under sediments ranging in age from Early Cretaceous to Eocene. Kent (1974) reported a Shell announcement that ". . . direct seismic continuity of these typical 'oceanic' rocks with Early Cretaceous basalts of Mozambique" i.e. by implication continuity with the inland continental basalts. The age trend of the volcanics is continued in Madagascar where subaerial flows range from post-Aptian to Coniacian and Santonian. These ages overlap the ages of the offshore volcanics so that there is no obvious difference between the volcanic history of the Mozambique Channel and Madagascar.

During Barremian times, the sea advanced over the weathered basalts and beds of the Maputo Formation, which rest with slight unconformity upon the volcanics, were deposited in the south near the South African border. Five hundred kilometers to the north only transgressive Campanian and Maastrichtian sediments rest upon Karroo volcanics. These beds are lateral equivalents of the Sena Formation, a thick sequence of deltaic sandstones and conglomerates (see Table XII). In common with such deposits the Sena Formation is diachronous and may rise to Maastrichtian. This, and the Domo Formation are evidence of uplift which engendered the erosion leading to the production of these detrital sediments. Open-water conditions thus really date from the time of deposition of the Grudja Formation (Senonian).

The eastward diminution in thickness seen in the Maputo Formation, the easterly source of the Domo Formation points to the existence of a positive area

TABLE XII

Summary of the Mesozoic lithologies in Mozambique (from Flores, 1973; Blant, 1973; Du Toit, 1954)

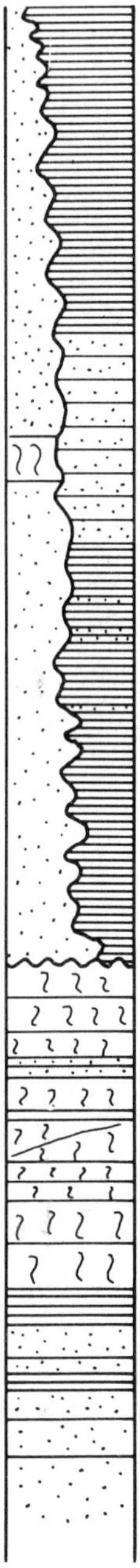

Senonian	Grudja Formation: a neritic argillaceous deposit at outcrop, in subsurface a deeper-water environment is continuous up to Lower Eocene; an important break in the succession south of Sul do Salve, and here the Grudja Formation transgresses over continental sandstones; formation thickest in vicinity of the present coast and thins to the east of Beira
Cenomanian–Turonian	Domo Formation: consists of arenaceous sediments derived from the east and deposited in a north–south trough; these sediments confirm the existence of a positive axis in Mozambique
Neocomian–Albian	Maputo Formation: littoral to neritic argillaceous marine facies with sandy marls, glauconitic sandstones, occasionally sandy limestones and layers of calcareous; like to Domo and Grudja Formations it is laterally equivalent to parts of the Sena Formation; age ranges from Neocomian (Barremian) to end-Albian–mid-Cenomanian; thickness 0–600 m
Neocomian–Maastrichtian	Sena Formation: 2,500 m of fluviatile–torrential sandy and conglomeratic deposits, surface often laterized; essentially an enormous detrital talus deposited in deltaic conditions; some similarities to the Adigrat sandstone; a diachronous facies which may range up to Maastrichtian according to location
Jurassic–Cretaceous	Volcanic Sequence: consisting of 1,500 m of amygdaloidal basalts, some andesites, agglomerates and a horizon of rhyolites and rhyolite intrusions near the top and greyish sandstones and grits with plant remains overlain by red "desert" sandstones with dinosaur bone fragments; the age of the volcanics ranges from Jurassic to Early Cretaceous, flows are progressively younger from west to east
	Mpiusa Shales: probably equivalent to the Beaufort of South Africa, a sequence of greyish grits, red and purple marls and sandstones with occasional calcareous concretions and layers, fossil wood and molluscs
	Tete Sandstone

in the region of the present Mozambique Channel, and comparison may be drawn with southwestern Somalia.

Madagascar. The Mesozoic beds in Madagascar occupy a narrow strip along the western seaboard. The lower part of the succession, assigned to the Karroo Supergroup, was deposited in a trough whose axis lay close to the eastern margin, and elsewhere on a penecontemporaneously block-faulted zone (Cliquet, 1957, referred to by Kent, 1974). The beds thin westwards, a further indication that the Mozambique Channel cannot be regarded as a former broad Karroo basin as Dixey (1956) proposed.

The Karroo sequence consists of three units, the Sakoa, Sakamena and Isalo Groups. The Paleozoic–Mesozoic boundary is marked by an unconformity between the lower and upper members of the Sakamena Group (Table XIII). In the northern part of the island beds of the Upper Sakamena Group contain an ammonite fauna which shows Australian affinities. In the southern part of the island the marine horizon is represented by *Estheria* beds. Following the Early Triassic a thick continental sequence, the Isalo Formation, was deposited, with no marine horizon until the Late Liassic (Toarcian). This transgression as with the earlier ones came from the east, but by the beginning of the Dogger a significant inversion had occurred with the Callovian transgression reaching Madagascar from the west (Blant, 1973). It suggests the collapse of the arch which had formerly occupied the Mozambique Basin, and the westward tilting of the island, and perhaps the first

TABLE XIII

Summary of the Mesozoic lithologies of Madagascar (largely after Besairie, 1957)

Upper Cretaceous		in north succession begins with a thick cross-bedded sandstone which at the tip of the island contains marine horizons; transgression begins with Upper Santonian marls; continental sandstones recur in the Maastrichtian; south of Cap Saint André, marine Santonian occurs in the south, but further north first marine horizon is Middle Campanian; Santonian marine beds give way to continental deposits Maastrichtian is known at several points on the east coast
Middle Cretaceous		For convenience the Aptian–Coniacian succession is grouped together and in the north this sequence is continuous from Middle Albian, including Cenomanian marly and pyritous beds; the higher part is sandy; south, towards Cap Saint André the succession consists of largely unfossiliferous red sandstones with intercalated ammonitic marls in the north generally marine in continuity with Jurassic at base, giving way later to continental sandstones; sometimes followed by an Aptian transgression
Lower Cretaceous–Upper Jurassic		succession begins with marine transgression; north of Cap Saint André Upper Jurassic consists entirely of marine limestones (often oolitic) Callovian, marly and frequently gypsiferous Oxfordian and higher stages; south of Cap Saint André section is faulted and incomplete, with mixed continental and marine facies in the extreme south continental intercalations occur in the lower part, and fully marine conditions established only in the Oxfordian; the Portlandian is absent
	Isalo III	base synchronous with Bajocian marine transgression, continental facies similar to Isalo II; in extreme southwest almost entirely marine with limestones and coral reefs
Middle and Lower Jurassic	Isalo II	characterized by thick lenses of red shale, soft sandstones finer in grain than Isalo I; cross-bedding common, also found thick torrential deposits; silicified wood is abundant and locally fish teeth (*Ceratodus*) and phytosaur dermal scales are present; some lenses of bituminous sandstone occur; thickness may exceed 2,000 m; marine intercalations are found in the north
Triassic	Isalo I	coarse often conglomeratic, poorly cemented sandstones with large-scale cross-bedding, and torrential type deposits in middle; feldspar grains are kaolinized; thickness ranges from a few hundred metres to over 3,000 m; in north, the succession is capped by a Triassic marine horizon
Eo-Trias	Sakamena Formation	transgressive and discordant upon the Sakoa Formation; the upper part of the Sakamena begins with shales containing Eo-Triassic fish followed by sandstones, shales and red argillaceous beds; total thickness 350–1,000 m

phase of the faulting leading to the collapse which produced the elevations from which the continental Isalo sediments were derived. The sea invaded from both north and south.

The history of the isolation of Madagascar can be traced in the faunal succession. The reptilian fauna of the Trias suggests continuity between Africa and Madagascar but with the establishment of marine faunas of Indian aspect along the west coast of Madagascar, in the Callovian there are affinities with South Africa, and by the end of the Oxfordian faunal affinities are with the east coast of Africa and India. During the Cretaceous there are two noteworthy features: the appearance of a European influence marking a good link with Tethys, and the development of a peculiar endemic fauna in Turonian times (Collignon, 1973) marking a progressive isolation of Madagascar. The latter may be related to contemporary tectonic movements indicated by volcanic activity, which can be traced into the Somali Basin.

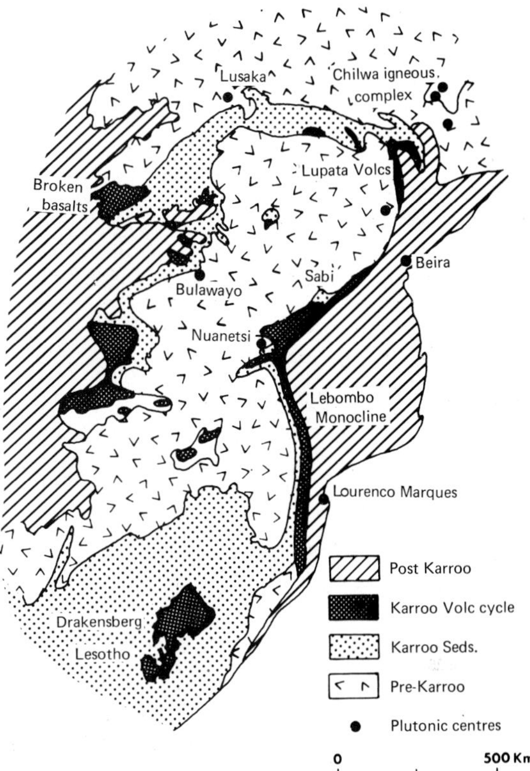

Fig. 18. Distribution of Karroo cycle, igneous rocks and of volcanic centers in southeastern Africa (from Cox, 1972).

IGNEOUS ACTIVITY

The termination of the Karroo Supergroup times was marked by the extrusion of extensive series of basalts accompanied by swarms of dolerite sills and dykes in southern Africa. In South Africa these volcanics of Late Triassic to Early Jurassic age (Fitch and Miller, 1971) have been described by Du Toit (1954 and earlier papers) and detailed petrographic studies are reported by Walker and Poldervaart (1949). Although the principal outcrops are in South Africa, volcanic rocks are widely distributed in eastern Botswana, Southern Rhodesia and Mozambique, but do not extend north of Mozambique (see Fig. 18).

Cox (1972) considered the concept of a Karroo volcanic cycle, to include volcanic rocks which in a strict sense are too young to be included in the Karroo Supergroup. This was linked with the attempt to associate all the Mesozoic rocks with a single petrogenic (thermo-tectonic) event, the event postulated being the disruption of Gondwana. While in the context of South Africa, there is much to recommend the idea, whether or not it be associated with the disruption of Gondwana. When all of the east coast of Africa is considered, it is difficult to maintain for in Egypt and the Sudan there are Cretaceous igneous rocks coeval with the Cretaceous igneous activity of eastern and southern Africa, but no extensive earlier phase of volcanic activity. There are, however, in Egypt some ring complexes which date back to Triassic and some dykes are as old as Carboniferous (El Shazly, 1977). What is clear is the significant Cretaceous igneous event, which although associated with a relatively small volume of volcanic rocks (e.g. the Wadi Natash volcanics of Egypt, the Lupata volcanics of Mozambique, the Cretaceous volcanics of Madagascar) was accompanied by the emplacement of ultrabasic rocks from Angola to Egypt. These deep intrusives include the kimberlite pipes of South Africa, carbonatites in Tanzania, and ring complexes in the Southeastern Desert of Egypt. A summary of the radiometric ages for some of the more important units is given in Table XIV.

The timing of the phases of the Karroo volcanic cycle given by Fitch and Miller (1971) is reproduced below (Table XV). The table suggests a younging of igneous activity towards the northeast, and this trend can be followed further in the subsurface in Mozambique (see p. 360, and Förster, 1975). The volcanic ac-

TABLE XIV

Partial list of the published radiometric ages of the Mesozoic igneous rocks

Rock unit	Location	Mineral	Age (m.y.)	Source
Carbonatite	Museni Hill, Mbeya, Tanzania	biotite	96 ± 9	1
		biotite	101 ± 12	1
Feldspathised cataclastic rock	Mbeya Range Tanzania	potash feldspar	100 ± 10	1
Fareig basalts	Sudan, about 200 km south of Wadi Halfa	whole rock	100 ± 15	2
		whole rock	88 ± 8	2
Alkaline volcanics	Lupata Volcanic Series, Mozambique		110.5	3
		anorthoclase	106	3
Alkaline lava	Lupata lava, Mozambique	whole rock	115 ± 10	4
Carbonatite	Panda Hill, Mbeya, Tanzania	phlogopite	113 ± 5	1
Chambe pluton perthosite	southern Malawi	biotite	116 ± 6	1
Chaone Mountain pulaskite	Kasupe district, southern Malawi	biotite	116 ± 6	1
Kangankunde Hill Carbonatite	southern Malawi	phlogopite	123 ± 6	1
Mlanje syenite	southern Malawi	biotite with hornblende and augite	128 ± 6	1
Basalt flow	Khor Shinab area, Sudan		148 ± 30	2
Lupata rhyolites	Mozambique	whole rock	166 ± 10	4
Granite Mateka Hills	Southern Rhodesia		167 ± 7	5
Karroo dolerite sub. max.	Lesotho and widespread basalts		~172	5
Karroo dolerite max.	central S. Africa		~187 ± 7	5
Nuanetsi and Lebombo rhyolites			190–194 ± 12	5
Nuanetsi and Lebombo dolerites			190–195	5
Marangudzi ring complex	Southern Rhodesia	average	~195	5
Biotite–gabbro		biotite	192 ± 10	
Biotite–foyaite		biotite	182 ± 10	
Malignite		biotite	196 ± 10	
Gabbro		biotite	187 ± 10	
Pulaskite granite		hornblende	195 ± 10	
		whole rock	186 ± 10	
Shawa ijolite	Shawa carbonatite ring complex, southern Rhodesia	Rb/Sr of biotite	197 ± 15	6

Sources: 1, Cahen and Snelling (1966); 2, Whiteman (1971); 3, McDougall, in Gough et al. (1964); 4, Flores (1964); 5, Fitch and Miller (1971), in Cox (1972) and in Gough et al. (1964); 6, Nicholaysen et al. (1962), and Gough and Brock (1964).

tivity in Madagascar is also consistent with this pattern.

In general Mesozoic extrusive igneous rocks are not recorded in East Africa, although basic to ultrabasic intrusives are known in southwest Tanzania (see Table XIV). In the Congo, kimberlites from Kundelungu, Bushimaie and Kasai-Lundu are reported in Cahen (1954). None has been dated with any accuracy, the best that can be said is the Bushimaie kimberlite is post-Jurassic–pre-Pleistocene in age and Cahen (1954) regards it as contemporaneous with the South African kimberlites. Similarly he regards the undated Kasengo basalts as possibly contemporaneous

TABLE XV

The Karroo Volcanic Cycle (from Fitch and Miller, 1971)

Waning phase		
subcycle	Lupata alkali basalts	105–115 ± 10 m.y.
subcycle	Karroo dolerite magmatism and metasomatism subsidiary maximum	~155 to 161 m.y.
subcycle	Lupata rhyolites	166 ± 10 m.y.
subcycle	Karroo dolerite magmatism and metasomatism subsidiary maximum	~172 m.y.
Culminating phase		
subcycle	Drakensberg lavas of Lesotho and flood basalts of central southern Africa	~184 ± 7 m.y.
subcycle	Nuanetsi and Lebombo rhyolites	190–194 ± 12 m.y.
Waxing phase		
subcycle	lower alkaline and tholeiitic basalts of Nuanetsi and Lebombo	~190–195 m.y.
subcycle	Marangudzi igneous complex	~195 m.y.
subcycle	Shawa igneous complex	~197 m.y.

with the Late Triassic–Early Jurassic Stormberg basalts.

Late Cretaceous volcanics are widespread in the Eastern and Western Deserts of Egypt particularly in the south where alkaline ring complexes, volcanic plugs, alkali olivine basalt sheets and also small granite bodies occur (El Shazly, 1977). Interbedded within the base of the local *Nubian Sandstone* are the Wadi Natash volcanics which range in composition from alkali basalt to trachyte and acidic rhyolites. The paleomagnetic data from these rocks are consistent with results from the Cretaceous of Southern Africa.

GEOLOGICAL HISTORY OF NORTH AND EAST AFRICA

The geological history of northern and eastern Africa during the Mesozoic was dominated by vertical movements, there is very little evidence for translational motion. The extent of vertical movement is considerable with displacements in excess of 6,000 m recorded. Movements along these faults result in the form of Africa as known today, and the intervening stages can be followed in the paleogeography. While the vertical tectonic movement does not require the intervention of plate motion (continental drift) to explain what is observed, plate motion must necessarily receive some attention for both represent lithosphere reaction to underlying processes and are thus part of a single system.

Even though the trend of the East African Rift System so closely parallels earlier trends, the spectacular recent movements have tended to obscure their underlying unity with the overall structural pattern. Both Dixey (1956) and McConnell (1951) have indicated that the rifts tend to parallel lines of ancient structures, superposed on zones of intense shearing or mylonitization between ancient shields, lines which originated at the close of "Basement System" times (McConnell, 1951). Dixey (1956) also related the Lebombo monocline, which northwards becomes more anticlinal, to the East African Rift bulge and the coastal fault system and even considers the Mozambique Channel as simply one of the more complex rifts. The latter idea however requires considerable modification following the work of Green (1972) and Ludwig et al. (1968). Dixey also remarked upon the tendency of carbonatites to follow alignments parallel to the main course of the rifts.

This great fracture system has been active upon a number of occasions, for it can be seen to affect Karroo sedimentation and as a result of latter reactivation affected the deposition of the Mesozoic. Clearly not all the faults were necessarily active simultaneously. During the Mesozoic, however, a general agreement between periods of tectonic activity between northern and eastern Africa exists even if the principal activity of the one was Jurassic and the other Cretaceous. It is the attempt to find a mechanism operating on this scale that the "new global tectonics" is a valuable unifying hypothesis, and it comes as no surprise that where continental fragmentation is projected it occurs following old lines of structural weakness though from place to place the trend may change, just as the apparently older trend through the Somali Mandara north–south depression changed in mid-Jurassic times to a direction parallel to the present coast.

The event dominating the Mesozoic is the proposed disruption of Gondwana, and in particular the separation of east and west Gondwana. The geological history of East Africa, however, places severe limits on the possible original positions. The thick wedge of Jurassic and Cretaceous sediments in the Somali Basin (and Triassic may also be present) effectively excludes locating Madagascar close to the Tanzanian coast (see Kent, 1974), and this is confirmed by seismic profiles which indicate continental crust extending offshore (Fig. 14) and by the magnetic profiles (Green, 1972). Whether Madagascar can be placed against the Natal–Mozambique coast is equivocal. Green (1972) did indicate that in the southern part of the Mozambique Channel magnetic anomalies have the amplitude and wavelength of typical linear oceanic anomalies. It was not however possible to demonstrate symmetry across the Mozambique Ridge, although Ludwig et al. (1968) have indicated typical oceanic crust exists on either side. Förster (1975) and others have argued that there are few geological arguments for proposing any other position of Madagascar but the present. Certainly there is no reason to suppose any separation more than a few tens of kilometers (i.e. comparable with East African or Red Sea rifts). The profiles do indicate a possible short gap in the centre (see Förster, 1975).

The extension observed is more like the extension of an expandable bracelet, that is in the zones of weakness, than separation by the production of new oceanic crust; with the possible exception of the Mozambique Ridge.

It may be a mistake to have included the Seychelles as part of the southeastern margin of the Seychelles Basin, given the uncertainty of its origin, particularly with a possible subduction zone proposed between it and Madagascar. However, this does not affect the interpretations favoured here.

Broad tectonic conclusions of the above type are not easily made for the Mediterranean coastal area, although there is a marked similarity in the geological histories. This may be because the final, Cenozoic phase, of development was associated with compressional tectonics of plate collision.

As far as the stratigraphic history of the region is concerned, the feature of note was the development of the East African coastline. In general the continental area remained continental during the Mesozoic. For most of the time the region must have remained close to sea level, and on two occasions the sea penetrated into the Congo Basin, and once a seaway was established from the Gulf of Gabes to the Gulf of Guinea. Periodic uplift occurred off-setting the erosion levelling, although it must be pointed out that there existed sub-Karroo and sub-Adigrat topography measured in hundreds of metres. The present elevation of Africa may be regarded as just one such period.

The most obvious effect of the vertical movements on the continental facies is the continual removal of previously formed deposits. Thus continental deposits of Mesozoic age, from whatever region, tend to be thin, and often isolated, the isolation of the Congo Basin being perhaps the most spectacular example. Only in troughs are relatively thick sequences preserved. However, as these troughs tend to be synsedimentation features, the thickness is little clue as to what the thickness generally may have been. Their paleogeographic history however indicates that the pre-existing topography was submerged and hence deposits were formerly more widely spread.

As is true with most continental deposits dating is extremely difficult. Fossils are far from common and many of those found tend to be of little value in chronostratigraphy. This has resulted in the widespread application of terms such as "Karroo" in southern and eastern Africa, and "Nubian" in northern Africa based upon lithological similarities which may often be more apparent than real. The most glaring example of this misuse of terms, discussed earlier, is the Nubian Sandstone. Even in the region where the term was established, four lithologies, all continental and admittedly interrelated, are now recognized, while in adjoining regions marine deposits were assigned to the same system. To this must be added the application of the term to deposits ranging in age from Carboniferous to Cretaceous. The same is true of the misuse of the term Karroo, and although the errors are not as great, the assignments are inappropriate. However, from

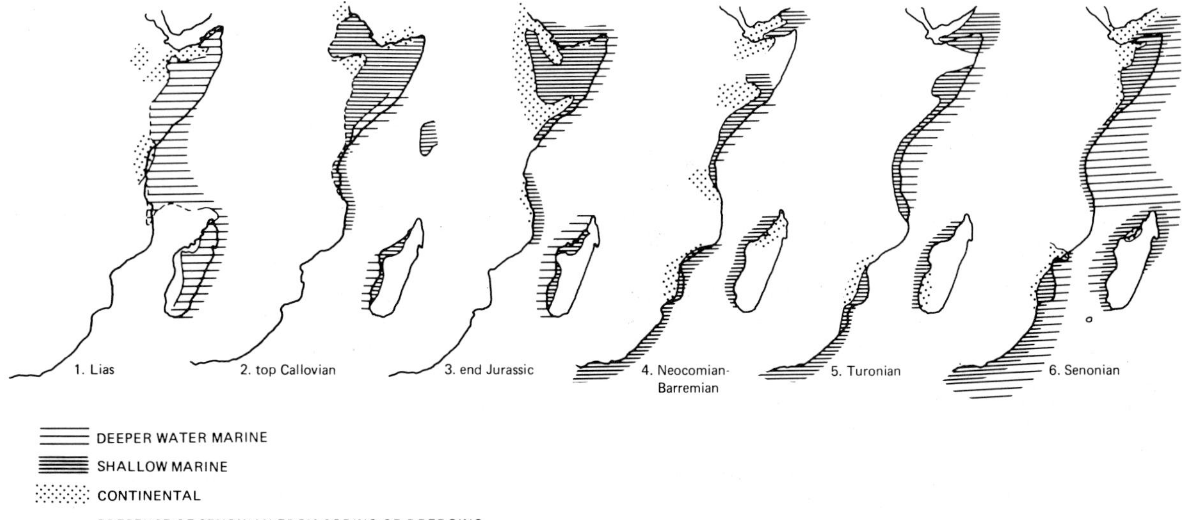

Fig. 19. Summary of the Mesozoic facies distribution along the east coast of Africa (after Blant, 1973).

these errors arise one point of note, and that is that the environmental conditions remained relatively uniform over a long period of time. The most striking example, one not within the Mesozoic, is the general occurrence of Ecca coals over the southern half of Africa, other environmental conditions in northern Africa during the Late Jurassic–Cretaceous are reflected by the *Nubian Sandstone,* while the widespread desert conditions of the Late Triassic in the southern half of Africa forms another example.

The Mesozoic era saw the modern shape of Africa taking form. Along the east coast, during the Jurassic, a shoreline in the Somali Basin not very different from the modern one was established. It was, however, separated from the gulf occupying the Mozambique Basin, and not until Cretaceous times was there a continuous seaway along the eastern seaboard of Africa. However, it must be reiterated that the evidence of the Mandawa evaporites as well as the more general, if short-lived Permian and Triassic transgressions, indicate that marine links were possible even earlier. An excellent survey of this development is given by Blant (1973) from whom Fig. 19 was derived.

In northern Africa some of the details of the marine history are well established, in particular in Tripolitania where lay the eastern margin of the Algero-Tunisian evaporite basin. The Tripolitanian coastal zone appears to have been a nearshore region throughout most of the Mesozoic, with the evaporite basin occupying a region with a long history as a zone of depression extending into the present offshore. When the results of offshore drilling become available this region will become better known, but at the moment it may be considered as a former horst graben platform, of which the eastern Tunisian Platform and Sicily form parts. The modern Egyptian coastline however lies far to the north of the earlier coasts, and in consequence information has to be derived from boreholes. Isopach maps generated from well data show a number of highs, but are not sufficiently precise to indicate whether these truly reflect horst structures. Faulting activity in the north seems to indicate major movements later than those in East Africa. While the East African coast is a Middle Jurassic feature and in Tripolitania Jurassic activity also occurred, the onset of the development of the Sirte graben dates from Cenomanian.

Clearly the final word on the Mesozoic geology of northern and eastern Africa has not yet been written. Yet, if an enormous amount of work remains to be done, certain broad lines appear to be emerging which may make that work more stimulating.

REFERENCES

Abdallah, A.M., El Addani, A. and Fahmy, N., 1963. Stratigraphy of the Lower Mesozoic rocks western side of Gulf of Suez, Egypt. *Geol. Surv. Min. Res. Dep., Pap.* 27: 23.

Abdel-Khalek, M.L., 1973. Characteristics and utilization potential of a Jurassic coal from the Sinai Peninsula, Egypt. *Egypt. J. Geol.,* 17: 71–84.

Akaad, M.K. and El-Ramly, M.F., 1960. Geological history and classification of the basement rocks of the central eastern desert of Egypt. *Geol. Surv. Min. Res. Dep., Pap.*, 9: 24.

Assereto, R. and Benelli, F., 1971. Sedimentology of the pre-Cenomanian formations of the Jebel Gharian, Libya. In: C. Gray (Editor), *Symposium on the Geology of Libya.* University of Libya, Tripoli, pp. 37–85.

Ayers, F.M., 1952. *Geology of the Wajir–Mandera District, North-east Kenya.* Geol. Survey Kenya, Nairobi, 36 pp.

Baker, B.H. and Miller, J.A., 1963. Geology and geochronology of the Seychelles Islands and structure of the floor of the Arabian Sea. *Nature,* 199: 346–348.

Barbieri, F., Civetta, L., Gasparini, P., Innocenti, F., Scandone, R. and Villari, L., 1974. Evolution of a section of the Africa–Europe plate boundary: paleomagnetic and vulcanological evidence from Sicily. *Earth Planet. Sci. Lett.,* 22: 123–132.

Barr, F.T. and Weegar, A.A., 1972. *Stratigraphic Nomenclature of the Sirte Basin, Libya.* Pet. Explor. Soc. Libya, Tripoli, 179 pp.

Beltrandi, M.D. and Pyre, A., 1973. Geological evolution of southwest Somali. In: G. Blant (Editor), *Sedimentary Basins of the African Coasts.* Assoc. Afr. Geol. Surv., Paris, pp. 159–178.

Besairie, H., 1957. *La géologie de Madagascar en 1957.* Haut Comm. Madagascar Depend. Serv. Géol., Tananarive, 159 pp.

Beydoun, Z.R. and Bichan, H.R., 1970. The geology of Socotra Island, Gulf of Aden. *Q. J. Geol. Soc. Lond.,* 125: 413–441.

Beyth, M., 1973. The Paleozoic–Mesozoic sedimentary basin of the Mekele outlier – Northern Ethiopia. In: G. Blant (Editor), *Sedimentary Basins of the African Coasts, 2. South and East Coasts.* Assoc. Afr. Geol. Surv., Paris, pp. 179–191.

Blant, G., 1973. Structure et paléogéographie du littoral méridional et oriental de l'Afrique. In: G. Blant (Editor), *Sedimentary Basins of the African Coasts, 2. South and East Coasts.* Assoc. Afr. Geol. Surv., Paris, pp. 193–231.

Bond, G. and Bromley, K., 1970. Sediments with remains of dinosaurs near Gokwe, Rhodesia. *Palaeogeogr., Palaeoclimatol., Palaeoecol.,* 8: 313–327.

Burollet, P., 1967. General geology of Tunisia. In: *Guidebook to the Geology and History of Tunisia.* Pet. Explor. Soc. Libya, Tripoli, pp. 51–58.

Burollet, P., 1963. Discussion of Libyan stratigraphy. *Rev. Inst. Fr. Pet.,* 18: 1323–1325.

Burollet, P.F., Mugniot, J.M. and Sweeny, P., 1977. The geology of the Pelagian Block: A. The margins and basins off southern Tunisia and Tripolitania. In: W.H. Kanes, A.E.M. Nairn and F.G. Stehli (Editors), *Ocean Basins and Margins 4, The Mediterranean Sea.* Plenum Press, New York, N.Y.

Busson, G., 1967. *Le Mésozoïque Saharien, première partie: l'Extrême-sud tunisien.* Centre Natl. Rech. Sci., Paris, 185 pp.

Busson, G., 1970. *Le Mésozoïque saharien: deuxième partie: Essai de synthèse des données des sondages Algéro-Tunisiens.* Centre Natl. Rech. Sci., Paris, 811 pp. (vols. 1 and 2).

Cahen, L., 1954. *Géologie du Congo Belge.* Vaillant-Carmaune, Liège, 577 pp.

Cahen, L. and Snelling, N.J., 1966. *The Geochronology of Equatorial Africa.* North-Holland, Amsterdam, 195 pp.

Caire, A., 1977. The Central Mediterranean mountain chains in the Alpine Orogenic Environment. In: W.H. Kanes, A.E.M. Nairn and F.G. Stehli (Editors), *The Ocean Basins and Margins 4. The Mediterranean Sea.* Plenum Press, New York, N.Y.

Christie, A.M., 1955. *Geology of the Garian area.* U.N.T.A.P., New York, N.Y., 60 pp.

Cliquet, P.L., 1957. *La tectonique profonde du Sud Bassin de Morondawa Madagascar.* C.C.T.A. Com. Géol. Centre-Est et Sud, Tananarive.

Conant, L.C. and Goudzari, G.H., 1967. Stratigraphic and tectonic framework of Libya. *Bull. Am. Assoc. Pet. Geol.,* 51: 719–730.

Collignon, M., 1973. Le bassin côtier du Golfe du Ménabe (Madagascar), caractère endémique de sa faune. In: G. Blant (Editor), *Sedimentary Basins of the African Coasts, 2. South and East Coasts.* Assoc. Afr. Geol. Surv., Paris, pp. 69–79.

Cox, K.G., 1972. The Karroo Volcanic Cycle. *J. Geol. Soc. Lond.,* 128: 311–336.

D'Argenio, B., De Castro, P., Emiliani, C. and Simone, L., 1975. Bahamian and Apenninic limestones of identical lithofacies and age. *Bull. Am. Assoc. Pet. Geol.,* 59: 524–530.

Davies, D. and Francis, T.J.G., 1964. The crustal structure of the Seychelles Bank. *Deep-Sea Res.,* 11: 921–927.

De Lapparent, A.F. and Lelubre, M., 1948. Interprétation stratigraphique des séries continentales au nord d'Edjeleh (Sahara central). *C.R. Acad. Sci. Paris,* 241: 2399–2402.

Desio, A., Rossi-Ronchetti, C., Pozzi, R., Clerici, F., Invernizzi, G., Pisoni, C. and Vigano, P.L., 1963. Stratigraphic studies in the Tripolitanian Jebel (Libya). *Riv. Ital. Paleontol. Stratigr. Mem.,* 9: 126 pp.

Dietz, R.S., 1973. Morphologic fits of North America/Africa and Gondwana: A review. In: D.H. Tarling and S.K. Runcorn (Editors), *Implications of Continental Drift to the Earth Sciences, 2.* Academic Press, New York, N.Y., pp. 865–871.

Dingle, R.V. and Klinger, H.C., 1971. Significance of Upper Jurassic sediments in the Knysna outlier (Cape Province) for timing of the break-up of Gondwanaland. *Nature, Phys. Sci.,* 232: 37–38.

Dixey, F., 1956. The East African Rift System. *Bull. Colon. Geol. Min. Resour., Suppl.,* 1: 76 pp.

Dixey, F., 1960. The geology and geomorphology of Madagascar and a comparison with Eastern Africa. *Q. J. Geol. Soc. Lond.,* 114: 255–268.

Du Toit, A.L., 1954. *The Geology of South Africa.* Oliver and Boyd, Edinburgh, 3rd ed., 611 pp.

El Shazly, E.H., 1977. The coast of Egypt. In: W.H. Kanes, A.E.M. Nairn and F.G. Stehli (Editors), *Ocean Basins and Margins 4. The Mediterranean Sea.* Plenum Press, New York, N.Y.

Fitch, F.J. and Miller, J.A., 1971. Potassium–argon radio-ages of Karroo Volcanic Rocks from Lesotho. *Bull. Volcanol.,* 35: 64–84.

Flores, G., 1964. On the age of the Lupata rocks, Lower

Zambezi River, Mozambique. *Trans. Geol. Soc. S. Afr.*, 67.

Flores, G., 1973. The Cretaceous and Tertiary sedimentary basins of Mozambique and Zululand. In: G. Blant (Editor), *Sedimentary Basins of the African Coasts, 2. South and East Coasts.* Assoc. Afr. Geol. Surv., Paris, pp. 81–110.

Förster, R., 1975. The geological history of the sedimentary basin of southern Mozambique, and some aspects of the origin of the Mozambique Channel. *Palaeogeogr., Palaeoclimatol., Palaeoecol.*, 17: 267–287.

Francis, T.J.G., Davies, D. and Hill, M.N., 1966. Crustal structure between Kenya and the Seychelles. *Philos. Trans. R. Soc.*, A 259: 240–261.

Furon, R., 1963. *Geology of Africa.* Hafner, New York, N.Y., 377 pp.

Goudzari, G.H., 1970. Geology and mineral resources of Libya – a reconnaisance. *U.S. Geol. Surv., Prof. Pap.*, 660: 104 pp.

Gough, D.I. and Brock, A., 1964. The palaeomagnetism of the Shawa ijolite. *J. Geophys. Res.*, 69: 2489–2493.

Gough, D.I., Brock, A., Jones, D.L. and Opdyke, N.D., 1964. The palaeomagnetism of the ring complexes at Marangudzi and the Mateke Hills. *J. Geophys. Res.*, 69: 2499–2507.

Gray, C., 1971. Structure and origin of the Garian Domes. In: C. Gray (Editor), *Symposium on the Geology of Libya.* University of Libya, Tripoli, pp. 307–319.

Green, A.G., 1972. Seafloor spreading in the Mozambique Channel. *Nature, Phys. Sci.*, 236: 19–21.

Green, D., 1966. The Karroo System in Bechuanaland. *Geol. Surv. Dep. Bechuanaland Bull.*, No. 2: 74 pp.

Gregor, C.B., Nairn, A.E.M. and Negendank, J.F.W., 1975. Paleomagnetic investigations of the Tertiary and Quaternary rocks: IX. The Pliocene of southeast Sicily and some Cretaceous rocks from Capo Passero. *Geol. Rundsch.*, 64: 948–958.

Hammuda, O.S., 1971. Nature and significance of the Lower Cretaceous unconformity in Jebel Nefusa, northwest Libya. In: C. Gray (Editor), *Symposium on the Geology of Libya.* University of Libya, Tripoli, pp. 87–97.

Hea, J.P., 1971. Petrography of the Paleozoic–Mesozoic sandstones of the southern Sirte Basin, Libya. In: C. Gray (Editor), *Symposium on the Geology of Libya.* University of Libya, Tripoli, pp. 107–125.

Heirtzler, J.R. and Burroughs, R.H., 1971. Madagascar's paleoposition: New data from the Mozambique Channel. *Science*, 174: 488–490.

Hinnawy, M.El. and Cheshilev, G., 1975. *Explanatory Booklet: Geological Map of Libya 1 : 250,000 Sheet Tarabulus N 1 33-13.* Ind. Res. Centre, Tripoli.

Kennedy, W.Q., 1965. The influence of basement structure on the evolution of the coastal (Mesozoic and Tertiary) basins of Africa. In: *Salt Basins around Africa.* Institute of Petroleum, London, pp. 7–15.

Kent, P.E., 1965. An evaporite basin in southern Tanzania. In: *Salt Basins around Africa.* Institute of Petroleum, London, pp. 41–54.

Kent, P.E., 1972. Mesozoic history of the East Coast of Africa. *Nature*, 238: 147–148.

Kent, P.E., 1974. Continental margin of East Africa – A region of vertical movements. In: C.A. Burk and C. Drake (Editors), *Geology of the Continental Margins.* Springer-Verlag, Heidelberg, pp. 313–320.

Kent, P. and Perry, J.T.O'B., 1973. The development of the Indian Ocean margin in Tanzania. In: G. Blant (Editor), *Sedimentary Basins of the African coasts, 2. South and East Coasts.* Assoc. Afr. Geol. Surv. Paris, pp. 113–130.

King, L.C., 1958. Basic palaeogeography of Gondwanaland during late Palaeozoic and Mesozoic eras. *Q. J. Geol. Soc. Lond.*, 114: 47–70.

Klitzsch, E., 1963. Geology of the northeast flank of the Murzuk basin. *Rev. Inst. Fr. Pét.*, 18: 97–113.

Klitzsch, E., 1971. The structural development of parts of North Africa since Cambrian time. In: C. Gray (Editor), *Symposium on the Geology of Libya.* University of Libya, Tripoli, pp. 253–263.

Laughton, A.S., McKenzie, D.P. and Sclater, J.G., 1973. The structure and evolution of the Indian Ocean. In: D.H. Tarling and S.K. Runcorn (Editors), *Implications of Continental Drift to the Earth Sciences, I.* Academic Press, New York, N.Y., pp. 203–212.

Lefranc, J.P., 1959a. Les séries continentales intercalaires du Fezzan nord-occidental (Libye), leur âge et leurs correlations. *C.R. Acad. Sci. Paris*, 249: 1685–1689.

Lefranc, J.P., 1959b. Existence, au Fezzan nord-occidental (Libye) de lacunes et discordances dans les séries du Continental intercalaire. *C.R. Acad. Sci. Paris*, 249: 2345–2347.

Lehmann, E.P., 1964. Tertiary–Cretaceous boundary facies in the Sirte Basin, Libya. *Int. Geol. Congr., 22nd, New Delhi, 1964, pt. 2, sect. 3*, pp. 56–73.

Lelubre, M., 1952. Aperçu sur la géologie du Fezzan, Algérie. *Serv. Cart Géol. Bull. Travaux Récents Collab.*, pt. 3: 109–148.

Leriche, M., 1938. L'état actuel de nos connaissances sur la paléontologie du Congo. *Ann. Soc. R. Belge*, 69: 139–150.

Ludwig, W.J., Nafe, J.E., Simpson, E.S.W. and Sacks, S., 1968. Seismic-refraction measurements of the southeast African continental margin. *J. Geophys. Res.*, 73: 3707–3719.

McConnell, R.B., 1951. Rift and shield structures in East Africa. *Rep. Int. Geol. Congr., 18th, London*, pt. 14: 199–209.

MacGregor, A.M., 1941. Geology of the Mafungabusi Gold Belt. *Geol. Surv. S. Rhod.*, No. 35: 1–26.

McGregor, A.M., 1947. An outline of the geological history of southern Rhodesia. *S. Rhod. Geol. Surv. Bull.*, 38: 73 pp.

McKenzie, D.P., 1972. Active tectonics of the Mediterranean region. *Geophys. J.*, 30: 109–185.

McKenzie, D.P., Davies, D. and Molnar, P., 1970. Plate tectonics of the Red Sea and East Africa. *Nature*, 226: 243–248.

McKinlay, A.C.M., 1954. Geology of the Ketewaka–Mchuchuma Coalfield Njombe District. *Tanganyika Geol. Surv. Dep. Bull.*, 21: 46 pp.

Matthews, D.H., 1963. A major fault scarp under the Arabian Sea displacing the Carlsberg Ridge near Socotra. *Nature*, 198: 950–952.

Mohr, P.A., 1961. *The Geology of Ethiopia.* Univ. Coll. Addis Ababa Press, 268 pp.

Nairn, A.E.M., 1964. Palaeomagnetic measurements on Karroo and post-Karroo rocks; a second progress report. *Bull. Overseas Geol. Min. Res.,* 9: 302–320.

Nicholaysen, L.O., Burger, A.J. and Johnson, R.L., 1962. The age of the Shawa Carbonatite complex. *Trans. Geol. Soc. S. Afr.,* 65: 293–294.

Papazachos, B.C. and Cominakis, P.E., 1971. Geophysical and tectonic features of the Aegean Arc. *J. Geophys. Res.,* 76: 8517–8533.

Pomeyrol, R., 1968. Nubian Sandstone. *Bull. Am. Assoc. Pet. Geol.,* 52: 537–590.

Quennell, A.M., McKinlay, A.C.M. and Aitken, W.G., 1956. Summary of the geology of Tanganyika pt 1, Introduction and Stratigraphy. *Tanganyika Geol. Surv. Dep. Mem.,* 1: 264 pp.

Rigassi, D.A., 1970. Comments to Discussions. *Bull. Am. Assoc. Pet. Geol.,* 54: 531–532.

Said, R., 1962. *The Geology of Egypt.* Elsevier, Amsterdam, 377 pp.

Saint Seine, P., 1953. Poissons fossiles de la cuvette congolaise. *C.R. Somm. Soc. Géol. Fr.,* No. 16: 343–345.

Salah, H.A., 1975. Stratigraphic nomenclature of the Sirte Basin. *Nat. Oil Co. Geol. Rep.,* 54: 23 pp.

Salaj, J., 1977. The Geology of the Pelagian Block: B. The Eastern Tunisian Platform. In: W.H. Kanes, A.E.M. Nairn and F.G. Stehli (Editors), *Ocean Basins and Margins 4. The Mediterranean Sea.* Plenum Press, New York, N.Y.

Schult, A., 1973. Palaeomagnetism of Upper Cretaceous volcanic rocks in Sicily. *Earth Planet. Sci. Lett.,* 19: 97–100.

Shukri, N.M. and Said, R., 1944. Contribution to the geology of the Nubian Sandstone, pt. 1. Field observations and mechanical analysis. *Fac. Sci. Bull. Cairo Univ.,* 25: 149–172.

Smith, A.G. and Hallam, A., 1970. The fit of the southern continents. *Nature,* 225: 139–144.

Spence, J., 1954. The geology of the Galula Coalfield. *Bull. Geol. Surv. Tanganyika,* No. 25.

Stockley, G.M. and Oates, F., 1931. The Ruhuhu Coalfields, Tanganyika Territory. *Min. Mag.,* 45: 73–91.

Talwani, M., 1962. Gravity measurements on HMS Acheron in South Atlantic and Indian Oceans. *Geol. Soc. Am. Bull.,* 73: 1171–1182.

Veevers, J.J., Jones, J.G. and Talent, J.A., 1971. Indo-Australia stratigraphy and the configuration and dispersal of Gondwanaland. *Nature,* 229: 383–388.

Walker, F. and Poldervaart, A., 1949. Karroo dolerites of the Union of South Africa. *Geol. Soc. Am. Bull.,* 60: 591–706.

Walters, R. and Linton, R.E., 1973. The sedimentary basin of coastal Kenya. In: A. Blant (Editor), *Sedimentary Basins of the African Coasts, 2. South and East Coasts.* Assoc. Afr. Geol. Surv., Paris, pp. 133–158.

Wegener, A., 1922. *Die Entstehung der Kontinente und Ozeane.* Vieweg, Braunschweig (3rd ed.: 1974).

Weissbrod, T., 1970. Nubian Sandstone: Discussion. *Bull. Am. Assoc. Pet. Geol.,* 54: 526–529.

Whiteman, A.J., 1970. Nubia Group: Origin and status. *Bull. Am. Assoc. Pet. Geol.,* 54: 522–526.

Whiteman, A.J., 1971. *The Geology of the Sudan Republic.* Clarendon Press, Oxford, 290 pp.

Chapter 10

THE WEST AFRICAN SEDIMENTARY BASINS

I. DE KLASZ

INTRODUCTION

The West African sedimentary basins are mostly "open" type coastal basins, the existence of which is attributed to the separation of Africa and South America.

The separation of the two continents has been the subject of numerous publications, especially during the last fifteen or twenty years. Many of these papers have been published in volumes consecrated to sea-floor spreading, to the marginal structure of the continents, etc. (such as Blackett et al., 1965; Tarling and Runcorn, 1973; Nairn and Stehli, 1973; Burk and Drake, 1974; etc.), while others are dispersed in various periodicals.

The object of this chapter is to provide an up-to-date description of the sedimentary sequence of the West African basins from the Rio de Oro (southwesternmost Morocco) to Angola (Fig. 1). Whereas in some cases the task is relatively easy, either because of personal knowledge of the basins, or because there is general agreement in the published literature, in other cases a choice has to be made between often contradictory data.

The most recent reviews of the geology of these basins are Franks and Nairn (1973) for the part stretching from Angola to Cameroun and by Machens in the same volume for the basins from northern Cameroun and Nigeria to the Ivory Coast.

Bibliographic quotations are restricted as far as possible to the newest information available. However, in some cases older literature contained more relevant information. The illustrations are mostly taken from already published papers.

The stratigraphic subdivisions in the various basins are not always uniform. While in most cases lithostratigraphic subdivisions are used, in some places only chronostratigraphic units have been adopted. In some basins no names corresponding to the international rules of stratigraphic nomenclature exist and the various units have only been designated by their age (e.g., Albian, Cenomanian, etc.).

Finally, in one or two basins surface formations have not been identified in borehole sections. This makes the homogeneous presentation of the stratigraphic subdivisions practically impossible. For this reason we opted for a summary of the stratigraphic subdivisions basin by basin.

REGIONAL STRATIGRAPHY

Rio de Oro Basin

This basin is situated in the extreme southwest of Morocco and in Rio de Oro (formerly also called Spanish Sahara). It stretches from south of Agadir to around Villa Cisneros, with a maximum thickness of Mesozoic sediments in the region of Tarfaya–El Aaiun, often referred to in the literature as Tarfaya and/or Aaiun "basins". The basin is thus nearly 1,000 km long with a width reaching up to 200 km (Hourcq, 1966a).

The earliest secondary rocks according to Querol (1966) are probably of Triassic age consisting of "evaporitic and continental . . . facies" encountered in several oil exploration wells in the region of El Aaiun [northwest Rio de Oro (Spanish Sahara)]. They are not known in outcrop. Geophysical evidence indicates active salt tectonics both in the region of El Aaiun and offshore on the continental shelf between Villa Cisneros and Cabo Bojador (R.E. King in discussion after Querol's 1966, paper), but the age of these beds is not yet certain.

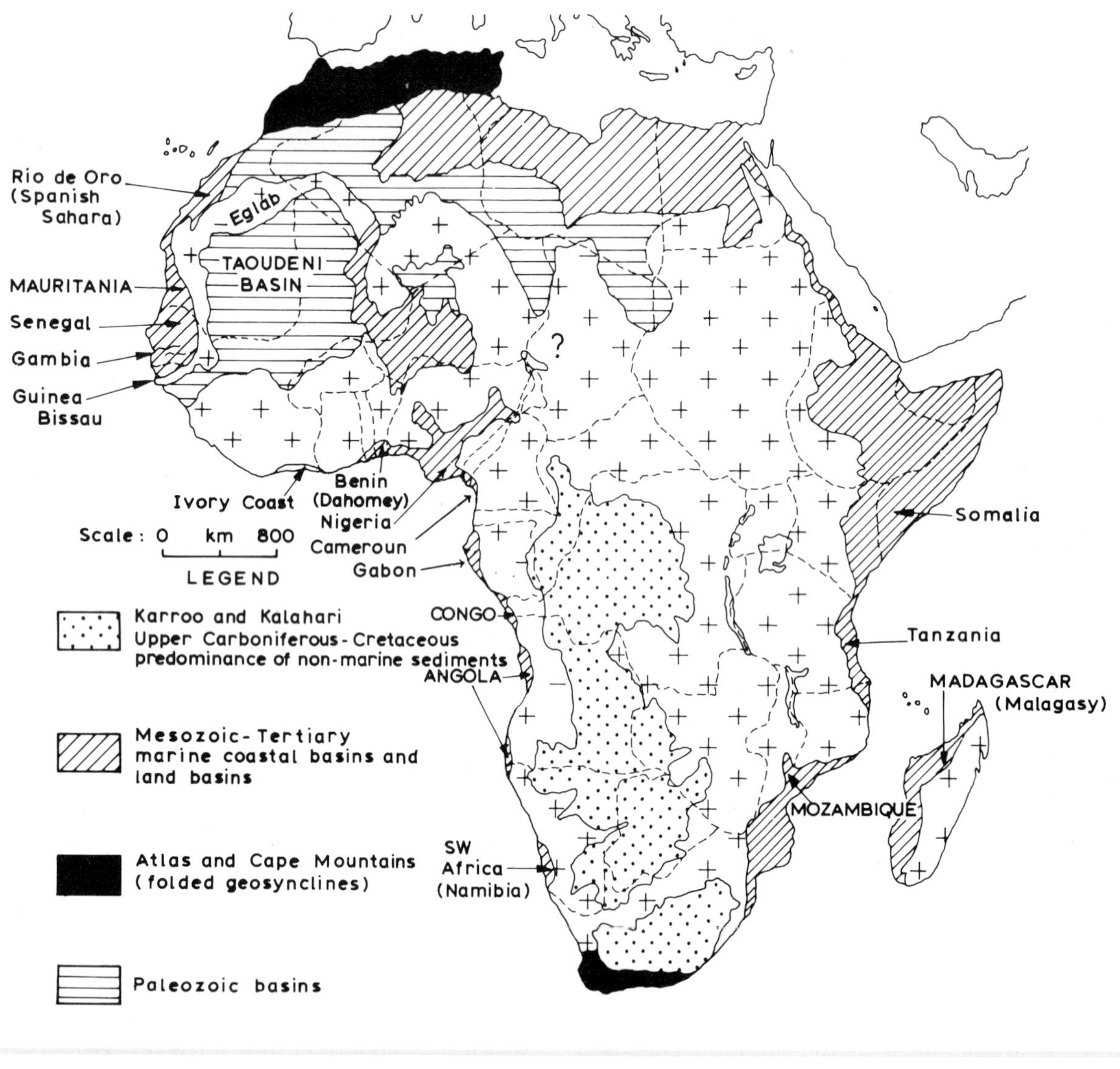

Fig. 1. Structural sketch map of Africa, showing the location of the marginal basins.

Jurassic has also not been found in outcrop in the Rio de Oro Basin, the oldest Jurassic beds again occurring in boreholes in the Tarfaya–El Aaiun area. Querol (1966) is of the opinion that the Jurassic deposits result from a transgression from the west. He mentions continental and lagoonal or evaporitic environments alternating with neritic facies in the Early Jurassic.

In the well Puerto Cansado I in the Tarfaya "basin" Viotti (1965) found marine facies of Liassic (Domerian) age, dated by larger foraminifera, especially *Pseudocyclammina* sp., similar to the species known elsewhere in the Domerian of Morocco (see also Martinis and Visintin, 1966). The Lias and Dogger are apparently not easily separated.

Calcareous rocks predominate during the Jurassic. In the lower part interbedded shales and sands are common, often with anhydrite in the lowermost part. The transgression reached its peak during the Late Jurassic, when 2,000 to 2,500 m of mostly calcareous deposits were formed. The sediments are characterised by the presence of the larger foraminifera *Pseudocyclammina jaccardi* and *Iberina lusitanica*, important as horizon markers. In the south and east some red shales and sands interfinger in the limestone, indicating the start of a new regression.

This regression foreshadowed during the Late Jurassic had its full extension during the Early Cretaceous with the deposition of littoral and continental sandstones and variegated shales. In the Puerto Cansado well of the Tarfaya "basin" some gravel beds and conglomerates with argillaceous cement have been found.

Marine Early Cretaceous (limestones associated with reefs) is known only close to the present northwest coastline in the El Aaiun region.

According to Querol, an unconformity within the Lower Cretaceous is suggested if not proved, by seismic and well electric logs. The thickness of the lowermost Cretaceous (below the probable unconformity) gradually increases from east to west with more than 1,500 m at the coast. It is actually believed that there was some tectonic activity during the Early Cretaceous with a subsidence in the west and erosion of uplifted sediments in the east.

The upper part of the Lower Cretaceous may reach 1000 m in thickness in the El Aaiun "basin", near the hinge line; it becomes thinner towards the west.

Widespread transgression was repeated during the Late Cretaceous. The sediments are well exposed in outcrops in the Aaiun "basin". The basal beds consist of dolomites and shales, sometimes with anhydrite. They are covered with foraminiferal calcareous shales and siltstones. In the south of the Aaiun "basin" the sediments represent a more shallow-water facies, with several sand intercalations.

In the Tarfaya "basin" according to Martinis and Visintin (1966), the Upper Cretaceous Aguidir Limestone Formation can be subdivided into three members. Details concerning the faunas of these strata have been given by numerous authors (Viotti, 1965; Hottinger, 1965; Lehmann, 1965; Masoli, 1965; Martinis and Visintin, 1966; Freneix, 1972; etc.). The greatest thickness of Upper Cretaceous strata (750 m) occurs in the proximity of the northwestern coast of Rio de Oro.

As Querol states, Paleocene–Eocene sediments generally unconformably overlie the Cretaceous, although towards the northwest of the Aaiun "basin" the unconformity or erosional break is questionable.

Senegal Basin

This basin, generally referred to in the literature as the Senegal Basin, reaches from the southern part of Rio de Oro (Villa Cisneros) in the north across Mauritania, Senegal and Gambia to Guinea Bissau in the south. It is the largest of the West African basins, with a total land area of about 340,000 km^2, to which is to be added the very large offshore segment of the basin. The exact extent of the latter is not known. It reaches at least to the Cape Verde Islands, where, on Maio Island, Jurassic sediments have been found. The largest width on land is about 560 km. In appearance at least, it is once again an "open basin", monoclinal as a whole. The eastern part of the basin is tectonically uncomplicated and rather more complicated in the west (Fig. 2). In the east the coastal basin is separated from the Taoudeni Basin by an elevated hinge zone created during the Hercynian orogeny. The pre-Mesozoic basement, more or less faulted, dips gradually toward west and around Dakar most probably exceeds 6,000 m.

Most of the basin is covered with Pliocene and Quaternary sands and outcrops are generally rare, occurring mostly not far from the present coastline. While the uppermost Cretaceous and Tertiary are well known from outcrops and from numerous waterboreholes, the older levels can only be studied in petroleum exploration deep wells. Good summaries of the geological history of the basin have been published by Castelain (1965) and De Spengler et al. (1966).

The oldest Mesozoic levels reached by oil boreholes are of Late Jurassic age; recently, offshore, reportedly Paleozoic (?Devonian) has been found below the Jurassic, but no detailed results have been published. From the Late Jurassic to the Maastrichtian there was a more or less continous transgression from west to east, followed by a generalised regression at the end of the Cretaceous (Fig. 3).

Onshore the deepest levels penetrated without reaching the base are of Late Jurassic age. Some 760 m of marine sediments: dolomitic limestones, often oolitic, without sandy elements and rich in organic components (algae, bryozoa, echinoderms, gastropods, ostracodes, foraminifera) were penetrated. The latter include *Pseudocyclammina jaccardi* (Schrodt) and *Iberina lusitanica* (Egger), the former a proof of the Jurassic age of these levels (Dogger–Sequanian–Early Kimmeridgian).

W
E
RUFISQUE
TIENABA
Atlantic Ocean
DAKAR
NDIASS
PALEOCENE and EOCENE
PALEOZOIC
0 m
0 m
5000 m
10,000 m
Oligocene-Eocene
Senonian
Turonian-Albian
Neocomian s.l.
Jurassic
hinge zone
SCALE
(Approx.)
5000m
0
100 km
LEGEND
Shale
Limestone
Sandy limestone
Dolomitic limestone
Sand or sandstone
Alternating sand-shale
Volcanic rocks
Crystalline and volcanic rocks

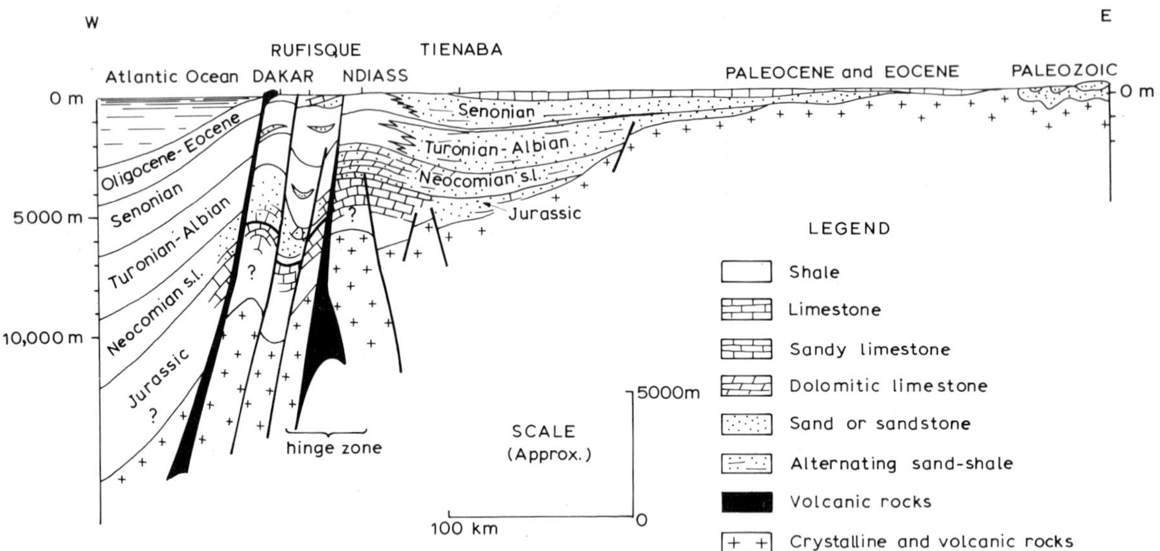

Fig. 2. Schematic cross-section through the Senegal Basin (from De Spengler et al., 1966).

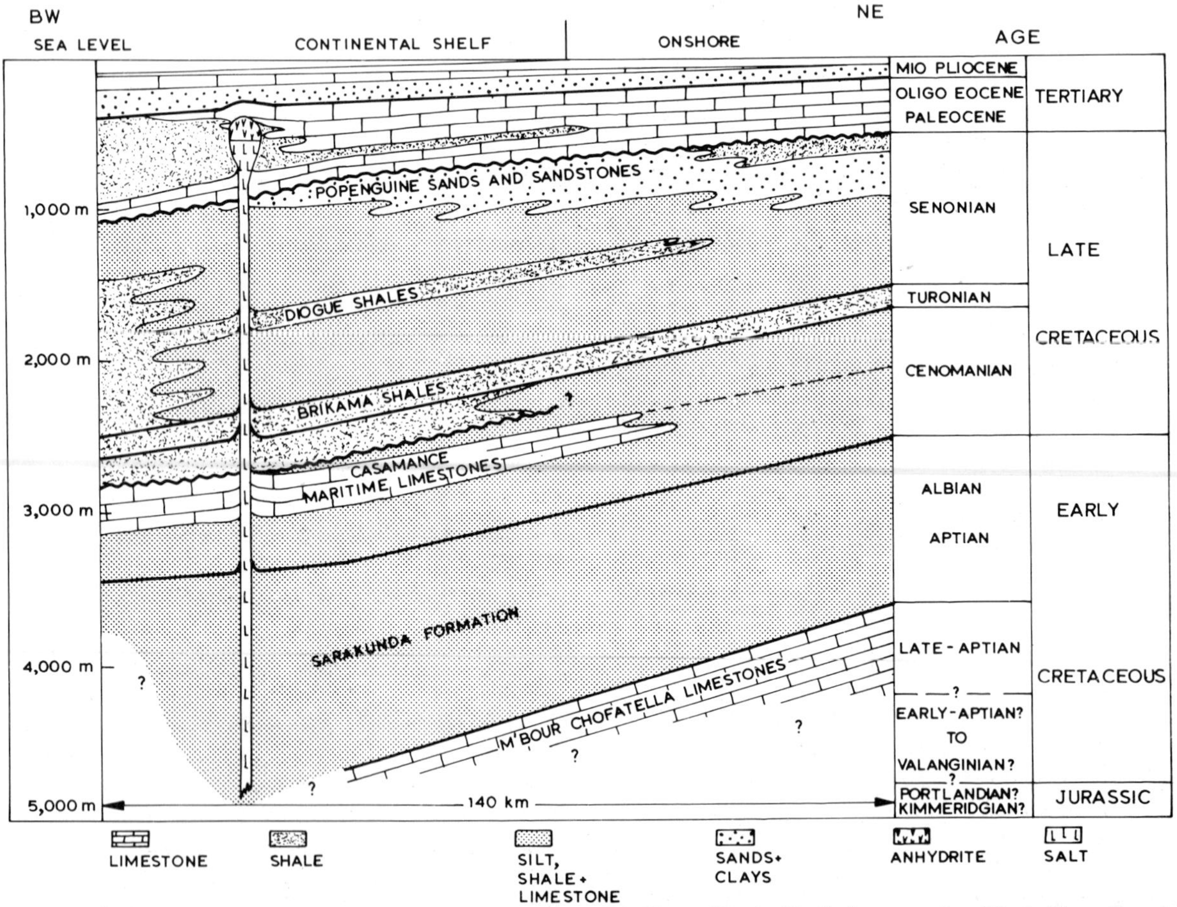

BW
NE
SEA LEVEL
CONTINENTAL SHELF
ONSHORE
AGE
MIO PLIOCENE
OLIGO EOCENE
PALEOCENE
TERTIARY
POPENGUINE SANDS AND SANDSTONES
SENONIAN
LATE
TURONIAN
DIOGUE SHALES
CRETACEOUS
CENOMANIAN
BRIKAMA SHALES
CASAMANCE
MARITIME LIMESTONES
ALBIAN
APTIAN
EARLY
SARAKUNDA FORMATION
LATE-APTIAN
CRETACEOUS
EARLY-APTIAN?
TO
VALANGINIAN?
M'BOUR CHOFATELLA LIMESTONES
PORTLANDIAN?
KIMMERIDGIAN?
JURASSIC
1,000 m
2,000 m
3,000 m
4,000 m
5,000 m
140 km
LIMESTONE
SHALE
SILT, SHALE + LIMESTONE
SANDS + CLAYS
ANHYDRITE
SALT

Fig. 3. Generalised geological section through the southern part of the Senegal Basin. Vertical exaggeration 25 : 1. (From Templeton, 1971.) (NERC copyright. Reproduced by permission of the Director, Institute of Geological Science, London SW7.)

Above the Jurassic, in the Lower Cretaceous, appear the first detritic layers, although in the west the carbonate sedimentation is still predominant. A layer of 1320 m Neocomian s.l. (Valanginian–Aptian) has been penetrated. In the upper part (Barremian–Aptian), *Choffatella decipiens* Schlumberger, associated at the top of the Aptian with *Pseudocyclammina hedbergi* Maync is common. In the Upper Aptian–Lower Albian of the borehole of Baladine Freneix (1966) describes a pelecypod fauna.

The Albian to Lower Cenomanian sequence is made up of shaly-quartzitic and calcareo-dolomitic layers as well as quartzitic sandstones in the lower part, becoming shaly or shaly–sandy in the upper part. Its maximum thickness is 1,500 m. In the lower part (Lower Albian) *Orbitolina texana* (Roemer) has been found in the west and arenaceous foraminifera dominated by *Reophax (Haplostiche?) texana* (Conrad) in the east. The upper part (Late Albian–Early Cenomanian) has been dated with the help of a planktonic microfauna including among others *Favusella washitensis* (Carsey) and *Hedbergella planispira* (Tappan).

The Upper Cenomanian–Turonian has a thickness varying between 450 and 700 m. In the west it is completely shaly, in the east shaly and sandy. Its age has been determined by both microfauna and microflora. The microfauna includes many planktonic foraminifera, mainly Heterohelicids and *Hedbergella.* In the Cenomanian part of this sequence the foraminifer *Thomasinella aegyptia* Omara has been found, in the Turonian *Gabonita levis* (De Klasz, Marie and Rerat), a species characteristic elsewhere (Gabon, Egypt, etc.) of the same levels.

The Lower Senonian–Campanian is 400–930 m thick, shaly or shaly–sandy. The limit between the Lower Senonian and Campanian has been determined by the appearance of a rich planktonic microfauna in the latter. In the east the Lower Senonian contains many ostracodes while the Campanian is sandy and contains no fossils. In the lower part of this sequence an ammonite, *Texanites* aff. *bourgeoisi* De Gross of Middle Coniacian age has been found.

The Maastrichtian shows considerable variation both in thickness and facies. While it is only 100 m thick near Dakar, it reaches a thickness of more than 2,000 m only 20 or 25 km distant and is again thicker in a sandy facies further to the east, where it represents the terminal part of the "aquiferous sands" of Senegal.

The Maastrichtian can be subdivided into a lower part, with numerous specimens of *Afrobolovina afra* Reyment, a very characteristic African foraminiferal species, and a thicker upper part, with a typical Maastrichtian pelagic foraminiferal fauna and many Buliminidae. The top of the Maastrichtian crops out south of Dakar, where some ammonites have been found: *Daradiceras gignouxi* Sornay and Tessier and *Sphenodiscus corroyi* Sornay and Tessier.

Toward the end of the Maastrichtian a generalised regression took place, followed by intense erosion. On the Cape Verde Peninsula (i.e. Dakar), more than 1,000 m of sediments are missing.

Ivory Coast Basin

The Ivory Coast Basin stretches from Fresco (Ivory Coast) to Axim (Ghana). It is about 400 km long with a maximum width onshore of only 35 km and covers about 8,000 km^2. It is in fact only the landward edge of an offshore basin of which little has been published. A "resumé" of the geological history of the basin has been published by De Spengler and Delteil (1966). There are notes on the Cretaceous palynostratigraphy by Jardiné and Magloire (1965).

The narrow onshore basin is divided lengthwise by a fault with a throw of several thousand metres into two very distinct zones (Fig. 4): a northern part, where the sedimentary cover rarely reaches 300 m and a southern one where the basement at the coast is 4,000–5,000 m below the surface.

North of the fault the sediments are mostly of Mio-Pliocene age and of continental origin. There are however two outcrops of older, marine rocks: one at Fresco at the western end of the basin, where well-dated Paleocene crops out, and one at Anwiafutu in Ghana in the east, where Cenomanian is known. South of the great fault which played such an important role in the sedimentation, lagoons, swamps or Quaternary sediments cover the older strata with the exception of the outcrop of Eboinda and Eboko (Maastrichtian bituminous sands and asphaltic limestones with numerous shell fragments of Paleocene age, respectively). In Ghana sands similar to those of Eboinda crop out at Bokakre and shelly limestones and shales of the Upper Cretaceous are known at Nandiuli.

The deeper part of the basin is only known through drilling. Subsidence started at the beginning of the Cretaceous or during the Late Jurassic, in the onshore

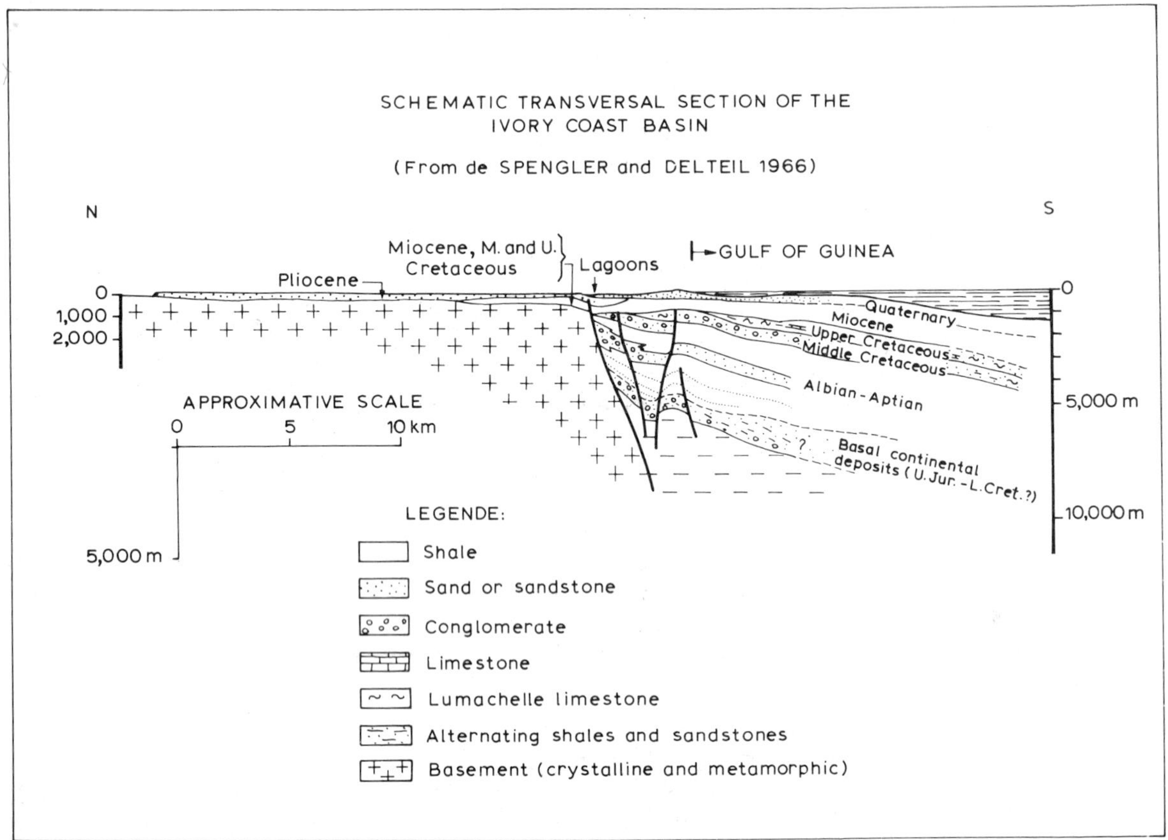

Fig. 4. Schematic transversal section of the Ivory Coast Basin (from De Spengler and Delteil, 1966).

part of the basin; no data on the offshore basin have been published.

The basal beds are essentially continental, consisting of variegated sands, sandstones, conglomerates and shales with a few intercalations of black shales. In the west these rocks directly overlie the basement (Fig. 5). In the east, the basement has never been reached, but it can reasonably be assumed that the situation is similar. This basal continental sequence is over 2,000 m thick in Ghana and 470 m in the west of the basin (borehole Adiadon I).

The age of this sequence is imperfectly known. Its top corresponds, according to palynological studies, to the marine Aptian–Albian of the centre of the basin. However, molluscs and some non-marine ostracodes in boreholes in Ghana indicate (according to H.H. Renz in De Spengler and Delteil, 1966) and older i.e. Late Jurassic–earliest Cretaceous age. Krömmelbein (1968) has described two small ostracod assemblages separated from each other by 900 m of barren sediments from a borehole of Ghana. He thinks that these two faunules (which he received from H.H. Renz) are of Cretaceous age. The limit Aptian–Albian could not be determined, but several arguments are in favour of the presence of Late Aptian.

The first marine transgression probably took place during the Aptian. The Aptian–Albian sequence is thus transgressive and probably unconformable. In the centre of the basin its thickness is more than 2,600 m. Its lower part is made up of black shales with more or less important sand intercalations. The upper part shows rapid lateral facies change: in the centre of the basin it is composed mainly of sandstones, marls and fissile shales, the latter with a typically Middle–Late Albian fauna [*Elobiceras, Dipoloceras, Neophlycticeras, Lyelliceras, Favusella washitensis* (Carsey), *Hedbergella planispira* (Tappan)]. In the upper part there are also very coarse levels with conglomerates containing elements from the basement. Hydrocarbon shows have been observed in boreholes.

The Cenomanian is probably represented by a regressive sequence of more or less coarse sediments.

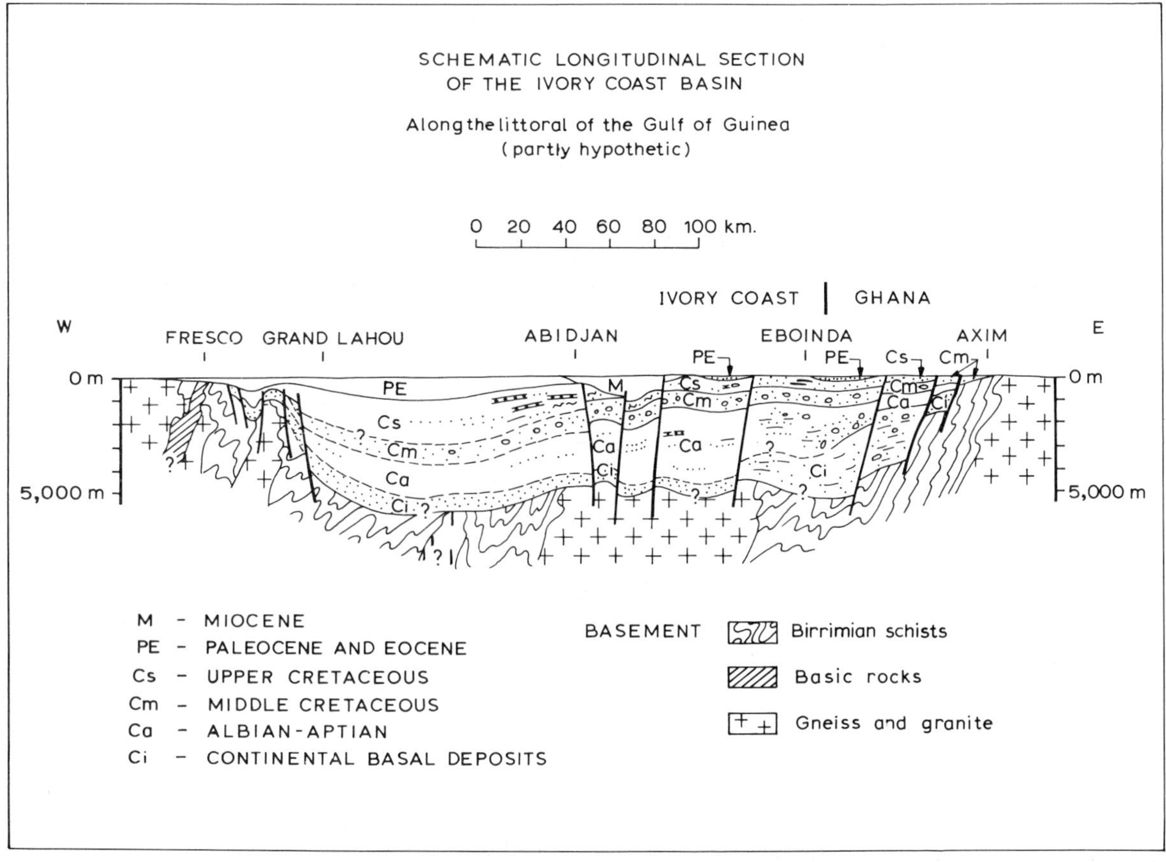

Fig. 5. Schematic longitudinal section of the Ivory Coast Basin along the coast of the Gulf of Guinea (partly hypothetical).

These are conglomerates and sands of often fluviatile origin, as well as sandy shales with in the centre of the basin sometimes sandy, dolomitic limestones. Its thickness, remarkably constant, is around 600–700 m. The rate of subsidence is less than that found in the Albian. The limits of the Cenomanian have been fixed somewhat arbitrarily.

The Turonian–Senonian–Maastrichtian sequence is again transgressive, without having such a rapid subsidence as the Aptian–Albian. In the west it overlies various parts of the continental basal rocks, or the Aptian–Albian in the centre. In the east there is an epicontinental facies with zoogenous limestones, shell banks and sands. The microfauna has its maximum development in the upper part of this unit, especially in the western part of the basin. The microflora shows marked similarities with the flora of the other coastal basins of West Africa (Senegal and Gabon). The Turonian period is defined from palynological studies, while the Coniacian–Maastrichtian interval can be dated with the help of the microfauna. Cox (1952) described a macrofauna including *Texanites* cf. *stangeri* (Baily) and numerous other molluscs of a probably Campanian age in the Nandiuli limestones.

A generalised emersion occurred at the end of the Maastrichtian.

Upper Cretaceous is sometimes found north of the great fault, always in a very thin and sandy (N̍'Zida, Lokodjo, Eboko) facies.

The Benin Basin

The Benin Basin is a group of sub-basins, defined by Hourcq (1966a) which extend from southeastern Ghana in the west across Togo, the Benin Republic (formerly Dahomey) and Nigeria to Cameroun in the east (Fig. 6). From the Niger Valley in the northwest and the Benue Trough in the northeast it extends to a

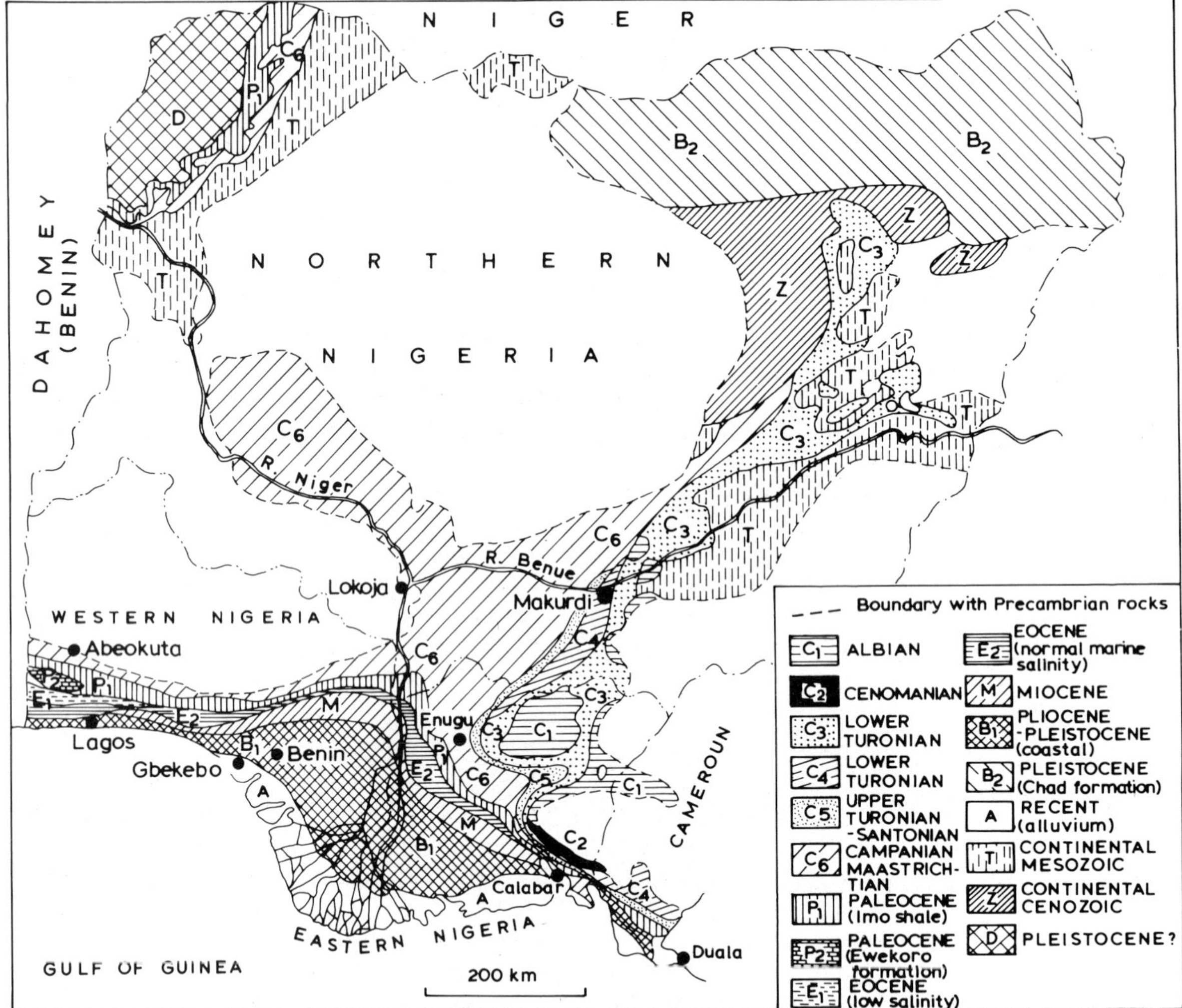

Fig. 6. Geological sketch map of Nigeria (from Reyment, 1956).

considerable distance offshore in the south. The surface area is about 350,000 km^2 on land with 1,200 km coastline.

The "Sokoto Basin" in the extreme northwest of Nigeria and the northern Benin Republic (ex-Dahomey) belongs geologically to the Iullemeden Basin, which covers a considerable part of the interior of West Africa. The northeast part of Nigeria is covered by sediments of the Lake Chad Basin.

The sub-basins are:

(1) *West Nigerian "Basin"*, often also called "*Dahomey Embayment*", which extends from southeastern Ghana in the west through southern Togo and southern Benin Republic (Dahomey) to southwestern Nigeria. It is separated in the east from the Niger Delta by a sub-surface basement high, known as the *Okitipupa Ridge*. The offshore limit is not well defined.

(2) The *Niger Delta* is the most important sub-basin from the point of view of size, thickness of sediments and economics, for the oil reserves found there provide a large part of Nigeria's national income. The apex of the Niger Delta is situated southeast of the confluence of the Niger and Benue rivers. In east–west direction it extends from the Okitipupa Ridge to southwestern Cameroun.

(3) The *Mid-Niger Valley* sub-basin reaches from the confluence of the Niger and Benue to the Kainji reservoir, where it is separated by basement rocks from the southeast part of the Iullemeden Basin.

(4) The *Benue Valley* stretches northeast from the northern Niger Delta along the Benue River to northern Cameroun. There are two large sedimentary basins within Nigeria which played an important role in the geological history of the Benin Basin, and which they are only partially separated.

(5) The Nigerian part of the Iullemeden Basin, often called *Sokoto Basin*, occupies the northwestern corner of Nigeria [and the northern Benin Republic (ex-Dahomey)].

(6) The northeastern part of the country is covered by a sedimentary sequence, part of a much larger basin covering large areas of Central Africa. It is separated from the Benue Trough by the sub-surface *Zambuk Ridge.*

These sub-basins have their own, partly independent history, and are becoming better known as a result of numerous contributions published in recent years.

The West-Nigerian "Basin" ("Dahomey Embayment")

This basin is comprehensively reviewed by Slansky (1962), especially as far as the Benin (Dahomey) part is concerned. See also numerous other papers by Reyment (1965), Dessauvagie (1975) as well as Dessauvagie and Whiteman (1972) and Kogbe (1976a).

In the western region, particularly in Togo and the Benin Republic (Dahomey) the oldest beds, on land, are of Maastrichtian age (Slansky, 1962). Offshore deep wells have shown that considerably older sediments are also present (Billman, 1976). The oldest sediments are pre-Albian, unfossiliferous folded rocks of unknown thickness and age. These are unconformably overlain by about 430 m of unnamed, predominantly non-marine sandstones of Early Cretaceous (Early to Late Albian) age, followed unconformably by 600 m of massive, marginally marine to non-marine sandstones. Near the coast these sandstones rest directly on crystalline basement rocks, seaward on the older Cretaceous sediments and are dated palynologically as Turonian. This sandstone is conformably overlain by a unit of interbedded sandy shale and sandstone, calcareous shaly sand and thin limestone bands, max. 200 m thick. It has been deposited in marginally marine conditions and is dated as Turonian to Early Senonian and has been compared with the Awgu Shales of the Niger Delta. A regression occurred above which rest about 180 m of black, fossiliferous shale with abundant Maastrichtian microfossils marking the top of the Cretaceous. This formation has been compared with the Nkporo Shales of the Niger Delta.

Onshore the Maastrichtian, about 200 m thick, rests on the basement with non-fossiliferous sands, followed by a sequence of coals, clays and marls, and capped by a fossiliferous horizon containing *Roudairea drui* and smaller Foraminifera.

The Cretaceous can be divided into two geographic zones. In the northern zone, the sequence is simple: lower sandy deposits passing gradually to upper clayey beds. Away from the interior basin margin some lignitic horizons and sometimes shales with leaves appear. Thick shelled lamellibranchs, like *Roudairea,* and ostracodes seem to indicate a shallow-water facies in this area.

The southern zone has a more complicated stratigraphy. The upper part of the Maastrichtian, entirely argillaceous further north, passes into marly beds with limestone levels. This is consistent with increasing distance from the basin margin, but does not explain the sandy horizons and sandstones absent in the north.

In addition to a north–south facies change there are west–east changes in the basal sandy beds which are far less developed in Togo than in the Benin Republic (ex-Dahomey).

From southwestern Nigeria there are no published sub-surface Cretaceous data. Presumably, the sedimentary sequence is similar to that described above for the Benin Republic.

In onshore exploration wells drilled in the border area, some thin Aptian–Albian to Cenomanian beds dated by palynology are overlain by a thick sequence of sands, sometimes with shaly intercalations. The lower and thicker part of these sands, which may exceed 1,000 m in thickness, contain mostly rare, arenaceous foraminifera, internal casts of lamellibranchs, gastropods and fish debris. It is dated as Senonian palynologically. The top part of the sequence contains a characteristic Maastrichtian benthonic microfauna, separated from the overlying Paleogene deposits by an unconformity or a depositional gap, for the base of the Paleocene is missing.

In the outcrop the Cretaceous sediments are referred to the *Abeokuta Formation* which was deposited directly on the crystalline basement. The formation begins with a basal conglomerate (1–3 m thick) overlain by coarse-grained, micaceous sandstone, often cross-bedded, with silty, clayey and mudstone intercalations. Glauconite-rich horizons are common toward the top. Over a considerable area east of Ijebu Ode, the Abeokuta Formation is impregnated with bitumen, at least in part. The rock is generally soft and friable except where locally cemented with ferruginous or siliceous matrix (Kogbe, 1974; Dessauvagie, 1975).

The formation name, Abeokuta Formation, is unfortunate, for the town of Abeokuta lies on basement rocks 16 km north of the "type locality". The type section has been compiled from data derived from correlations of three different boreholes, and is, therefore, invalid according to the rules of stratigraphic nomenclature.

In one of the three boreholes a 50 cm thick shale intercalated in the upper part of the section has been dated as Maastrichtian using foraminifera and ostracodes (Reyment, 1965). It has therefore, been assumed that the upper part of the known section (max. thickness 240 m near the Benin Republic border) is of Maastrichtian age, with the lower part possibly Senonian [1]. This formation is probably equivalent to the thick sub-surface sands found in deep exploration wells drilled further south. Here the uppermost, more shaly, part is dated Maastrichtian. A detailed palynological study of the non-weathered material, which contains a very rich microflora, should assist in establishing the relationships between the outcrop area and the offshore sections.

The Abeokuta Formation can probably be correlated with the onshore Maastrichtian of the Benin Republic and Togo. The comparison with the offshore section in that area is less certain in the absence of detailed studies.

The Niger Delta

The geological history of the Niger delta is considerably more complex than the Nigerian part of the "Dahomey Embayment". The fact that many publications contain conflicting information makes interpretation more complicated. Principal sources are Reyment (1965), Short and Stauble (1967), Murat (1972), Dessauvagie (1975), and the volumes edited by Dessauvagie and Whiteman (1972) and Kogbe (1976a).

The Niger miogeocline is thought to be underlain by oceanic crust (Reyment, 1969: Burke et al., 1972) as a result of a triple rift junction. Murat (1972) demonstrated that the Cretaceous and Early Tertiary geological history is controlled by a few major tectonic features. These are fracture zones or "hinge lines" formed and then reactivated during three major tectonic phases, the Albian, the Late Santonian–Early Campanian and the Late Eocene, when the axis of the main basin was displaced.

Mesozoic sediments are only known in the northern part of the Niger Delta. In the south, if present, they are beyond the reach of economical drilling. According to Merki (1972) Cenozoic sediments are 10,000–13,000 m thick.

While most authors consider the dominantly Albian *Asu River Group* as the oldest sediments of the Niger Delta region, Murat (1972) indicates the presence of a Lower Cretaceous pre-Albian "*Basal Sandstone*" in this area. The sandstone is separated from the Asu River Group by an unconformity and the long period of non-deposition probably encompasses most of the Aptian and the Early Albian. Unfortunately little is known about these "Basal Sandstones" although their importance in the geological history of the whole Bight of Benin region is considerable. They may be time-equivalents of offshore unfossiliferous pre-Albian rocks of the Benin Republic.

The *Asu River Group* is said to be more than 3,000 m thick (Dessauvagie, 1975). It is made up of olive brown sandy shale, fine-grained, micaceous sandstone and mudstone, poorly bedded shale with occasional lenses of sandstone and sandy limestone. The group has not been studied in detail, but ammonites indicate that it is in part of Albian age. Reyment (1965) considers that the uppermost beds are younger, comparable with the "Grès de la Cross" of Cameroun dated as Albian–Cenomanian.

The *Odukpani Formation* near Calabar in the east is a sequence of shales, sandy shales, sandstones and limestones with a coarse basal sandstone resting unconformably upon the basement. Its total thickness is about 750 m (Dessauvagie, 1972, 1975). A Cenomanian age is assigned to the lower (and larger) part of the Odukpani Formation based on micro- and macrofossil evidence. The suggestion of an Albian age for its basal sandstone is based on comparisons with similar formations in Cameroun (Diebold, 1960). The upper part of the formation has yielded Early Turonian ammonites (Reyment, 1965).

The *Eze Aku Shale* deposited in the "Abakaliki Trough" consists of hard grey or black calcareous

[1] Recently (1976) both Cretaceous and Paleocene foraminifera and palynomorphs have also been found in the upper part of the Abeokuta Formation.

shale, limestone and siltstone containing some horizons with frequent impressions of *Inoceramus.* Locally the shales pass laterally into sandstones (*Amasiri Sandstone* near Afikpo and the *Makurdi Sandstone* at Makurdi). The maximum thickness is unknown but is estimated to exceed 1,200 m (Dessauvagie, 1975). The age of this formation is Early Turonian. According to Reyment (1965), the beds were deposited in a shallow sea.

The overlying unit is the *Awgu Shale* which consists of about 900 m of bluish-grey, bedded shale with some fine-grained carbonaceous limestone beds. Locally sandstones can be predominant, as in the case of the Agbani Sandstone. Ammonites in the lower part of the formation indicate a Turonian age, whereas at the top Coniacian molluscs have been found.

The greater part of the Senonian and Maastrichtian of southern Nigeria is represented by the *Nkporo Shale*, a sequence of bluish to dark-grey shale and mudstone, with occasional thin sandy shale and sandstone intercalations. Rare thin bands of shelly limestone are also found. Laterally a change of facies takes place as the shale passes into the *Owelli Sandstone* and the *Afikpo Sandstone.*

In the region of Enugu the Upper Cretaceous is more differentiated. The lower part is shaly with sandstones, sandy shales and coal (*Enugu Shale*). Laterally the shale passes into the *Otobi* and *Owelli Sandstones.* The thickness of the formation is estimated to be 1,000 m. The age is said to be Campanian–Maastrichtian and considered as paralic equivalent of the marine Nkporo Shale, but the paleontological evidence is meagre.

The *Mamu Formation* is a paralic sequence of well-bedded fine to medium sandstone, yellow and white in colour, alternating with grey shale, mudstone and sandy shale with several coal seams. Its thickness varies from 70 to 600 m (Dessauvagie, 1975). The Maastrichtian age of the Mamu Formation is deduced more from its stratigraphic position than based on paleontological evidence. Beds of similar lithology occur in the upper Enugu Shale Sequence and in the lower part of the overlying Ajali Sandstone Formation.

The *Ajali Sandstone* is a continental sequence of 10–460 m of poorly sorted white or pinkish sandstone and siltstone with light-coloured mudstone and shale. Cross-bedding is common. The palaeontological evidence for the Maastrichtian age which has been attributed to this formation is scarce.

The highest Mesozoic formation in this area, one which extends into the Paleocene according to most authors, is the *Nsukka Formation.* It consists of fine-grained, well-bedded sandstone alternating with dark shale, sandy shale and various coal horizons. Lithologically it is similar to the Mamu Formation. At the top thin limestone layers occur locally. It is reported to reach 400 m thickness although no maximum thickness data are available.

The Mid-Niger Valley

This sub-basin has apparently a relatively simple geological history.

The most recent review is by Adeleye (1976), but other papers contain important data concerning the thickness of the sediments in the area. Adeleye (1976) describes the Mid-Niger Valley as a "shallow trough" with only 300 m of sediments, yet Adeleye and Dessauvagie (1972) and Dessauvagie (1975) mention up to 3,000 m of sediments based on airborne magnetometric data and Ojo and Ajakaiye (1976) estimate a maximum thickness of ca. 1,000 m (and never exceeding 2,000 m) on the basis of gravimetric interpretation.

According to Adeleye "the epeirogenesis responsible for the basin genesis seems closely connected with the crustal movements of the Santonian orogeny in southeastern Nigeria and the nearby Benue Valley." He suggests a rugged basement topography overlain by "rounded to subrounded coarse conglomerates, clay–sand–pebble admixtures and cross-stratified sandstones locally with scattered pebbles, cobbles and boulders." These basal strata are possibly of alluvial fan origin.

The conglomerates are conformably overlain by a sequence dominated by sandstones, with some claystones, fine conglomerates and siltstones as well as two oolitic and pisolitic ironstone levels. The whole sequence has been called *Nupe Group,* within which several formation names have been proposed (e.g. *Bida Sandstone, Sakpe Ironstone, Enagi Siltstone* and *Batati Ironstone* in the region of Bida; *Lokoja Sandstone* and *Patti Formation* near the confluence of Niger and Benue). Although the larger part of the clastic sequence is fluviatile to deltaic, brief marine transgressions occurred, indicated by the occurrence of marine fossils. The oolitic ironstones may also indicate shallow marine incursions.

The generic determination of marine molluscs indicates only a Late Cretaceous age but on the basis of plant remains in the southern part and sporomorphs further north, a Campanian–Maastrichtian age for these strata can be assigned.

The Benue Valley

In addition to the works of Reyment (1965), Dessauvagie (1975) and Offodile (1976), the origin of the Benue Trough has been discussed by Burke et al. (1972) and Wright (1976).

It is generally accepted that the Benue Trough is a rift valley, possibly a branch of a triple junction system. The extent of the rifting is controversial. Burke et al. (1972) advocate considerable sea-floor spreading (200 km), but according to Wright (1976) there was crustal stretching of no more than a few kilometres. Subsequently in Santonian times the Benue Valley narrowed as a consequence of compressional movements resulting in considerable longitudinal folding of the sedimentary fill and apparently simultaneous deformation of the underlying basement. Further folding episode occurred after the Maastrichtian. The superimposed effects of these two tectonic movements are clearly visible in the Middle Benue region.

The same tectonic events which controlled the geological history of the Niger Delta have also been dominant in the Lower Benue Valley. Here the sedimentary sequence is similar to that described from the northern, Mesozoic part of the Niger Delta.

In the Benue Valley the sedimentary fill is mainly the product of three major sedimentary cycles recording a succession of transgressions and regressions (Murat, 1972):

Coniacian–Santonian: regression (partial)
Late Turonian–Early Santonian: transgression
Early Turonian: regression
Early Turonian: transgression
Late Albian–Cenomanian: regression
Middle Albian: transgression
Pre-Aptian: transgression then regression

The succession of the Middle Benue Valley is less clear. As Offodile (1976) points out, "in this part of the Benue Valley very little is known about the geology".

The oldest sediments are again the "*Basal Sandstones*" of pre-Aptian age (Murat, 1972) and are separated from the overlying *Keana Formation* by an unconformity. The latter is often quoted as "Keana Sandstone"; it includes however also an extensive shale facies (called by Murat "Intermediate Group", and partially correlated with the Asu River Group) and therefore the name Keana Formation is probably more adequate. It is composed of "coarse-grained, feldspathic sandstone, micaceous sandstone, siltstone and shale" (Dessauvagie, 1975). The thickness has not been determined.

The formation has tentatively been dated Albian–Cenomanian, as further up-river (around Kumberi) it is overlain by unnamed marine strata with Early Turonian ammonites.

The Keana Formation is followed upwards by the "*Wukari Group*". This *informal* name has been proposed by Dessauvagie (1975) for a sequence of poorly exposed marine shales, siltstones and "transitional" sandstones and shales, which have not been systematically mapped.

No thickness data are known and the tentative age indication is "pre-Santonian".

The youngest Mesozoic sediments in the area are made up by the "*Lafia Sandstone*", another informal name for "a succession of coarse to fine-grained sandstones with occasional mudstones and black coal" deposits, for which Murat (1969) has suggested a Late Senonian age. Current palynological investigations seem to confirm this age determination. Cratchley and Jones (1965) considered these sandstones as equivalents of the above-mentioned Mamu, Ajali and Nsukka Formations and of the Gombe Sandstone of the Upper Benue Valley and Tchad Basin.

The upper Benue region has again been more studied and has more detailed biostratigraphic subdivisions.

The stratigraphic succession starts in this area with the *Bima Sandstone* unconformably overlying the basement complex. Ir varies considerably in thickness from 300 to 3,000 m. These are thick feldspathic sandstones, often cross-bedded, with somewhat finer-grained layers, as well as dark red and whitish siltstones and mudstones. In the Lamurde anticline, where the formation reaches its greatest thickness (3,000 m) it has been subdivided into three members. Carter et al. (1963) attributed an Albian–Cenomanian age to the Bima Sandstone, although few fossils have been found (sporadic fossil wood, lamellibranch fragments) and Reyment (1965) regards it as "probably older than Turonian".

The following unit, called *Yolde Formation*, marks the transition from continental to marine sedimentation (Carter et al., 1963). It consists of a variable sequence of thinly bedded sandstones, sandy mudstones and shelly limestones. The formation is 70–300 m thick. Most authors consider it Cenomanian–Turonian, basing their age assignment on lithological correlations with other, more fossiliferous beds (e.g., Gongila Formation).

The lower part of the *Gongila Formation* is a lateral equivalent of the Yolde Formation, the upper part of the Pindiga Formation. The lower, carbonaceous part of the sequence resting on the Bima Sandstone is followed by an alternating succession of sandstones and shales. The limestones contain ammonites of the earliest Turonian. The thickness of the formation is estimated as approximately 400 m.

The *Pindiga Formation* found in the northern part of the Benue Valley is composed of an alternation of fossiliferous limestones and shales with a few intercalations of siltstones and sandstones. To the south it is replaced by the *Dukul, Jessu,* and *Sekule Formations.* The first of these consists of shales and thin limestones, the second of shales and mudstones, the third of shales and limestones. The Gulani Sandstone is considered as a sandy facies of the Pindiga Formation.

The age of the Pindiga Formation is, according to Reyment (1965), Early Turonian to Maastrichtian, as both vascoceratids of the earliest Turonian and *Libycoceras* of the Early Maastrichtian have been found in it. No maximum thickness data are available.

The *Numanha Shale* (130–300 m shales with occasional bands of sandstone, nodular mudstone and limestone) conformably overlies the Sekule Formation, but has a restricted distribution. It contains a Santonian fauna. Above follows a 50 m succession of sandstones, shales and thin coal seams, called *Lamja Sandstone*, dated by spores and fish remains as Santonian. It is folded with the underlying pre-Campanian, mostly marine rocks.

The next formation is, according to Dessauvagie (1975), the *Fika Shale*, a blue-grey, occasionally gypsiferous shale, which "spills over" the Zambuk Ridge into the Chad Basin. The age assigned is based on fish and reptile remains and can only be regarded as tentative since it stratigraphically overlies the Pindiga Formation and underlies the Gombe Sandstone. The best estimate of its age is "high Upper Cretaceous (possibly Maastrichtian)" (Reyment, 1965).

Maximum thickness has been measured in a borehole in the Chad Basin (430 m).

The *Gombe Sandstone* is an estuarine and deltaic sandstone and siltstone (in outcrop often ferruginous), with some shales and locally with thin coal seams. The macrofauna consists of indeterminable lamellibranchs. The coals have been dated by palynomorphs as Late Senonian–Maastrichtian. Since it is stratigraphically above the Lower Maastrichtian Upper Pindiga Formation and is in turn underlain by the Fika Shale, a Maastrichtian age is the more probable.

Throughout the deposition of the Benue Trough sediments there was minor *volcanic activity* for thin tuffs and possibly lavas are intercalated in sediments from the Bima Sandstone to the Numanha Shale (Carter et al., 1963). The source vents are generally not known. The intrusions were accompanied by lead–zinc mineralisation.

The Chad Basin

About 10% of the Chad Basin lies in northeastern Nigeria. The geology of this basin, the largest area of inland drainage in Africa, has recently been reviewed by Matheis (1976).

The western limit of the Chad Basin is formed by the divide between the Niger and the Chad drainage systems. The southern limit is formed by the divide between the Chad and Benue systems. (In sub-surface the Chad Basin is separated from the Benue Trough by the Zambuk Ridge.)

The Cretaceous in-filling of the basin is practically identical with that of the uppermost Benue Valley.

Sediments from the latter "spill over" into the Chad Basin but are mostly covered by Tertiary to Recent strata.

The Cretaceous stratigraphic sequence is summerised in Table I.

According to Matheis (1976) the deposition of the Bima Sandstone probably began in the latest Albian. Over 1,000 m were deposited unconformably on the Precambrian basement, especially in the southwestern part of the basin (Barber, 1965).

Maximum transgression occurred during the Turonian and the Gongila Formation (a partial equivalent of the Yolde Formation) was deposited. This limestone/shale sequence is overlain by over 500 m of marine shales of the Fika Formation of Senonian–Maastrichtian age. The maximum thickness of the latter has been measured in the Chad Basin in a borehole near Maiduguri. During the later Maastrichtian the estuarine–deltaic strata of the Gombe Sandstone were deposited. According to Matheis this formation does not extend far to the east in the Chad Basin and

TABLE I

Cretaceous stratigraphic sequence of the Chad Basin

Age	Formation	Environment
Maastrichtian	unconformity Gombe Sandstone	estuarine–deltaic
Senonian–Maastrichtian	Fika Shales	marine
Turonian	Gongila Formation	marine–estuarine
Albian–Cenomanian	Bima Sandstone	continental
Pre-Albian	unconformity basement complex	

its thickness near Maiduguri is very restricted (Barber and Jones, 1960).

At the end of the Cretaceous the sediments of the Chad Basin were folded into a series of anticlines and synclines and were later partly eroded producing an erosional unconformity at the top of the Cretaceous. Cratchley (1960) suggests on geophysical grounds that the Cretaceous beds fill broad troughs within the basement. One major WSW–ENE trough has a probable depth of over 3,500 m with an estimated thickness of 3,000 m of Cretaceous. Along the flanks of the basin the Cretaceous is generally absent, the basement being directly overlain by more recent strata deposited in smaller troughs. These troughs in the Chad Basin have a similar trend to structures in the Benue Valley and may have formed either by rift faulting or folding (Turner, 1971).

While the above data give the most important traits of the Cretaceous geological history of the Nigerian part of the Chad Basin, much remains to be learned about the details. Given the recent sedimentary cover of a large part of the basin this can only be achieved with the help of deep-drilling data.

The "Sokoto Basin"

The term "Sokoto Basin" is often used in the literature for what is in fact the Nigerian part of the Iullemeden Basin of West Africa. It covers a considerable area in the northwestern part of Nigeria. It is separated from the Mid-Niger Valley by Precambrian rocks in the region of the Kainji Dam Lake on the River Niger. The geology of the region has been reviewed by Kogbe (1976b) and Dessauvagie (1975).

The sedimentary sequence consists of two continental units separated by a marine interval.

The lower unit consists of two sedimentary formations, the Gundumi and Illo Formations. The *Gundumi Formation* consists of 300 m of false-bedded, coarse-grained sandstones, arkoses, clayey sandstones, clays and shales with lignites. There is a 3 m basal conglomerate, which contains silicified wood.

Dessauvagie (1975) assigned a Turonian age to the Gundumi Formation by analogy with similar deposits in the Niger Republic. However, recent unpublished studies of fossil wood suggest a considerably older age, Early Cretaceous or possibly even Late Jurassic. Greater precision requires additional study, especially palynological.

The *Illo Formation* consists of arkosic, coarse to fine-grained current-bedded sandstones and mudstones, locally exceeding 210 m in thickness. In older works the coarser beds were frequently termed grits. The formation is divided in two by a thin, non-continuous layer (3–10 m) of pisolitic and nodular clay which has a high alumina content.

Based on lithological comparison with deposits found in the Benin Republic (ex-Dahomey) and with the Nupe Sandstone, a Senonian age has been assigned to the Illo Formation. Recent field investigations (Kogbe, 1976b) indicate however that the Gundumi and Illo Formations may be contemporaneous lateral equivalents.

The Gundumi and Illo Formations are unconformably overlain by the *Rima Group* made up of the Taloka, Dukamaje and Wurno Formations.

Dessauvagie (1975) describes the *Taloka Formation* as composed of white, fine-grained, friable sandstones with intercalations of mudstone and carbonaceous mudstone. Gypsum and ferruginous material is also present. The maximum thickness of 210 m was recorded in a borehole near the type area.

Mosasaurian bones and fish teeth suggest a Maastrichtian age, but until the sporomorph content is studied no more precise age of these beds can be assigned.

The *Dukamaje Formation* overlying the Taloka Formation consists of a dark, fossiliferous shale with white, spheroidal nodules, some thin limestone bands, and locally gypsum. The maximum thickness is only about 30 m. Where it thins out it is impossible to map the limit of the underlying Taloka and overlying

Wurno Formations with any certainty (Kogbe, 1976b).

The presence of *Mosasaurus nigeriensis*, several species of fishes and turtle fragments in a bone-bed as well as the presence of ammonites including *Lybicoceras* makes it possible to date these levels as Maastrichtian.

The overlying *Wurno Formation*, also considered Maastrichtian, is made up of friable, fine-grained sandstones with intercalations of soft mudstone and shale. It locally contains carbonaceous matter and iron sulfide, suggesting deposition under stagnant conditions (Dessauvagie, 1975). The maximum thickness, found in a borehole, is 50 m (Kogbe, 1976b).

The top of the Wurno Formation marks the end of the Mesozoic in the "Sokoto Basin". The sediments of the Rima Group are conformably overlain by deposits of the equally marine Paleogene Sokoto Group, which, however, marks a distinct change of sedimentary environment.

The Douala Basin

Whereas part of Cameroun's sedimentary basin lying west of the volcanic trend which stretches across Mount Cameroun, Fernando Poo, Ilha do Principe and São Tomé (i.e., the delta of the Rio del Rey and the offshore area opposite to it) is simply the prolongation of the Nigerian Basin, the Douala (or Wouri) Basin has a different geological history and a different stratigraphy. Summaries of data concerning the stratigraphy have been published by Belmonte (1966) and Reyre (1966a).

This is a small, more or less triangular basin situated between 3 and 5°N. Its land surface is approximately 7,000 km^2, but the basin is in fact larger, as, like the other "open" basins of the coast, it is enlarged by an offshore area, the continental platform having an average width of 25 km.

The only Aptian sediment known is rock-salt near Kribi in the extreme south, where a thick layer as been encountered in a borehole. Marine Albian has been found offshore off the Sanaga River and it is quite likely that part of the basal detrital sediments of the interior is also Albian. These basal detrital sediments seem to reach higher and higher in the sequence from north to south. The marine Albian contains mainly nannofossils and some sporomorphs.

Cenomanian sediments, identified by palynology, are only known in offshore exploration wells. Onshore the major part of the Cenomanian is probably represented together with the Albian by the basal detritic sediments called "Grès de base", which can reach a thickness of about 600 m. These detrital rocks of continental to fluviatile origin, according to Reyre (1966), occupy a similar position as the Bima of the Middle Benue Valley.

While in the region of Calabar the Turonian section is complete, according to Reyment and Barber (1956) (*fide* Reyre, 1966a), in the Douala Basin its lowermost part (vascoceratid zone) is missing or is represented by the basal detrital facies. In offshore boreholes the Turonian has been formally identified with the help of palynology and there are Turonian outcrops with ammonites and other molluscs in the region of the Mungo River well known in the literature (Lower Turonian according to Wright and Reyment). Reyment (1956) compared this fauna from the proximity of Ediki with that of the Wadatta Limestone Member of the Makurdi Formation of the Lower Benue Valley of Nigeria. Undifferentiated Turonian–Coniacian has been penetrated nearly 1,000 m at Logbaba near Douala.

The Senonian (including Maastrichtian) rocks make up the most important part of the sedimentary sequence of Cameroun. These are mainly shaly deposits with occasional sandy and very rare limestone intercalations. The thickness of this unit can be considerable; at Logbaba (near Douala) for instance it is at least 2,750 m, but 10 km to the east this thickness is reduced to 1,400 m. This is due, according to Belmonte (1966), to the Douala flexure, which has also influenced the distribution of facies. Upwards of the flexure clastic layers, often carbonate-cemented, are well developed, but the maximum development of sandy intercalations is located in the proximity of the flexure; there is a reduction of this phenomenon toward the west. The thickness of the sandy layers is very irregular. At Logbaba for example in two wells, only 500 m apart from each other, the sand thickness varies from 9 to 25 m.

Most of the Senonian and Maastrichtian is relatively well dated. The most complete outcrops are at the Mungo River, where the macrofauna of ammonites and lamellibranchs has been studied by various authors since the end of the nineteenth century (see Reyment, 1956). While the microfauna does not make possible a good delimitation of the Turonian and Coniacian, it is fairly characteristic in the Santonian, Campanian and the Maastrichtian.

The top of the Cretaceous is marked by a generalised regression.

The Gabon Basin (Table II)

The final stage of the separation of Africa and South America is nowhere reflected more clearly than in the Gabon Basin and the corresponding regions of Brazil. This is also a relatively well-known basin as, thanks to the understanding of the management of the oil company longest active in the area, much information which is usually confidential can be made public.

Among the numerous papers summing up the geological and stratigraphical results during the last two decades in addition to some very important internal reports and many specialised papers are: Freneix (1959, 1966), Belmonte et al. (1965), De Klasz and Gageonnet (1965), S.P.A.F.E./Reyre et al. (1966), Krömmelbein and Wenger (1966), Hudeley and Belmonte (1970), De Klasz and Micholet (1972), and Brink (1974).

Excluding some Karroo remnants of Paleozoic age nearly the whole coastal sedimentary basin of Gabon is in fact a product of the Mesozoic rifting and subsequent separation of the two continents. It extends onshore from Campo in the extreme southeast of Cameroun to Mayumba in southern Gabon. It is widest at the latitude of Port Gentil and the Ogooué River Valley, where it reaches 200 km.

According to Reyre (1966), the land area covers about 50,000 km^2. However, as recent information has demonstrated this is only part of a vast geological ensemble.

The Gabon Basin is partially divided into two unequal parts by a crystalline ridge, which crops out between Lambaréné and Chinchoua, although it continues below the surface at both ends. The two parts are generally called for convenience the Oriental "basin" and Atlantic "basin" although they are of course not independent basins but parts of a whole. The Atlantic "basin" is in turn divided by some flexure zones of which especially the so-called "Atlantic flexure" (or "hinge belt" of Brink, 1974) controlled the sedimentation to a large extent since Cenomanian times if not as early as the Albian.

Onshore Brink (1974) uses most of the local stratigraphic unit names mentioned in this chapter as formation names and not in the accepted chronostratigraphic sense proposed by De Klasz and Micholet (1972). In the present text local stratigraphic names refer to time-stratigraphic units if not explicitly stated otherwise.

The first Mesozoic sediments, separated from the Permian Upper Karroo by a long period of non-deposition, are restricted to the Oriental "basin". These are polygenic conglomerates filling old ravines, followed by red sandstones and finally by red to violet continental or fresh-water shales. This unit, called *M'Vone*, has a recorded thickness of 70 to about 300 m. Its age, based on some ostracode internal casts and some palynomorphs in the upper, shaly part, is believed to be Jurassic.

Another short sedimentation break separates the M'Vone from the overlying sequence of fluviatile, detritic sediments (*N'Dombo*). These coarse, feldspar-

TABLE II

Chronostratigraphic table for the Gabon Basin

LOCAL SUBDIVISIONS		GENERAL SCALE		
ALEWANA	AKOSSO	POST - MIOCENE		QUATERNARY
			NEOGENE	CENOZOIC
	N'TCHENQUÉ	MIOCENE		
	M'BÉGA			
	MANDOROVÉ			
		OLIGOCENE	PALEOGENE	
MANDJI	N'GOLA	EOCENE		
	ANIMBA			
	OZOURI			
	IKANDO	PALEOCENE		
ASSEWE	EWONGUÉ	MAASTRICHTIAN	UPPER CRETACEOUS	MESOZOIC
	POINTE CLAIRETTE	CAMPANIAN (SENONIAN)		
	ANGUILLE	SANTONIAN CONIACIAN		
	AZILÉ	TURONIAN		
REMBO N'KOMI	CAP LOPEZ	CENOMANIAN		
	MADIELA	ALBIAN	LOWER CRETACEOUS	
COCOBEACH	EZANGA (SEL)	APTIAN		
	N'ZÉMÉ ASSO			
	N'TOUM	NEOCOMIAN-BARREMIAN ("WEALDEN")		
	REMBOUÉ			
	KANGO			
KOUGOULEU	N'DOMBO			
	M'VONE		JURASSIC-TRIAS	
KARROO	AGOULA	PERMIAN	PERMO-CARBONIFEROUS	PALEOZOIC
	N'KHOM			
NOYA				PRE-CAMBRIAN?

rich, often conglomeratic deposits can be followed in the Oriental "basin" from the Ogooué River to the coast. They range in thickness from 150 to about 350 m. Only two samples have yielded a microflora of latest Jurassic to earliest Cretaceous age. Contrary to what can be seen in the Oriental "basin", these levels are not represented in the Atlantic Basin or in the adjacent Congo Basin.

The following complex is the *Cocobeach.* This name encompasses in fact two very different entities not separable for nomenclature reasons. The lower part is composed of three chronostratigraphic units:

(3) *N'Toum*
(2) *Remboue*
(1) *Kango*

These subdivisions are based on microfaunal and palynological interpretation. The fauna consists mainly of fresh-water ostracodes, occasional *Estheria*, as well as rare protoconchs of lamellibranchs and gastropods. In one well a fresh-water lamellibranch (*Unio*) has been found and in some levels fish remains [among others of *Lepidotes* sp., *Leptolepis congolensis* Arambourg and Schneegans, *Prochanos aethiopicus* (Weiler), *Chirocentrites guinensis* Weiler] are abundant.

This very thick sequence of fluviatile and – mainly – lacustrine sediments has been deposited in a huge rift-valley system, subdivided by various horsts and grabens, regarded as the original phase in the final separation of Africa and South America.

The upper part of Cocobeach (*N'Zémé-Asso* and *Ezanga* units) were deposited following a major unconformity, which affected a considerable part of the South Atlantic. The first signs of marine incursion are pseudomorphoses of salt crystals in the N'Zémé-Asso. This was followed by the precipitation of salt deposits many hundreds of metres thick (Ezanga) when saturation was reached. These huge salt masses played an important role in the geological development of extensive areas of the basin, as the active salt tectonics lasted until the Miocene. The lithology of the Cocobeach starts with a basal sandstone overlying the basement, becoming later essentially shaly with some sandy intercalations.

The pre-evaporitic part of the Cocobeach can reach a thickness of 6,000 m in some graben areas. Tectonic movements are of corresponding magnitude: the N'Toum fault has a throw of about 5,000 m.

The age of the Cocobeach is earliest Cretaceous to Middle or Late Aptian, the salt being of Middle to Late Aptian age.

The similarities of lithology, microflora and especially the microfauna with equivalent parts of Brazil is truly amazing: the same fresh-water ostracode species appear in the same order as in Gabon in the Reconcavo–Tucano and Sergipe–Alagoas Basins of Brazil (Krömmelbein, 1966; Krömmelbein and Wenger, 1966; Viana, 1966; Grékoff and Krömmelbein, 1967, etc.)

During the Late Aptian a transgression and flooding of the basin by the sea occurred, and during the Albian a considerable thickness of sediments accumulated in the Atlantic "basin"; much less – not more than 200 m – in the Oriental "basin". The maximum thickness of the Atlantic "basin" sediments is estimated at 2,900 m. These are mainly carbonaceous or shaly, and generally of shallow-water origin. In the west some deeper-water facies are developed and towards the eastern or landward margin continental facies are found. The age of this uppermost Aptian and Albian *Madiéla* sequence has been determined with the help of ammonites (including such genera as *Deshayesites* and *Elobiceras*). Other elements of the macrofauna include some lamellibranchs and gastropods. The microfauna is often rather poor, especially in the limestones. It can be composed of foraminifers or ostracodes or both, with a sprinkling of radiolaria, protoconchs of gastropods and of lamellibranchs.

The Cenomanian overlying the Albian exists only in the Atlantic "basin"; it is represented by the *Cap Lopez Stage.*

The regression which began during the Late Albian continued with the rapid progression of continental facies over a large part of the platform situated east of the "Atlantic flexure". West of the flexure the sedimentation remains marine. Between the two facies a littoral, or in some places lagoonal, transition zone exists. The end of this unit was marked by the onset of another general transgression.

The Cap Lopez Stage overlies the Late Albian carbonates dated by ammonities; its upper limit, defined by palynology, is just above the base of the transgression, which reached its peak during the Turonian. The marine sediments can be subdivided into the Lower and Upper Cap Lopez using mostly benthonic foraminifera, both calcareous and arenaceous, and in the littoral transition area with the help of ostracodes. Among the stratigraphically significant foraminifera, *Favusella washitensis* (Carsey) and *Schackoina cenomana* (Schako) should be noted; among the palynomorphs *Triorites africaensis* (Jardiné and Magloire) is the most important form of the assemblage characteristic of this unit and generally one of the best palynological markers of the Cretaceous of West Africa (Jardiné et al., 1972).

The following sequence, the *Azile Stage*, corresponds grosso modo to the Turonian with the transgression begun during the latest Cap Lopez continuing. Carbonate facies are common and may reach a thickness varying from 200 to more than 1,000 m. In some places there is a relatively rich, if not very well-preserved, macrofauna with such typically African ammonite genera as *Bauchioceras* and *Benueites* and also lamellibranchs and gastropods. The microfauna is of a transitory type: in the lower part of the Azile Stage it is rather similar to that of the Cap Lopez Stage fauna, whereas in the upper part, when it is not missing, it has Senonian affinities. It is often difficult to identify; here again palynological studies are extremely helpful.

Two local subdivisions are considered as Senonian:
(2) *Pointe Clairette (Campanian)*
(1) *Anguille (Santonian–Coniacian)*

It was a period of repeated transgressions and regressions. In the region centered on Port Gentil and further offshore a thick deltaic sequence is developed, and is the main oil producer of the country. The sedimentation is mostly shaly and silty with sandy intercalations. The thickness is very variable. It can reach well over 1,000 m.

The macrofauna of the Early Senonian is well known. South of Port Gentil, at Milango and Komandji there are numerous though mostly poorly preserved ammonites and lamellibranchs (*Inoceramus*) sometimes reaching gigantic sizes. The microfauna, often, but by no means always, poor and uncharacteristic in the Early Senonian, is generally richer in the Campanian (Pointe Clairette). There is also a characteristic microflora and nannofossils.

At the end of the Pointe Clairette period an important lowering of the sea level resulted in the deposition of detritic sediments in the deeper parts. Then with the return of the sea these sands have been buried below a shaly cover. The cumulative thickness varies between 260 and more than 750 m.

Rare ammonites and *Venericardia ameliae* (Peron), as well as some planktonic Foraminifera permitted a Maastrichtian age determination of this unit, the *Ewongué*. In the east the lateral equivalents have a more littoral type of sedimentation, though not continental.

The Congolese Coastal Basin

This basin stretches from Mayumba in southernmost Gabon across the Congo, the enclave of Cabinda, Zaïre and across the Congo River south to the region south of Ambrizete in Angola (Fig. 7). Its surface area is approximately 17,000 km^2 with a maximal width of about 100 km. Although petroleum exploration resulted in extensive studies, relatively little information has been published about this basin. The best summary is Hourcq's posthumous paper (1966b). Belmonte et al. (1965) briefly discussed certain aspects of the geology of this region, while Freneix (1959, 1966) studied the macrofauna. Krömmelbein

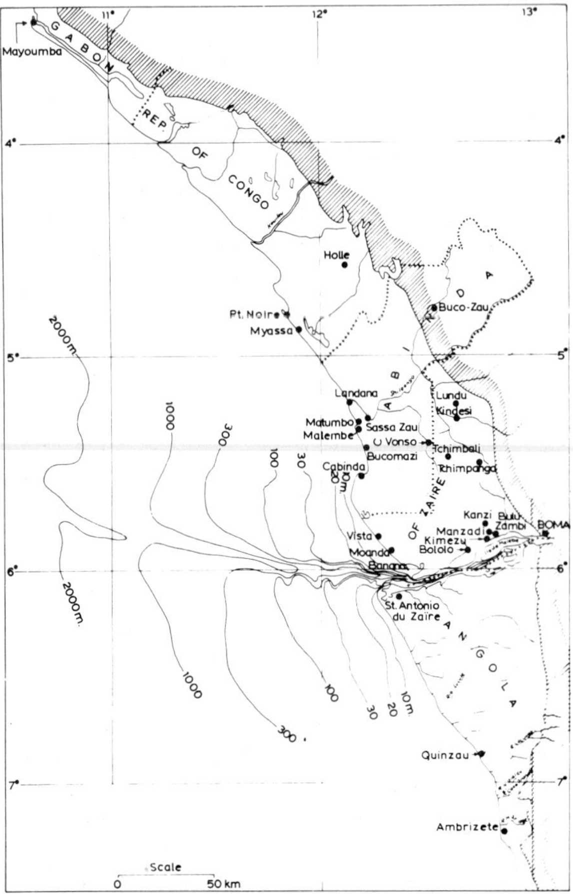

Fig. 7. The Congolese Coastal Basin.

as well as Grosdidier described in various papers some fresh-water ostracodes from the Lower Cretaceous.

The oldest sediments in outcrop occur in the Lower Congo River region of Zaïre. These, the so-called "Sublittoral Sandstones", can be divided into two units according to Cahen (see Hourcq, 1966):

(2) More or less marly argillites, violet and yellow sandstones with some cherts with plant remains.
(1) Red and pink arkosic sandstones, shaly sandstones and argillites with plant remains.

The plants have been determined as *Pagiophyllum*, a genus related to *Sphenolepidium* (Jurassic to Early Cretaceous, where it is the most abundant).

There are lateral facies variations. A greenish equivalent is called Lucula Sandstone or Loango Sandstone or Basal Sandstone.

The deposition took place on an irregular surface filling depressions, sometimes following the basement relief. Some of Cabinda's oil is produced from these basal sandstones. The contact with the basement is a normal transgressive one, often starting with a conglomerate or a conglomeratic arkose in the lower part.

Above these basal detrital beds follows a thick sequence, which is the equivalent of the Cocobeach of Gabon and is actually called so. As mentioned above, these fluvio-lacustrine sediments of the Early Cretaceous were deposited in a rift-valley system before the separation of Africa and South America. The sequence is here, however, only half as thick as in Gabon (about 2,000–2,500 m in a given place).

The part corresponding to the N'Toum of Gabon is generally missing ("South Atlantic unconformity"), but below the salt there exists a thin sequence of sandstones, conglomerates, lumachelles and shales, called the *Chela Formation*, which has yielded a small oil production near Pointe Noire. This is the equivalent of the equally oil-producing N'Zémé-Asso sequence of Gabon. Similar facies exist also in Angola in identical position.

As in Gabon, this is overlain by an evaporitic horizon, which is in some places rich in sylvinite. The salt tectonics is less intensive than in Gabon. In fact in most places the salt did not suffer any deformation, no doubt because of the low overburden. Salt domes proper are rare if not absent. The thickness of the salt is about 300–900 m, depending on the distance of the landward border of the basin. The salt sequence can be divided into rhythmically repeated sedimentary cycles. The evaporitic part is made up of alternating rock-salt and potassic salt layers. At the top there is generally a massive layer of anhydrite.

North of the Congo River and offshore the Albian has been identified with the help of palynology. Off the Congo River it has been reported to be oil bearing. South of the Congo River its existence is not proved, but can be represented by continental and fluvio-continental deposits.

North of the Congo River the Cenomanian is generally represented by sands and sandstones with thin dolomitic intercalations, which can have a reddish, more or less continental facies in some areas. Hydrocarbons have been found offshore at the top of the Cenomanian.

Non-differentiated Cenomanian–Turonian is known at the Lower Congo River. The molluscan fauna collected at Bulu–Zambi indicates a Turonian age for that part of the series. Turonian has also been encountered in the Lower Congo River region in boreholes with planktonic Foraminifera. A detrital facies crops out south of Pointe Noire and has been penetrated by offshore boreholes. No macrofauna has been found there and the microfauna is rather poor. There is however a normal microflora, making a dating possible.

In the Congo Senonian (including Maastrichtian) is known in very limited areas only offshore, around Pointe Noire and the lower Kouilou River. The outcrops at Pointe Noire belong to the Lower Senonian; they are possibly slightly younger at the lower Kouilou. In offshore wells a very thin layer of Maastrichtian overlying Lower Senonian is known.

In the Lower Congo River area of Zaïre and in Cabinda well-dated Senonian (including Maastrichtian) is known in boreholes as well as in outcrops.

The Cuanza Basin

Onshore this basin forms a narrow coastal strip from Mossulo (south of Ambrizete, Angola) to the mouth of the Rio Dande, where it becomes much larger to reach a maximum width of about 150 km (Fig. 8). It becomes again much narrower south of Porto Amboim, and extends as another narrow coastal band as far as Cabo Santa Maria. It is continued offshore, where the width of the continental platform varies between a few kilometres and 50 km (Fig. 9). Its land surface is about 22,000 km^2 (Brognon and Verrier, 1966).

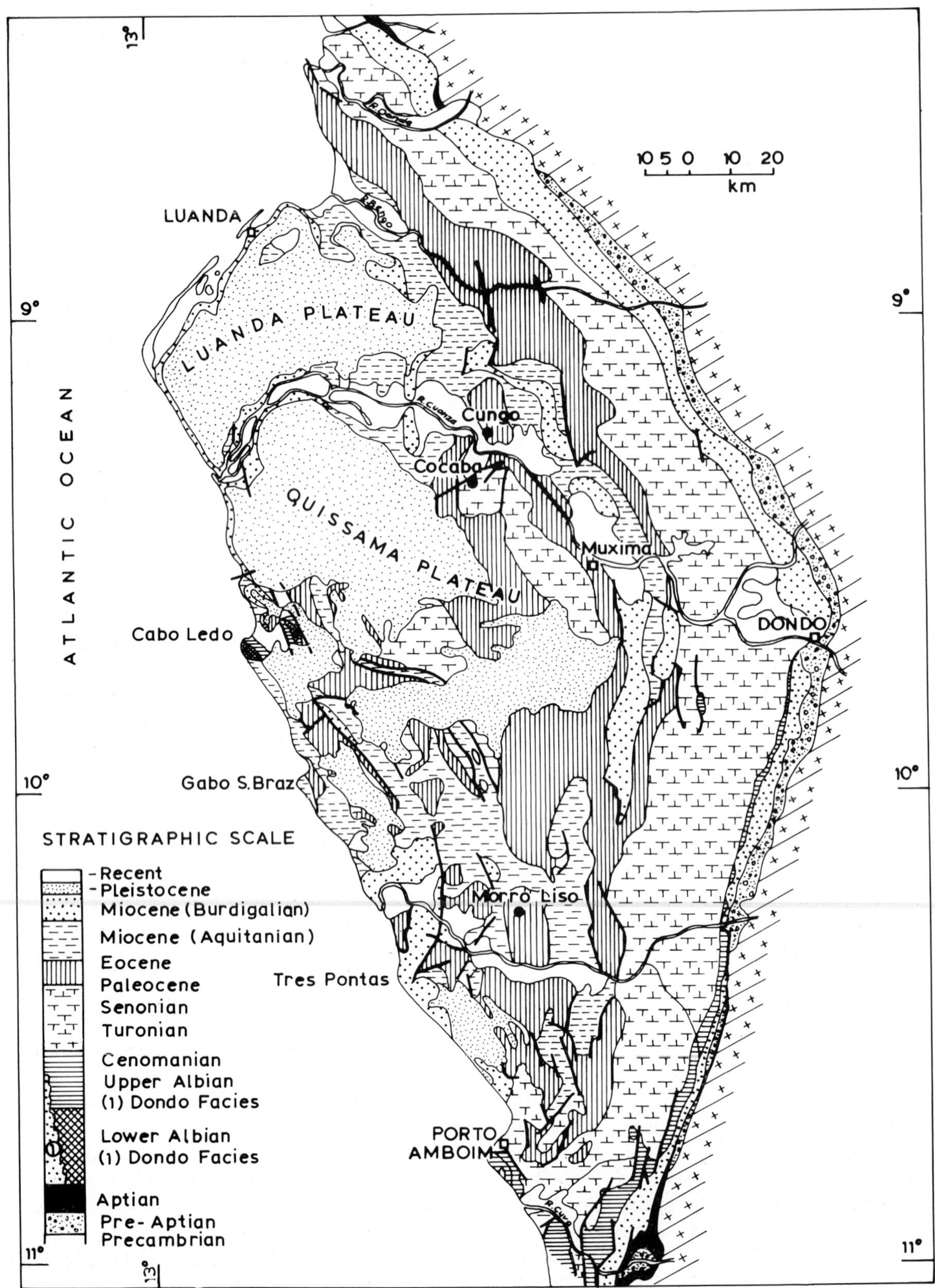

Fig. 8. Geological sketch map of the Cuanza Basin (from Brognon and Verrier, 1966).

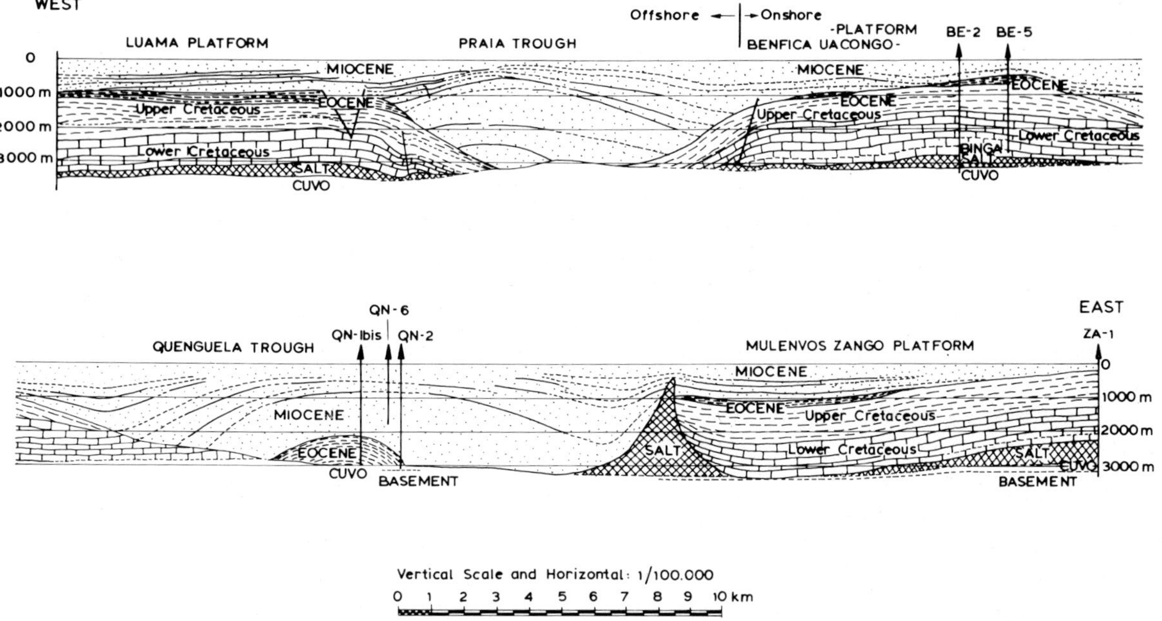

Fig. 9. Generalised cross-section through the north Cuanza Basin (Angola) (from Brognon, 1971). (NERC copyright. Reproduced by permission of the Director, Institute of Geological Sciences, London SW7.)

Several geological reviews have been published about this basin, one of the most recent by Brognon and Verrier (1966). The macrofauna has been studied by Freneix (1959, 1966) as well as by Darteville and Freneix (1957) and Freneix (in collaboration with Cahen and Hourcq, 1959). There exist also several shorter papers. Bibliographies of micropaleontological papers have been compiled by Rocha (1968, 1971). A short synthesis of micropaleontological results has been published by Hanse (1965) and more recent data presented (but not published) by Dufaure (1970).

The basic pattern of the geological history of the Cuanza Basin shows marked similarities with that of the Gabon Basin.

The sequence begins with terrigenous sediments of the *Cuvo Formation*. These have been variously deposited on the crystalline basement or on basalts and dolerites, which often cover the basement. The lower part of the *Cuvo Formation* appears in a continental facies, with coarse, red sandstones and conglomerates (sometimes interbedded with volcanic ashes). The upper Cuvo is composed mainly of sandstones. There are however also intercalations of shelly limestone often rich in ostracodes. (Unfortunately these have not been compared with Gabonese and Brazilian faunas, although they could yield interesting regional palaeogeographic clues.) The age of the Cuvo is Early Cretaceous (pre-Aptian to Early Aptian). These sediments can probably be equated to the lower and middle part of the Cocobeach of Gabon and Congo, although they.are not so thick.

Upwards follow evaporitic rocks with dolomites, anhydrite and salt deposited during the Aptian and Albian. The main salt cycle, called "sel massif" or rock salt, can reach 300–400 m in thickness, or even more. It is transgressive over the sediments of the lower or upper Cuvo or on to the gneisses of the basement or (at Morro Liso and Cabo Ledo) on dolerites. At Cabo Ledo a lateral transition of the rock salt into a carbonaceous reef facies (bioclastic and oolitic reef facies associated with algal limestones) can be observed. These sediments correspond to the Upper Cocobeach (Ezanga) of Gabon and the transgression follows the "South Atlantic unconformity" of Wenger. The halokinesis of the salt deposits considerably influenced the later geologic history of the basin.

The first two evaporitic cycles above the salt massif are easily distinguished. They are the *Quianga Formation* and the *Binga Formation*, both consisting of anhydrite and carbonates. Brognon and Verrier (1966) assigned an Aptian age for the rock salt as well as for the Quianga and Binga Formations. The other evaporitic and carbonate cycles above these layers are grouped into the *Tuenza Formation*, which is subdivided in three parts: saliferous, anhydritic and dolomitic Tuenza, the age of which is given as "Lower Albian" (Brognon and Verrier, 1966).

Overlying the Tuenza is the *Catumbela Formation*, made up of oolitic and bioclastic calcarenites and biogenous limestones. The youngest formation of this group is a unit of shaly and marly limestones called *Quissonde.* This is marine and well dated by ammonites and foraminifera. It is the result of the opening-up of the evaporitic basin. Both the Catumbela and the Quissonde are of Late Albian age.

The Upper Cretaceous and the Lower Tertiary of the Cuanza Basin are essentially represented by a bathyal to neritic sequence covering the whole present basin, thus more than the Albian. There was a continuous marine sedimentation practically from the Late Albian to the Eocene (or possibly even to the Miocene) over the major part. The littoral, deltaic or lagoon facies are restricted to the eastern edge of the basin, where several stratigraphic gaps can be observed. The age determinations during the Late Cretaceous are essentially based on micropaleontological investigations, corroborated especially in later years by new macropaleontological data.

The Cenomanian *Cabo Ledo Formation* is characterised by species of the foraminiferal genus *Rotalipora.* This is a period of generalised transgression (in contrast to the history of most other West African basins at this time). The lithology is mostly marly with thin layers of foraminiferal limestones. The following Turonian–Coniacian formation, *Itombe*, is also mainly marly. The Itombe is overlain by a unit of light-coloured marls, the *N'Golome Formation,* containing numerous Santonian Foraminifera. It is remarkably homogeneous throughout the basin and no doubt represents the largest extension of truly marine facies. This Cenomanian to Santonian period starts and finishes with well-individualised transgressive phases separated by a more or less regressive interval during the Turonian.

The sedimentation was continuing throughout the Late Senonian (including Maastrichtian) and Paleocene without major geological events. The lithology is mostly marly–silty and rather monotonous, containing numerous, often planktonic, Foraminifera, as well as *Inoceramus* fragments. The Campanian and Maastrichtian are represented by the *Teba Formation*, the lower and middle Teba corresponding to the Campanian, the upper Teba to the Maastrichtian.

The Cretaceous–Tertiary boundary is marked by a sudden change in the foraminiferal fauna.

DISCUSSION OF THE "ORIGIN" OF THE SOUTH ATLANTIC

The Mesozoic evolution of the West African coastal sedimentary basins usually shows the same basic sequence: continental or fluvio-lacustrine sediments followed by evaporites as salt water penetrated the basin, and ending with normal marine sedimentation with a pattern of transgressions and regressions. This sequence has been well described by Delteil et al. (1975) and probably fits most of the West African basins for which sufficient data are available, with possible exception for the Benin Basin. As indicated earlier, this history has been closely linked to the origin of the South Atlantic.

In some cases there may exist a Paleozoic section beneath the Mesozoic. In the Gabon Basin Early Permian and (?) Late Carboniferous remnants have been recognised in the continental succession underlying the Mesozoic, and the Lower Cocobeach contains reworked Devonian acritarchs from an unknown source. Meyerhoff and Meyerhoff (1974) indicate that marine Silurian occurs in the "Rio Muni" part of equatorial Guinea. Along the Ivory Coast pre-Mesozoic rocks are beyond the limits of economic drilling; however, Devonian rocks are known on the Ghana coast. Paleozoic rocks are known in Guinea and Sierra Leone (Templeton, 1971) and a thickness of about 1,000 m of Cambrian and possible Early Denovian is projected in coastal and offshore Liberia. In Senegal, Sierra Leone and Liberia these older rocks have been intruded by volcanic rocks. The age of the folded pre-Albian offshore the Benin Republic is not known, nor whether there are still older sediments (Billman, 1976). Although the only known borehole to penetrate the Mesozoic of Rio de Oro, near Tarfaya, passed into Precambrian below (?) Triassic, the presence of Paleozoic sediments is possible for they cover a large part of the basin. In Senegal Paleozoic (Cambrian–Devonian) is known, and Devonian (and older?) beds have been encountered below the Mesozoic offshore. Aymé (1965, fig. 3) estimates they may reach there a thickness of 1,000–2,000 m.

In the Congo Basin and the Cuanza Basin of Angola no Paleozoic is known.

The existence of the Paleozoic sediments suggests

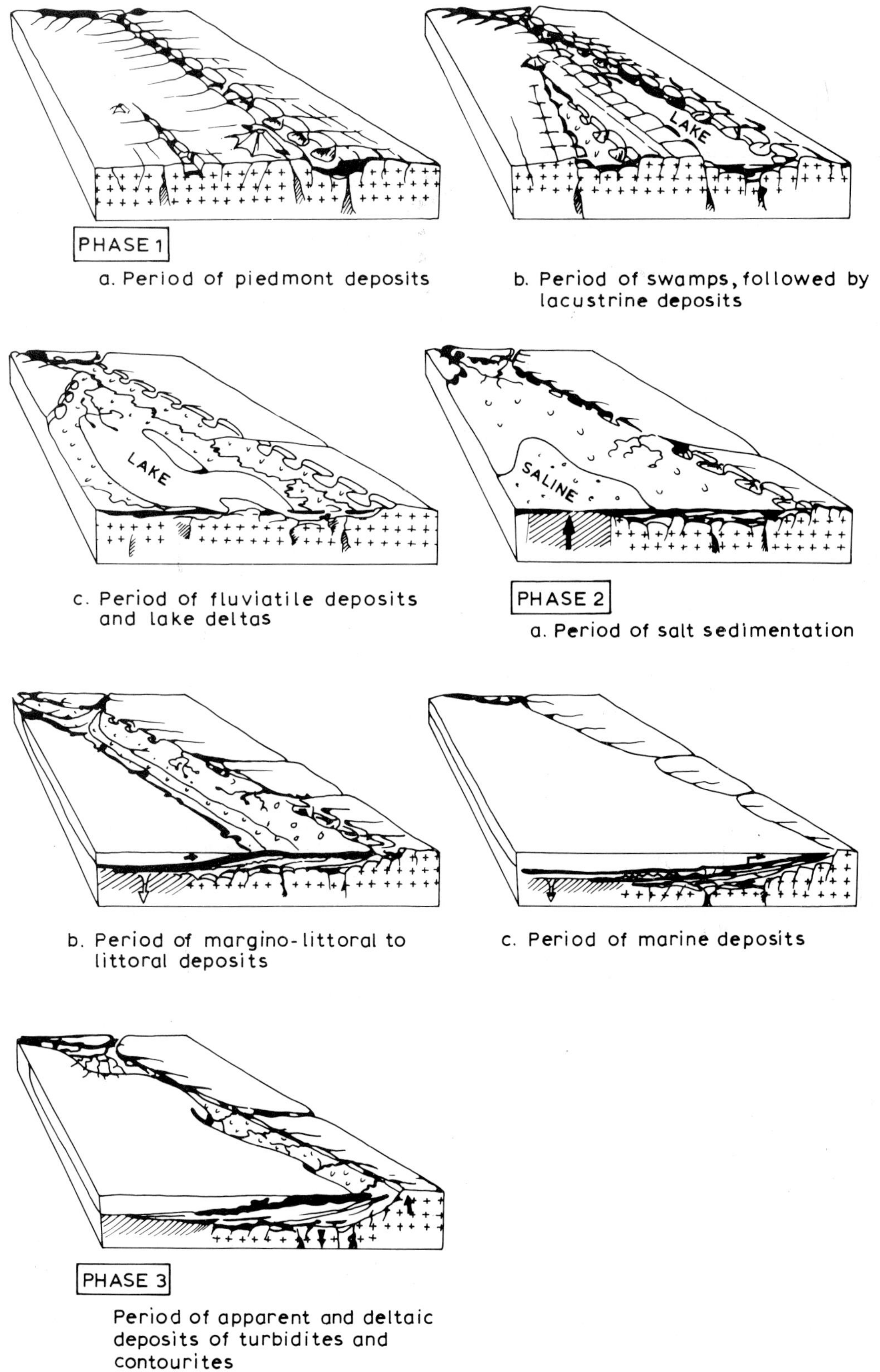

Fig. 10. Schematic interpretation of the sedimentary evolution of the Gulf of Guinea continental margin.

that, at least in the case of the basins north and west of Gabon, the Mesozoic basins occupy persistent zones of weakness which existed along what are the present day continental margins. Unfortunately, due to the sporadic nature of the information available, and the extensive erosion of the cratonic area, it is difficult to generalise about the nature of the troughs in which the Paleozoic sediments accumulated. It would be extremely interesting to know whether they were indeed elongated fault basins.

The zone of Paleozoic deposition is the one where there is the clearest indication of Mesozoic time and facies change. In Rio de Oro the earliest, red continental sediments associated with evaporites are of probable Triassic age and are followed by marine Jurassic beds. In Senegal the salt deposits are possibly Liassic whereas off Liberia the oldest Mesozoic sediments are Early Cretaceous (Albian or older). In the eastern part of the Ivory Coast Basin, thick (in excess of 2,000 m) continental red beds occur below the marine beds, and the upper part of these continental deposits may be correlated with the Aptian of the central part of the basin. The lower part is presumably pre-Aptian and may be in part Jurassic.

No sediments older than Cretaceous are known in the Benue Basin, although the folded non-marine pre-Albian rocks, of which 230 m were penetrated offshore the Benin Republic (Billman, 1976) have not been dated. In those parts of the basin where the oldest Cretaceous beds are exposed, they are present in continental or deltaic facies (e.g., Basal Sandstone and Bima Sandstone in the Benue Valley). Evaporites have not been found in the Benin Basin although saline springs and lakes do occur. As the oldest sediments are Cretaceous the source of the salt is presumably not older than Early Cretaceous. Mascle et al. (1973) report diapiric structures below the continental slope and rise about 100 km south of the Niger Delta, which are considered as possible Lower Cretaceous salt diapirs. Although Tertiary clay domes are known in the Niger Delta since the pillar-shaped structures originate in the lower part of a sedimentary section at least 4,000 m thick and penetrate much of it, Mascle's salt diapir interpretation cannot be entirely dismissed.

Salt has been found off the southernmost part of the Douala Basin and the existence of thick evaporites in the Gabon, Congo and in the Cuanza Basins is well known. The age of the evaporites is mostly Middle to Late Aptian.

This stratigraphic information can be fitted into a dynamic history of the South Atlantic which involves a long period of gradual motion resulting in rifting as the result of vertical movements deep within the crust. The process culminated with the opening of the rifts to the sea and the subsequent separation of the two continents. In the north the opening proceeded from the Rio de Oro towards the Benin Basin, seen in the progressively younger age of the oldest Mesozoic marine sediments. In the southern part of the South Atlantic rifting progressed from south to north according to Wenger (1972). The final separation may have occurred in the centre of the Gulf of Guinea, the Benin Basin. Based upon faunal links Reyment and Tait (1972) believe this occurred during Turonian times when the break occurred at the Ivory Coast Basin. The presence of a thick Aptian–Cenomanian marine section (in excess of 3,000 m) rules out this hypothesis. The date of separation, so far as it can be established geologically was in Early Albian times.

A paleogeographic model developed by Delteil et al. (1975) for the region stretching from Angola to Cameroun (Fig. 10) is probably also applicable to the northern part of the West African continental margin. Simple depressions ("Buttonholes" of Delteil et al.) opened up, into which alluvial fans descended. The central parts of the depression, undrained, became swampy or developed into lakes. The lakes filled and as fault movements continued, a drainage system developed by rivers flowing along the rifts, connecting the lakes there generating deltas.

Fluviatile erosion continued along the rift margins, eroding the higher regions and gradually filling up the rifts. Thus, there was sedimentation contemporaneous with the faulting, and in the case of the Lambaréné horst, it can be seen to be buried to a large extent in its own debris.

At the top of these non-marine deposits there is a hiatus, important enough to prompt Wenger (1972) to call it the "South Atlantic unconformity". At this time the rift stretched from the Walvis Ridge in the south to southern Cameroun, and although there is no confirmation of it, Mascle et al. (1973) suggest that the diapirs southeast of the Niger Delta may have resulted from salt deposited in the continuation of this southern basin. This hiatus is also recorded on the other side of the Atlantic.

The sea invaded the southern part of the rift. The inflow was obviously restricted, for evaporites developed. This may have been the time of the first appearance of oceanic crust. At this stage the basin stretched a distance of 1,900 km over which distance evaporites were forming, a size greater than the present Red Sea. It was also wider, and if the widths of the evaporite zones are combined, the resulting figure is of the order of 500 km.

The Aptian evaporite episode, which reached into the Albian in some basins, as in Angola, was replaced by normal marine carbonate sedimentation, implying the progressive widening of the rift, leading to a free exchange of seawater which stopped the evaporite formation. Two features however should not go unrecorded. First, that some pre-concentration of marine water occurred, for in the evaporites the sulphates are normally absent, and second, the occurrence of primary polyhalite in the Congo which implies an extremely high natural salinity.

The epicontinental carbonates of the Albian are represented by formations such as the Binga, Tuenza, Catumbela and Quissonde Formations of the Cuanza Basin, the Madiéla in Gabon or the Riachuelo Formation of the Sergipe–Alagoas Basin (Lange, 1975). Following this the ocean developed in a strict sense and its widening effectively separated the basins along the margins. The African and South American basins have from this time on their own separate history. The pericratonic half-basins created show a general down-warping oceanward. This down-warping continued with some fluctuation due to eustatic movements until the Early Tertiary. At the beginning of the Neogene a rather brusque decrease in down-warping occurred. The sea level was stabilised or even slightly lowered. Detritic sediments accumulated since the end of the evaporitic period, have built up accretional sedimentary bodies, which make up the present continental platform and the larger deltas, like the Niger Delta in Nigeria or the Ogooué Delta in Gabon. This phenomenon lead to a stabilisation of the continental margin, which reacts now only in an elastic way below the sedimentary burden, without more dramatic, deep-seated events (Delteil et al., 1975).

The above described model could possibly be applied also to the northern part of the South Atlantic, where as we have seen, the opening-up started earlier and progressed from north to south. Unfortunately, the amount of information published about the evaporites in the northern part of the South Atlantic is rather scarce. Even the geographic extent of the evaporitic deposition is practically unknown. Rona (1971) indicated the presence of diapirism north of the Cape Verde Archipelago. These diapirs have deformed sediments below the lower continental rise (4,600–5,600 m below sea level). They have been interpreted as salt diapirs of Late Triassic–Jurassic age and compared with evaporitic deposits of the Rio de Oro and Senegal Basins. The continuity of these strata (and that of the Jurassic ones found on the Cape Verde Islands) is not proved up to now. [This would mean a 1,500–2,000 km wide evaporitic (then marine) basin during Late Triassic–Jurassic times. If the Paleozoic fit of Africa and North America as proposed by Bullard et al. (1965) is admitted and the distribution of evaporitic deposits both in the above-mentioned area as well as in Florida and the Gulf of Mexico is considered, this hypothesis does not seem shocking (Templeton, 1971, figs. 2 and 3). Uchupi et al. (1976) believe however for various reasons that these diapir-like structures are in fact basement pinnacles. An important argument for this is that DSDP borehole 141 drilled above one of these structures encountered about 1.3 m of strongly weathered basalt under a few metres of undated, mottled clay underlying young Tertiary (Hayes et al., 1972). The presence of the basalt itself is however not an *absolute* proof: it must be remembered that the salt of the Rio de Oro Basin is Triassic or Early Jurassic, and that the Jurassic has been a time of widespread volcanism in Africa from the Kaokoveld lavas of Southwest Africa through the Younger Granites of Nigeria to the Liberian offshore, where the Early Cretaceous is underlain by a basalt of considerable thickness (Schlee et al., 1974).]

The continuity of the sedimentary basins south of the Senegal Basin is not known, or in many cases not published, even where oil prospection wells have been drilled. It is, however, most likely that offshore the continuity of sedimentary areas is far more extensive than indicated on most geological maps.

Whereas Lower Cretaceous non-marine ostracode faunas show a remarkable similarity from Angola to Equatorial Guinea (and also with faunas of the Reconcavo–Tucan and Sergipe–Alagoas Basins of Brazil), a faunule studied by Krömmelbein (1968) from Ghana did not show any species common with the above-mentioned areas. (This can however possibly be due to local conditions and is not an absolute proof of geographic separation.)

Sea connection across Africa, especially between the South Atlantic and the Mediterranean during the Mesozoic is also an aspect to be mentioned. The earliest connection between the Mediterranean and the Atlantic (Gulf of Guinea region) is a seaway, which opened up probably during the Early Turonian. This connection existed – possibly with occasional local interruptions – till the Maastrichtian. The environment was almost invariably shallow marine. It made possible, however, a considerable faunal interchange. Various authors (Reyment, 1965; Reyre, 1966; Machens, 1973; Kogbe, 1976a, etc.) thought that this seaway (or at least its western branch) existed till the Paleogene. However, based on faunal and micropaleontological evidence Petters (1977) convincingly demonstrated that no connection existed between the Nigerian part of the Iullemeden Basin ("Sokoto Basin") and southern Nigeria during the Maastrichtian and Paleocene, in spite of the presence of marginally marine strata in northwest Nigeria at that time.

The large amount of information derived mainly from petroleum exploration since the Second World War and especially as far as offshore exploration is concerned during the last fifteen or twenty years has probably convinced most of the doubters of the land connection of Africa and South America during the Mesozoic. It is now largely proved that the drifting apart of the two continents occurred to a considerable extent during the Cretaceous. Campanian sediments found in the South Atlantic indicate that the ocean basin had a considerable width already at that time (Nairn and Stehli, 1973b).

REFERENCES

Adeleye, D.R., 1976. The geology of the Middle Niger Basin. In: C.A. Kogbe, (Editor), *Geology of Nigeria. Proc. Conf. Geol. Nigeria, Ile-Ife 1974.* Elizabethan Publ. Co., Lagos, pp. 283–287.

Adeleye, D.R. and Dessauvagie, T.F.J., 1972. Stratigraphy of the Niger Embayment near Bida, Nigeria. In: T.F.J. Dessauvagie and A.J. Whiteman (Editors), *African Geology. Proc. Conf. Afr. Geol., Ibadan, 1970.* Dep. Geol. Univ. Ibadan, Ibadan, pp. 181–186.

Aymé, J.M., 1965. The Senegal salt basin. In: *Salt basins Around Africa.* Institute of Petroleum, London, pp. 83–90.

Barber, W., 1965. Pressure water in the Chad Formation of Bornu and Dikwa Emirates, North-Eastern Nigeria. *Geol. Surv. Niger. Bull.*, No. 35: 138 pp.

Barber, W. and Jones, D.G., 1960. The geology and hydrology of Maiduguri, Bornu Province. *Rec. Geol. Surv. Niger.*, 1958: 5–20.

Belmonte, Y.C., 1966. Stratigraphie du basin sédimentair du Cameroun. In: *Proc. 2nd W. Afr. Micropaleontol. Colloq., Ibadan 1965.* E.J. Brill, Leiden, pp. 7–23.

Belmonte, Y., Hirtz, P. and Wenger, R., 1965. The salt basins of the Gabon and the Congo (Brazzaville). In: *Salt Basins Around Africa.* Institute of Petroleum, London, pp. 55–74.

Billman, H.G., 1976. Offshore stratigraphy and paleontology of the Dahomey Embayment, West Africa. In: *Proc. 7th Afr. Micropaleontol. Colloq. Ile-Ife,* in press.

Blackett, P.M.S., Bullard, E. and Runcorn, S.K. (Editors), 1965. A symposium on continental drift. *Philos. Trans. R. Soc., Ser. A,* 258: 329 pp.

Brink, A.H., 1974. Petroleum geology of the Gabon basin. *Am. Assoc. Pet. Geol. Bull.*, 58 (2): 216–235.

Brognon, G. and Verrier, G., 1966. Tectonique et sédimentation dans le bassin du Cuanza (Angola). In: D. Reyre (Editor), *Sedimentary Basins of the African Coasts, 1. Atlantic Coast.* Assoc. Afr. Geol. Surv., Paris, pp. 207–252.

Bullard, E.C., Everett, J.E. and Smith, A.G., 1965. The fit of continents around the Atlantic. *Philos. Trans. R. Soc., Ser. A,* 258 (1088): pp. 41–51.

Burk, C.A. and Drake, C.L. (Editors), 1974. *The Geology of Continental Margins.* Springer-Verlag, Heidelberg, 1009 pp.

Burke, K.C., Dessauvagie, T.F.J. and Whiteman, A.J., 1972. Geological history of the Benue Valley and adjacent areas. In: T.F.J. Dessauvagie and A.J. Whiteman (Editors), *African Geology. Proc. Conf. Afr. Geol., Ibadan, 1970.* Dep. Geol., Univ. Ibadan, Ibadan, pp. 187–205.

Carter, J.D., Barber, W., Tait, E.A. and Jones, G.P., 1963. The geology of parts of Adamawa, Bauchi and Bornu Provinces in North-eastern Nigeria. *Geol. Surv. Niger. Bull.*, No. 30: 1–108.

Castelain, J., 1965. Aperçú stratigraphique et micropaléontologique du bassin du Sénégal. Historique de la découverte paléontologique. In: *1er Colloq. Int. Micropaléontol. Ouest Afr. Dakar, 1963. Mém. Bur. Rech. Géol. Min.*, No. 32: 135–159.

Cox, L.P., 1952. Cretaceous and Eocene fossils from the Gold Goast. *Geol. Surv. Gold Coast Bull.*, No. 17: 68 pp.

Cratchley, C.R., 1960. Geophysical survey of the Southwestern part of the Chad Basin. *C.C.T.A. Publ.*, No. 31. (*fide* Matheis, 1976).

Cratchley, C.R. and Jones, G.P., 1965. An interpretation of the geology and gravity anomalies of the Benue Valley, Nigeria. *Overseas Geol. Surv. Geophys. Pap.*, No. 1: 1–26.

Darteville, E. and Freneix, S., 1957. Lamellibranches fossiles du Crétacé de la côte occidentale d'Afrique (du Cameroun à l'Angola). *Ann. Mus. R. Congo Belge, Sér. in 8e, Sci. Géol.*, 20: 271 pp. (Inoceramidae by J. Sornay, pp. 56–61).

De Klasz, I. (with R. Gageonnet), 1965. Biostratigraphie du bassin gabonais. In: *1er Colloq. Int. Micropaléontol. Ouest-Afr., Dakar, 1963. Mém. Bur. Rech. Geol. Min.*, No. 32: 277–303.

De Klasz, I. and Micholet, J., 1972. Eléments nouveaux concernant la biostratigraphie du bassin gabonais. *IV Colloq. Afr. Micropaléontol., Act. Congr.* pp. 119–143.

Delteil, J.R., Le Fournier, J. and Micholet, J., 1975. Schéma dévolution sédimentaire d'une marge continental stable:

exemple type du Golfe de Guinée, de l'Angola au Cameroun. *IXe Congr. Int. Sédimentol., Nice, 1975, Thème 4,* pp. 91–98.

De Spengler, A. and Delteil, J.R., 1966. Le bassin secondaire–tertiaire de Côte d'Ivoire (Afrique Occidentale). In: D. Reyre (Editor), *Sedimentary Basins of the African Coasts, 1. Atlantic Coast.* Assoc. Afr. Geol. Surv., Paris., pp. 99–113.

De Spengler, A., Castelain, J., Cauvin, J. and Leroy, M., 1966. Le bassin secondaire–tertiaire du Sénégal. In: D. Reyre (Editor), *Sedimentary Basins of the African Coasts, 1, Atlantic Coast.* Assoc. Afr. Geol. Surv., Paris, pp. 80–94.

Dessauvagie, T.F.J., 1972. Biostratigraphy of the Odukpani (Cretaceous) type section, Nigeria. In: T.F.J. Dessauvagie and A.J. Whiteman (Editors) *African Geology. Proc. Conf. Afr. Geol., Ibadan, 1970* Dep. Geol. Univ. Ibadan, Ibadan, pp. 207–218.

Dessauvagie, T.F.J., 1975. Explanatory note to the geological map of Nigeria, scale 1 : 1,000,000. *J. Min. Geol.,* 9 (1972): 3–28 (2 geol. maps 1 : 1,000,000).

Dessauvagie, T.F.J. and Whiteman, A.J. (Editors), 1972. *African Geology. Proc. Conf. Afr. Geol., Ibadan, 1970.* Dep. Geol., Univ. Ibadan, Ibadan, 668 pp.

Diebold, P., 1960. Notes on the geology of southern Cameroons. *Afr. West Cent. Reg. Comm. Geol., 3rd Meet., Publ.* No. 55: 16–17.

Franks, S. and Nairn, A.E.M., 1973. The equatorial marginal basins of West Africa. In: A.E.M. Nairn and F.G. Stehli (Editors), *The Ocean Basins and Margins, 1. The South Atlantic.* Plenum Press, New York, N.Y., pp. 301–350.

Freneix, S. (with Hourcq, V. and Cahen, L.), 1959. Mollusques fossiles du Crétacé de la côte occidentale d'Afrique du Cameroun à l'Angola. III. Conclusions stratigraphiques et paléontologiques. *Ann. Mus. Congo Belge, Sér. in 8°, Sci. Géol.,* 24: I–XV; 1–126.

Freneix, S., 1966. Faunes de bivalves et corrélations des formations marines du Crétacé des bassins côtiers de l'Ouest Africain. In: D. Reyre (Editor), *Sedimentary Basins of the African Coasts, 1. Atlantic Coast.* Assoc. Afr. Geol. Surv., Paris, pp. 52–78.

Freneix, S., 1972. Les Mollusques bivalves crétacés du bassin côtier de Tarfaya (Maroc Méridional). *Notes Mém. Serv. Géol. Maroc,* No. 228: 49–255.

Grékoff, N. and Krömmelbein, K., 1967. Des Ostracodes de la série de Cocobeach, Gabon, Afrique Equatoriale, et de la série de Bahia, Brésil. *Rev. Inst. Fr. Pét. (et Ann. Combust. Liquides),* 22: 1307–1353.

Hanse, A., 1965. Les microfaunes en Angola. In: *1er Colloq. Int. Micropaléontol. Ouest-Afr., Dakar, 1963. Mém. Bur. Rech. Géol. Min.,* No. 32: 327–334.

Hayes, D.E. et al., 1972. 7. Site 140. In: *Initial Reports of the Deep Sea Drilling Project, XIV.* U.S. Government Printing Office, Washington, D.C., pp. 179–215.

Hottinger, L., 1965. Evolution et variation morphologique des *Palmula* et *Flabellinella* du Coniacien et du Santonien de Tarfaya (Maroc Méridional). In: *1er Colloq. Int. Micropaléontol. Ouest-Afr. Dakar, 1963. Mém. Bur. Rech. Géol. Min.,* No. 32: 101–111.

Hourcq, V., 1966a. Enumération des bassins sédimentaires côtiers ouest-africains. In: D. Reyre (Editor), *Sedimentary Basins of the African Coasts, 1. Atlantic Coast.* Assoc. Afr. Geol. Surv., Paris, pp. 1–3.

Hourcq, V., 1966b. Le bassin côtier congolais. In: D. Reyre (Editor), *Sedimentary Basins of the African Coasts, 1. Atlantic Coast.* Assoc. Afr. Geol. Surv., Paris, pp. 197–206.

Hudeley, H. (with Belmonte, Y.C.), 1970. Carte géologique de la République Gabonaise au 1/1.000.000. *Mém. Bur. Rech. Géol. Min.,* No. 72.

Jardiné, S. and Magloire, L., 1965. Palynologie et stratigraphie du Crétacé des bassins de Sénégal et de Côte d'Ivoire. In: *1er Colloq. Int. Micropaléontol. Ouest-Afr., Dakar, 1963. Mém. Bur. Rech. Géol. Min.,* No. 32: 187–245.

Jardiné, S., Doerenkamp, A. and Legoux, O., 1972. Le genre *Hexaporotricolpites* Boltenhagen 1967. Morphologie, systématique, stratigraphie et extension géographique. *IV. Coll. Afr. Micropaléontol. Act. Congr.,* p. 175–194.

Kogbe, C.A., 1974. The Upper Cretaceous sediments of South-Western Nigeria. *The Nigerian Field,* XXXIX (4): 169–179.

Kogbe, C.A. (ed.), 1976a. *Geology of Nigeria. Proc. Conf. Geol. Nigeria, Ile-Ife, 1974.* Elizabethan Publ. Co., Lagos, 436 pp.

Kogbe, C.A., 1976b. Outline of the geology of the Iullemeden Basin in North-Western Nigeria. In: C.A. Kogbe (Editor), *Geology of Nigeria. Proc. Conf. Geol. Nigeria, Ile-Ife, 1974.* Elizabethan Publ. Co., Lagos, pp. 331–338.

Kogbe, C.A., 1976c. Palaeogeographic history of Nigeria from Albian times. In: C.A. Kogbe (Editor), *Geology of Nigeria. Proc. Conf. Geol. Nigeria, Ile-Ife, 1974.* Elizabethan Publ. Co., Lagos, pp. 331–338.

Krömmelbein, K., 1966. On "Gondwana-Wealden" Ostracoda from NE Brazil and West Africa. *Proc. 2nd W. Afr. Micropaléontol. Colloq., Ibadan, 1965.* E.J. Brill, Leiden, pp. 113–118.

Krömmelbein, K., 1968. The first non-marine Cretaceous Ostracods from Ghana, West Africa. *Paleontology,* 11 (pt.2): 259–263.

Krömmelbein, K. and Wenger, R., 1966. Sur quelques analogies remarquables dans les microfaunes crétacées du Gabon et du Brésil Oriental (Bahia et Sergipe). In: D. Reyre (Editor), *Sedimentary Basins of the African Coasts, 1. Atlantic Coast.* Assoc. Afr. Geol. Surv., Paris, pp. 193–196.

Lange, F.W., 1975. Stratigraphy of the Cretaceous sedimentary basins of Brazil. *Proc. Vth Afr. Colloq. Micropaléontol., Addis Ababa, 1972,* pp. 565–621.

Lehmann, R., 1965. Résultats d'une étude des Globotruncanidés du Crétacé Supérieur de la Province de Tarfaya (Maroc Méridional). In: *1er Colloq. Int. Micropaléontol. Ouest-Afr., Dakar, 1963. Mém. Bur. Rech. Géol. Min.,* No. 32: 113–117.

Machens, E., 1973. The geologic history of the marginal basins along the North shore of the Gulf of Guinea. In: A.E.M. Nairn and F.G. Stehli (Editors), *The Ocean Basins and Margins, 7. The South Atlantic.* Plenum Press, New York, N.Y., pp. 351–390.

Martinis, B. and Visintin, V., 1966. Données géologiques sur le bassin sédimentaire côtier de Tarfaya (Maroc Méridional). In: D. Reyre (Editor), *Sedimentary Basins of the African Coasts, 1. Atlantic Coast.* Assoc. Afr. Geol. Surv., Paris, pp. 13–26.

Mascle, J.R., Bornhold, B.D. and Renard, V., 1973. Diapiric structures off Niger Delta. *Am. Assoc. Pet. Geol. Bull.*, 57 (9): 1672–1678.

Masoli, M., 1965. Sur quelques Ostracodes fossiles mésozoiques (Crétacé) du bassin côtier de Tarfaya (Maroc Méridonial). In: *1er Colloq. Int. Micropaléontol. Ouest-Afr. Dakar, 1963. Mém. Bur. Rech. Géol. Min.* No. 32: 119–134.

Matheis, G., 1976. Short review of the geology of the Chad Basin in Nigeria. In: C.A. Kogbe (Editor), *Geology of Nigeria. Proc. Conf. Geol. Nigeria, Ile-Ife, 1974.* Elizabethan Publ. Co., Lagos, pp. 289–294.

Merki, P., 1972. Structural geology of the Cenozoic Niger Delta. In: T.F.J. Dessauvagie and A.J. Whiteman (Editors), *African Geology. Proc. Conf. Afr. Geol., Ibadan, 1970*, pp. 635–646.

Meyerhoff, A.A. and Meyerhoff, H.A., 1974. Test of plate tectonics. In: *Plate Tectonics – Assessments and Reassessments. Am. Assoc. Pet. Geol. Mem.*, No. 23: 43–145.

Murat, R.C., 1969. *Geological Map of the Southern Nigerian Sedimentary Basin. Scale 1 : 1,000,000.* Shell–BP Pet. Devel. Co. of Nigeria Ltd.

Murat, R.C., 1972. Stratigraphy and palaeogeography of the Cretaceous and Lower Tertiary in Southern Nigeria. In: T.F.J. Dessauvagie and A.J. Whiteman (Editors), *African Geology. Proc. Conf. Afr. Geol., Ibadan, 1970*, pp. 251–266.

Nairn, A.E.M. and Stehli, F.G. (Editors), 1973a. *The Ocean-Basins and Margins, 1. The South Atlantic.* Plenum Press, New York, N.Y. 583 pp.

Nairn, A.E.M. and Stehli, F.G., 1973b. A model for the South Atlantic. In: A.E.M. Nairn and F.G. Stehli (Editors), *The Ocean Basins and Margins, 1. The South Atlantic.* Plenum Press, New York, N.Y., pp. 1–24.

Offodile, M.E., 1976. A review of the geology of the Cretaceous of the Benue Valley. In: C.A. Kogbe (Editor), *Geology of Nigeria. Proc. Conf. Geol. Nigeria, Ile-Ife, 1974.* Elizabethan Publ. Co., Lagos, pp. 319–330.

Ojo, S.B. and Ajakaiye, D.E., 1976. Preliminary interpretation of gravity measurements in the Middle Niger basin area, Nigeria. In: C.A. Kogbe (Editor), *Geology of Nigeria. Proc. Conf. Geol. Nigeria, Ile-Ife, 1974.* Elizabethan Publ. Co., Lagos, pp. 295–307.

Petters, S.W., 1977. Ancient seaway across the Sahara. *The Nigerian Field*, 42 (part 1): 22–30.

Querol, R., 1966. Regional geology of the Spanish Sahara. In: D. Reyre (Editor), *Sedimentary Basins of the African Coasts, 1. Atlantic Coast.* Assoc. Afr. Geol. Surv., Paris, pp. 27–39.

Reyment, R.A., 1956. On the stratigraphy and paleontology of the Cretaceous of Nigeria and the Cameroons, British West Africa. *Geol. Fören. Stockh. För.*, 78: 17–96.

Reyment, R.A., 1965. *Aspects of the Geology of Nigeria.* Ibadan Univ. Press, Ibadan, 145 pp.

Reyment, R.A., 1969. Ammonite biostratigraphy, continental drift and oscillatory transgressions. *Nature*, 224 (5215): 137–140.

Reyment, R.A. and Barber, W., 1956. Nigeria and Cameroons. In: *Lexique Stratigraphique International, 4. Afrique*, pp. 35–59.

Reyment, R.A. and Tait, E.A., 1972. Biostratigraphical dating of the early history of the South Atlantic Ocean. *Philos. Trans. R. Soc., Ser. B. Biol. Sci.*, 264 (858): 55–95.

Reyre, D., 1966a. Histoire géologique du bassin de Douala (Cameroun). In: D. Reyre (Editor), *Sedimentary Basins of the African Coasts, 1. Atlantic Coast.* Assoc. Afr. Geol. Surv., Paris, pp. 143–161.

Reyre, D., 1966b. Particularités géologiques des bassins côtiers de l'Ouest Africain. In: D. Reyre (Editor), *Sedimentary Basins of the African Coasts, 1. Atlantic Coast.* Assoc. Afr. Geol. Surv., Paris, pp. 253–301.

Rocha, A.T., 1968. Contribuição para o conhecimento da bibliografia micropaleontológica de Portugal. *Bol. Inst. Invest. Cient. Angola*, 5 (1): 67–92.

Rocha, A.T., 1971. Contribuição (2ª) para o conhecimento da bibliografia micropaleontológica de Portugal. *Bol. Inst. Invest. Cient. Angola*, 8 (1): 1–15.

Rona, P.A., 1971. The continental margin between the Canary and Cape Verde Islands and symmetries with eastern North America. In: F.M. Delany (Editor), *The Geology of the East Atlantic Continental Margin (ICSU/SCOR Working Party 31 Symp., Cambridge 1970), 4. Africa. Nat. Environ. Res. Couc., Inst. Geol. Sci. Rep.*, No. 70/16: 37–42.

Schlee, J., Behrendt, J.C. and Robb, J.M., 1974. Shallow structure and stratigraphy of the Liberian continental margin. *Am. Assoc. Pet. Geol. Bull.*, 58 (4): 708–728.

Short, K.C. and Stauble, A.J., 1967. Outline of geology of the Niger Delta. *Am. Assoc. Pet. Geol. Bull.*, 51 (5): 761–779.

Slansky, M., 1962. Contribution à l'étude géologique du bassin sédimentaire côtier du Dahomey et du Togo. *Mém. Bur. Rech. Géol. Min.*, No. 11: 1–270.

S.P.A.F.E. (Reyre, D. with Belmonte, Y., Derumaux, F. and Wenger, R.), 1966. Evolution géologique du bassin gabonais. In: Reyre, D. (Editor), *Sedimentary Basins of the African Coasts, 1. Atlantic Coast.* Assoc. Afr. Geol. Surv., Paris, pp. 171–191.

Tarling, D.H. and Runcorn, S.K. (Editors), 1973. *Implications of Continental Drift to the Earth Sciences.* Academic Press, London, 1184 pp. (2 vols.).

Templeton, R.S.M., 1971. Geology of the continental margin between Dakar and Cape Palmas. In: F.M. Delany (Editor), *The Geology of the East Atlantic Continental Margin (ICSU/SCOR Working Party 31 Symp., Cambridge, 1970), 4. Africa. Nat. Environ. Res. Counc., Inst. Geol. Sci. Rep.*, No. 70/16: 47–60.

Turner, D.C., 1971. The sedimentary cover in Nigeria. In: *Tectonics in Africa.* UNESCO Publ., Paris, pp. 403–405.

Uchupi, E., Emery, K.O., Bowin, C.O. and Phillips, J.D., 1976. Continental margin off Western Africa: Senegal to Portugal. *Am. Assoc. Pet. Geol. Bull.*, 60 (5): 809–878.

Viana, C.F., 1966. Stratigraphic distribution of Ostracoda in the Bahia Supergroup (Brazil). *Proc. 2nd W. Afr. Micropaleontol. Colloq., Ibadan, 1965.* E.J. Brill, Leiden, pp. 240–256.

Viotti, C., 1965. Microfaunes et microfaciès du sondage

Puerto Cansado 1, (Maroc Méridional, Province de Tarfaya). In: *1er Colloq. Int. Micropaléontol. Ouest-Afr., Dakar, 1963. Mém. Bur. Rech. Géol. Min.*, No. 32: 29–59.

Wenger, R., 1972. *Géologie du bassin sédimentaire côtier gabonais.* Internal Report ELF R.E. (unpublished).

Wright, T.B., 1976. Origins of the Benue Trough – a critical review. In: C.A. Kogbe (Editor), *Geology of Nigeria. Proc. Conf. Geol. Nigeria, Ile-Ife, 1974.* Elizabethan Publ. Co., Lagos, pp. 309–317.

Chapter 11

SOUTH AFRICA

R.V. DINGLE

INTRODUCTION

The Mesozoic sediments of South Africa comprise two lithologically and geographically distinct successions: the Triassic to Lower Jurassic sequence of the Karroo and Zambezi Basins; and the Jurassic to Cretaceous sequences along the continental margin and coastal basins (Fig. 1). The former are continental deposits and are well-exposed in the central and northern parts of the country, where they have been extensively studied since the mid-nineteenth century. The latter are mainly marine, with continental and paralic facies only in the lower parts, and although the Algoa (southeast Cape Province) and Zululand/Mozambique coastal Mesozoic basins had been described in the nineteenth century, it was not until academic and commercial exploration of the continental shelf began in 1967, that thick, widespread Mesozoic sediments were proved offshore. Relatively little is still known about the smaller basins on the continental shelf off the east coast.

In southern Africa, the Palaeozoic/Mesozoic boundary falls within the Karroo Supergroup: a well-defined lithostratigraphic subdivision which has, until recently, been used by South African geologists in the sense of a system. It comprises four groups (Dwyka, Ecca, Beaufort and Stormberg), and ranges in age from Early Carboniferous to Jurassic. A biostratigraphic zonation has been established only for the Beaufort Group (Permo-Triassic), where subdivision is based on evolutionary trends within the prolific reptilian assemblages. Although no direct correlation can be achieved with the standard international molluscan biozones, good comparisons can be made with vertebrate faunas in North America, Europe and Asia. Most recent zonations (e.g., C.B. Cox, 1967; Kitching, 1970) place the Permian/Triassic boundary between the Lower and Middle divisions of the Beaufort Group (i.e. between the Middleton and Balfour Formations), in contrast to earlier classifications where the boundary was placed within, or at the top of the Middle Beaufort (e.g., Du Toit, 1954).

Since the breakup of Gondwanaland the Triassic Beaufort and Stormberg sediments and overlying Triassic/Jurassic lavas have occupied an elevated, intracontinental position. This has lead to the vigorous erosion, diminution and isolation of their outcrop, with the result that only where they are flexured across the "Lebombo Line", are the Triassic and Jurassic sediments and lavas seen to underlie Lower Cretaceous sediments. Since Jurassic times, the continental margin basins have, to a large extent, been involved in downward epeirogenic movements so that although the late Mesozoic sequences are relatively complete, they are overlain by extensive Cainozoic cover. The complex geomorphological relationships that have been produced by the relative vertical movements of the interior and marginal areas have been studied by King in a series of classic publications (summarised in 1967). Recently his views have been challenged or modified (e.g., De Swart and Bennett, 1974), but the reality of several well-defined Mesozoic erosional surfaces (Gondwana, post-Gondwana, and African) is generally accepted.

The most comprehensive general account of onshore South African Mesozoic stratigraphy remains that by Du Toit (1954) although Haughton (1969) and Truswell (1970) have updated various aspects. A wealth of new data on Triassic stratigraphy and palaeontology has been published in the Proceedings of the three Gondwanaland Symposia (individual authors cited below), and recent revision has been made to Cretaceous molluscan zonations in Zululand by Kennedy and Klinger (1971, 1975). Prelimi-

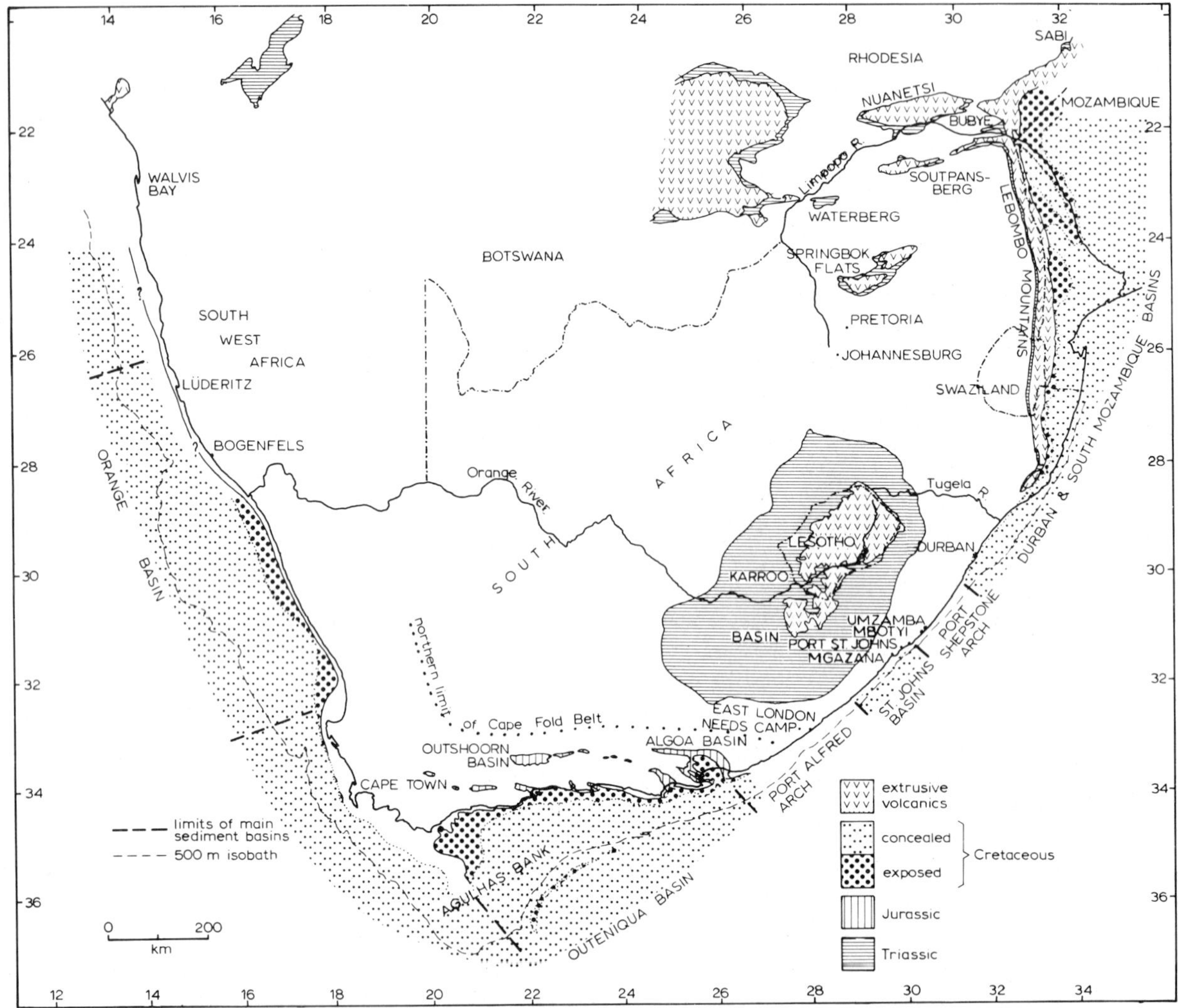

Fig. 1. Distribution of Mesozoic rocks in South Africa. After Du Toit (1954), Haughton (1969), Kitching (1970) and Dingle and Siesser (1977).

nary accounts of offshore Mesozoic stratigraphy have been given by Dingle (1973a) and Du Toit (1976).

TRIASSIC

Triassic rocks crop out in the Karroo Basin (centred on Lesotho and the northeast Cape Province) and in a series of erosional outliers to the north, which formed part of the Zambezi Basin (Rust, 1975) (Figs. 1, 3 and 4).

In southern Africa, the base of the Mesozoic cannot be accurately located. On the basis of reptilian faunas in the Beaufort Group, the Permian/Triassic boundary is placed at the junction of the Lower and Middle divisions of the Beaufort Group in the Karroo Basin (type area: northeast Cape Province) (e.g., Kitching, 1970; Raynor, 1970; Cooper, 1974) (Fig. 2). Because of the possibly diachronous character of this junction, and the unfossiliferous or non-diagnostic nature of Beaufort sediments away from the main Karroo Basin (for instance in the northern Transvaal, and in the Lebombo area, Van Eeden, 1973; Cooper, 1974) it is frequently not possible to confidently assign the Beaufort beds to any of the three classic litho-stratigraphic units (Lower, Middle, or Upper), nor to place the beds in a biostratigraphic

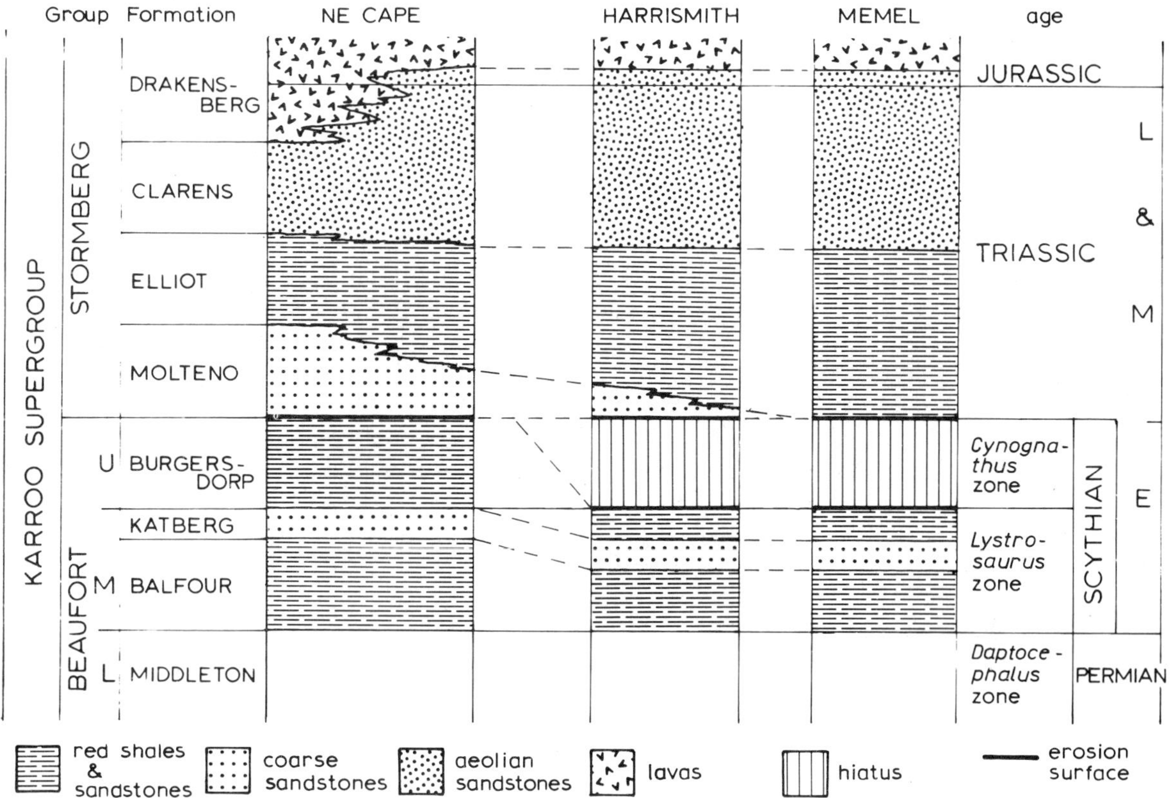

Fig. 2. Classification of Triassic rocks in the Karroo Basin. Based on Kitching (1970), Cooper (1974), Turner (1975), and Johnson et al. (1977).

scheme. In still other areas (for example, the Sabi coalfield, and eastern Botswana), the Beaufort is apparently missing, and the basal representatives of the Stormberg Group rest directly on Permian Ecca Group.

It is equally difficult to define the Triassic/Jurassic boundary in the Karroo Basin. In the northeastern Cape, where the succession is best developed, the uppermost Stormberg sediments (Clarens Formation) are overlain by the Drakensberg lava sequence, the lower part of which frequently contains sediment lenses. The age of the main Lesotho basalts is ~187 ± 7 m.y. (Fitch, 1972) which suggests that the Triassic/Jurassic boundary lies within the Clarens Formation (Cooper, 1974) although no biostratigraphic zonation exists whereby it can be accurately located.

Succession

Lower Triassic (Balfour, Katberg, and Burgersdorp formations)

In the Karroo Basin, the Lower Triassic (Scythian) is represented by the Middle and Upper divisions of the Beaufort Group (the classic subdivisions, as used, for instance by Du Toit, 1954), which have recently been re-classified into formations (Johnson et al., 1977). Although this classification is still provisional, we have utilized it, and Figs. 2 and 3 show a summary of the litho- and biostratigraphical relationships of the South African Triassic succession.

The Permian/Triassic boundary is placed at the first appearance of the therapsid reptile *Lystrosaurus* (C.B. Cox, 1967; Kitching, 1970) and in the northeastern Cape, this event is used to define the junction between the Middle and Lower divisions of the Beaufort Group (i.e. Middleton and Balfour Formation boundary), although it does not coincide with any marked change in lithology. Both the upper Middleton and lower Balfour Formations consist of alternating bright red, maroon, purple, and greenish mudstones, with red calcareous concretions, together with bluish-green fine sandstone horizons. The concretions commonly contain reptilian bone fragments

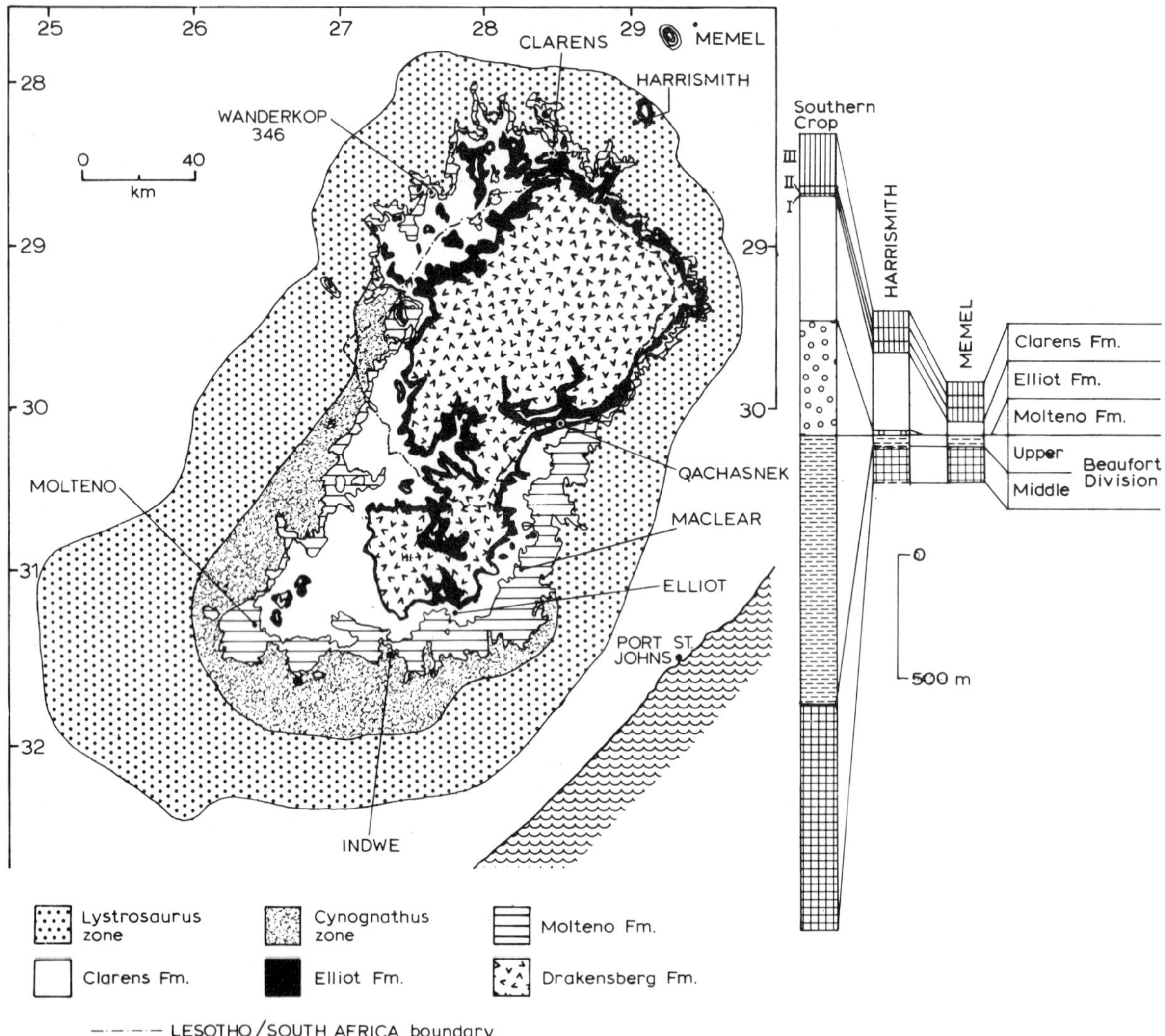

Fig. 3. Distribution and variations in thickness of Triassic rocks in the Karroo Basin. After Beukes (1970), Kitching (1970); Turner (1971 and 1975).

and skulls. Lateral facies changes are rapid and the succession is characterized by local unconformities, washouts, and clay pellet conglomerates. The Balfour Formation is overlain by a massive, yellowish, feldspathic sandstone: Katberg Formation, above which, in the northeastern Cape, *Lystrosaurus* is not found (Du Toit, 1954; Kitching, 1970). It contains pebbles of igneous and metamorphic rocks, possibly derived from the southeast. In northwestern Natal and southern Transvaal, however, *Cynognathus,* and other reptiles typical of Kitching's (1970) upper zone, do not occur in the shales above this massive sandstone, so that although the Burgersdorp Formation has been recorded from the area between Harrismith and Memel by Du Toit (1954, p. 294), Kitching (1970) does not recognize the *Cynognathus* zone north of about 28.5°S. This demonstrates the diachronous nature of the boundary between the Middle and Upper Beaufort lithostratigraphic divisions (Fig. 2).

The outcrop of the Middle division of the Beaufort Group (i.e. Balfour and Katberg Formations) is an approximately elliptical erosional remnant surrounding Lesotho, and the succession thins from a maximum of about 900 m along the southern crop, to about 150 m in the north (Fig. 3). It is overlain,

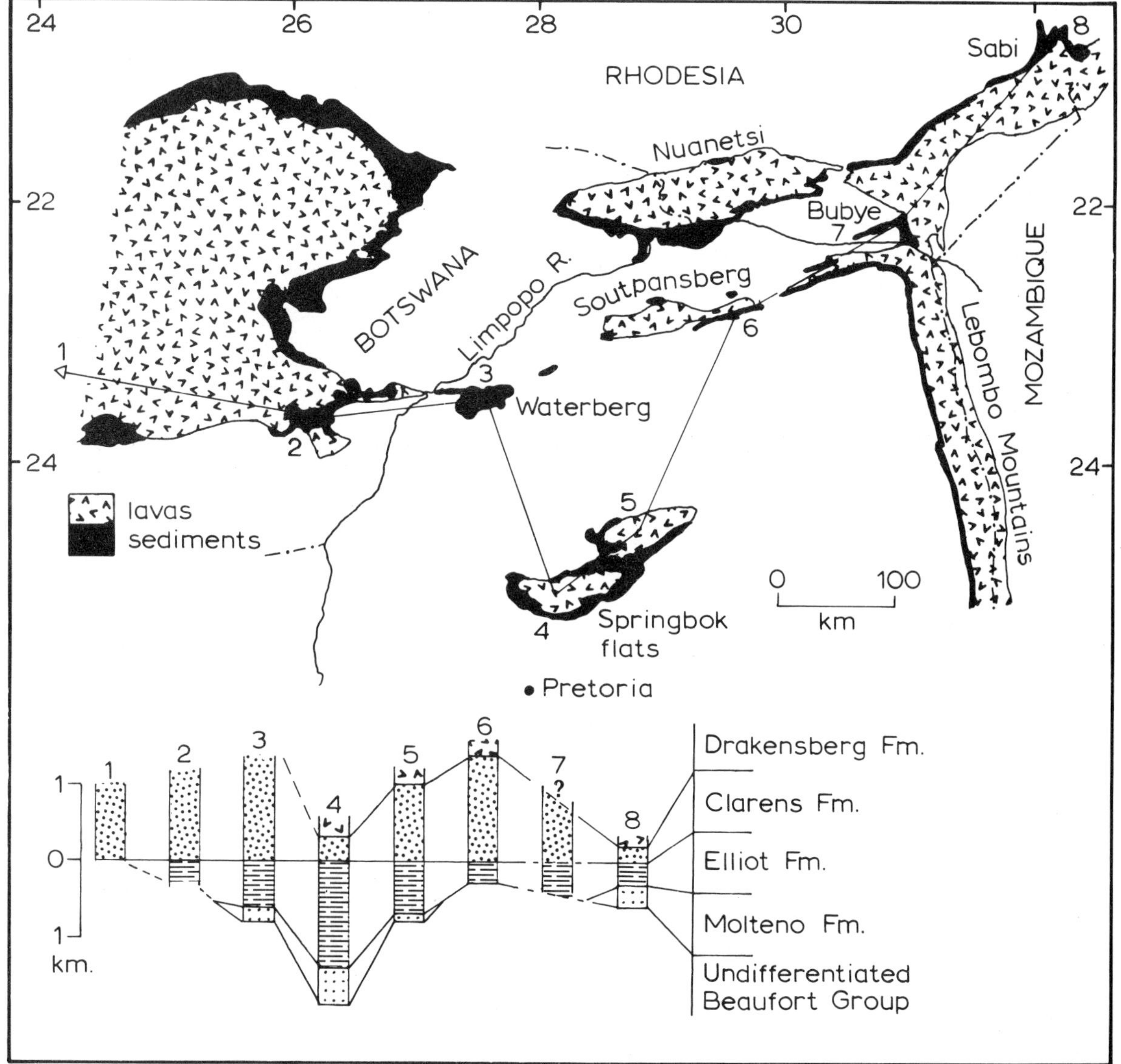

Fig. 4. Distribution and variations in thickness of Triassic rocks in the southern part of the Zambezi Basin. After Du Toit (1954), Haughton (1969), Van Eeden (1973), and Cooper (1974).

apparently conformably, by the Burgersdorp Formation (Figs. 2 and 3) which is lithologically similar to the underlying sequence: red and green variegated mudstones, with abundant red carbonate nodules, and widely spaced fine to medium, frequently feldspathic sandstones. Haughton (1969) considers that the red shales of this formation are more prominent than in the underlying Balfour Formation. Its outcrop centres on Lesotho and the succession thins from about 1070 m in the northeastern Cape, to about 46 m along the northern crop (Fig. 3).

Sediments ascribed to the Beaufort Group have been recorded from several large outliers in the Transvaal, Rhodesia, Botswana, and along the foot of the Lebombo Mountains (Fig. 4). These constitute remnants of the Zambezi Basin (Rust, 1975) but so far, they have not been palaeontologically zoned, nor have they been confidently assigned to any of the lithostratigraphic subdivisions of the Beaufort Group (see Du Toit, 1954; Haughton, 1969; Van Eeden, 1973). In all these areas, the total thickness for the so-called Beaufort Group is thin compared to the type areas, although in the Waterberg, Springbok Flats and Nuanetsi outliers and in the Lebombo outcrops this may be due to erosion (Du Toit, 1954, p. 312; Haughton, 1969, p. 378). Lithological descriptions (mostly from boreholes) indicate bluish-grey, red and variegated shales and mudstones, with occasional sandstones and carbonaceous shales. Lack of palaeontological data from these northern outcrops preclude the identification of the Permian/Triassic boundary outside the Karroo Basin.

Middle and Upper Triassic (Stormberg Group)

The Stormberg Group consists of continental clastic sediments overlain by thick lavas. Subdivision within the group is on lithostratigraphic criteria, and closely follows the scheme originally proposed by Dunn (1878) of Molteno, Red Beds, Cave Sandstone, overlain by Drakensberg lavas. These have recently been renamed as the following formations: Molteno, Elliot, Clarens and Drakensberg, respectively (see Falcon, 1975). The base of the Group is usually a well-defined erosional surface overlain by Molteno or Elliot Formations. Vertebrate fossils are restricted to the Elliot and Clarens Formations, and are predominantly diapsid types (compared to synapsid in the Beaufort Group). They have been dated as Late Triassic by comparison with Northern Hemisphere taxa (Kitching, 1970; C.B. Cox, 1973). Middle Triassic rocks may be represented by the Molteno Formation.

The group is more widespread in areas north of the main Karroo Basin than are the underlying Middle and Upper divisions of the Beaufort, although the possibility of diachronism within it, and lack of definitive palaeontological data preclude the recognition of biostratigraphic zones in these northern outcrops.

Triassic sedimentation was locally terminated by extrusion of the Drakensberg lavas.

Molteno Formation. The formation consists of sandstones and shales with subordinate conglomerates, siltstones, and coals. Maximum development is at Maclear (460 m) and the formation thins northwards to 3.5 m at Harrismith, and does not occur at Memel, where the overlying Elliot Formation rests directly on the Beaufort Group (Turner, 1972b) (Fig. 3). The basal beds rest on an erosion surface cut across the Beaufort Group (*Cynognathus* zone in the south and *Lystrosaurus* zone in the north). Turner (1970 and 1971) considers the Molteno Formation to represent four upward thinning sedimentary cycles, each consisting of four lithofacies. These can be summarized (base A, top D):

D. Carbonaceous shale and lenticular coal seams, with thin beds of fine sandstone and siltstone. Montmorillonite is common in the shales. The four main coal seams (Indwe, Guba, Molteno, and Gubenxa) thin rapidly northwards away from the southern crop, and seams greater than 1 m thick are restricted to the Indwe district.

B. Pale grey to yellowish-grey, coarse to medium sandstones, fining upwards to fine sandstones and siltstones. Trough and planar cross bedding and ripple marks are characteristic.

C. Flat-bedded siltstones and silty shales, with shale chips and carbonaceous debris along some bedding planes.

A. Basal pebble horizon or conglomerate (cobbles and boulders) containing poorly sorted quartz clasts [derived from the Witteberg Formation (Lower Carboniferous) in the Cape Fold Belt to the south, which formed the main provenance area for the Stormberg Group sediments], together with clay pellets and plant debris.

The sandstones are feldspathic, and secondary silica overgrowths give rise to a characteristic sparkling appearance in many of the outcrops (Turner, 1972a).

No reptilian fossils have been recovered from the Molteno Formation (Turner, 1972b) which is tentatively placed in the Middle–Upper Triassic on floral evidence (*Dicroidium* flora).

A prominent, whitish sandstone, correlated with the Molteno Formation, occurs in several of the northern outliers and in the Lebombo area (Fig. 4) (Du Toit, 1954; Haughton, 1969; Van Eeden, 1973). It lies between reddish shales and sandstones (Elliot Formation), and the local equivalent of the Beaufort Group, and in the Sabi coalfield it contains workable coals (Haughton, 1969, p. 375). Thicknesses range from 30 m in the west (Waterberg) to over 100 m in the Sabi coalfield (Fig. 4). About 70 m occur in Swaziland. In the northern Transvaal and southern Rhodesia, the Molteno Formation is missing.

A small, downfaulted outlier of Molteno Formation on the coast near Port St. Johns contains 490 m of pebbly, "glittering" sandstones, and shales.

Elliot Formation (formerly known as the Red Beds). In the main Karroo Basin, the Molteno Formation passes conformably upwards (and possibly laterally) into red and purple mudstones and shales with abundant calcareous nodules, together with less common red and yellow sandstones. These red beds locally have a basal clay pellet conglomerate. About

490 m of Elliot Formation occurs along the southern crop in the Cape Province, but it thins northwards: 90 m at Harrismith and 50 m at Memel. At the latter locality, Elliot Formation rests directly on *Lystrosaurus* zone strata (which at this locality occur in the Burgersdorp Formation, see Fig. 2), from which it can be distinguished by the presence of saurischian dinosaur remains (especially prosauropods). These vertebrate faunas indicate a Late Triassic age for the Elliot Formation (Haughton, 1969; Raynor, 1970).

Thick sequences of red shales and sandstones, correlated on lithological and faunal grounds with the Elliot Formation, occur in all the outliers in the southern Zambezi Basin (Fig. 4). They appear to reach their maximum development (360 m) in the north (Sabi coalfield), but thin westwards into Botswana, where Van Eeden (1973) does not recognize the subdivision west of about 25°E (Fig. 4).

Clarens Formation (formerly known as the Cave Sandstone). Beukes (1970) has provided a comprehensive description of the regional variations in lithofacies for the youngest sediments in the Stormberg Group. He defines the boundary between the Elliot and Clarens Formations as the base of the lower "massive" sandstone units at Qachasnek and Wanderkop 346 (Fig. 3) and discards alternative criteria (such as variations in grain size, colouration, presence of nodules) as unsatisfactory because of their inconsistent development. According to Haughton (1969) the base is frequently marked by an erosion surface.

The formation consists of massive (i.e. apparently structureless) or cross-bedded light brown, light red, and yellowish sandstones with minor lenticular reddish siltstones, conglomerates, clay pellet horizons and calcareous nodules. Beukes (1970) recognizes three lithofacies (base I, top III):

III. Thick-bedded, massive, very fine grained sandstone. In the southeast, this lithofacies grades laterally into thick-bedded silts and silty sandstones.

II. Alternating cross-bedded, laminated and massive fine, to very fine grained sandstones.

I. Massive basal sandstone with small lenticles of shallow-water-lain sand with ripple marks, sun cracks and clay pellets. Calcareous nodules are common, and locally lumps and thin bands of chert occur.

Lithofacies I and II generally increase in thickness towards the northeast, whereas lithofacies III retains an approximately constant thickness over the whole outcrop, except in the very south, where it thickens rapidly southwards (see Fig. 3). It is the Clarens Formation, and particularly lithofacies I and II that gives rise to the impressive cliffs with caves in the north-eastern Cape and Orange Free State. The reptilian fauna of the Clarens Formation is predominantly saurischian and Haughton (1969) has suggested that the heavier limbed dinosaurs are more common in the lower part of the formation and that they give way to lighter limbed varieties higher up. By comparison with vertebrate faunas in the Northern Hemisphere, the bulk of the Clarens Formation is considered to be Late Triassic in age (Haughton, 1969; Raynor, 1970), whilst the youngest sediments (top of lithofacies III), which interfinger with the Drakensberg lavas, are Early Jurassic (Cooper, 1974).

Thick, "massive" or dune-bedded, whitish sandstones correlated with the Clarens Formation occur in all the northern outliers and along the Lebombo line (see Fig. 4), suggesting that the formation had the greatest original extent of all the South African Triassic sedimentary sequences.

Triassic sedimentation was locally brought to a close by the eruption of the Drakensberg lavas. The lowest part of the volcanic sequence locally contains numerous thin, laterally impersistent, sediment lenses, which are mostly well-bedded and shaley. Locally, the Clarens Formation was eroded prior to these eruptions, and in the south and southeast, the sediment intercalations are associated with bedded tuffs and agglomerates. Clearly, early volcanic activity and sedimentation occurred simultaneously, although sediment lens do not occur more than about 300 m above the local basal lavas.

In all the southern Zambezi Basin outliers, the Clarens Formation is overlain by thick basaltic lavas, which in the Nuanetsi outlier are associated with rhyolites and alkali basalts (see K.G. Cox, 1972, for a summary of extra-South African distribution in southern Africa). Kent (1974) also considers that the Mozambique Channel is underlain by "Karroo" lavas which lie on top of an extensive "Karroo" sediment sequence.

Triassic palaeontology

Following Bain's (1845) discovery of reptilian fossils in Karroo strata, a large number of publications dealing with the varied and well-preserved vertebrate fauna has appeared (e.g., Broom, 1932; Haughton, 1954). These faunas display an almost unrivalled record of terrestrial vertebrate evolution during the Late Palaeozoic and Early Mesozoic, and are particularly important for the insight they allow into the early development of the mammals and dinosaurs. Excellent summaries have been given by Du Toit (1954), C.B. Cox (1967, 1973) and Haughton (1969). Kitching (1970) has recently revised earlier reptilian zonal schemes, which form the current basis for biostratigraphical correlation in the upper Karroo Supergroup. Here, we will mention only the stratigraphically more important reptilian taxa, and Table I shows some of the significant relationships to which reference is made in the text.

Reptiles appear in the Lower division of the Beaufort Group (Permian) and although all three sub-classes are represented in the succeeding Triassic succession, it is the Synapsida which are numerically more abundant, and stratigraphically more useful. A significant biostratigraphic event is the appearance of advanced mammal-like reptiles and the earliest dinosaurs in the Elliot Formation (Upper Triassic), which is separated from the Lower Triassic Burgersdorp Formation by the barren Molteno Formation (Middle Triassic).

Scythian faunas (*Lystrosaurus* and *Cynognathus* zones) are dominated by therapsid types, particularly the Theriodonta, which progressively developed mammal-like characteristics (especially the double palate). This trend is best displayed by the Cynodonta, small, carnivorous "dog tooth" types which gave rise to the primitive sub-mammals in the Late Triassic. Of the related taxa, Gorgonopsia, the least mammal-like of the Theriodonta, did not persist beyond the *Lystrosaurus* zone, whereas the Therocephalia developed mammal-like tendencies, which became most prominent in the *Cynognathus* zone. They do not occur in the Middle and Upper Triassic.

Within the other therapsid sub-orders, only the Anomodontia survived the Permian. They generally increase in size from the Late Permian, and in the Scythian became the dominant herbivorous reptiles, of which *Lystrosaurus* (possibly a hippopotamus-like aquatic type) was predominant in the lower zone.

Diapsid reptiles occur only in small numbers through the Early Triassic, but in the Elliot and Clarens Formations (Late Triassic—Early Jurassic) gave rise to the saurischian-forerunners of the dinosaurs, as well as early crocodiles.

Amphibian remains are confined to the Lower Triassic, and consist entirely of varieties of the labyrinthodonts. They occur in relatively small numbers and are known mainly from skull remains.

TABLE I
Ranges of some important South African Triassic tetrapods

Subclass	Orders and suborders	Upper Permian	Scythian		Middle and Upper Triassic
			Lystrosaurus zone	*Cynognathus* zone	
ANAPSIDA					
SYNAPSIDA	Therapsida				
	Dinocephalia				
	Anomodontia			dominant herbivores	
	Theriodonta (mammal-like reptiles)		*Lystrosaurus*		
	Gorgonopsia	common, least mammal-like			
	Therocephalia				
	Cynodontia	dominant mammal-like reptiles			advanced types
				Cynognathus	
DIAPSIDA	(give rise to dinosaurs, lizards, birds, crocodiles)				earliest dinosaurs

The affinities of south and east African Triassic tetrapods have been the subject of a preliminary assessment by C.B. Cox (1973), who shows a close relationship (% similarity at family level) with all other regions except Europe: Europe (58%), South America (74%), India (75%), North America (75%) and Asia (89%). Although, as Cox points out, these figures are susceptible to distortion by unusually rich finds in particular areas and time zones, the apparently closer relationship between Africa and Asia than between Africa and South America is difficult to explain in terms of Gondwanaland reconstructions, and may be the result of comparing somewhat different environments.

Rare fish have been recorded from all lithostratigraphic units, except the Molteno Formation, and together with rare molluscs and crustacea are freshwater species.

Plant remains are common at various horizons throughout the Lower and Middle Triassic strata, and have been used to produce biostratigraphic zonal schemes [macroflora: Plumstead, 1970; microflora: Hart, 1969 (Lower Triassic only)]. Charred tree trunks near the contact of the Clarens Formation and Drakensberg lavas (Beukes, 1970) are the only plants so far recorded above the Molteno Formation. Lower Triassic floras are dominated by various species of *Glossopteris* and associated types, although they become impoverished and less diverse up the succession. *Dicroidium* makes its appearance near the top of the *Cynognathus* zone, and during the Middle Triassic (especially in the Molteno coal seams) is the zonal type of a distinctive and varied flora containing cycads, ginkgoes and ferns. Falcon (1975) has suggested that this floral change is not necessarily an evolutionary phenomenon, but is more likely to have been occasioned by environmental changes (such as an alteration in the level of the water table).

No palynological zonation has been published for the post-*Lystrosaurus* Triassic succession, but one is urgently needed: the macroflora are not present in post-Molteno sediments, and no reptilian zonation is available for post-Beaufort strata (the Molteno is apparently barren). In addition, it would reflect the evolutionary trends in the total flora more faithfully than the facies-controlled macroflora.

The abundance of plant life during the whole of the Triassic cannot be doubted (even in the so-called barren Elliot and Clarens Formations) in view of the large number of herbivorous and dependent carnivorous reptiles.

JURASSIC

During the Jurassic, the main sediment depocentres shifted from the intra-continental Karroo and Zambezi Basins into the newly established basins around the margins of what was to become the African continent. Several of these basins had been established in Permian times, but tectonic activity in the southern Cape in the mid-Triassic led to further small rift and graben-type structures and intermontane basins. Onshore these basins remained active depocentres only during the Mesozoic, whereas on the Agulhas Bank the Outeniqua Basin contains thick marine Cainozoic deposits. The Outeniqua Basin is subdivided by several basement ridges (extensions of Cape Fold Belt structures) over which the thickness of total Mesozoic sediments is usually less than 2500 m whereas in the intervening basins it is usually greater than 3000 m, a with a maximum of about 5700 m (Dingle, 1973a).

Small, poorly known basins along the east coast (see Simpson and Dingle, 1973), *may* contain sediments as old as Jurassic.

Geological activity in the interior consisted of continued volcanism, and erosion of Triassic lavas and sediments. Shallow seas, which had temporarily penetrated central Gondwanaland as early as the Permian, became established in eastern Africa by Early and Middle Jurassic times, and had reached South Africa and the Falkland Plateau on at least one occasion by latest Jurassic times.

The diachronous nature of these marine transgressions, together with continued volcanism and localized, tectonically-controlled continental sedimentation, means that identifiable basal Jurassic strata vary greatly in age from place to place. Similarly, the passage from Jurassic to Cretaceous can nowhere be unequivocally identified, although complete sequences across this boundary possibly occur on the continental shelf.

Succession

Jurassic sediments are thickest and most extensively developed in the southern Cape and in the Outeniqua Basin, where they infill intermontane, partially fault-bounded basins in the Cape Fold Belt. Onshore, erosion has produced small, isolated outliers, the largest of which is the Algoa Basin, whereas offshore, the infill has buried basement topography to produce a large sediment basin which has been completely covered by Cretaceous and Cainozoic sediments (Fig. 6). Elsewhere, suspected Jurassic sediment outcrops are confined to small, fault-bounded outliers on the east coast and intercalated lenses in the Lebombo lavas in southwest Mozambique.

Because lithofacies are generally continental and non-marine, with only rare marine interfingers, and it has not been possible to confidently date much of the succession, our understanding of the Jurassic succession of South Africa is unsatisfactory. Although it is generally accepted that the major lithofacies are strongly diachronous and probably span the Jurassic/ Cretaceous boundary (e.g., Du Toit, 1976; McLachlan et al., 1977a) there is no unanimity on the dating of the marine horizons. The following account is, therefore, tentative and liable to revision as more commercial data are analyzed and published. Fig. 5 summarizes possible bio- and lithostratigraphic relationships.

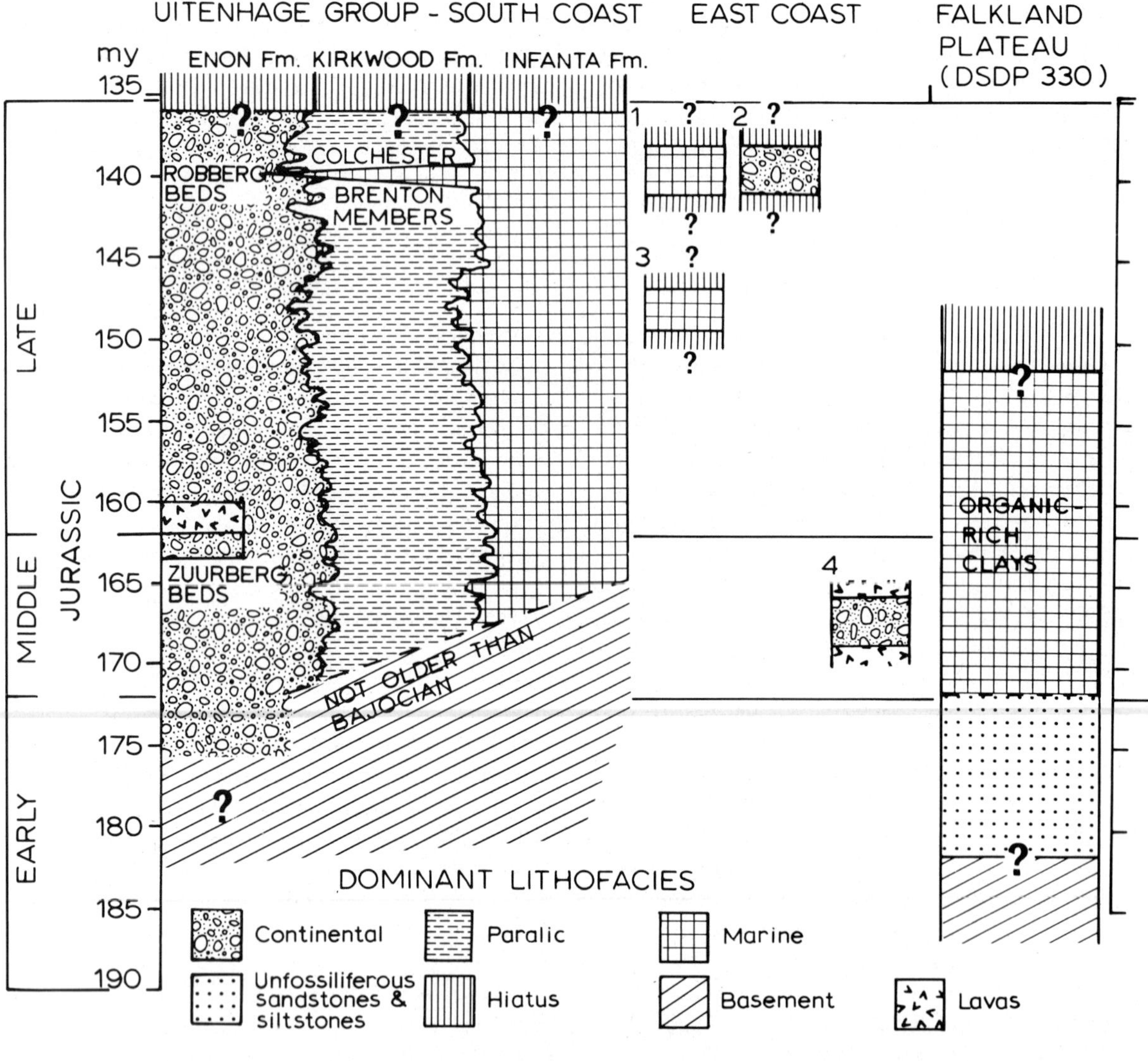

Fig. 5. Tentative classification, and dominant lithofacies of Jurassic sediments in South Africa and on the Falkland Plateau. After Dingle (1973c), Winter (1973), Barker et al. (1974) and Du Toit (1976). Note that the main lithofacies in the Algoa Basin probably range into the Lower Cretaceous.

Jurassic volcanic rocks occur extensively over the whole of the Karroo and Zambezi Basins, in the Lebombo range, and locally in the southern Cape.

Uitenhage Group

The type area for Jurassic rocks in southern South Africa is the Algoa Basin, where the generalized stratigraphy was first described by Atherstone (1857) (Fig. 7). Du Toit (1954) has summarized the earlier surface geological and palaeontological work, but most of our knowledge of litho- and bio-stratigraphic relationships comes from recent oil company activity, which has been published by Winter (1973), and Lock et al. (1975). Similarly, the succession in the Outeniqua Basin has been established on seismic and borehole evidence (Dingle, 1973a; Du Toit, 1976).

The whole of the Jurassic succession has been assigned to the Uitenhage Group (Winter, 1973), with many of the type sections in Southern Oil Exploration Corporation boreholes. In the Algoa Basin, Winter (1973) subdivided the Uitenhage Group into three formations, and in the Outeniqua Basin, Du Toit (1976) subdivided the group into five formations. Our classification is a combination of these two schemes (Fig. 5).

Enon Formation. This is a strongly diachronous lithofacies which usually forms the base of the Mesozoic

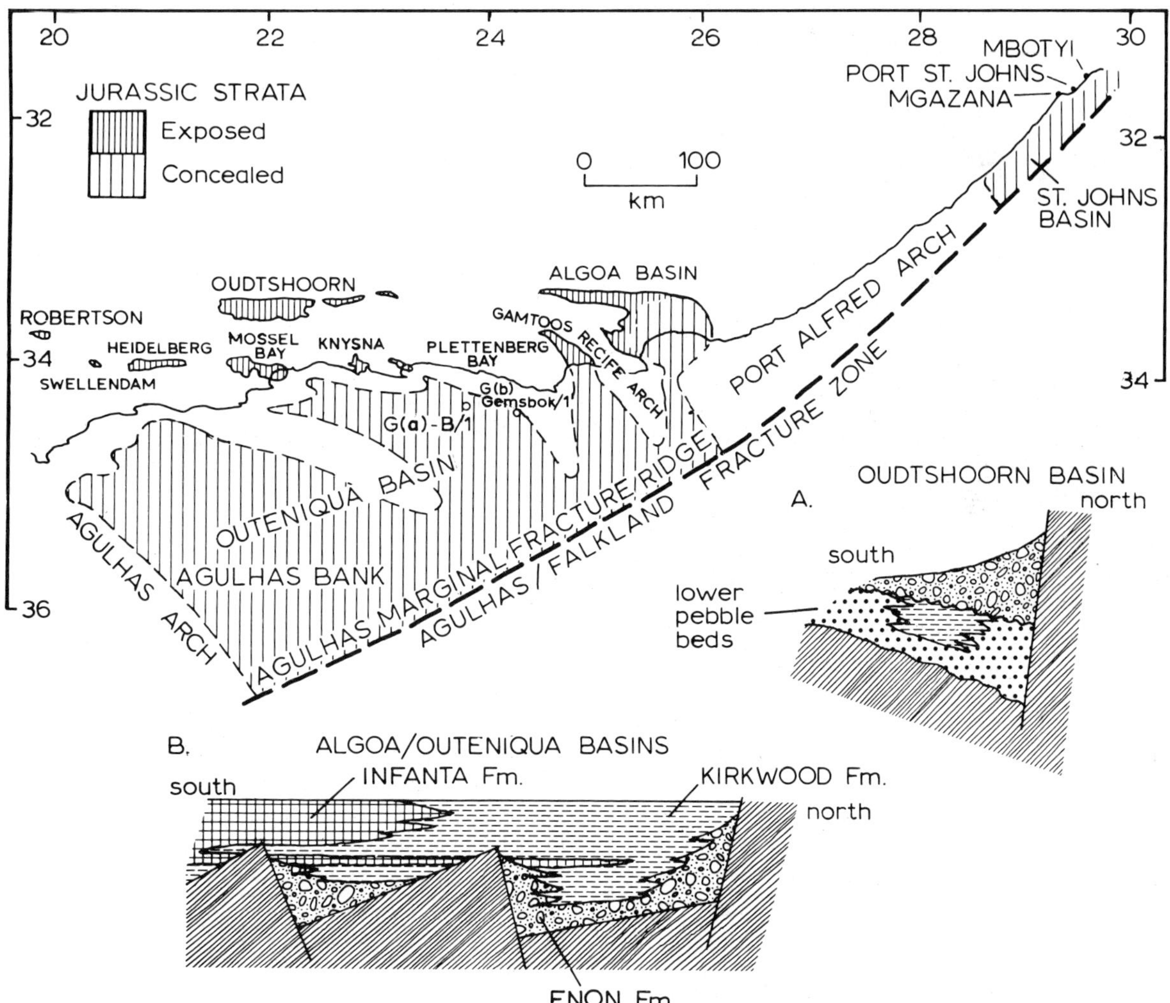

Fig. 6. Jurassic sediments on the south coast. After Dingle (1973c) and Du Toit (1976). A. Schematic section across Oudtshoorn Basin (after Lock et al., 1975). B. Schematic section across Algoa/Outeniqua Basins (after Winter, 1973; Lock et al., 1975; Du Toit, 1976).

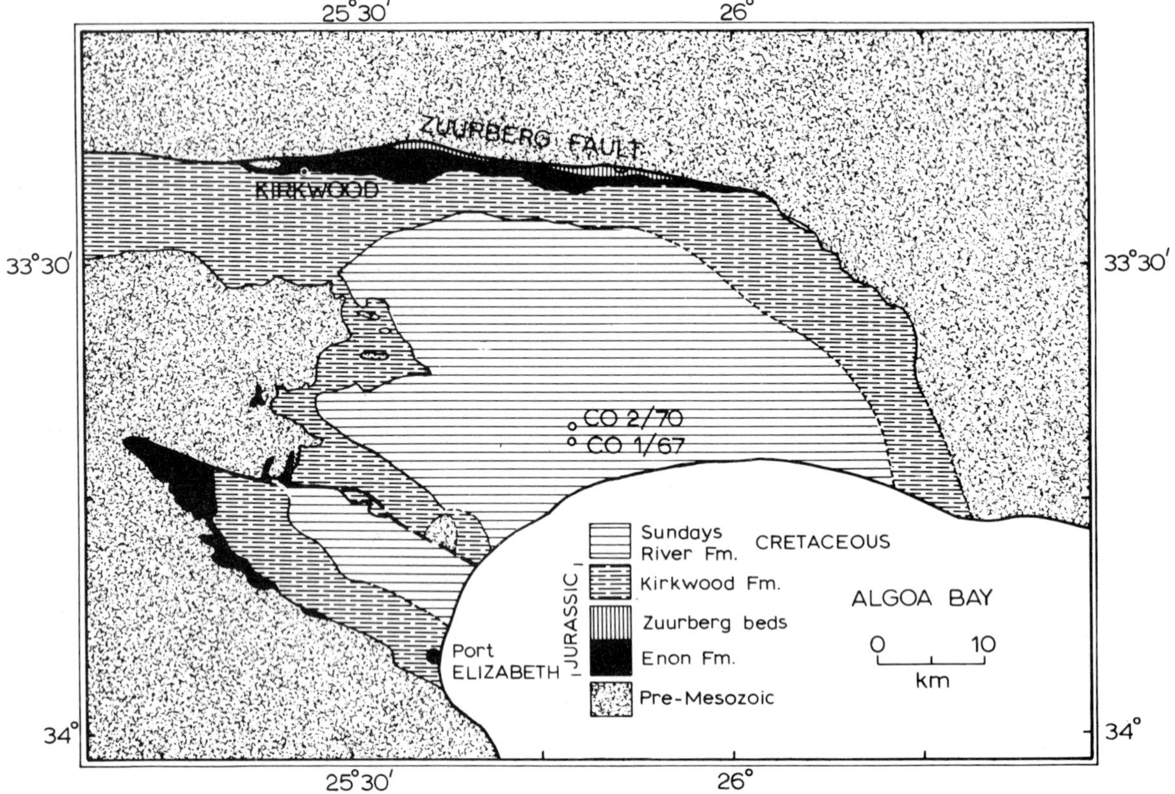

Fig. 7. Sketch map of the Algoa Basin (after Winter, 1973).

in the deepest parts of the southern Cape and Outeniqua Basins and interfingers with the overlying Kirkwood Formation. Although Dingle (1973b) has suggested that the Enon Formation ranges in age from Triassic to Early Cretaceous, no fossils older than Jurassic have so far been recovered. Its unsuitable lithology militates against fossil preservation, and in most instances its age is inferred from stratigraphical relationships.

In the type area (Figs. 6 and 7) it consists of coarse conglomerates interbedded with red and yellow sandstones and reddish brown shales and siltstones. Sorting in the conglomerates is poor, and the composition of clasts varies with the local bedrock: Table Mountain quartzites, and Bokkeveld and Dwyka shales. Lateral facies and thickness changes are rapid, and the sequence probably represents high-energy, alluvial fans and piedmont deposits. Lock et al. (1975) have demonstrated the strongly diachronous nature of the boundary of the Enon and overlying Kirkwood Formations by a study of the Oudtshoorn Basin. Here a lower pebble conglomerate sequence occurs on both sides of the sediment basin, and grades laterally and vertically into the Kirkwood Formation. Reactivation of the northern boundary fault scarp resulted in the formation of coarse fanglomerates, with large angular boulders, which fine upwards and southwards (Fig. 6). Subsequent reactivation of this fault produced continuous piedmont fan construction, possibly well into Cretaceous times.

The Enon Formation occurs in all the other Jurassic outliers in the southern Cape, and in most of them lateral and vertical gradation into the Kirkwood Formation is apparent. Marine intercalations in the Enon Formation are seen at two localities: Knysna and Plettenberg Bay. In the former a shallow-marine sandy clay is thought to interfinger with red conglomerates, whereas at Plettenberg Bay the situation is less clear. Poorly preserved marine fossils occur in a silicified pebbly sandstone which is overlain by more normal Enon facies. It has been suggested that the fossiliferous sequence is a separate stratigraphical unit (see Lock et al., 1975), but it seems more likely to be a marine intercalation (beach deposit) within the terrestrial Enon facies (Figs. 5 and 6).

Adjacent to the Zuurberg Fault, along the northern edge of the Algoa Basin, lavas and tuffs dated at 162 m.y. (Du Toit, 1976) overlie 65 m of breccias, conglomerates and sandstones intercalated with tuffs (Figs. 5 and 7). These sediments are here tentatively assigned to the Enon Formation, younger representatives of which overlie the upper lavas and tuffs.

In the Outeniqua Basin, Du Toit (1976) has identified the Enon Formation in a number of boreholes and, like the other Jurassic sediments in the basin, it is thickest in the grabens and rarely present over the basement highs, onto which it overlaps (Fig. 6).

Because of strong tectonic controls, the Enon Formation varies considerably in thickness: Oudtshoorn, >3000 m (Du Toit, 1954); Gamtoos, 1500 m (Lock et al., 1975); type locality (just east of Kirkwood), 300 m (Winter, 1973); Outeniqua Basin, up to 2000 m (Dingle, 1973a).

Kirkwood Formation. In the Mesozoic outliers of the southern Cape, the coarse Enon Formation passes laterally and vertically into the predominantly fine-grained Kirkwood Formation (Fig. 6). In the type locality, the formation consists of reddish-brown variegated silty mudstones (with greenish reduction blotches), with subordinate grey shales and sands. The latter are silty (reddish-yellow to yellow) or yellow, white, or green coarse, cross-bedded, with thin pebbly layers and abundant carbonaceous fragments. It is a fluviatile deposit, laid down on valley floors at the foot of Enon facies piedmont and alluvial fans. From boreholes in the distal parts of the Algoa Basin two additional lithologic units were defined by Winter (1973): Colchester and Swartkops Members (Fig. 5). The former will be discussed in the next section, as it represents a marine intercalation of the Infanta Formation. The Swartkops Member is a fine-grained, poorly sorted sandstone with interbedded brownish and dark grey shales which occurs near the base of the Kirkwood Formation. According to Du Toit (1976) it locally rests directly on the Enon Formation or on Palaeozoic basement in the Outeniqua Basin, where it has been intersected by several boreholes.

In the Algoa Basin the Kirkwood Formation has a thickness of 1314 m in the type section borehole, which increases to 2210 m 2 km to the southeast. In the Outeniqua Basin Du Toit (1976) records 1100 m from borehole G(a)B/1 (south of Plettenberg Bay), and on seismic evidence Lock et al. (1975) indicate 1800 m in the Gamtoos Basin.

The Kirkwood Formation has been recognized in the following outliers: Algoa, Gamtoos, Oudtshoorn, Mossel Bay, and Heidelberg. All contain abundant plant material, and at Heidelberg fresh-water molluscs, fish and crustaceans have been found. Palynological studies on borehole material from the Outeniqua Basin indicate that the Kirkwood Formation is not older than Bajocian (Du Toit, 1976).

Infanta Formation. The Kirkwood Formation grades laterally, and in some places vertically, into the Infanta Formation (Fig. 6), and Du Toit (1976) defines the boundary as the change from predominantly red beds (Kirkwood) to greyish sediments (Infanta).

The Infanta Formation does not occur in the Algoa Basin, and has been defined from borehole sections in the Outeniqua Basin (Du Toit, 1976). It is a relatively homogeneous sequence of poorly fossiliferous, light to medium, occasionally dark-grey siltstones, mudstones, and shales, and is thought to have been deposited in estuarine or shallow-water marine environments (Dingle, 1973b; Du Toit, 1976). It therefore represents a more distal basinal facies than the Kirkwood Formation, into which it grades laterally, and in some sections it is underlain by Kirkwood-type red beds, and locally by the Swartkops Member. In the type section borehole (G(b)-Gemsbok/1) the Infanta Formation is about 1300 m thick, but it thins rapidly towards the basin's edge.

Near the base of the Infanta Formation, a tongue of grey shales extends laterally northwards into the Algoa Basin, representing a short-lived, shallow marine transgression which interfingers with the Kirkwood facies: the Colchester Member (Winter, 1973). This is a grey shale, with a 1-m band of dark grey, waxy, organic carbon-rich shale towards the base. The Colchester Member has a sparse fauna of agglutinated foraminifera and hyposaline ostracods, and thickens to over 400 m in the distal parts of the Algoa Basin (Winter, 1973). A similar tongue occurs within the Kirkwood Formation over much of the Outeniqua Basin and probably correlates with the shallow-marine Brenton Beds in the Knysna outliers (Figs. 5 and 6). These are irregularly bedded greenish sandy clays, silty sands, sands and conglomerates with a marine molluscan fauna. Glauconite and carbonaceous debris are common. The proximity of the Brenton bed outcrops to cliffs of Enon conglomerate suggests that the marine facies rapidly passes laterally into the conglomerate and that a silicified pebbly sandstone with marine molluscs at Plettenberg Bay is a beach deposit of the same transgression (Robberg Formation of Rigassi and Dixon, 1972).

Dingle and Klinger (1972) considered the marine ostracod faunas from the Brenton beds to be similar to assemblages from Madagascar and assigned them a Late Jurassic age, a date which is apparently corroborated by the presence of one specimen of the ammonite *Hybonoticeras* (generic range Middle Kimmeridgian to Early Tithonian) (Klinger et al., 1972). More recent micropalaeontological work has questioned this date (Stapleton and Beer, 1976; McLachlan et al., 1977a) and has attempted a correlation with the Valanginian/Hauterivian lower Sundays River Formation (see below). However, until the precise local and regional age ranges of the benthos utilized for these correlations have been established, and until the anomaly of the ammonite has been clarified, it seems prudent to provionally retain a Late Jurassic assignment for the Brenton Beds. This is particularly so in view of the fact that the recent work has assigned a ?Late Jurassic to Early Valanginian age to the Colchester Member (Soekor, 1976).

Outliers on the east coast and Mozambique

Two small, fault-bounded outliers occur on the Transkei coast on a broad basement arch on which Cretaceous and younger sediments are generally absent or thin: the Transkei Swell (Simpson and Dingle, 1973). At Mbotyi a conglomerate containing boulders of dolerite, sandstone and shale, with a few sandy layers containing wood fragments rests on Beaufort Group sediments. At Mgazana, 40 km away, a conglomerate containing dolerite, sandstone, amygdaloidal basalt, and agates rests on the Molteno Formation. It contains layers of coarse pebbly sand and a limestone band, which has a marine molluscan fauna, including *Bochianites* (H.C. Klinger, pers. comm., 1976) together with plant fragments (including *Otozamites*). The Mgazana beds are probably a beach deposit.

The ages of these outliers are equivocal. McLachlan et al. (1977b) have described the benthic microfauna from Mgazana and have correlated the ostracods faunas with assemblages from the lower Sundays River Formation (Valanginian/Hauterivian) in the Algoa Basin (see below). However this link seems weak – only five species were identified, none of whose age ranges are known, in addition to which, several types were mentioned as being close to Kimmeridgian forms described by Bate (1975) from Tanzania. On the grounds of the latter associations we have tentatively assigned the Mgazana beds (and hence also the Mbotyi outlier, purely on lithological similarities) a Tithonian age. This uncertainty will, presumably, be clarified after the ammonite faunas have been studied.

No marine Jurassic sediments have been reported in Mozambique south of Nacala (14°S), where Henriques da Silva (1966) records a shallow-water Kimmeridgian fauna. A sandstone intercalation in the Lebombo rhyolites (165 m.y., Bathonian) at Namahacha contains fresh-water molluscs and *Otozamites*.

In the southern Cape, Jurassic sediments are overlain by Lower Cretaceous grey-coloured beds: in the Algoa Basin the boundary is between red beds and grey shale whereas in the more distal parts of the Outeniqua Basin the boundary is an apparent unconformity between two grey shaley sequences. Both these boundaries are lithostratigraphic and nominal, there being no palaeontological framework within which to precisely define the Jurassic/Cretaceous contact.

Jurassic palaeontology

In strong contrast to the Triassic, which has a rich terrestrial vertebrate fauna, and the overlying Cretaceous, with its abundant marine invertebrates, the Jurassic is characterized by thick barren, or sparsely fossiliferous, non-marine or hyposaline, water-lain deposits. Marine fossils occur only rarely, and the most abundant fossil group are plants, which occur predominantly in the Kirkwood Formation. No biostratigraphic zonation has been established for the South African Jurassic, although a palynological scheme would seem most feasible.

Vertebrate remains have been reported from three basins, all within the Kirkwood Formation: reptile teeth at Oudtshoorn, dinosaur bones (including *Algoasaurus baini* Broom) at Algoa (two localities), and fresh-water fish scales at Heidelberg.

Fresh-water molluscs (*Unio*) and crustaceans (*Estheria*) occur in the Kirkwood Formation at

Heidelberg, and fresh-water and hyposaline molluscs (*Modiola, Melania, Viviparus* and *Cyrena*) in a more distal facies of the Kirkwood Formation in the Algoa Basin (Du Toit, 1954). Although the stratigraphical relationships are not clear, this fauna may be related to the marine transgression represented by the Colchester Member, which contains unidentified arenaceous foraminifera and hyposaline ostracods (Winter, 1973). *Unio* and *Viviparus* occur in the sandstone intercalations in rhyolite at Namahacha (165 m.y.).

Bearing in mind the equivocal datings discussed above, marine fossils have been reported from Mgazana, Plettenberg Bay and Knysna (Brenton), and a summary of their faunas is given in Table II. To date only certain microfaunas (especially ostracods) are well-known. The Brenton ostracod assemblages are rich in numbers (up to 20%) and species (5) of the subgenus *Progonocythere* (*Majungaella*) which is so typical of Upper Jurassic to Neocomian strata of the western Indian Ocean area (Tanzania, Madagascar, South Africa). Ostracods from the Mgazana beds are more restricted: no representative of *Progonocythere* (*Majungaella*) have been reported, and none of its cytheracean fauna has unequivocally been identified from other localities. *Hybonoticeras* and *Bochianites* have been identified from Brenton and Mgazana, respectively, and McLachlan et al. (1977b) imply that a more diverse ammonite fauna awaits identification from Mgazana.

Only rare and fragmentary plant remains have been found in the Enon Formation (Plettenberg Bay: Du Toit, 1954). The Kirkwood Formation, however, contains abundant plant material including lignitic masses, tree trunks and well-preserved leaves and stems. Initial studies (for example Seward, 1908) claimed this to be a "Wealden" flora, intimating an age, as well as a facies, connotation (i.e. Lower Cretaceous). Although the Kirkwood plants probably do represent a Wealden facies flora (i.e. paralic/lacustrine/fluvial), the formation probably ranges in age from Middle or Late Jurassic to Valanginian (McLachlan et al., 1977b). The flora is varied and consists of ferns (6 species), conifers (4 species) and cycads (10 species). (The numbers of species are minimal values and based on Seward, 1908).

TABLE II

Summary of marine Jurassic fossils from South Africa

	Brenton	Plettenberg Bay	Mgazana
Molluscs			
Perna thesensi Schwarz 1915	X		
P. brentonensis Schwarz 1915	X		
Trigonia kitchini Schwarz 1915	X		
T. sp.		X	X
?*Ptychoma* sp.	X		
?*Pseudomelania* sp.			X
Hybonoticeras aff. *H. hildebrandti* (Beyrich)	X		
Bochianites sp.			X

Ostracoda – dominant genera and number of species

Number of specimens belonging to dominant genera expressed as % total ostracod population

	Mgazana (Sample 4838 of McLachlan et al., 1977 b)	Brenton (Sample 3 of Dingle and Klinger, 1972)
Argilloecia sp.	26%	
Bairdiacypris sp.	13%	
Procytherura sp.	11%	
Cytherella spp.	39% (4 spp.)	72% (2 spp.)
Progonocythere (*Majungaella*) spp.		23% (2 spp.)

Plant fragments also occur in the Mgazana outlier and in the Namahacha intercalations. In the former, ferns (2 species) and cycads (3 species) are found, but none of the species appears to be the same as those found in the Kirkwood Formation in the southern Cape. Both east coast outcrops contain the cycad genus *Otozamites* (range Rhaetic to Early Cretaceous; Alvin et al., 1967), which has not been recorded from the Kirkwood Formation.

CRETACEOUS

Following the breakup of West Gondwana at 127 m.y. (Valanginian), an additional sediment depocentre was created on the rifted continental margin of southwestern Africa: the Orange Basin (Fig. 1). In addition, the established basins (Outeniqua, and south Mozambique/Natal Valley) continued to receive sediment, although in the east much of the terrigenous input bypassed the continental margin and found its way into the adjacent deep ocean basins.

Most of the Cretaceous sediments were laid down under marine conditions and, consequently, carry a varied and easily datable invertebrate fauna. For this reason, stratigraphical correlation is more satisfactory in the Cretaceous rocks than for either of the earlier Mesozoic systems. Despite this, no section has yet been described in which the Jurassic/Cretaceous boundary can be identified. With the possible exception of distal localities in the Outeniqua and Mozambique Basins, this boundary is probably disconformable or unconformable in all the continental margin basins. The demonstrably oldest Cretaceous sediments are Valanginian, and occur in the south (Algoa and Outeniqua Basins) whereas on the east coast no sediments older than Barremian have been identified.

The succession across the Mesozoic/Cainozoic boundary (i.e. Maastrichtian/Danian) is probably complete on the Natal continental shelf (JC-1 borehole), and on the continental rise off the southwestern Cape (DSDP site 361), but elsewhere (onshore, and in the proximal parts of the continental margin basins) seems to be marked by a hiatus and, locally, a strong unconformity. Fig. 8 summarizes the stratigraphy of South African Cretaceous sediments, and Figs. 9 and 10 show details of outcrops on the east and south coasts.

Succession

It is convenient to discuss the South African Cretaceous strata by reference to two standard successions: Zululand/Natal, and southern Cape. Although the former is less complete, it is better known and, being exposed onshore, likely to be more useful as a standard for correlating macro- and microfossil zonation schemes. The Cretaceous succession off the west coast is thick, but poorly known and can only be described in general terms.

Zululand, Natal, and southern Mozambique

Since Garden (1855) gave the first descriptions of Cretaceous beds from the northeast Cape coast, numerous workers have described the sediments and rich invertebrate faunas in Natal and Zululand. For a summary of earlier work, and for details of the revised stratigraphy and ammonite faunas the reader is referred to Kennedy and Klinger (1975), and Du Toit (1954).

The succession in Natal and Zululand thickens eastwards and northwards and consists of three transgressive formations which progressively overlap southwestwards (Fig. 9). Offshore (borehole JC-1), the Upper Cretaceous section is probably complete, but south of Durban only Santonian–Campanian strata occur (Fig. 8). Exposures are generally poor and dips low to the east. The area in the north is mostly covered by Cainozoic sediments and, locally, dense bush. Most exposures are confined to small river or lagoon banks and man-made cuts.

Makatini Formation (Barremian–Aptian). The oldest Cretaceous rocks onshore in eastern South Africa crop out in northern Zululand and are Barremian piedmont fan and fluviatile sands and conglomerates, which rest on deeply weathered Lebombo lavas where they pass under the Mozambique coastal plain (Kennedy and Klinger, 1975; Förster, 1975a). Because of the transgressive nature of the sequence, Kennedy and Klinger (1975) suspect that older Cretaceous sediments occur further east, and seismic records shown in Du Toit and Leith (1974) indicate that there are certainly thick Lower Cretaceous sequences on the continental shelf off Natal (at least 1050 m, 0.35 sec at 2993 m/sec near borehole JC-1). The earliest marine sediments are Late Barremian sands and conglomerates which are overlain by Barremian and Aptian glauconitic sands and silts.

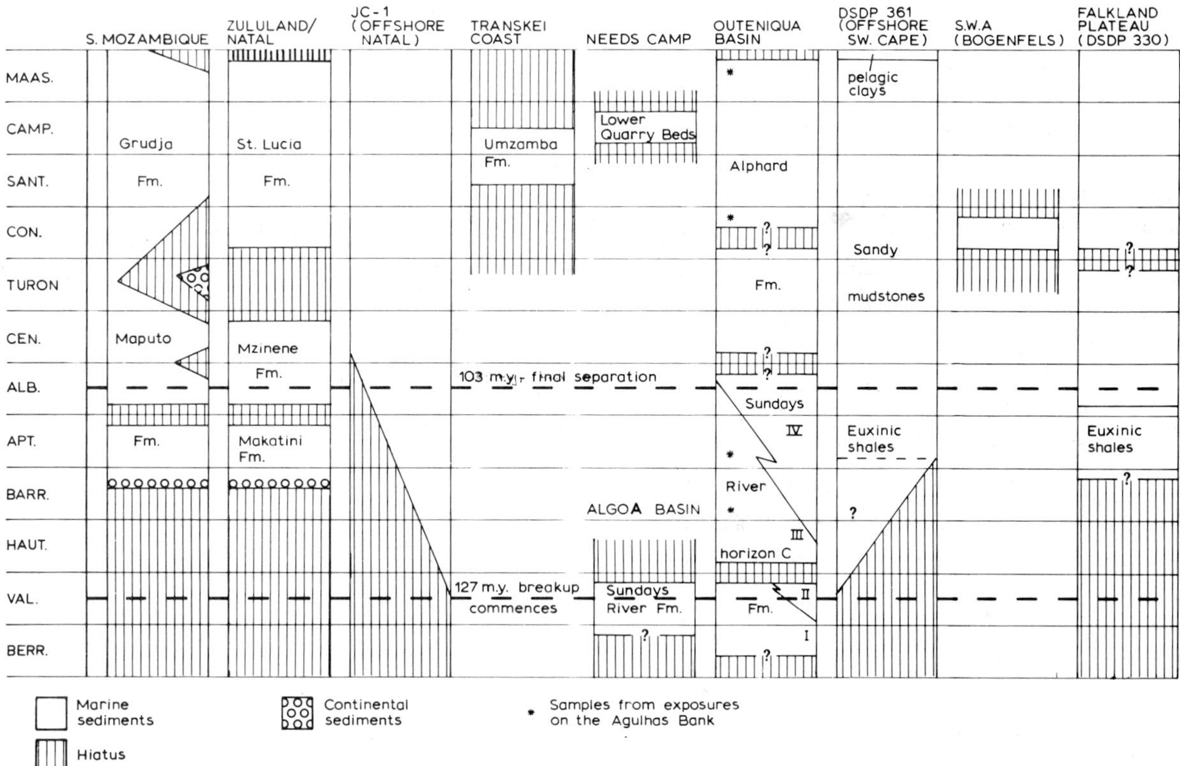

Fig. 8. Stratigraphical classification of Cretaceous sediments in South Africa. Based on Haughton (1930), King (1972), Dingle (1973a), Winter (1973), Barker et al. (1974), Du Toit and Leith (1974), Bolli et al. (1975), Förster (1975a), Kennedy and Klinger (1975), Du Toit (1976) and H.C. Klinger (pers. comm., 1976).

Carbonaceous fragments are common, and locally logs and large wood fragments occur. In the Aptian, higher-energy environments are represented by minor erosion surfaces with bored bivalves and reworked fossils. Towards the south of the outcrop, Aptian marine sediments overstep onto Lebombo lavas. The uppermost Aptian is missing.

Makatini Formation outcrops are confined to the northern part of Zululand, and are overstepped southwards by younger formations (Fig. 9).

Mzinene Formation (Lower Albian–Upper Cenomanian). The Aptian/Albian boundary is a non-sequence which represents latest Aptian and earliest Albian times. It is marked by a hardground disconformity of bored and rolled fossils overlain by a Lower Albian conglomerate, and can be traced for 175 km southward from the Mozambique border. Albian sediments are shallow-marine silts and sands, with shelly horizons, and small-scale sediment cycles. In the lower part several erosional surfaces with bored bivalves and reworked fossils occur locally. Lower, Middle and Upper Cenomanian silty glauconitic sands, with a rich marine fauna follow conformably. A major regression terminated deposition of the Mzinene Formation, and the uppermost Cenomanian is missing.

St. Lucia Formation (Lower Coniacian–Maastrichtian). Coniacian sediments are separated from underlying strata by a hiatus which represents latest Cenomanian, all Turonian, and Early Coniacian time. It is marked by a hardground of bored and eroded fossils below basal Coniacian conglomerates and concretions. In the Lake St. Lucia area a complete sequence of Lower Coniacian through Maastrichtian is seen, which consists of shelly glauconitic silts with occasional concretions and limestones. Basal Coniacian strata become younger towards the south, illustrating the overstepping nature of the transgression (Fig. 9),

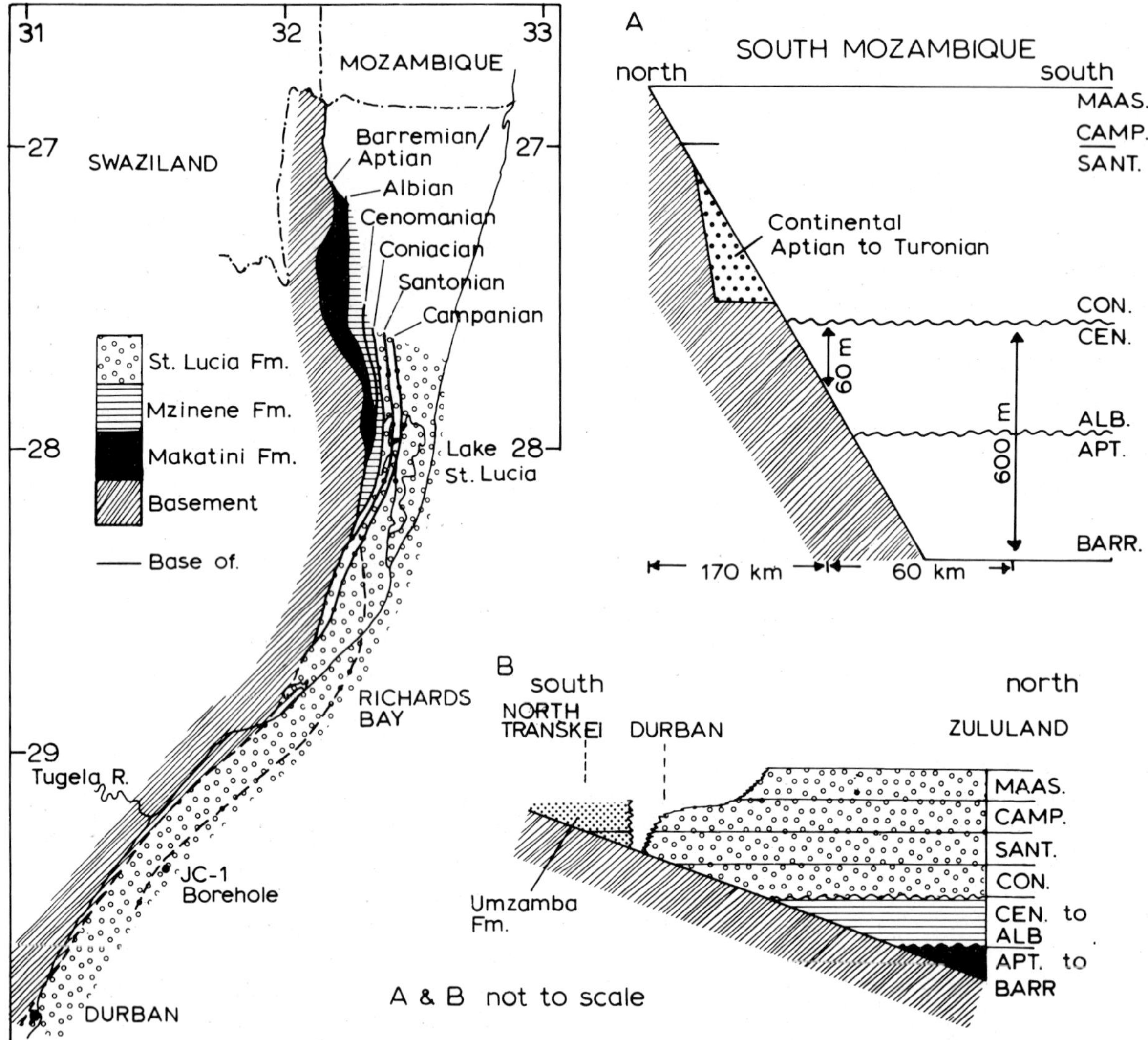

Fig. 9. Distribution of Cretaceous sediments in Zululand/Natal and southern Mozambique (after Du Toit and Leith, 1974; Kennedy and Klinger, 1975; Förster, 1975a). A and B show overlap of successive Cretaceous stages onto basement in south Mozambique and Zululand/Natal, respectively.

and in the vicinity of Hluhluwe rest on the Makatini Formation and finally on Lebombo lavas and basement granites.

Boreholes in the coastal area indicate an hiatus between Maastrichtian and Danian sediments (Stapleton, 1975).

As a whole, the Cretaceous succession in Zululand/Natal has been intensely bioturbated and lacks small-scale sedimentary structures. Plant fragments are very common in the Barremian to Lower Campanian succession.

Santonian and Campanian sediments occur at Durban, and to the south only Upper Cretaceous strata occur: Umzamba Formation (Middle Santonian to Lower Campanian: Klinger, 1977). Here a basal conglomerate resting on Palaeozoic rocks is overlain by shelly sands and dark glauconitic sandy clays. The Umzamba Formation is rich in plant fragments and is wholly marine. Reptilian remains (including *Tylosaurus*) have been recorded by Broom (for example, 1912).

In the Mozambique coastal basin, oil company boreholes have penetrated thick Cretaceous sequences above Jurassic lavas (Flores, 1970, 1973). Förster

(1975a, b) has recently described the Barremian to Cenomanian ammonites from southern Mozambique and a summary of the succession in this area is shown in Figs. 8 and 9. The sequence is subdivided by an Upper Cenomanian to Coniacian hiatus into two major formations (Maputo: Barremian to Cenomanian; and Grudja: Coniacian to Maastrichtian). Locally an intercalation of Turonian continental and fluviatile sediments occurs in the south (up to 100 m thick) which is equivalent to thick (>2500 m) 'mid'-Cretaceous sequences in the Zambezi Valley area where continental sandstones are intercalated with lavas dated at 115 ± 10 m.y. (Flores, 1973). These deposits form the base of the Cretaceous in the north-western and central parts of the south Mozambique coastal basin.

The oldest marine sediments in south Mozambique are Barremian in age and a marked Aptian/Albian non-sequence within the Maputo Formation corresponds to the hiatus between the Makatini and Mzinene Formations in Zululand. Marine Turonian strata have been proved in the eastern, more distal parts of the basin, where a complete Upper Cretaceous to Lower Tertiary succession occurs. The transgressive nature of Cretaceous sedimentation in southern Mozambique is illustrated in Fig. 9, where a steady northward onlap of progressively younger Cretaceous rocks is seen.

On the continental shelf north of Durban, marine Cretaceous sediments were penetrated by SOEKOR borehole JC-1 (Du Toit and Leith, 1974). Cenomanian sandy shales rest unconformably on a diamictite of unknown age, and are overlain by an apparently complete marine Upper Cretaceous succession of grey clays and siltstone, with minor fine-grained sandstones. Thin hard limestone bands occur in the Cenomanian and Turonian part of the sequence. These sediments are continuous into the Palaeocene.

Because of poor exposure and low dips, estimates of thicknesses are tentative for Zululand. Kennedy and Klinger (1975) estimate at least 1000 m for the whole of the Cretaceous in the St. Lucia area. In Mozambique the succession thickens eastwards and reaches a maximum of about 4000 m in the coastal areas in the vicinity of 23°S (Flores, 1973). Offshore, the seismic profile in Du Toit and Leith's (1974) paper shows the Cretaceous succession thickening from about 520 m 2 km offshore, through 748 m in the JC-1 borehole (23 km offshore) to about 1570 m 45 km offshore.

Southern Cape

The succession in the southern Cape consists of two formations separated by an Albian hiatus (Dingle, 1973a; Du Toit, 1976) (Figs. 8 and 10).

Sundays River Formation (Berriasian–Lower Albian). Du Toit (1976) has defined the lithostratigraphy of the Lower Cretaceous in the southern Cape from boreholes in the Outeniqua Basin. Although the type section for the Sundays River Formation was defined in the Algoa Basin (Winter, 1973), only the lower part of the formation is present in this area. The sequence consists of two prograding shale/sandstone deltaic/marine sequences, separated by an erosion surface or unconformity (horizon C of Du Toit, 1976) (see Figs. 8 and 10). The succession can be summarized (base I, top IV):

IV. Medium-grained, porous light grey sandstone with lignitic and shaley layers towards the base.
III. Dark grey mudstones with subordinate sands and thin dolomitic layers. Lignite common.
Erosion surface
II. Medium- to fine-grained, grey shaley sandstones.
I. Medium to dark grey shale with lignite.

The lower member (I) occurs in the Algoa Basin, where the sequence has a rich, shallow marine fauna with occasional ammonites. On the basis of species of *Eodesmoceras* and *Rogersites*, Spath (1930) dated the Algoa Basin Sundays River sediments as Valanginian, but in the more distal parts of the basin Berriasian sediments may be present. The base rests both conformably and locally unconformably upon various members of the Uitenhage Group, and locally upon

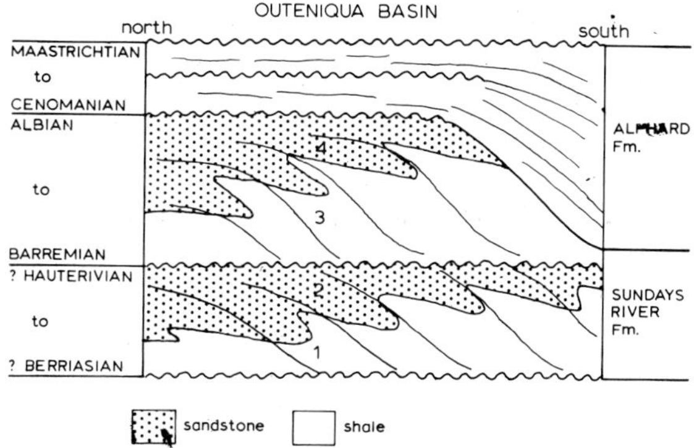

Fig. 10. Schematic relationship of Cretaceous sediments in the Outeniqua Basin (after Dingle, 1973a; Du Toit, 1976).

Palaeozoic basement. The top is not younger than Albian (Du Toit, 1976).

The Sundays River Formation crops out only in the nearshore areas of the Outeniqua Basin where glauconitic, dark grey marine shales dated as Barremian and Aptian have been sampled (Dingle, 1971). Thicknesses in the formation are variable: at least 1800 m in the Algoa Basin (Winter, 1973), and 1600 m in the Outeniqua Basin (Du Toit, 1976).

By Late Albian times the basement irregularities in the Outeniqua Basin had been buried, and the various sub-basins had lost their identities.

Alphard Formation (Cenomanian–Maastrichtian). Upper Cretaceous sediments occur only in the Outeniqua Basin. They are separated from the Sundays River Formation by an erosion surface, and they prograded southwards from sediment supply points which lay in the vicinity of the present-day coastline. Apart from a possible Coniacian hiatus (Du Toit, 1976), the Upper Cretaceous succession is probably complete, and is separated from the overlying Eocene by a strong unconformity. The Alphard Formation consists of soft, medium grey and greenish grey, marine clays, with subordinate thin limestones and sandstones. Glauconitic, pyrite-rich Maastrichtian clayey sands with a rich microfauna have been recovered from the distal parts of the basin on the present-day upper continental shelf (Dingle, 1971).

The formation crops out intermittently in a nearshore zone parallel to the south Cape coast, and forms extensive outcrops southeast of Cape Agulhas. Specimens of *Yabeiceras* and *Eubaculites* have been dredged from these areas. In the central Outeniqua Basin it is hidden by Cainozoic sediments.

Du Toit (1976) records a thickness of at least 1100 m for the Alphard Formation in the Outeniqua Basin (borehole G(b)-Gemsbok/1) and on seismic evidence Dingle (1973a) suggests at least 1400 m from the central part of the basin.

The only marine Cretaceous sediments preserved onshore in the southern Cape occur near East London (Needs Camp). Here a soft, whitish calcarenite is exposed in the "lower" quarry (at 344 m). It rests on a "Karroo" dolerite sill and contains a Campanian fauna dominated by polyzoan fragments (Lang, 1908; Woods, 1908; McGowan and Moore, 1971; King, 1972). A reptile tooth (Plesiosauroidea) has been recorded by McGowan and Moore (1971). The nearby "upper" quarry (at 366 m) exposes cream-coloured crystalline Eocene limestones with reworked Upper Cretaceous fossils (King, 1972).

West coast

Following continental separation in the South Atlantic, Cretaceous sediments were deposited on the continental margin and adjacent oceanic crust off southwestern Africa. The Orange River provided the bulk of the terrigenous sediments and constructed a large prograding sediment wedge: Orange Basin. Although the thickest part of the basin was drilled in 1972, no details have so far been released. The Cretaceous sediments are at least 3500 m thick (Dingle, 1973c), but stratigraphical data are available from only the distal parts of the basin at DSDP site 361 (Bolli et al., 1975; Siesser, 1977). Here, about 1000 m of marine Cretaceous strata were proved in an apparently complete sequence from Lower Aptian to Maastrichtian. The base of the sediments was not reached. The lowest sediments cored (Aptian) consisted of highly carbonaceous (~10%) shales, sandy mudstones and sandstones, which pass upwards into alternating dark grey and reddish shales. This lithology continues to the top of the Maastrichtian, where it is overlain by a deep-water pelagic clay. A characteristic of the Upper Cretaceous sequence is the abundance of parallel and cross-bedded fine sand and silt laminae (Siesser, 1977).

Cretaceous sediments have been recorded from two small outcrops near the coast of southwestern Africa. Just south of the Orange River mouth dinosaur bones (*Kangnasaurus coetzeei* Haughton, 1915) and silicified wood have been found in an argillaceous, lignitic sand. The outcrop is in a small buried valley and the assemblage has been dated as Late Cretaceous (Cluver, pers. comm., 1976). Further north, in the vicinity of Bogenfels, a small erosional outlier of massive limestone containing abundant *Rhynchostreon* lies at 60 m above sea level. A single ammonite specimen, *Proplacenticeras merenskyi* (Haughton) 1930, suggests a Coniacian age.

Suggestions (for example Du Toit, 1954, p. 408) that the lower parts of the continental Kalahari Sand, which cover large areas of central southern Africa, may be of Cretaceous age remain purely speculative.

Cretaceous palaeontology

The bulk of Cretaceous sediments in South Africa are marine, and generally contain a rich invertebrate fauna, particularly mollusca. Although large faunal lists are available for many of these groups, comparison with other areas is severely hampered by the lack of modern taxonomic studies. Notable exceptions are ammonites, and to a lesser extent bivalves and ostracods.

Through Mesozoic time southeast African faunas belonged to a well-defined proto-Indian Ocean province which, for most taxa, included Mozambique, Madagascar, East Africa, South India and parts of Australasia. This is to be expected in view of Early Mesozoic Gondwanaland palaeogeographies. A loosening of ties with these areas is evident in the Late Mesozoic as Gondwanaland fragmented.

East African Scythian ammonoids and bivalves belonged to the Tethyan Realm (Kummel, 1973), whose seas extended as far south as Madagascar, and which stretched from southern Europe, across northern India to China. By Early Jurassic times a more restricted province can be defined by the distribution of the ammonite *Bouleiceras*: East Africa, Madagascar and eastern Tethys (Howarth, 1973). This foreshadowed an even smaller, and more clearly defined "Indo-Madagascay" ammonite province (East Africa, India, Indonesia) which persisted from Callovian to Portlandian times (Cariou, 1973; Enay, 1973). Many of the genera were endemic to the province, which may have included parts of Argentina.

The distribution of belemnites suggests to Stevens (1973) even stricter limits on the size of the province in Kimmeridgian times. He considers the East African and South Indian (Ethiopian) province to be distinct from the Mediterranean–Himalayan–Indonesian province (both of which fall within the Tethyan Realm). On the basis of marine ostracods, we would extend Stevens' Ethiopian province as far as South Africa (see Fig. 11). As the Gondwana fragments became more widely separated during Cretaceous times, and in particular when West Gondwana broke up, the Ethiopian province began to lose its identity as it received immigrants and provided emigrants to the South Atlantic and Pacific areas. Nevertheless a high degree of endemism persisted in certain taxa. We will briefly review a few of the better-known Cretaceous groups in South Africa, although it must be stressed that so far, very little is known about Cretaceous faunas from the west coast, and that most of the following discussion applies to the south and east coast areas.

Mollusca: bivalves

Kauffman (1973) has summarized worldwide Cretaceous bivalve distribution. He recognized a clearly defined East Africa province (southeast Africa, Madagascar and India) with a warm temperate fauna having a high degree of endemism. Characteristic types include: *Pleurotrigonia, Sphenotrigonia, Herzogina, Isotancredia* and *Megacucullacea* (Lower Cretaceous); and *Trigonocallista* (Upper Cretaceous). "Trigonia" types are common in the Lower Cretaceous of northern Zululand and in the Sundays River Formation, where nine species have been recorded (Du Toit, 1954). The decline in endemism in the East African province during the Cretaceous is well illustrated by Kauffman's analyses (1973, figure 5) (% genera and subgenera endemic): 27% (Berriasian-Barremian), 16–13% (Aptian–Albian), and 10–8% (Cenomanian–Maastrichtian). In fact, by Kauffman's definition, the East Africa–India area forms only a sub-province in post-Barremian times.

Mollusca: ammonites

Matsumoto (1973) has summarized worldwide Late Cretaceous ammonite distribution and places southeast Africa in a circum-Indian Ocean province which appears to confirm Kauffman's (1973) conclusions that the Ethiopian or East African province had lost its identity by the end of Early Cretaceous times.

Unfortunately no modern taxonomic studies have been made of South African Early Cretaceous ammonite faunas. Du Toit (1954) lists sixteen species in six genera from the Valanginian lower Sundays River Formation, in the Algoa Basin: *Bochianites* (2), *Distoloceras* (1), *Eodesmoceras* (1), *Leopoldia* (1), *Phylloceras* (1), and *Rogersites* (10). None of these genera are endemic to the western Indian Ocean area, although most of the species are probably so.

Kennedy and Klinger (1975), and Förster (1975a) have recently revised the taxonomy of southern African ammonites. We have extracted details on endemism (for the area including South and East Africa, Madagascar, India, Australia, New Zealand and Antarctica) from their results using the *Treatise* (Moore, 1957). These are summarized in Tables III, IV and V. Unfortunately, no species data were to hand for the Zululand/Natal succession, but Förster's work (1975a) shows that the Albian "high" in endemic genera is mirrored by an increase in the number of endemic species (Table IV). It is interesting to note that the three endemic Albian genera in Mozambique contributed only 9 of the 22 endemic species, which indicates development of "geographical species" within genera which migrated into, or out of, the Indian Ocean province. Taxonomic distribution of abundant and/or diverse taxa in relation to generic endemism (Table V) shows that throughout the Cretaceous, families containing endemic genera are not abundantly represented [exceptions being the Brancoceratidae (Albian) and Collignoniceratidae (Campanian)], and that in post-Albian times endemism was restricted to three families.

The only ammonite recorded from the west coast, *Proplacenticeras merenskyi* (Haughton) 1930, which was dated as "probably Coniacian" (Haughton, 1930), belongs to one of the abundantly represented Coniacian families from Zululand (Table V).

TABLE III

Ammonite genera from Barremian to Maastrichtian strata in Zululand/Natal (after Kennedy and Klinger, 1975)

Barremian (7)

"*Emericiceras*" *
"*Acrioceras*"
Heteroceras
Sanmartinoceras
Phylloceras
Eulytoceras
Colchidites (C.) *

Aptian (13)

Procheloniceras *
Tropaeum *
Ancyloceras
Cheloniceras *
Valdedorsella
Australiceras
?*Rossalites*
Acanthohoplites *
Diadochoceras
Phylloceras
?*Protanisoceras*
?*Tonohamites*
Lytoceras

Albian (37)

Douvilleiceras *
Damesites
Lyelliceras *
Neosilesites
Hypophylloceras
Beudanticeras
"*Cleoniceras*"
"*Sonneratia*"
Rossalites
Ammonoceratites
Anagaudryceras *
Eubrancoceras *
Oxytropidoceras *
Manuaniceras *,**
Tarfayites
desmoceratids
Hysteroceras
Diploceras
Mortoniceras *
Deiradoceras
Erioliceras **
Arestoceras **
Cainoceras **
Puzosia
Bhimaites **
Desmoceras
Gaudryceras
Tetragonites
Hamites *
Anisoceras *
Labeceras **
Myloceras **
Durnovarites
Stoliczkaia
Idiohamites *
Lechites
Mariella
Hypengonoceras **

Cenomanian (21)

Sharpeiceras *
Mariella *
Desmoceras
Sciponoceras
Ostlingoceras *
Hypoturrilites *
Mantelliceras *
Scaphites
Tetragonites
Forbesiceras
Turrilites *
Acanthoceras *
Calycoceras *
Hypophylloceras
Borissjakoceras
Anisoceras
Stomohamites
Sciponoceras
Puzosia
Bhimaites **
Eucalycoceras

Coniacian (27)

Proplacenticeras *
Kossmaticeras
Bostrychoceras
Pachydesmoceras
Peroniceras
Forresteria *
"*Eedenoceras*" **
Basseoceras
Puzosia
Lewesiceras
Yabeiceras
Pseudoxybeloceras
Hyphantoceras
Allocrioceras
Baculites *
Scaphites
Zuluiceras **
Protexanites
Miotexanites
Paratexanites
?*Praemuniericeras*
Zuluites **
?*Gauthiericeras*
"*Fluminites*"
"*Hluhluweoceras*"
"*Andersonites*"
Pseudoschloenbachia

Santonian (9)

Texanites *
Plesiotexanites *
Hauericeras *
Pseudoschloenbachia
Pseudophyllites
?*Karapidites* **
?*Eupachydiscus*
Gaudryceras
Hyphantoceras
diplomoceratids

Campanian (19)

Submortoniceras *
Bevahites **
Menabites *
Hauericeras
Pseudoschloenbachia
Bostrychoceras *
Australiella *,**
Baculites *
Anapachydiscus
Pachydiscus
Hoplitoplacenticeras
Maorites **
Neogaudryceras
Gaudryceras
Eupachydiscus *
Saghalinites *
Gunnarites **
Nostoceras
Neodesmoceras

Maastrichtian (6)

Eubaculites *
Saghalinites
Neodesmoceras
Menuites
"*Epiphylloceras*"
Hopoloscaphites

* Abundant or diverse. ** Endemic to Indo-Malagasy–Australian province.

Ostracoda

South African Cretaceous ostracod faunas are incompletely known, but Dingle (1969a, b, and 1971) and Brenner and Oertli (1976) have described Valanginian, Barremian, Aptian, Coniacian/Santonian, Campanian, and Maastrichtian assemblages. Table VI lists the cytheracean genera known to occur in the South African Cretaceous, and shows that endemism within a western Indian Ocean (Ethiopian or East African) province is high. The Lower Cretaceous faunas are very close to those from Madagascar, being characterized by a *Sondagella/Majungaella/Rostrocytheridea* assemblage. The latter two genera occur in the Australian Upper Cretaceous (Bate, 1972), but are probably endemic to the western Indian Ocean during the Early Cretaceous. Of the Upper Cretaceous genera so far recognized 35% appear to be endemic to the western Indian Ocean (e.g., Sigal, 1974). *Haughtonileberis* (which ranges into the Eocene) and *Agulhasina* both have close relatives in the Australian Santonian/Campanian (Bate, 1972).

Contrary to the conclusions drawn by Bertels (1975), the South African (south and east coast) Cretaceous ostracod faunas are not close to those from Argentina. The latter have strong affinities with West African assemblages which are characterized by abundant Buntoniinae, and genera such as *Veenia, Ovocytheridea,* and numerous species belonging to *Brachycythere.* Although the latter is poorly represented in South Africa, it has not so far been recorded from Australia. The other genera are absent. At the specific level, shallow-water ostracods are prone to local endemism, but at the generic level are good indicators of provincial endemism, and the South African Cretaceous faunas clearly indicate the persistence of an Indian Ocean ostracodal province that can be traced from at least early Late Jurassic times, and which was still identifiable as late as the Eocene (Dingle, 1976b).

TABLE IV

Southern African Cretaceous ammonites endemic to an Indo-Malagasy–Australian province

Age	Number	Endemic	%
Zululand and Natal (after Kennedy and Klinger, 1975)			
Maastrichtian	6 (genera)	none	–
Campanian	19	4	21
Santonian	9	1	11
Coniacian	27	3	11
(Several uncertain types not cited in Moore, 1957)			
Cenomanian	21 (genera)	1	5
Albian	37	8	22
Aptian	13	none	–
Barremian	7	none	–
Mozambique (after Förster, 1975a)			
Cenomanian	17 (species)	9	53
Albian	32	22	69
Aptian	25	9	36
Barremian	1	1	100
Endemic genera were identified only in Albian sediments			
Albian	18 (genera)	3	17

TABLE V

Relationship between abundance and diversity, and endemism in Zululand/Natal Cretaceous ammonites at family level

Age	Family status of abundant and/or diverse genera	Family status of endemic genera
Maastrichtian	Baculitidae	none
Campanian	Collignoniceratidae Baculitidae Pachydiscidae Nostoceratidae	Collignoniceratidae Kossmaticeratidae
Santonian	Collignoniceratidae Desmoceratidae	Kossmaticeratidae
Coniacian	Placenticeratidae Collignoniceratidae Baculitidae	three genera of uncertain taxonomic status
Cenomanian	Turrilitidae Acanthoceratidae	Desmoceratidae
Albian	Douvilleiceratidae Tetragonitidae Brancoceratidae Anisoceratidae Hamitidae	Placenticeratidae Labeceratidae Brancoceratidae Desmoceratidae
Aptian	Douvilleiceratidae Ancyloceratidae	none
Barremian	Ancyloceratidae Heteroceratidae	none

MESOZOIC IGNEOUS ACTIVITY

Two main phases of activity can be identified in time and space (Table VII and Fig. 11). The classic

TABLE VI
South African Cretaceous cytheracean ostracod genera and subgenera

Maastrichtian
?*Acrocythere* (1)
Phacorhabdotus (1)
Trachyleberis (1)
Aguhasina * (1)
Paraplatycosta * (1)
Bairdia (2)
Krithe (1)
Campanian/Santonian
Pondoina * (1)
Brachycythere (3)
Paraphysocythere * (1)
Veenia (1)
Amphicytherura (1)
?*Cnestocythere* (1)
Haughtonileberis * (2)
Gibberleberis * (1)
Cythereis (1)
Acanthocythereis (1)
Coniacian/Santonian
?*Amphicytherura* (1)
Brachycythere (1)
Aptian
Majungaella (1)
Cythereis (1)
Barremian
Arculicythere * (1)
Sondagella * (1)
Isocythereis (1)
Majungaella (1)
?*Cytheropteron* (1)
Valanginian
Majungaella (3)
Rostrocytheridea (2)
Sondagella * (1)
?*Orthonotocythere* (1)
?*Fastigatocythere* (1)
Infracytheropteron (1)
Metacytheropteron (1)
?*Annosacythere* (1)
?*Bythocypris* (2)
Acrocythere (2)
Procytherura (3)
Vesticytherura (1)
Mandelstamia (1)
Pontocyprella (1)
?*Argilloecia* (1)
Saxellacythere (1)
?*Oligocythereis* (1)

* Endemic to western Indian Ocean area. ( ) number of species.

Karroo volcanics are Triassic–Jurassic in age and occur in the east and south of South Africa. They are associated with vertical tectonics and continental breakup in the western Indian Ocean area. Cretaceous volcanism occurs predominantly in the central and western areas and is associated with rifting in the South Atlantic.

Triassic–Jurassic

The youngest Mesozoic deposits in the Karroo Basin are the Drakensberg volcanics. They are widely distributed over southern Africa, and in particular form the spectacular precipices along the Natal/Lesotho border, and all the high ground (>1800 m) in Lesotho. K.G. Cox (1972) has discussed the Karroo volcanic cycle, which is associated with vertical tectonics and the breakup of Gondwanaland in the western Indian Ocean area (Kent, 1974; Dingle, 1976a), and Fitch (1972) has proposed a time sequence for Karroo volcanic activity based on K–Ar dating. His data, plus other radiometric dates relating to South Africa are shown on Table VII. The main types of Triassic–Jurassic volcanic activity were:

Lebombo lavas and intrusions. The earliest activity was in the east, where it persisted for nearly 60 m.y. Up to 13 km of lavas overlie the Clarens Formation, and both lavas and underlying sediments dip steeply eastwards across the so-called Lebombo Monocline, beyond which they plunge under younger Mesozoic sediments which dip eastwards at about 5° into the Mozambique coastal basin. The Lebombo line runs for about 700 km and marks the western edge of the area affected by Mesozoic vertical tectonics (Figs. 11 and 12). The sequence consists of extensive flows of amygdaloidal tholeiite (plus a few olivine basalts), intruded by augite granophyre in Swaziland, overlain by rhyolitic lavas, which are in turn overlain by a sequence of basalts and intercalated rhyolites. The lower rhyolite unit is particularly resistant to weathering and gives rise to the striking Lebombo mountain range. It consists of a series of flows, usually 10–500 m thick each of which has a thin basal tuff overlain by flow banded and brecciated lava (Wachendorf, 1973). An eastward extension of this lava sequence has been found in southern Mozambique at depths ranging from 1830 to 2200 m (Flores, 1970), and according to Kent (1974) it may extend under the Mozambique Channel (Fig. 11A).

TABLE VII

Mesozoic igneous activity in southern Africa

	m.y.	Lebombo	Keetmans-hoop silts	Karroo flood basalts	Karroo intru-sives	Hoachanas basalts	Zuur-berg lavas and tuffs	Kaoko-veld lavas	Alkaline intrusives	Kimberlites
Cretaceous	65 –									
	70 –									
	–									83
	80 –									*
	–									
	90 –									*
	–									90
	100 –									
	–									
	110 –							114		
	–							*		
	120 –								127	127
	–								*	*
	130 –									
	–	137						136	136	
Jurassic	140 –	*						*	* Cape Cross	
	–									*
	150 –				155					147
	–	160			*					
	160 –	*				161	*			
		Upper					162		164	
	–	Rhyolites				*			*	
	170 –								Okonjeje	
	–		178			*				
	180 –		*			173				
	–			*	*				*	
	190 –			187 ± 7	187				190 ± 8	
Triassic	–	*	*						Klein	
	200 –	195	198						Spitzkop	

Jurassic tuffs of uncertain age occur at: Plettenberg Bay (?M.–U. Jur.), Oudtshoorn, Heidelberg, Robertson and Algoa.

Diatremes. Circular and elongate vents up to 2 km across pierce the Stormberg sediments in an area of south Lesotho and northeast Cape, covering about 120,000 km^2. They are filled with basaltic lava, agglomerate, tuff and fragmented sediments (mostly Stormberg lithologies), and fail to penetrate the overlying main flood basalts for which they seem not to have acted as eruptive centres. Over 240 diatremes have so far been recognized, but presumably many more lie under the lava sheets of Lesotho. So far these intrusives have not been dated, but must be older than 187 m.y.

Main flood basalts. These form the high ground of Lesotho, and are composed of thick (up to 50 m), sub-horizontal sheets which can often be traced for more than 100 km. Their maximum development was at least 1370 m, but they have been heavily eroded since extrusion. The main rock types are tholeiites, with occasional olivine basalts. Each flow is typically amygdaloidal at its base, but weathering of the upper surface is very rare, indicating rapid accumulation of free-flowing lava. Dates available so far indicate an Early Jurassic age (187 m.y.). Extrusion is thought to have been from feeder dykes (fissure eruptions), which are sometimes seen to cut through the whole sequence.

Intrusions in the Karroo Basin. Except in the Cape Fold Belt, the entire main Karroo Basin sediment sequence (Dwyka to Beaufort Groups) is riddled with dykes (usually <10 m) and sills (usually 20–200 m). The main rock types are tholeiitic- and olivine-dolerites, and their resistance to weathering is responsible for the characteristic flat-topped mesa and butte morphology of the Karroo landscape. It is assumed that these intrusions were feeders for an extension of the flood basalt covering now preserved only around Lesotho. Activity continued from Early Jurassic (187 m.y.) to Late Jurassic (155 m.y.).

Zuurberg and southern Cape. A narrow, 160 km long outcrop of volcanic rocks occurs along the northern edge of the Algoa Basin, at the foot of the Zuurberg Mountains (Figs. 7 and 11). Several hundred metres of tholeiitic lavas overlie sands, conglomerates and tuffs, and are themselves overlain by the Enon Formation, which contains basalt boulders. These lavas are of Middle Jurassic age (162 m.y.).

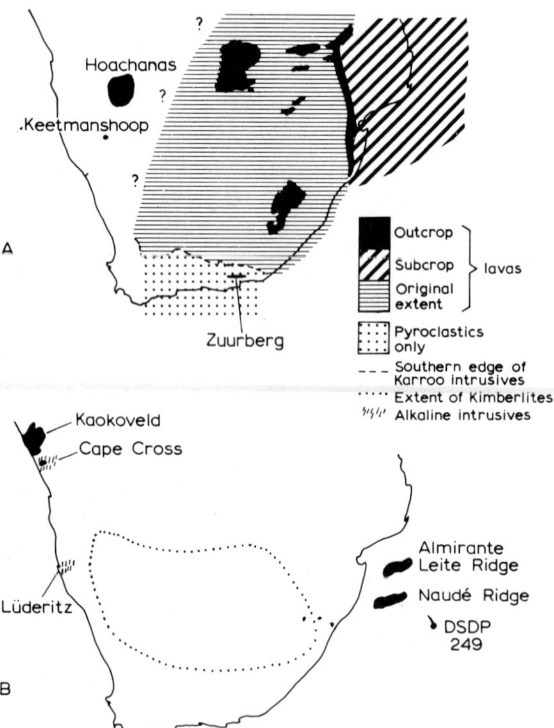

Fig. 11. Distribution of Mesozoic igneous rocks in South Africa (after numerous authors). A. Triassic to Jurassic (associated with vertical tectonics and continental breakup in the southwest Indian Ocean). B. Cretaceous (associated with continental breakup in the southeast Atlantic Ocean).

Pyroclastics (tuffs and occasional bentonites) have been found associated within the Enon Formation at Robertson, Heidelberg and Oudtshoorn (Lock et al., 1975), in the silicified beach deposits at Plettenberg Bay (Du Toit, 1954), and in the lower Kirkwood Formation in the northern Algoa Basin (Du Toit, 1954).

Sills and lavas in South West Africa. An outcrop of basaltic lava flows (at least 360 m in total thickness) overlies Dwyka strata at Hoachanas in south-eastern South West Africa. About 150 km to the south two small dolerite sills occur near Keetmanshoop. These two volcanic sequences have been dated at 161–173 m.y. and 178–198 m.y., respectively (Gidskehaug et al., 1975).

Cretaceous

Upper Mesozoic volcanic activity occurred in the west and central parts of southern Africa. It was of two distinct types (alkaline intrusives, and tholeiitic extrusives) and took place during two distinct phases (Lower and Upper Cretaceous) (Table VII and Fig. 11).

Lower Cretaceous

Intrusives. Two belts of alkaline intrusives occur in South West Africa, whose origins have been related by Marsh (1973) to igneous activity along continental extensions of marginal offsets.

The Cape Cross line (21°S) is an ENE series of intrusive plugs containing alkaline granites, gabbros, anorthosites, syenites, foyaites, essexites and carbonatites. They have been dated at 136–127 m.y. (Portlandian–Valanginian), although the Okonjeje complex has an age of 164 m.y. (Middle Jurassic) and Klein Spitzkop 190 ± 8 m.y. (Late Triassic).

A further set of plugs, with a similar trend occurs immediately south of Lüderitz. Here foyaites are the dominant rock type, one of which (Granietberg) has been dated at 130 ± 2 m.y. (Berriasian–Valanginian) (Marsh, 1973).

Extrusives. Lava flows occur along the northern coastal region of South West Africa: the Kaokoveld. Petrologically they are similar to the Triassic–Jurassic

Karroo flood basalts (tholeiites), but have been dated at 136 m.y.–114 m.y. (Portlandian–Barremian), with a peak of activity around 125 m.y. (Valanginian). They overlie sediments correlated on lithological grounds with the Stormberg Group in South Africa, and with the Etjo sandstone outcrops 250 km to the southeast. In the latter area conglomerates and coarse sandstones are overlain by red mudstones and shales, topped by thick yellow and red massive and aeolian bedded sandstones. The sequence is 600 m thick and contains dinosaur footprints (see Du Toit, 1954, for a full description).

The Kaoko lavas are considered to be the product of volcanic activity which accompanied continental breakup in the South Atlantic (127 m.y.) and as such are analogous to the Triassic–Jurassic activity in eastern Africa which coincided with the breakup of East Gondwana.

Upper Cretaceous

Kimberlite pipes (ultrabasic rocks composed of olivine, enstatite, diopside, phlogopite, Mg-rich garnet, ilmenite, and occasionally diamond, with xenoliths of lherzolite, saxonite and eclogite all set in a groundmass characteristically rich in $CaCO_3$ and serpentine) occur over a wide area in west and central South Africa (Fig. 11). Several of these have been dated, and most have a Late Cretaceous age (range 90–83 ± 4 m.y., Turonian to Coniacian). Individuals not falling within this group have been dated as Precambrian, Late Jurassic (147 m.y.) and Valanginian (127 m.y.) (Moore, 1976).

The pipes tend to occur in clusters and are frequently aligned. Because of their potential economic importance, their mode of origin and distribution in time and space have been extensively discussed, although not until recently within a plate-tectonic context. Moore (1976) has attempted to explain the kimberlite intrusions, together with alkaline volcanism in southwestern Africa (Lower Cretaceous to Oligocene) in terms of crustal flexuring at continental margins, which is in some way related to epeirogenic movements. Fractures and lines of weakness resulting from lithospheric stress tap magma sources at depths which vary according to their position relative to the axis of flexure. Although the scheme has only been outlined in broad terms, it may prove to be a more fruitful line of thought than extant alternatives (for example, stationary mantle plumes or hot spots).

In the following section, the various volcanic phenomena discussed above will be related to plate movements around southern Africa.

PALAEOGEOGRAPHICAL RECONSTRUCTIONS

Triassic

Prior to the break up of Gondwanaland (which started around 180 m.y., e.g., Scrutton, 1973), large-scale vertical movements had taken place in East Africa and the Mozambique/Madagascar area, where fault-bounded basins accumulated thick Late Palaeozoic and Early Mesozoic sediments. During Permo-Triassic times shallow epi-continental seas penetrated into these basins (Fig. 12A), but apart from a possible temporary marine incursion during deposition of the Permian middle Ecca Group in Natal (Rillett, 1963; Hart, 1964), they were not widespread in South Africa and did not persist into the Triassic.

Triassic sedimentation therefore took place in a mid-continental environment, with the accumulation of fluviatile, lacustrine and aeolian deposits. The Karroo Supergroup was deposited in two large basins: Karroo and Zambezi (Rust, 1975), which were separated by a positive area over which formations thin or are absent [Du Toit's (1954) northern and southern facies] (Fig. 12A). The Karroo Basin axis lay approximately east–west, just to the north of the present Cape Fold Belt, and Karroo Supergroup sedimentation can be considered as the last phase in the silting up of the Agulhas–Clanwilliam trough, which also contains rocks of the Cape Supergroup (Cambrian to Devonian) (Theron, 1970). The axis of sedimentation in this trough migrated northwards as it filled and the lower Beaufort Group (Upper Permian) sediments were the last of the Karroo Supergroup to be laid down before uplift of the thick pile began, which culminated in the formation of the Cape Fold Belt. Earth movements possibly occurred as early as Late Carboniferous, but the main phase started in the Early Triassic and reached a climax in the Early–Middle Triassic. The reader is referred to Newton (1975) for a full description of structures and tectonism in the Cape Fold Belt. Lower Triassic (Balfour, Katberg and Burgersdorp Formations) sediments were, therefore, deposited in a shrinking sedimentary basin which lay to the north of the rising

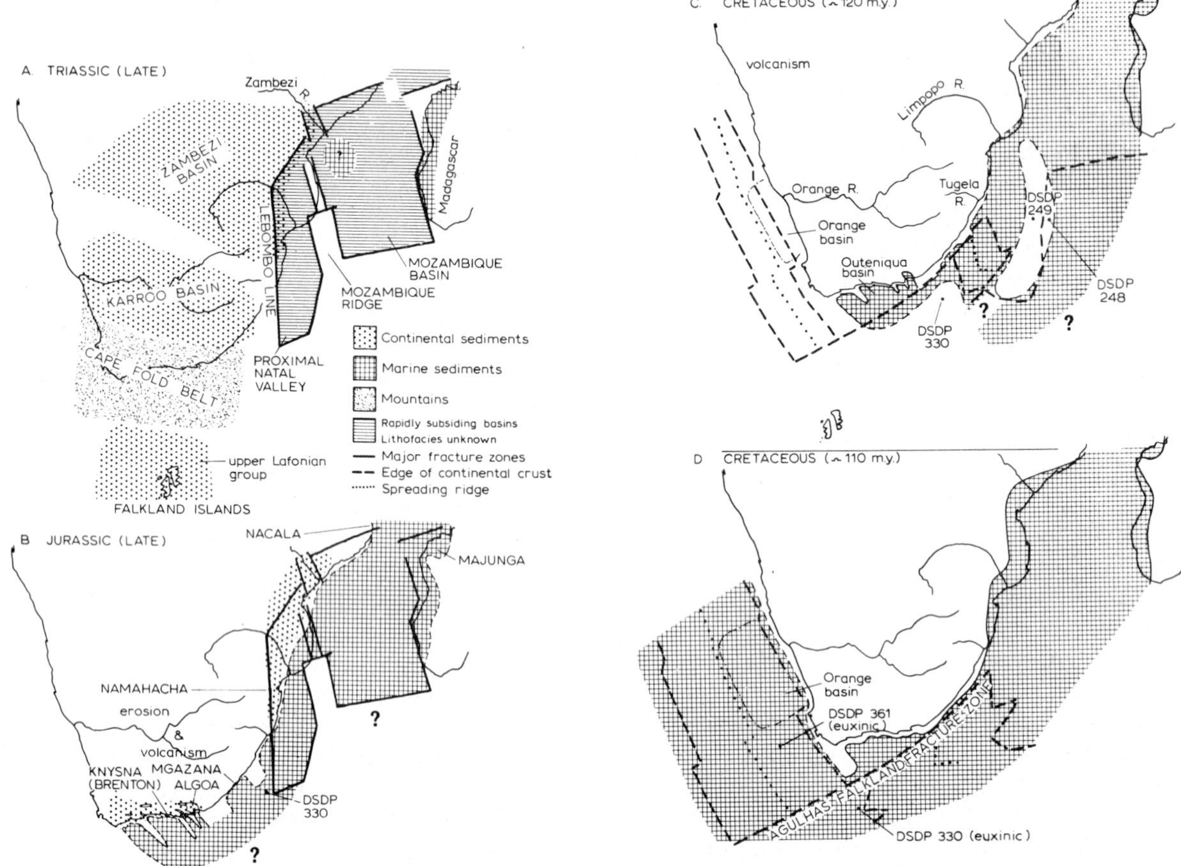

Fig. 12. Schematic palaeogeographic maps of South Africa (after numerous authors). Note that thinned continental crust is thought to underlie the Proximal Natal Valley and Mozambique Basin. These areas, together with adjacent parts of Mozambique and Madagascar and the Mozambique Ridge, are characterized by vertical tectonics. Lithofacies in these basins are unknown, but are probably shallow marine. Restricted marine conditions possibly obtained in the early southeast Atlantic Ocean (including the Orange Basin) in post-Valanginian to pre-Aptian times. Post-Aptian to Maastrichtian conditions on the South African continental margin were shallow, open marine (see Dingle, 1976a, for discussion of Mesozoic ocean basin sedimentary environments around southern Africa).

Cape Fold Belt (Fig. 12A). Deposition took place mainly in a low-energy environment of almost featureless mud flats crossed by slow, shifting streams with occasional sheet flood deposits of fine sand. Raynor (1970) points out that although the sediments are predominantly water-lain, most of the tetrapods were dry land types (the most important exception being *Lystrosaurus*). The abundance of herbivores attests to a plentiful supply of vegetation, but the relative lack of plant remains in the Lower Triassic precludes large numbers of "woody" types. Horizons such as the Katberg Formation, with its content of coarse allochthonous clasts probably reflect pulses in the rise of the Cape Fold Belt which gave rise to temporary increases in the energy of the environment of deposition.

The main folding was completed after the deposition of the *Cynognathus* zone sediments (i.e. post-Scythian) and the Molteno Formation represents coarse detrital fans emanating from the base of the newly formed mountains. The cyclic nature of Molteno sedimentation possibly indicates late tectonic pulses, with each cycle fining vertically and northwards and being laid in braided river channels wandering across a flood plain which enlarged northwards. Coal seams in the Molteno cycles attests to the abundance of vegetation and at least periodic wetness of the climate.

The Elliot Formation red beds probably indicate a return to conditions similar to those which obtained during Scythian times, and although reptile remains are common, plant remains are scarse. During the course of deposition of the Elliot Formation (Ladinian) aridity probably intensified, giving rise to the increasing importance of lighter-limbed saurischian reptiles. The change from water-lain Elliot to semi-arid aeolian Clarens zone I sediments took place gradually. Only the middle (zone II) of the Clarens Formation is a purely aeolian deposit, the upper and lower units (I and III) showing evidence of redeposition by water (periodic heavy storms).

Triassic sediments are represented in the Falkland Islands by the Upper Lafonian Group which is lithologically similar to the Middle and Upper divisions of the Beaufort Group. This suggests that a basin lay to the south of the Cape Fold Belt in which conditions were similar to those that obtained in the Karroo Basin (Fig. 12A).

Jurassic

Horizontal and vertical crustal movements in eastern Africa, and in the Madagascar/India area accompanied the fragmentation of East Gondwana. Although the vertical movements had been under way since Permian times, continental separation probably did not take place until the Early Jurassic (i.e. about 180 m.y.). Karroo volcanism preceded this event by about 15 m.y., and soon terminated accumulation of sediments in the Karroo (and Zambezi) Basin. It reached its maximum intensity just before breakup (~180 ± 7 m.y.), and persisted until Late Jurassic times, although there is no evidence for large-scale volcanicity in the Karroo Basin after about 155 m.y. (Table VII, Fig. 11).

Mid-Mesozoic sedimenation took place in a series of depocentres around the margins of the old lava-covered, Karroo and Zambezi Basins, which may have been uplifted during the Jurassic (Fig. 12B). Denudation this high hinterland gave rise to the Gondwana erosion surface (King, 1967). The earliest local, Jurassic, marine incursions became progressively younger southwards (Fig. 12B): Toarcian (north Madagascar); Bajocian (western Madagascar, and DSDP site 330); Upper Jurassic (Outeniqua Basin). Marine sedimentation persisted off the Natal coast (DSDP site 330) until Oxfordian times, and possibly in the central parts of the Outeniqua Basin throughout much of Late Jurassic time, where the Colchester Member/Brenton Bed tongues represent far-reaching, but temporary, incursions. Evidence from southern Africa is too sparse to accurately chronicle these transgressions, but data from East Africa suggests they took place in three stages (Early Jurassic, Middle Jurassic, and Callovian: Kent, 1974). The similarity of the ostracod faunas indicates a shallow-marine connections between the basins of Outeniqua and Majunga (Madagascar) (Grekoff, 1963) during Late Jurassic time.

Sediment accumulation in the southern Cape, and probably on the adjacent parts of the Falkland Plateau, took place between basement ridges of the Cape Fold Belt. In the proximal basins only torrential piedmont and alluvial fans, and fluviatile valley flat sedimentary environments obtained (Enon and Kirkwood Formations) but in the distal basins these facies passed laterally into hyposaline estuary (distal Kirkwood Formation) and normal or hyposaline shallow-marine conditions (Infanta Formation). As sedimentation progressed, the basement ridges were buried, depocentres coalesced, and marine conditions moved closer to the present coast.

Subsidence of the Outeniqua Basin, was partly facilitated by contemporaneous large-scale movement along the boundary faults of the intermontane half grabens (mostly on the northern sides). As pointed out by Lock et al. (1975) much of this normal, down-to-the-south faulting, which seems to have been in an opposite sense to the original northward-moving thrusts, took place after the deposition of the Enon facies, and may have continued on an extensive scale until Early Cretaceous times. Recent earthquake activity indicates that some of these faults are still active (for example, the Worcester Fault, which has a total throw exceeding 6000 m). Subsidence of the proximal Natal Valley was probably accompanied by faulting in coastal Natal, where Maud (1961) has dated the youngest faults as "end Jurassic", and in the south Mozambique coastal basin by subsidence to the east of the Lebombo fracture line, where draped volcanics show that it was active from Late Triassic to Late Jurassic times. Kent (1974) discusses similar, normal large-scale Jurassic faulting in East Africa, which he considers was the final stage of the Karroo phase of faulting that had been responsible for the deep coastal sediment basins, stretching from Somalia to northern Mozambique and Madagascar.

Continental sedimentation proceeded intermittently in southwestern Mozambique and Madagascar (Isalo Formation), but Karroo volcanism continued in the Lebombo area and possibly Mozambique Channel (Kent, 1974). The ash bands in the southern Cape basins are attributed to this activity.

Cretaceous

The separation of Africa and South America has been dated at 127 m.y. (Valanginian) by Larson and Ladd (1973). This event was probably preceded and accompanied by differential crustal movements along the whole South African coast (commencement of the post-Gondwana erosion cycle of King, 1967). Except in the deep-water sedimentary basins, from which we have no stratigraphical data, there is a hiatus in marine sedimentation between Upper Jurassic and Barremian in the east (Mozambique and Madagascar), and Upper Jurassic and Valanginian in the south (Algoa Basin). Continental separation was accompanied in South West Africa by extensive lava extrusions (Kaokoveld: 136–114 m.y.) and linear alkaline volcanism, the latter possibly along onshore lines of weakness associated with marginal offsets (Fig. 11).

Early Cretaceous marine sediments are known only from the Outeniqua Basin (see Fig. 8). Shallow-water Valanginian deltaic sands and clays prograded southwards (Figs. 10 and 12C, D) filling the last irregularities caused by the pre-Jurassic basement topography (sequences I and II of the Sundays River Formation). The renewed marine invasion of the Outeniqua Basin at this time may have been due to subsidence activated by the first movements along the Agulhas/Falkland fracture zone, which truncated the original basin, and along which South America and southern Africa separated (Dingle and Scrutton, 1974).

By analogy with other rifted continental margins, sedimentation probably commenced in the southeast Atlantic rift zone prior to Valanginian times (i.e. latest Jurassic or earliest Cretaceous), and the basement blocks can be expected to have been draped with locally thick continental clastics, pyroclastics and lavas, with fresh-water sediments in low-lying areas. These conditions may have persisted until Aptian times (earliest marine sediments recorded at DSDP site 361) when restricted entry of the sea was achieved, although in view of the proximity of Valanginian marine sedimentation in Outeniqua Basin, local marine incursions into the southeast Atlantic may have occurred earlier. An intra-formational hiatus occurs within the Sundays River Formation, and renewed progradation of the (Barremian to Aptian) Upper Sundays River Formation deltaic sands and clays (sequences III and IV) may correlate with the regional Barremian transgression recognized in Zululand and Mozambique. During Late Barremian to Aptian times, marine conditions prevailed over most of the South African continental margin (Fig. 12C). In Zululand and Mozambique the marine Makatini/lower Maputo sediments were laid down on an eroded surface of Lebombo lavas, and locally continental Barremian clastic deposits. Open ocean connections are indicated by the rich molluscan (especially ammonites and lamellibranchs) faunas.

Recent seismic work in the northern Natal Valley (between the Lebombo line and the Mozambique Ridge) (Dingle et al., 1977) indicates that sedimentation commenced here prior to Cenomanian times. A prominent mid-Cretaceous reflecting horizon, which has been recognized over most of the area, is underlain on the Tugela Cone by at least 0.75 sec (~1500 m) of sediments. These thin eastwards to about 0.20 sec (~400 m) in the centre of the valley (Fig. 13A). Rapid sediment accumulation continued on the cone during the remainder of Late Cretaceous time, but sediment dispersion throughout the rest of the basin was strongly influenced by two NE–SW basement ridges. These ridges are at least as old as Neocomian, and their strong magnetic signature suggests that they are volcanic in origin. Jurassic and Early Cretaceous volcanism in southeast Africa may, therefore, include not only the north–south Lebombo and Mozambique ridges, but also the intrabasinal Almirante Leite and Naudé ridges (Figs. 11 and 13).

Throughout Late Neocomian times, the Falkland Plateau and the continental margin of southeast Africa moved adjacent to each other, and by Aptian times (110 m.y.) the tip of the Falkland Plateau lay off the southern Cape (Fig. 12D). In the expanding Orange Basin a huge sediment cone was rapidly accumulating off the Orange River in deep stagnant marine conditions (Fig. 13B). Poorly sorted sandy shales and sands containing large quantities of wood and other carbonaceous matter were dumped on the continental slope which prograded rapidly oceanwards.

Final separation between the Falkland Plateau and

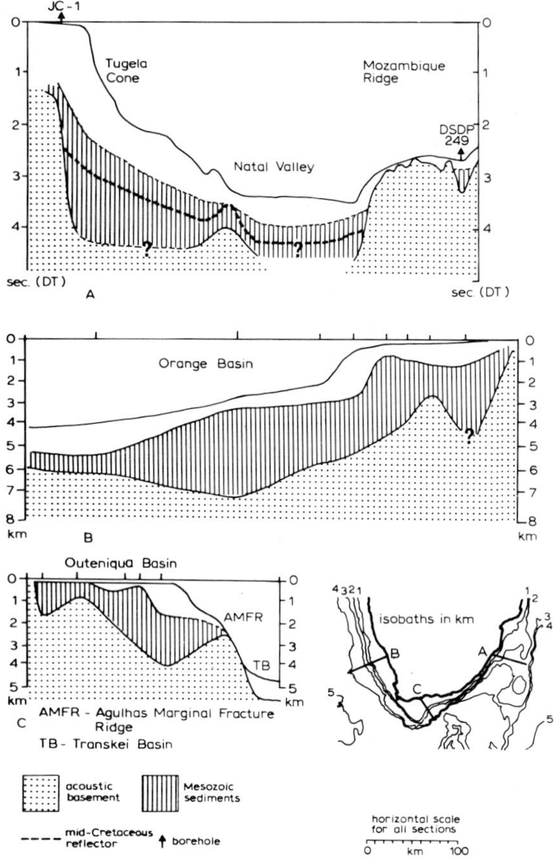

Fig. 13. Thicknesses of Mesozoic sediments on the continental margins around South Africa. A. East coast (Tugela Cone, Natal Valley, Mozambique Ridge). B. West coast (Orange Basin). C. South coast (Outeniqua Basin, Agulhas marginal fracture ridge and western Transkei Basin). A is after Dingle et al. (1977), B and C are mainly after Dingle (1976a). Note that the vertical scale of A is in seconds (two-way time), whereas B and C are in km. Vertical marks along tops of each profile are seismic refraction data sites used in the construction of the sections.

South Africa took place at about 103 m.y. (Dingle and Scrutton, 1974). There was a hiatus in all the sedimentary basins about this time, but as can be seen from Fig. 8 it was not synchronous. In the Outeniqua Basin the lower Alphard Formation rests unconformably on the Sundays River Formation and here the break apparently coincided with the separation event. In Zululand the mid-Cretaceous hiatus is earlier (Late Albian to Early Aptian) and is of a similar age in Mozambique. There may not be a common cause for these breaks in the succession. Following separation, open-ocean, well-oxygenated conditions replaced the stagnant environments in the South Atlantic (Bolli et al., 1975) and after this change a homogeneous shale sequence was deposited during Albian to Maastrichtian time at DSDP site 361.

A major marine regression commenced in the Late Cenomanian. Its effects were pronounced in near-shore areas (such as Natal and Zululand), whereas in the more distal parts of the Outeniqua Basin it interrupted sedimentation for only a short period during Coniacian times, producing a disconformity in the middle of the Alphard Formation (Du Toit, 1976) (Figs. 8 and 10). As pointed out by Moore (1976) this regression coincided with the main phase of kimberlite pipe emplacement (90–83 m.y.), which suggested to him that the regression had an epeirogenic cause. Although this may be true, a possible Turonian/Coniacian hiatus on the Falkland Plateau (which by this time lay well to the southwest) suggests that the epeirogenism coincided with a eustatic sea-level lowering.

The following transgression is Coniacian in age (diachronous from north to south: Early to Late Coniacian respectively in Zululand/Natal) and although the sea receded somewhat from a Late Coniacian high level, sedimentation was continuous in all the marginal basins until the latter part of the Maastrichtian (the upper Umzamba beds may have been eroded). In Zululand the fine-grained lithofacies and abundance of plant remains in the St. Lucia Formation suggest a relatively nearshore, quiet, shallow-water environment throughout the Senonian, and off the Natal coast, borehole JC-1 indicates a nearshore, hyposaline environment in a lateral shelf facies of the Tugela River delta. Here there is a complete sequence across the Maastrichtian/Danian boundary, where the sedimentary environment remained unchanged until well into the Palaeogene (Du Toit and Leith, 1974). A disconformity marks the Mesozoic/Cainozoic boundary in Zululand and southern Mozambique. In the Outeniqua Basin oceanward prograding post-Coniacian sediments were dammed behind the Agulhas marginal fracture ridge (Fig. 13C), which was topped only at the end of Cretaceous times (Dingle and Scrutton, 1974). Here, and in the proximal Orange Basin, the Mesozoic is separated from the Cainozoic by a break in the succession and, locally, an Early Tertiary period of folding and volcanism (Dingle and Gentle, 1972).

REFERENCES

Alvin, K.L., Barnard, P.D.W., Harris, T.M., Hughes, N.F., Wagner, R.H. and Wesley, A., 1967. Gymnospermophyta. In W.B. Harland et al. (Editors), *The Fossil Record.* Geological Society, London, pp. 247–268.

Atherstone, W.G., 1857. Geology of Uitenhage. *Eastern Prov. Mon. Mag., Grahamstown,* 1(11): 580–595.

Bain, A.G., 1845. On the discovery of the fossil remains of bidental and other reptiles in South Africa. *Trans. Geol. Soc. Lond.,* 2.

Barker, P.F., Dalziel, I.W.D., Elliot, D.H., Von der Borch, C.C., Thompson, R.W., Plafker, G., Tjalsma, R.C., Wise, S.W., Dinkelman, M.G., Gombos, A.M., Lonardi, A. and Tarney, J., 1974. South Western Atlantic. *Geotimes,* November: 16–18.

Bate, R.H., 1972. Upper Cretaceous Ostracoda from the Carnavon Basin, Western Australia. *Spec. Pap. in Palaeontol., 10.* Palaeontol. Assoc., London, 85 pp.

Bate, R.H., 1975. Ostracods from Callovian to Tithonian sediments of Tanzania, East Africa. *Bull. Br. Mus. (Nat. Hist.), Geol.,* 26(5): 163–223.

Bertels, A., 1975. Upper Cretaceous (middle Maastrichtian) ostracodes of Argentina. *Micropaleontology,* 21(1): 97–130.

Beukes, N.J., 1970. Stratigraphy and sedimentology of the Cave Sandstone stage, Karroo System. In: S.H. Haughton (Editor), *Proc. 2nd I.U.G.S. Gondwanaland Symp.* CSIR, Pretoria, pp. 321–341.

Bolli, H.M., Ryan, W.B.F., McKnight, B.K., Kagami, H., Melguen, M., Siesser, W.G., Natland, J., Longoria, J.F., Proto-Decima, F., Foresman, J.B. and Hottman, W.B., 1975. Basins and margins of the eastern South Atlantic. *Geotimes,* 20(6): 22–24.

Brenner, P. and Oertli, H.J., 1976. Lower Cretaceous ostracodes (Valanginian to Hauterivian) from the Sundays River Formation, Algoa Basin, South Africa. *Bull. Centre Rech. Pau – SNPA,* 10: 471–533.

Broom, R., 1912. On a species of *Tylosaurus* from the Upper Cretaceous beds of Pondoland. *Ann. S. Afr. Mus.,* 7: 332–333.

Broom, R., 1932. *The Mammal-Like Reptiles of South Africa and the Origin of Mammals.* Witherby, London, 376 pp.

Cariou, E., 1973. Ammonites of the Callovian and Oxfordian. In: A. Hallam (Editor), *Atlas of Palaeobiogeography.* Elsevier, Amsterdam, pp. 287–295.

Cooper, M.R., 1974. Discussion of: The correlation of the subdivisions of the Karroo system by O.R. Van Eeden. *Trans. Geol. Soc. S. Afr.,* 77: 377–380.

Cox, C.B., 1967. Changes in terrestrial vertebrate faunas during the Mesozoic. In: W.B. Harland et al. (Editors), *The Fossil Record.* Geological Society, London, pp. 77–89.

Cox, C.B., 1973. Triassic terapods. In: A. Hallam (Editor), *Atlas of Palaeobiogeography.* Elsevier, Amsterdam, pp. 213–223.

Cox, K.G., 1972. The Karroo Volcanic Cycle. *J. Geol. Soc. Lond.,* 128: 311–336.

De Swart, A.M.J. and Bennett, G., 1974. Structural and physiographic development of Natal since the late Jurassic. *Trans. Geol. Soc. S. Afr.,* 77: 309–322.

Dingle, R.V., 1969a. Marine Neocomian ostracoda from South Africa. *Trans. Roy. Soc. S. Afr.,* 38: 139–164.

Dingle, R.V., 1969b. Upper Senonian ostracods from the coast of Pondoland, South Africa. *Trans. Roy. Soc. S. Afr.,* 38: 347–385.

Dingle, R.V., 1971. Some Cretaceous ostracodal assemblages from the Agulhas Bank (South African continental margin). *Trans. Roy. Soc. S. Afr.,* 39: 393–418.

Dingle, R.V., 1973a. Post-Palaeozoic stratigraphy of the eastern Agulhas Bank, South African continental margin. *Mar. Geol.,* 15: 1–23.

Dingle, R.V., 1973b. Mesozoic palaeogeography of the southern Cape, South Africa. *Palaeogeogr., Palaeoclimatol., Palaeoecol.,* 13: 203–213.

Dingle, R.V., 1973c. Regional distribution and thickness of post-Palaeozoic sediments on the continental margin of southern Africa. *Geol. Mag.,* 110: 97–102.

Dingle, R.V., 1976a. A review of the sedimentary history of some post-Permian continental margins of Atlantic-type. *Ann. Brazil Acad. Sci.,* 48 (Supplement): 67–80.

Dingle, R.V., 1976b. Palaeogene ostracods from the Continental Shelf off Natal, South Africa. *Trans. Roy. Soc. S. Afr.,* 42: 35–79.

Dingle, R.V. and Gentle, R.I., 1972. Early Tertiary volcanic rocks on the Agulhas Bank, South African continental shelf. *Geol. Mag.,* 109: 127–136.

Dingle, R.V. and Klinger, H.C., 1972. The stratigraphy and ostracod fauna of the Upper Jurassic sediments from Brenton, in the Knysna Outlier, Cape Province. *Trans. Roy. Soc. S. Afr.,* 40: 279–298.

Dingle, R.V. and Scrutton, R.A., 1974. Continental breakup and the development of post-Permian sedimentary basins around southern Africa. *Geol. Soc. Am. Bull.,* 85: 1467–1474.

Dingle, R.V. and Siesser, W.G., 1976. *Geology of the Continental Margin between Walvis Bay and Ponta do Ouro.* Geol. Surv. S. Afr., map, in press.

Dingle, R.V., Goodlad, S.W. and Martin, A.K., 1977. The bathymetry and stratigraphy of the northern Natal Valley (S.W. Indian Ocean) – a preliminary account. *Mar. Geol.,* in press.

Dunn, E.J., 1878. *Geological Report of the Stormberg Coalfield.* Parliament Rep. G.4, Cape Twon, 36 pp.

Du Toit, A.L., 1954. *The Geology of South Africa.* Oliver and Boyd, Edinburgh, 2nd ed., 611 pp.

Du Toit, S.R., 1977. *The Mesozoic Geology of the Agulhas Bank, South Africa.* Thesis, University of Cape Town (unpublished).

Du Toit, S.R. and Leith, M.J., 1974. The J(C)-1 bore-hole on the continental shelf near Stanger, Natal. *Trans. Geol. Soc. S. Afr.,* 77: 247–252.

Enay, R., 1973. Upper Jurassic (Tithonian) Ammonites. In: A. Hallam (Editor), *Atlas of Palaeobiogeography.* Elsevier, Amsterdam, pp. 297–307.

Falcon, R.M.S., 1975. Application of palynology in subdividing the coalbearing formations of the Karroo sequence in Southern Africa. *S. Afr. J. Sci.,* 71: 336–344.

Fitch, F.J., 1972. Discussion of: The Karroo Volcanic Cycle.

J. Geol. Soc. Lond., 128: 311–336.

Flores, G., 1970. Suggested origin of the Mozambique Channel. *Trans. Geol. Soc. S. Afr.*, 73: 1–16.

Flores, G., 1973. The Cretaceous and Tertiary sedimentary basins of Mozambique and Zululand. In: G. Blant (Editor), *Sedimentary Basins of the African Coasts, 2. South and East Coasts.* Assoc. Afr. Geol. Surv., Paris, pp. 81–111.

Förster, H.D., 1975a. Die geologische Entwicklung von Süd-Mozambique seit der Unterkreide und die Ammoniten-Fauna von Unterkreide und Cenoman. *Geol. Jahrb.*, 12: 3–324.

Förster, H.D., 1975b. The geological history of the sedimentary basin of southern Mozambique, and some aspects of the origin of the Mozambique Channel. *Palaeogeogr., Palaeoclimatol., Palaeoecol.*, 17: 267–287.

Garden, R.J., 1855. Notice of some Cretaceous rocks near Natal, South Africa. *Q.J. Geol. Soc. Lond.*, 11: 453.

Gidskehaug, A., Creer, K.M. and Mitchell, J.M., 1975. Palaeomagnetism and K–Ar ages of the South-West African basalts and their bearing on the time of initial rifting of the South Atlantic Ocean. *Geophys. J. R. Astron. Soc.*, 42: 1–20.

Grékoff, N., 1963. Contribution à l'étude des ostracodes du Mésozoïque moyen (Bathonien–Valanginien) du bassin de Majunga. *Rev. Inst. Fr. Pet.*, 18: 1709–1762.

Hart, G.F., 1964. Where was the lower Karroo Sea? *Sci. S. Afr.*, 1: 289–290.

Hart, G.F., 1969. The stratigraphic subdivision and equivalents of the Karroo sequence as suggested by palynology. *Gondwana Stratigr., Earth Sci.*, 2: 23–25 (UNESCO, Paris).

Haughton, S.H., 1915. On some dinosaur remains from Bushmanland. *Trans. Roy. Soc. S. Afr.*, 5: 259–264.

Haughton, S.H., 1930. On the occurrence of Upper Cretaceous marine fossils near Bogenfels, S.W. Africa. *Trans. Roy. Soc. S. Afr.*, 18: 361–365.

Haughton, S.H., 1954. Gondwanaland and the distribution of early reptiles. *Trans. Geol. Soc. S. Afr.*, (annexure), 56: 1–30.

Haughton, S.H., 1969. *Geological History of South Africa.* Geol. Soc. S. Afr., Cape Town, 535 pp.

Henriques da Silva, G., 1966. Sobre a Ocorrencia do Jurássico marinho no norte de Mozambique. *Est. Gerais Univ. Mozambique,* III(II): 61–68.

Howarth, M.K., 1973. Lower Jurassic (Pliensbachian and Toarcian) ammonites. In: A. Hallam (Editor), *Atlas of Palaeobiogeography.* Elsevier, Amsterdam, pp. 275–282.

Johnson, M.R., Botha, B.J.V., Hugo, P.J., Keyser, A.W., Turner, B.R. and Winter, H. de la R., 1977. Preliminary report on stratigraphic nomenclature in the Karroo Sequence. *Rep. S. Afr. Comm. Stratigr. Nomencl.,* in press.

Kauffman, E.G., 1973. Cretaceous Bivalvia. In: A. Hallam (Editor), *Atlas of Palaeobiogeography.* Elsevier, Amsterdam, pp. 353–383.

Kennedy, W.J. and Klinger, H.C., 1971. A major intra-Cretaceous unconformity in eastern South Africa. *J. Geol. Soc. Lond.*, 127: 183–186.

Kennedy, W.J. and Klinger, H.C., 1975. Cretaceous faunas from Zululand and Natal, South Africa, Introduction and Stratigraphy. *Bull. Br. Mus. (Nat. Hist.), Geol.*, 25(4): 265–315.

Kent, P.E., 1974. Continental margin of East Africa – a region of vertical movements. In: C.A. Burk and C.L. Drake (Editors), *The Geology of Continental Margins.* Springer-Verlag, Berlin, pp. 313–320.

King, L.C., 1967. *Scenery of South Africa.* Oliver and Boyd, Edinburgh, 2nd ed., 308 pp.

King, L.C., 1972. Geomorphic significance of the Late Cretaceous limestones at Needs Camp, near East London. *Trans. Geol. Soc. S. Afr.*, pp. 1–3.

Kitching, J.W., 1970. A short review of the Beaufort zoning in South Africa. In: S.H. Haughton (Editor). *Proc. 2nd I.U.G.S. Gondwanaland Symp.* CSIR, Pretoria, pp. 309–312.

Klinger, H.C., 1977 The Umzamba Formation at its type section. *Ann. Geol. Surv. S. Afr.*, in press.

Klinger, H.C., Kennedy, W.J. and Dingle, R.V., 1972. A Jurassic ammonite from South Africa. *Neues Jahrb. Geol. Paläontol., Monatsh.*, 11: 653–659.

Kummel, B., 1973. Lower Triassic (Scythian) molluscs. In: A. Hallam (Editor), *Atlas of Palaeobiogeography.* Elsevier, Amsterdam, pp. 225–233.

Lang, W.D., 1908. Polyzoa and Anthozoa from the Upper Cretaceous limestones of Needs Camp, Buffalo River. *Ann. S. Afr. Mus.*, 7: 1–11.

Larson, R.L. and Ladd, J.W., 1973. Evidence for the opening of the South Atlantic in the early Cretaceous. *Nature,* 246: 209–212.

Lock, B.E., Shore, R. and Coates, A.T., 1975. Mesozoic Newark type sedimentary basins within the Cape Fold Belt of Southern Africa. *Proc. 9th Int. Congr. Sedimentol., Nice, 1975.* Preprint.

McGowan, B. and Moore, A.C., 1971. A reptilian tooth and Upper Cretaceous microfossils from the Lower Quarry at Needs Camp, South Africa. *Trans. Geol. Soc. S. Afr.*, 74: 103–105.

McLachlan, I.R., Brenner, P.W. and McMillan, I.K., 1977a. The stratigraphy and micropalaeontology of the Brenton Formation and PB-A/1 well near Knysna, Cape Province. *Trans. Geol. Soc. S. Afr.*, 79: 341–370.

McLachlan, I.R., McMillan, I.K. and Brenner, P.W., 1977b. Micropalaeontology of the Cretaceous beds at Mbotyi and Mnganzana, Transkei, South Africa. *Trans. Geol. Soc. S. Afr.*, 79: 321–340.

Marsh, J.S., 1973. Relationships between transform directions and alkaline igneous rock lineaments in Africa and South America. *Earth Planet. Sci. Lett.*, 18: 317–322.

Matsumoto, T., 1973. Late Cretaceous ammonoidea. In: A. Hallam (Editor), *Atlas of Palaeobiogeography.* Elsevier, Amsterdam, pp. 421–429.

Maud, R.R., 1961. A preliminary review of the structure of coastal Natal. *Trans. Geol. Soc. S. Afr.*, 64: 247–256.

Moore, A.E., 1976. Controls of post-Gondwanaland alkaline volcanism in southern Africa. *Earth Planet. Sci. Lett.*, 31: 291–296.

Moore, R.C. (Editor) 1957. *Treatise on Invertebrate Palaeontology, Part I, Mollusca 4, Cephalopoda, Ammonoidea.* Kansas University Press, Lawrence, Kansas, 490 pp.

Newton, A.R., 1975. *Structural Applications of Remote*

Sensing in the Cape Province, and a New Model for the Evolution of the Cape Fold Belt. Thesis, University of Cape Town (unpublished).

Plumstead, E.P., 1970. A review of contributions to the knowledge of Gondwana mega-plant fossils and floras of Africa published since 1950. In: *Reviews prepared for the first symposium on Gondwana stratigraphy (1967).* IUGS, Haarlem, pp. 139–148.

Raynor, D.H., 1970. Data on the environment and preservation of late Palaeozoic tetrapods. *Proc. Yorks. Geol. Soc.,* 38(4): 437–495.

Rigassi, D.A. and Dixon, G.E., 1972. Cretaceous of the Cape Province, Republic of South Africa. *Proc. Conf. Afr. Geol., 1970,* pp. 513–527.

Rillett, M.H.P., 1963. A fossil cephalopod from the Middle Ecca beds in the Klip River coal field near Dundee, Natal. *Trans. Roy. Soc. S. Afr.,* 37: 73–74.

Rust, I.C., 1975. Tectonic and sedimentary framework of Gondwana Basins in southern Africa. In: K.S.W. Campbell (Editor), *Gondwana Stratigraphy.* Australian National University Press, Canberra, A.C.T., pp. 537–564.

Scrutton, R.A., 1973. The age relationships of igneous activity and continental breakup. *Geol. Mag.,* 110: 227–234.

Seward, A.C., 1908. Fossil flora of Cape Colony. *Ann. S. Afr. Mus.,* 28: 131–158.

Siesser, W.G., 1977. Leg 40 results in relation to continental shelf and onshore geology. *Init. Rep. Deep Sea Drilling Proj.,* Leg 40, in press.

Sigal, J., 1974. Comments on Leg 25 sites in relation to the Cretaceous and Paleogene stratigraphy in the eastern and southeastern Africa coast and Madagascar regional setting. In. E.S.W. Simpson et al., *Initial Reports of the Deep Sea Drilling Project, 25.* U.S. Government Printing Office, Washington, D.C., pp. 687–725.

Simpson, E.S.W. and Dingle, R.V., 1973. Offshore sedimentary basins on the southeastern continental margin of South Africa. In: G. Blant (Editor), *Sedimentary Basins of the African Coasts,2. South and East Coasts.* Assoc. Afr. Geol. Surv., Paris, pp. 63–68.

Soekor, 1976. *The Structure of the Mesozoic Succession of the Agulhas Bank.* Technical Department, Southern Oil Exploration Corporation, Johannesburg, 13 pp.

Spath, L.F., 1930. On the Cephalopoda of the Uitenhage Beds. *Ann. S. Afr. Mus.,* 28: 131–158.

Stapleton, R.P., 1977. Planktonic foraminifera and calcareous nannofossils at the Cretaceous–Tertiary Contact in Zululand. *Palaeontol. Afr.,* Johannesburg, 18: 53–69.

Stapleton, R.P. and Beer, E.M., 1976. Upper Jurassic sediments of southern Africa. *Nature,* 264: 49.

Stevens, G.R., 1973. Cretaceous belemnites. In: A. Hallam (Editor), *Atlas of Palaeobiogeography.* Elsevier, Amsterdam, pp. 385–401.

Theron, J.N., 1970. A stratigraphic study of the Bokkeveld Group (Series). In: S.H. Haughton (Editor), *Proc. 2nd I.U.G.S. Gondwanaland Symp.* CSIR, Pretoria, pp. 197–204.

Truswell, J.F., 1970. *An Introduction to the Historical Geology of South Africa.* Purnell, Cape Town, 157 pp.

Turner, B.R., 1970. Facies analysis of the Molteno sedimentary cycle. In: S.H. Haughton (Editor), *Proc. 2nd I.U.G.S. Gondwanaland Symp.* CSIR, Pretoria, pp. 313–319.

Turner, B.R., 1971. The geology and coal reserves of the north-eastern Cape Province. *Bull. Geol. Surv. S. Afr.,* 52: 74 pp.

Turner, B.R., 1972a. Silica diagenesis in the Molteno sandstone. *Trans. Geol. Soc. S. Afr.,* 75: 55–66.

Turner, B.R., 1972b. Revision of the stratigraphic position of Cynodonts from the upper part of the Karroo (Gondwana) System in Lesotho. *Geol. Mag.,* 109: 349–360.

Turner, B.R., 1975. *The Stratigraphy and Sedimentary Histories of the Molteno Formation in the Main Karroo Basin of South Africa and Lesotho.* Thesis, University of Witwatersrand, Johannesburg, 314 pp. (unpublished).

Van Eeden, O.R., 1973. The correlation of the subdivisions of the Karroo System. *Trans. Geol. Soc. S. Afr.,* 76: 201–206.

Wachendorf, H., 1973. The rhyolitic lava flows of the Lebombos (SE-Africa). *Bull. Volcanol.,* 37(4): 515–529.

Winter, H. de la R., 1973. Geology of the Algoa Basin. In: G. Blant (Editor), *Sedimentary Basins of the African Coasts, 2. South and East Coasts.* Assoc. Afr. Geol. Surv., Paris, pp. 17–48.

Woods, H., 1908. Echinoidea, Brachiopoda and Lamellibranchia from the Upper Cretaceous limestone of Needs Camp, Buffalo River. *Ann. S. Afr. Mus.,* 7: 13–20.

Chapter 12

ARABIAN PENINSULA

P. SAINT-MARC

INTRODUCTION

The Arabian Peninsula is bordered to the west by the Mediterranean and Red Seas, to the south by the Arabian Sea, to the east by the Persian Gulf and to the north and northeast by the Taurus and Zagros Mountains (Fig. 1). The peninsula is a vast Precambrian platform covered by non-metamorphic sediments ranging in age from Middle Cambrian to Pleistocene. Basaltic lavas, the youngest of which are Neogene and Quaternary in age, cover large areas in Syria and Saudi Arabia.

The basement crops out along the Red Sea coast. Eastwards about the longitude of Riyadh it disappears under a sedimentary cover. The basement and the overlying sediments dip gently and regularly towards the Persian Gulf. The age of the rocks exposed at the surface becomes progressively younger in the same direction. In the vicinity of the gulf the sedimentary sequence may reach a thickness in excess of 10,000 m.

The southeastern part of the Arabian Peninsula is occupied by the Rub al Khali Desert, which conceals below recent dunes a vast syncline whose margin to the south is formed by the Hadhramut Arch and in the east by the mountains of Oman. Along the shore of the Arabian Sea structural uplift is marked by sporadic outcrops of basement rocks.

To the north the basement crops out in the southern part of Sinai and along the Gulf of Aqaba. It dips regularly towards the north and northeast under a cover of sediments of both continental and marine origin. Still further to the north the surface of the basement is more irregular (troughs and uplifts) and the structural trends become more variable, from northeast–southwest in the Persian Gulf and in the east of Iraq (Tigris Valley) they swing to the west in the region of Mosul and back to southwest close to the Mediterranean Sea. The Mediterranean margin of the Arabian Shield is broken by the faults of the Gulf of Aqaba and the Dead Sea and their extensions which are responsible for uplifted massifs.

Mesozoic outcrops are of limited extent and essentially confined to the western part of the Arabian Shield where they mantle the basement. In the east they are confined to the Oman Mountains and in northeastern Iraq, but much subsurface information is available from well data.

The nature of sedimentation at the beginning of the Mesozoic, the extent of transgression, etc., were influenced by the paleorelief inherited from earlier orogenic activity. This paleorelief, progressively modified throughout the Paleozoic, was dominated by the Arabian Shield, an extension of the African Shield. Formed of a rigid mass of ancient rocks, the Arabian Shield has remained a relatively stable region since the Cambrian, generally emergent. After prolonged peneplanation it was gently tilted northeastwards at the beginning of the Paleozoic towards the ancestral trough of the Tethys. General sinking produced a vast epicontinental platform upon which from Cambrian times onward sediments were deposited. The most extensive transgression was that of the Early Ordovician which covered most of the Arabian Shield and left as witness graptolitic shales.

At the end of the Paleozoic (post-Early Devonian to pre-latest Permian) central and southern Arabia were affected by orogenic movements resulting in the formation of the east–west striking Central Arabian Arch (at the latitude of Riyadh) and the Hadhramut Arch (extending probably as far as Dhufar) bordering each side of the Rub al Khali depression. The Central Arabian Arch was a slightly positive structure towards the end of the Paleozoic but

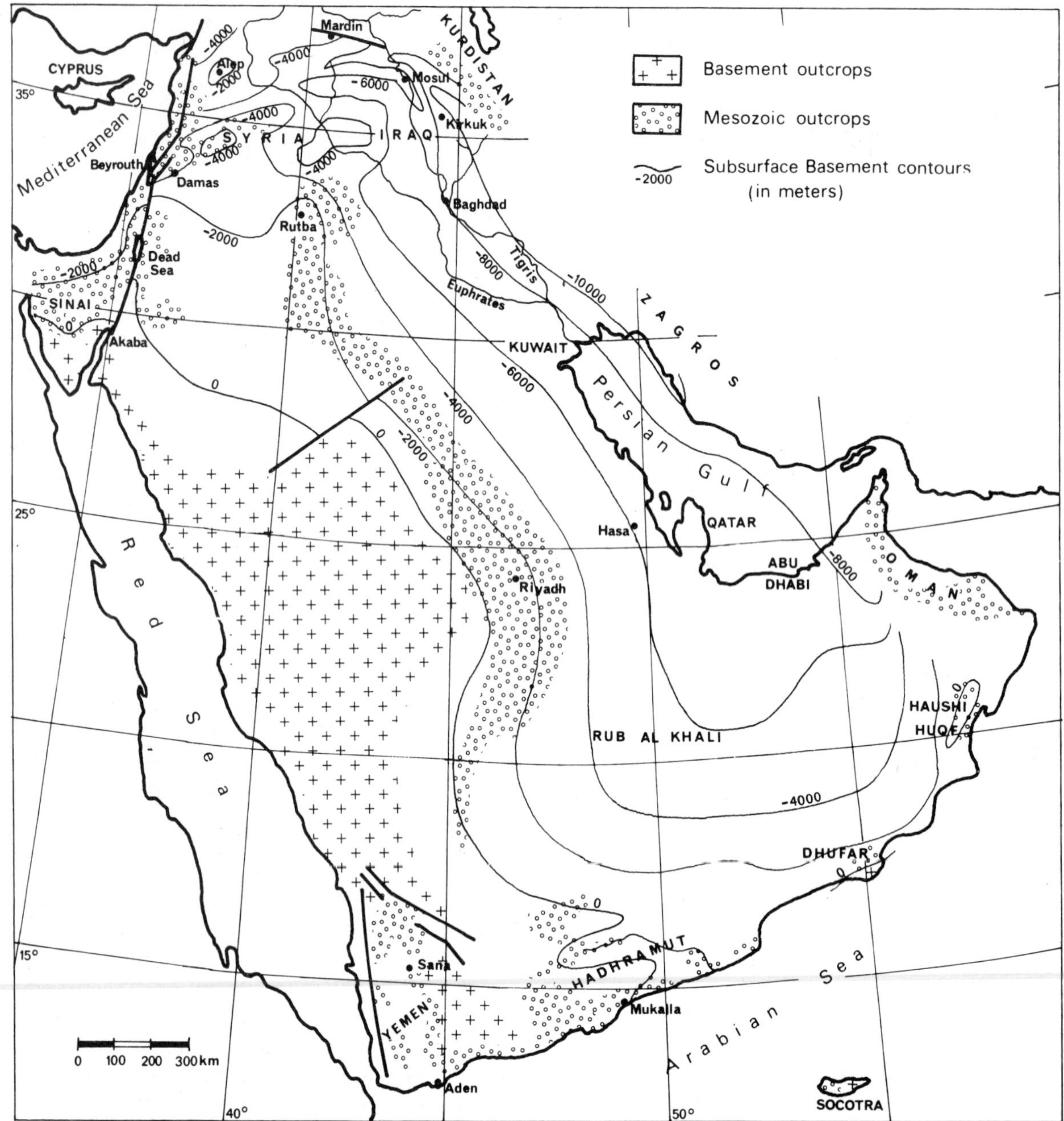

Fig. 1. Structural outline of the Arabian Peninsula (after Dubertret and André, 1969) indicating the area of Mesozoic outcrops.

prior to the Late Permian, became a divide separating two regions with distinctive sedimentation patterns. To the south in the Rub al Khali depression detrital continental sediments (cross-bedded sandstones, conglomerates and shales) together with lacustrine dolomites accumulated between Early Devonian and Early Permian times (Wajid Sandstone). This region was open to the sea in the east for Early Permian marine horizons are present. Still predominantly clastic in the region of Haushi-Huqf, the Lower Permian becomes dolomitic and calcareous in the moutains of Oman. To the north and east of the Central Arabian Arch sedimentation consisted essentially of epicontinental marine carbonates with regular facies varia-

Fig. 2. Reconstruction of the continents into the universal landmass of Pangaea as at the end of the Permian (after Dietz and Holden, 1970).

tions and an eastwards thickening of the section towards the Zagros Mountains.

The Syrian Platform, which forms the northern margin of the Arabian Peninsula, shows some special features. While during the Early Paleozoic it was submerged by the vast transgression which covered most of the Arabian Shield, orogenic movements between the end of the Devonian and Early Permian led to the formation in the north of an E–W arch, the Mardin Arch. The arch was probably emergent, and subsequently became a source of clastic sediments barring marine access to the region to the south. The Mardin Arch is continued westwards, but with a change in trend to NE–SW, by the Amanus Arch. Thus the Syrian Platform is characterized at the top of the Paleozoic by continuous sedimentation of marine clastics in some regions, but discontinuous in others.

Arabia, which formed part of the margin of the African Shield, was then the margin of the ancestral Tethys, and from this latter region periodic transgressions during the Mesozoic covered much of it. Fig. 2 shows the reconstruction of Pangaea during the Permian (Dietz and Holden, 1970). The unusual forms result from distortion due to the projection and to the use of the 1,000-m isobath rather than the shoreline. To aid in recognition, the Gulf of Aden, which did not form until the Cenozoic, is shown open.

The separation of the various plates during the course of the Mesozoic and their movement towards the Eurasian block lead to the progressive closure of the Tethys in this region. It is with this history that the present article is largely concerned.

MESOZOIC SEDIMENTATION

Upper Permian and Triassic

Stratigraphy

In the Middle East it is generally impossible to separate the Triassic from the Upper Permian. The two form part of a sedimentary cycle beginning with transgression at the beginning of the Late Permian followed by progressive regression up to the end of the Triassic interrupted by a minor transgressive phase during the Middle Triassic.

In the centre of the Arabian Peninsula (Jabal Tuwayq) there is a broad band of Permo-Triassic beds with a low easterly dip, stretching for nearly 1,000 km. Four formations have been recognized: Khuff Formation (base of the Upper Permian), Sudair Shale Formation (top Upper Permian–basal Triassic), Jihl Formation (Lower and Middle Triassic) and Minjur Sandstone (Upper Triassic and ? Liassic). The succession is concordant and generally well-dated by marine fossils (ammonites, molluscs, etc.) or by pollen and spores. The Permo-Triassic of this region, well studied by Powers et al. (1966), has been chosen as a type section. Not only has it a wide outcrop, but it is well known in subsurface. In addition, it was sufficiently close to the old sea margin to reflect clearly in the sedimentary sequence transgression and regression but sufficiently distant so that continental influx does not hide these movements.

The base of the Upper Permian. In central Arabia north of 23°N the Knuff Formation, the base of the Upper Permian, is composed primarily of neritic limestones with shale intercalations. The limestones are often dolomitized and the gypsum beds are common. South of that latitude there is a progressive appearance of clastics (first shales, then sandstones) which increase in thickness southwards at the expense of the limestone. At 19°N the formation is entirely arenaceous.

Towards the east in subsurface (Khurais, Qatar, Bahrein) the limestones were deposited at shallow depths and horizons of dolomite and anhydrite occur. The Permian limestones of Zagros, on the eastern side of the Persian Gulf, show the same facies. The formation increases in thickness from 0–300 m in the west to 500–900 m in the east. Towards the south (Rub al Khali) in subsurface the limit between the carbonates and the clastics occurs at about 20°N, but the carbonates (found in wells Al'Ubaylah and ST-3) consist rather of dolomite alternating with anhydrite horizons. Further to the south, the arch of Hadhramut and Dhufar were emergent thus limiting the transgression in the south.

In Oman, if only the outcrops regarded as autochthonous are considered, the succession is calcaro-dolomitic, but in the Fahud well the formation (200 m) begins with sandy discordant limestones. The same discordance is also seen at Jabal Akhdar and south of Saih Hatat where the sequence is dolomitic and thicker (500–800 m). Finally to the northwest of Oman (Ru'us al Jibal) the Upper Permian consists of thick neritic limestones concordant with the base of the Permian.

In the northern part of the Arabian Peninsula, in northern Iraq, the Upper Permian Chia Zairi Limestone Formation rests directly without discordance (?) on Lower Carboniferous. The uppermost Carboniferous beds at the contact with the Permian have a ferruginous crust. The Upper Permian appears to be similar to that of central Arabia and is composed of two thick neritic calcareous units, occasionally with reefs, bounding an evaporitic (dolomitic) unit.

On the Syrian platform, the Upper Permian consists essentially of clastics (shales and sandstones) with limestone intercalations. It has only been found in a few wells and is lithologically indistinguishable from the beds above and below. It is assigned to a clastic sequence (Doubayat Group) with an age range from Devonian (?) to Early Triassic. Along the Dead Sea graben and in Sinai, the Upper Permian is arenaceous at outcrops, but to the west well data show that shales and then limestones (Yamin Formation) appear.

Top of the Permian and Triassic. Beds of latest Permian and Triassic age outcrop over a distance of 800 km in central Arabia where they are seen to rest concordantly upon Upper Permian. The basal beds, the Sudair Shale Formation, of latest Permian to earliest Triassic age are composed of presumed continental red and green shales with intercalations of siltstones, sandstones and gypsum. To the north (27°N) and south (20°N) lenses of carbonate appear. Towards the east in subsurface (Suwei Formation of Qatar) shales are still abundant but alternate with bands of dolomite and anhydrite. The thickness of the formation is variable, 120 m in southern outcrops, 200 m in the north and the same to the east in subsurface.

At the top of the Lower Triassic and in the Middle Triassic marked lateral facies changes are evident in the Jihl Formation. It changes from continental sandstones (130 m) in the south (22°N) to 320 m of sandstones, marine shales and carbonates in the central area of outcrop (24°N) and to limestones 250 m thick in the north (27°N). The formation passes eastwards into a 160 m alternation of dolomite and anhydrite identified in subsurface under the name Gulailah Formation (Qatar). The formation is thus calcareous in the north and thins from the central area of outcrop to the north, south and east.

The same feature persists in the Minjur Sandstone which forms the Upper Triassic – (?)base of the Lias. During this period however, the sequence is continental along the whole length of the outcrop, being formed essentially of sandstones with conglomeratic lenses and red and green shales. Ripple marks, mudcracks, dune bedding and plant fossils all indicate its sub-aerial origin. In subsurface, between the outcrops and the Persian Gulf, it still consists of arenaceous continental beds (Umm Al Jamajim and Khurais wells). Further to the east in Qatar there is a break in the depositional sequence for the Middle Jurassic Izhara Formation rests disconformably directly on the Middle Triassic Gulailah Formation.

In southern Arabia, southern Yemen, the Hadhramut Arch and Dhufar were emergent as during the Late Permian. There is little data from the Rub al Khali region but well ST-3 (20°16′N 46°23′E) indicates that the top of the Permian and Lower Triassic are represented by red and green shales, and the Middle Triassic, which consists of marine sandstones reduced in thickness, is directly and discordantly overlain by Middle Jurassic sandstones. Further to the east, in the Al Ubaylah well (21°59′N 50°57′E), dolomites and red and green shales form the top of the Permian and Lower Triassic overlain by dolomites and marine shales of the Middle Triassic. Once again the Upper Triassic and Lower Liassic are absent.

At Haushi-Huqf the Upper Permian and Trias are absent with the Middle Jurassic resting discordantly upon Lower Permian. The same succession is found in the wells drilled westwards in the vicinity (Haima, Al Ghubar, Ghaba, Afar).

In Oman, the Triassic outcrops generally accepted as autochthonous consist of thick neritic limestones. At Jabal Akhdar they consist of detrital or pelleted dolomitic limestones. At Ru'us al Jibal the Triassic is a calcareous dolomite with the uppermost beds of sandstones, marls and shales conformably overlain by Liassic limestones. To the west however, in the Fahud well, the Triassic is represented by a dolomitic succession with marly and anhydritic horizons in the upper part. This is discordantly overlain by a Jurassic sequence beginning with a sandstone.

The outcrops in northern Arabia of rocks belonging to this time interval are few and restricted to Iraq (Kurdistan and the Ga'ara depression) and along the Dead Sea graben. More complete data are available from the subsurface.

In northern Iraq (Kurdistan) the Lower Triassic consists of the Mirga Mir Formation (shales and argillaceous limestones) overlain by the marls and shales of the Beduh Shale Formation. These beds are concordantly overlain by the Middle Triassic Geli Khana Formation which consists of limestones and dolomites alternating with shales. The top of the formation is marked by ferruginous and silicified dolomites with a ferruginous crust containing limonitic pisolites. This horizon represents an erosional unconformity. The Upper Triassic rests upon these beds without obvious angular discordance. The lower part consists of thick limestones, shales and dolomites of the Kurra Chine Formation overlain by the Baluti Shale Formation which contains shales and marls with intercalations of evaporitic carbonates and anhydrite.

The outcrops in the Ga'ara depression in southwestern Iraq, because of proximity of the ancient shoreline, show a different facies. The oldest beds of Middle Triassic age are the shales and marls of the probably continental Nijili Formation, followed by variegated sandstones of the Ga'ara Sandstone Formation. The Upper Triassic consists of two formations, the lower Mulussa Formation, an alternation of massive, dolomitized limestones and marls, and the upper shales and gypsiferous marls alternating with limestones of the Zor Hauran Formation. In this area several unconformities (non-depositional or erosional) are recognized by the occurrence of conglomeratic horizons and of ferruginous crusts. The most distinct is that separating the Middle and Upper Triassic.

In central Syria the Triassic, not clearly differentiated from either the Permian or the Jurassic, is known only in boreholes where it forms the top of the Doubayat Group and the base of the Dolaa Group. The Lower Triassic, sometimes absent, is essentially composed of shales. The Middle Triassic is calcareous while the Upper Triassic is formed of an alternation of limestones and shales with thin bands of anhydrite. In southern Syria (south of the Palmyrides) the Triassic is made up of sandstones, sometimes ferruginous, alternating with red and green shales, similar to those of Ga'ara (southwest Iraq), and is evidence of the nearness to the shoreline. In Bassit (northwest Syria) blocks of sandy limestones and fine-grained limestones with an Upper Triassic fauna are found associated with ophiolites in a complex which includes sedimentary rocks of ages ranging from Paleozoic to Cretaceous. The origin and date of emplacement of the ophiolites varies according to the author (cf. p. 457).

Along the Dead Sea graben and in Sinai the exact position of the shoreline is known as the result of many studies. The great variety of facies, continental, deltaic, littoral, etc., due to the proximity of the continental source leads to the introduction of numerous local formations. East of this shoreline stretching SSW–NNE in Sinai and N–S along the Dead Sea the continental facies of sandstones and siltstones is assigned to the Nubian Sandstone. To the west marine facies predominate with gradual diminution of the clastics with respect to the carbonates as the distance from the shoreline increases. In the transition zone in the Negev, lithological variations in the Triassic reflect oscillations in the shoreline locations. The transition from Permian to Triassic, the Lower Triassic and the base of the Middle Triassic are formed by sandstones and shales of the Zafir Formation and the Sandy Sequence. The top of the Middle Triassic and the base of the Upper Triassic (Calcareous Sequence) are made up of limestones with some marls and shales while the top of the Triassic is formed by the Evaporitic Sequence consisting of a thick gypsum overlain by limestones and anhydritic, bird's eye and mud-cracked dolomites.

Paleogeographic and stratigraphic synthesis

The Permo-Triassic formations found in the various regions of Arabia and their correlation are shown in Table I. Exact equivalence is often hard to establish because of the poor dating of some formations. However, it is clear that the Arabian Peninsula was subjected to sedimentary and orogenic phenomena some of which affected the whole region, some only parts.

From an analysis of the stratigraphy it appears that the Late Permian was marked by a vast transgression which drove the coastline far to the west in Arabia. The map of the Upper Permian facies distribution (Fig. 3) gives an impression of the position of the

TABLE I

Correlation chart of the Permian, Triassic and basal Liassic formations of the Arabian Peninsula

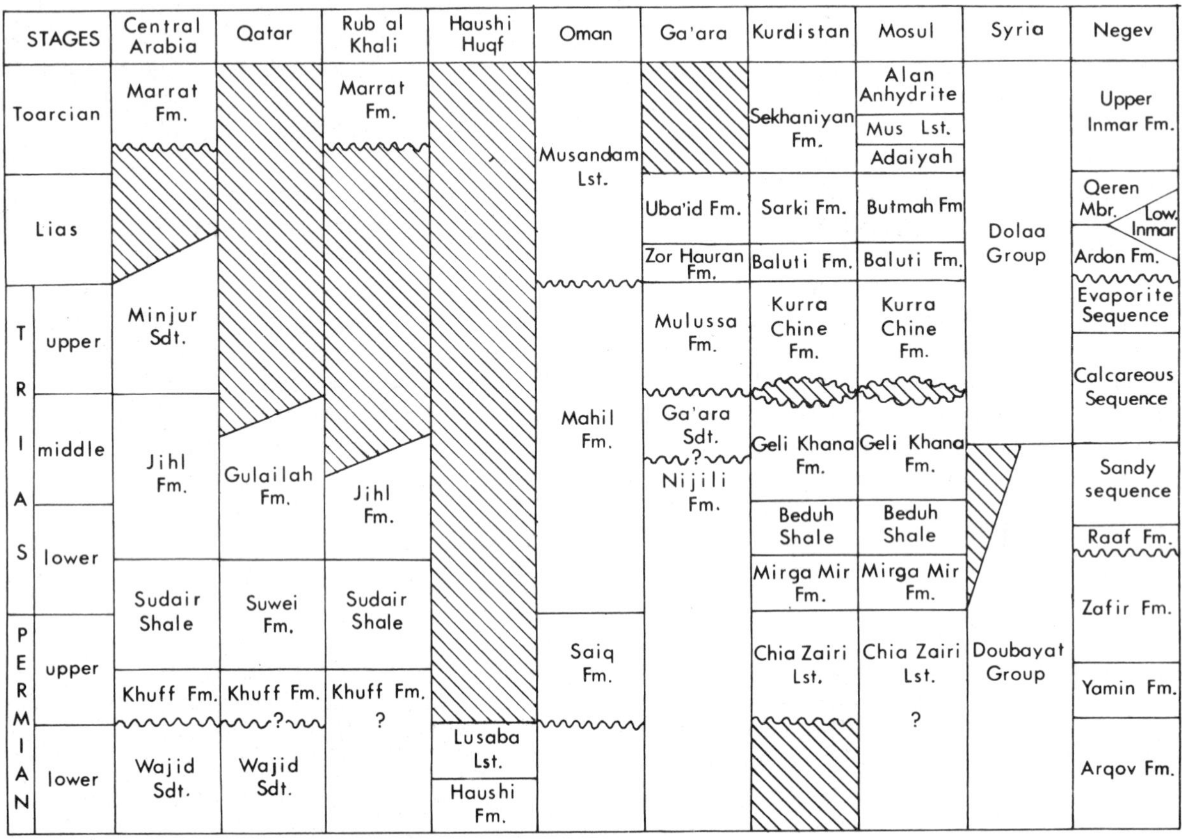

STAGES		Central Arabia	Qatar	Rub al Khali	Haushi Huqf	Oman	Ga'ara	Kurdistan	Mosul	Syria	Negev
Toarcian		Marrat Fm.		Marrat Fm.		Musandam Lst.		Sekhaniyan Fm.	Alan Anhydrite Mus Lst. Adaiyah		Upper Inmar Fm.
Lias							Uba'id Fm. Zor Hauran Fm.	Sarki Fm. Baluti Fm.	Butmah Fm Baluti Fm.	Dolaa Group	Qeren Mbr. Low. Inmar Ardon Fm.
TRIASSIC	upper	Minjur Sdt.				Mahil Fm.	Mulussa Fm.	Kurra Chine Fm.	Kurra Chine Fm.		Evaporite Sequence Calcareous Sequence
	middle	Jihl Fm.	Gulailah Fm.				Ga'ara Sdt. ? Nijili Fm.	Geli Khana Fm.	Geli Khana Fm.		Sandy sequence
	lower			Jihl Fm.				Beduh Shale Mirga Mir Fm.	Beduh Shale Mirga Mir Fm.		Raaf Fm. Zafir Fm.
PERMIAN	upper	Sudair Shale Khuff Fm.	Suwei Fm. Khuff Fm.	Sudair Shale Khuff Fm. ?		Saiq Fm.		Chia Zairi Lst.	Chia Zairi Lst. ?	Doubayat Group	Yamin Fm.
	lower	Wajid Sdt.	Wajid Sdt.		Lusaba Lst. Haushi Fm.						Arqov Fm.

emergent zone, bordered by a vast epicontinental platform opening northeastwards to the Tethys. In central and southern Arabia, the continent is encircled by a belt of clastic sediments, derived from the exposed area. Beyond that lays a zone of shallow-water carbonates. The presence of nodules or horizons of anhydrite associated with the carbonates, from the Rub al Khali to Qatar, gives evidence that in this region the sea was somewhat restricted and hence subject to evaporitic conditions. This may be explained by a positive structure created before the Late Permian transgression and forming a barrier between the region and the open sea. Between Early and Late Permian, orogenic movements locally produced a discordance between beds of the two ages. This disconformity, hardly detectable in central and southwest Arabia (Yemen), becomes clear in the central mountains of Oman (Jabal Akhdar, south of Saih Hatat) and in wells sunk south of here (Fahud) for the base of the Upper Permian is often detrital (conglomerates, sandstones and sandy limestones). These movements do not seem to be recorded at the northwestern tip of Oman (Ru'us al Jibal) where a calcareous sequence is continuous throughout the Permian.

In the area of Haushi-Huqf and in the wells drilled to the north the Dogger and Malm lie discordantly upon Lower Permian. The Upper Permian, Triassic and basal Jurassic were presumably deposited but eroded prior to Middle Jurassic times. The hypothesis of the existence of a positive structure before Late Permian times, perhaps emergent in the region of Haushi-Huqf and trending NNW–SSE to just east of Qatar, seems consistent with the Permian isopach map (Kamen-Kaye, 1970). Along the eastern margin of this swell (or horst?), which was probably not very pronounced and for the most part submarine, the Upper Permian is clearly discordant (Fahud, Jabal

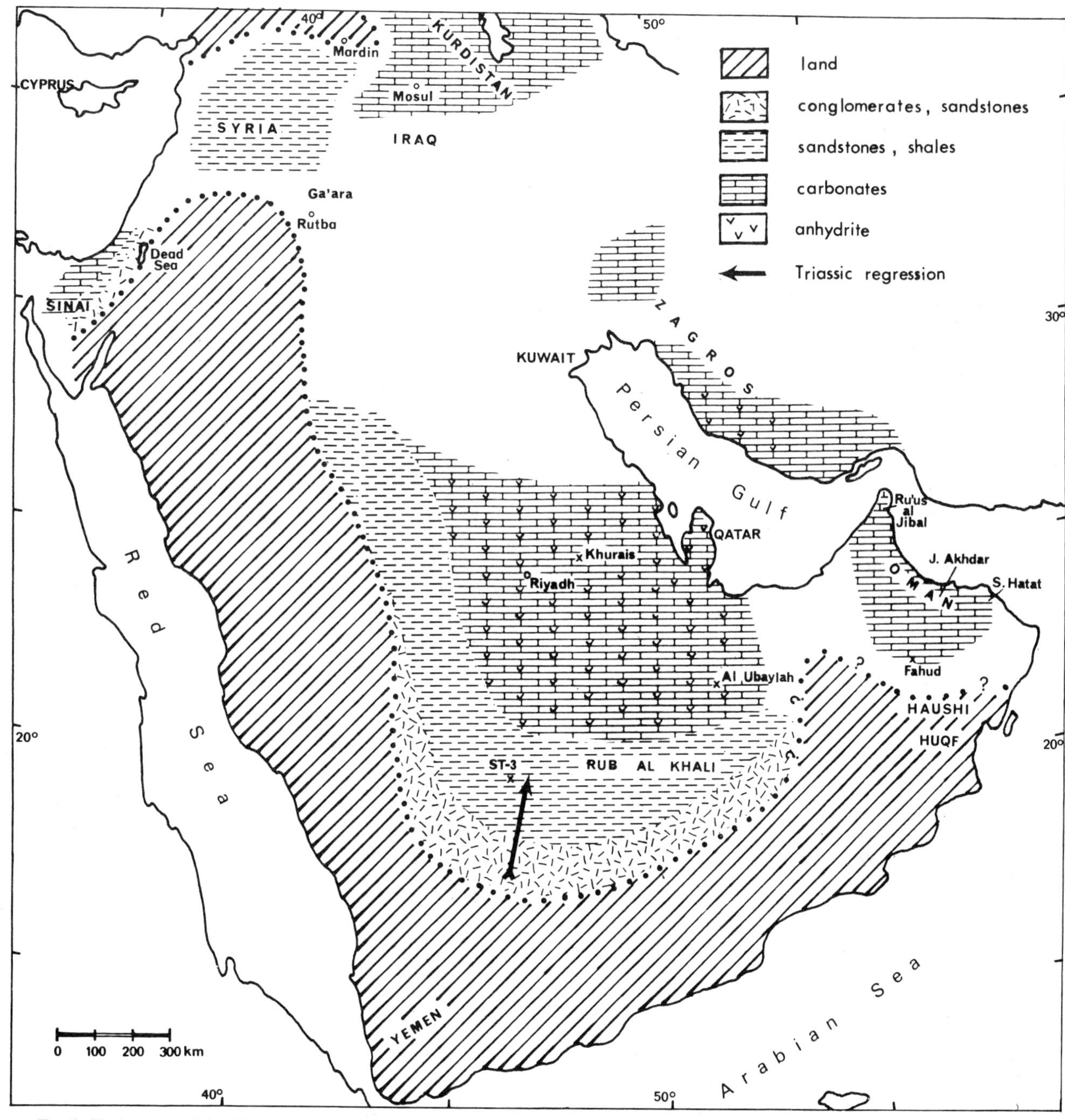

Fig. 3. Facies map of the Upper Permian deposits of the Arabian Peninsula.

Akhdar). The discordance is still visible northeastwards as far as the southern part of Saih Hatat. On the western side the absence of outcrop and the rarity of boreholes (Rub al Khali) prevent any observation being made. Further to the west (Jabal Tuwayq) the discordance has never been conclusively demonstrated. The orientation of the proposed structure is quite different from the E–W trend of earlier structures but has the same direction as the Jurassic structures responsible for the localization of the rich petroleum deposits along the Persian Gulf.

In northern Arabia the Late Permian transgression

can be observed in Iraq (Kurdistan) and along the Dead Sea graben where it is also marked by calcareous deposits. In contrast, on the Syrian Platform there is an essentially continuous clastic marine sequence from the Carboniferous to Lower Triassic. The origin of the sediments lay in part to the south, the Rutba area of the Arabian Shield, and in part from the shoals of Mardin and Amanus to the north.

Following the Late Permian transgression the Triassic was a period of regression, but this was not regular for in some regions, e.g., central Arabia and the Dead Sea, a slight Middle Triassic transgression is recorded. Further in both Iraq and the Dead Sea region numerous unconformities (? emersion, erosional) without visible discordance are known. The most important, as mentioned early, lies at the end of the Middle Triassic.

At the end of the Triassic (? or at the beginning of the Lias) orogenic movements affected central and southern Arabia (Oman) as well as the area of the Dead Sea. Here the transgressive Jurassic beds are discordant. The discordance is not marked in Iraq and Syria where the Jurassic follows the Triassic concordantly without a break in the sedimentation pattern.

Lower and Middle Lias (pre-Toarcian)

Stratigraphy

Marine Lower and Middle Liassic beds, including Domerian, have been recorded in the region of the Dead Sea, in Syria, Iraq, Kuwait, Iran and Oman. Elsewhere, in central and southern Arabia, deposits of this age are unknown.

In Israel the basal marine Jurassic beds lie discordantly upon the Triassic. Outcrops are rare but borings in the Negev and Coastal Plain permit a reconstruction of the depositional environment. The Jurassic transgression in Early and Middle Liassic times began with initial dolomites and anhydrite giving way to fine-grained limestones and dolomites. The beds were deposited on a shallow continental shelf, in a supratidal to lagoonal environment, with the proximity of the shoreline indicated by fluviodeltaic sediments. To the east and southeast (Dead Sea, Jordan and the south of Sinai) continental sandstones appear.

In southern Syria (Palmyra area) the Lower and Middle Lias is concordant upon the Triassic and is formed of calcareous shales and thick beds of anhydrite. Limestone horizons are rare. This is the middle part of the Dolaa Group. Limestones increase in importance further to the north in Syria but are always associated with clastics and anhydrite. In the extreme north against the Turkish–Syrian border (Mardin) the Lias is absent.

In Iraq the base of the Jurassic is likewise concordant. The succession in the northeast in Kurdistan consists of an alternation of oolitic and evaporitic limestones with argillaceous intercalations. In subsurface from Mosul in the north to Kuwait in the south it consists of a heterogeneous mixture of calcareous–argillaceous beds with evaporites. The lithology varies from one region to the next and it is often difficult to distinguish the top of the Triassic from the Lower Liassic. Sandy, silty and argillaceous intercalations are known in the middle of the succession at Kuwait and in the vicinity of Baghdad. In the outcrops in the Wadi Hauran (the Ga'ara depression in southwest Iraq), the Liassic deposits are again concordant. They are dominantly argillaceous and evaporitic at the base (Zor Hauran Formation) passing up into calcaro-dolomitic beds with chert at the top (Uba'id Formation).

Throughout central Arabia (Jabal Tuwayq) no outcrops of beds of this age are known and none have been recognized in wells drilled as far east as Qatar. Equally in the Rub al Khali there is a break in the sequence at this time. The same is true for the whole of southern Arabia from the Yemen to Haushi-Huqf.

In Iran, on the eastern shores of the Persian Gulf (Fars), the Lower and Middle Lias (Neyriz Formation) is present and concordantly overlies the Triassic (Khaneh Kat Formation). It consists of dolomites and dolomitic limestones, occasionally argillaceous with intercalations of silty shales, sandstones and siltstones. Further to the northwest (Lurestan, Khuzestan) the facies found in northern Iraq, dolomites and gypsum, reappear.

In Oman, in the accepted autochthonous regions, the basal Jurassic is represented by a ferruginous sandstone or by a sandy limestone lying with slight discordance on the Triassic. This is overlain by a thick limestone sequence, the Musandam Limestone, of which the lower part is Early and Middle Liassic age (the base of the "*Orbitopsella* Limestones" unit). These are platform limestones, often dolomitic or sandy, and formed at shallow depths. The same succession of an argillaceous basal sandstone discordant on the Triassic and overlain by massive limestone has been drilled at Fahud, south of the mountains of Oman.

Paleogeographic and stratigraphic synthesis

It is apparent from the preceding and from Table I that the marine Lower and Middle Lias is restricted to northern Arabia (Israel, Syria, Iraq) and to the eastern border of central Arabia (Iran, Oman). At the beginning of the Jurassic the sea was less extensive than during the Triassic, and only in Syria and Iraq are the deposits concordant with the Triassic. In Israel and Oman the Jurassic is discordant upon the Triassic and this along with the absence of deposits in the central and southern parts of the Arabian Peninsula is interpreted as evidence of orogenic movements at the end of the Triassic which presumably led to the emergence of most of central and Southern Arabia. The northern part of Arabia was either unaffected or very little affected by these movements.

As a result of the uplift and the limited extension of the Liassic sea, the truly open marine facies linked with the Tethys are only seen in the eastern area (Iran: Lurestan and Khuzestan). In Iraq, Kuwait and along the eastern margin of the Persian Gulf (Fars) the carbonates are always associated with clastics and evaporites. On the Syrian Platform the pattern of sedimentation is very close to that observed in Iraq differing only in the presence of greater quantities of clastics derived from the Mardin Arch to the north. This E–W arch, formed prior to the Early Permian, probably still emergent, constituted, as during the Triassic, a barrier to free contact with the open sea to the north as well as provided a source of clastic sediments. In the region of the Dead Sea as in Oman, the facies although littoral are clearly marine implying direct connection with the open sea.

Toarcian to Bathonian

Stratigraphy

The Toarcian marked the onset of a major transgression which extended over wide areas of the Arabian Peninsula, an area which increased during the Bajocian–Bathonian. This marine incursion is particularly marked in southern Arabia.

In central Arabia the Toarcian Marrat Formation and the Bajocian–Bathonian Dhruma Formation crop out along the long band of outcrops of Jabal Tuwayq, from 19°N to 27°N. Discordantly overlying the Triassic, they are represented west of Riyadh (24°20′N) by 500 m of shallow-water shales and carbonates. Towards the south the series thins and the carbonates and shales are progressively replaced, first by marine and then by continental sandstones. This change is first apparent in the basal beds at 24°N, the uppermost beds only becoming totally arenaceous south of 21°N. This gives a clear picture of the marine transgression to the south. To the north of Riyadh, along the Jabal Tuwayq, the replacement of the carbonates by shales and then by continental sandstones suggests the presence of a shoreline to the west.

Towards the Persian Gulf a lateral passage to shallow-water platform carbonates and a progressive diminution in the importance of shale horizons can be followed in subsurface. At Qatar and to the north, gaps in the sequence are evident along the Persian Gulf. At Qatar the limestones and dolomites of the Bajocian Izhara Formation rest discordantly upon the Middle Triassic, further north at Safaniya (28°13′N 48°49′E) the Upper Toarcian is absent with the Bajocian resting directly on the basal Toarcian. These gaps presumably reflect the existence of shoals which were progressively submerged during the course of the transgression.

In the southwestern Arabian Peninsula (Yemen, Hadhramut, Dhufar) continental sedimentation occurred in the emergent region. Here the Kohlan Series consists of conglomerates, sandstones, sometimes ferruginous and with current bedding, shales and marls with plants which were deposited in subsiding basins (up to 500 m in the Yemen). At the top the formation becomes progressively marine and there is continuity of sedimentation up into the Upper Jurassic. The latter is marine and transgresses widely in this region.

The exact limits of the transgression are not known in the Rub al Khali. It is however known that the Toarcian extension to the south was less than that of the Bajocian–Bathonian. In the west (well ST-3: 20°16′N 46°23′E) the Dogger consists of marine sandstone while further to the east at Al Ubaylah (21°59′N 50°57′E) it is composed of limestone. In the region of Haushi-Huqf the Middle Jurassic limestones rest unconformably upon the Lower Permian. Though not so thick they resemble the facies found in the mountains of Oman. In the latter region the Toarcian, Bajocian and Bathonian form part of a thick carbonate series of the autochthon. They were deposited without break in relatively shallow water (platform facies) from the Liassic to the top of the Jurassic (Lower Musandam Limestone). The thickness of the deposits increases from SW to NE.

In northern Arabia, in northeastern Iraq (Kurdistan), the Toarcian Sekhaniyan Formation is concordant over the Lias. It consists essentially of dolomites of evaporitic origin but contains a limestone in the middle of the unit. In subsurface from Mosul to Kuwait this calcareous horizon, the Mus Limestone, is found between evaporitic horizons (Alan Anhydrite, Adaiyah Anhydrite) which are the lateral equivalents of the Kurdistan dolomites. The limestone forms a transgressive horizon representing more open water conditions and appears to be the equivalent of the Marrat Formation at the base of the Jurassic sequence in central Arabia. In the Ga'ara region (southwest Iraq) the Toarcian is absent (gap or erosion?). The Bajocian–Bathonian Sargelu Formation is characterized by pronounced lithological changes. It consists of shales and limestones with *Posidonia* and radiolarians (sediments of deep-water origin) deposited in a subsiding zone whose axis parallels the trace of the Tigris River, from Mosul in the northeast to Kuwait in the southwest. To the west and southwest, from this trough to Ga'ara, subsidence was not marked, and the beds formed consist of neritic limestones, locally oolitic and sandy, indicating the closeness of a shoreline.

On the Syrian platform and in Lebanon, the Toarcian and Middle Jurassic concordant with the Lias show varying thickness and facies according to locality. A thinning from Lebanon in the west (1200 m) to the Euphrates in the east (100 m) can be observed and the facies can be grouped into three types. To the south between the emergent Arabian Shield and Palmyra the facies are essentially clastic reflecting a near littoral environment with conglomerates, sandstones, neritic limestones (oolitic and sandy), dolomites and marls. North of Palmyra and in eastern Lebanon clays, limestones and dolomites with evaporite (gypsum) intercalations make their appearance. In Lebanon and northwestern Syria a thick series of limestones and dolomites which had direct contact with the open sea were laid down. In the proximity of the Iraq–Syria border the absence of deposits in the Deir ez Zor region implies the existence of an emergent region.

In Jordan, Israel and Sinai emergent regions were close and these areas provided an important source of detrital clastic sediments. However, the passage from a littoral facies in the east and southeast (Jordan and south of Sinai) to the open marine facies in the west and northwest (Negev and Israeli Coastal Plain) was relatively rapid. All the facies from continental to deep-water marine can be recognized. Their location is related to the movements of transgression and regression. Further there is an increase in sedimentary thickness of these deposits from east to west.

Paleogeographic and stratigraphic synthesis

This part of the Jurassic was characterized by a widespread transgression which took place from Toarcian to Bathonian times and expanded further in the Late Jurassic. Central and southern Arabia which was emergent during the Lias was covered by a shallow sea in the Toarcian which did not seem to have extended beyond 22°N. In northern Arabia (Iraq and Israel) the Toarcian, in sedimentary continuity with the Lias, appears to have been a time of a regressive tendency bounded by two transgressive periods, the Domerian and Bajocian.

During the Bajocian (Fig. 4) and Bathonian the area covered by the sea was greater than during the Toarcian. In Syria sedimentation was directly affected by the emergent zones, the Arabian Shield to the south, the Mardin Arch to the north and the Deir ez Zor swell to the east. Thus clastic and evaporitic sedimentation dominates to the north and east, clastics in the south becoming calcareous to the west where the influence of open sea prevailed. In Iraq along the line of the present Tigris Valley a subsiding basin formed which received deep-water sediments. Along the Persian Gulf from Kuwait to Qatar the absence of part or all of the Toarcian may be explained by the existence of shoals (? emergent) formed prior to the Toarcian (? end of the Triassic) which may have been locally rejuvenated before the Bajocian. By Middle Jurassic however they were all submerged. To the west (Jabal Tuwayq) and south (Rub al Khali) there is a progressive passage from marine to continental facies. In southern Arabia the Hadhramut Arch was emergent. The sea did not extend as far as the Yemen, but this was a region of subsidence in which continental deposits accumulated, and in the Late Jurassic provided a passage between the Arabian Sea and the African Sea (in Somalia and Ethiopia).

From a tectonic stand-point, the period was one of calm. The concordant series in the north show a continuity of sedimentation. In central and southwestern Arabia the discordance between the Toarcian (or Bajocian) and the Upper Triassic (or older beds) was

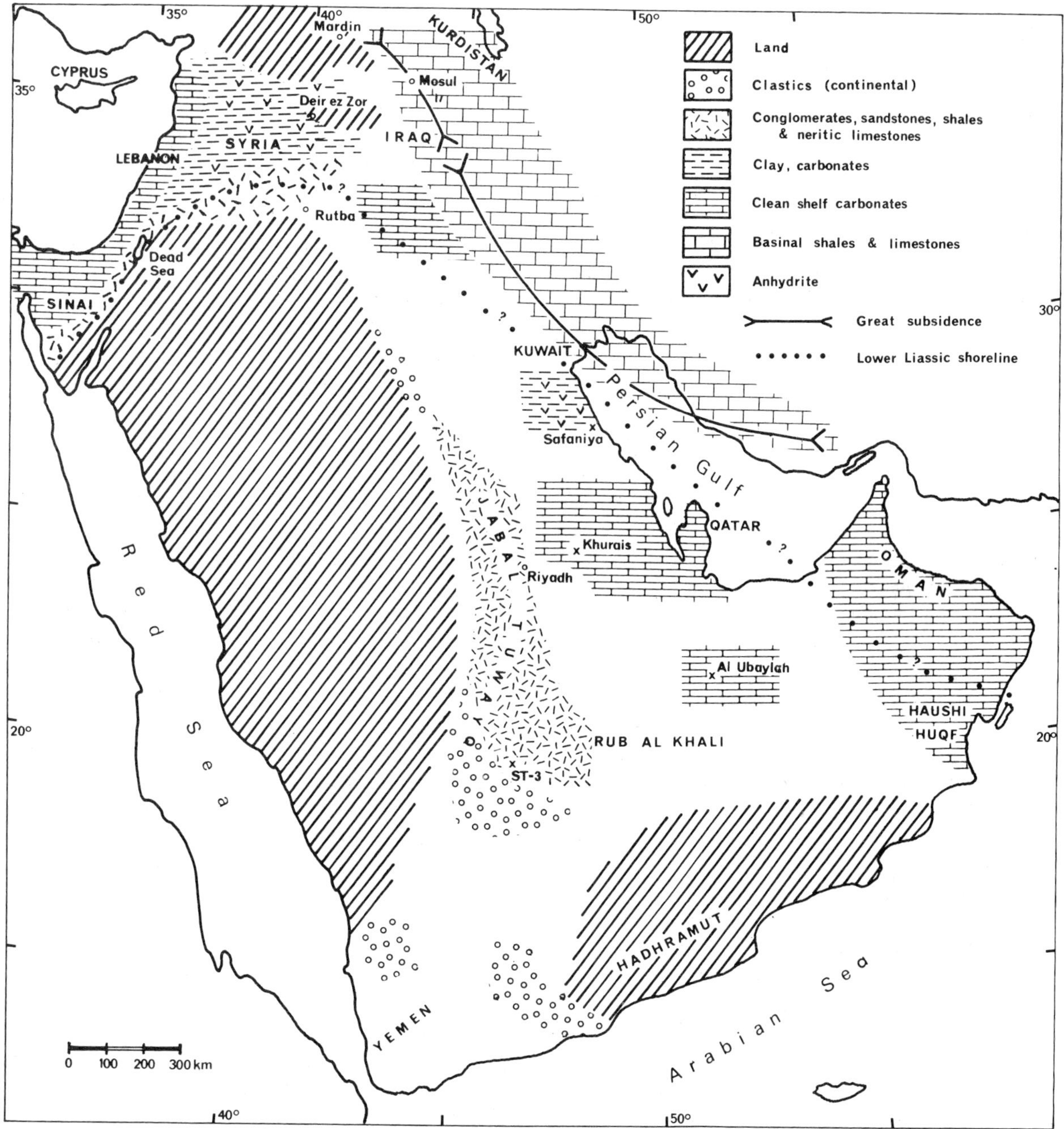

Fig. 4. Facies map of the Bajocian deposits of the Arabian Peninsula.

linked to orogenic movements at the end of the Triassic which had elevated all this region above sea level during the Early Jurassic.

Upper Jurassic–basal Lower Cretaceous

The Upper Jurrassic and basal Lower Cretaceous rocks crop out extensively in the Middle East. In general they are present in the form of shallow-water carbonates with a widespread evaporitic episode at the top of the Kimmeridgian. The most remarkable outcrops in central Arabia are those of Jabal Tuwayq which have a lateral extent of about 1,000 km. The Arabian Shield was largely covered by a transgressive sea which extended to the Yemen where it was in continuity with the East African Sea.

Stratigraphy

In central Arabia the Callovian–Oxfordian Tuwayq Mountain Limestone Formation and the basal Lower Kimmeridgian Hanifa Formation and Jubaila Formation form a continuous arcuate outcrop (from 17°30′N to 27°30′N) bordering the Arabian Shield. The Callovian and Oxfordian which lie unconformably on the Bathonian Dhruma Formation consist of pure shelf limestones with a lithographic or clastic texture and sometimes contain reefs with corals in growth positions. At the base of the Kimmeridgian the limestones are much more clastic than during the Callovian–Oxfordian.

The conditions of sedimentation change during the Middle Kimmeridgian (Arab Formation) which was a period of transition characterized by the deposition of an alternation of neritic limestones, dolomites and anhydrite horizons. At the end of the Kimmeridgian and beginning of the Tithonian sedimentation became almost totally evaporitic with the deposition of the Hith Anhydrite, though there were nonetheless a few minor intercalations of detritic limestone and dolomite.

The Tithonian–Berriasian represented by the Sulaiy Formation and the Yamama Formation of the Lower Valanginian are marked by the deposition of pure fine-grained or detrital limestones rich in molluscs and foraminifera. Immediately on top of these lies the Buwaib Formation of Late Hauterivian age for Upper Valanginian and Lower Hauterivian are missing.

To the east (Khurais, Qatar, Dammam) the same formations are present in subsurface but are generally impure (clay) and finer in texture. Further the Kimmeridgian evaporites are, with respect to the carbonates, poorer in anhydrite than in central Arabia. Finally along the border of the Persian Gulf the Callovian–Oxfordian is commonly absent but there is a continuity of sedimentation up to the Aptian with neither a gap nor discordance between the Valanginian and the Hauterivian.

Southern Arabia appears to have been largely covered with Upper Jurassic marine deposits, possibly only the region of Dhufar remained emergent. Wells drilled in Rub al Khali show that the Callovian, Oxfordian and Lower Kimmeridgian are formed of shelf carbonates of shallow-water origin with the Upper Kimmeridgian evaporitic, a sequence which is directly comparable to that in central Arabia. Above the latter horizon in the east (Al Ubaylah) essentially carbonate sedimentation continues through the Tithonian to Aptian whereas in the west (well ST-3) the basal Cretaceous beds have been removed by erosion.

In southeastern Arabia (Abu Dhabi, Oman, Haushi-Huqf) correlation of the beds by benthonic foraminifera and algae is difficult. For Abu Dhabi the age assignments of Dunnington (1967) are adopted. Sedimentation of shallow-water carbonates continued uninterruptedly from Callovian to Aptian, although at the top of the Kimmeridgian an evaporitic horizon consisting of an alternation of dolomite and anhydrite appears.

In Oman the Callovian, Oxfordian and Lower Kimmeridgian deposits belong to the top of the Lower Musandam Limestone and are concordant with the Bathonian. The rocks consist of highly fossiliferous (corals, stromatoporoids and benthonic foraminifera) shallow-water shelf limestones of varying textures including fine-grained, oolitic, detrital and pelletal. Between the Lower and Upper Musandam Limestone (? Upper Jurassic to Aptian) there is no angular discordance but it is possible that the Upper Kimmeridgian is absent. In fact the Upper Musandam Limestone begins with a breccia containing porcellanous limestones which perhaps form, in part, the Upper Kimmeridgian (= evaporitic facies of central Arabia), but which most probably represent the Tithonian for the limestones contain calpionellids. The overlying beds are porcellanous limestones with radiolaria and calpionellids of Berriasian and ? Valanginian age. To the southwest the lithologies in subsurface are identical, and the transition from the Jurassic to the Cretaceous is marked by a silty, argillaceous limestone with pyritic pellets. In the region of Haushi-Huqf the Upper Jurassic and basal Cretaceous is also represented by shallow-water limestones formed in a near-coastal environment. These beds are much thinner than beds of the same age found in subsurface in Oman.

In southwestern Arabia (Yemen, Hadhramut) the deltaic and lagoonal clastic sediments of the Kohlan Formation are considered to be ? Triassic to Middle Jurassic in age. However at the top marine horizons (calcareous sandstones and sandy limestones) appear and there is a gradational passage to the Upper Jurassic Shuqra Formation. When this is examined within the general framework of the Arabian Peninsula it seems that the transgressive facies of the uppermost Kohlan Formation correspond to the transgressive facies of Late Jurassic age in central Arabia (Tuwayq Mountain Limestones). There is however one difference for in the Yemen the transgression came from the south or southeast. From the Callovian to the Lower Kimmeridgian (Shuqra Formation and base of the Amran Series) the sequence over the region is marine and consists of neritic beds, an alternation of detrital or oolitic limestones, occasionally sandy, and marls. The top of the Kimmeridgian and the ? base of the Tithonian in the centre and northeastern part of the area consist of evaporites with intercalations of shale and sandstone (Sab'atayn Formation and Transition Beds). To the south and east the facies are marine (Madbi Formation) consisting of marls, shales and limestones with beds of gypsum. During the Tithonian and ? at the beginning of the Berriasian (Naifa Formation) the whole region was covered by marine deposits, but important orogenic movements during the course of the Berriasian and which probably began during the Tithonian (intraformational breaks in the Naifa Formation) were responsible for conditions which led to the erosion of the greater part of the formation. The overlying Qishn Formation (Barremian–Aptian) is discordant. Upper Jurassic beds are found in Somalia and Ethiopia with identical facies. This similarity is explained by the link across the Yemen of the seas covering the Arabian Peninsula and East Africa.

In northeastern Iraq (east Kurdistan) the Callovian, Oxfordian and basal Kimmeridgian (Naokelekan Formation) which rest with apparent conformity on the Bathonian Sargelu Formation are represented by a condensed sequence a few meters thick of deep-water limestones and shales rich in ammonites and radiolaria. The top of the Kimmeridgian is also a condensed sequence but marked by an evaporitic tendency for the Barsarin Formation is made up of an alternation of finely crystalline limestones and dolomites and anhydrite.

This zone of condensed sedimentation, passing through Rowanduz, is effectively parallel to the strongly subsiding zone stretching NNE–SSW along the line of the Tigris. The thickness increases from northeast (Kurdistan) to south (Tigris) accompanied by a progressive change in facies and the early appearance of evaporites. Between Mosul and Baghdad the Callovian to Lower Kimmeridgian Najmah Formation consists of thick shallow-water carbonates (oolitic, detrital) and the Middle and Upper Kimmeridgian Gotnia Anhydrite is a thick sequence of massively bedded anhydrite with some detrital limestone and shale intercalations. In Kuwait further to the south the same formations appear but the evaporitic facies begins in the Lower Kimmeridgian (Gotnia Formation = Arab + Hith Formations of Saudi Arabia). To the northwest (Jabal Sinjar) and southwest (Ga'ara) the basal beds of the Upper Jurassic were probably eroded during the Neocomian.

The Tithonian and Berriasian are quite similar to the preceding time interval, with the strongly subsiding zone just east of the Tigris Valley. In eastern Kurdistan relatively thin, deep-water radiolarian shales and finely bedded limestones with ammonites and calpionellids of the Chia Gara Formation were formed.The Berriasian is absent however, and the Valanginian (base of the Balambo Formation) rests directly on the Tithonian.

From Kurdistan southwestwards a progressive change in facies can be observed. Thus in the region of Kirkuk, the zone of maximum subsidence, the sequence consists of a basal Tithonian identical to that in Kurdistan overlain by calcareous mudstones of the Karunia Mudstone Formation (Upper Tithonian–basal Berriasian). The Valanginian and Hauterivian Sarmod Formation overlies it unconformably, for the Upper Berriasian is missing, and consists of glauconitic marly limestones. Still further southwest in the Tigris Valley, the Tithonian is represented by argillaceous limestones and calcareous mudstones with rare intercalations of detrital limestone. Here the Berriasian is absent and the Tithonian is locally reduced by erosion. This erosion, linked to tectonic movements during the Early Berriasian, is more important in northern Iraq.

In the Euphrates Valley the Tithonian–basal Berriasian Makhul Formation in the Awasil well consists of radiolarian shales and calpionellid limestones with sand and silt intercalations. In the Upper Berriasian the whole of the area between the Tigris and Euphrates was characterized by the deposition of detrital limestones associated with calcareous mudstones, dolomites and calcareous shales (Zangura Formation). They are overlain without break by the concordant beds of the Valanginian Garagu Formation. In the extreme southwest of Iraq (Ga'ara) the Tithonian has probably been eroded away.

In the northwestern part of the Arabian Peninsula, the Upper Jurassic follows the Middle Jurassic without break, but the succession, though thick, is often incomplete. The base of the Lower Cretaceous is found only in Syria.

In the Lebanese massif the "calcaires de Kesrouane" represent the Callovian, Oxfordian and Lower Kimmeridgian. They consist of a sequence of limestones and dolomites capped by a volcanic complex (niveau de Bhannès) in which basalts, tuffs and ash layers are associated with oolitic and detrital limestones and marls. This is followed by limestones and dolomites with a rich fauna of stromatoporoids, corals and *Nerinea* (falaise de Bikfaya) and an alternation of oolitic limestones, marls and shales (calcaire de Salima). The age of the latter, given as ? Portlandian, is not well established. In the Anti-Lebanon (Hermon) the Upper Jurassic overlies a thick limestone and dolomite sequence ranging from Bajocian to Lower Callovian. In the Upper Callovian and Oxfordian limestones and dark marls with ammonites alternate. These were products of a shallow, restricted marine environment. The Kimmeridgian– ? Portlandian sequence consists of a basal reef limestone (= falaise de Bikfaya) capped by a marly, occasionally oolitic, limestone (= calcaire de Salima).

The orogenic movements which led to the emergence of these massifs at the end of the Jurassic had their beginning during the Kimmeridgian, a time when faulting allowed the emplacement of basalts. Deep erosion which followed emergence at the beginning of the Cretaceous resulted in the discordance of ? Barremian–Aptian sandstones on various horizons of the Jurassic down to Bathonian.

The Upper Jurassic is only known from the western part of Syria, although the base of the Cretaceous crops out widely. The most complete sequence is seen in the Dolaa (35°15′40″N 38°20′10″E) and Cherrifé (34°30′N 37°27′E) wells in the vicinity of Palmyra. The top of the Dolaa Group (Upper Jurassic) consists of limestones and dolomitic limestones, occasionally argillaceous with bands of shale and anhydrite. There is no discernable break between it and the overlying Cherrifé Shale Formation (Upper Jurassic –basal Cretaceous) in central Syria. However, there may be a short break in sedimentation whose duration increases towards northern and northeastern Syria. The Cherrifé Shale Formation is made up of reddish, hematitic sandy shales with occasional intercalations of limestones, dolomites and calcareous sandstones. Sometimes it may include weathered basalts and ash layers. The formation becomes sandy in northern Syria.

It is probable that Syria, like the Lebanon, was subjected to orogenic movements at the end of the Jurassic leading to strong erosion of the Jurassic beds. This is particularly apparent in eastern Syria where the Upper Jurassic is absent. The Lower Cretaceous is widely distributed for to this age are assigned the clastic beds of northeastern (Hassetché) and northwestern (Kurd Dagh) Syria. Correlation of these formations with those of Lebanon and Iraq is difficult in the absence of a characteristic fauna in the former.

In Jordan there are no marine Upper Jurassic beds. The end of the Jurassic was here too marked by an orogenic phase. Sandstones, often coarse grained, of the base of the Hathira Sandstone Formation (? Upper Jurassic–Lower Cretaceous) rest discordantly upon an erosion surface cutting Middle Jurassic beds.

In Israel and Sinai, the Upper Jurassic, though concordant, is represented by only the Callovian and Oxfordian. Lower Cretaceous sandstones truncate beds of Late Oxfordian age. However, in the Bay of Haifa and north of Galilea fine-grained algal limestones of Kimmeridgian age appear, separated from the Oxfordian by marls and volcanic rocks. In the west (Coastal Plain) the Callovian and Oxfordian are represented by massif reef limestones followed by fine-grained (lagoonal) and detrital limestones, marls and pyritic shales (= Anti-Lebanon facies), while in the east and southeast (Dead Sea and south Negev) the beds are of terriginous origin. Open water sediments and dark shales with reef debris are found offshore west of the present coastline.

Paleogeographic and stratigraphic synthesis

The correlations of Upper Jurassic and Lower Cretaceous beds are shown in Table II.

From the beginning of the Callovian to the Early Kimmeridgian, the transgression which had already begun in Middle Jurassic times expanded widely over the Arabian Peninsula (Fig. 5). The emergent region was of much less importance than during the preceding periods. Southern Arabia (? with the exception of Dhufar) was covered by deposits laid down in a shallow sea, which connected directly with the sea over East Africa. An actively subsiding NNW–SSE zone in Iraq, also active in the Middle Jurassic, extended to the Zagros and was the location of deep-water sedimentation; on its northeastern margin lay a zone in which an identical, but condensed, sequence was deposited, while to the west there was a progres-

TABLE II

Correlation chart of the Jurassic and basal Cretaceous formations of the Arabian Peninsula

STAGES	Central Arabia	Qatar	Kuwait	Iraq: Baghdad	Iraq: Kirkuk	Iraq: Kurdistan E	Lebanon	Israel	Oman	Oman	Hadhramut	Yemen
Hauterivian	Buwaib Fm.	Buwaib Fm.	Ratawi Fm.	Zubair Fm.	Qamchuqa	Balambo Fm			j-n	Upper Musandam Lst.		
Valanginian	Yamama Fm.	Yamama Fm.	Yamama Fm	Garagu Fm.	Sarmord Fm				i			
Berriasian – M. Tithonian	Sulay Fm.	Sulay Fm.	Sulay Fm.	Zangura Fm / Makhul Fm.	Karunia / Chia Gara	Chia Gara Fm.	c. Salima? / F. Bikfaya?		h / g		Naifa Fm.	Naifa Fm.
not dated ? Upper & Middle Kimmeridgian	Hith Fm. / Arab Fm.	Hith Fm. / Qatar Fm. / Fahalil Fm.	Gotnia Fm.	Gotnia Fm. / ?	Barsarin	Barsarin Fm.	Niveau de Bhannès / ?		?		Sabatain / Madbi Fm.	Madbi Fm. / Trans. beds / Amran
Lower Kimmeridgian	Jubaila Fm. / Hanifa Fm.	Darb Fm.	Najmah Fm.	Najmah Fm	Najmah Fm. / Naokelekan Fm.	Naokelekan Fm.	Calcaire	c. à Algues	f	Lower	Shuqra Fm.	Shuqra Fm. / Series
Callovian Oxfordian	Tuwayq Lst.	Diyab Fm.				?		Halutza / Beer Sheba / Kidod	d-e		?	
Bathonian passing L. Callovian at top	Dhruma Fm.	Araej Fm.	Sargelu Fm.	Sargelu Fm.	Sargelu Fm.	Sargelu Fm.	de	Zohar Fm. / Sherif Fm.	c	Musandam	Kohlan Series	Kohlan Series
Bajocian		Izhara Fm.						Daya Fm.	b	Lst.		
Toarcian	Marrat Fm.		Alan Anhy. / Mus Lst. / Adaiyah A.	Alan Anhy. / Mus Lst. / Adaiyah A.	Alan Anhy. / Mus Lst. / Adaiyah A.	Sekhaniyan Fm.	Kesrouane	Rosh Pina / Upper Inmar	a			
Lower Liassic			Butmah Fm.	Butmah Fm.	Butmah Fm	Sarki Fm.		Qeren Mbr. / Low Inmar / Ardon Fm.			?	?

sive transition to a shallow-water shelf carbonate sequence, identical with that found in central Arabia. The Syrian Platform and the western part of Iraq were presumably covered by the sea, but any deposits were eroded away during the Early Cretaceous. Towards the west the region opened to a sea with the sediments at first littoral then of deeper-water type. In central Arabia, along the Persian Gulf and in Rub al Khali and Oman the sediments are shallow-water carbonates, but to the southwest, towards Yemen, they become more detrital.

The passage from the Middle to the Upper Jurassic, while generally concordant and gradual, is marked by a slight discordance in central Arabia and throughout the northeastern part of the Arabian Peninsula from Qatar to Iraq, i.e. in the area bounding the zone of subsidence and the subsident zone itself. This discordance is probably more due to local subsidence than to true orogenic movements. In addition the subsidence appears to have been compensated by the uplift of adjoining areas along faults as seen west of Mosul, recognized by either the absence of deposits or by erosion of the older beds (in Qatar).

The period of transgression was followed by regression during the Late Kimmeridgian which culminated in the deposition of evaporites over nearly the whole region, with the exception of Oman (? break in deposition) and the northwestern part of the peninsula. In the latter region orogenic movements, in the form of faulting and extrusion of basalt flows, occurred. The movements, equally felt in Yemen, were precursors of the main orogenic phase at the end of the Jurassic (? or Middle Berriasian) which affected not only the Arabian Peninsula but also the bordering areas of Turkey, central Iran, Ethiopia and Somalia. This phase of block-faulting is apparent everywhere except in the subsiding zone and the margins where there is continuity of sedimentation with local breaks (in eastern Kurdistan).

Following the orogenic phase, the uplifted and emergent Jurassic beds in the northwestern and southwestern parts of the Arabian Peninsula were subjected to intense erosion. The products of this erosion provided the material for the discordant clastic sediments of the basal Cretaceous. Over the rest of the region marine sedimentation conditions

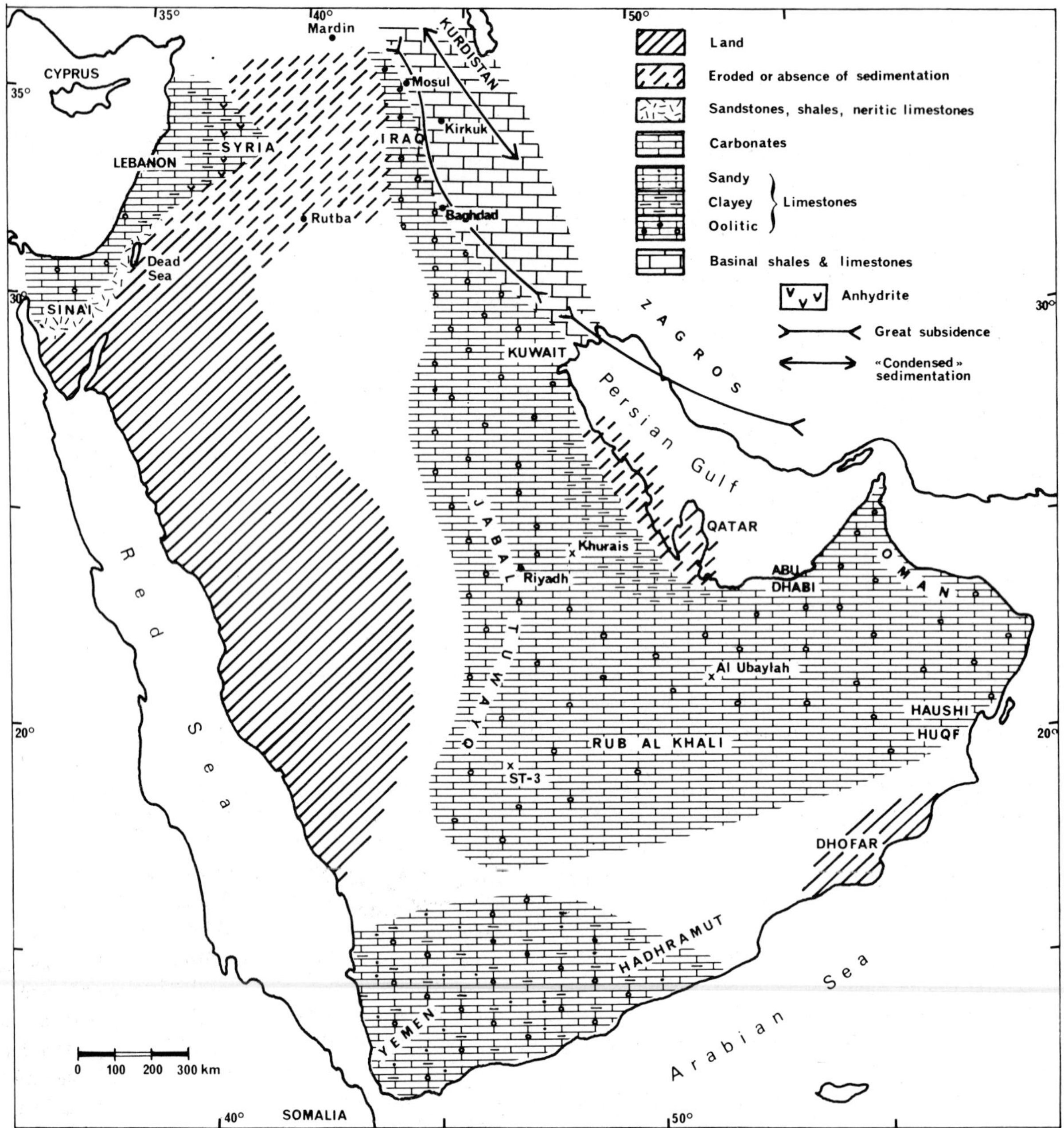

Fig. 5. Facies map of the Callovian–Oxfordian deposits of the Arabian Peninsula.

continued without major break into the Lower Cretaceous, with a facies pattern essentially the same as that of the Upper Jurassic.

Top Lower Cretaceous and Middle Cretaceous

Stratigraphy

In general this was a time of continental and shallow marine sedimentation reflecting regressive and transgressive cycles.

In central Arabia the deposits of this period seen in outcrop are essentially clastic, however, at the beginning of the Hauterivian, there is a thin detrital foraminiferal limestone (Buwaib Formation) resting discordantly upon Valanginian beds. Towards the top, sandstones are intercalated and the series passes progressively to continental (Biyadh Sandstone). A twofold division can be made of these continental beds, a lower sequence of cross-bedded sandstones, variegated shales and conglomerates covering the Barremian and basal Lower Aptian, which after a break comprising nearly all of the Aptian and Albian is overlain by a Vraconian (top Upper Albian) clastic sequence in which is intercalated a thin carbonate bed with marine molluscs. The immediately overlying Upper Cenomanian Wasia Formation is a marine alternation of shales and sandstones, in turn discordantly overlain by the Campanian–Maastrichtian Aruma Formation.

Towards the Persian Gulf in the east, there is a progressive passage into a marine facies, the sequence becoming more continuous and without break. The basal Hauterivian limestone passes rapidly into shales concordant with the Valanginian beds. The Barremian and the base of the Lower Aptian (Biyadh Sandstone) retains its continental sandy facies as far as 49°E, from Rub al Khali to Safaniya. From 49° to 51°E the sequence becomes marine with an alternation of shales, sandstones and limestones, while beyond 51°E only neritic limestones are found. A carbonate horizon with orbitolinids, the Shu'aiba Formation of Aptian age, covers a vast area, from Rub al Khali to Iraq, with almost constant thickness, thus separating the Barremian–basal Lower Aptian from the Upper Aptian–Cenomanian. While probably removed by erosion to the west, it is dolomitic at Khurais and a calcareous, sometimes dolomitic, shale along the Persian Gulf and in the Rub al Khali. It is overlain by the marine Wasia Group in which by means of benthonic foraminifera seven members have been recognized. Although the Turonian is absent, in part, the sequence is covered with apparent conformity by the ? Coniacian–Maastrichtian Aruma Formation.

In southern Arabia from Yemen to Dhufar, due either to non-deposition or erosion, there are no basal Cretaceous deposits. The top of the Lower Cretaceous and Middle Cretaceous which rest unconformably on the Jurassic are marine in the east (Mahra Group) and continental in the west (Tawilah Group). Open sea lay to the east.

The Barremian–Aptian Qishn Formation consists of a marine sequence of limestones and marls in Dhufar and Mahra in the east and of limestones with shales and sandstones in the Hadhramut; still further to the west in Yemen this time interval is represented by continental sandstones, conglomerates and shales forming the base of the Tawilah Group. The transition from marine to continental facies is gradual but the presence of N–S structures formed at the end of the Jurassic results in significant thickness variations.

The same basic pattern persists in the Albian and Cenomanian with marine facies in Dhufar and Mahra passing westwards into continental beds. In the east the Albian and Cenomanian Fartaq Formation consists of richly fossiliferous limestones, marly limestones and marls, and it is the presence of *Praealveolina* which indicates that they do not extend up into the Turonian. In the west, in eastern Hadhramut there are two distinguishable sequences, the lower Albian Harshiyat Formation, a clastic but marine sequence of sandstones, shales and marls with oysters and orbitolinids, and an upper, also called Fartaq Formation, essentially of carbonates with marl and shale intercalations. Still further to the west the carbonate facies disappears and the Albian–Cenomanian (? and Turonian) sequence becomes entirely continental with sandstones containing shale and marl intercalations.

The examination of the Barremian to Cenomanian succession shows two deep incursions to the west from a sea lying to the east, the first during the Aptian (top of the Qishn Formation), the second at the top of the Cenomanian. The earlier transgression was associated with the formation of a highly fossiliferous limestone with a fauna and facies similar to that found at the same period in Lebanon and Syria (Falaise de Blanche) as well as in Iraq and along the Persian Gulf. Above the Middle Cretaceous a sandstone of the Mukalla Formation of presumed Senonian age was deposited in apparent concordance.

In southeastern Arabia in Rub al Khali, Abu Dhabi and Oman, Hauterivian deposits follow the Valanginian concordantly. They consist of fine-grained limestones with radiolaria and *Nannoconus*, a facies which persists in the Barremian to the tip of Oman (Jabal Habhab) while in other regions such as Haushi-Huqf and Oman Mountains (Sayh Hatat and Jabal Akhdar) the Barremian is represented by shallow-water, occasionally argillaceous, limestones. The Aptian over the whole region is represented by a massive, orbitolinid, limestone. The uppermost Aptian and Lower Albian are known in the Fahud and Lekhward wells where they are represented by shales with thin intercalations of argillaceous limestones (the Nahr Umr Shale Formation) whilst to the northeast in Jabal Akhdar massive shelf limestones occur.

The Wasia Group of Upper Albian and Cenomanian age consists of shallow-water shelf carbonates. The Turonian and base of Senonian are absent, and on top of a deeply eroded Middle Cretaceous the ? Santonian to Campanian Muti Formation of marls, shales, lenses of conglomeratic limestones and flyschoid turbidites lie unconformably. The beds contain planktonic foraminifera as well as reworked microfaunas.

In northeastern Arabia (Kuwait, Iraq) during the Early and Middle Cretaceous, there are numerous rapid variations in thickness and facies and breaks in sedimentation. The faulting which accompanied the earth movements at the end of the Jurassic resulted in the formation of a number of subsiding zones receiving heavy sedimentation interspersed with highs, which when exposed underwent intense erosion. Thus the principal transgressive and regressive events are not readily discernable.

In Kuwait and southwestern Iraq up to the River Tigris, the succession found is identical with that of central Arabia. At Burgan the Valanginian is conformably overlain by a thin Hauterivian oolitic limestone of the Ratawi Formation followed by a thick series of sandstones of the Barremian–basal Lower Aptian Zubair Formation. The sandstones are transgressed by an Aptian dolomitic limestone of the Shu'aiba Formation. The Upper Aptian and Albian Nahr Umr Formation consists of sandstones and sands with lignite and amber. This regressive facies has been partly or totally removed by erosion to the west of Baghdad (well Awasil 5). In the Upper Albian and in the Cenomanian the sequence is similar to that of central Arabia. The top of the Cenomanian, the Magwa Formation (= Rumaila + Mishrif Formation of central Arabia), is often profoundly eroded and covered discordantly by the Gudair Formation (Upper Campanian).

Because of emersion following the terminal Jurassic earth movements west and southwest of Mosul, the Valanginian and Hautcrivian are not found. The tabular relief was only submerged during the Barremian–Aptian, at which period marls and neritic marly limestones, sometimes sandy or silty, of the Sarmord Formation were deposited. The Lower Albian deposits (Qamchuqa Formation), directly overlain by Campanian–Maastrichtian, consist of limestones west of Mosul and marly limestones with anhydrite southwest of Mosul.

The region east of the Tigris continued to be strongly subsiding and in the region of Kirkuk a thick series of reef limestones assigned to the Qamchuqa Formation (Barremian–Lower Albian) formed. They are covered conformably by fine-grained cherty limestones dated from planktonic foraminifera as Late Turonian–? Coniacian (Kometan Formation). Further to the east the Valanginian to Turonian is represented by the deep-water deposits of the Balambo Limestone Formation, a series of marls, shales and fine-grained limestones with radiolaria, ammonites and planktonic foraminifera.

It should be remarked that north of Mosul, in the vicinity of Jabal Sinjar, a narrow, strongly subsiding zone, trending WNW–ESE (Barremian–Aptian) to NW–SE (Albian–Cenomanian) existed which was probably fault-controlled. In it were deposited deep-water facies of Barremian to Aptian age identical with those found in southeastern Iraq. Evidently less subsident during the Albian–Cenomanian it contains neritic reef limestones.

Unlike central Arabia where the Turonian is absent (? by erosion) there are in Iraq Turonian rocks in the southeastern part of the country, along the lower course of the Tigris and Euphrates and along the border with Iran. In this area sedimentation was continuous from Cenomanian to Senonian with Turonian present as a deep-water facies of fine-grained limestones and marls (top of the Balambo Formation). In central Iraq more or less important erosional gaps separate it from the underlying beds, and then they rest directly without angular discordance (?) on eroded Albian or Cenomanian. In other parts of Iraq the Turonian beds were eroded during the Senonian, but it may be supposed that the facies were probably neritic resembling the rudist reef limestones (Mergi Limestone) in the extreme north of Iraq.

In northwestern Arabia, during the Barremian, sandstones (Grès de base) were deposited in Lebanon, in which were intercalated beds of lignitic shale, of basalt and of tuff along with thin beds of limestones or marls with oysters. The sandstones become less pronounced and alternate with marls and neritic limestones at the base of the Aptian. The thickness of these transgressive beds varies from one region to the next depending upon the surface of deposition which resulted from the terminal Jurassic earth movements. The marine transgression to the east was progressive for the arenaceous marine facies appears later in the Anti-Lebanon and Syria (in Aptian times). Throughout this period and up until the Albian, volcanic horizons (basalt) occur within the sequence.

In the vicinity of the Dead Sea the basal Cretaceous succession is the same. The sandy, continental facies are found along the eastern bank of the Jordan and extend to southern Sinai. In both Israel and northern Sinai the facies become progressively more marine (the Lebanon facies) and in the coastal plain consist of neritic limestones, calcareous shales with sandstone intercalations.

The top of the Lower Aptian is marked by the deposition of a massive limestone with rudists and orbitolinids (Falaise de Blanche), a facies which is widely spread throughout the Mediterranean littoral from Kurd Dagh in the north to Israel in the south. Towards the east it passes laterally into an arenaceous facies or was not deposited. At the top of the Upper Aptian the facies indicate the onset of regression (shales, sandstones, a few bands of neritic limestone). From Albian to Cenomanian times the pattern reversed with a major transgression towards the east. In Lebanon, Israel and in northern Sinai the Lower and Middle Albian (couches à *Knemiceras*) consist of marls and marly limestones; these are followed by essentially shelf limestones of the top of the Albian and Cenomanian. The varieties found consist of neritic, often reef limestones, fine-grained cherty limestones, marly limestones, dolomites and some marl intercalations. The beds are highly fossiliferous. Within the sequence the Vraconian (uppermost Albian) and Upper Cenomanian beds seem to be of slightly deeper water origin corresponding to minor transgressive peaks.

At the end of the Cenomanian epeirogenic movement lead to the formation of basins and highs trending NE–SW in Israel and NNE–SSW in Lebanon. At the beginning of the Turonian the basins received heavy sedimentation with fine-grained limestones and marls with planktonic foraminifera and ammonites, while the highs, occasionally emergent, developed a coral reef facies. The end of the Turonian was marked by the partial filling of the basins with the deposition of neritic limestones.

The carbonate facies at the top of the Middle Cretaceous extend towards the east and communicated directly with the seas over Iraq. However, sills (? emergent) probably existed in the vicinity of Deir ez Zor and Jabal Abd el Aziz for the Aptian and Albian are locally evaporitic. To the north the Mardin Arch, which during earlier periods barred the seas over Turkey access to the Syrian Platform, was submerged during the Middle Cretaceous.

Senonian deposits of variable age (? Coniacian to Campanian) rest upon the Middle Cretaceous with apparent concordance or are discordant where erosion has removed the Upper Turonian and where the lower members of the Senonian are absent.

Paleogeographic and stratigraphic synthesis

The ages and correlations of the principal formations of the Lower and Middle Cretaceous discussed above are shown in Table III. As a result of important orogenic movements between the Jurassic and Cretaceous the whole of the western part of Arabia was uplifted and heavily eroded. The products of this erosion are found in the clastic series bordering the region. Subsequent movements, linked probably to the reactivation of faults, led to the extrusion of basalts up until Albian times.

The approximate position of the coastline during Barremian–Early Aptian times is shown in Fig. 6. Purely carbonate rocks are restricted to eastern Iraq and southeastern Arabia in which regions there is sedimentary continuity with the Jurassic. Elsewhere the deposits rest discordantly upon a more or less eroded Jurassic surface. These deposits are essentially clastic and there is a progressive passage from marine to continental facies as the shoreline is approached.

From the Barremian up to the end of the Cenomanian the sea transgressed over the emergent area interrupted by a minor regression in the Late Aptian. Two major marine incursions over the continental area appear to be general, that of the Aptian (Falaise de Blanche in Lebanon, Shu'aiba Formation in Iraq and the Persian Gulf, top of the Qishn Formation in southwestern Arabia) and that of the Upper Cenomanian (the presumed location of this shoreline is shown in Fig. 6). The transgression of the Vraconian is less distinct probably because of the difficulty of precise dating.

Between the Cenomanian and Turonian epeirogenic movements resulted in northwestern Arabia in the formation of basins and highs resulting in turn in a marked facies differentiation. Over the rest of the Arabian Peninsula, excluding only the deep-water deposits found in eastern Iraq, Turonian deposits are absent. This absence is presumably due to their removal by erosion following orogenic movements at the beginning of the Senonian which profoundly modified the submarine topography.

TABLE III

Correlation chart of the Cretaceous formations of the Arabian Peninsula

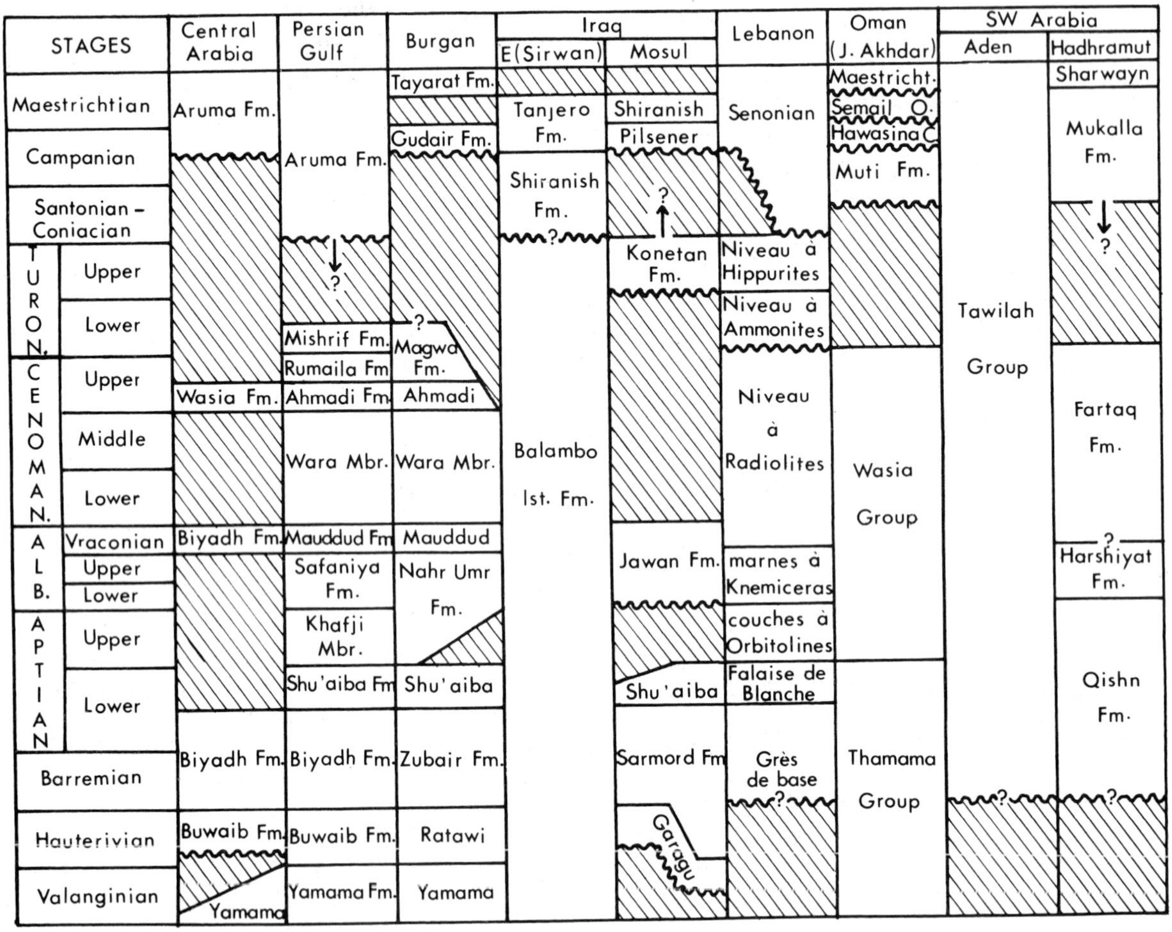

STAGES		Central Arabia	Persian Gulf	Burgan	Iraq E(Sirwan)	Iraq Mosul	Lebanon	Oman (J. Akhdar)	SW Arabia Aden	SW Arabia Hadhramut
				Tayarat Fm.				Maestricht.		Sharwayn
Maestrichtian		Aruma Fm.			Tanjero Fm.	Shiranish	Senonian	Semail O.		Mukalla Fm.
Campanian			Aruma Fm.	Gudair Fm.		Pilsener		Hawasina C. / Muti Fm.		
Santonian – Coniacian					Shiranish Fm. ?	?				?
TURON.	Upper		?			Konetan Fm.	Niveau à Hippurites			
	Lower		Mishrif Fm.	? Magwa Fm.			Niveau à Ammonites		Tawilah Group	
CENOMAN.	Upper	Wasia Fm.	Rumaila Fm / Ahmadi Fm.	Ahmadi			Niveau à Radiolites			Fartaq Fm.
	Middle		Wara Mbr.	Wara Mbr.	Balambo lst. Fm.			Wasia Group		
	Lower									
ALB.	Vraconian	Biyadh Fm.	Mauddud Fm	Mauddud						?
	Upper		Safaniya Fm.	Nahr Umr Fm.		Jawan Fm.	marnes à Knemiceras			Harshiyat Fm.
	Lower									
APTIAN	Upper		Khafji Mbr.				couches à Orbitolines			Qishn Fm.
	Lower		Shu'aiba Fm	Shu'aiba		Shu'aiba	Falaise de Blanche			
Barremian		Biyadh Fm.	Biyadh Fm.	Zubair Fm.		Sarmord Fm	Grès de base ?	Thamama Group	?	?
Hauterivian		Buwaib Fm.	Buwaib Fm.	Ratawi		Garagu				
Valanginian		Yamama	Yamama Fm.	Yamama						

Upper Cretaceous

Stratigraphy

During the Late Cretaceous there are numerous gaps and discordances within the stratigraphic succession in central Arabia and Rub al Khali (Aruma Formation) which reflect the orogenic movements which began in the Turonian and continued until the Maastrichtian.

At the beginning of the Late Cretaceous the whole western part of central Arabia as far as Khurais in the east was emergent, and along the Persian Gulf there existed numerous highs some of which may have been emergent. The ? Turonian and the base of the Upper Cretaceous (Coniacian, Santonian and ? Lower Campanian) lying transgressively and discordantly upon the Cenomanian are known only in depressions, in certain areas bordering the Persian Gulf (Hasa, Dammam) and in the central part of the Rub al Khali (in the Al Ubaylah Well).

During the Late Campanian and particularly during the Maastrichtian a marine transgression left a large part of central Arabia covered by carbonate deposits. Shallow-water deposits, as shales and limestones with *Globotruncana,* are known in subsurface in the vicinity of Ghawar and in the eastern part of Rub al Khali (Abu Dhabi and the wells between Haushi-Huqf and the Oman Mountains). Shelf carbonates with benthonic foraminifera and rudists are found in the central part of Rub al Khali and in wells in central Arabia from the Khurais well up to the outcrops in the west. This belt of outcrops extending from 24° to 30°N consists almost entirely of carbonates, however in the north at Sakakah (30°N

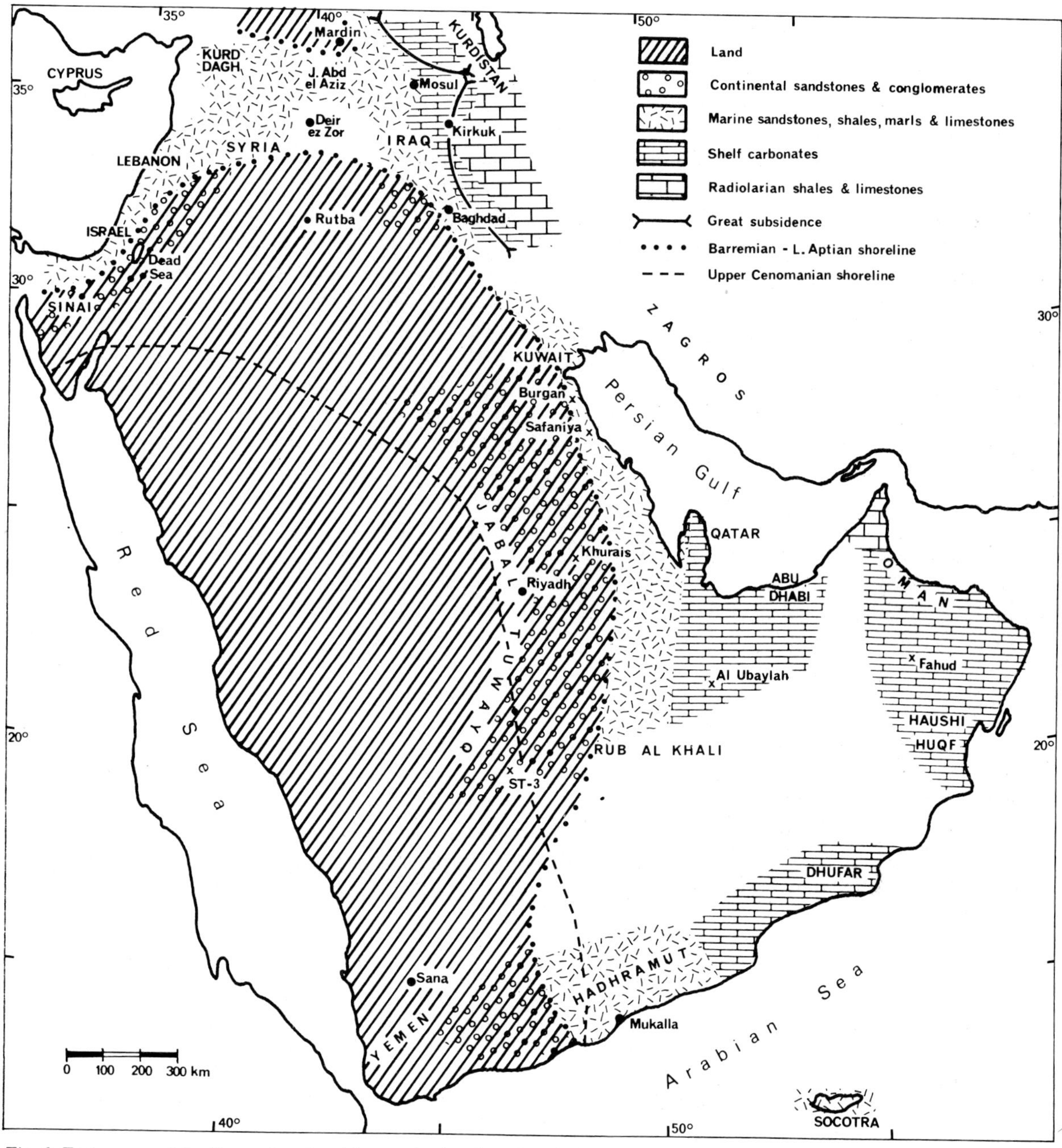

Fig. 6. Facies map of the Barremian–basal Lower Aptian deposits of the Arabian Peninsula.

40°E), the intercalation of sandstones indicates the proximity of the shoreline. In a similar manner south of 24°N arenaceous bands appear within the limestone and south of 22°N the formation is represented entirely by continental sandstones. The continental facies can apparently be traced in the western part of Rub al Khali and correlated with the continental sandstone facies of southwestern Arabia (see below).

During the Late Cretaceous the greater part of southwestern Arabia was also emergent and clastic, continental facies predominate. In Yemen the Upper Cretaceous is represented by the continental sandstones of the top of the Tawilah Group. Towards the east, that is in the direction of Hadhramut and

Dhufar and the location of open sea, the base of the Upper Cretaceous, the Mukalla Formation, is still composed of continental clastics but the upper part, the Sharwayn Formation of Maastrichtian age, is marine consisting of neritic limestones, marls and shales with benthonic foraminifera and rudists. Thus, it would seem that only during the Maastrichtian did marine conditions penetrate into the region, the period of its maximum penetration in central Arabia. In the west, in Aden, beginning in Senonian times active volcanicity commenced and continued through to the beginning of the Tertiary. Although dominantly basaltic the Aden Trap Series also contains andesites, trachytes and rhyolites.

In Oman the end of the Late Cretaceous was marked by important orogenic movements which seem to have begun also during the Turonian and continued to the Campanian–Maastrichtian where they reached their maximum intensity. The magnitude of these movements has been variously interpreted depending upon whether the outcrops of the Hawasina Group and Semail Ophiolites are considered as autochthonous or allochthonous. The rocks in the Mountains of Oman have been divided into five major units superposed one upon the other as follows:

1. The sialic basement (Pre-Permian) consisting of granites and folded and partially metamorphosed sediments and volcanic rocks.
2. Sediments ranging in age from Late Permian to Campanian, mostly limestones formed under shallow-water conditions and belonging to the Arabian Platform. These rocks are universally considered autochthonous and have been incorporated in the paleogeographic reconstructions attempted here. In the particular case of the Senonian, in the southwestern part of the Mountains of Oman in the Fahud, Natih and other wells, planktonic foraminiferal shales (Aruma Shale Formation) pass northeastwards at Jabal Akhdar into flyschoid deposits of the Muti Formation. Their age is from Late Santonian to Campanian; the Lower Senonian and Turonian are absent.
3. A complex mixture of highly folded rocks, the Hawasina Complex, formed in deep water and consisting principally of radiolarites, turbidites, calcilutites, pillow lavas and tuffs. The age of the identifiable fossils (reworked according to some) ranges between ? Late Permian and Middle Cretaceous. Within the sequence olistoliths of large dimensions and consisting of shallow-water Permo-Triassic dolomites and limestones are found.
4. An ophiolitic sequence, the Semail Ophiolites, in which an orderly progression from coarsely crystalline plutonic rocks to effusive rocks can be traced, with peridotites (at the base), gabbros, dolerites and pillow lavas (at the top).
5. Autochthonous marine sediments of Maastrichtian and Tertiary age, transgressive and discordant.

The autochthonous hypothesis for the origin of this sequence suggests that all the units are concordant (Morton 1959; Tschopp, 1967a; Wilson, 1969). The greater part of the Hawasina fauna must thus be reworked. According to this hypothesis Wilson assumes that from the Santonian to the beginning of the Maastrichtian there was a deepening from the Arabian Platform with its shallow-water carbonates passing into an intracratonic trough lying where the Oman Mountains now stand. At the margin of the trough the foraminiferal shales of the Aruma Shale Formation formed a transition facies while in the trough itself radiolarites represented the deep-water facies. The trough was bordered to the northeast beyond the actual shoreline by a submerged shoal, which as the result of faulting became the site of submarine volcanism. Turbidity currents were responsible for erosion of the rocks on the shoal and for their transport into the trench to the southwest. This mixture of reworked rocks and deep-water deposits constitutes the Hawasina Formation. At the beginning of the Maastrichtian after the sedimentation of the Hawasina Formation the shoal was subjected to strong faulting which resulted in the gravity sliding of exotic blocks of its Permo-Triassic cover. The continuation of this faulting equally resulted in the eruption of the Semail ultra-basic magma in the trough where they covered the Hawasina beds.

The second hypothesis favours an allochthonous origin with the thrusting of the Hawasina Formation and the Semail Ophiolites (Lees, 1928; Allemann and Peters, 1972; Glennie et al., 1973). The origin of the latter formations is sited to the northeast well beyond the present coast of Oman. According to Glennie et al., their formation and emplacement is linked to the expansion of an oceanic ridge from Late Permian times with the creation of Tethys by the separation of the supercontinent of Pangaea into Laurasia and Gondwana. The oceanic ridge brought up peridotites and gabbros and allowed the submarine volcanicity, these materials subsequently forming the Semail nappe. The rocks of the Hawasina Formation from the Permian to Cenomanian were deposited in a deep basin northeast of Oman between the Arabian Platform and the oceanic ridge. According to Glennie et al., the Hawasina Formation represents a succession of nappes, each constituting a lithologic unit with the lowest nappes originally deposited closer to the Arabian foreland and with a lesser tectonic displacement than the upper which originated closer to the oceanic ridge. It is thus possible to develop the sedimentological evolution of the basin from the Permian to the Cenomanian. The Semail and Hawasina nappes were emplaced at the end of the Campanian by the combined effects of sea-floor expansion and the displacement of the Arabian Peninsula to the east and northeast, the two acting in opposite directions.

In northern Arabia, Kuwait was probably under water during the Turonian and into the Early Senonian, although no deposits of this age occur. The contact between the lower beds of the Upper Cretaceous Aruma Formation and the Cenomanian Wasia Group is discordant and marked sedimentologically by a concentration of limonitic oolites. From the Campanian to the beginning of the Maastrichtian the area lay at the limit between neritic facies to the southwest and pelagic to the north and east. At Burgan in the southwest the Gudair Formation consists of detrital neritic limestones with benthonic foraminifera and shales. To the north and east, in the Zubair wells, deep-water chalky and marly limestones with planktonic foraminifera are found. Within the deep-water assemblage however are found intercalations of neritic detrital limestones and shales with a mixture of benthonic and pelagic foraminifera. The facies of the Tayarat Formation at the top of the Maastrichtian, shallow-water reef limestones, crystalline dolomitic limestones sometimes containing anhydrite, and shales, are found uniformly over the whole region. The beds appear to be conformably overlain by the basal Tertiary Radhuma Formation.

In Iraq Lower Senonian beds are known only in the southeast in the vicinity of Baghdad where the Mushorah Formation is made up of marls containing planktonic foraminifera, and in the northwest where at Jabal Sinjar the base of the Shiranish Formation consists of cherty, marly limestones with *Pithonella*. In the first case the beds are in continuity with the Turonian, in the second they transgress over eroded Albian limestones.

The transgression of the (? Upper) Campanian–Maastrichtian covered the whole of Iraq. West of the Tigris an E–W facies distribution can be found, with basin deposits in troughs (Shiranish Formation) and shelf carbonates on the bordering highs (Pilsener Formation). The basins and highs are presumed to be due to the activity of deep faults. East of the Tigris a further basin aligned NE–SW also received deep-water deposits with globotruncanids. Eastwards a sandy marly flysch facies occurs, the Tanjero Formation, in which are found reworked radiolarites and ophiolites. Still further to the east, in Kurdistan, discordantly above the latter formation occur successively radiolarites, massive limestones (of ? Permian to Cenomanian age), then gabbros and dolerites. According to some the sequence is autochthonous with the radiolarites being of Late Cretaceous age while the igneous rocks extruded *in situ* occasionally carrying with them large blocks of limestone. The second, allochthonous, hypothesis assumes tectonic emplacement during the Maastrichtian of beds formed far distant from their present location (cf. Oman).

The Mediterranean margin of the Arabian Peninsula is characterized during the Senonian by the deposition of a chalky or marly limestone facies with a variable amount of flint. Locally however there may be considerable variations in thickness and some facies variations. The presence of an emergent region to the southeast led to a differentiation between open-sea deposits of chalk to the west and northwest and a littoral facies sandy chalk rich in flint to the east and southeast. In addition orogenic movements during the Turonian and earliest Senonian resulted in a submarine topography broken up into basins and highs and the transgressive beds of the base of the Upper Cretaceous (Coniacian, Santonian, Lower Campanian) are known only in the basins; the highs were either emergent or areas of non-deposition. Between Early and Late Campanian epeirogenic movements caused a slight uplift in Jordan and Israel. The Upper Campanian and Maastrichtian deposits expand over most of the region over a relief largely smoothed by the infilling of the basins by Santonian and Lower Campanian sediments, but some facies change (a passage from neritic to deep-water or from high-energy to low-energy environments) suggests there was still some submarine relief.

In northwest Syria and the Turkish province of Hatay there are significant outcrops of ophiolites and radiolarites within the Upper Cretaceous succession. The Middle Cretaceous limestones are directly overlain by limestones or glauconitic sandstones followed by marls and limestones of ? Late Campanian–Maastrichtian age. Thus, the Coniacian, Santonian and ? Lower Campanian are not represented. The beds are successively overlain in apparent conformity by peridotites, gabbros, dolerites, pillow lavas and then by radiolarites with blocks of *Halobia* limestone (Triassic in age). Locally blocks of metamorphic rocks are also included. The succession may also vary slightly. The sequence is terminated by neritic limestones with benthonic foraminifera which disconformably overlie the preceding.

As in both Oman and Kurdistan the ophiolites and radiolarites are considered allochthonous by some, autochthonous by others. Dubertret (1955) regards the rocks as autochthonous with the magma extruded during the Maastrichtian carrying with it xenolithic blocks of substratum (Triassic limestones). Continued extrusion below the lavas allowed slow cooling and the more crystalline members of the sequence. The radiolarites are regarded as penecontemporaneous

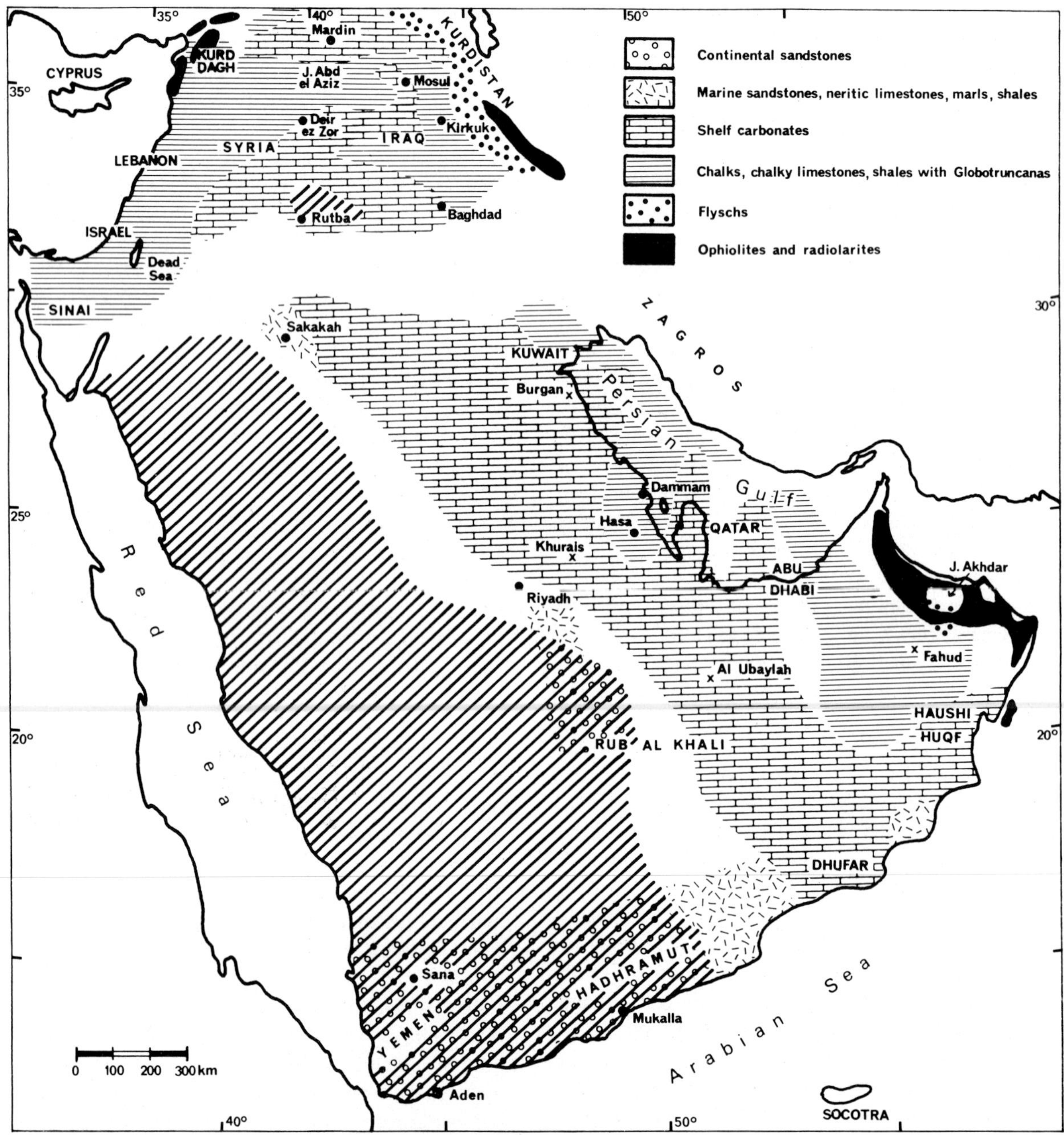

Fig. 7. Facies map of the Upper Campanian–Lower Maastrichtian deposits of the Arabian Peninsula.

with the effusives. According to Ponikarov et al. (1967) the effusive rocks were extruded from Triassic times until the beginning of the Cretaceous in association with radiolarites in a deep basin separated from the Arabian Platform by faulting. During the Maastrichtian orogenic movements resulted in their emplacement over the margin of the carbonate zone. The allochthonous hypothesis (Kober, 1915; Ricou, 1971) considers that the ophiolitic–radiolaritic suite was tectonically emplaced as a result of important tangential movements during the course of the Maastrichtian, the suite being older than the time of its emplacement.

Paleogeographic and stratigraphic synthesis

The Senonian was a time of transgression which expanded and became more pronounced from the Coniacian to Maastrichtian, and at its maximum the transgression covered the greater part of the Arabian Shield. Tectonic movements during the Turonian and continuing into the Early Senonian resulted in a diversified submarine topography of basins and highs which influenced the nature of sedimentation over the Arabian Platform.

At the base of the Senonian the sea encroached to only a limited extent over the Arabian Shield. Along the northern and eastern borders of the shield the structural form determined the nature and distribution of the facies. Sedimentation was continuous in the deep basins with characteristic deep-water deposits; on the highs deposits are generally absent for these were either emergent or current-swept (hard-ground). During the Late Campanian and Maastrichtian (Fig. 7) the marine transgression which covered most of the peninsula was marked by the deposition of shallow-water shelf carbonates, and similar deposits are found on the highs along the northern and eastern margins of the peninsula. These highs trend N–S in central Arabia, NW–SE in northeastern Iraq, and W–E in northern Arabia. They margin basins with deeper-water sediments, principally shales, marly limestones and chalk with globotruncanids. The shoreline is not easily defined in central Arabia, but must lie west of the Upper Cretaceous outcrops. The major tectonic phase at the end of the Cretaceous (Campanian–Maastrichtian) is clearly evidenced tectonically (? by nappes) and sedimentologically by flysch along the northern, northeastern and eastern margins of the peninsula. It was also responsible for the emplacement of the ophiolites and radiolarites. In central Arabia the effect of the movements is less clearly distinguishable and is shown essentially by breaks in the succession and local discordances. The source and age of the ophiolites and radiolarites vary according to whether they are considered as autochthonous (and hence Campanian–Maastrichtian) or allochthonous. In the latter case their age can be relatively early (? Permian to Cenomanian), their emplacement being due to tangential movements during the Maastrichtian and their origin located at a considerable distance from their present position.

CONCLUSIONS

During the Mesozoic the greater part of the Arabian Peninsula formed a vast, relatively stable platform across which transgressive seas advanced covering it more or less completely at certain times. The pattern of shallow-water sedimentation presents considerable variability due to the proximity of emergent areas. These were generally confined to the western part, that is the border of the present Red Sea. The open-sea (clean carbonates) facies and the deep-water facies lay along the northern, northeastern and eastern borders of the peninsula, and formed the transition zone between the epicontinental shelf and the Tethys. These bordering regions are marked by instability of the sea floor with the development of shoals and subsiding basins.

The analysis of the Mesozoic sequence allows the demonstration of sedimentary cycles of transgression and regression and of phases of orogenic activity.

As it has been previously indicated from a sedimentary standpoint the Upper Permian cannot be separated from the Triassic. The Late Permian was a time of a major marine transgression, followed by regression during the Triassic. The regression was interrupted by a slight transgression during Middle Triassic times. Between the Triassic and Jurassic important earth movements led to the emersion of the whole of central and southern Arabia. It follows that during the Liassic the sea was restricted to the northwest, that is the Syrian Platform and Mediterranean margin and to the east, that is to the subsiding Iraq Basin, the Persian Gulf and Oman. The Jurassic transgression, the onset of which began in Liassic times, reached its maximum extent in the Kimmeridgian at which time most of the Arabian Peninsula was under water. At that time a link existed between the sea covering the Arabian Peninsula and that existing in East Africa. The existence of a paleorelief created by orogenic movements at the end of the Triassic resulted in a facies differentiation although all were of shallow-water, shelf, origin. One strongly subsiding basin striking NW–SE did develop along the line of the present Tigris and the eastern shore of the Persian Gulf, and in it deep-water sedimentation occurred. At the end of the Kimmeridgian an important regression began, whose effects are seen in the formation of evaporites over almost the whole region with the exception only of Oman and the northwestern part of the peninsula. At this time in the western part of Arabia, in Syria, Lebanon, Israel and Yemen, the first signs, faulting and extrusion of lavas, of the major orogenic phase of the end of the Jurassic (? and the beginning of the Cretaceous) which affected not only the Arabian Peninsula but also the bordering zones (Turkey, central Iran, Ethiopia and Somalia) became apparent. They resulted in the emersion of central and western Arabia which then underwent erosion during the Early Cretaceous and became the source for the clastic sediments deposited around the margins. Late orogenic movements, presumably related to fault reactivation, occurred in northwestern Arabia up until Albian times.

Calcareous marine deposits were limited to the Lower Cretaceous of northeastern and southeastern Arabia, in which regions there is continuity of sedimentation with the Jurassic beds. From the Barremian to the latest Cenomanian the sea once more transgressed over the emergent land. This transgression was not regular or progressive being interrupted by slight regressions of which the most important occured during the Late Aptian. On two occasions, at the end of the Early Aptian and during the Late Cenomanian, the sea penetrated deep into the continent. The transgressive phase of the Vraconian (top Albian) was less distinct.

Between the Cenomanian and Turonian the region was again subjected to orogenic activity which resulted in the formation (? by faulting or folding) of basins and intervening highs. During the Early Turonian these highs, sometimes emergent, received either a neritic (reef) sedimentation or none at all, whilst in the basins deep-water sedimentation occurred. Further orogenic movements followed the Late Turonian regression at the beginning of the Senonian. The end of the Mesozoic was marked by a marine transgression which increased from the Coniacian to the Maastrichtian. The distribution of sediments was very much structurally controlled. During the Early Senonian the sea transgressed only slightly over the Arabian Shield, but during the Campanian to Maastrichtian the shoreline was displaced to the west. At the top of the Campanian and at the base of the Maastrichtian the effects of a major tectonic phase are clearly seen in the development of flysch and tectonically in the formation of nappes (?) along the northern, northeastern and eastern borders of the Arabian Peninsula. This was the period of emplacement of ophiolites and radiolarites. The origin of these formations is still under discussion for they are regarded as autochthonous by some and allochthonous by others.

Within the framework of plate tectonics a great number of hypotheses have been proposed to explain these sedimentary and orogenic phenomena. They offer tempting models, but, often contradictory, they leave unexplained important features.

The detailed studies of the sedimentary succession, particularly in the lesser known regions (Oman, Yemen), the results of paleomagnetic and other geophysical studies are still in the preliminary stages. The availability of complete records from petroleum exploration in particular would greatly aid in the analysis of data in the hope of developing a well-founded hypothesis.

REFERENCES

Allemann, F. and Peters, T., 1972. The ophiolite–radiolarite belt of North-Oman Mountains. *Eclog. Geol. Helv.*, 65: 657–697.

Al Naqib, K.M., 1967. Geology of the Arabian Peninsula. Southwestern Iraq. *Geol. Surv., Prof. Pap.*, 560G: 54 pp.

Aramco, 1959. Ghawar Oil Field, Saudi Arabia. *Bull. Am. Assoc. Pet. Geol.*, 43(2): 434–454.

Arkell, F.R.S., 1952. Jurassic Ammonites from Jebel Tuwayq, Central Arabia. *Philos. Trans. R. Soc. Lond.*, 633(236) 241–313.

Arkin, Y. and Hamaoui, M., 1967. The Judea Group (Upper Cretaceous) in central and southern Israel. *Geol. Surv. Isr.*, 42: 17 pp.

Baker, N.E. and Henson, F.R.S., 1952. Geological conditions of oil occurence in Middle East fields. *Bull. Am. Assoc. Pet. Geol.*, 36(10): 1185–1901.

Banner, F.T. and Wood, G.V., 1964. Lower Cretaceous–Upper Jurassic stratigraphy of Umm Shaif Field, Abu Dhabi marine areas, Trucial Coast, Arabia. *Bull. Am. Assoc. Pet. Geol.*, 48(2). 191–206.

Beauchamp, J. and Lemoigne, Y., 1974. Présence d'un bassin

de subsidence en Ethiopie centrale et essai de reconstitution paléogéographique de l'Ethiopie durant le Jurassique. *Bull. Soc. Géol. Fr.*, XVI(5). 563–569.

Bender, F., 1968. Geologie von Jordanien. In: H.J. Martini (Editor), *Beiträge zur regionalen Geologie der Erde.* Borntraeger, Berlin, 230 pp.

Bentor, Y.K., 1959. *Lexique stratigraphique international, vol. III, Asie, fasc. 10c2, Israel.* Centre National Recherche Scientifique, Paris.

Beydoun, Z.R., 1964. The stratigraphy and structure of the Eastern Aden Protectorate. *Overseas Geol. Min. Resour., Bull. Suppl.* 5: 107 pp.

Beydoun, Z.R., 1966. Geology of the Arabian Peninsula. Eastern Aden Protectorate and Part of Dhufar. *Geol. Surv. Prof. Pap.*, 560H: 49 pp.

Beydoun, Z.R., 1969. Note on the age of the Hadhramut Arch, Southern Arabia. *Overseas Geol. Min. Resour.*, 10(3): 236–240.

Beydoun, Z.R. and Bichan, H.R., 1970. The geology of Socotra Island, Gulf of Aden. *Q.J. Geol. Soc. Lond.*, 125: 413–446.

Beydoun, Z.R. and Greenwood, J.E.G.W., 1968. *Lexique stratigraphique international, vol. III, Asie, fasc. 10b2, Protectorat d'Aden et Dhufar.* Centre National Recherche Scientifique, Paris.

Clift, W., 1956. Sedimentary history of the Ogaden District, Ethiopia. *20th Congr. Geol. Int., Mexico,* I: 89–112.

Derin, B. and Gerry, E., 1972. Jurassic biostratigraphy and environments of deposition in Israel. *Isr. Insti. Petr.*, rep.2/72: 20 pp.

Derin, B. and Reiss, Z., 1966. Jurassic microfacies of Israel. *Isr. Inst. Pet., Spec Publ.*, 43 pp.

Dewey, J.F., Pitman, W.C., Ryan, W.B.F. and Bonnin, J., 1973. Plate tectonics and the evolution of the Alpine System. *Geol. Soc. Am. Bull.*, 84: 3137–3180.

Dietz, S.D. and Holden, J.C., 1970. Reconstruction of Pangaea: Breakup and dispersion of continents, Permian to Present. *J. Geophys. Res.*, 75: 4939–4956.

Dubertret, L., 1937. Le massif Alaouite. *Notes Mém. Syrie–Liban,* II. 9–42.

Dubertret, L., 1955. Géologie des roches vertes du Nord-Ouest de la Syrie et du Hatay (Turquie). *Notes Mém. Moyen-Orient,* VI: 1–180

Dubertret, L., 1959. Contribution à la stratigraphie et à la paléontologie du Crétacé et du Nummulitique de la marge NW de la péninsule Arabique. *Notes Mém Moyen-Orient,* VII. 193–220.

Dubertret, L., 1966. Liban, Syrie et bordure des pays voisins. Première partie. Tableau stratigraphique avec carte géologique au millionième. *Notes Mém. Moyen-Orient,* VIII 251–358.

Dubertret, L. and André, C., 1969. Carte orographique et cartes tectoniques de la péninsule Arabique. *Notes Mém. Moyen-Orient,* X: 285–318.

Dubertret, L. and Vautrin, H., 1937. Révision de la stratigraphie du Crétacé du Liban. *Notes Mém. Syrie–Liban,* II: 43–73.

Dubertret, L., Daniel, E.J. and Bender F., 1963. *Lexique stratigraphique international, vol. III, Asie, fasc. 10c1, Liban, Syrie, Jordanie.* Centre National Recherche Scientifique, Paris.

Dunnington, H.V., 1958. Generation, migration, accumulation and dissipation of oil in Northern Iraq. In: *Habitat of Oil.* Am. Assoc. Pet. Geol. Tulsa, Okla., pp. 1194–1251.

Dunnington, H.V., 1967. Stratigraphical distribution of oilfields in the Iraq–Iran–Arabia Basin. *J. Inst. Pet.*, 53(520) : 129–161.

El-Naggar, Z.R. and Al-Rifaiy, I.A., 1972. Stratigraphy and microfacies of type Magwa Formation of Kuwait, Arabia; pt. 1: Rumaila Limestone Member. *Bull. Am. Assoc. Pet. Geol.*, 56(8): 1464–1493.

El-Naggar, Z.R. and Al-Rifaiy, I.A., 1973. Stratigraphy and microfacies of type Magwa Formation of Kuwait, Arabia; pt. 2: Mishrif Limestone Member. *Bull. Am. Assoc. Pet. Geol.*, 57(11): 2263–2279.

Falcon, F.R.S., 1967. The geology of the north-east margin of the Arabian Basement Shield. *Adv. Sci.*, 24: 1–12.

Falcon, N.L., 1958. Position of oil fields of southwest Iran with respect to relevant sedimentary basins. In: *Habitat of Oil.* Am. Assoc. Pet. Geol., Tulsa, Okla., pp. 1279–1293.

Flexer, A., 1968. Stratigraphy and facies development of Mount Scopus Group (Senonian–Paleocene) in Israel and adjacent countries. *Isr. J. Earth-Sci.*, 17: 85–94.

Flexer, A., 1971. Late Cretaceous paleogeography of Northern Israel and its significance for the Levant geology. *Palaeogeogr., Palaeoclimatol., Palaeoecol.*, 10: 293–316.

Freund, R., 1959. On the stratigraphy and tectonics of the Upper Cretaceous in Western Galilee. *Bull. Res. Counc. Isr.*, G8(1) 43–50.

Freund, R., 1961. Distribution of Lower Turonian ammonites in Israel and the neighbouring countries. *Bull. Res. Counc. Isr.*, 10G(1–2). 79–100.

Freund, R., 1965. Upper Cretaceous reefs in Northern Israel. *Isr. J. Earth-Sci.*, 14: 108–121.

Geukens, F., 1966. Geology of the Arabian Peninsula. Yemen. *Geol. Surv., Prof. Pap.*, 560B: 23 pp.

Glennie, K.W., Boeuf, M.G.A., Clarke, M.W.H., Moody-Stuart, M., Pilaar, W.F.H. and Reinhardt, B.M., 1973. Late Cretaceous nappes in Oman Mountains and their geologic evolution. *Bull. Am. Assoc. Pet. Geol.*, 57: 5–27.

Greenwood, J.E.G.W. and Bleackly, D., 1967. Geology of the Arabian Peninsula. Aden Protectorate. *Geol. Surv., Prof. Pap.*, 560C: 96 pp.

Greig, D.A., 1958. Oil Horizons in the Middle East. In: *Habitat of Oil.* Am. Assoc. Pet. Geol., Tulsa, Okla., pp. 1182–1193.

Hajash, G.M., 1967. The Abu Sheikhdom – The onshore oilfields history of exploration and development. *7th World Pet. Congr., Proc., Mexico,* 2: 129–139.

Henson, F.R.S., 1950. Cretaceous and Tertiary reef formations and associated sediments in Middle East. *Bull. Am. Assoc. Pet. Geol.*, 34(2): 215–238.

Henson, F.R.S., 1951. Observations on the geology and petroleum occurrences of the Middle East. *3rd World Pet. Congr., Proc., The Hague,* 1: 118–140.

Hudson, R.G.S., 1960. The Permian and Trias of the Oman Peninsula, Arabia. *Geol. Mag.*, 47(4): 299–308.

Hudson, R.G.S. and Chatton, M., 1959. The Musandam Limestone (Jurassic to Lower Cretaceous) of Oman, Arabia. *Notes Mém. Moyen-Orient,* VII: 69–93.

Hudson, R.G.S., McGugan, A. and Morton, D.M., 1954. The

structure of the Jebel Hagab Area, Trucial Oman. *Q.J. Geol. Soc. Lond.*, 110: 121–152.

James, G.A. and Wynd, J.G., 1965. Stratigraphic nomenclature of Iranian Oil Consortium agreement area. *Bull. Am. Assoc. Pet. Geol.*, 49(12): 2182–2245.

Kafri, U., 1972. Lithostratigraphy and environment of deposition Judea Group, Western and Central Galilee, Israel. *Geol. Surv. Isr.*, 54: 56 pp.

Kamen-Kaye, M., 1970. Geology and productivity of Persian Gulf Synclinorium. *Bull. Am. Assoc. Pet. Geol.*, 54(12): 2371–2394.

Kober, L., 1915. Geologische Forschungen in Vorderasien. I. Teil. A. Das Taurusgebirge. *Denkschr. Math.- Naturwiss. Kl. K. Akad. Wiss., Wien,* pp. 381–419.

Lees, G.M., 1928. The geology and tectonics of Oman and of parts of south-eastern Arabia. *Q.J. Geol. Soc. Lond.*, 84: 585–670.

McKenzie, D.P., 1970. Plate tectonics of the Mediterranean region. *Nature,* 226: 239–243.

Maync, W., 1966. Microbiostratigraphy of the Jurassic of Israel. *Geol. Surv. Isr.*, 40: 45 pp.

Mina, P., Razaghnia, M.T. and Paran, Y., 1967. Geological and Geophysical studies and exploratory drilling of the Iranian continental shelf – Persian Gulf. *7th World Pet. Congr., Proc., Mexico,* 2: 871–903.

Morton, D.M., 1959. The geology of Oman. *5th World Pet. Congr., Proc.,* 1, paper 14: 1–14.

Owen, R.M.S. and Nasr, S.N., 1958. Stratigraphy of the Kuwait-Basra Area. In: *Habitat of Oil.* Am. Assoc. Pet. Geol., Tulsa, Okla, pp. 1252–1278.

Parnes, A., 1962. Triassic ammonites from Israel. *Geol. Surv. Isr.*, 33: 76 pp.

Ponikarov, V.P., Kazmin, V.G., Mikhailov, I.A., Razvaliayev, A.V., Krasheninnikov, V.A., Kozlov, V.V., Soulidi-Kondratiyev, E.D., Mikhailov, K.Ya., Kulakov, V.V., Faradzhev, V.A. and Mirzayev, K.M., 1967. *The Geology of Syria. Explanatory Notes on the Geological Map of Syria. Part I: Stratigraphy, Igneous Rocks and Tectonics.* Ministry of Industry, Syrian Arab Republic, 230 pp.

Powers, R.W., 1968. *Lexique stratigraphique international, vol. III, Asie, fasc. 10b1, Saudi Arabia.* Centre National Recherche Scientifique, Paris.

Powers, R.W., Ramirez L.F., Redmond, C.D. and Elberg, E.L., 1966. Geology of the Arabian Peninsula. Sedimentary geology of Saudi Arabia. *Geol. Surv., Prof. Pap.,* 560D: 147 pp.

Reinhardt, B.M., 1969. On the genesis and emplacement of ophiolites in the Oman Mountains Geosyncline. *Schw. Min. Petrogr. Mitt.* 49(1): 1–30.

Renouard, G., 1955. Oil Prospects of Lebanon. *Bull. Am. Assoc. Pet. Geol.*, 39(11): 2125–2169.

Ricou, L.E., 1971. Le croissant ophiolitique péri-arabe, une ceinture de nappes mises en place au Crétacé supérieur. *Rev. Géogr. Phys. Géol. Dyn.*, 13(4): 327–350.

Rigo de Righi, M. and Cortesini, A., 1964. Gravity tectonics in foothills; structure belt of south-east Turkey. *Bull. Am. Assoc. Pet. Geol.*, 48(12): 1911–1937.

Saint-Marc, P., 1970. Contribution à la connaissance du Crétacé basal au Liban. *Rev. Micropaléontol.*, 12(4): 224–233.

Saint-Marc, P., 1974. Etude stratigraphique et micropaléontologique de l'Albien, du Cénomanien et du Turonien du Liban. *Notes Mém. Moyen-Orient,* XIII: 342 pp.

Steineke, M., Bramkamp, R.A. and Sander, N.J. 1958. Stratigraphic relations of Arabian Jurassic Oil. In: *Habitat of Oil.* Am. Assoc. Pet. Geol., Tulsa, Okla., pp. 1294–1329.

Stocklin, J., 1968. Structural history and tectonics of Iran: a review. *Bull. Am. Assoc. Pet. Geol.*, 52(7): 1229–1258.

Stocklin, J. and Setudehnia, A.O., 1972. *Lexique stratigraphique international vol. III, Asie, fasc. 9b, Iran.* Centre National Recherche Scientifique, Paris.

Stoneley, R., 1975. On the origin of ophiolite complexes in the southern Tethys region. *Tectonophysics,* 25: 303–322.

Sudgen, W., and Standring, A.J., 1975. *Lexique stratigraphique international, vol. III, Asie, Qatar.* Centre National Recherche Scientifique, Paris.

Tschopp. R.H., 1967a. The general geology of Oman. *7th World Pet. Congr., Proc., Mexico,* 2: 231–242.

Tschopp, R.H., 1967b. Development of the Fahud field. *7th World Pet. Congr., Proc., Mexico,* 2: 243–250.

Van Bellen, R.C., Dunnington, H.V., Wetzel, R. and Morton, D.M., 1959. *Lexique stratigraphique international, vol. III, Asie, fasc. 10a, Iraq.* Centre National Recherche Scientifique, Paris.

Weber, H., 1963. Ergebnisse erdölgeologischer Aufschlussarbeiten der DEA in Nordost-Syrien. Teil I. Schichtfolge, Fazies und Tektonik in der Haute Djésireh. *Erdöl Kohle,* 16: 669–682.

Weber, H., 1964. Ergebnisse Erdölgeologischer Aufschlussarbeiten der DEA in Nordost-Syrien. Teil II. Geophysikalische Untersuchungen und Tiefbohrungen in der Haute Djésireh. *Erdöl Kohle,* 17: 249–261.

Weissbrod, T., 1969. The Paleozoic of Israel and adjacent countries. *Geol. Surv. Isr.*, 47: 34 pp.; 48: 32 pp.

Wetzel, R. and Morton, D.M., 1959. Contribution á la géologie de la Transjordanie. *Notes Mém. Moyen-Orient,* VII: 95–191.

Wilson, H.H., 1969. Late Cretaceous eugeosynclinal sedimentation, gravity tectonics, and ophiolite emplacement in Oman Mountains, Southeast Arabia. *Bull. Am. Assoc. Pet. Geol.*, 53(3): 626–671.

Wolfart, R., 1967. Geologie von Syrien und dem Libanon. In: H.J. Martini (Editor), *Beiträge zur regionalen Geologie der Erde.* Borntraeger, Berlin, 326 pp.

AUTHOR INDEX

GENERAL INDEX